TRENDS IN RESEARCH IN EPHEMEROPTERA AND PLECOPTERA

TRENDS IN RESEARCH IN EPHEMEROPTERA AND PLECOPTERA

Edited by

Eduardo Domínguez

National University of Tucumán
Tucumán, Argentina

Kluwer Academic / Plenum Publishers
New York, Boston, Dordrecht, London, Moscow

Library of Congress Cataloging-in-Publication Data

Trends in research in Ephemeroptera and Plecoptera/edited by Eduardo Dominguez.
p. cm.
"Proceedings of the IX International Conference on Ephemeroptera and XIII International Symposium on Plecoptera, held August 16–21, 1998, and August 20–23, 1998, respectively, in Tafí de Valle, Tucumán, Argentina"—T.p. verso.
Includes bibliographical references and index.
ISBN 0-306-46544-2
1. Mayflies—Congresses. 2. Stoneflies——Congresses. I. Dominguez, Eduardo, 1958– II. International Conference on Ephemeroptera (9th: 1998: Tucumán, Argentina) III. International Symposium on Plecoptera (13th: 1998: Tucumán, Argentina)

QL505 .T74 2001
595.7′34—dc21

00-067112

Proceedings of the IX International Conference on Ephemeroptera and XIII International Symposium on Plecoptera, held August 16–21, 1998, and August 20–23, 1998, respectively, in Tafí de Valle, Tucumán, Argentina

ISBN 0-306-46544-2

233 Spring Street, New York, New York 10013

http://www.wkap.nl/

10 9 8 7 6 5 4 3 2 1

A C.I.P. record for this book is available from the Library of Congress

Printed in the United States of America

DEDICATION

This volume is dedicated to our two dear colleagues and friends: "Mick" Gillies and "Bill" Peters. Their important contributions to our knowledge of Ephemeroptera, and above all their enthusiasm and warm friendship will remain always in our memory.

Dr. **William Lee Peters**
(1939 -2000)

Dr. **Michael Thomas Gillies**
(1920-1999)

PREFACE

The International Conferences on Ephemeroptera (Mayflies) and Symposia on Plecoptera (Stoneflies) are held every three years, in different parts of the world. These events allow specialists from different countries to interact and present the results of their latest investigations.

The IX International Conference on Ephemeroptera and XIII International Symposium on Plecoptera, were held August 16-21, 1998, and August 20-23, 1998, respectively, in Tafí del Valle, Tucumán, Argentina, with a joint symposium on August 20, 1998. These events were hosted by the "Instituto Superior de Entomología" Facultad de Ciencias Naturales, from the National University of Tucumán, and attended by approximately 80 specialists from 25 countries.

At present, when the biodiversity crisis and the pressures on fresh-water environments and their inhabitants are worse than ever before, the information, discussions and guidelines coming out of events of this kind are becoming more and more important.

The 54 papers included in this volume are among those presented during the meetings, and accepted after peer review by international specialists.

The papers are grouped in five loosely defined sections (except the first that corresponds to a panel discussion), although several of the papers could exceed the subject boundaries where they are located.

I would like to thank the many individuals and institutions that helped with both this book and the organization of the events, namely:

The members of the organizing committee: H. R. Fernández, M. G. Cuezzo, F. Romero, C. Molineri and C. Nieto. Also collaborating were M. Ceraolo, J. Chocobar, M. Guzmán de Tomé, S. Moro, M. Orce, V. Manzo and many volunteers and students too numerous to detail here. Without their untiring efforts, the events simply would not have been possible.

The following persons acted as manuscript reviewers: J. Alba-Tercedor, J. V. Arnekleiv, R. Baumann, J. E. Brittain, I. C. Campbell, J. M. Elouard, J. F. Flannagan, R. W. Flowers, C. Froehlich, E. Gaino, M. T. Gillies, P. Goloboff, P. M. Grant, M. D. Hubbard, Y. Isobe, J. Jackson, N. N. Kapoor, P. Landolt, W. P. McCafferty, I. McLellan, R. Nelson, W. L. Peters, G. Pritchard, M. Sartori, I. Sivec, J. Stanford, K. W. Stewart, D. Studemann, B. Stark, P. Suter, B. Sweeney, S. Szczytko, and P. Zwick.

These institutions provided their institutional and/or economic support: Facultad de Ciencias Naturales e Instituto Miguel Lillo, Universidad Nacional de Tucumán; Consejo Nacional de Investigaciones Científicas y Técnicas (CONICET); Consejo de Investigaciones, Universidad Nacional de Tucumán (CIUNT); Fundación Antorchas, Fundación Miguel Lillo; Dirección de Turismo, Municipalidad de Tafí del Valle.

The Permanent Committee on Ephemeroptera Conferences provided partial financial support for fellowships for students from the Czech Republic, China and Slovenia; and the North American Benthological Society for students from Bolivia and Venezuela.

The completion of this book would not have been possible without the invaluable effort of Gustavo Sánchez, who made the digital work of the originals and the cover design, C. Molineri and C. Nieto who patiently reviewed the final copies, and valuable advice from Mary Ann McCarra and Robert Wheeler. The cover illustrations were done by S. Roig Juñent (Ephemeroptera) and A. Dupuy (Plecoptera).

Eduardo Domínguez
Convenor and Editor
August, 2000

CONTENTS

STATUS OF THE KNOWLEDGE OF EPHEMEROPTERA IN THE WORLD

ECOLOGY AND BEHAVIOUR

BIOGEOGRAPHY AND DISTRIBUTION

PHYLOGENY AND SYSTEMATICS

MORPHOLOGY AND ULTRASTRUCTURE

INTRODUCTION TO THE PANEL DISCUSSION "THE STATUS OF THE KNOWLEDGE OF EPHEMEROPTERA"

Javier Alba-Tercedor

Departamento de Biología Animal y Ecología
Facultad de Ciencias, Universidad de Granada
18071, Granada, Spain

It gives me great pleasure to introduce this section of the Proceedings. It represents the efforts of several authors to develop a panel on the "Status of the Knowledge of Ephemeroptera." When Eduardo Domínguez, as convenor, asked me to coordinate this panel, I was very enthusiastic and totally convinced of the importance of the topic. I contacted colleagues with geographical interests from around the world who made my work easy, thanks to the magnificent cooperation of every proposed participant. We never intended an exhaustive revision but only a summary of the present status of mayfly taxonomy. For personal reasons, Yeon Bae (Seoul Women's University) could not travel to Argentina, but his contribution is included herein.

Brief guidelines were given to each author indicating major topics to be included in their presentations. Every panelist did his best, or perhaps what could be done on short notice, and the following pages give the results. Presentations were followed by an animated discussion that is briefly summarized by four major points:

- The worldwide Ephemeroptera fauna is not well known from a faunistic or taxonomic point of view, and there is little knowledge on the biology and ecological requirements of many species. In fact, the mayflies of geographical areas such as North America or Europe, which are considered well studied, are not well known. Many areas in North America and Europe remain uncollected and many taxonomic revisions of various genera are needed. The fauna of other continents (e.g., Africa) is practically unknown, and many of the few described species are known only from one developmental stage.

- Despite this, taxonomy is considered to be "old-fashioned" in developed countries and therefore it is difficult to obtain funds to conduct research. Some participants expressed their concern for the future, commenting that graduate students refuse to conduct thesis research on taxonomic topics due to the difficulty of finding future employment. In the future, some existing collections may disappear without adequate care, and it is not clear that curators' positions will be maintained.

Trends in Research in Ephemeroptera and Plecoptera
Edited by E. Dominguez, Kluwer Academic/Plenum Publishers, 2001

- Despite these pessimistic opinions, there were many positive comments concerning the future of mayfly taxonomy. These suggested the assimilation of taxonomy with historical change, biodiversity, and conservation especially in developing countries. Such studies can be funded if they include water quality assessment and bio-monitoring, genetics, or biology of evolution.

- We need to continue advancing "global information," using the new computer media such as that started as West Lafayette (Purdue University) by Patrick McCafferty and Tallahassee (Florida A&M University) by Michael Hubbard and William Peters.

STATUS OF THE KNOWLEDGE OF EPHEMEROPTERA IN NORTHEAST ASIA AND GUIDELINES FOR FUTURE RESEARCH

Yeon Jae Bae

Department of Biology
Seoul Women's University
Seoul 139-774, Korea

ABSTRACT

Mayfly studies in Northeast Asia are summarized from a historical perspective. The number of known Ephemeroptera taxa, major bibliographic sources, and current mayfly workers are provided. Future research guidelines are additionally discussed.

OVERVIEW

Northeast Asia may be defined by the region encompassed by the Baykal lake, Mongolian dry lands, and the Huang river in China, including Northeastern China, Russian Far East, Korean peninsula, and Japanese islands (Bae, 1997). Systematists, biogeographers, and ecologists have being attracted to this fauna rich region since long time ago.

Systematic studies of Ephemeroptera in this region have been started by Japanese workers early in this century although there were some pioneer investigations by European workers (e.g., McLachlan, 1875). Ueno (e.g., 1928) and Imanishi (1930-1938) have made major systematic works in Japan. Gose (1979-1980, 1985) prepared larval and adult keys to Japanese species of mayflies, which are still in use in Japan. There have been 101 species of mayflies in 12 families known in Japan (Table 1).

In Far East Russia (FE Russia), Tshernova (e.g., 1952), Bajkova (e.g., 1972), and Kluge (e.g., 1983) have dealt with FE Russian fauna. Tshernova et al. (1986) and Kluge (1997) prepared adult and larval keys to FE Russian mayflies, respectively. So far, 149 species of mayflies in 17 families have been known in FE Russia (Table 1).

Korean mayfly fauna has been investigated mainly by Yoon and Bae (e.g., 1988); and a larval key (Yoon, 1995) has been available. More taxonomic investigations on the Baetidae of Korea (Park et al., 1996; Bae and Park, 1997, 1998; Bae et al., 1998) and North Korean mayfly fauna (Bae and Soldan, 1997; Bae and Andrikovics, 1997) have been carried out relatively recently. There have been 75 species of mayflies in 12 families known in Korea (Table 1).

Trends in Research in Ephemeroptera and Plecoptera
Edited by E. Dominguez, Kluwer Academic/Plenum Publishers, 2001

Table 1. Number of Ephemeroptera taxa recorded from Northeast Asian countries

Family	Japan	FE Russia	Korea	NE China
Acanthametropodinae	0	1	0	0
Ameletidae	8	7	2	2
Ametropodidae	0	1	0	0
Baetidae	23	30	15	4
Behningiidae	0	2	0	0
Caenidae	1	12	0	0
Ephemerellidae	25	20	16	10
Ephemeridae	4	9	4	4
Heptageniidae	24	45	24	11
Isonychiidae	1	2	2	0
Leptophlebiidae	6	6	2	1
Metretopodidae	0	2	1	0
Neoephemeridae	0	1	1	1
Oligoneuriidae	1	1	0	0
Polymitarcyidae	3	2	1	1
Potamanthidae	2	2	4	1
Siphlonuridae	3	6	3	3
Species total	101	149	75	38
Genus total	32	42	32	18
Family total	12	17	12	10
Species grand total				232
Genus grand total				47
Family grand total				17

Mayfly fauna of Northeastern China is currently under investigation by Bae and Liu (1999). So far, 38 species of mayflies in 10 families have been recognized from the area, but more species will be described in the near future.

It has been argued that taxonomic comparisons are necessary between Northeast Asian countries (e.g., Bae et al., 1998). Regional revisions are being conducted in some groups of mayflies (e.g., Chun and Bae: Heptageniidae; Park and Bae: Baetidae). Bae (1997) reviewed mayfly systematics in Northeast Asia from a historical perspective, which contained comprehensive bibliographic sources. Bae et al. (1999) reviewed Imanishi's (1940) report on larval Ephemeroptera which is the first monographic work covering almost all the areas in Northeast Asia.

From the ecological and environmental perspectives, larval mayflies have been the major interest not only in the analysis of communities of aquatic insects or benthic macroinvertebrates but also in the use of biological indicators in stream ecosystems (see, Bae, 1992, 1996). Hundreds of streams and rivers have been investigated for environmental purposes especially in Japan and Korea (see Bae, 1996).

It must be, however, notify that population level ecological approaches, e.g., life history and habitat adaptation, trophic and behavioral ecology, etc., are very wanting although some important groups such as *Ephemera* (e.g., Takemon, 1990; Lee et al., 1999), *Potamanthus* (e.g., Watanabe, 1989), and *Ephoron* (e.g., Watanabe et al., 1999), are recently under investigation.

Mayfly workers in Northeast Asian region are listed in Table 2. More mayfly scientists in diverse fields are needed in the future.

Table 2. Current mayfly workers who reside in Northeast Asia

Name	Institution	Status	Area interest*
Japan			
K. Gose		Retired	1
N. C. Watanabe	Kagawa Univ.	Professor	2, 3
S. I. Ishiwata	Kanagawa Env.Res.Center	Researcher	1, 3
Y. Takemon	Osaka Pref. Univ.	Professor	2, 3
N. Kobayashi	Asahi Tech. Inst.	Researcher	1, 3
FE Russia			
O. Ja. Bajkova		Retired	1
T. Tiunova	Russian Acad. Sci., Vladivostok	Researcher	1, 2
Korea			
I. B. Yoon	Korea Univ.	Professor	1, 2, 3
Y. J. Bae	Seoul Women's Univ.	Professor	1, 2, 3
D. J. Chun	Korean Entomol. Inst.	Researcher	1
S. J. Lee	Korea Univ.	PhD student	2
S. Y. Park	Seoul Women's Univ.	Researcher	1
J. M. Hwang	Seoul Women's Univ.	MS student	1
NE China			
G. C. Liu	Shenyang Agr. Univ.	Professor	1, 3
Y. T. Quan	Harbin Agr. Univ.	Professor	1

*1: Systematics, 2: Ecology, 3: Environment.

ACKNOWLEDGMENTS

This work was supported by GRANT No 961-0585-063-2 from the Korea Science and Engineering Foundation.

REFERENCES

Bae, Y. J. 1992. Current research and methods in aquatic entomology with special reference to mayfly studies. '92 Symp. Proc. Korean Soc. Limnol. p. 9-32. (in Korean).

Bae, Y. J. 1996. Current status and problems of aquatic insect research in Korea. '96 Symp. Proc. Korean Soc. Limnol. p. 63-71. (in Korean).

Bae, Y. J. 1997. A historical review of Ephemeroptera systematics in Northeast Asia. p. 405-417 In: P. Landolt and M. Sartori (eds). Ephemeroptera & Plecoptera: Biology-Ecology-Systematics. MTL, Fribourg, Switzerland.

Bae, Y. J. and S. Andrikovics. 1997. Mayfly (Ephemeroptera) fauna of North Korea (2). Insecta Koreana 14: 153-160.

Bae, Y. J., N. Ju. Kluge and D. J. Chun. 1998. New synonymy and new distributional data of mayflies from Korea and Russian Far East (Ephemeroptera). Zoosyst. Rossica. 7: 89-94.

Bae, Y. J., J. E. Lee and I. B. Yoon. 1999. Review of Imanishi's (1940) report on the Ephemeroptera in Northeast Asia. Ent. Sci. (Submitted.)

Bae, Y. J. and G. C. Liu. 1999. Mayflies (Ephemeroptera) from Zhiangbeisan area in Northeast Asia. Ent. Res. Bull. (KEI), Seoul. (In press.)

Bae, Y. J. and S. Y. Park. 1997. Taxonomy of *Cloeon* and *Procloeon* (Ephemeroptera: Baetidae) in Korea. Korean J. Syst. Zool. 13: 303-314.

Bae, Y. J. and S. Y. Park. 1998. *Alainites*, *Baetis*, *Labiobaetis*, and *Nigrobaetis* (Ephemeroptera: Baetidae) in Korea. Korean J. Syst. Zool. 14: 1-12.

Bae, Y. J. and T. Soldan. 1997. Mayfly (Ephemeroptera) fauna of North Korea (1). Insecta Koreana 14: 137-152.

Bajkova, O. Ja. 1972. On the mayflies of the Amur basin: I. Imagines (Ephemeroptera: Ephemerellidae). Izv. Tikhookean. Nauchn. Issled Inst. Rybn. Khoz. Okeanogr. 77: 178-206. (in Russian).

Gose, K. 1979-80. The mayflies of Japan. Aquabiol. (Japan) (Reference sources various). (in Japanese).

Gose, K. 1985. Ephemeroptera. In: Kawai, T (ed.) An illustrated book of aquatic insects of Japan, pp. 7-32, Tokai Univ. Press, Tokyo. (in Japanese).

Imanishi, K. 1930-38. Mayflies from Japanese torrents. I-IX. (Reference sources various).

Imanishi, K. 1940. Ephemeroptera of Manchoukuo, Inner Mongolia, and Chosen. pp. 169-263 In: Kawamura, T. (Ed.) Report of the limnological survey of Kwantung and Manchoukuo. Kyoto. (in Japanese).

Kluge, N. Ju. 1983. New and little known mayflies of the Far East SSSR. Gen. *Ecdyonurus* Etn. (Ephemeroptera, Heptageniidae). Ekol. Sist. Presn. Organ. Dal. Vost. pp. 27-36. (in Russian).

Kluge, N. Ju. 1997. Order Ephemeroptera. p. 116-220 In: S. J. Tshalolikhin (ed.) Key to freshwater invertebrates of Russia and adjacent lands. St. Petersburg.

Lee, S. J., Y. J. Bae, I. B. Yoon and N. C. Watanabe. 1999. Comparisons of temperature related life histories in two ephemerid mayflies (*Ephemera separigata* and *E. strigata*: Ephemeridae, Ephemeroptera, Insecta) from a mountain stream in Korea. Korean J. Limnol. (in press).

McLachlan, R. 1875. A sketch of our present knowledge of the neuropterous fauna of Japan (excluding Odonata and Trichoptera). Trans. ent. Soc. London. pp. 167-190.

Park, S. Y., Y. J. Bae and I. B. Yoon. 1996. Revision of the Baetidae (Ephemeroptera) of Korea. (1) Historical review, *Acentrella* Bengtsson and *Baetiella* Ueno. Ent. Res. Bull., KEI, Seoul 22: 55-66.

Takemon, Y. 1990. Timing and synchronicity of the emergence of *Ephemera strigata*. pp. 61-70 In: I. C. Campbell (Ed.) Mayflies and stoneflies. Kluwer Academic Publ.

Tshernova, O. A. 1952. Mayflies (Ephemeroptera) of the Amur river basin and adjacent waters and their role in the nutrition of Amur fishes. Tr. Amurskoy Ichtiol. Ekspeditsii 3: 229-360. (in Russian).

Tshernova, O. A., N. Ju. Kluge, N. D. Sinitshenkova and V. V. Belov. 1986. Order Ephemeroptera. pp. 99-142 In: Ler, P.A. (ed.) Identification of insects of Far East USSR. Vol. 1, Leningrad Press, Leningrad. (in Russian).

Ueno, M. 1928. Some Japanese mayfly nymphs. Mem. Coll. Sci., Kyoto Imp. Univ., Kyoto 4: 19-63.

Watanabe, N. C. 1989. Seasonal and diurnal changes in emergence of *Potamanthodes kamonis* in a stream of central Japan (Ephemeroptera: Potamanthidae). Jap. J. Limnol. 50: 157-161.

Watanabe, N. C., I. Mori and I. Yoshitaka. 1999. Effect of water temperature on the mass emergence of the mayfly, *Ephoron shigae*, in a Japanese river (Ephemeroptera: Polymitarcyidae). Freshw. Biol. 41: 537-541.

Yoon, I. B. and Y. J. Bae. 1988. I. Order Ephemeroptera. pp. 95-184 In: Yoon, I.B. Illustrated encyclopaedia of animals and plants of Korea. Vol. 30. Animals (Aquatic Insects), Ministry of Education, Korea. (in Korean).

Yoon, I. B. 1995. An Illustrated Key Book of Aquatic Insects of Korea. Jung Hang Sa, Seoul. (in Korean).

THE CURRENT STATUS OF EPHEMEROPTERA BIOLOGY IN AUSTRALIA

Ian C. Campbell

Department of Biological Sciences
Monash University, Clayton 3168, Australia

ABSTRACT

Since the most recent comprehensive review of the biology of Australian Ephemeroptera by Peters and Campbell in 1991, knowledge has been rapidly increasing. In formal taxonomy there have been new genera and/or species described within the Caenidae, Leptophlebiidae and Prosopistomatidae, but there has also been substantial progress in informal taxonomy with the publication of several keys for identification. In ecology there have been studies published on egg development and effects of river regulation and there are papers in preparation on mayfly production. Some of the ecological studies have shed light on some apparent conundrums relating to the distribution of some southeastern Australian taxa such as *Coloburiscoides*, and there have been additional useful biogeographical studies on previously little known taxa such as *Ephemerellina*.

INTRODUCTION

There have been several comprehensive reviews of the biology of Australian Ephemeroptera, the earliest being that of Tillyard (1926) and the most recent being those prepared for the two editions of the CSIRO book "The Insects of Australia" (Riek 1970, Peters and Campbell 1991). Two catalogues of Australian Ephemeroptera have also been published (Campbell 1988, Hubbard and Campbell 1996). This paper reviews progress in research on Australian ephemeropteran biology since Peters and Campbell (1991).

TAXONOMIC STUDIES

There has been considerable progress in Australian mayfly taxonomy since 1991. Table 1 documents new taxa described since that time. Within the Leptophlebiidae, which is the largest Australian family, two new genera *Kalbaybaria* and *Tillyardophlebia*, have been described together with 9 new species from those two genera, *Atalomicria* and *Austrophlebioides*.

Trends in Research in Ephemeroptera and Plecoptera
Edited by E. Dominguez, Kluwer Academic/Plenum Publishers, 2001

There have also been significant taxonomic additions to the Australian Caenidae with two new genera and 6 new species described (Suter 1993, 1999), and the only Australian prosopistomatid has been described as a new species (Campbell and Hubbard 1998). In addition there has been an informal key to the genera published by Dean and Suter (1996). Within the Baetidae Lugo-Ortiz and McCafferty (1998a, b, 1999) have described two new genera and two new species, and the first record of the genus *Cloeodes* from Australia. Dean and Suter also noted the presence of *Platybaetis* in northern Australia and suggest that neither true *Baetis* not *Pseudocloeon* actually occur in Australia.

The most bizarre taxonomic addition has been the record by Lugo-Ortiz and McCafferty (1998c) of a species of *Siphlaenigma* (Siphlaenigmatidae) from Australia, a record almost certainly based on an erroneous data label.

Taxonomic piracy is a continuing problem for Australia as it is also for many developing countries. There continue to be frequent cases of foreign taxonomists obtaining small collections of Australian Ephemeroptera and using these to describe new genera and species with types lodged in foreign museums where they are inaccessible to Australian taxonomists. Frequently the published descriptions are poor and incomplete, and usually they are made without reference to the remainder of the Australian fauna. The motivation is apparently solely the acquisition of publications, certainly they make little contribution to the progress of Australian taxonomy since the material requires later redescription by Australian taxonomists before it can be used. Frequently the poor quality of the work requires more than the usual taxonomic work to correct the errors.

ECOLOGICAL STUDIES

It is not possible in the context of this paper to review all the ecological data which has been produced since 1991, because much of that data resides in references to Ephemeroptera in studies of stream invertebrates, often only available as unpublished theses. For the purposes of this paper I have restricted consideration to studies of which Ephemeroptera are the principal subject of the study.

Hearnden and Pearson (1991) investigated habitat partitioning among 13 mayfly species in Yuccabine Creek, a tropical semi-permanent stream in tropical north Queensland. Regrettably the taxonomy used in the paper is poor, they referred to two *Atalophlebioides* species, presumably in error for *Austrophlebioides*, and *Atalonella* presumably in error for either *Nousia, Koorrnonga* or *Neboissophlebia.* The study was carried out using multiple regression of field data consisting of mayfly numbers and a series of environmental variable measured at the sampling site including current velocity, leaf litter dry mass and rock size. They also carried out an experiment using replicate trays with four different substrate types with and without added leaf litter placed in riffles and pools. They found no one controlling factor, but only 2 of the 13 species preferred riffles to pools, and two of the pool species showed a preference for leaf litter presence. Two species, a *Tasmanocoenis* species and an "*Atalonella*" had rather narrow niches as measured by the Simpson-Yule index but the others all demonstrated broad niche preferences. One difficulty with the study was the rather coarse methods used to assess environmental factors. In particular the scale of current measurements was fairly large particularly in the context of the extensive discussions in the literature on possible invertebrate responses to hydrodynamics (e.g. see Davis 1986, Davis and Barmuta 1989).

A further study of tropical Australian mayflies was a description of the life cycles of three *Jappa* species from two streams in far north Queensland (Campbell 1991). Not surprisingly, in view of the high water temperatures, the life cycles were all quite rapid and very aseasonal with emergence, egg hatching and growth throughout the year.

Pardo et al. (1998) investigated the influence of an irrigation reservoir on mayfly assemblages and life histories in two southeastern Australian streams. They found a richer assemblage of 17 species in the unregulated stream, compared with only 11 in the regulated

Table 1. A list of the genera and species of Australian Ephemeroptera described since 1991.

Taxon	Reference
Leptophlebiidae	
Kalbaybaria	Campbell 1993
Kalbaybaria doantrangae	Campbell 1993
Atalomicria banjdjalama	Campbell and Peters 1993
Atalomicria bifasciata	Campbell and Peters 1993
Atalomicria chessmani	Campbell and Peters 1993
Atalomicria dalagara	Campbell and Peters 1993
Austrophlebioides booloumbi	Parnrong and Campbell 1997
Austophlebioides marchanti	Parnrong and Campbell 1997
Tillyardophlebia	Dean 1997
Tillyardophlebia rufosa	Dean 1997
Tillyardophlebia alpina	Dean 1997
Caenidae	
Wundacaenis	Suter 1993
Wundacaenis angulata	Suter 1993
Wundacaenis dostini	Suter 1993
Wundacaenis flabellum	Suter 1993
Irpacaenis	Suter 1999
Irpacaenis deani	Suter 1999
Irpacaenis kaapi	Suter 1999
Irpacaenis coolooli	Suter 1999
Prosopistomatidae	
Prosopistoma pearsonorum	Campbell and Hubbard 1998
Baetidae	
Offadens	Lugo-Ortiz and McCafferty (1998a).
Offadens sobrinus	Lugo-Ortiz and McCafferty (1998a).
Cloeodes fustipalpus	Lugo-Ortiz and McCafferty (1998b).
Cloeodes illiesi	Lugo-Ortiz and McCafferty (1998b).
Edmundsiops	Lugo-Ortiz and McCafferty (1999).
Edmundsiops instigatus	Lugo-Ortiz and McCafferty (1999).
Siphlaenigmatidae	Lugo-Ortiz and McCafferty (1998c).
Siphlaenigma edmundsi	Lugo-Ortiz and McCafferty (1998c).

Mitta Mitta River. The pattern of flow in the regulated stream included a summer release of cold water which resulted in the highest water levels of the year occurring during a normally low water period, and temperatures 5-10°C lower than would normally be expected at that time of year. The two most abundant mayflies – a species they identified as belonging to *Baetis* and a species of *Coloburiscoides*, both declined numerically during the summer water release, and the period of cold water appeared to delay nymphal hatching of *Coloburiscoides*.

An earlier study of egg development of *Coloburiscoides* by Brittain and Campbell (1991) found that egg hatching rates were related to water temperature between 10° and 25°C but that no eggs hatched at temperatures of 5°C or 30°. Some of the streams in which *Coloburiscoides* occurs have temperatures below 5° for a significant part of the year, so this data supported the earlier suggestion by Campbell (1986) that *Coloburiscoides* display an egg diapause in some streams near Mt Kosciusko.

The most substantial, although as yet incomplete, study of an Australian mayfly is that of *Austrophlebioides marchanti* being carried out by Parnrong (pers. comm). She has investigated the egg development time, life cycle and production of the species at four sites, two in forest streams and two in farmland streams. Life cycles of *Austrophlebioides* species had previously been described based purely on field samples (Marchant 1986, Campbell et al.

1990), but the growth is asynchronous making it very difficult to ascertain growth rates or to accurately determine voltinism. As part of her study Parnrong grew nymphs in cages in the field, allowing a much more accurate determination of nymphal growth rate. Previous authors had suggested that *Austrophlebioides* was univoltine (Campbell et al. 1990) or bivoltine (Marchant 1986). Parnrong has demonstrated that *Austrophlebioides marchanti* has 2.5 to 3 generations per year.

The final body of substantial work on Australian Ephemeroptera is the work on trichomycete fungi which occur in the guts of many aquatic insects, and have been collected from the guts of a number of Australian mayflies by Lichtwardt and Williams (1990, 1992a,b,c). The main hosts seem to be baetids, including *Centroptilum* nymphs (Lichtwardt and Williams 1990, 1992a) but some species have also been recorded from the guts of *Nousia* (*Australonousia*) (Leptophlebiidae) (Lichtwardt and Williams 1992b). The ecological significance of Trichomycetes either as parasites or as endosymbionts is unclear.

BIOGEOGRAPHICAL STUDIES

As yet there have been few biogeographic studies completed or published on Australian mayflies. As noted above Brittain and Campbell (1991) suggested that the requirement of *Coloburiscoides* eggs for water temperatures higher than 5°C for development may explain the absence of that genus from Tasmania, and it is likely that the same explanation will be true for *Mirawara* which extends into the tropics on the Australian mainland.

Dean and Cartwright (1992) published survey data from the Pelion Valley from Tasmania which included the first record of the genus *Austrophlebioides* from Tasmania. They collected two species from the genus, both of which had previously been recorded from the mainland of Australia. They also collected the species "*Jappa tristis*" and suggested that it does not in fact belong to *Jappa*, and that true *Jappa* do not appear to occur in Tasmania.

One of the most interesting biogeographical publications is that of Chessman and Boulton (in press), documenting new localities for the sole Australian ephemerellid, *Ephemerellina* (*Austremerella*) *picta* (Riek). This species was previously known from a single locality in southeastern Queensland, and was definitely known not to occur in southeastern Australia, as might be expected for a Gondwanian relict, or in tropical northern Australia (as might be expected for a recent tropical arrival). One possibility was that it was an exotic species introduced by Europeans (Campbell 1990). Chessman and Boulton record the species from 18 new localities in northern New South Wales making it extremely unlikely that it is a recent exotic species, but its geographic restriction and its isolation from related species, which seem to occur in Africa (Allen 1965) still pose a biogeographic conundrum.

CURRENT RESEARCH

Taxonomic studies of Australian Ephemeroptera are continuing. Kyla Finlay, a PhD student at Monash University is working on the taxonomy and biogeography of selected southeastern Australian leptophlebiid genera, and Faye Christidis, a PhD. student at James Cook University in North Queensland is working on the taxonomy of tropical Leptophlebiidae. The National River Health initiative has provided funding to support some graduate student work on formal taxonomy and is also supporting the production of keys and voucher collections.

ACKNOWLEDGMENTS

This paper could not have been written without the assistance of Kyla Finlay and Natalie Lloyd who obtained references and passed them to me while I was based in Thailand.

REFERENCES

Allen, R. K. 1965. A review of the subfamilies of the Epemerellidae (Ephemeroptera). J. Kans. ent. Soc. 38: 262-266.

Brittain, J. E. and I. C. Campbell. 1991 The effect of temperature on egg development of the Australian mayfly genus *Coloburiscoides* (Ephemeroptera: Coloburiscidae) and its relationship to distribution and life history. J. Biogeogr. 18: 231-235.

Campbell, I. C. 1986 Life histories of some Australian siphlonurid and oligoneuriid mayflies (Insecta: Ephemeroptera). Aust. J. Mar. Freshw. Res., 37: 261-288.

Campbell, I. C. 1988 Ephemeroptera. In: Zoological Catalogue of Australia. Volume 6. ABRS, Canberra.

Campbell, I. C. 1990. The Australian mayfly fauna : composition, distribution and convergance. In: I. C. Campbell (ed.). Mayflies and Stoneflies, Life Histories and Biology. Kluwer Academic Press, Dordrecht, Netherlands.

Campbell, I. C. 1991. Size allometry in some Australian mayfly nymphs (Insecta: Ephemeroptera). Aquat. Insects 13: 79-86.

Campbell, I. C. 1993. A new genus and species of leptophlebiid mayfly (Ephemeroptera: Leptophlebiidae: Atalophlebiinae) from tropical Australia. Aquat. Insects 15:159-167.

Campbell, I. C. 1995. The life histories of three tropical species of *Jappa* Harker (Ephemeroptera: Leptophlebiidae) in the Mitchell River system, Queensland, Australia, pp. 197-206. In: J. Ciborowski and L. Corkum (eds.). Current Directions in Research on Ephemeroptera. Canadian Scholars Press, Toronto.

Campbell, I. C., M. J. Duncan and K. M. Swadlin. 1990. Life histories of some Ephemeroptera from Victoria, Australia. In: I. C. Campbell (ed.). Mayflies and Stoneflies, Life Histories and Biology. Kluwer Academic, Dordrecht, Netherlands.

Campbell, I. C. and M. D. Hubbard. 1998. A new species of *Prosopistoma* (Ephemeroptera: Prosopistomatidae) from Australia. Aquat. Insects 20: 141-8.

Campbell, I. C. and W. L. Peters. 1993. A revision of the Australian Ephemeroptera genus *Atalomicria* Harker (Leptophlebiidae: Atalophlebiinae). Aquat. Insects 15: 89 -117.

Campbell, I. C. and P. J. Suter. 1988. Three new genera, a new subgenus and a new species of Leptophlebiidae (Ephemeroptera) from Australia. J. Aust. Ent. Soc. 27: 259-273.

Chessman, B. C. and A. J. Boulton. Occurrence of the mayfly family Ephemerellidae in northern New South Wales. Mar. Freshw. Res. In press.

Davis, J. A. 1986. Boundary layers, flow microenvironments and stream benthos. In: P. de Dekker and W. D. Williams (eds.). Limnology in Australia. Junk, Dordrecht. pp. 291,12.

Davis, J. A. and L. A. Barmuta. 1989. An ecologically useful classification of mean and near-bed flows in streams and rivers. Freshw. Biol. 21: 271-82.

Dean, J. C. 1997. Descriptions of new Leptophlebiidae (Insecta: Ephemeroptera) from Australia. 1. *Tillyardophlebia* Gen. Nov. Mem. Mus. Vict. 56: 83-89.

Dean, J. C. and D. I. Cartwright. 1992. Plecoptera, Ephemeroptera and Trichoptera of the Pelion Valley, Tasmanian World Heritage Area. Occ. Papers Mus. Vict. 5: 73-79.

Dean, J. C. and P. J. Suter. 1996. Mayfly Nymphs of Australia. A guide to genera. Cooperative Research Centre for Freshwater Ecology. Identification Guide Number 7. 76 pp.

Hearnden, M. N. and Pearson, R. G. 1991. Habitat partitioning among the mayfly species (Ephemeroptera) of Yuccabine Creek, a tropical Australian stream. Oecologia 87:91-101.

Hubbard, M. D. and I. C. Campbell. 1996. A checklist of the Australian Ephemeroptera (Mayflies). Australian Society for Limnology, Special Publication.

Lichtwardt, R. W. and M. C. Williams, M. C. 1990. Trichomycete gut fungi in Australian aquatic insect larvae. Can. J. Bot. 68: 1057-1074.

Lichtwardt, R. W. and M. C. Williams. 1992a. Western Australian species of *Smittium* and other trichmycetes in aquatic insect larvae. Mycologia 84: 392-398.

Lichtwardt, R. W. and M. C. Williams. 1992b. Tasmanian trichomycete gut fungi in aquatic insect larvae. Mycologia 84: 384-391.

Lichtwardt, R. W. and M. C. Williams. 1990c. Two new Australasian species of Amoebidiales associated with aquatic insect larvae, and comments on their biogeography. Mycologia 84: 376-383.

Lugo-Ortiz, C. R. and W. P. McCafferty. 1998a. *Offadens*, a new genus of small minnow mayflies (Ephemeroptera: Baetidae) from Australia. Proc. ent. Soc. Wash. 100: 306-309.

Lugo-Ortiz, C. R. and W. P. McCafferty. 1998b. First report and a new species of the genus *Cloeodes* (Ephemeroptera: Baetidae) from Australia. Ent. News 109: 122-128.

Lugo-Ortiz, C. R. and W. P. McCafferty. 1998c. First report of the genus *Siphlaenigma* Penniket and the family Siphlaenigmatidae (Ephemeroptera) from Australia. Proc. ent. Soc. Wash. 100: 209-213.

Lugo-Ortiz, C. R. and W. P. McCafferty. 1999. *Edmundsiops instigatus*: a new genus and species of small minnow mayflies (Ephemeroptera: Baetidae) from Australia. Ent. News 110: 65-69.

Marchant, R. 1986. Estimates of annual production for some aquatic insects from the LaTrobe River, Victoria. Aust. J. Mar. Freshw. Res. 37: 113-120.

Pardo, I., I. C. Campbell and J. E. Brittain. 1998. Influence of dam operation on mayfly assemblage structure and life histories in two southeastern Australian streams. Regulated Rivers, Res Manag. 14: 285-295.

Parnrong, S. and I. C. Campbell. 1997. Two new species of *Austrophlebioides* Campbell and Suter (Ephemeroptera: Leptophlebiidae) with notes on the genus. Aust. J. ent. 36: 121-127.

Peters, W. L. and I. C. Campbell. 1991 Ephemeroptera. In : CSIRO (ed.) The Insects of Australia, A textbook for students and research workers. 2nd Edition. Melbourne University Press, Carlton.

Riek, E. F. 1970. Ephemeroptera. In CSIRO (eds). The Insects of Australia. CSIRO/ Melbourne University Press, Carlton.

Suter, P. J. 1993. *Wundacaenis*, a new genus of Caenidae (Insecta: Ephemeroptera) from Australia. Invert. Taxonomy 7: 787-803.

Suter, P. J. 1999. *Irpacaenis*, a new genus of Caenidae (Insecta: Ephemeroptera) from Australia. Aust. J. ent. 38: 159-167.

Tillyard, R. J. 1926. Order Plectoptera (or Ephemeroptera) (May-flies). In: R. J. Tillyard (ed.). The Insects of Australia and New Zealand. Angus and Robertson, Sydney.

KNOWLEDGE OF THE AFRICAN-MALAGASY MAYFLIES

J. M. Elouard

IRD, BP5045
F-34032 Monpellier cedex 1, France

ABSTRACT

Actually, African Mayflies are depauperate and few diversified (81 genera and 368 species). The study of genus and species numbers described during each decades since the last century till now, associated to the number of authors working on that order and, the setting up of a map of prospected areas, proof that the poverty of Mayfly in Africa is rather due to a poor knowledge than to a real poverty.

INTRODUCTION

The African-Malagasy continent contains about 81 genera and 368 mayfly species. Taking into account the size of the continent and the richness of the biota, one could suppose that either the mayfly fauna is very depauperate or that it is poorly known.

The second proposition, which seems more realistic, leads to the two following questions:

- Are there still undiscovered genera and species?
- Are there many areas of the African continent which have been, for mayflies, investigated poorly or not at all?

Answering these two questions should indicate the degree of our knowledge of the African-Malagasy mayfly fauna.

Are there still undiscovered genera and species?

It is difficult to answer this question directly. By studying the frequency of descriptions of genera and species in the past we might gain insight into the probability of discovering new taxa.

OVERVIEW

Thirteen genera were cited from Africa during the 19th century, but only five of them were new; others were European or cosmopolitan, because the first African species discovered were tied to previously known genera. This assertion is confirmed by the fact that the first species description occurred in 1833 whereas the first true new African genus was described in 1881, that is to say nearly half a century later.

Table 1. Numbers of new genera and species and numbers and percentage of renamed species per decade.

Years	New Genera	New spec.	nb. renamed species	% renamed species
19th cent.	13	15	11	73
1900-09	1	2	2	100
1910-19	6	17	11	65
1920-29	5	17	9	53
1930-39	5	54	25	46
1940-49	2	26	15	58
1950-59	6	28	15	54
1960-69	5	37	6	16
1970-79	8	24	5	21
1980-89	8	33	11	33
1990-98	22	115	14	12
Sum	81	368	124	34

During the 1900-1989 period, 54 new genera were established, which corresponds to a mean of six new genera per decade, with a minimum during the 1900-1909 (1 genus) and 1940-1949 (2 genera) periods. The decades of maximum descriptions were 1910-1919 (7 genera), 1970-1979 (8 genera) and 1980-1989 (8 genera). During the same 1900-1989 period, the number of new species described was around 26 per decade, which corresponds to a mean of two to three per year. Among them, 99/237 (41%) were renamed. This is because of the creation of endemic African genera and the reclassification of known species into them.

During the 1990-1998 period 22 new genera and 115 new species were described. These numbers correspond respectively to 29% of the total known African genera and 31% of the African species. The newly established genera were largely due to the reclassification of previously known species, particularly in the Baetidae family , and partly to the discovery of new taxa, mainly in West Africa and Madagascar. However, in spite of the establishment of numerous genera, only 14% of the 115 new species described during those 8 years were renamed (figs. 1 and 2).

Authors

- During the 19th century, the main descriptions were done by Vayssiere (1890, 1891, 1893, 1895), Eaton (Eaton, 1868, 1871, 1881, 1883-88), McLachlan (1868) and Latreille (1833) who described mayflies from throughout the world.

- During the 1900-1919 period, Ulmer (1916), Esben-Petersen (1913), Navas (1909, 1911, 1912, 1913, 1915 a, b, c) and Eaton (1912, 1913 a, b, c) described almost all species while Navas (1922 a, b, 1926, 1927, 1929) and Lestage (1923 a, b, 1924 a, b) were the main mayfly describers between 1920-1929 (see also Ulmer, 1920).

- Barnard (1932, 1937) did the major part of the work between 1930 and 1939 (see also Ulmer, 1932; Kimmins, 1937; Lestage, 1939) and Crass (1947) and Harrison (1943, 1949 a and b) and Kimmins (1949) between 1940-1949 (see also Barnard, 1940, Lestage, 1945). The descriptions during the fifties were done mainly by Kimmins (1956, 1957), Demoulin (1952, 1955 a and b), and Gillies (1954, 1957)(see also Edmunds, 1953). The describers during the 1960-1969 period were more diversified. Demoulin (1965, 1966 a and b, 1967), Kimmins and Gillies (1960) continued to publish new species but several authors such as Agnew (1961 a, b, 1962), Schoonbe (1968) and the american teams also contributed (Allen and Edmunds, 1963; Peters and Edmunds, 1964, Peters et al., 1964; McCafferty, 1968). Gillies (1977) and Demoulin (1970, 1973) dominated during the 1970-1979 period assisted by Puthz (1971), McCafferty (1971) and Agnew (1973). During the 1980-1989

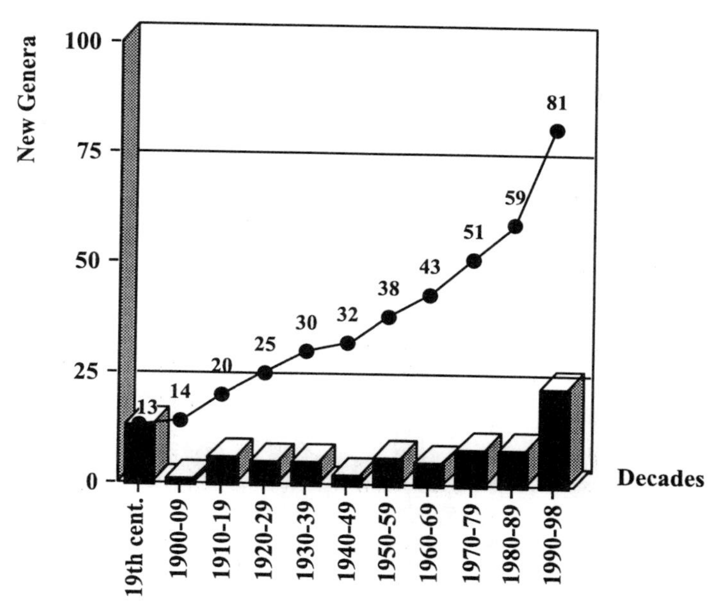

Fig. 1. Number of new genera described per decade and cumulative descriptions.

period, authors were more diversified and although Gillies (1980 a, b, 1982, 1984, 1985) and Kopelke were the main authors, Malzacher (1987) and Elouard (Elouard and Gillies, 1989) made some contributions. During the 1990-1998 period, 115 new species and 22 new genera were described. Descriptions of new genera were done principally by two teams: one from Purdue University: mainly McCafferty, Waltz and Lugo-Ortiz (Waltz and McCafferty, 1994; McCafferty and Wang, 1995; Povonsha and McCafferty, 1995; Edmunds and McCafferty, 1996, Lugo-Ortiz and McCafferty, 1996 a and b, 1997 a, b and c, 1998; Barber-James and McCafferty, 1997; McCafferty et al., 1997; Lugo-Ortiz et al. 1999) and the Orstom team associated with the Musée Zoologique de Lausanne mainly: Elouard, Wuillot, Oliarinony, Sartori, Gattolliat (Elouard et al. 1990; Elouard and Oliarinony, 1997; Elouard and Sartori, 1997; Oliarinony and Elouard, 1997; Gattolliat et al., 1999. The descriptions of most new species were done by researchers of these two teams, but three independent researchers, Gillies, Malzacher and Wuillot, also contributed greatly (Gillies, 1990; 1991, Gillies and Elouard, 1990; Gillies et al., 1990; Malzacher, 1990, 1993, 1995; Wuillot and Gillies, 1993 a, b).

Two remarks could be made:

- the more people (who) work on mayflies, the more new taxa are discovered and more taxonomic problems are solved;

- the abundance of descriptions during the 1980-1998 period point out that the African mayfly fauna was previously very poorly known.

Are mayflies well investigated in Africa and Madagascar?

We have outlined on the map (fig. 3), in dark gray those countries where the mayfly fauna is well known and in light gray countries where the mayfly fauna is only partly known. Outside of these areas, there still exist some places were a few mayflies have been recorded, but the mayfly fauna for these countries is for the large part unknown. One can see that less than one third of Africa has been really prospected.

Some areas, such as the great lakes landscape (east Zaïre, Rwanda, Burundi, Uganda), present an abundant variety of aquatic biota but still very few mayfly taxa are known from those areas.

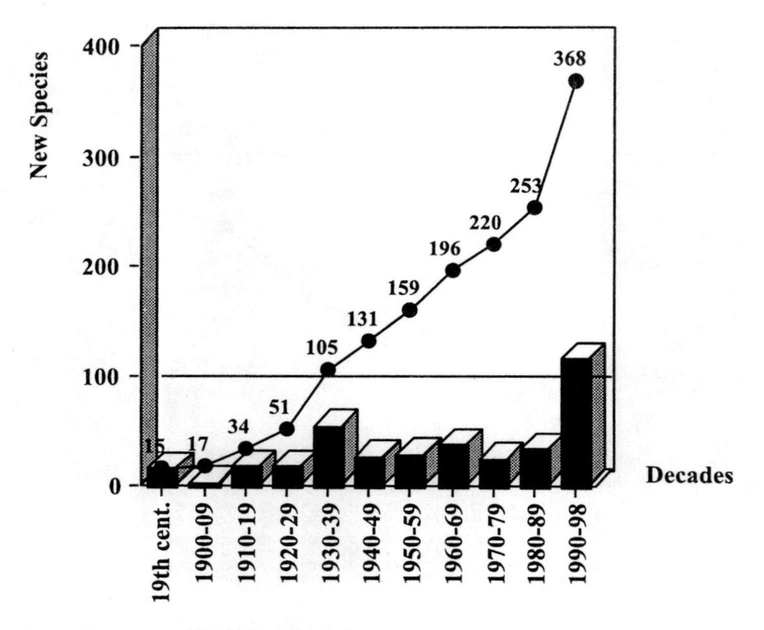

Fig. 2. Number of new species described per decade and cumulative descriptions.

Some other areas, such as West Africa and Madagascar, have been well investigated by the ORSTOM team but the majority of taxa are not yet described in the literature. We estimate that more than 250 new species are recorded and still unpublished.

To this lack of descriptions is added the lack of knowledge of the distribution of taxa. Most are known only from the type locality or for two or three stations. The South Malagasy and South African mayflies are exceptions to this situation.

CONCLUSION

The fact that well investigated areas contain numerous new genera and species indicates that the general African mayfly fauna is not depauperate but rather largely unknown.

The lack of knowledge concerns not only the cryptic or microendemic species but also large or medium-sized species, often with a more or less wide distribution. The discovery in Madagascar of two *Eatonica* and seven *Proboscidoplocia* illustrates this.

One can also note that for the majority of African taxa, only one ecophase is known, nymph or adult, and sometimes only one instar, often multiplying the number of described species (i.e.: *Machadorythus maculatus*).

In conclusion we think that the African-Malagasy mayfly fauna is largely unknown even though much progress has been made during the last two decades. To alleviate this situation, we think it is necessary to:

- intensify efforts to describe previously recorded mayflies;
- sample seriously the uninvestigated areas;
- generate distribution maps from systematic records;
- and ultimately increase rearing efforts in order to establish correspondence between nymphs and imagoes.

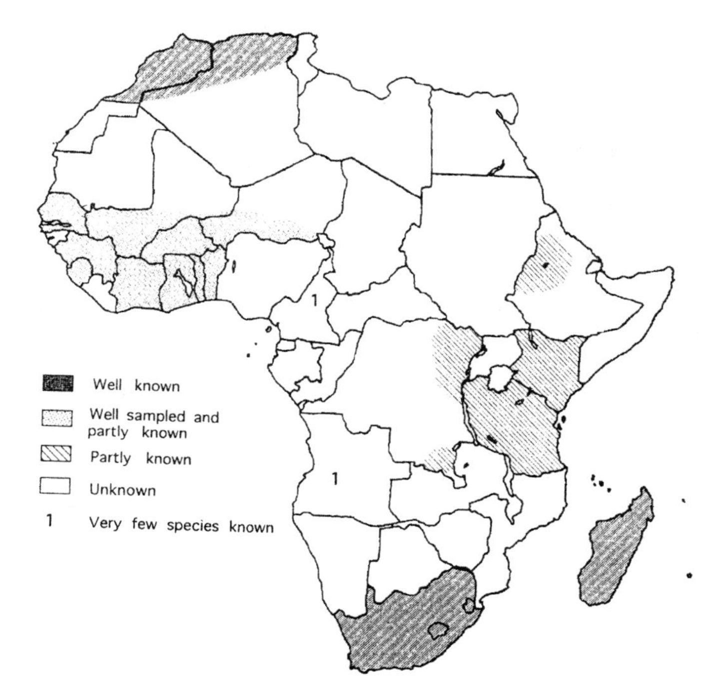

Fig. 3. Actual status of the knowledge on Ephemeroptera in Africa.

REFERENCES

Agnew, J. D. 1961a. New Transvaal Leptophlebiidae (Ephem). Novos Taxa ent., 26: 3-9.

Agnew, J. D. 1961b. New Baetidae (Ephem.) from South Africa. Novos Taxa ent., 25: 3-18.

Agnew, J. D. 1962. New Leptophlebiidae (Ephem.) from the Transvaal. Arch. Hydrobiol., 58 (3): 358-366.

Agnew, J. D. 1973. Two new species of *Oligoneuriopsis* Crass from the republic of South Africa (Oligoneuriidae: Ephemeroptera). Proc. of the First Int. Conf. on Ephemeroptera. W. L. Peters and J. G. Peters (eds.). E. J. Brill, Leiden., 114-121.

Allen, R. K. and G. F. Edmunds. 1963. New and little known Ephemerellidae from Southern Asia, Africa and Madagascar. Pac. Insects, 5 (1): 11-22.

Barber-James, H. M. and W. P. McCafferty. 1997. Review and a new species of the African genus *Acanthiops* (Ephemeroptera: Baetidae). Ann. Limnol., 33 (2): 85-92.

Barnard, K. H. 1932. South African may-flies (Ephemeroptera). Trans. R. Soc. S. Afr., 20 (3): 201-259.

Barnard, K. H. 1937. A new may-fly from Natal (Ephemeroptera). Ann. Natal Mus., 8 (2): 275-278.

Barnard, K. H. 1940. Additionnal Records, and descriptions of new species of South African Alder-flies (Megaloptera), May-flies (Ephemeroptera), Caddis-flies (Trichoptera), Stone-flies (Perlaria) and Dragon-flies (Odonata). Annls. S. Afr. Mus., 32 (6): 609-661.

Crass, R. S. 1947. The may-flies (Ephemeroptera) of Natal and the Eastern Cape. Ann. Natal Mus., 11 (1): 37-110.

Demoulin, G. 1952. Contribution à l'étude des Ephoronidae Euthyplociinae (Insectes Ephéméroptères). Bull. Inst. r. Sci. nat. Belg., 28 (45): 1-22.

Demoulin, G. 1955a. *Atalophlebiodes inequalis* sp. nov., Epheméroptères Leptophlebiidae nouveau de Madagascar. Bull. Inst. r. Sci. nat. Belg., 31 (15): 1-4.

Demoulin, G. 1955b. *Afromera* gen. nov., Ephemeridae de la faune éthiopienne (Ephemeroptera). Bull. Ann. Soc. r. ent. Belg., 91 (11-12): 291-295.

Demoulin, G. 1965. Contribution à l'étude des Palingeniidae (Insecta, Ephéméroptères). Nova Guinea, Zool., 33: 305-344.

Demoulin, G. 1966a. Contribution à l'étude des Euthyplociidae. (Ephemeroptera). IV - Un nouveau genre de Madagascar. Ann. Soc. ent. Fr. (N.S.), 2 (4): 941-949.

Demoulin, G. 1966b. Quelques Ephéméroptères nouveaux de Madagascar. Ann. Soc. Ent. Fr. (N.S.), 2 (3): 711-717.

Demoulin, G. 1967. Description de deux larves atypiques de Baetidae (Ins. Ephemeroptera). Bull. Ann. Soc. r. ent. Belg., 103: 226-233.

Demoulin, G. 1970. Ephemeroptera des faunes éthiopienne et malgache. S. Afr. anim. Life, 14: 24-170.

Demoulin, G. 1973. Ephéméroptères de Madagascar. III. Bull. Inst. r. Sci. nat. Belg., 49 (7): 1-20.

Eaton, A. E. 1868. An outline of re-arrangement of the genera of Ephemeridae. Ent. mon. Mag., 5: 82-91.

Eaton, A. E. 1871. A monograph of the Ephemeridae. Trans. ent. Soc. London., 1: 1-164.

Eaton, A. E. 1881. An announcement of new genera of Ephemeridae. Ent. mon. Mag., 17: 191-197.

Eaton, A. E. 1883-1888. A revisional monograph of recent Ephemeridae or mayflies. Trans. Linn. Soc. London, 2nd Ser. Zool., 3: 1-352.

Eaton, A. E. 1912. On a new species of *Oligoneuria* (Ephemeridae) from British East Africa. Ann. Mag. nat. Hist., 8 (10): 243-244.

Eaton, A. E. 1913a. Ephemeridae from tropical Africa. Ann. Mag. nat. Hist., 8 (12): 271-278.

Eaton, A. E. 1913b. The Percy Sladen Trust Expedition to the Indian Ocean in 1905. IV. 26. Ephemeridae. Trans. Linn. Soc. London, 2nd Ser. Zoology, 15: 433-434.

Eaton, A. E. 1913c. Ephemeridae of the Seychelles. Trans. Linn. Soc. London, 2nd Ser. Zoology, 15: 433.

Edmunds, G. F. 1953. Taxonomic notes on the genus Adenophlebiodes Ulmer (Ephemeroptera: Leptophlebiidae). Revue Zool. Bot. afr., 43: 1-2.

Edmunds, G. F. and W. P. McCafferty. 1996. New field observations on burrowing in Ephemeroptera from around the world. Ent. News, 107 (2): 68-76.

Elouard, J. M. 1986. Ephémères d'Afrique de l'Ouest. Le genre *Eatonica* (Ephemeridae). Rev. Hydrobiol. Trop., 19 (2): 87-92.

Elouard, P. and Elouard J. M. 1991. Mayflies of West Africa: the adults of the subgenus *Adenophlebiodes s.s.* (Ephemeroptera: Leptophlebiidae). Aquat. Insects., 13 (1): 1-18.

Elouard, J. M. and M. T. Gillies. 1989. West African Ephemeroptera. The genus *Machadorythus* (Tricorythidae). Aquat. Insects., 10 (3): 1-10.

Elouard, J. M., M. T. Gillies and J. Wuillot. 1990. Ephemeroptera from West Africa: The genus *Pseudopannota* (Baetidae). Rev. Hydrobiol. Trop., 23 (1): 27-39.

Elouard, J. M. and R. Oliarinony. 1997. Biodiversité aquatique de Madagascar. 6- *Madecassorythus* un nouveau genre de Tricorythidae définissant la nouvelle sous-famille des Madecassorythinae (Ephemeroptera, Pannota). Bull. Soc. ent. Fr., 102 (3): 225-232.

Elouard, J. M. and M. Sartori. 1997. Aquatic biodiversity of Madagascar, *Proboscidoplocia* a singular plural (Ephemeroptera: Polymitarcyidae: Euthiplociinae), pp. 439-447. In: P. Landolt and M. Sartori (eds.). Ephemeroptera and Plecoptera - Biology, Ecology, Systematics. MTL-MAURON-TINGUEM and LACHAT S.A., Fribourg.

Esben-Petersen, P. 1913. Ephemeridae from South Africa. Annls S. Afr. Mus., 10: 177-187.

Gattolliat, J. L., M. Sartori and J. M. Elouard. 1999. Three new species of Baetidae (Ephemeroptera) from the Réserve Naturelle Intégrale d'Andohahela, Madagascar. Fieldiana, 94 (7): 115-124.

Gillies M.T. 1954. The adult stages of *Prosopistoma* Latreille (Ephemeroptera) with descriptions of two new species from Africa. Trans. R. Ent. Soc. London., 105 (15): 355-372.

Gillies M.T. 1957. New records and species of *Euthraulus* Barnard (Ephemeroptera) from East Africa and the Oriental region. Proc. R. ent. Soc. Lond., 26 B: 43-48.

Gillies M.T. 1960. A new genus of Tricorythidae (Ephemeroptera) from East Africa. Proc. R. ent. Soc. Lond., 29 (3-4) B: 35-40.

Gillies M.T. 1977. A new genus of Caenidae (Ephemeroptera) from East Africa. J. nat. Hist. (G.B.), 11: 451-455.

Gillies M.T. 1980a. An introduction to the study of *Cloëon* Leach (Baetidae, Ephemeroptera) in West Africa. Bull. I.F.A.N., 42 (A, 1): 135-156.

Gillies M.T. 1980b. The African Euthyplociidae (Ephemeroptera) (Exeuthyplociinae subfam. n.). Aquat. Insects., 2 (4): 217-224.

Gillies M.T. 1982. A second large-eyed genus of Caenidae (Ephemeroptera) from Africa. J. nat. Hist., 16: 15-22.

Gillies M.T. 1984. On the synonymy of *Notonurus* Crass with *Compsoneuriella* Ulmer (Heptageniidae). Proc. IVth Intern. Confer. Ephemeroptera. V. Landa *et al.* (eds.), 21-25.

Gillies M.T. 1985. A preliminary account of the est african species of *Cloeon* Leach and *Rhithrocloeon* gen. n. (Ephemeroptera). Aquat. Insects, 7 (1): 1-17.

Gillies M.T. 1990. A revision of the African species of *Centroptilum* Eaton (Baetidae, Ephemeroptera). Aquat. Insects, 12 (2): 97-128.

Gillies M.T. 1991. A new species of *Afroptilum (Afroptiloides)* from East Africa (Ephem., Baetidae). Ent. mon. Mag., 127: 109-116.

Gillies M.T. and J. M. Elouard. 1990. The Mayfly mussel association, a new example from the river Niger basin, pp. 289-297. In: I. C. Campbell (ed.). Mayflies and stoneflies: life histories and biology. Kluwer Acad Publ., (London).

Gillies, M. T., J. M. Elouard and J. Wuillot. 1990. Ephemeroptera from West Africa: the genus *Ophelmatostoma* (Baetidae). Rev. Hydrobiol. Trop., 23 (2): 115-120.

Harrison A. C. 1943. Cape May-flies. I. The family Leptophlebiidae. Piscator, 2 (49): 117-120.

Harrison A. C. 1949a. Cape may-flies. Part II. Piscator, 2 (49): 48-53.

Harrison A. C. 1949b. Cape may-flies. III. The family Leptophlebiidae. Piscator, 2 (49): 83-87.

Kimmins, D. E.. 1937. Some new Ephemeroptera. Ann. Mag. nat. Hist., (10) 19: 430-440.

Kimmins, D. E., 1949. Ephemeroptera from Nyasaland, with descriptions of new species. Ann. Mag. nat. Hist., 12 (1): 825-836.

Kimmins, D. E., 1956. New species of Ephemeroptera from Uganda. Bull. br. Mus. (nat. Hist.), 4: 71-87 .

Kimmins, D. E. 1957. New species of the genus *Dicercomyzon* Demoulin (Ephemeroptera, fam. Tricorythidae). Bull. br. Mus. (nat. Hist.), 6 (5): 129-136.

Latreille, P. A. 1833. Description d'un nouveau genre de Crustacés. Ann. Mus. Hist. nat. Paris, 3e Sér., 2: 23-34.

Lestage, J. A. 1923a. Etude des Ephémères du Congo belge. I. Notes sur *Eatonica schoutedeni* Nav. Revue Zool. Bot. Afr., 11 (3): 301-307.

Lestage, J. A. 1923b. Les *Cloëon* africains (Ephémères) et description d'une espèce nouvelle du Congo Belge. Rev. Zool. Bot. afr., 11 (2): 192-195.

Lestage, J. A. 1924a. Etudes sur les Ephémères du Congo Belge. II. Un nouveau *Cloëon* du Katanga (*Cloëon smaeleni* n. sp.). Revue Zool. Bot. afr., 12 (4): 426-428.

Lestage, J. A. 1924b - Les Ephémères de l'Afrique du Sud. Catalogue critique et systématique des espèces connues et description de trois genres nouveaux et de sept espèces nouvelles. Revue Zool. Bot. afr., 12 (3): 316-352.

Lestage, J. A. 1939 - Contribution à l'étude des Ephéméroptères. XXIII. Les Polymitarcyidae de la faune africaine et description d'un genre nouveau du Natal. Bull. Ann. Soc. r. ent. Belg., 79: 135-138.

Lestage, J. A. 1945. Contribution à l'étude des Ephéméroptères. XXVI. Étude critique de quelques genres de la faune éthiopienne. Bull. Ann. Soc. Roy. Ent. Belg., 81: 81-89.

Lugo-Ortiz, C. R. and W. P. McCafferty. 1996a. *Crassabwa*: a new genus of small minnov mayflies (Ephemeroptera: Baetidae) from Africa. Ann. Limnol., 32 (4): 235-240.

Lugo-Ortiz, C. R. and W. P. McCafferty. 1996b. The *Bugilliesia* complex of African Baetidae (Ephemeroptera). Trans. Amer. ent . Soc., 122 (4): 175-197.

Lugo-Ortiz, C. R. and W. P. McCafferty. 1997a. A new genus and redescriptions for African species previously placed in *Acentrella* (Ephemeroptera: Baetidae). Proc. ent. Soc. Wash., 99 (3): 429-439.

Lugo-Ortiz, C. R. and W. P. McCafferty. 1997b. Contribution to the systematics of the genus *Cheleocloeon* (Ephemeroptera: Baetidae). Ent. News, 108 (4): 283-289.

Lugo-Ortiz, C. R. and W. P. McCafferty. 1997c. *Maliqua*: a new genus of Baetidae (Ephemeroptera) for a species previously assigned to *Afroptilum*. ent. News, 108 (5): 367-371.

Lugo-Ortiz, C. R. and W. P. McCafferty. 1998. The *Centroptiloides* complex of Afrotropical small minnow mayflies (Ephemeroptera: Baetidae). Ann. ent. Soc. Amer., 91 (1): 1-26.

Lugo-Ortiz, C. R., W. P. McCafferty. and J. L. Gattolliat. 1999. The small minnow mayfly genus *Cloeodes* Traver (Ephemeroptera: Baetidae) in Madagascar. Proc. ent. Soc. Wash., 101 (1): 208-211.

Malzacher, P. 1987. Eine neue Caeniden-Gattung *Afrocercus* gen. nov. und Bemerkungen zu *Tasmanocaenis tillyardi* (Insecta: Ephemeroptera). Stuttg. Beitr. Naturk., A (407): 1-10.

Malzacher, P. 1990. Caenidae der äthiopsichen Region (Insecta: Ephemeroptera). Teil 1. Beschreibung neuer Arten. Stuttg. Beitr. Naturk., ser. A, 454 (28): 1-28.

Malzacher, P. 1993. Caenidae der äthiopischen region (Insecta: Ephemeroptera). Teil 2. Systematische Zusammenstellung aller bisher bekannten Arten. Mitt. Schweiz. Ent. Ges., 66: 379-416.

Malzacher, P. 1995. Caenidae from Madagascar (Insecta, Ephemeroptera). Stuttg. Beitr. Naturk., n° 530: 1-12.

McCafferty, W. P. 1968. A new genus and species of Ephemeridae (Ephemeroptera) from Madagascar. Ent. Records, 80: 293.

McCafferty, W. P. 1971. New burrowing mayflies from Africa (Ephemeroptera: Ephemeridae). J. ent. Soc. S. Afr., 34 (1): 57-62.

McCafferty, W. P., C. R. Lugo-Ortiz and H. M. Barber-James. 1997. *Micksiops*, a new genus of small minnow mayflies (Ephemeroptera: Baetidae) from Africa. Ent. News, 108 (5): 363-366.

McCafferty, W. P. and T. Q. Wang. 1995. A new genus and species of Tricorythidae (Ephemeroptera: Pannota) from Madagascar. Ann. Limnol., 31 (3): 179-183.

McLachlan, R. 1868. On a new species belonging to the ephemerideous genus Oligoneuria (*O. trimeniana*). Ent. mon. Mag., 4: 177-178.

Navas, L. 1909. Neurôpteres de Zambo (Africa Orientale, Mozambique). Broteria, Zool., 8: 106.

Navas, L. 1911. Deux Ephemeridae (Ins-Neur.) nouveaux du Congo Belge. Annls. Soc. Sci., 35: 221-224.
Navas, L. 1912. Notes sur quelques Névroptères d'Afrique. Revue Zool. Bot. afr., 1: 401-410.
Navas, L. 1913. Algunos órganos de las alas de los insectos. Trans. 2nd. Int. Congr. Ent., Oxford, 2: 178-186.
Navas, L. 1915a. Notes sur quelques Névroptères du Congo Belge. III. Revue Zool. afr., 4: 172-182.
Navas, L. 1915b. Neuropteros nuevos o poco Conocidos (4a serie). Mems R. Acad. Cienc. Artes. Barcelona, 11 (3): 373-398.
Navas, L. 1915c. Neuroptera nova Africana. VI Series. Mém. Acad. pont. Nuovi Lincei, 33: 30-39.
Navas, L. 1922a - Insecto de Fernando Poo. Treb. Mus. Ci. Nat. Barcelone, 4: 109-116.
Navas, L. 1922b. Voyage de M. le Baron Maurice de Rotschild en Ethiopie et en Afrique Orientale Anglaise (1904-1905) Nevroptères. Publication MNHN, Paris, Imprimerie Nationale, 329-332.
Navas, L. 1926. Algunos Insectos del Museo de Paris. Broteria, Zool., 23 (3): 95-115.
Navas, L. 1927. Insectos de la Somalia Italiana. Mem. Soc. ent. Ital., 6: 85-93.
Navas, L. 1929. Insectes du Congo Belge. Série III. Revue Zool. Bot. afr., 18: 1-21.
Oliarinony, R. and J. M. Elouard. 1997. Biodiversité aquatique de Madagascar. 7. *Ranorythus* un nouveau genre de Tricorythidae définissant la nouvelle sous-famille des Ranorythinae (Ephemeroptera Pannota). Bull. Soc. ent. Fr., 102 (5): 439-447.
Peters, W. L. and G. F. Edmunds. 1964. A revision of the generic classification of the Ethiopian Leptophlebiidae (Ephemeridae). Trans. R. ent. Soc. London., 116 (10): 225-253.
Peters, W. L., M. T. Gillies and G. F. Edmunds. 1964. Two new genera of mayflies from the Ethiopian and Oriental region (Ephemeroptera: Leptophlebiidae). Proc. R. ent. Soc. Lond., 33 (7-8) B: 117-124.
Puthz, V. 1971. Über zwei *Afronurus*-Arten von Mount Elgon (Insecta, Ephemeroptera). Ent. Ts. Arg., 92 (3-4): 178-182.
Provonsha, A. V. and W. P. McCafferty. 1995. New brushlegged caenid mayflies from South Africa (Ephemeroptera: Caenidae). Aquat. Insects., 17 (4): 241-251.
Schoonbee, H. J. 1968. A revision of the genus *Afronurus* Lestage (Ephemeroptera: Heptageniidae) in south Africa. Mem. ent. Soc. S. Afr., 10: 1-60.
Ulmer, G. 1932. Bemerkungen über die seit 1920 neu aufgestellten Gattungen der Ephemeropteren. Stettin. ent. Ztg., 93: 204-219.
Ulmer, G. 1916. Ephemeropteren von aequatorial-Afrika. Arch. Naturgesch., 81 (A), 7 (1915): 1-19.
Ulmer, G. 1920. Neue Ephemeropteren. Arch. Naturgesch., 85 (A), 11 (1919): 1-80.
Vayssiere, M. A. 1890. Sur le *Prosopistoma variegatum* de Madagascar. C. r. Acad. Sci. Paris, 110: 95-96.
Vayssiere, M. A. 1891. Observations sur l'*Euthyplocia sikoraï*, type d'Ephéméridé de grande taille, provenant de l'île de Madagascar. C.r. Ass. fr. Avanc. Sci., 20: 243.
Vayssiere, M. A. 1893. Note sur l'existence au Sénégal d'une espèce nouvelle de *Prosopistoma*. Ann. Sci. Nat., Zool., 15 (7): 337-342.
Vayssiere, M. A. 1895. Description zoologique de l'*Euthyplocia Sikoraï*, nouvelle espèce d'Ephéméridé de Madagascar. Ann. Soc. ent. Fr. (N.S.), 64: 297-306.
Waltz, R. D. and W. P. McCafferty. 1994. *Cloeodes* (Ephemeroptera: Baetidae) in Africa. Aquatic Insects., 16 (3): 165-169.
Wuillot, J. and M. T. Gillies. 1993a. New species of *Afroptilum* (Baetidae, Ephemeroptera) from West Africa. Rev. Hydrobiol. Trop., 26 (4): 269-277.
Wuillot, J. and M. T. Gillies. 1993b. *Cheleocloeon*, a new genus of Baetidae (Ephemeroptera) from West Africa. Rev. Hydrobiol. Trop., 26 (3): 213-217.

THE GENTLE QUEST: 200 YEARS IN SEARCH OF NORTH AMERICAN MAYFLIES

W. P. McCafferty

Dep. Entomology, Purdue University
West Lafayette, IN 47907, USA

ABSTRACT

The first 200 years of mayfly faunistics in North America are reviewed. After meager beginnings for most of the Nineteenth Century, a burst of descriptive activity occurred in the 1920's and 1930's, the Golden Age of ephemeropterology, when over half of the currently known valid species were described, mainly by John McDunnough and Jay Traver. Thirteen of the 77 authors of North American mayflies species in the past 200 years have contributed over 80% of the known fauna, and represent all time periods and regions. The North American fauna currently consists of close to 700 species, with only moderate increases expected in the next century. Growth of the known fauna over time, contributions by decade, changes in family and generic classification, regional inventories, pertinent revisions, and major researchers are reviewed. Current status and recommendations for future research are also outlined.

OVERVIEW

As the Twentieth Century comes to a close, it is most befitting to review the historical development of our knowledge of the Ephemeroptera fauna of North America. I refer to the activity of discovering and describing mayflies, and refining our knowledge of them, as the "gentle quest" because of the dedication and labor of love that workers have long demonstrated in their searches for these delicate, exquisite, and challenging aquatic insects.

Because of the relative recent colonization of North America (N. A. hereafter) by Europeans and the subsequent scientific explorations of its vast territories and environments, it was not until the Nineteenth Century (and by far the latter half) that naturalists began to make significant inroads in describing mayflies from the "New World." Initially, small collections from various expeditions in N. A. were being sent to European museums and specialists, with only a few early pioneering American and Canadian naturalists able to study and describe mayflies. It was in the Twentieth Century that home-grown workers took over the task of describing the fauna and that a comprehensive knowledge of its diversity and distribution

Trends in Research in Ephemeroptera and Plecoptera
Edited by E. Dominguez, Kluwer Academic/Plenum Publishers, 2001

was acquired. In fact, by the end of the Twentieth Century, the major vast diversity of N. A. mayflies will have been discovered.

In the future, remaining voids in N. A. faunistics (especially species ranges) will be filled at an accelerated rate and necessary systematic revisions will continue to be performed. The following historical review of ephemeropterology in N. A. will hopefully provide both perspective and direction for the Ephemeroptera workers of the Twenty-First Century. Clearly, in the Twenty-First Century the outstanding priority of mayfly faunistics and biodiversity will be its dynamic synthesis, including the electronic management and dissemination of data for not only the purposes of deducing the evolutionary history of the fauna and its many ramifications, but predicting the future of the fauna in the face of its special ecological requirements and any impending or potential natural or unnatural alterations to the environment. The rudiments of such efforts are already present in these waning years of the Twentieth Century.

Considerable data are referenced in this historical review, but the numerous particulars concerning the nomenclature of the Mayflies of N. A. that cannot be given in the short space allotted here may be found at *Mayfly Central* at <http://www.entm.purdue.edu/entomology/research/mayfly/mayfly.html>. Descriptive Ephemeroptera literature not cited herein may be found in Needham et al. (1935), McCafferty and Randolph (1998), and revisionary and faunistic works that are referenced herein. Species counts are given principally for their comparative value, considering that exact counts of valid species are subject to revision and thus transitory. Unpublished reports or theses are not referenced or accounted for in the data sets.

The Nineteenth Century

No N. A. mayflies were to be found in the 1758 edition of Linnaeus' «Systema Naturae.» The first recognizable mayfly species known in N. A. was *Ephoron leukon* Williamson, described from New Jersey in 1802. *Siphlonurus noveboracana* (Lichtenstein), which was described from New York in 1796, represents the only description of a mayfly originally taken in N. A. previous to the Nineteenth Century; however, it remains an unresolved nomen dubium. In all, 96 valid species were described from N. A. in the Nineteenth Century.

European specialists who described species in the Nineteenth Century from N. A. included foremost the Reverend Alfred E. Eaton. He authored 42 valid N. A. species from 1871 to 1892, including the renaming of six species. Two of the 42 Eaton species were originally described from elsewhere and only subsequently discovered in N. A. by other workers. Eaton's (1883-88) world monograph was seminal to much of the N. A. research well into the Twentieth Century.

Early Nineteenth Century N. A. descriptions by Europeans included an obscure species described in H. C. C. Burmeister's 1832 "Handbuch der Entomologie." Also, however, Jean Serville, in Guérin-Méneville's 1829 "Iconographie du Règne Animal de C. Cuvier . . ." described what today is arguably the best known species in N. A., the common burrower *Hexagenia limbata* (Serville), often called the Giant Michigan Mayfly. François Pictet had access to certain N. A. materials, and three of his 1843 N. A. species remain valid, although *Stenonema flaveolum* (Pictet) must be regarded as a nomen dubium (McCafferty and Bae, 1992). In 1853, Francis Walker (who in his lifetime proposed a record ca. 20,000 insect names) described 17 Canadian mayflies housed in the British Museum (only seven remain valid). While still in Prussia, Herman Hagen described seven valid new species from N. A. Four homonymic Hagen species were later renamed by Eaton. *Ephemera compar* Hagen, described in 1875 from the Colorado Territory (see Edmunds and McCafferty, 1984) after Hagen moved to America, is now considered extinct (McCafferty, 1996a).

Of the N. A. naturalists who contributed to the faunistics of mayflies in the Nineteenth Century, Williamson's single pioneering contribution has already been mentioned. The famous American naturalist Thomas Say made significant early contributions in the 1820's and 30's with 10 valid species. The less famous, but important Benjamin Walsh described 15 currently valid species from Illinois and added significantly to generic classification. The

Table 1. Notable reviews and revisions since Edmunds et al. (1976). See References for complete citations of works. Numbers of new species, new synonyms, and new combinations are applicable to North American species only. Some earlier reviews are cited in the narrative.

Taxon	Reference	n. sp.	n. syn.	n. comb.
Ameletus	Zloty (1996)	7	9	0
Acerpenna s.l.	Lugo-Ortiz & McCafferty (1994)	2	0	0
Americabaetis	Lugo-Ortiz & McCafferty (1996a)	0	0	2
Ametropus	Allen & Edmunds (1976)	1	0	0
Anthopotamus	Bae & McCafferty (1991)	0	4	0
Baetis s.l.	Morihara & McCafferty (1979a)	0	13	0
	Waltz & McCafferty (1987a)	0	0	9
	McCafferty & Waltz (1990)	2	3	8
	Lugo-Ortiz & McCafferty (1998)	0	0	10
Baetisca	Pescador & Berner (1981)	0	2	0
Barbaetis s.l.	Lugo-Ortiz & McCafferty (1998)	0	0	1
Brachycercus	Soldán (1986)	5	0	0
Caenis	Provonsha (1990)	1	5	0
Callibaetis	McCafferty & Waltz (1990)	0	10	0
	Lugo-Ortiz & McCafferty (1996b)	1	1	0
Camelobaetidius	Lugo-Ortiz & McCafferty (1995c)	1	5	0
Centroptilum s.l.	McCafferty & Waltz (1990)	0	0	19
Choroterpes	Burian (1995)	0	3	0
Cloeodes	Waltz & McCafferty (1987b)	3	0	0
Cloeon s.l.	McCafferty & Waltz (1990)	0	0	11
Ephemerella s.l.	McCafferty (1977)	0	0	1
	Allen (1980)	0	0	55
Eurylophella	Funk & Sweeney (1994)	3	0	0
	McCafferty & Wang (1994)	0	0	1
Fallceon	Lugo-Ortiz, McCafferty & Waltz (1994)	0	0	1
Farrodes	Domínguez, Molineri & Peters (1996)	1	0	0
Heptagenia s.l.	Flowers (1980)	0	0	20
Heterocloeon	Morihara & McCafferty (1979b)	0	0	1
Homoeoneuria	Pescador & Peters (1980)	2	0	0
Labiobaetis	Morihara & McCafferty (1979c)	1	3	0
Isonychia	Kondratieff & Voshell (1984)	2	11	0
Macdunnoa	Flowers (1982)	1	0	2
Metretopus	Berner (1978)	0	0	0
Moribaetis	Waltz & McCafferty (1985)	1	0	1
Neoephemera	Bae & McCafferty (1998)	0	0	0
Neochoroterpes	Henry (1993)	1	2	3
Pseudiron	Pescador (1985)	0	1	0
Pseudocloeon s.l.	McCafferty & Waltz (1990)	0	3	17
Siphloplecton	Berner (1978)	3	1	0
Stenonema	Bednarik & McCafferty (1979)	0	9	0
Timpanoga	McCafferty & Wang (1994)	0	1	3

demise of Walsh's types in the great Chicago fire of 1871 and subsequent actions to validate the species were discussed by Burks (1953). The considerable descriptive work in the 1860's depicted in Fig. 1 is attributable to Walsh along with Hagen. The French Canadian Léon Provancher studied mayflies from Quebec, which he referred to as the "Petit Faune." The three species he described in 1876 and 1878 are still recognized, although one is dubious. In the final year of the Nineteenth Century, the American Nathan Banks (1900) added four species in the first of his several works that would include mayflies.

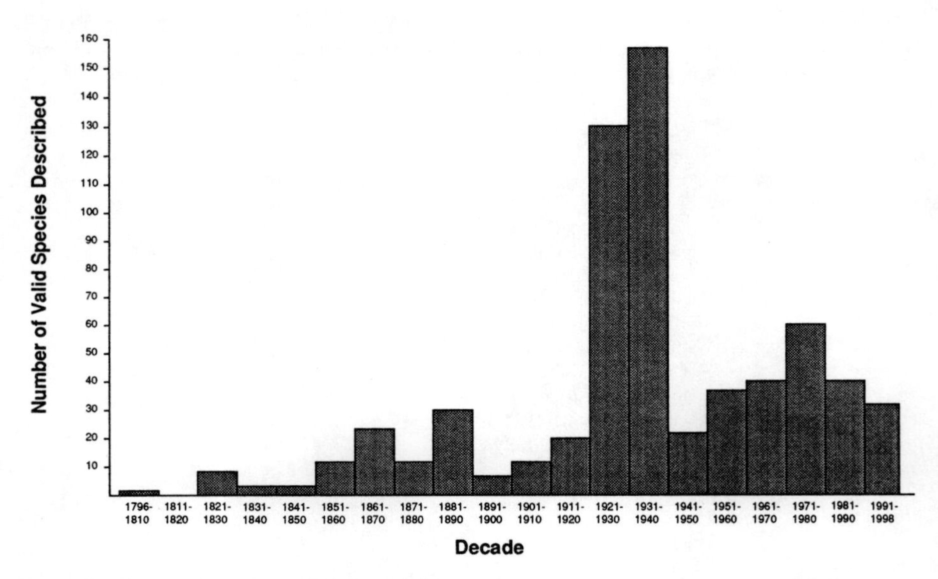

Fig. 1. Relative numbers of mayfly species described from N. A. by decade from 1796 to 1998. Only junior synonyms not counted (homonyms are counted in the decade of original description). The 1990 decade is modified for an additional five new species known to be in preparation.

The Twentieth Century

In the first two decades of the Twentieth Century, 33 valid species were added to the N. A. mayfly fauna. This was a somewhat inauspicious beginning to the century, given the past activity in the late Nineteenth Century and that which was to follow in the 1920's and 30's (Figs. 1 and 2). Contributions to the known fauna were by Banks; by W. A. Clemens; by one of the forefathers of aquatic entomology, James Needham and one of his protégés, Anna Haven Morgan; and by Georg Ulmer. In one sense, N. A. entomology was truly coming of age in this period, because only one species was described by a European (Ulmer).

What surely must be considered the *Golden Age* of N. A. ephemeropterology in terms of discovery and description of new species occurred in the third and fourth decades of the Twentieth Century. This Golden Age is due primarily to the outstanding descriptive works of two of the most prolific mayfly taxonomists the world has ever known: the Canadian John McDunnough and the American Jay Traver. McDunnough proposed 208 species names from 1921 to 1943, and of those, 161 remain valid at the species level; three are now recognized as subspecies only, two are homonymic and have been renamed, and 42 have proven to be junior synonyms. Nearly all of McDunnough's species were described from Canada, and all but one valid species was described in the 1920's and 30's. A complete review of McDunnough's descriptive literature can be found in McCafferty and Randolph (1998). Traver proposed 124 species names in the 1930's. Of those, 84 remain valid and 40 have fallen as synonyms. Although McDunnough essentially finished his Ephemeroptera work by the early 1940's, Traver continued to describe N. A. mayflies through 1968, resulting in 106 valid N. A. species authored by her.

Besides the contribution of over 240 descriptions of valid N. A. species during the 1920's and 30's by McDunnough and Traver, these workers also provided important faunal accounts for N. A. McDunnough (1925) gave a first substantial list for Canada and its provinces, and Traver (1935) reviewed 517 species recognized in N. A. north of Mexico at that

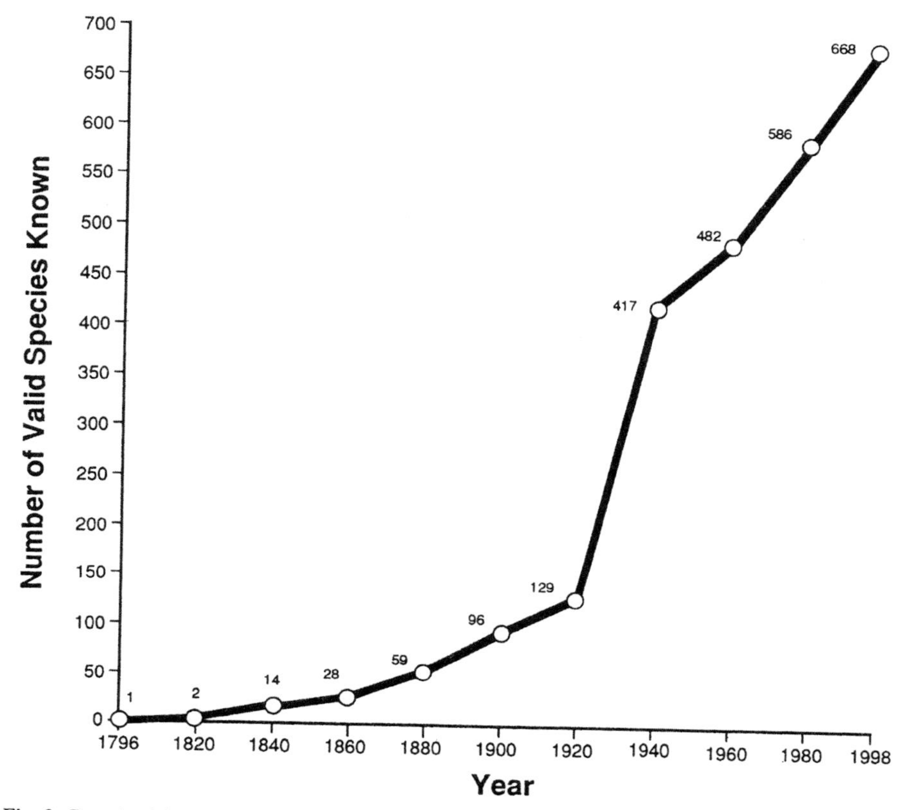

Fig. 2. Growth of the known valid mayfly fauna of N. A., accounting for species originally described from N. A. and subsequently discovered species originally described elsewhere.

time. Traver's synopsis was a remarkable achievement and continued to be of value for the remainder of the Twentieth Century. It should be noted, however, especially for the purposes of tracking faunistic developments, that based on subsequent research, there were actually 137 junior synonyms, two homonyms, and at least one misidentification treated as species, making in actuality a currently valid 370 species treated by Traver (1935). Still, this number represented about a ten-fold increase of N. A. species known since Eaton's monograph. In addition, by 1935 there were an additional 10 N. A. species known that were restricted to Mexico.

Besides the valid McDunnough and Traver species described during the Golden Age, another 31 valid N. A. species were described by the Americans Banks, Lewis Berner, G. S. Dodds, Justin Leonard, Velma Knox Mayo, Needham, Theresa Seemann, and Herman Spieth, and the Canadians Clemens and Fred Ide. These workers tended to be more regional in their scope; for example, Berner worked in Florida, Dodds in Colorado, Ide in Ontario, and Seemann in California. The Englishman D. E. Kimmins described another two species from Mexico, and the Spanish Jesuit P. Longinos Navás described one species from Mexico during this period. New descriptions along with discoveries of European species in N. A during the period resulted in a total of 417 valid species known from N. A. by the end of 1940. The steep growth curve for N. A. known species during the Golden Age is most evident from the graph shown in Fig. 2.

Table 2. Notable North American regional faunistic reports published since Edmunds *et al.* (1976). See References for complete citations of works. Earlier faunistics works are cited in the narrative.

Region	Author	Year
Alabama	Kondratieff & Harris	1986
Alaska	McCafferty	1985
Alberta	McCafferty & Randolph	1998
Arizona	Lugo-Ortiz & McCafferty	1995a
Arkansas	McCafferty & Provonsha	1978
British Columbia	McCafferty & Randolph	1998
Colorado	McCafferty, Durfee & Kondratieff	1993
Connecticut	Burian & Bednarik	1994
Florida	Berner & Pescador	1988
Illinois	Randolph & McCafferty	1998
Indiana	Randolph & McCafferty	1998
Iowa	Klubertanz	1995
Kentucky	Randolph & McCafferty	1998
Maine	Burian & Gibbs	1991
Manitoba	Flannagan & Flannagan	1982
	McCafferty & Randolph	1998
Mexico	McCafferty & Lugo-Ortiz	1996
Michigan	Randolph & McCafferty	1998
Missouri	Sarver & Kondratieff	1997
New Brunswick	Petersen	1989
	McCafferty & Randolph	1998
Newfoundland	Larson & Colbo	1983
	McCafferty & Randolph	1998
New Mexico	McCafferty, Lugo-Ortiz & Jacobi	1997
North Carolina	Unzicker & Carlson	1982
Northern Canada	Harper & Harper	1981
	Cobb & Flannagan	1980
	McCafferty & Randolph	1998
Nova Scotia	Petersen	1989
	McCafferty & Randolph	1998
Ohio	Randolph & McCafferty	1998
Oklahoma	McCafferty, Heth & Waltz	1997
Ontario	McCafferty & Randolph	1998
Saskatchewan	Lehmkuhl	1976
	McCafferty & Randolph	1998
South Carolina	Unzicker & Carlson	1982
South Dakota	McCafferty	1990
Quebec	Dulude	1992
	McCafferty & Randolph	1998
Tennessee	Long & Kondratieff	1997
Texas	Lugo-Ortiz & McCafferty	1995b
Virginia	Kondratieff & Voshell	1983
Wisconsin	Hilsenhoff	1995
	Randolph & McCafferty	1998
West Virginia	Faulkner & Tarter	1977
Yukon	Harper & Harper	1997
	McCafferty & Randolph	1998

Far fewer valid N. A. species were described in the 1940's and 50's (Fig. 1), perhaps because many of the common species had already been described, but also because of the precedence of World War II in the 1940's and its aftermath. The major contributors from the

Table 3. Major historical authors of the North American mayfly fauna (with 10 or more currently valid species). Periods are those in which species counted were named. Species counts are of currently valid North American species.

Worker	Period	Species
Say, T.	1823 - 1839	10
Walsh, B. D.	1862 - 1863	15
Eaton, A. E.	1871 - 1892	42
Banks, N.	1900 - 1924	13
Needham, J. G.	1903 - 1932	13
McDunnough, J.	1921 - 1943	161
Traver, J. R.	1931 - 1968	106
Mayo, V. K.	1939 - 1973	11
Berner, L.	1940 - 1978	22
Day, W. C.	1952 - 1954	12
Edmunds, G. F.	1957 - 1976	23
Allen, R. K.	1957 - 1988	72
McCafferty, W. P.	1977 - 1998	33

period tended to be regional in their scope and included Berner (Southeast), B. D. Burks (Illinois), Richard Daggy (Minnesota), Bill Day (California), and Mayo (West and Southwest). Traver also described eight valid species during this period. Minor descriptive contributions were made by Dick Allen, George Edmunds, Ide, McDunnough, Bill Peters, and Ulmer (the latter being the only non-N. A. species author in the period).

The 1950's did, however, witness the first of regional checklists or more comprehensive regional faunal accounts of mayflies in N. A. Two of the works, in particular, were noteworthy because of the wealth of specific information they contained: the Florida fauna was treated extensively by Berner (1950) and the Illinois fauna extensively by Burks (1953). A work on California by Day (1956) contained keys but was quite preliminary, and checklists were given for Utah by Edmunds (1954) and Oregon by Allen and Edmunds (1956).

Dick Allen led the way in descriptive taxonomy in the next two decades in authoring or co-authoring 58 of the 99 valid N. A. species described between 1961 and 1980. A good deal of the species discoveries were from the relatively poorly known southwestern and Mexican regions. In particular, this involved major contributions to the knowledge of the austral genera *Baetodes* by Cohen and Allen (1972), *Thraulodes* by Traver and Edmunds (1967), *Camelobaetidius* by Traver and Edmunds (1968), and *Leptohyphes* by Allen and Brusca (1973). Many of Allen's singular descriptions were also from this area, for example, from his work on *Tricorythodes* (Allen, 1967). In addition to the authors mentioned above, minor descriptive contributions were made by Berner, F. L. Carle, E. S. M. Chao, D. L. Collins, Françoise Harper, Steve Jensen, J. I. Kilgore, Ralph Kirchner, Dick Koss, Dennis Lehmkuhl, Phil Lewis, Mayo, Pat McCafferty, Dennis Morihara, Chad Murvosh, Ingrid Müller-Liebenau (the only non-N. A. species author in this period), Manny Pescador, Peters, R. F. Schneider, and Donald Tarter. The few contributions to regional faunas published during this period involved Michigan (Leonard and Leonard, 1962), British Columbia (Scudder, 1975), and Arkansas (McCafferty and Provonsha, 1978).

The advent of co-authorship of taxonomic descriptions also became apparent in this time period, although one species had already been co-authored by Clemens and Leonard earlier. The renowned American ephemeropterist George Edmunds can be credited as much as anyone with this new trend. In the 1960's he co-authored descriptions by Edmunds and Allen; Edmunds and Traver; Edmunds, Berner and Traver; and Allen and Edmunds (see McCafferty, 1995). Such co-authorship marked the beginning of a new era of team work and schools among mayfly taxonomists. Beginning in the 1970's, Allen co-authored species with

Table 4. Evolution of family classification of the North American mayfly fauna in the Twentieth Century. Listing order of modern families does not necessarily indicate relationships.

Needham *et al.* (1935)	Burks (1953)	Edmunds *et al.* (1976)	McCafferty (1997)
Baetidae	=Baetidae	=Baetidae	=Baetidae
		+Siphlonuridae	=Siphlonuridae
			+Acanthametropodidae[1]
			+Ameletidae
			+Isonychiidae
	+Ametropodidae	=Ametropodidae	=Ametropodidae
		+Metretopodidae	=Metretopodidae
		+Heptageniidae partim	=Pseudironidae
	+Baetiscidae	=Baetiscidae	=Baetiscidae
	+Caenidae	=Caenidae	=Caenidae
		+Tricorythidae	=Leptohyphidae
	+Ephemerellidae	=Ephemerellidae	=Ephemerellidae
	+Leptophlebiidae	=Leptophlebiidae	=Leptophlebiidae
	+Oligoneuriidae[2]	=Oligoneuriidae	=Oligoneuriidae
Heptageniidae	=Heptageniidae	=Heptageniidae partim	=Heptageniidae
			+Arthropleidae
Ephemeridae	=Ephemeridae	=Ephemeridae	=Ephemeridae
		+Behningiidae[1]	=Behningiidae
		+Polymitarcyidae	=Polymitarcyidae
		+Potamanthidae	=Potamanthidae
	+Neoephemeridae	=Neoephemeridae	=Neoephemeridae

[1] Taxa not known in North America in 1935 or 1953.
[2] Taxa not known in North America in 1935.

seven different workers. Such collaborative efforts were commonplace for the remainder of the Twentieth Century.

Edmunds (1962) provided a useful account of the type localities of all of the N. A. species recognized at that time. The pivotal "The Mayflies of North and Central America," by Edmunds, Jensen, and Berner, appeared in 1976. This work concentrated on generic treatments, but also listed a total of 657 N. A. species. Also, of considerable significance in the 1960's was the first comprehensive restudy and review of an entire large group of N. A. mayflies: the outstanding series dealing with the family Ephemerellidae by Allen and Edmunds (1959, 1961ab, 1962ab, 1963ab, 1965). Both new species and new synonymies were included in the latter, but perhaps their most useful aspects were extensive species keys for what are now considered seven genera.

Near the end of the period, major revisionary works were published for the large and problematic genera *Baetis* by Morihara and McCafferty (1979a) and *Stenonema* by Bednarik and McCafferty (1979). These revisions were the first to demonstrate that there were a large number of synonyms amongst the N. A. species names, thus indicating that a tendency toward typology had been common among many early workers. For example, of those 657 N. A. species listed by Edmunds et al. (1976), actually only 535 are currently considered valid. Other revisionary works in the final two decades of the Twentieth Century (see Table 1), such as that of Kondratieff and Voshell (1984) on *Isonychia* and McCafferty and Waltz (1990) on Baetidae in general would continue to demonstrate that the fauna had been over-described. Table 1 indicates that among combined revisionary works, synonyms have largely outnumbered new species in the last quarter of the century. Additions subsequent to the publication of Edmunds et al. (1976) with respect to new descriptions and to discoveries of species originally described from the Palearctic or Neotropics are greater than might be suggested in Fig. 2. This is because considerable synonymyzing of previously proposed species names also took place after 1976.

Table 5. Growth of genera as recognized in North America, based on current, valid genera discovered or described from North America, raised to rank or changed in name by the interval dates indicated. For authors of genera, see *Mayfly Central*.

Known by 1900	Added 1901-1935	Added 1936-1976	Added 1977-1997
Ameletus	*Ametropus*	*Acanthametropus*	*Acanthomola*
Baetis	*Anepeorus*	*Analetris*	*Acentrella*
Baetisca	*Arthroplea*	*Apobaetis*	*Acerpenna*
Brachycercus	*Centroptilum*	*Baetodes*	*Amercaenis*
Caenis	*Cinygmula*	*Cloeon*	*Americabaetis*
Callibaetis	*Habrophlebia*	*Dolania*	*Anthopotamus*
Campsurus	*Habrophlebiodes*	*Edmundsius*	*Attenella*
Choroterpes	*Heterocloeon*	*Epeorus*	*Barbaetis*
Cinygma	*Ironodes*	*Homoeoneuria*	*Camelobaetidius*
Ephemera	*Metretopus*	*Lachlania*	*Caudatella*
Ephemerella	*Neoephemera*	*Leptohyphes*	*Caurinella*
Ephoron	*Paraleptophlebia*	*Litobrancha*	*Cercobrachys*
Euthyplocia	*Parameletus*	*Paracloeodes*	*Cloeodes*
Heptagenia	*Pseudiron*	*Stenacron*	*Diphetor*
Hexagenia	*Siphlonisca*	*Tortopus*	*Drunella*
Isonychia	*Siphloplecton*	*Traverella*	*Eurylophella*
Leptophlebia	*Stenonema*		*Fallceon*
Pentagenia	*Thraulodes*		*Farrodes*
Rhithrogena	*Tricorythodes*		*Hydrosmilodon*
Siphlonurus			*Labiobaetis*
			Leucrocuta
			Macdunnoa
			Moribaetis
			Neochoroterpes
			Nixe
			Procloeon
			Pseudocentroptiloides
			Raptoheptagenia
			Serratella
			Timpanoga

Some 69 valid N. A. species were described between 1981 and the time of this writing (1998). I know of at least five more that will be published by the end of the century. McCafferty authored or co-authored 29 species in this period. His co-authors included Carlos Lugo-Ortiz, Arwin Provonsha, Bob Waltz, and Nick Wiersema, all of whom were major contributors in their own right (see e.g., Tables 1 and 2). The other 40 species were authored by an array of workers, with Thomas Soldán from the Czech Republic and Jack Zloty from Alberta each contributing seven species. More minor contributions were made by Javier Alba-Tercedor (the only non-N. A. species author besides Soldán in the period), R. K. (Dick, above) Allen, R. T. Allen, W. F. Botts, W. L. Burrows, J. R. Davis, Richard Durfee, J. L. Evans, John Flannagan, Wills Flowers, D. H. Funk, F. Harper, P. Harper, Brad Henry, Boris Kondràtieff, J. H. Kennedy, Lehmkuhl, R. G. Lowen, Murvosh, G. Roemhild, J. R. Voshell, and Eric Whiting.

There have been major advances in our knowledge of distributional ranges of N. A. species in the last two decades of the Twentieth Century. These data have been contributed via numerous small published reports of individual species records as well as from a variety of faunistic studies and inventories (Table 2), ranging from basic checklists to extensive faunal analyses (some partial state inventories are not shown). The most extensive study is that of Randolph and McCafferty (1998) for the U. S. states of Kentucky, Illinois, Indiana, Michigan, Ohio, and Wisconsin, where not only were extensive habitat, drainage system,

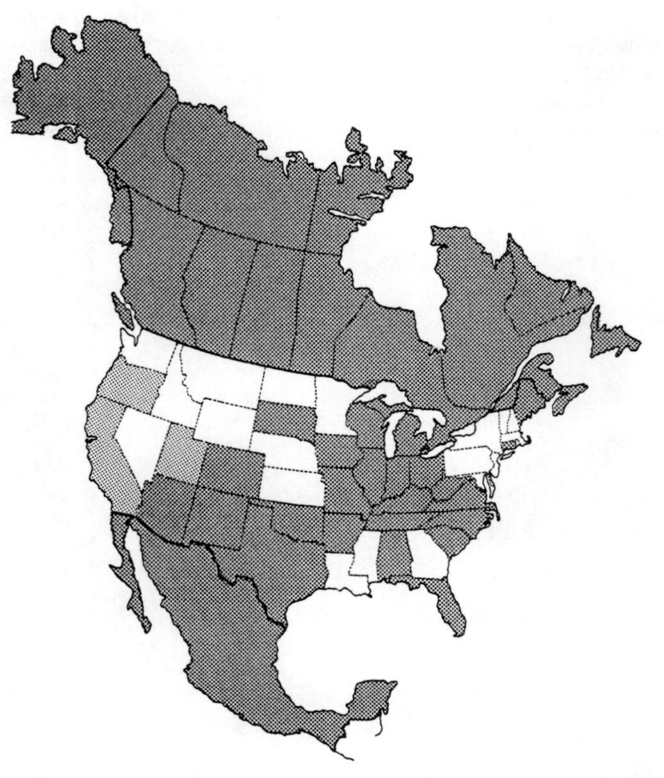

Fig. 3. North American subregions with significant inventories (those inventoried only previous to 1976 shown by slight shading; those since 1976 are darkly shaded; unpublished surveys and some partial inventories not shown).

historical geology, and fine geography detailed, but also risk analyses of species included. Studies of the Maine and New Mexico faunas are also especially notable (see Table 2). During the 1990's, comprehensive documentation of the entire faunas of Canada (McCafferty and Randolph, 1998), including provincial and territorial distributions, and Mexico (McCafferty and Lugo-Ortiz, 1996) were provided for the first time. Regions that have been surveyed are highlighted in Fig. 3. Those studied only prior to 1976 undoubtedly require considerable updating as do some others.

As of this writing, there are close to 700 valid species of mayflies in N. A. The net growth of the known fauna over time can be visualized from Fig. 2. The growth has been relatively steady since the Golden Age in the 1920's and 30's (see also Fig. 1). Eventually when lesser known subregions such as Alaska and Mexico are more fully documented, I expect the growth curve (Fig. 2) to flatten out, with species discoveries more or less balanced by new synonyms. Also as of this writing, some 321 species are currently known from Canada (McCafferty and Randolph, 1998), 124 species from Mexico (McCafferty and Lugo-Ortiz, 1996; McCafferty et al., 1997; McCafferty, unpublished), 564 species from the conterminous United States (McCafferty, unpublished), and only 19 species from Alaska (McCafferty, 1985, 1994). Updates of these numbers may be found on the WWW at *Mayfly Central*. The 13 primary authors of N. A. mayflies and their contributions of 543 valid species, or over 80% of the known fauna mentioned above, are summarized in Table 3.

Nine N. A. species were originally described from Europe and later discovered in N. A. (similarly, five species originally described from N. A. were later discovered to occur in the Palearctic) (Flowers, 1989; McCafferty, 1985, 1994; McCafferty and Randolph, 1998). Also several species were originally described from Central America, South America, or the Antilles and only subsequently discovered in Mexico. The dynamic biogeographic interchange of lineages between North and South America was detailed by McCafferty (1998). *Euthyplocia hecuba* (Hagen), *Hexagenia albivitta* (Walker), and *H. mexicana* Eaton are the only species known to co-occur in North and South America. The only mayfly known from every subregion of N. A. (including Alaska and the Yukon) is *Baetis tricaudatus* Dodds; interestingly, however, it is not known beyond N. A.

This review has dealt essentially with species discoveries; however, higher classification has also developed immensely during the Twentieth Century. Transitions reflect not only a change from a phenetic to an evolutionary to a phylogenetic basis (see Edmunds and Traver, 1954; McCafferty and Edmunds, 1979; McCafferty, 1991), but also reflect the intent to impart substantial information content with more finite taxa, such as genera. Tables 4 and 5 illustrate the major transitions and advancements in familial and generic classification, respectively, as they have pertained to the N. A. fauna.

The last published checklist for the entire N. A. continent was given by McCafferty (1996a). However, the rate of change in nomenclature instigated by on going revisionary work can and often does necessitate modifications to such lists before they appear in print. To alleviate this problem and to provide immediately available, up to date tracking of the fauna and its nomenclature, "The Mayflies of North America" has existed on the WWW as part of *Mayfly Central* since 1995 (McCafferty 1996b). It provides a full listing of species and facilitates searching any name (synonym, homonym, misspelling, generic combination, etc.) applicable to any species. It also provides regional distribution data for each species. Such data management and presentation are previews of the kind of utilitarian information that is in store for the Twenty-First Century as the gentle quest continues.

The Future

A status report for the systematics of N. A. Ephemeroptera was given by McCafferty et al. (1990), and many recommendations for future priorities of research indicated at that time still hold. The fauna of special habitats of large rivers continue to require study, as does the general fauna of several geographical regions that were indicated. Regional priorities are best judged from Table 2 and Fig. 3. Certainly, there remain genera that require revision. This is particularly true for many of the heptageniid genera. The most needed revisionary work in the Baetidae involves the genera *Centroptilum* and *Procloeon*, and they are now being comprehensively studied. The leptohyphid genera *Leptohyphes* and *Tricorythodes* require revision because they were probably largely over-described as species and surely involve several additional genera. The caenid genus *Brachycercus* also remains problematic. The large leptophlebiid genus *Paraleptophlebia* and siphlonurid genus *Siphlonurus* may appear to require revision; however, that remains to be seen because I have found that most current species have morphological integrity as adults. The challenge will be to delineate the species of these genera in the larval stage.

As our knowledge of species improves appreciably, species keys, now unavailable for many genera, must be provided. Larval and adult stages need to be associated for the many species known only as one or the other. Distributional work needs to be taken to the next level by plotting species against numerous parameters at a finer level, and incorporating modern GIS (geographic information systems) analysis to help explain past and present distributions. Perhaps most importantly, data should be used to determine species at risk from present, impending, or potential environmental degradation.

ACKNOWLEDGMENTS

Thanks to Arwin Provonsha, who aided in the preparation of graphics and to Pat Randolph and Luke Jacobus for their assistance. This paper has been assigned Purdue Agricultural Research Program Journal No. 15782.

REFERENCES

Allen, R. K. 1967. New species of New World Leptohyphinae (Ephemeroptera: Tricorythidae). Can. Ent. 99: 350-375.

Allen, R. K. 1980. Geographic distribution and reclassification of the subfamily Ephemerellinae (Ephemeroptera: Ephemerellidae), pp. 71-91. *In*: J. F. Flannagan and K. E. Marshall (eds.). Advances in Ephemeroptera biology. Plenum, New York.

Allen, R. K. and R. C. Brusca. 1973. New species of Leptohyphidae from Mexico and Central America (Ephemeroptera: Tricorythidae). Can. Ent. 105: 83-95.

Allen, R. K. and G. F. Edmunds, Jr. 1956. A list of the mayflies of Oregon. Utah Acad. Proc. 33: 85-87.

Allen, R. K. and G. F. Edmunds, Jr. 1959. A revision of the genus *Ephemerella* (Ephemeroptera: Ephemerellidae). I. The subgenus *Timpanoga*. Can. Ent. 91: 51-58.

Allen, R. K. and G. F. Edmunds, Jr. 1961a. A revision of the genus *Ephemerella* (Ephemeroptera: Ephemerellidae). II. The subgenus *Caudatella*. Ann. ent. Soc. Amer. 54: 603-612.

Allen, R. K. and G. F. Edmunds, Jr. 1961b. A revision of the genus *Ephemerella* (Ephemeroptera: Ephemerellidae). III. The subgenus *Attenuatella*. J. Kansas ent. Soc. 34: 161-173.

Allen, R. K. and G. F. Edmunds, Jr. 1962a. A revision of the genus *Ephemerella* (Ephemeroptera: Ephemerellidae). IV. The subgenus *Dannella*. J. Kansas ent. Soc. 35: 333-338.

Allen, R. K. and G. F. Edmunds, Jr. 1962b. A revision of the genus *Ephemerella* (Ephemeroptera: Ephemerellidae). V. The subgenus *Drunella* in North America. Misc. Publ. ent. Soc. Amer. 3: 147-179.

Allen, R. K. and G. F. Edmunds, Jr. 1963a. A revision of the genus *Ephemerella* (Ephemeroptera: Ephemerellidae). VI. The subgenus *Serratella* in North America. Ann. ent. Soc. Amer. 56: 583-600.

Allen, R. K. and G. F. Edmunds, Jr. 1963b. A revision of the genus *Ephemerella* (Ephemeroptera: Ephemerellidae). VII. The subgenus *Eurylophella*. Can. Ent. 95: 597-623.

Allen, R. K. and G. F. Edmunds, Jr. 1965. A revision of the genus *Ephemerella* (Ephemeroptera: Ephemerellidae). VIII. The subgenus *Ephemerella* in North America. Misc. Publ. ent. Soc. Amer. 4: 243-282.

Allen, R. K. and G. F. Edmunds, Jr. 1976. A revision of the genus *Ametropus* in North America (Ephemeroptera: Ametropodidae). J. Kansas ent. Soc. 49: 625-635.

Bae, Y. J. and W. P. McCafferty. 1991. Phylogenetic systematics of the Potamanthidae (Ephemeroptera). Trans. Amer. ent. Soc. 117: 1-143.

Bae, Y. J. and W. P. McCafferty. 1998. Phylogenetic systematics and biogeography of the Neoephemeridae (Ephemeroptera: Pannota). Aquat. Insects 20: 35-68.

Banks, N. 1900. New genera and species of Nearctic neuropteroid insects. Trans. Amer. ent. Soc. 26: 239-259.

Bednarik, A. F. and W. P. McCafferty. 1979. Biosystematic revision of the genus *Stenonema* (Ephemeroptera: Heptageniidae). Can. Bull. Fish. Aquat. Sci. 201: 1-73.

Berner, L. 1950. The mayflies of Florida. Fla. Univ. Stud. Ser. 4: xii + 1-267.

Berner, L. 1978. A review of the mayfly family Metretopodidae. Trans. Amer. Ent. Soc. 104: 91-137.

Berner, L. and M. L. Pescador. 1988. The mayflies of Florida, revised edition. Univ. Presses Florida, Gainesville.

Burks, B. D. 1953. The mayflies or Ephemeroptera of Illinois. Ill. Nat. Hist. Surv. Bull. 26: 1-216.

Burian, S. K. 1995. Taxonomy of the eastern Nearctic species of *Choroterpes* Eaton (Ephemeroptera: Leptophlebiidae), pp. 433-453. *In*: L. D. Corkum and J. J. H. Ciborowski (eds.). Current directions in research on Ephemeroptera. Can. Scholars' Press, Toronto.

Burian, S. K. and A. F. Bednarik. 1994. The mayflies (Ephemeroptera) of Connecticut: an initial faunal survey. Ent. News 105: 204-216.

Burian, S. K. and K. E. Gibbs. 1991. Mayflies of Maine: an annotated faunal list. Maine Agr. Exper. Stat. Tech. Bull. 142: 1-109.

Cobb, D. G. and J. F. Flannagan. 1980. The distribution of Ephemeroptera in northern Canada, pp. 155-166. *In*: J. F. Flannagan and K. E. Marshall (eds.). Advances in Ephemeroptera biology. Plenum, New York.

Cohen, S. D. and R. K. Allen. 1972. New species of *Baetodes* from Mexico and Central America. Pan-Pac. Ent. 48: 123-135.

Day, W. C. 1956. Ephemeroptera, pp. 79-105. *In*: R. L. Usinger (ed.). Aquatic insects of California. Univ. Calif. Press, Berkeley.

Domínguez, E., C. Molineri, and W. L. Peters. 1996. Ephemeroptera from Central and South America: New species of the *Farrodes bimaculatus* group with a key for the males. Stud. Neotrop. Fauna Environ. 31: 87-101.

Dulude, Y. 1992. Les éphémères du pêcheur Québécois. Les Éditions de l'Homme, Montréal.

Eaton, A. E. 1883-88. A revisional monograph of recent Ephemeridae or mayflies. Trans. Linn. Soc. Lond., Second Ser. Zool. 3: 1-352.

Edmunds, G. F., Jr. 1954. The mayflies of Utah. Proc. Utah Acad. Sci. Arts Lett. 31: 64-66.

Edmunds, G. F., Jr. 1962. The type localities of the Ephemeroptera of North America north of Mexico. Univ. Utah Biol. Ser. 12: iii + 1-39.

Edmunds, G. F., Jr., S. L. Jensen, and L. Berner. 1976. The mayflies of North and Central America. Univ. Minnesota Press, Minneapolis.

Edmunds, G. F., Jr. and W. P. McCafferty. 1984. *Ephemera compar*: an obscure Colorado burrowing mayfly (Ephemeroptera: Ephemeridae). Ent. News 95: 186-188.

Edmunds, G. F. , Jr. and J. R. Traver. 1954. An outline of reclassification of the Ephemeroptera. Proc. ent. Soc. Wash. 56: 236-240.

Faulkner, G. M. and D. C. Tarter. 1977. Mayflies, or Ephemeroptera, of West Virginia with emphasis on the nymphal stage. Ent. News 88: 202-206.

Flannagan, P. M. and J. F. Flannagan. 1982. Present distribution and the post-glacial origin of the Ephemeroptera, Plecoptera and Trichoptera of Manitoba. Man. Dept. Nat. Res. Fish. Tech. Rept. No. 82-1: ii + 1-79.

Flowers, R. W. 1980. Two new genera of Nearctic Heptageniidae (Ephemeroptera). Florida Entomol. 63: 296-307.

Flowers, R. W. 1982. Review of the genus *Macdunnoa* (Ephemeroptera: Heptageniidae) with descriptions of a new species from Florida. Gr. Lakes Ent. 15: 432-444.

Flowers, R. W. 1989. Holarctic distribution of three taxa of Heptageniidae (Ephemeroptera). Ent. News 97: 193-197.

Funk, D. H. and B. W. Sweeney. 1994. The larvae of eastern North American *Eurylophella* Tiensuu (Ephemeroptera: Ephemerellidae). Trans. Amer. ent. Soc. 120: 209-286.

Harper, F. and P. P. Harper. 1981. Northern Canadian mayflies (Insecta; Ephemeroptera), records and descriptions. Can. J. Zool. 59: 1784-1789.

Harper, P. P. and F. Harper. 1997. Mayflies (Ephemeroptera) of the Yukon, pp. 151-167. *In*: H. V. Danks and J. A. Downes (eds.). Insects of the Yukon. Biol. Surv. Can. (Terr. Arthropods), Ottawa.

Henry, B. C., Jr. 1993. A revision of *Neochoroterpes* (Ephemeroptera: Leptophlebiidae) new status. Trans. Amer. ent. Soc. 119: 317-333.

Hilsenhoff, W. L 1995. Aquatic insects of Wisconsin. Nat. Hist. Council Univ. Wisconsin-Madison 3: 1-79.

Klubertanz, T. H. 1995. Survey of Iowa Mayflies (Ephemeroptera). J. Kansas ent. Soc. 68: 20-26.

Kondratieff, B. C. and S. C. Harris. 1986. Preliminary checklist of the mayflies (Ephemeroptera) of Alabama. Ent. News 97: 230-236.

Kondratieff, B. C. and J. R. Voshell, Jr. 1983. A checklist of the mayflies (Ephemeroptera) of Virginia, with a review of pertinent taxonomic literature. J. Georgia ent. Soc. 18: 273-279.

Kondratieff, B. C. and J. R. Voshell, Jr. 1984. The North and Central American species of *Isonychia* (Ephemeroptera: Oligoneuriidae). Trans. Amer. ent. Soc. 110: 129-244.

Larson, D. J. and M. H. Colbo. 1983. The aquatic insects: biogeographic considerations, pp. 593-677. *In*: G. R. South (ed.). Biogeography and ecology of the island of Newfoundland. Junk, The Hague.

Lehmkuhl, D. M. 1976. Mayflies. Blue Jay 34: 70-81.

Leonard, J. W. and F. A. Leonard. 1962. Mayflies of Michigan trout streams. Cranbrook Inst. Sci. Bull. 43: x + 1-139.

Long, L. S. and B. C. Kondratieff. 1997. The mayflies (Ephemeroptera) of Tennessee, with a review of the possibly threatened species occurring within the state. Gr. Lakes Ent. 29: 171-182.

Lugo-Ortiz, C. R. and W. P. McCafferty. 1994. The mayfly genus *Acerpenna* (Insecta: Ephemeroptera: Baetidae) in Latin America. Stud. Neotrop. Fauna Environ. 29: 65-74.

Lugo-Ortiz, C. R. and W. P. McCafferty. 1995a. An annotated inventory of the mayflies (Ephemeroptera) of Arizona. Ent. News 106: 131-140.

Lugo-Ortiz, C. R. and W. P. McCafferty. 1995b. The mayflies (Ephemeroptera) of Texas and their biogeographic affinities, pp. 151-169. *In*: L. Corkum and J. Ciborowski (eds.). Current directions in research on Ephemeroptera. Can. Scholars' Press, Toronto.

Lugo-Ortiz, C. R. and W. P. McCafferty. 1995c. Taxonomy of the North and Central American species of *Camelobaetidius* (Ephemeroptera: Baetidae). Ent. News 106: 178-192.

Lugo-Ortiz, C. R. and W. P. McCafferty. 1996a. Taxonomy of the Neotropical genus *Americabaetis*, new status (Insecta: Ephemeroptera: Baetidae). Stud. Neotrop. Fauna Environ. 31: 156-169.

Lugo-Ortiz, C. R. and W. P. McCafferty. 1996b. Contribution to the taxonomy of *Callibaetis* (Ephemeroptera: Baetidae) in southwestern North America and Middle America. Aquat. Insects 18: 1-9.

Lugo-Ortiz, C. R. and W. P. McCafferty. 1998. A new North American genus of Baetidae (Ephemeroptera) and key to *Baetis* complex genera. Ent. News 109: 354-356.

Lugo-Ortiz, C. R., W. P. McCafferty, and R. D. Waltz. 1994. Review of the Panamerican genus *Fallceon* (Ephemeroptera: Baetidae). J. N. Y. ent. Soc. 102: 460-475.

McCafferty, W. P. 1977. Biosystematics of *Dannella* and related subgenera of *Ephemerella* (Ephemeroptera: Ephemerellidae). Ann. ent. Soc. Amer. 70: 881-889.

McCafferty, W. P. 1985. The Ephemeroptera of Alaska. Proc. ent. Soc. Wash. 87: 381-386.

McCafferty, W. P. 1990. Biogeographic affinities of the Ephemeroptera of the Black Hills, South Dakota. Ent. News 101: 193-199.

McCafferty, W. P. 1991. Toward a phylogenetic classification of the Ephemeroptera (Insecta): a commentary on systematics. Ann. ent. Soc. Am. 84: 343-360.

McCafferty, W. P. 1994. Additions and corrections to the Ephemeroptera of Alaska. Proc. ent. Soc. Wash. 96: 177.

McCafferty, W. P. 1995. George Edmunds, ephemeropterist par excellence, pp. 3-18. *In*: L. Corkum and J. Ciborowski (eds.). Current directions in research on Ephemeroptera. Can. Scholars' Press, Toronto.

McCafferty, W. P. 1996a. The Ephemeroptera species of North America and index to their complete nomenclature. Trans. Amer. ent. Soc. 122: 1-54.

McCafferty, W. P. 1996b. The mayflies (Ephemeroptera) of North America online. Ent. News 107: 61-63.

McCafferty, W. P. 1997. Ephemeroptera, pp. 89-117. *In*: R. W. Poole and P. Gentili (eds.). Nomina Insecta Nearctica, a check list of the insects of North America. Vol. 4: Non-holometabolous orders. Entomol. Inform. Serv., Rockville, Maryland.

McCafferty, W. P. 1998. Ephemeroptera and the great American interchange. J. N. Am. Benthol. Soc. 17: 1-20.

McCafferty, W. P. and Y. J. Bae. 1992. Taxonomic status of historically confused species of Potamanthidae and Heptageniidae (Ephemeroptera). Proc. ent. Soc. Wash. 94: 169-171.

McCafferty, W. P. and G. F. Edmunds, Jr. 1979. The higher classification of the Ephemeroptera and its evolutionary basis. Ann. ent. Soc. Amer. 72: 5-12.

McCafferty, W. P. and C. R. Lugo-Ortiz. 1996. Ephemeroptera, pp. 133-145. *In*: J. E. Llorente-Bousquets, A. N. García-Aldrete, and E. González-Soriano (eds.). Biodiversidad, taxonomía y biogeografía de atrópodos de México: hacea una síntesis de su conocimiento. Univ. Nacional Autónoma de México, México D.F.

McCafferty, W. P. and A. V. Provonsha. 1978. The Ephemeroptera of mountainous Arkansas. J. Kansas ent. Soc. 51: 360-379.

McCafferty, W. P. and R. P. Randolph. 1998. Canada mayflies: a faunistic compendium. Proc. ent. Soc. Ontario 129: 47-97.

McCafferty, W. P. and R. D. Waltz. 1990. Revisionary synopsis of the Baetidae (Ephemeroptera) of North and Middle America. Trans. Amer. ent. Soc. 116: 769-799.

McCafferty, W. P. and T.-Q. Wang. 1994. Phylogenetics and the classification of the *Timpanoga* complex (Ephemeroptera: Ephemerellidae). J. N. Amer. Benthol. Soc. 13: 569-579.

McCafferty, W. P., R. S. Durfee, and B. C. Kondratieff. 1993. Colorado mayflies (Ephemeroptera): an annotated inventory. Southwest Natural. 38: 252-274.

McCafferty, W. P., R. K. Heth, and R. D. Waltz. 1997. The Ephemeroptera of Spring Creek, Oklahoma, with remarks on notable records. Ent. News 108: 193-200.

McCafferty, W. P., C. R. Lugo-Ortiz, and G. Z. Jacobi. 1997. Mayfly fauna of New Mexico. Gr. Basin Natural. 57: 283-314.

McCafferty, W. P., C. R. Lugo-Ortiz, A. V. Provonsha, and T.-Q. Wang. 1997. Los efemerópteros de México: I. Clasificación superior, diagnosis de familias y composición. Dugesiana 4: 1-29.

McCafferty, W. P., B. P. Stark, and A. V. Provonsha. 1990. Ephemeroptera, Plecoptera, and Odonata, pp. 43-58. *In*: M. Kosztarab and C. W. Schaefer (eds.). Systematics of the North American insects and arachnids: status and needs. Virginia Agr. Exper. Stat. Inform. Ser. 90-1, Virginia Polytechnic Inst. Univ., Blacksburg.

McDunnough, J. 1925. Ephemeroptera. pp. 104-106. *In*: N. Criddle. The entomological record, 1924. Ann. Rept. ent. Soc. Ontario 55: 89-106.

Morihara, D. K. and W. P. McCafferty. 1979a. The *Baetis* larvae of North America (Ephemeroptera: Baetidae). Trans. Amer. ent. Soc. 105: 139-221.

Morihara, D. K. and W. P. McCafferty. 1979b. The evolution of *Heterocloeon*, with the first larval description of *Heterocloeon frivolus* comb. n. (Ephemeroptera: Baetidae). Aquat. Insects 1: 225-231.

Morihara, D. K. and W. P. McCafferty. 1979c. Systematics of the *propinquus* group of *Baetis* species (Ephemeroptera: Baetidae). Ann. ent. Soc. Amer. 72: 130-135.

Needham, J. G., J. R. Traver, and Y.-C. Hsu. 1935. The biology of mayflies. Comstock, Ithaca.

Pescador, M. L. 1985. Systematics of the Nearctic genus *Pseudiron* (Ephemeroptera: Heptageniidae: Pseudironinae). Florida Ent. 68: 432-444.

Pescador, M. L. and L. Berner. 1981. The mayfly family Baetiscidae (Ephemeroptera), Part II. Biosystematics of the genus *Baetisca*. Trans. Amer. ent. Soc. 107: 163-228.

Pescador, M. L. and W. L. Peters. 1980. A review of the genus *Homoeoneuria* (Ephemeroptera: Oligoneuriidae). Trans. Amer. ent. Soc. 106: 367-393.

Petersen, R. H. 1989. Species distribution of mayfly (Ephemeroptera) nymphs in three stream systems in New Brunswick and Nova Scotia with notes on identification. Can. Tech. Rept. Fish. Aquat. Sci. No. 1685: iii + 1-14.

Provonsha, A. V. 1990. A revision of the genus *Caenis* in North America (Ephemeroptera: Caenidae). Trans. Amer. ent. Soc. 116: 801-884.

Randolph, R. P. and W. P. McCafferty. 1998. Diversity and distribution of the mayflies (Ephemeroptera) of Illinois, Indiana, Kentucky, Michigan, Ohio, and Wisconsin. Ohio Biol. Surv. Bull. New Ser. 13 (1): vii + 188pp.

Sarver, R. and B. C. Kondratieff. 1997. Survey of Missouri mayflies with the first description of adults of *Stenonema bednariki* (Ephemeroptera: Heptageniidae). J. Kans ent. Soc. 70: 132-140.

Scudder, G. G. E. 1975. An annotated checklist of the Ephemeroptera (Insecta) of British Columbia. Syesis 8: 311-315.

Soldán, T. 1986. A revision of the Caenidae with ocellar tubercles in the nymphal stage (Ephemeroptera). Acta Univ. Carolinae, Biol. 5-6, 1982-1984: 289-362.

Traver, J. R. 1935. Part II systematic. North American mayflies order Ephemeroptera, pp. 237-739. *In*: J. G. Needham, J. R. Traver and Y.-C. Hsu. The biology of mayflies. Comstock, Ithaca, New York.

Traver, J. R. and G. F. Edmunds, Jr. 1967. A revision of the genus *Thraulodes* (Ephemeroptera: Leptophlebiidae). Misc. Publ. ent. Soc. Amer. 5: 349-395.

Traver, J. R. and G. F. Edmunds, Jr. 1968. A revision of the Baetidae with spatulate clawed nymphs (Ephemeroptera). Pac. Insects 10: 629-677.

Unzicker, J. D. and P. H. Carlson. 1982. Ephemeroptera, pp. 3.1-3.97. *In*: A. R. Brigham, W. U. Brigham and A. Gnilka (eds.). Aquatic insects and oligochaetes of North and South Carolina. Midwest Aquat. Enterprises, Mahomet , Illinois.

Waltz, R. D. and W. P. McCafferty. 1985. *Moribaetis*, a new genus of Neotropical Baetidae (Ephemeroptera). Proc. ent. Soc. Wash. 87: 239-251.

Waltz, R. D. and W. P. McCafferty. 1987a. New genera of Baetidae for some Nearctic species previously included in *Baetis* Leach (Ephemeroptera). Ann. ent. Soc. Amer. 80: 667-670.

Waltz, R. D. and W. P. McCafferty. 1987b. Revision of the genus *Cloeodes* Traver (Ephemeroptera: Baetidae). Ann. ent. Soc. Amer. 80: 191-207.

Zloty, J. 1996. A revision of the Nearctic *Ameletus* mayflies based on adult males, with descriptions of seven new species (Ephemeroptera: Ameletidae). Can. Ent. 128: 293-346.

THE STATUS OF THE TAXONOMY OF THE MAYFLY (EPHEMEROPTERA) FAUNA OF SOUTH AMERICA

Manuel L. Pescador[1], Michael D. Hubbard[1], and María del Carmen Zúñiga[2]

[1] Entomology, Center for Water Quality
Florida A&M University
Tallahassee, Florida 32307, USA
[2] Departamento de Procesos Químicos y Biológicos
Universidad del Valle, Apartado Aéreo 25360
Cali, Colombia

ABSTRACT

A total of 375 nominal species representing 91 genera in 13 families are presently known in South America. The family Leptophlebiidae is presently the most diverse group representing 38 % of the mayfly genera and 30 % of the species in the region. Brazil and Argentina have the highest number of mayfly taxa known followed distantly by Peru and Chile. Approximately 60 % of the South American genera and 80 % of the species are endemic to the region. A high percentage of taxa that are known from only one life stage, disparity of the fauna known from different countries, and lack of communication and information exchange among mayfly workers are some of the problems with mayfly taxonomic studies of the region.

INTRODUCTION

It has been roughly two centuries since Weber (1801) described the first South American mayfly species, *Ephemera atrostoma.* This first species has an interesting taxonomic history. Pictet (1843) transferred it to *Palingenia* without any explanation, but Eaton (1871) thought that the species belonged to the genus *Hexagenia* Walsh. However, Eaton later (1883) transferred the species back to *Palingenia.* Lestage (1931) referred to this species as enigmatic because *Palingenia* does not occur in South America and nobody has seen the type specimen of the species. Today it is considered a *nomen dubium.*

Knowledge of the mayfly fauna of South America has been slow in coming. It took almost a century after Weber's first description of *Ephemera atrostoma* before a considerable number of taxonomic studies of South American mayflies began to appear in various publi-

Trends in Research in Ephemeroptera and Plecoptera
Edited by E. Dominguez, Kluwer Academic/Plenum Publishers, 2001

cations. Figure 1 shows a remarkable increase in the number of taxonomic papers on mayflies from 1901 to 1975. The majority of the taxonomic papers that were published during this period included those of Demoulin, Edmunds, Lestage, Mayo, Navás, Traver, and Ulmer. An even more dramatic increase in taxonomic studies of the mayfly fauna in the region started in the mid 1970's because of the efforts of a newer generation of Ephemeropterists (Domínguez, Flowers, Hubbard, Lugo-Ortiz, McCafferty, Pescador, Peters, and Savage, whose works account for about one third of the over 253 taxonomic papers on mayflies of the region (see Hubbard 1982, Hubbard and Peters 1977, 1981).

TAXON RICHNESS OF THE MAYFLY FAUNA IN THE REGION

We are just beginning to recognize the diversity of the South American mayfly fauna. Following McCafferty's (1991) recent higher classification of the Ephemeroptera in which 24 extant families are recognized worldwide, 13 families are presently represented in South America including the enigmatic monotypic genus *Melanemerella,* whose familial placement, however, remains unresolved, as some consider it an ephemerellid while others think it is a leptophlebiid. Of the approximately 2 500 mayfly species known worldwide, 375 species are presently known to occur in South America, almost half of the number of species known from America north of Mexico (~700 species). The South American mayfly fauna is presently represented by 91 described genera, roughly 23% of the 330 genera known worldwide. Approximately sixty and eighty percent of the South American mayfly genera and species, respectively, are endemic to the region.

MAYFLY FAUNA OF THE VARIOUS SOUTH AMERICAN COUNTRIES

Comparing the currently known mayfly taxa of the various countries in South America, Brazil and Argentina have the highest number of genera (59% and 38% respectively), distantly followed by Peru and Chile with 28% and 23% of the total number of currently known genera, respectively (Fig. 2). A similar pattern occurs in species richness; Brazil and Argentina again have the highest number (38% and 30% respectively) of species in the region (Fig. 3). We believe that the enormous gap in faunal composition reflects more the history of collecting in certain countries rather than the actual taxon richness in these countries. It is hard to imagine Chile, for instance, a small country, and with comparatively less diverse habitats, having more mayfly taxa than the other countries in the region except for Argentina, Brazil, and Peru. Based on the number of recently published taxonomic studies on mayflies, it is obvious there have been more concerted collecting efforts in Argentina, Brazil and especially Chile than in the other countries of South America.

GENERIC AND SPECIES DIVERSITY OF MAYFLY FAMILIES IN THE REGION

The generic composition of the different mayfly families in the region ranges from monogeneric [Coloburiscidae (*Murphyella*), Ephemeridae (*Hexagenia*), Metamoniidae (*Metamonius*), Oniscigastridae (*Siphlonella*), and Ephemerellidae (*Melanemerella*)] to as many as 25 and 35 genera in the Baetidae and Leptophlebiidae, respectively (Fig. 4). Based on the number of genera and species (35 and 118, respectively), the family Leptophlebiidae is the largest mayfly group representing 38% and 30% of the 91 genera and 375 species of mayflies in the region (Figs. 4 and 5). The taxonomy of this family is also comparatively well known. The leptophlebiid fauna of southern South America (Chile and Andean region of Argentina including the Magallanes Islands) is very much identifiable to species, and a good number of species in the area have the nymph and adults associated. Indeed, there is a good possibility that the leptophlebiids could very well be the most diverse mayfly group in

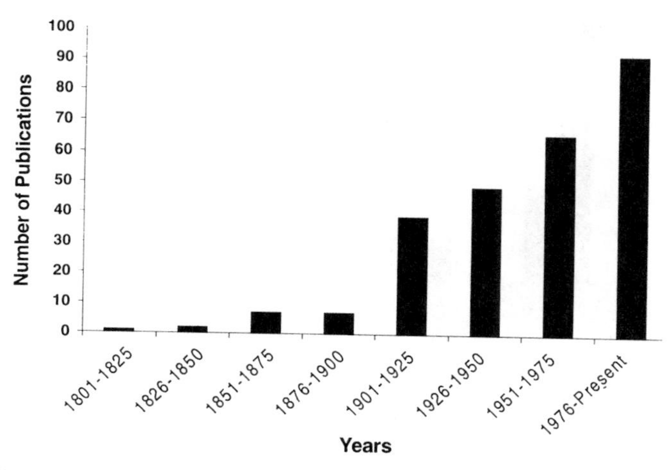

Fig. 1. Publications dealing with South American mayflies from 1801 to present (total N° of publications: 263).

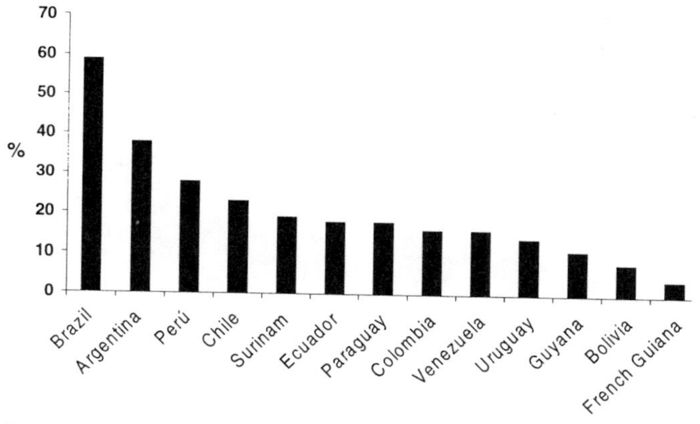

Fig. 2. Percentage of mayfly genera in each country of South America (total N° of genera: 91).

South America but human factors should not be ignored. For example, Savage in his recent taxonomic works on the Leptophlebiidae from Brazil, Domínguez and Pescador in Argentina, and Pescador and Peters in Chile have contributed considerably to the tremendous increase of described leptophlebiid taxa in the region. The family Baetidae represents the second largest mayfly group with approximately 92 species currently known representing 25 genera. Recent papers by Flowers, McCafferty, Lugo-Ortíz, and Waltz on the baetid fauna in the region are significant additions to the earlier works of Traver and Edmunds.

The family Leptohyphidae and Oligoneuriidae each have seven genera and are represented by 53 and 15 species, respectively. The speciose *Leptohyphes* has thirty-four of the 53 known leptophyphid species in the region. The families Euthyplociidae and Polymitarcyidae each have three genera and are represented by 6 and 52 species, respectively. Interestingly, approximately 81% of the 52 known polymitarcyid species in the region belong to the genus *Campsurus*.

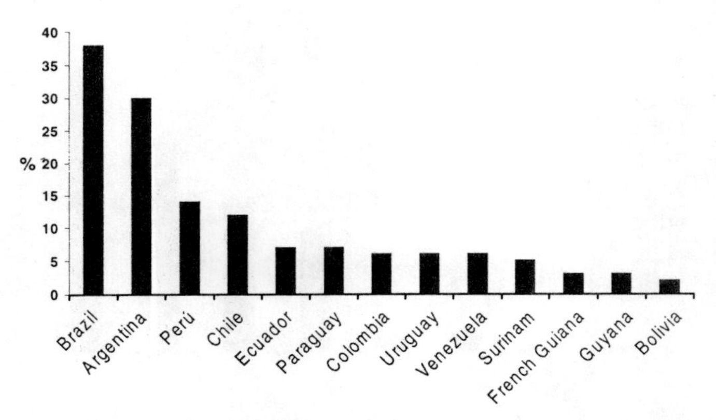

Fig. 3. Percentage of mayfly species in each country of South America (total N° of species: 375).

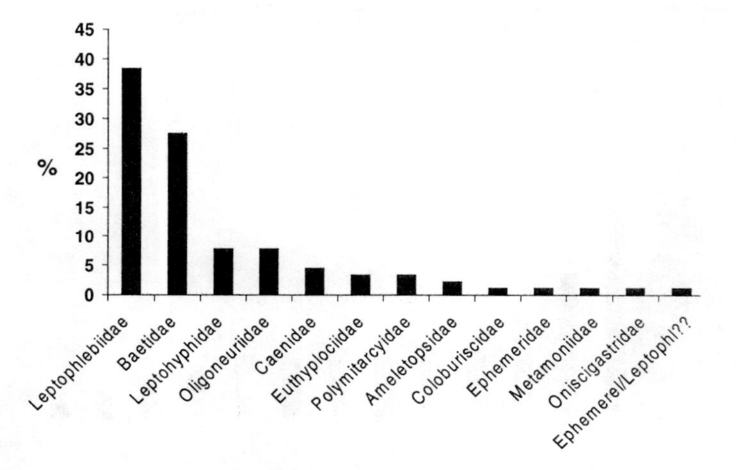

Fig. 4. Generic composition of the mayfly fauna of South America.

In addition to the genus *Brasilocaenis* that is endemic to Brazil, the family Caenidae is represented in the region by three other geographically widespread genera (*Brachycercus, Caenis, Cercobrachys*). The four caenid genera occurring in the region contain 25 species, approximately 7% of the mayfly species in the region (Fig. 5).

PROBLEMS ASSOCIATED WITH TAXONOMIC STUDIES OF THE REGION

A high percentage of taxa are known or described from only one life stage. Approximately 53% of the species in the region are known from the alate forms while only 11 % are known from both the adults and nymph. Most species of leptohyphids and baetids are known only from the nymphal stage. Conversely, all species of *Campsurus* (Polymitarcyidae) are known from adults; some of the descriptions are based only on females and a number of subimagos.

Disparity in the portion of the fauna known from various countries in South America. Present data appear to be more a reflection of collecting effort than anything else. More

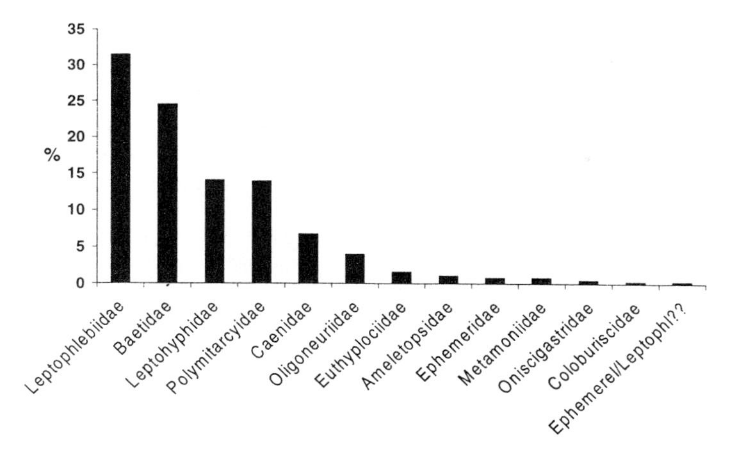

Fig. 5. Species composition of the mayfly fauna of South America.

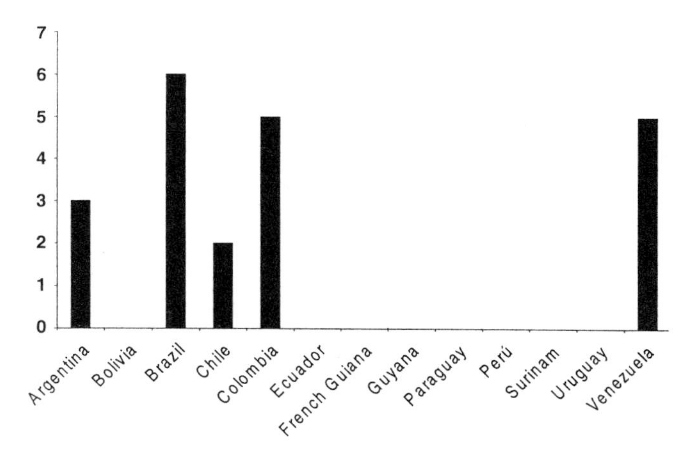

Fig. 6. Estimate of number of mayfly taxonomists.

collection and study must be done in countries where the mayfly fauna is virtually unknown. Figure 6 lists the approximate number of workers in the various countries of South America presently engaged in taxonomic activities with mayflies. It is obvious that the paucity of endemic workers contributes to the lack of knowledge of various faunas.

Lack of knowledge of the location where many type specimens are deposited. This is a particular problem with the species of Navás. Additionally, difficulties involved in borrowing type specimens and lack of financial support to visit museums and examine types contribute to taxonomic difficulties in South America.

Lack of communication and information exchange. There appears to be less than ideal communication among mayfly workers in the region and exchange of publications is often limited. Workshops, seminars, conferences, and visitations are all excellent methods of enhancing communication.

CONCLUDING REMARKS

Our knowledge of the South American mayflies though still fairly inadequate, is steadily improving. The quality of taxonomic papers which have been published recently is also steadily improving and many have included the descriptions of eggs, nymphs, and adults of some taxa. Additionally, interpretations of the phylogenetic relationships and zoogeography of some taxonomic groups have also been included in many of the mayfly papers dealing with the region.

We must, however, accelerate our efforts to collect the fauna in many areas of the region before the undescribed taxa become eliminated by environmental perturbations (e.g., deforestation, urbanization, industrialization, and agricultural activities). We, as systematists, must find a way to educate funding agencies of the importance of systematic research in the region and convince them to support our research efforts.

REFERENCES

(Selected references; a complete list of references to the South American fauna is beyond the space limitations of this paper)

Eaton, A. E. 1871. A monograph on the Ephemeridae. Trans. ent. Soc. London 1871: 1-164, pl. 1-6.

Eaton, A. E. 1883-1888. A revisional monograph of recent Ephemeridae or mayflies. Trans. Linn. Soc. Lond, Zool., 3: 1-352, pl. 1-65.

Hubbard, M. D. 1982. Catálogo abreviado de Ephemeroptera da América do Sul. Pap. Avul. de Zool. 34: 257-282.

Hubbard, M. D. and W. L. Peters. 1977. Ephemeroptera, pp. 165-169. In: S. H. Hurlbert (ed.). Biota acuática de Sudamérica austral. San Diego State University, San Diego, California.

Hubbard, M. D. and W. L. Peters. 1981. Ephemeroptera, pp. 55-63 In: S.H. Hurlbert, G. Rodríguez and N. D. Santos (eds.). Aquatic biota of tropical South America, Part 1: Arthropoda. San Diego State University, San Diego, California.

Lestage, J. A. 1931. Contribution à l'étude des larves des Ephéméroptères. VII. Le groupe Potamanthidien. Mém. Soc. ent. Belg. 23: 73-146.

McCafferty, W. P. 1991. Toward a phylogenetic classification of the Ephemeroptera (Insecta): A · commentary on systematics. Ann. ent. Soc. Amer. 84: 343-360.

Pictet, F. J. 1843-1845. Histoire naturrelle générale et particulière des insectes néuroptères. Famille des éphémérines. Chez J. Kessmann et Ab. Cherbuliz, Geneva. 300 pp., xix + 47 pl.

Weber, F. 1801. Observations entomologicae, pp. 99-100.

THE EPHEMEROPTERA OF NEW ZEALAND AND NEW CALEDONIA

William L. Peters

Entomology, Orr Drive
Florida A&M University, Tallahassee
FL 32307, USA

ABSTRACT

The status of our systematic knowledge of the Ephemeroptera of New Zealand and New Caledonia is discussed. In New Zealand the Leptophlebiidae are the largest group, while in New Caledonia the Leptophlebiidae are the only family except for one report of Baetidae. In both cases modern systematic studies are available for the Leptophlebiidae. All species and genera are endemic to New Zealand and New Caledonia, and all genera are closely related to those in Australia, southern South America, South Africa, Madagascar, the Seychelles, Sri Lanka, Southern India and Sulawesia. The remaining genera in other families occurring in New Zealand are listed.

OVERVIEW

New Zealand

The first mayflies recorded from New Zealand were *Palingenia humeralis* and *Baetis remota* (both now referred to *Coloburiscus humeralis*) and *Baetis scita* (now referred to *Neozephlebia*) based on descriptions by Walker (1853). Gradually, other species were added to this list so that by 1899, Eaton could count 11 described species and one undescribed species from New Zealand.

Since the introduction of freshwater trout to New Zealand in the 1860's (McDowell 1994), the mayflies and other aquatic insects of New Zealand have been studied by many fishermen and photographers, but unfortunately few systematic papers have ever been published on the mayflies.

For many years both amateurs and scientists alike had to rely on the works of Phillips (1930a,b; 1931) to identify New Zealand mayflies. Since Phillips' revisions of the New Zealand Ephemeroptera, Penniket published additional studies and these included Penniket (1961) on the Leptophlebiidae, Penniket (1962a,b) on *Oniscigaster* and *Siphlaenigma*, and Penniket (1966) on *Rallidens*. Towns and Peters (1978, 1979a,b) and Towns (1983) began

Trends in Research in Ephemeroptera and Plecoptera
Edited by E. Dominguez, Kluwer Academic/Plenum Publishers, 2001

to revise the genera of New Zealand Leptophlebiidae, and these results were included in the 1996 monograph on the Leptophlebiidae of New Zealand. Thirty species in 12 genera are represented, all endemic to New Zealand. Closely related genera occur in New Caledonia, Australia, southern South America, South Africa, Madagascar, the Seychelles, Sri Lanka, Southern India and Sulawesia. The revision includes descriptions of all genera and species and illustrated keys to the imagos and nymphs of all taxa.

The imaginal and nymphal keys also include all other mayfly genera in New Zealand. These are: Siphlaenigmatidae (*Siphlaenigma*, 1 sp.), Coloburiscidae (*Coloburiscus*, 1 sp.), Nesameletidae (*Nesameletus*, 2 spp.), Rallidentidae (*Rallidens*, 1 sp.), Ameletopsidae (*Ameletopsis*, 1 sp.), Oniscigastridae (*Oniscigaster*, 3 spp.), and Ephemeridae (*Ichthybotus*, 1 sp.). All genera are endemic and with the exception of *Ichthybotus*, all are closely related to genera in Australia and southern South America. Currently, Mr. Terry Hitchings of the Canterbury Museum is reevaluating Penniket's original collection in the Canterbury Museum, especially Penniket's unpublished study on *Nesameletus*.

New Caledonia

Kimmins (1953) recorded the first mayfly from New Caledonia as "? *Atalophlebia* sp." based on two male imagos and one female subimago collected in Nouméa. In June, 1965, Dr. George F. Edmunds, Jr., University of Utah, received from Prof. Dr. F. Starmühlner, 1. Zoologischen Institutes der Universität Wien, a collection of mayfly nymphs from New Caledonia. All specimens were Leptophlebiidae and none could be assigned to known species and genera in the world.

This collection prompted me to request from the National Geographic Society, Washington, D.C., funds to collect and rear mayflies in New Caledonia. All mayfly specimens obtained on this trip were again Leptophlebiidae, except for two subimaginal specimens of Baetidae from Rivière Bleue upstream of a reservoir in southern New Caledonia. Results of this collecting trip are published in six systematic papers: Peters and Peters (1979-1980, 1981a,b) and Peters, Peters and Edmunds (1978, 1990, 1994); two remaining systematic parts and keys to the entire fauna are yet to be published. When completed, these papers will describe 46 species arranged in 19 genera. All taxa are endemic with sister group relations in New Zealand, except for one genus (*Kariona*) which appears most closely related to a genus in Madagascar.

Currently, Nathalie Mary is collecting mayflies and other aquatic insects in New Caledonia for a biotic index study sponsored by the Université Paul-Sabatier in Toulouse. Her extensive collection of nymphal mayflies does contain several new species and at least one new genus.

Although the present state of knowledge of mayflies of New Zealand and New Caledonia appears quite good, it is clear from studies by both amateurs and professionals in New Zealand and from those of Nathalie Mary in New Caledonia that there are still species to be discovered. All Leptophlebiidae in this region belong to the subfamily Atalophlebiinae, but a complete study of the phylogenetic relationships of southern hemisphere genera will require more knowledge of genera from other regions, particularly Australia and Madagascar.

REFERENCES

Eaton, A. E. 1899. An annotated list of the Ephemeridae of New Zealand. Trans. ent. Soc. Lond. 1899: 285-293.

Kimmins, D. E. 1953. XXII.–Miss L. E. Cheesman's expedition to New Caledonia, 1949–Orders Odonata, Ephemeroptera, Neuroptera and Trichoptera. Ann. Mag. Nat. Hist., Ser. 12, 6 (64): 241-257.

McDowell, R. M. 1994. Gamekeepers for the Nation: the story of New Zealand's acclimatisation societies 1861-1990. Canterbury Univ. Press, Christchurch.

Penniket, J. G. 1961. Notes on New Zealand Ephemeroptera. I. The affinities with Chile and Australia and remarks on *Atalophlebia* Eaton (Leptophlebiidae). N. Z. Ent. 2 (6): 1-11.

Penniket, J. G. 1962a. Notes on New Zealand Ephemeroptera. II. A preliminary account of *Oniscigaster wakefieldi* McLachlan, recently rediscovered (Siphlonuridae). Rec. Canterbury Mus. 7:375-388.
Penniket, J. G. 1962b. Notes on New Zealand Ephemeroptera. III. A new family, genus and species. Rec. Canterbury Mus. 7:389-398.
Penniket, J. G. 1966. Notes on New Zealand Ephemeroptera. IV. A new siphlonurid subfamily: Rallidentinae. Rec. Canterbury Mus. 8: 163-175.
Peters, W. L. and J. G. Peters. 1979-1980. The Leptophlebiidae of New Caledonia (Ephemeroptera). Part II – Systematics. Cah. ORSTOM, sér. Hydrobiol. 13: 61-82.
Peters, W. L. and J. G. Peters. 1981a. The Leptophlebiidae: Atalophlebiinae of New Caledonia (Ephemeroptera). Part III – Systematics. Rev. Hydrobiol. Trop. 14: 233-243.
Peters, W. L. and J. G. Peters. 1981b. The Leptophlebiidae: Atalophlebiinae of New Caledonia (Ephemeroptera). Part IV – Systematics. Rev. Hydrobiol. Trop. 14: 245-252.
Peters, W. L., J. G. Peters and G. F. Edmunds, Jr. 1978. The Leptophlebiidae of New Caledonia (Ephemeroptera). Part I – Introduction and systematics. Cah. ORSTOM, sér. Hydrobiol. 12: 97-117.
Peters, W. L., J. G. Peters and G. F. Edmunds, Jr. 1990. The Leptophlebiidae: Atalophlebiinae of New Caledonia (Ephemeroptera). Part V – Systematics. Rev. Hydrobiol. Trop. 23: 121-140.
Peters, W. L., J. G. Peters and G. F. Edmunds, Jr. 1996. The Leptophlebiidae: Atalophlebiinae of New Caledonia (Ephemeroptera). Part VI – Systematics. Rev. Hydrobiol. Trop. (1994) 27: 97-105.
Phillips, J. S. 1930a. A revision of New Zealand Ephemeroptera. Part 1. Trans. Proc. N. Z. Inst. 61: 271-334. Plates 50-60.
Phillips, J. S. 1930b. A revision of New Zealand Ephemeroptera. Part 2. Trans. Proc. N. Z. Inst. 61: 335-390. Plates 61-67.
Phillips, J. S. 1931. Studies of New Zealand mayfly nymphs. Trans. ent. Soc. Lond. 79: 399-422. Plates XVI-XXII.
Towns, D. R. 1983. A revision of the genus *Zephlebia* (Ephemeroptera: Leptophlebiidae). N. Z. J. Zool. 6: 409-419.
Towns, D. R. and W. L. Peters. 1978. A revision of genus *Atalophlebioides* (Ephemeroptera: Leptophlebiidae). N. Z. J. Zool. 5: 607-614.
Towns, D. R. and W. L. Peters. 1979a. Three new genera of Leptophlebiidae (Ephemeroptera) from New Zealand. N. Z. J. Zool. 6: 213-235.
Towns, D. R. and W. L. Peters. 1979b. New genera and species of Leptophlebiidae (Ephemeroptera) from New Zealand. N. Z. J. Zool. 6: 439-452.
Towns, D. R. and W. L. Peters. 1996. Leptophlebiidae (Insecta: Ephemeroptera). Fauna of New Zealand 36. Manaaki Whenua Press.
Walker, F. 1853. Ephemeridae. pp. 533-585, IN: List of the specimens of neuropterous insects in the collection of the British Museum, 3.

CURRENT KNOWLEDGE OF MAYFLY RESEARCH IN EUROPE (EPHEMEROPTERA)

Michel Sartori

Museum of Zoology
P.O. Box 448, CH-1000 Lausanne 17
Switzerland

ABSTRACT

The current status of mayfly research in Europe and North Africa (West Palaearctic area) is briefly reviewed and focused on three main topics: systematics, faunistics, and current research areas.

INTRODUCTION

It is obvious that Ephemeroptera have been studied for a long time in Europe. The pioneer work by Linnaeus (1758) is the root of our actual binomial nomenclature, and was mainly based on European animals and plants. He recognized eleven mayfly species placed in the single genus *Ephemera* within the Neuroptera. Things changed considerably during the 19th century, with the works of Pictet (1843) and, especially, Eaton (1882-1888). The number of European mayfly species continuously increased during the first half of the 20th century, with works such as those of Bengtsson, Esben-Petersen, Klapálek, Mikulski, Schönemund, and Ulmer. After World War II, other scientists (such as Grandi, Jacob, Kimmins, Landa, Macan, Müller-Liebenau, Puthz, Soldán, Sowa, and Thomas, among many others) published major works on European mayflies.

In this small review, only subjects relating directly to mayfly research, i.e. systematics, faunistics, autecology, as well as anatomy, physiology and behaviour will be discussed. Hydrobiological themes involving mayflies as a part of the benthic community will not be treated.

SYSTEMATICS

In the first edition of the "Limnofauna Europaea", Illies (1967) mentioned approximately 200 species (excluding non European Mediterranean countries). In the second edition, Puthz (1978) recorded about 220 species.

Trends in Research in Ephemeroptera and Plecoptera
Edited by E. Dominguez, Kluwer Academic/Plenum Publishers, 2001

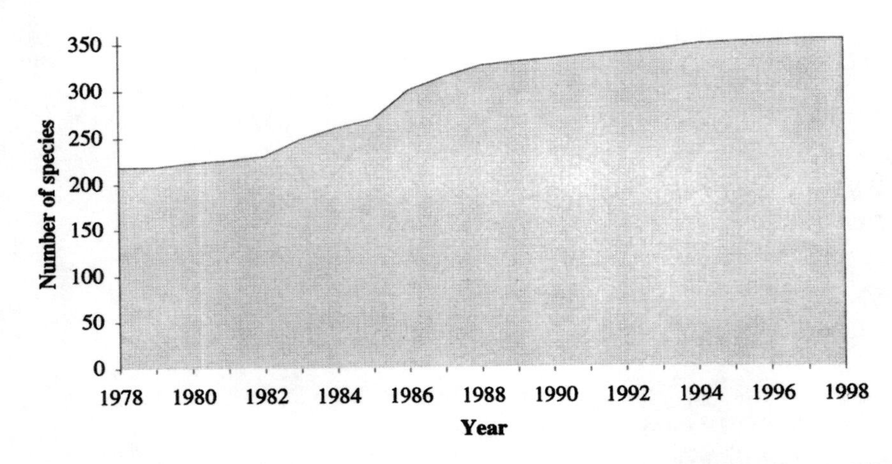

Fig. 1. Evolution of the number of mayfly species in the West Palaearctic from 1978 until 1998.

Figure 1 illustrates the evolution of the number of European mayfly species between 1978 and 1998. A little more than 350 species are actually known, i.e. that during the last 20 years, our taxonomic knowledge increased by 60%. As the shape of the curve indicates (Fig. 1) most of these new taxa were described between 1982 and 1988, with the maximum peak reached in 1986 when the descriptions of 31 new species were published. Since the beginning of the 1990's, less than half a dozen new species have been described each year. Two reasons can be put forward: first, because the European fauna is now pretty well known; second, because systematic activities obey a strange cycle in which periods of "low activity" (i.e., periods with low description rates) follow periods of "high activity" (i.e., with high description rates). The next years will tell us which is the right one.

Speaking about the evolution of mayfly species number in general is a little bit gross since taxa are not involved in the same manner. On one hand, the last European species described in the genus *Ephemera*, among the six actually known, was *E. hellenica* by Demoulin in 1955. On the other hand, the number of species in Baetidae and *Rhithrogena* drastically increased during the last twenty years (fig. 2). The European Baetidae actually reach ca. 80 species (+ 43% since 1978) whereas the heptageniid genus *Rhithrogena* has ca. 75 species (+ 180% since 1978!). The main works on these taxa were those of Alba-Tercedor, Belfiore, Soldán, Sowa, and Thomas.

FAUNISTICS

For literature references used in this chapter, consult the list established by Sartori and Landolt (1999) or Landolt and Sartori, this volume.

From a faunistic point of view, there are important discrepancies among West Palaearctic countries. I will split West Palaearctic countries in four categories with respect to the degree of knowledge in their mayfly fauna (alphabetically listed):

1) Countries with well known mayfly fauna, i.e., for which at least 95% of the supposed fauna is recorded: Austria, Belgium, Bulgaria, Czech Republic, Denmark, Finland, France, Ireland, Luxembourg, Netherlands, Norway, Poland, Slovakia, Sweden, Switzerland, United Kingdom.

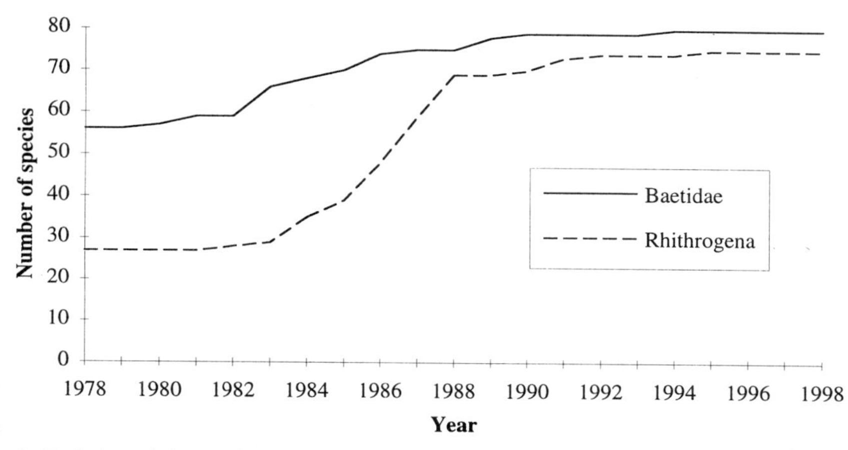

Fig. 2. Evolution of the number of species within the family Baetidae and the genus *Rhithrogena* from 1978 until 1998.

2) Countries with numerous mayfly records, but in which some geographic areas or some taxonomic groups lack recent data: Algeria, Germany, Hungary, Israel, Italy, Macedonia, Morocco, Portugal, Romania, Slovenia, Spain, Tunisia.

3) Countries with only some mayfly records, for which no actual checklists are available: Croatia, Estonia, Greece, Latvia, Lebanon, Lithuania, Syria, Turkey.

4) Countries poorly known, with some isolated records or frankly "terra incognita" concerning mayflies: Albania, Belorussia, Bosnia, Egypt, Libya, Moldovia, Montenegro, Serbia, Ukraine.

These categories perhaps are roughly made, but they highlight places where much has to be done: the Balkans, the former USSR states, as well as the Eastern Mediterranean area. The advancement of the mayfly knowledge in these areas will certainly also lead to the discovery of new species, and thus, will also increase the systematic count in Europe.

Countries mentioned in category 1, as well as some of category 2, have been analyzed further, in order to see if the notion of "well known mayfly fauna" had some realistic basis. In other words, would it be possible to extrapolate the expected number of mayfly species to be found in such countries?

Figure 3 shows the relationship between the number of known species and the size of the investigated country. As it can be seen, there is a semi-logarithmic relationship between the variables. But it is clearly different between Central - South and North European countries. Latitude, climate, landscape homogeneity and insularity are responsible for these differences. Netherlands lies somewhat in-between, whereas Austria is the country with the highest diversity, considering the country size. Italy, Germany and Spain lies below the regression curve, suggesting that additional species could be found.

OTHER FIELDS INVESTIGATED

The anatomy of mayflies has been relatively poorly studied in Europe, except for the pioneer works by Marta Grandi. More recently, researches led by Landa and Soldán (1985) on the anatomy of mayfly larvae with its phylogenetic implications have to be mentioned. The

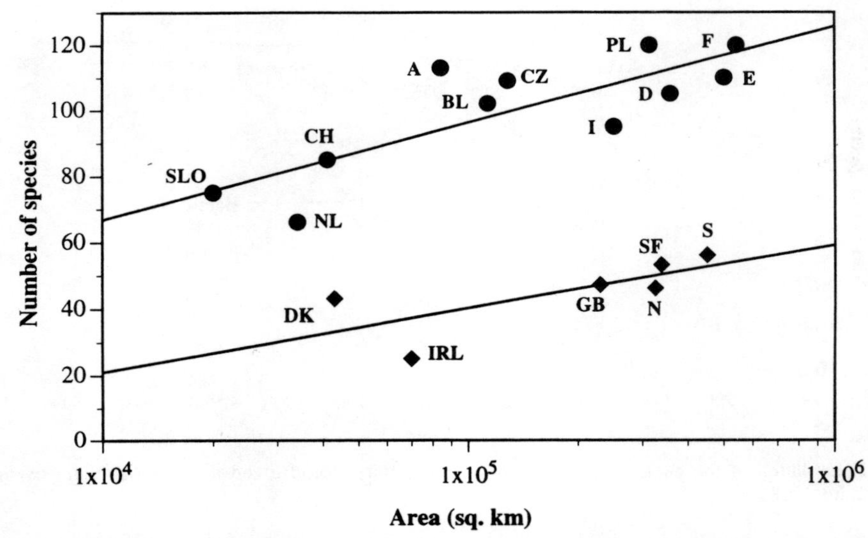

Fig. 3. Relationship between area of the country and the number of mayfly species actually known. Upper curve: Central and South European countries (regression coefficient $r^2 = 0.66$; $p<0.001$); lower curve North European countries (regression coefficient $r^2 = 0.53$; $p<0.005$). A: Austria, BL: Bulgaria, CH: Switzerland, CZ: former Czechoslovakia, D: Germany, DK: Denmark, E: Spain, F: France, GB: Great Britain, I: Italy, IRL: Ireland, N: Norway, NL: Netherland, PL: Poland, S: Sweden, SLO: Slovenia, SF: Finland. (Redrawn from Sartori and Landolt, 1999, courtesy of the Centre Suisse de Cartographie de la Faune, Neuchatel).

study of the ultrastructural organization in mayflies is almost entirely the property of Elda Gaino, with numerous works on egg chorionic structure, as well as ovarioles, fat body, and sensilla, for instance. Functional morphology, especially of the mouthparts, is the domain of Austrian and German scientists, with works such as those of Elpers, Schönemann, Staniczek, and Strenger. Autecological studies have received considerable attention since the pioneer work by Degrange (1960), leading to numerous researches, involving life cycle strategies, such as those of Brittain, Elliott, Humpesch, and Macan.

CURRENT RESEARCH AREAS

A compilation of available literature for three years is presented in fig. 4. Faunistic and ecological studies are the dominant fields investigated. Applied aspects involving mayflies, such as biomonitoring, and human impacts on the environment are still underrepresented compared to other areas (e.g. North America). Autecological studies, mainly life histories, need to be encouraged since we lack data on a great number of species.

Thanks to the increased knowledge in systematics, more and more scientists each year are working on mayflies, leading to the diversification of the investigated themes.

NEW RESEARCH AREAS

Mayflies constitute a unique and fascinating material for basic research such as behavioural ecology. The study of particular cases can shed new light on our understanding of general biological processes. Surprisingly, little has been done in this field. Some recent

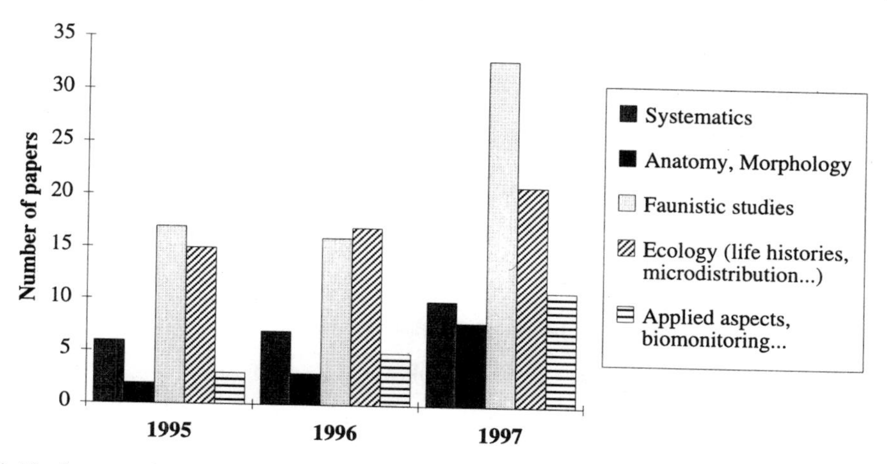

Fig. 4. Mayfly research main topics in Europe during three years (1995-1997). Source: NABS current and selected bibliographies on benthic biology.

attempts have proven that mayfly study can bring some answers of interest to all biologists: mating behaviour and competition (e.g., Flecker et al., 1988; Allan and Flecker, 1989; Harker, 1992), effect of parasites on physiology and predation (Vance, 1996; Vance and Peckarsky, 1997), asphalt as a trap for polarization-sensitive aquatic insects (Kriska et al., 1998), or flightlessness as a response to predator removal (Ruffieux et al., 1998). These studies have to be encouraged since they contribute to making mayflies better known to the scientific community, as well as to the public.

Surprisingly, new molecular techniques (e.g. DNA sequences) are still not used in mayfly research, despite their important contribution to systematics, phylogeny, and biogeography (Landolt, 1991). This field will certainly be developed all over the world in the next years.

CONCLUSIONS

Mayfly knowledge in Europe has increased considerably during the last ten years. On a systematic and faunistic point of view, more information is needed from eastern and southeastern Europe. This situation will be improved only when economic and political stability again occur. We need autecological data on numerous mayfly species. This basic research is a necessary condition to understanding which factors affect populations in order to protect them. Then, the use of mayflies in applied aspects of hydrobiology will be more efficient.

ACKNOWLEDGMENTS

I sincerely thank all the colleagues who help me in providing numerous unpublished data or information used in this paper. Special thanks to Javier Alba-Tercedor (Granada) who pushed me to complete an earlier draft for publication, and to Jean-Luc Gattolliat for helpful comments. Despite all this help, I am solely responsible for errors that could still occur.

REFERENCES

Allan, J.D. and Flecker, A.S. 1989. The mating biology of a mass-swarming mayfly. Anim. Behav. 37: 361-371.

Degrange, Ch. 1960. Recherches sur la reproduction des Ephéméroptères. Trav. Lab. Hydrobiol. Pisc. Univ. Grenoble. 50/51: 7-193.

Eaton, A.E. 1883-1888. A revisional monograph of recent Ephemeridae or Mayflies. Trans. Linn. Soc. London 3: 1-352.

Flecker, A.S., Allan, J.D. and McClintock, N.L. 1988. Male body size and mating success in swarms of the mayfly *Epeorus longimanus*. Holarct. Ecol. 11: 280-285.

Harker, J.E. 1992. Swarm behaviour and mate competition in mayflies (Ephemeroptera). J. Zool., Lond. 228: 571-587.

Illies, J. 1967. Ephemeroptera. In: J. Illies ed., Limnofauna Europaea, pp. 212-219, Gustav Fischer Verlag, Stuttgart.

Kriska, G., Horvath, G. and Andricovics, S. 1998. Why do mayflies lay their eggs en masse on dry asphalt roads? Water-imitating polarized light reflected from asphalt attracts Ephemeroptera. J. Exp. Biol. 201: 2273-2286.

Landa, V. and Soldán, T. 1985. Phylogeny and higher classification of the order Ephemeroptera: a discussion from the comparative anatomical point of view. Academia, Praha, 121 p.

Landolt. P. 1991. An approach to the application of molecular biological methods to solve taxonomical and phylogenetic problems of the Ephemeroptera, pp.3-14. In: J. Alba-Tercedor and A. Sanchez-Ortega (eds.). Overview and strategies of Ephemeroptera and Plecoptera, pp. 3-14, Sandhill Crane Press, Gainsville.

Landolt, P. and M. Sartori. 2000. The mayfly fauna of Switzerland. In: E. Dominguez (ed.). Trends in Research in Ephemeroptera and Plecoptera. Kluwer-Plenum Pub.

Pictet, F.-J. 1843-1845. Histoire naturelle générale et particulière des insectes Névroptères. Famille des Ephémérines. Kessmann and Cherbuliez, Geneve, 300 p. + 49 plates.

Puthz, V. 1978. Ephemeroptera. In: J. Illies ed., Limnofauna Europaea, 2nd edition, pp. 256-263, Gustav Fischer Verlag, Stuttgart.

Ruffieux, L., Elouard, J.-M. and Sartori, M. 1998. Flightlessness in mayflies and its relevance to hypotheses on the origin of insect flight. Proc. R. Soc. London, Ser. B 265: 2135-2140.

Sartori, M. and Landolt, P. 1999. Atlas de distribution des Ephemeres de Suisse (Insecta, Ephemeroptera). Fauna helvetica 3, 214 p., SEG-CSCF ed., Neuchâtel.

Vance, S.A. 1996. Morphological and behavioural sex reversal in nermithid-infected mayflies. Proc. R. Soc Lond., Ser. B 263: 907-912.

Vance, S.A. and Peckarsky, B.L. 1997. The effect of nermithid parasitism on predation of nymphal *Baetis bicaudatus* (Ephemeroptera) by invertebrates. Oecologia 110: 147-152.

STATUS OF THE SYSTEMATIC KNOWLEDGE AND PRIORITIES IN EPHEMEROPTERA STUDIES: THE ORIENTAL REGION

Tomáš Soldán

Institute of Entomology
Academy of Sciences of the Czech Republic
31 Branišovská
CZ - 370 05 České Budějovice, Czech Republic

ABSTRACT

Of the 28 families containing about 400 genera and approximately 3,000 species of extant Ephemeroptera, 19 families, 104 genera (47 of these endemic) and more than 535 species have been described from the Oriental region including its transition zones (Afghanistan, Himalaya, China, and Papua-New Guinea). The families Baetidae, Heptageniidae and Leptophlebiidae exhibit the greatest diversity in this region. Mayfly diversity and distribution are briefly reviewed at the generic level based on the literature. However, our knowledge of Oriental mayflies seems to be fragmentary since most species are known only in either the larval (about 35%) or adult stage (about 39%); the total number of species is estimated to be at least 3-4 times higher. With few exceptions (e.g. Sri Lanka, Hong Kong, and the Sunda Islands), there are no detailed data on local faunas, biogeography, ecology and life cycles. Further taxonomic research, association of larvae and adults previously described separately, revision of available type material, detailed revisions of genera or species-groups, studies of faunas of defined regions (however small), biogeographic analysis including definition of the limits of the Oriental region, and intensive study of mayfly biology are the main research priorities discussed in this paper.

INTRODUCTION

Although considerable attention has been paid to mayflies of the Oriental region, our present knowledge is very fragmentary. Besides original descriptions by earlier authors (e.g. Pictet, Eaton, Walker, Hagen, and others) and the extensive monographs concerning the Sunda Islands (Ulmer, 1939, 1940), our knowledge is mostly based on isolated or occasional descriptions of new species and genera collected mostly by expeditions directed to other problems, or by general collectors. Furthermore, many types are lost (cf. Hubbard, 1984a, Alba-Tercedor and Peters, 1985) and subsequent taxonomic treatment of some groups is difficult. Only a few higher taxa, i.e. the Leptophlebiidae, Potamanthidae and Neoephemeridae, have been treated in detail (Peters & Edmunds, 1970; Bae & McCafferty, 1991, 1998) and

Trends in Research in Ephemeroptera and Plecoptera
Edited by E. Dominguez, Kluwer Academic/Plenum Publishers, 2001

catalogs exist for species of some areas, e.g. Tonkin (Lestage, 1921, 1924), the Indian subcontinent (Hubbard & Peters, 1978), the Philippines (Hubbard & Pescador, 1978), Hong Kong (Hubbard, 1986) or China (Gui, 1985; You & Gui, 1995). Larvae of mayflies of Taiwan are being studied intensively at the present time (Kang & Yang, 1994a-f, 1995, 1996). From the biogeographical point view, only Sri Lanka (Hubbard & Peters, 1984) and the Malaysia Peninsula and the Sunda Islands (Edmunds & Polhemus, 1990) have been discussed and Edmunds (1972, 1975, 1979) gave an extensive discussion of the origin of most Oriental genera.

The principal objective of this review is to briefly summarize knowledge of the diversity of mayflies of the Oriental region emphasizing which taxa should be urgently treated and to suggest the priorities of mayfly studies in this geographical area.

DIVERSITY AND DISTRIBUTION OF THE ORIENTAL EPHEMEROPTERA

In the following discussion, the mayfly taxa discussed are arranged according to the catalog of the families and genera of the world by Hubbard (1990) although this list principally follows the higher classification by McCafferty (1991a). In evaluating individual taxa, I have attempted to follow their original meaning when established to emphasize the needs for further revision. The Oriental region (sometimes called the Indomalayan region), although usually limited by the Thar Desert in Pakistan, the Himalaya's main mountain ridge, southern China and the Wallace's line, is considered here to include transition zones and/or adjacent subareas of neighbor biogeographic regions, i.e. Afghanistan, Pakistan, most of China, Cele-

Table 1. Families, genera and approximate number of species of Ephemeroptera in the Oriental region (* endemic taxa; the genus *Teloganella* of unclear family status not included)

Family	N° of genera	N° of species	Larval stage only	Adult stage only	Limited distribution or endemic taxa
Baetidae	23	152	78	55	9 genera*, Tab. 2
Ameletidae	1	5	3	1	Himalaya, Taiwan
Siphlonuridae	1	2	-	2	*Siphluriscus**, China
Oligoneuriidae	2	3	1	1	Chromarcyinae*
Isonychiidae	1	7	2	3	
Heptageniidae	19	114	37	48	7 genera*, Tab. 3
Leptophlebiidae	23	65	10	16	16 genera*, Tab. 4
Behningiidae	1	1+	-	-	Thailand
Euthyplociidae	1	3	-	2	*Polyplocia**
Ephemeridae	2	42	2	24	*Eatonigenia**, subg. *Aethephemera**
Palingeniidae	3	22	-	19	*Plethogenesia**
Polymitarcyidae	2	13	3	7	subg. *Languidipes**
Potamanthidae	2	17	3	6	*Rhoenanthus** and subg. *Stygifloris**
Ephemerellidae	10	39	19	7	4 genera*
Teloganodidae	3	11	3	7	*Teloganodes**, *Macafertiella**
Tricorythidae	1	1+	-	-	
Neoephemeridae	1	5	2	-	*Potamanthellus**
Caenidae	6+	29	17	10	*Caenoculis**
Prosopistomatidae	1	9+	8	1	
Total (19 families + *Teloganella**)	103 (+ 1)	536 (+ 1)	188 (35%)	209 (39%)	141 spp. in both stages (26 %)

bes, Papua-New Guinea and northernmost Australia. The total number of families and genera and the approximate number of species described are listed in Table 1 and reviewed below.

Baetidae. This family represents the most diversified mayfly group within the Oriental region. The approximate number of species described for individual genera are apparent from Table 2. Fourteen genera are classified within the subfamily Baetinae, 7 genera within the Cloeoninae; however, the generic classification of the Baetidae is not still fully understood. Some genera are classified to subgeneric rank (e.g. *Acentrella* and *Baetiella* are considered to be subgenera of *Baetis*) and intergeneric taxonomic shifts are frequent in contemporary literature. For instance, an examination of *Pseudocloeon*-like species showed that only the type species, *P. kraepelini* Klapálek (larva unknown), could be included in the species-rich genus *Pseudocloeon* (cf. Waltz & McCafferty, 1987a); however, the most recent study (Lugo-Ortiz et al., 1999) considered *Pseudocloeon* congeneric with *Labiobaetis* and all species previously assigned to *Labiobaetis* were reassigned to *Pseudocloeon*. Consequently, there are 21 species of *Pseudocloeon* occurring in the Oriental region which were originally described in *Acentrella, Baetis, Labiobaetis* and *Pseudocloeon*, 8 of them known only in the adult stage and 12 described only as larvae. The only species known in both larval and adult stages is *P. fulmeki* (Ulmer).

Detailed study of *Cloeodes*-like species described as *Centroptella* (Braasch & Soldán, 1980) showed that these Oriental species actually belonged to *Cloeodes* (originally considered Neotropical) and *Chopralla* which is endemic to the Oriental region (Waltz & McCafferty, 1987b, c). Distributions of endemic genera are given in Table 2. Larvae of the peculiar genus *Symbiocloeon* live in bivalve molluscs (Müller-Liebenau, 1979). Adults of most endemic species have not been described, except for the adult female of *Echinobaetis* (Lugo-Ortiz & McCafferty, 1997).

The nearly cosmopolitan genera *Baetis* and *Cloeon* are very diverse comprising 43 and 25 species, respectively, in the Oriental region. However, these numbers are probably inaccurate. On one hand, many new species are yet to be described (there are at least 10 unnamed species of *Baetis* in Vietnam); on the other hand some species are probably synonyms named from different life cycle stages. For instance, all species of *Baetis* described from the Philippines (Müller-Liebenau, 1982), West Malaysia and Sabah (Müller-Liebenau, 1984a, b), Taiwan (Müller-Liebenau, 1985, Kang & Yang, 1994a, 1995a) and Sri Lanka (Müller-Liebenau & Hubbard, 1985) are known solely in the larval stage. Species known as both larvae and adults are very rare within *Baetis* (e.g. *B. lahauensis* Kaul & Dubey, *B. javanicus* Ulmer, altogether only 3 species) or *Cloeon* (e.g. *C. marginale* Hagen, *C. fluviatile* Ulmer or *C. exiguum* Navás, altogether 7 species). In some species, revision is difficult because the species is known only from the adult female (e.g., *Baetis seragruis* Dubey).

Predominantly Palaearctic genera extending into the Oriental region include: *Alainites* from China, Malaysia and Taiwan; *Nigrobaetis* from West Malaysia (one species) and Taiwan (8 species); *Acentrella* from Taiwan; *Procloeon* from India, Sri Lanka, Thailand, Malaysia, Hong Kong and Taiwan; and *Baetiella* from the Himalaya and Taiwan. *Alainites pekingensis* (Ulmer), *Baetiella bispinosa* (Gose), and *B. ladakae* Traver are the only species of these genera known in the adult stage. *Centroptilum* occurs in India and China and larvae of an undescribed species have been found in Vietnam. The Nearctic genus *Fallceon* is represented by a single species, *F. candidus* (Kang & Yang, 1995a) in Taiwan.

Ameletidae. Only the genus *Ameletus* of this exclusively Holarctic family extends to the Oriental region or, more precisely, to its transition zones. Three species known only in the larval stage have been recently described from Taiwan (Kang & Yang, 1994a). *Ameletus primitivus* Traver (larva and female) was described from Kashmir (Kyam) (Traver, 1939) and larvae closely related to the Central Asian species *A. alexandrae* Brodsky were collected in Nepal Himalaya (Braasch, pers. comm.). A new subgeneric classification of *Ameletus* is presented in these proceedings (Zloty, in press).

Table 2. Genera and approximate number of species of the family Baetidae in the Oriental region (* presumably endemic genera/subgenera, § adults unknown).

Genus	N° of species	Larval stage only	Adult stage only	Distribution of endemic genera
Acentrella	1	-	-	
Alainites	4	3	1	
Baetiella	9	7	1	
Baetis	43	19	22	
Baetopus subg. *Raptobaetopus**	1	1	-	Malaysia
Centroptilum	4	-	4	
*Chopralla**§	5	-	-	Sri Lanka, Malaysia
Cloeodes	3	3	-	
Cloeon	25	2	17	
*Gratia**§	1	1	-	Thailand
*Echinobaetis**	1	-	-	Sulawesi
Fallceon	1	1	-	
*Indobaetis**§	2	2	-	Sri Lanka
*Indocloeon**§	1	1	-	Sri Lanka
*Jubabaetis**§	1	1	-	Philippines
*Liebebiella**§	9	9	-	Nepal, Sri Lanka, Malaysia, Philippines
Nigrobaetis	8	8	-	
*Platybaetis**	4	3	-	Nepal, Malaysia, Borneo, Vietnam, Philippines
Procloeon	4	3	-	
Pseudocentroptiloides subg. *Psamonella* *§	1	1	-	Sri Lanka
Pseudocloeon	21	12	8	
*Symbiocloeon**§	1	1	-	Thailand
Takobia	2	-	2	
Total (23 genera)	152	78	55	19 spp. in both stages

Siphlonuridae. This predominantly Holarctic family is represented in the Oriental region by the genus *Siphluriscus*. Originally assigned to Acanthametropodidae (cf. McCafferty & Wang, 1994), it is found in the Tsayin Mts., China (*Siphluriscus chinensis* Ulmer, known only from adult male and female subimago). *Siphluriscus davidi* Navás described from Chusan (China) represents an unclear species and the type material is probably lost (cf. Alba-Tercedor & Peters, 1985).

Oligoneuriidae. Except for *Oligoneuriella kashmirensis* (Ali) (Oligoneuriinae) known in the larval stage from the transition zone in Pakistan (Ali, 1971) and Afghanistan, only the monotypic Chromarcyinae are present in the Oriental region. *Chromarcys feuerborni* (Ulmer) is known from Sumatra (Ulmer, 1939, 1940); other species of this genus were found in Thailand, Srí Lanka (Hubbard & Peters, 1984), and Vietnam. *Chromarcys magnifica* Navás is known from the female adult from China (Demoulin 1967).

Isonychiidae. This monotypic family widespread in the Holarctic region includes 7 described species from the Oriental region. Only *I. formosana* (Ulmer) (Japan and Taiwan) and *I. kiangsinensis* Hsu (China) are known in both larval and adult stages; larvae of other species from the Greater Sunda Islands [*I. grandis* (Ulmer), *I. sumatrana* (Navás), *I. winkleri* Ulmer] and Hainan (*I. hainanensis* She & You) remain unknown (Ulmer, 1940, Zhou, et al., 1997). Larvae of several species have been found in Indochina (e.g. in Laos, Vietnam and Thailand). *Isonychia khyberensis* Ali, known only in the adult stage, occurs in Pakistan (Ali, 1971).

Heptageniidae. Genera and species of the subfamily Heptageniinae from the Oriental region are listed in Table 3. The generic classification of this family is still not satisfactory, although individual genera or species groups mostly representing natural clades have been defined at different (and sometimes arbitrary) levels by various authors. There is little doubt that the Oriental region is the most important center of biodiversity for this family, and revision of genera in this region should resolve many of the problems in the rest of the world. The former genera *Ororotsia* and *Nixe* are generally considered to be congeneric with *Cinygmula* and *Ecdyonurus*, respectively, or to be subgenera. Similarly, the genera *Belovius* and *Iron* are frequently considered to be subgenera of *Epeorus*. Undoubtedly natural, monophyletic groups of *Ecdyonurus*-like genera consist of closely related taxa with weakly defined differential diagnoses and/or distributional types (McCafferty, 1991b). Consequently, original genera are often subjected to taxonomic shifts and Kluge (1988) even synonymized the genera *Thalerosphyrus, Afronurus, Cinygmina, Notacanthurus, Electrogena,* and *Asionurus* with *Ecdyonurus*.

Endemic genera and their distribution are given in Table 3. At the generic level, larval stages are known for all genera except *Epeorella.* Adults of *Asionurus* and *Trichogenia* have not been described. Description of larval stage and revision of the genus *Atopopus* were published by Wang & McCafferty (1995).

As with the Baetidae, Palaearctic and Holarctic genera are found mostly within transition zones. *Belovius* reaches the southern slopes of the Himalaya and China. The Holarctic genus *Iron* has a similar distribution with the southernmost limits in India (Punjab) or Nepal; the only exception is *I. martius* Braasch & Soldán from Vietnam. *Epeorus* reaches to Vietnam and West Malaysia (Gombak Riv.). The predominantly Holarctic *Rhithrogena* (only 8 species described in the Oriental region, more than 120 Holarctic species) also reaches its southern limits in Nepal and Kashmir, with an isolated distribution of *R. diehliana* Braasch & Soldán in Sumatra (Braasch & Soldán, 1986a). *Cinygmula* enters the Oriental region only

Table 3. Genera and approximate number of species of the family Heptageniidae in the Oriental region (* presumably endemic genera/subgenera, § adults unknown, §§ larvae unknown).

Genus	N° of species	Larval stage only	Adult stage only	Distribution of endemic genera
Afronurus	10	3	4	
*Asionurus**§	2	2	-	Vietnam
*Atopopus**	3	-	-	Borneo
Belovius	6	2	2	
Cinygma	3	-	-	
*Cinygmina**	9	3	4	India, Vietnam, China
Cinygmula	2	1	1	
incl. *Ororotsia** and	1	-	1	Himalaya
Compsoneuria	5	1	2	
Ecdyonurus	23	1	15	
incl. *Nixe* and	3	3	-	
Notacanthurus	2	2	-	
Electrogena	1	1	-	
Epeorus	5	4	1	
*Epeorella**§§	1	-	1	Borneo
Heptagenia	11	2	8	
Iron	8	3	2	
*Paegniodes**	1	-	-	China
Rhithrogena	8	4	3	
*Rhithrogeniella**	2	1	1	Sumatra, Tonkin
Thalerosphyrus	7	2	4	
*Trichogenia**§	1+	1	-	Vietnam, Sulawesi
Total (19 genera)	114	36	49	29 spp. in both stages

in India and southern China. *Cinygmina* is endemic to the Oriental region with 3 species described from the Indian subcontinent and 6 species from Vietnam and China.

There are two groups of genera that are widely distributed, but no genus is cosmopolitan. The genera *Compsoneuria* and *Afronurus* have an Ethiopian-Oriental distribution. *Compsonueria cingulata* (Navás) is described from Chusan Prov., China (type material lost), and 4 species are known from Southeast Asia (Braasch & Soldán, 1986b). Eleven species of *Afronurus* are described from the Oriental region (India to the Philippines); *Thalerosphyrus* displays a similar distribution (India and China to the Sunda Islands).

Twenty-three species of the predominantly Palaearctic *Ecdyonurus* have been reported from the Oriental region, most of them from the foothills of the Himalaya and China. Southernmost records include *E. illotus* Navás (the Philippines) and 3 other species described by Navás (e.g., *E. pallescens, E. pichoni, E. radialis*); none of the type material of any of these species has been studied. Three species of *Nixe* are reported from Taiwan (Kang & Yang, 1994b) and 2 species of *Notacanthurus* are described from the Himalaya in the larval stage (Braasch 1986). The material of *Electrogena fracta* Kang & Yang appears to belong to some other genus (cf. Kang & Yang, 1994c). The genus *Heptagenia* has the largest distribution within the Oriental and Nearctic regions with 8 species from the Sunda Islands, one from China (*H. limbata* Navás) and one from Assam, India (*H. nubila* Kimmins).

Leptophlebiidae. Two subfamilies of Leptophlebiidae are known from the Oriental region. Genera and the approximate number of species are listed in Table 4. Contrary to the families Baetidae and Heptageniidae, the generic classification of Leptophlebiidae (or, more precisely, that of the Eastern Hemisphere genera) seems to be well-understood thanks to the extensive revision by Peters & Edmunds (1970). The subfamily Leptophlebiinae includes 5 genera. Species of the Nearctic *Leptophlebia* and *Paraleptophlebia* occur in China, northern India, Nepal and Taiwan. The Nearctic and Oriental genus *Habrophlebiodes* contains 4 species from Sumatra, Java, China, and Hong Kong. The monotypic *Dipterophlebiodes sarawacensis* Demoulin is restricted to Borneo and West Malaysia and *Gilliesia hindustanica* (Gillies) seems to be endemic to northern India (Peters & Edmunds, 1970).

Most genera of the Oriental Leptophlebiidae belong to the subfamily Atalophlebiinae. The genus *Choroterpes* appears to be African in origin and *Choroterpes* s. str. may have given rise to the subgenus *Euthraulus;* both are widespread throughout the Oriental and Ethiopian regions and 18 species are described from tropical Asia. *Choroterpides* is known from Nepal, Thailand, Indochina, Java and Sumatra. One species of *Cryptopenella* known from Hong Kong may be a synonym of *Choroterpes* (Kluge, 1984). Species of *Thraulus* are known from India, Thailand, Vietnam, Hong Kong, and Taiwan. Other endemic genera are listed in the Table 4. The genera *Kimminsula*, *Petersula*, and *Nathanella* (southern India and Sri Lanka) represent a true Gondwanian derivative and share a common ancestry with genera living in Madagascar; *Sulawesia haema* Peters & Edmunds occurs in Sulawesi (Celebes) and is a member of the ancient Gondwanian *Atalophlebioides* lineage which is widespread in Australia, New Zealand, South America, Madagascar, Srí Lanka and Southern India (Peters & Edmunds, 1990).

Behningiidae. Only one species of this largely Holarctic family, *Protobehningia merga* Peters & Gillies (male imago and larval exuviae collected in the river Kwâi in Thailand), is known from the Oriental region (Peters & Gillies, 1991; Hubbard, 1994). Based on the known distribution, *Protobehningia* seems to be Oriental in origin (Peters & Gillies, 1991). Another undescribed species of *Protobehningia* has been recently collected in Sumatra (Peters, pers. comm.).

Euthyplociidae. The genus *Polyplocia* is represented by 3 species in the Oriental region. Adults of *P. crassinervis* and *P. campylociella* were described by Ulmer (1939, 1940) from Borneo. Lestage (1921, 1924) described adults of *P. vitalisi* from Tonkin. Larvae of this genus were described by Demoulin (1966).

Table 4. Genera and approximate number of species of the family Leptophlebiidae in the Oriental region (* presumably endemic genera/subgenera, § adults unknown).

Genus	N° of species	Larval stage only	Adult stage only	Distribution of endemic genera
*Dipterophlebioides**	1	-	1	Borneo
*Gilliesia**	1	-	1	India (Assam)
Habrophlebiodes	4	1	-	
Leptophlebia	3	-	2	
Paraleptophlebia	2+	-	2	
*Barba**	1	-	-	Papua-New Guinea
Choroterpes	5	1	2	
Choroterpides	3	1	1	
*Cryptopenella**	2	-	-	Thailand, Hong Kong
Euthraulus	14	2	5	
*Indialis**	2	-	-	India (Kerala), China
*Isca** subg. *Isca*	1	-	-	India, Hong Kong
subg. *Minyphlebia*	1	-	-	Thailand
subg. *Tanycola*	1	-	-	Sri Lanka
*Kimminsula**	3	-	-	Sri Lanka
*Magnilobus**	1	-	1	Papua-New Guinea
*Megaglena**	1	-	-	Sri Lanka
*Nathanella**	2	-	-	South India
*Nonnullidens**	2	1	-	Papua-New Guinea
*Notophlebia**	1	-	-	India (Tamil Nadu)
*Petersula**	2	-	-	India (Tamil Nadu)
*Simothraulus**§	1	-	1	Borneo, Sabah
*Sulawesia**	1	-	-	Sulawesi
*Sulu**	2	-	-	Malaysia, Borneo, Mindanao
Thraulus	8	4	-	
Total (23 genera)	65	10	16	39 spp. in both stages

Ephemeridae. Only the genera *Eatonigenia* and *Ephemera* of this widely distributed family are recorded in the Oriental region. Previous records of *Ichthybotus* and *Hexagenia* from India seem to be obvious errors (cf. e.g., McCafferty, 1973a, b; Hubbard, 1990). There are 6 described species of the endemic genus *Eatonigenia*: *E. chaperi* (Navás) is known from Borneo, Java, Thailand and China (cf. Zhang, et al., 1995), larvae of *E. indica* (Chopra), *E. trirama* McCafferty (India), *E. seca* McCafferty (Thailand) and *E. philippina* (Navás) (Philippines) are unknown, and *E. chinei* (Dang) (Vietnam) is described only in the larval stage (McCafferty, 1973b, 1991a). More than 25 species of the genus *Ephemera* have been described from the Oriental region. *Ephemera nadinae* McCafferty & Edmunds (Thailand) belongs to the subgenus *Aethephemera* (McCafferty & Edmunds, 1973), but most species are placed in the subgenus *Ephemera* s.str. This genus is known from Taiwan, Sri Lanka, the Indian subcontinent and China. but evidently does not cross Wallace's line. It shows a relative high diversity here from whence it probably dispersed to the Ethiopian region (Edmunds, 1979). Many species are known only from the type series, and some types are lost. This genus requires a detailed revision.

Palingeniidae. Three genera of this family are known from the Oriental region, *Palingenia*, *Anagenesia* and *Plethogenesia*. The last two are endemic to the Oriental region while the distribution of the first includes the Palaearctic region as well. *Palingenia orientalis* Chopra is recorded from Iran and India. Besides the doubtful species *Anagenesia leucoptera* Navás recorded from Tonkin, 11 species of this genus have been described from the Sunda Islands; *A. lontona* Hafiz occurs in Burma and in Assam, and *A. robusta* (Eaton), *A. minor* (Eaton) and *A. birmanica* Navás were collected in India and Burma. All five species of *Plethogenesia* are endemic to Papua-New Guinea (Demoulin, 1965). With the exception of species of *Palingenia*, *Anagenesia robusta* (Eaton) and *Plethogenesia papuana* (Eaton) are the only species known in the larval stage.

Polymitarcyidae. Two genera of this family have been found in the Oriental region — *Ephoron* (Polymitarcyinae) and *Povilla* (Asthenopodinae). Of the former, 5 species are known from the Indian subcontinent including Sri Lanka, all described from adults. Recently *E. hainanensis* Gui & Zhang was described in adult stage from Hainan (Gui & Zhang, 1997). An uncertain record on the occurrence of the European species *E. virgo* from China was published by Hsu (1935). The genus *Povilla,* comprises a single Afrotropical species and 7 Oriental species, was revised by Hubbard (1984b). *P. andamanensis* Hubbard (South Andaman Island), *Povilla junki* Hubbard (Thailand), and *P.* (*Languidipes*) *taprobanes* Hubbard (Sri Lanka), are known in the larval stage only. Both genera have probably dispersed from the Oriental to the Ethiopian (Edmunds, 1975, 1979).

Potamanthidae. Detailed revision of this originally Laurasian family including a detailed analysis of distributional patterns has been completed by Bae & McCafferty (1991). All genera of this family except for the Nearctic *Anthopotamus* occur in the Oriental region. The subgenus *Rhoenanthus s. str.* comprises two allopatric species: *R. distafurcus* Bae & McCafferty (south India and South East Asia, larva unknown), and *R. speciosus* Eaton occurring in the Sunda Islands and southern Malaysian Peninsula. Species of *Rhoenanthus* (*Potamanthindus*) occur in Southeast Asia and China [*R. obscurus* Navás, *R. magnificus* Ulmer —larvae unknown, *R. youi* (Wu & You)]; only a single species [*R. coreanus* (Yoon & Bae)] extends past the borders of the Oriental region. Within *Potamanthus*, the monotypic subgenus *Stygifloris* seems to be endemic to Borneo (Sabah) and 10 species of the subgenus *Potamanthodes* are distributed in a relatively narrow belt-like area from central Japan to southernmost Malaysia. One doubtful species *P. subcostalis* Navás, originally described from adults, is recorded from southern India. Much of Southeast Asia is inhabited by the common *P. formosus* Eaton. On the other hand, *P. idiocerus* Bae & McCafferty seems to be endemic to Taiwan (Bae & McCafferty, 1991, Kang & Yang, 1994e) There are also several species of uncertain taxonomic position described on the basis of adult material from China (e.g., *P. macrophthalmus* You, *P. kwangsiensis* Hsu, *P. sangangensis* You, *P. yunnanensis* You et al.).

Ephemerellidae. Ten genera of this family inhabit the Oriental region. The endemic genus *Ephacerella* (=*Acerella* Allen 1971, nec Berlese, 1909 cf. Paclt, 1994) consists of three species occurring in Southeast Asia and Japan, *E. uenoi* (Allen & Edmunds) in India, and two species (*E. montana, E. glebosa*) described from Taiwan by Kang & Yang (1995b). Four species of *Crinitella* occur mostly in Kashmir, Nepal and Pakistan (a single species known from Malaysia), and most are known in the larval stage only. *Cincticostella* exhibits a disjunctive distribution in the Far East, Japan and the Southeast Asia (Allen, 1980). Three species (all in the subgenus *Rhionella*) occur in the latter area.; two additional species (*C. fusca, C. colossa*) have been recently described from Taiwan by Kang & Yang (1995b). The predominantly Holarctic *Drunella* is represented by 12 Oriental species classified in the subgenera *Drunella* s. str., *Myllonella*, and *Tribrochella*. These species occur mostly in montane streams of the Himalaya, Hindukush [e.g. *D. kabulensis* (Ali), *D. submontana* (Brodsky) and *D. serrata* Braasch] and China; species from China are known in the adult stage only (cf. Su & Gui, 1995). The southernmost extension of *Drunella* is represented by *D. soldani* Allen from southern Vietnam (Allen, 1986). Four species of *Serratella* have been described in adult stage from China (You & Gui, 1995). *Teloganopsis media* Ulmer is widely distributed in Java (Allen, 1980), *Torleya nepalica* (Allen & Edmunds) occurs in the Himalaya, another unnamed species lives in southern Vietnam, and *T. glareosa* Kang & Yang and *T. lutosa* Kang & Yang seem endemic to Taiwan. The monotypic genus *Eburella* (*E. brocha* Kang & Yang), known from the larval stage only, is also endemic to Taiwan, and the monotypic *Hyrtanella christinae* Allen & Edmunds may be endemic to Borneo. In addition a single species of *Ephemerella* is known from India in the larval stage (Kapur & Kripalani, 1963) and 3 species are described from China in the adult stage (You & Gui, 1995).

Teloganodidae. Three genera of this family (McCafferty & Wang,1997) are known to occur in the Oriental region. *Vietnamella* (subfamily Austremerellinae) is represented by 6 species in Vietnam and southern China. Some [e.g. *V. thani* Tshernova, *V. ornata* (Tshernova)] are known from larvae only; others [e.g. *V. sinensis* (Hsu)] as adults; and *V. dabieshanensis* You & Su from larvae and adults. Within the subfamily Teloganodinae, the monotypic *Macafertiella* is considered endemic to Sri Lanka; however, other species apparently live in south India. *Teloganodes* is widely distributed throughout the whole Oriental region showing relatively high diversity here although only 4 species are named. The most common is *T. tristis* (Hagen) occurring in the Sunda Islands, the Philippines, Southeast Asia, China and Indian subcontinent including Sri Lanka. Other species, like *T. dentata* Navás and *T. major* Eaton, are known only in the adult stage.

The monotypic genus *Teloganella* represents a taxon of uncertain systematic position. *T. umbrata* Ulmer, occurring in the Sunda Islands and Malaysia (Ulmer, 1939, Wang, et al., 1995), is classified either in the Teloganodidae or in the Tricorythidae (Peters & Peters, 1993, Wang, et al., 1995). Further, an unnamed species of the same or closely related genus occurs in southern India.

Tricorythidae. The only species of this family (subfamily Tricorythinae) known from the Oriental region is *Tricorythus jacobsoni* Ulmer. Adults were described by Ulmer in 1924 and larvae in 1940 from South Sumatra and West Java. The relationships of this species to other representatives of the genus from the Ethiopian region are not clear. There are several unnamed species closely related to *T. jacobsoni* collected throughout the Oriental region.

Neoephemeridae. Detailed revision of this family, the distributional patterns of which appear to be most similar to the Potamanthidae, was published by Bae & McCafferty (1998). The only Oriental genus is *Potamanthellus* (= *Neoephemeropsis*) with 4 described species. In addition to *P. chinensis* (Hsu) occurring in northwestern China and the Far East, *P. ganges* Bae & McCafferty (adult unknown) is described from northern India and the 3 other species [(*P. amabilis* (Eaton), *P. caenoides* (Ulmer) and *P. edmundsi* Bae & McCafferty, adults of the latter unknown] are sympatric in Southeast Asia. This genus does not cross Wallace´s line.

Caenidae. Five genera of this family have been reported from the Oriental region, i.e. *Brachycercus, Cercobrachys, Caenoculis, Caenis,* and *Clypeocaenis*. Only one species of *Brachycercus* (*B. gilliesi* from Sri Lanka) is described from the Oriental region (Soldán & Landa, 1990), representing an Oriental extension of this largely Holarctic genus. *Cercobrachys* is a widely distributed genus with 2 representatives in the Oriental region (Soldán 1986). On the other hand, *Caenoculis* seems to be endemic to this area; 2 species are known from Vietnam and one from Malaysia (Soldán, 1986) but other unnamed species undoubtedly occur in this region. *Clypeocaenis* is an Oriental-Ethiopian genus which seems to have its center of biodiversity in the Oriental region with 4 species described from India, Sri Lanka, Thailand, Vietnam; its sister-group taxon possessing brush-legged larvae occurs in South Africa. The taxonomy of *Caenis* in the Oriental region is not well studied and there are undoubtedly many species yet to be described. Of the19 species described, most are from the Indian subcontinent and Taiwan (Kang & Yang, 1994d).

Prosopistomatidae. Of the 16 species of this monotypic family, six are known from the Oriental region. In addition to *Prosopistoma wouterae* (Lieftinck) from Western Java, Peters (1967) described 5 further species from India and Sri Lanka (*P. indicum, P. lieftincki*), the Philippines (*P. boreus, P. palawana*) and Papua-New Guinea (*P. sedlaceki*). Two more species were described from Vietnam by Soldán & Braasch (1984) and one (*P. pearsonorum*) from far north Queensland in Australia (Campbell & Hubbard, 1998). The latter species is the only one known in adult stage. Distributional patterns of most species were treated by Koch (1988). However, there are a number of species yet to be described, at least 10 from Southeast Asia (Peters, pers. comm.).

CONCLUSIONS AND RESEARCH PRIORITIES

Of the total number of recent Ephemeroptera (28 families, about 400 genera and approximately 3000 species) altogether 19 families, 104 genera and approximately 537 species have been so far identified in the Oriental region. Contrary to a very high percentage of taxa at the family level, the number of known genera is relatively low. However, this is affected by the absence of most Gondwanian Leptophlebiidae, a highly diversified family. Despite a relatively large number of species described from the Oriental region, mayfly diversity is most probably 3-4 times higher than known since the total known number of species is comparable to that of the Nearctic or West Palaearctic faunas, areas with an original fauna considerably reduced by the last glaciation.

In addition to basic taxonomic research, research priorities of the Oriental Ephemeroptera can be summarized as follows:

(i) Revision of unclear or weakly defined genera, especially those in the families Baetidae and Heptageniidae.

(ii) Proper association of larvae and adults described separately. Some genera (e.g. *Epeorella*) are known in adult stage only and other genera (e.g. *Symbiocloeon, Trichogenia* and *Eburella*) only as larvae. Of about 43 species of the genus *Baetis* known from this area, approximately a half are known only in the adult stage (especially those of the Indian subcontinent) while many remaining species (mostly those in Southeastern Asia) are described in the larval stage.

(iii) Revision of type material scattered in different institutions; however about a third of the type material is lost or missing. Concentration on detailed generic or species-group revision and study of faunas of well defined areas however small.

(iv) Biogeographical analysis of species/genera origins. Considerable attention should be paid to transitory areas and to the definition of the Oriental limits. There is no doubt that the limits as defined on the basis of vertebrate distribution (e.g. Wallace's line) do not match the distribution of Ephemeroptera and invertebrates in general.

(v) Study of life cycle and ecology of individual species or mayfly taxocenes. Only very fragmentary data have been published on biology of the Oriental species. Apart from Ulmer's (1940) summary of seasonal occurrence of the Sunda Island species and Edmunds and Edmunds' (1980) study of adult and subimaginal adaptation, there are few detailed studies on mayfly seasonality and habits in this region. However, valuable data have been collected in the framework of limnological studies especially in Malaysia (Bishop, 1973), Hong Kong (Dudgeon, 1992); see this monograph for further references on mayfly biology), and the Indian part of the Himalaya (Pandit, 1999).

ACKNOWLEDGMENTS

I would like to sincerely thank Janice G. and William L. Peters and M. D. Hubbard, Florida A&M University, Tallahassee, for critical review of the manuscript. This manuscript was completed while I was conducting research at Florida A&M University. My thanks are due also to E. Dominguez who carefully revised the final version of the manuscript.

BIBLIOGRAPHY

Alba-Tercedor, J. and W. L. Peters. 1985. Types and additional specimens of Ephemeroptera studied by Longinos Navás in the Museo de Zoología del Ayuntamiento, Barcelona, Spain. Aquat. Insects, 7 215-227.

Ali, S. R. 1971. Certain mayfly nymphs (Order: Ephemeroptera) of Azad, Kashmir and Swat. Pakistan J. Sci., 23: 209-214.

Allen, R. K. 1980. Geographic distribution and reclassification of the subfamily Ephemerellinae (Ephemeroptera: Ephemerellidae), pp. 71-92. In: J. F. Flannagan and K. E. Marshall (eds.). Advances in Ephemeroptera biology. Plenum, New York.

Allen, R. K. 1986. Mayflies of Vietnam: *Acerella* and *Drunella* (Ephemeroptera, Ephemerellidae). Pan-Pac. Ent., 62: 301-302.

Bae, Y. J. and W. P. McCafferty. 1991. Phylogenetic systematics of the Potamanthidae (Ephemeroptera). Trans. Amer. ent. Soc., 3-4: 1-143.

Bae, Y. J. and W. P. McCafferty. 1998. Phylogenetic systematics and biogeography of the Neoephemeridae (Ephemeroptera: Pannota). Aquat. Insects, 20: 35-68.

Bishop, J. E. 1973. Limnology of a small Malayan river, Sungai Gombak. Monographiae Biologicae 22, Dr. W. Jung, The Hague, 485 pp.

Braasch, D. 1986. Zur Kenntnis der *Notacanthurus* Tshernova, 1974 aus dem Himalaya. Reichenbachia, 23:118-125.

Braasch, D. and T. Soldán. 1980. *Centroptella* n. gen., eine neue Gattung der Eintagsfliegen aus China. (Baëtidae, Ephemeroptera). Reichenbachia, 18: 123-127.

Braasch, D. and T. Soldán. 1986a. *Rhithrogena diehliana* n. sp. von Sumatra (Ephemeroptera, Heptageniidae). Reichenbachia, 24: 91-92.

Bräasch, D. and T. Soldán. 1986b. Die Heptageniidae des Gombak River in Malaysia (Ephemeroptera). Reichenbachia, 24: 42-52.

Campbell, I. C. and M. D. Hubbard. 1998. A new species of *Prosopistoma* (Ephemeroptera: Prosopistomatidae) from Australia. Aquat. Insects, 20: 141-148.

Demoulin, G. 1965. Contribution à l´étude des Palingeniidae (Insecta, Ephemeroptera). Nova Guinea (Zool.)., 33: 305-344.

Demoulin, G. 1966. Contribution à l´étude des Euthyplociidae 3 (Insectes, Ephéméroptčres). Zool. Meded., 41: 137-141.

Demoulin, G. 1967. Redescription de l´holotype ♀ de *Chromarcys magnifica* Navás et discussion des affinites phylétiques du genre Chromarcys Navás (Ephemeroptera, Chromarcyinae). Bull. Inst. r. Sci. Nat. Belg., 43 (31): 1-10.

Dudgeon, D. 1992 Patterns and processes in stream ecology. A synoptic review of Hong Kong running waters. G. Fischer, Stuttgart, Binnengewässer, 29, 267 pp.

Edmunds, G. F. 1972. Biogeography and evolution of Ephemeroptera. Ann. Rev. Ent., 49: 1-19.

Edmunds, G. F. 1975. Phylogenetic biogeography of mayflies. Ann. Mo. Bot. Gdn., 62:251-263.

Edmunds, G. F. 1979. Biogeographical relationships of the Oriental and Ethiopian mayfly faunas, pp. 11-14. In: K. Pasternak and R. Sowa (eds.). Proc. 2nd Intern. Confer. Ephemeroptera, Panstwowe Wydaw. Nauk., Krakow.

Edmunds, G. E. and C. H. Edmunds. 1980. Predation, climate and mating of mayflies, pp. 277-286. In: J. F. Flannagan and K. E. Marshall (eds.). Advances in Ephemeroptera biology. Plenum, New York.

Edmunds, G. F. and D. A. Polhemus. 1990. Zoogeographical patterns among mayflies (Ephemeroptera) in the Malay Archipelago, with special reference to Celebes, iv + 343 pp. In: W. J. Knight and J. D. Holloway (eds.). Insects and the rain forest of South East Asia. R. Entomol. Soc., London.

Gui, H. 1985. A catalog of the Ephemeroptera of China. J. Nanjing Normal Univ., 85 (4): 79-97.

Gui, H. and J. Zhang. 1997. A new species of Ephoron from China (Ephemeroptera: Polymitarcyidae). Acta Zootaxon. Sin., 22: 50-53.

Hsu, Y.-C. 1935-1936. New Chinese mayflies from Kiangsi Province (Ephemeroptera). Peking Nat. Hist. Bull., 10: 319-326.

Hubbard, M. D. 1984a. Ephemeroptera type-specimens in the Zoological Survey of India, Calcutta. Orient. Insects, 18: 1-4.

Hubbard, M. D. 1984b. A revision of the genus *Povilla* (Ephemeroptera: Polymitarcyidae). Aquat. Insects, 6: 17-35.

Hubbard, M. D. 1986. A catalog of the mayflies of Hong Kong. Insecta Mundi, 1: 247-254.

Hubbard, M. D. 1990. Mayflies of the world: a catalog of the family and genus group taxa (Insecta: Ephemeroptera). Sandhill Crane Press, Gainesville, Florida, 119 pp.

Hubbard, M. D. 1994. The mayfly family Behningiidae (Ephemeroptera: Ephemeroidea): keys to the recent species with a catalog of the family. Gr. Lakes Ent., 27: 161-168.

Hubbard, M. D. and M. L. Pescador. 1978. A catalog of the Ephemeroptera of the Philippines. Pac. Insects, 19: 91-99.

Hubbard, M. D. and W. L. Peters. 1978. A catalog of the Ephemeroptera of the Indian Subregion. Orient. Insects, Suppl., 9: 1-43.

Hubbard, M. D. and Peters W. L. (1984) Ephemeroptera of Sri Lanka: an introduction to their ecology and

biogeography, pp. 257-274. In: C. H. Fernando (ed.). Ecology and Biogeography in Sri Lanka. Dr. W. Junk Publ., The Hague.
Kang, S. Ch. and C. T. Yang. 1994a. A revision of the genus *Baetis* in Taiwan (Ephemeroptera, Baetidae). J. Taiwan Mus., 47: 9-44.
Kang, S. Ch. and C. T. Yang. 1994b. Three new species of the genus *Ameletus* from Taiwan (Ephemeroptera: Siphlonuridae). Chinese J. Ent., 14: 261-269.
Kang, S. Ch. and C. T. Yang. 1994c. Heptageniidae of Taiwan (Ephemeroptera). J. Taiwan Mus., 47: 5-36.
Kang, S. Ch. and C. T. Yang. 1994d. Leptophlebiidae of Taiwan (Ephemeroptera). J. Taiwan Mus., 47: 57-82.
Kang, S. Ch. and C. T. Yang. 1994e. Ephemeroidea of Taiwan (Ephemeroptera). Chinese J. Ent., 14: 391-399.
Kang, S. Ch. and C. T. Yang. 1994f. Caenidae of Taiwan (Ephemeroptera). Chinese J. Ent. 14: 93-113.
Kang, S. Ch. and C. T. Yang. 1995a. Two new species of *Baetis* Leach (Ephemeroptera: Baetidae) from Taiwan. Chinese J. Ent., 16: 61-66.
Kang, S. Ch. and C. T. Yang. 1995b. Ephemerellidae of Taiwan (Insecta, Ephemeroptera). Bull. Nat. Mus., Nat. Sci., 5: 95-116.
Kapur, A. P. and M. B. Kripalani. 1963. The mayflies (Ephemeroptera) from the Northwestern Hinalaya. Rec. Ind. Mus., 59: 183-221.
Kluge, N. Yu. 1984. Podenky podroda *Euthraulus* Barn. (Ephemeroptera, Leptophlebiidae, rod *Choroterpes*) fauny SSSR [Mayflies of the subgenus *Euthraulus* (Ephemeroptera, Leptophlebiidae, genus *Choroterpes* of the USSR fauna]. Entomol. Obozr., 63: 722-728.
Kluge, N. Yu. 1988. Reviziya rodov. sem. Heptageniidae (Ephemeroptera). I . Diagnozy trib, rodov i podrodov podsem. Heptageniinae [A revision of genera of the family Heptageniidae (Ephemeroptera). I. Diagnoses of tribes, genera and subgenera of the subfamily Heptageniinae.]. Entomol. Obozr., 67: 291-313.
Koch, S. 1988. Mayflies of Northern Levant (Insecta: Ephemeroptera). Zool. Middle East, 2: 89-112.
Lestage, J. A. 1921. Les Ephémères indo-chinoises. Ann. Soc. ent. Belg., 61: 211-222
Lestage, J. A. 1924. Faune entomologique de l'Indochine Française. Les Ephémères de l'Indochine Française. Opusc. Inst. Sci. Indochine, 3: 79-93.
Lugo-Ortiz, C. R. and W. P. McCafferty. 1997. First adult description of the unusual Baetid mayfly genus *Echinobaetis* (Ephemeroptera: Baetidae) Ent. News, 108: 113-116.
Lugo-Ortiz, C. R., W. P. McCafferty and R. D. Waltz. 1999. Definition and reorganization of the genus *Pseudocloeon* (Ephemeroptera: Baetidae) with new species descriptions and combinations. Trans. Amer. ent. Soc., 125: 1-37.
McCafferty, W. P. 1973a. Systematic and zoogeographic aspects of Asiatic Ephemeridae (Ephemeroptera). Orient. Insects, 7: 49-67.
McCafferty, W. P. 1973b. Commentary on the genus *Ichthybotus* (Ephemeridae) and the misplaced *I. dodecus* Dubey. Orient. Insects, 7: 351-352.
McCafferty, W. P. 1991a. Toward a phylogenetic classification of the Ephemeroptera: a commentary on systematics. Ann. ent. Soc. Amer., 84: 343-360.
McCafferty, W. P. 1991b. The cladistics, classification, and evolution of the Heptagenioidea (Ephemeroptera), pp. 87-102. In: I. Alba-Tercedor and A. Sanchez-Ortega (eds.). Overview and strategies of Ephemeroptera and Plecoptera. Sandhill Crane Press, Inc., Gainesville, Florida.
McCafferty, W. P. and G. F. Edmunds. 1973. Subgeneric classification of *Ephemera* (Ephemeroptera: Ephemeridae). Pan-Pac. Ent., 49: 300-307.
McCafferty, W. P. and T.-Q. Wang. 1994. Relationships of the genera *Acanthametropus*, *Analetris*, and *Siphluriscus*, and re-evaluation of their higher classification (Ephemeroptera: Pisciforma). Great Lakes Ent., 27: 209-215.
McCafferty, W. P. and T.-Q. Wang. 1997. Phylogenetic Systematics of the Family Teloganodinae (Ephemeroptera, Pannota). Ann. Cape Prov. Mus., Nat. Hist. 19: 387-437.
Müller-Liebenau, I. 1979. Symbiocloeon: A new genus of Baetidae from Thailand (Insecta, Ephemeroptera), pp. 57-66. In: K. Pasternak and R. Sowa (eds.). Proc. 2nd Intern. Confer. Ephemeroptera, Panstwowe Wydaw. Nauk., Krakow.
Müller-Liebenau, I. 1982. New species of the family Baetidae from the Philippines (Insecta, Ephemeroptera). Arch. Hydrobiol., 94: 70-82.
Müller-Liebenau, I. 1984a. Baetidae from Sabah (East Malaysia) (Ephemeroptera), pp. 85-99. In: V. Landa, T. Soldán and M. Tonner (eds.). Proc. 4th Int. Confer. Ephemeroptera, Czechoslovak Acad. Sci., České Budejovice.
Müller-Liebenau, I. 1984b. New genera and species of the family Baetidae from West Malaysia (River Gombak). Spixiana, 7: 253-284.

Müller-Liebenau, I. 1985. Baetidae from Taiwan with remarks on *Baetiella* Uéno, 1931 (Insecta, Ephemeroptera). Arch. Hydrobiol., 104: 93-110.

Müller-Liebenau, I. and M. D. Hubbard. 1985. Baetidae from Sri Lanka with some general remarks on the Baetidae of the Oriental Region (Insecta, Ephemeroptera). Florida Ent. 68: 537-561.

Paclt, J. 1994. *Ephacerella*, a replacement name for *Acerella* Allen, 1971 (Ephemeroptera), nec Berlese, 1909 (Protura). Ent. News, 105: 283-284.

Pandit, A. K. 1999. Freshwater ecosystems of the Himalaya. Parthenon Publ. Group, New York, 197 pp.

Peters, W. L. 1967. New species of *Prosopistoma* from the Oriental region (Prosopistomatoidea: Ephemeroptera). Tijdsch. Ent., 110: 207-222.

Peters, W. L. and G. F. Edmunds. 1970. Revision of the generic classification of the Eastern Hemisphere Leptophlebiidae (Ephemeroptera). Pac. Insects, 12: 157-240.

Peters, W. L. and G. F. Edmunds. 1990. A new genus and species of Leptophlebiidae: Atalophlebiinae from the Celebes (Sulawesi) (Ephemeroptera), pp. 327-335. In: I. C. Campbell (ed.). Mayflies and Stoneflies: Life Histories and Biology. Kluwer Academic, Dordrecht, Netherlands

Peters, W. L. and M. T. Gillies. 1991. The imago of *Protobehningia* Tshernova from Thailand (Ephemeroptera, Behningiidae), pp. 207-216. In: J. Alba-Tercedor and A. Sanchez-Ortega (eds.). Overview and strategies of Ephemeroptera and Plecoptera. Sandhill Crane Press, Gainesville, Florida.

Peters, W. L. and J. G. Peters. 1993. Status changes in Leptohyphidae and Tricorythidae (Ephemeroptera). Aquat. Insects, 15: 45-48.

Soldán, T. 1986. Revision of the Caenidae with ocellar tubercles in the nymphal stage (Ephemeroptera). Acta Univ. Carolinae, Biol., 1982-1984: 289-362.

Soldán, T. and D. Braasch. 1984. Two new species of the genus *Prosopistoma* (Ephemeroptera, Prosopistomatidae) from Vietnam. Acta ent. Bohemoslov., 81: 370-376.

Soldán, T. and V. Landa. 1990. Two new species of Caenidae (Ephemeroptera) from Sri Lanka, pp.235-244. In: J. Alba-Tercedor and A. Sanchez-Ortega (eds.). Overview and strategies of Ephemeroptera and Plecoptera. Sandhill Crane Press, Gainesville, Florida.

Su, C.-R. and H. Gui. 1995. The first record of the genus *Drunella* in China with description of a new species. Acta Zootaxon. Sin., 20: 451-453.

Traver, J. R. 1939. Himalayan mayflies (Ephemeroptera) Ann. Mag. Nat. Hist., 4: 32-56.

Ulmer, G. 1939. Eintagsfliegen (Ephemeropteren) von den Sunda-Inseln. Teil I. Einleitung und Systematik. Arch. Hydrobiol., Suppl., 16: 443-580.

Ulmer, G. 1940. Eintagsfliegen (Ephemeropteren) von den Sunda-Inseln. Teil II. Beschreibung von Nymphen. Teil III. Zusammenfassung. Arch. Hydrobiol., Suppl., 16: 581-692.

Waltz, R. D. and W. P. McCafferty. 1987a. Systematics of *Pseudocloeon*, *Acentrella*, *Baetiella*, and *Liebebiella*, new genus (Ephemeroptera: Baetidae). J.N.Y. ent. Soc., 95: 553-568.

Waltz, R. D. and W. P. McCafferty. 1987b. Revision of *Cloeodes* Traver (Ephemeroptera: Baetidae). Ann. ent. Soc. Amer., 80: 191-207.

Waltz, R. D. and W. P. McCafferty. 1987c. A generic revision of *Cloeodes* Traver and description of two new genera (Ephemeroptera: Baetidae). Proc. ent. Soc. Wash., 89: 177-184.

Wang, T.-Q., and W. P. McCafferty. 1995. First larval description, new species, and evaluation, of Southeast Asian genus *Atopopus* (Ephemeroptera, Heptageniidae). Bull. Soc. Hist. Nat. Toulouse, 131: 19-25.

Wang, T.-Q. and W. P. McCafferty. 1997. Phylogenetic systematics of the family Teloganodidae (Ephemeroptera, Pannota). Ann. Cape Prov. Mus., Nat Hist., 19: 387-437.

Wang, T.-Q., W. P. McCafferty and G. F. Edmunds, Jr. 1995. Larva and adult of *Teloganella* (Ephemeroptera: Pannota) and assessment of familial classification. Ann. ent. Soc. Amer., 88: 324-327.

You, D.-S. and H. Gui. 1995. Ephemeroptera. Economic Insect Fauna of China 48. vii + 152 pp.

Zhang, J., H. Gui and D. S. You. 1995. Studies on the Ephemeridae (Insecta: Ephemeroptera) of China. J. Nanjing Normal Univ., 18: 68-76.

Zhou, Ch., H. Gui and C. Su. 1997. Three new species of Ephemeroptera (Insecta) from Henan Province of China. Entomotaxonomia, 19: 268-272.

LIFE CYCLE AND ANNUAL PRODUCTION OF *CAENIS* SP (EPHEMEROPTERA, CAENIDAE) IN LAKE ESCONDIDO (BARILOCHE, ARGENTINA)

D. A. Añón Suárez and R. J. Albariño

Centro Regional Universitario Bariloche
Universidad Nacional del Comahue
Unidad postal Universidad
(8400) Bariloche, Argentina

ABSTRACT

Life cycle and secondary production of *Caenis* sp nymphs were analysed during a year of study in Lake Escondido, in relation to depth, water temperature and vegetation distribution. Higher densities of nymphs were found in the upper littoral zone with *Schoenoplectus californicus* (2665 ind m^{-2}), compared to lower littoral covered with *Potamogeton linguatus* (29 ind m^{-2}) and the profundal (4 ind m^{-2}) zones. During the study this mayfly was univoltine with one emergence in summer. The highest production (0.67 g m^{-2} yr^{-1}) was observed in the *S. californicus* zone, and it was related to the highest values of organic matter content and water temperature in the shallower area of the lake.

INTRODUCTION

The knowledge about the different aspects of life cycle of aquatic insects contributes greatly to explain the benthic community dynamics in lakes. The estimates of secondary production have outstanding ecological importance because this parameter integrates density, biomass and voltinism in a single figure, also indicating the community success and functional importance in the ecosystem (Benke, 1984, 1993). Among the factors affecting the magnitude of the secondary production and life cycles of aquatic insects, most authors indicate the temperature and the photoperiod together with the quantity and quality of food as the most relevant ones (Vannote and Sweeney, 1980; Ward and Stanford, 1982; Sweeney, 1984).

Caenis sp nymphs are the most common Ephemeroptera in Lake Escondido, and together with the Chironomids (Diptera) are numerically the most important taxa of the benthic macroinvertebrates (Añón Suárez, 1991, 1997).

In Argentina, Domínguez, et al. (1992) and Domínguez, et al. (1994) have studied taxonomic aspects of ephemeropterans, and there have been also studies carried out on distribution, life cycles and production in patagonian lakes and reservoirs (Kaisin, 1989; Añón Suárez, 1991).

This is the first contribution about the life cycle and production of *Caenis* sp in an Andean lake, therefore, there are no previous data for direct comparison. The main goal of this study was to determine the annual production of *Caenis* sp and also to analyse the life cycle parameters (density, biomass, size-classes distribution and voltinism) in different zones of Lake Escondido. The population studied was identified as *Caenis* sp reissi group (E. Domínguez, pers. com.) and is currently under study for complete identification.

STUDY AREA

Lake Escondido is situated at 30 km west of San Carlos de Bariloche (41°2' S; 71°4' W, Argentina), at 764 m a.s.l., with an area of 8 ha., a maximum depth of 8 m and a mean depth of 5.5 m (Fig.1). An evergreen mixed forest of *Nothofagus dombeyii* (Mirb) Blumeand, *Austrocedrus chilensis* (D.DON), Florin et Boutleje and also the deciduous *Nothofagus antarctica* (G. Foster) Oesterd. composes the riparian vegetation.

The littoral area of the lake is occupied by two macrophytes *Schoenoplectus californicus* (Meyer) Soják (=*Scirpus californicus*) and *Potamogeton linguatus* Hangström. The central deepest area is partially colonised by the Characeae *Nitella* sp, although in a very low density.

The thermal regime of Lake Escondido is warm monomictic with periods of complete mixing during autumn and winter and direct thermal stratification during late spring and early summer (Balseiro and Modenutti, 1990). However, during hard winters its surface often freezes behaving as dimictic, like in this study period.

Dissolved oxygen concentration is high throughout the year, with values near 100% of saturation in spring (Balseiro and Modenutti, 1990). The Secchi disc is visible at the maximum depth (8 m) and conductivity oscillates between 80 µS cm^{-1} (late summer) and 40 µS cm^{-1} (winter) (Balseiro and Modenutti, 1990). The pH is 7.02 and the chlorophyll *a* concentration is low (0.5 mg m^{-3} and 1.8 mg m^{-3}: winter and summer 1988 respectively) (Díaz and Pedrozo, 1993). These characteristics indicate the oligotrophic condition of this lake (Balseiro and Modenutti, 1990; Díaz and Pedrozo, 1993).

MATERIALS AND METHODS

Samples were obtained during April 1988 - May 1989, approximately monthly (autumn - winter) and biweekly (spring 1988 - autumn 1989), with an Ekman grab (225 cm^2) in three different zones: Station A (upper littoral with *S. californicus*, 50 cm mean depth), Station B (lower littoral with *P. linguatus*, 4m mean depth) and Station C (profundal, 8 m maximum depth). Samples could not be taken during July because the littoral zone of the lake remained cover by an ice layer in most of its surface. Each sampling date, four replicate samples were taken in Stations A and B and seven in Station C. Samples were filtered in situ by using a conic net (212 µm mesh size) and the material retained was fixed with formalin 5%.

Temperature was obtained on each sampling date in Stations A, B and C by using a thermometer inside a Ruttner bottle (at nearest 0.5°C). Also total accumulated degrees/day for the three stations and the difference of the degrees/day accumulated between littoral stations (A and B) with respect to Station C (profundal zone) were calculated. The granulometric and organic matter contents analysis was performed in all three zones according to Jackson (1964).

In the laboratory the nymphs were separated from the sediment and stored in formalin 4%. Cephalic width (interocular distance) was used to determine the size-classes (Snyder et al., 1991; Pritchard and Zloty, 1994: Sweeney et al., 1995). The density (ind m^{-2}) of each size-class was determined to construct size-frequency histograms. Then, individuals of each class (total n = 119) were dried at 60°C during 24 hr (Sweeney and Vannote, 1986) and weighed at nearest 0.01 mg to estimate mean dry weight of each size-class. The mean biomass for each size-class and the secondary production of *Caenis* sp (g m^{-2} yr^{-1}) were calculated by three methods: "Size-Frequency", "Allen Curve" and "Instantaneous Growth" (Waters and Crawford,

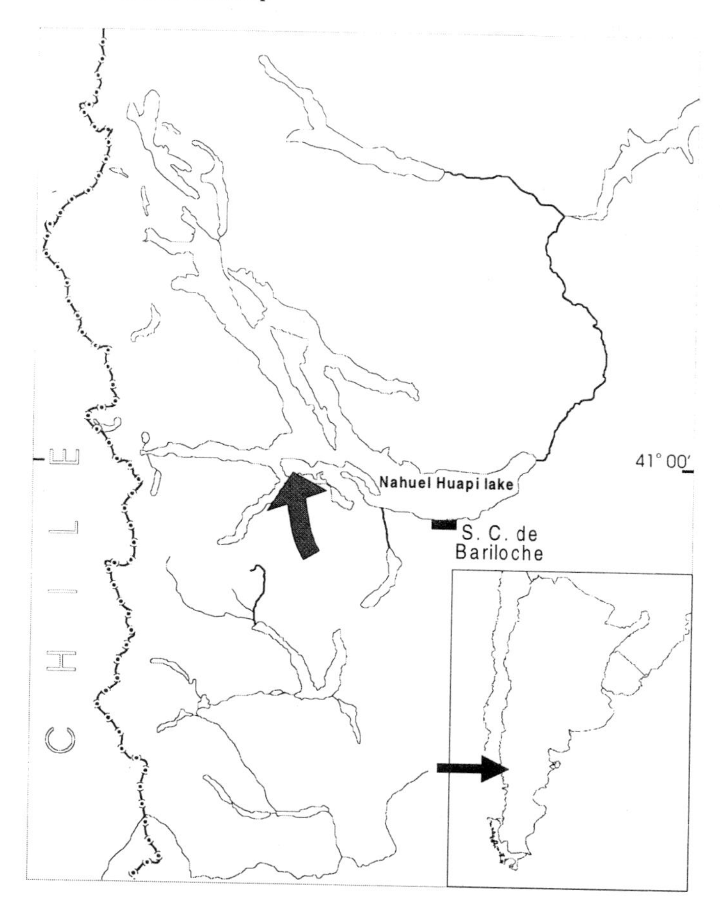

Fig. 1. Geographical position of Lake Escondido (grey arrow) and location of the study area (black arrow).

1973; Waters 1979; Benke, 1984, 1993). The former is used when cohorts are recognisable or not and the latter are recommended when cohorts are discernible.

For the Size Frecuency method the equation is expressed as follows:

$$P = i\sum_{j=1}^{i} \sqrt{(W_{j+1} \cdot W_j)} \cdot (N_{j+1} - N_j)$$

where P is the annual production, $(N_{j+1} - N_j)$ is difference in mean annual density of each size-class, $\sqrt{(W_{j+1}.W_j)}$ is the geometric mean of its respective dry weight and i is the number of size-classes.

For the Instantaneous Growth method the equation is:

$$P = \sum G.B$$

where P is the annual production, $G = ln\ (W_t + W_{t+\Delta t})$ is the instantaneous biomass growth rate and B is the biomass at a given time.

For the Allen Curve method, annual production (P) is the area under the curve $\overline{N}$ vs $\overline{W}$, where is the mean number during Δ_t and W is the mean individual biomass.

RESULTS

Lake Temperature

Temperature values presented differences between the three Stations (Fig. 2.a). The maximum temperature values were 22°C in early January and mid February 1989 at Station A, and 20°C in mid February 1989 at Stations B and C (Fig. 2.a). Average temperatures were 14.3°C, 13.0°C and 12.4°C for stations A, B and C respectively. Heat accumulation were calculated for these sampling stations (Fig. 2.b.) with maximum values of 4920.9 (Station A), 4539.4 (Station B) and 4349.8 (Station C) degrees/day accumulated at the end of the sampling period. The differences between the accumulation of heat (degrees-day) in the littoral zone were also calculated in relation to the coldest zone (Station C) (Fig. 2.c). In Station B, this accumulation mainly took place between October 1988 and January 1989, reaching a maximum value of 190 degrees-day (Fig. 2.c). The heat accumulation in Station A was more important, reaching a difference with Station C of 571 degrees-day accumulated at the end of the annual cycle (Fig. 2.c).

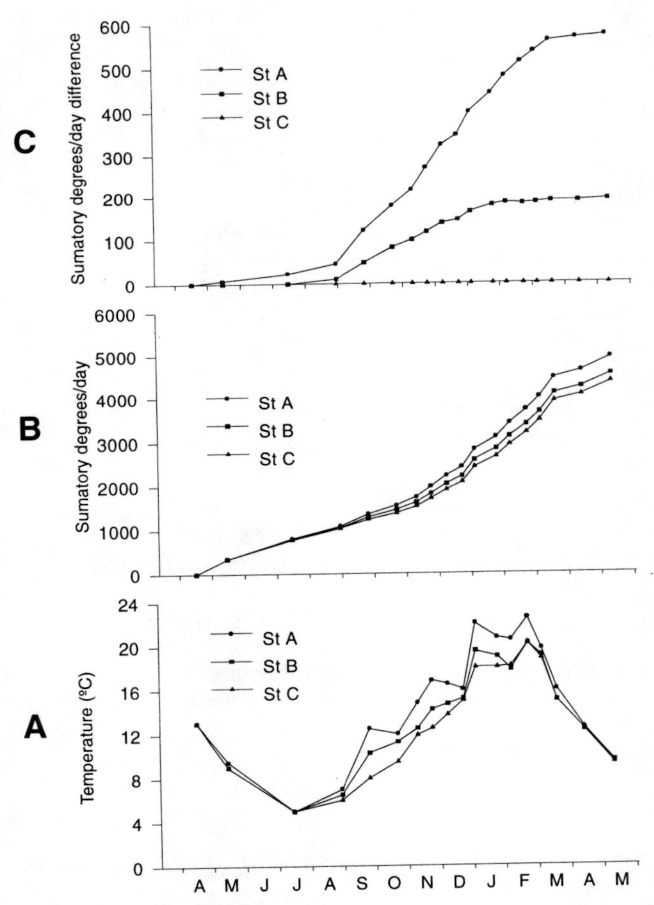

Fig. 2. (A) Monthly changes in temperature; (B) degrees/day accumulated and (C) Difference of degrees/day accumulated in Lake Escondido using Station C as reference, during April 1988 - May 1989.

Table 1. Granulometric analysis and organic matter content in Stations A, B and C of Lake Escondido.

Stations	sand (%)	silt (%)	clay (%)	organic matter (%)
A	52.0	45.0	3.0	36.0
B	88.0	10.5	1.5	14.5
C	85.0	11.0	4.0	13.5

The maximum differences registered in the degrees-day accumulated between the sampling stations during late spring and summer were related to the thermal stratification observed in Lake Escondido during this period.

Granulometry and Organic Matter Content

In Station A the percentages of sand and silt were similar, and that of clay was very low with respect to those components. In Stations B and C, high percentages of sand were observed in relation to the fractions of silt and clay. The high percentage of organic matter registered in Station A (36%) might be due to the incoming of allochthonous material originating in the surrounding vegetation and the presence of rhizomes of the boggy *S. californicus*. Finally, there were no outstanding differences in the content of organic matter between Stations B and C (Table 1).

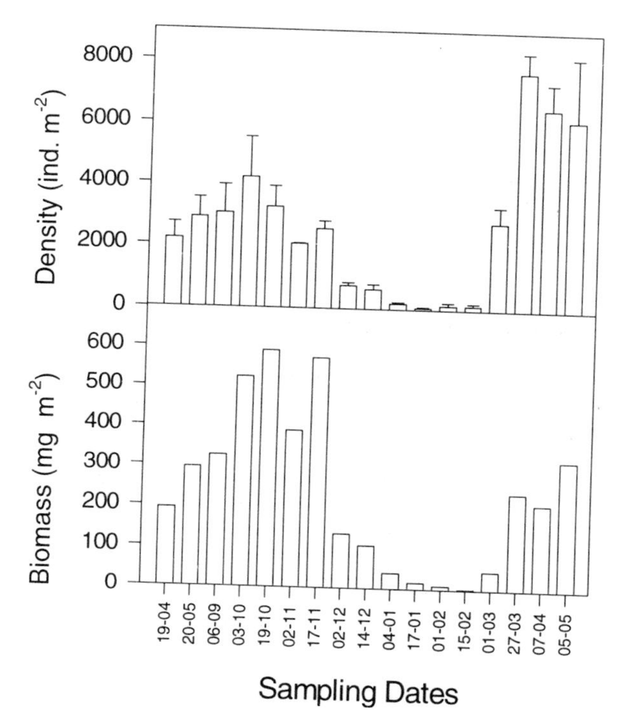

Fig. 3. Density and biomass variations of *Caenis* sp nymphs in Station A.

Life Cycle and Secondary Production

The density and mean biomass of *Caenis* sp showed evident differences between the area of *S. californicus* (Station A) and the deeper stations of the lake (B and C) (Table 2). In Station A the higher density values were observed from late March to early May 1989 (autumn) with a maximum of 7604 ind m^{-2} in early April (Fig. 3). The lowest densities were registered from December 1988 to mid February 1989 (summer) with a minimum of 67 ind m^{-2} in mid summer (Fig. 3).

The biomass presented the same pattern of density during this study (Fig. 3). Higher values were observed between October and November 1988, with a maximum of 591.90 mg m^{-2} in mid October (Fig. 3) and then, the biomass decreased with a minimum of 2.23 mg m^{-2} in mid February 1989 (Fig. 3).

The increment of body growth was evident during spring (Fig. 4). The greatest proportion of large size-classes was observed in mid January 1989; during February these classes showed a decreasing trend, and the individuals of the smallest ones, belonging to the new generation, began to increase (Fig. 4). The size-classes distribution, together with density and biomass variations (Fig. 3, 4), indicates that *Caenis* sp was univoltine, with one emergence period during summer.

Production estimates for Station A by the Allen Curve and Instantaneous Growth methods were similar while the Size Frequency method provided a higher estimate than the others (53 % and 83 % respectively) (Table2).

In Station A, the main habitat of *Caenis* sp population, secondary production and turnover ratio P/B were higher than Stations B and C (Table 2). In Stations B and C, the low number of organisms collected did not allow the calculation of the secondary production directly, so it was estimated multiplying the biomass of these stations by the turnover ratio of Station A (Waters, 1977,1979). By this way, secondary production in Station A was 2 to 3 orders in magnitude higher than that obtained in B and C (Table 2).

DISCUSSION

The high density, biomass and production of *Caenis* sp observed in Station A (Lake Escondido) (Table 2) corresponded to higher values of degrees-day accumulated and organic matter percent (Table 1, Fig. 2). Stations B and C presented lower values for the parameters mentioned above (Table 2). It is widely known that the temperature together with the quantity

Table 2. Population parameters of *Caenis* sp in Stations A, B and C of Lake Escondido. D= density (ind. m^{-2}); B= biomass (g m^{-2}); P= secondary production (g m^{-2} yr^{-1}) and P/B= turnover rate, by three methods of secondary production estimation.

Method	Stations	Density	B	P	P/B
Size Frequency	A	2665	0.24	1.029	4.3
	B	29	0.003	0.011	-
	C	4	0.001	0.003	-
Allen curve	A	2665	0.24	0.671	2.8
	B	29	0.003	0.008	-
	C	4	0.001	0.003	-
Instantaneous growth	A	2665	0.24	0.562	2.3
	B	29	0.003	0.007	-
	C	4	0.001	0.002	-

and quality of the food available are the main factors that affect the life history of aquatic insects (Benke, 1984; Ward, 1992). In this respect the genus *Caenis* has been mentioned by Edmunds and Waltz (1996) as sprawler on vegetated fine sediments, with feeding habits as scraper of algae and fine organic matter collector (Palmer et al., 1993).

We used the results obtained by the Allen Curve method to compare other freshwater systems because this is considered to give better estimates than other ones (Lindegaard, 1989). Waters and Crawford (1973) stressed that Size Frequency method overestimate se-

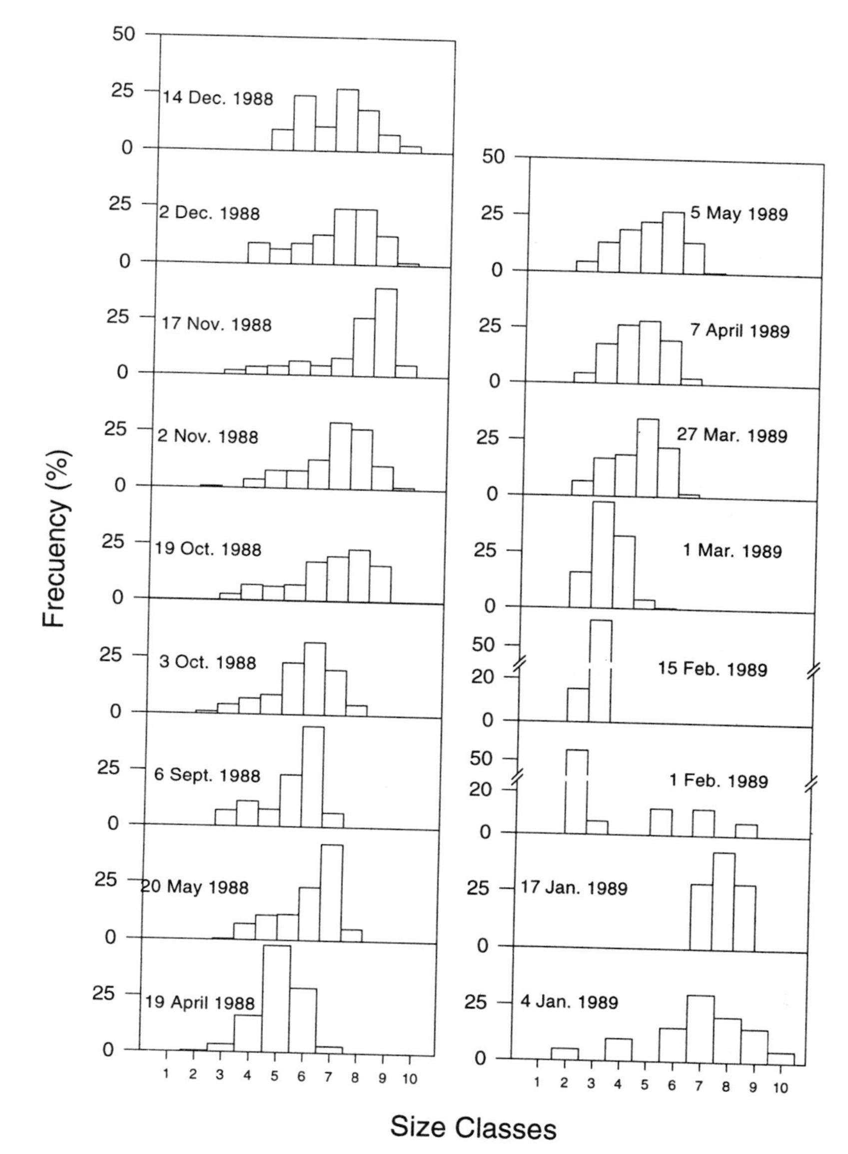

Fig. 4. Size-frequency histograms for *Caenis* sp nymphs in Station A of Lake Escondido during April 1988 - May 1989. X-axis (Size classes based on head-width, interocular distance, in mm): **1**= 0.14-0.21; **2**= 0.22-0.29; **3**= 0.30-0.37; **4**= 0.38-0.45; **5**= 0.46-0.53; **6**= 0.54-0.61; **7**= 0.62-0.69; **8**= 0.70-0.77; **9**= 0.78-0.85; **10**= 0.86-0.94.

condary production, because it is based on nymphal size it can wrongly consider more size classes due to different final sizes between male and female nymphs.

The early truncated survivorship curve in April-May 1988 (Fig. 4), may be due to missed small individuals or mortality related to biotic or abiotic factors. However, this loss would have a little effect in the estimation of secondary production.

Based on the Allen Curve method, the production in Station A is lower or similar compared with some waterbodies. For example, the maximum value obtained in Lake Escondido (0.67 g m^{-2} yr^{-1}) is lower than that estimated for *Caenis* sp in littoral zone of Ramos Mexía reservoir, North Patagonia (Argentina) (1.54 g m^{-2} yr^{-1}, Kaisin and Bosnia, 1987). A different situation was observed in the annual production of Chironomids of these two waterbodies. In this case, the values obtained in Lake Escondido for two genera common to both sites, were similar than those of the Ramos Mexía reservoir (Kaisin, 1989; Añón Suárez, 1997).

Other Caenids have showed higher or similar values than those obtained in Lake Escondido (Station A), for example *Caenis* sp (0.68 g m^{-2} yr^{-1}) (Rodgers, 1982) *Caenis macrura* (0.52 g m^{-2} yr^{-1}) (Zelinka, 1980) and *C. simulans* (0.75 g m^{-2} yr^{-1}) (Mac Farlane and Waters, 1982) in lotic systems, and *Brachycercus* sp (1.76 g m^{-2} yr^{-1}) in a small lake in Texas (Benson et al., 1980).

Production values for *Caenis* sp in Lake Escondido were also lower than was observed for species of other families. *Ephoron leucon* presented 2.860 g m^{-2} yr^{-1} (max.) (Snyder et al., 1991), *Ephemera spilosa* 0.037 g m^{-2} yr^{-1} (Dudgeon 1996a), and among five Heptagenid species the maximum production was 0.220 g m^{-2} yr^{-1} (Dudgeon, 1996b), as some examples of stream univoltine species. Waters and Crawford (1973) obtained higher values for the stream dweller *Ephemerella subvaria* (26.4 g m^{-2} yr^{-1}, larval wet weight; 4.4 g m^{-2} yr^{-1}, larval dry weight, see Waters, 1977).

On the other hand, the biomass turnover of *Caenis* sp in this study (P/B= 2.8) was lower than those included within the range indicated by Waters (1977) for univoltine species (P/B= 4 - 7). In Lake Escondido, this ratio was also lower than that obtained in the species above mentioned, as those show an average near 8.

It is difficult to explain the differences in the secondary production values from different waterbodies, because there are involved many factors interacting (food quantity and quality, temperature, salinity, habitat complexity and biologic interactions). However, in Lake Escondido, is clear that distribution of vegetation, high organic matter contents (that is food availability) and thermal pattern of the upper littoral zone (Station A), provide the most favourable conditions for the development of *Caenis* sp population.

ACKNOWLEDGMENTS

We want to thank the anonymous referee for the several suggestions and comments that helped us to improve significantly the manuscript. This study was supported by B-094 Grant to B. E. Modenutti and ANPCyT Grant 01-00000-01194 to E. G. Balseiro.

REFERENCES

Añón Suárez, D. A. 1991. Distribución del bentos del lago Escondido (Río Negro, Argentina) con especial énfasis en los quironómidos (Diptera, Chironomidae). Stud. Neotrop. Fauna Environ. 26: 149-157.

Añón Suárez, D. A. 1997. Estructura y dinámica de la taxocenosis Chironomidae (Diptera, Nematocera) de un lago andino. Doctoral Thesis. Facultad de Ciencias Naturales y Museo, Universidad Nacional de La Plata, 181 p.

Balseiro, E. G. and B. E. Modenutti. 1990. Zooplankton dynamics of lake Escondido (Río Negro, Argentina), with special reference to a population of *Boeckella gracillipes* (Copepoda, Calanoida). Int. Revue ges. Hydrobiol. 75: 475-491.

Benke, A. C. 1984. Secondary production of aquatic insects, pp. 289-322. In: The Ecology of Aquatic Insects, V. H. Resh and D. M. Rosenberg (eds.), Praeger, New York.

Benke, A. C. 1993. Concepts and patterns of invertebrate production in running waters. Edgardo Baldi Memorial Lecture. Verh. Internat. Verein. Limnol. 25: 15-38.

Benson, D. J., L. C. Fitzpatrick and W. D. Pearson. 1980. Production and energy flow in a benthic community of a Texas pond. Hydrobiologia 74: 81-93.

Díaz, M. M. and F. L. Pedrozo. 1993. Seasonal succesion of phytoplankton in a small Andean patagonian lake (Rep. Argentina) and some considerations about the PEG Model. Arch. Hydrobiol. 127: 167-184.

Domínguez, E., M. D. Hubbard and W. L. Peters. 1992. Clave para ninfas y adultos de las familias y géneros de Ephemeroptera (Insecta) sudamericanos. Biología Acuática 16: 1-32.

Domínguez, E., M. D. Hubbard and L. Pescador. 1994. Los Ephemeroptera en Argentina. Fauna de Agua Dulce de la República Argentina 33: 1-142.

Dudgeon, D. 1996a. The life history, secondary production and microdistribution of *Ephemera* spp. (Ephemeroptera: Ephemeridae) in a tropical forest stream. Arch. Hydrobiol. 135: 473-483.

Dudgeon, D. 1996b. Life histories, secondary production, and microdistribution of heptageniid mayflies (Ephemeroptera) in a tropical stream. J. Zool., Lond. 240: 341-361.

Edmunds, G. F. (Jr.) and R. D. Waltz. 1996. Ephemeroptera, pp. 126-163. In: An Introduction to the Aquatic Insects of North America, R.W. Merrit and K.W. Cummins (eds.), Kendall/Hunt Publishing Company, Iowa.

Jackson, M.L. 1964. Análisis químico de suelos. Omega, Barcelona, 662 p.

Kaisin, F. J. and A. S. Bosnia. 1987. Producción anual de *Caenis* sp (Ephemeroptera) en el embalse E. Ramos Mexía (Neuquen, Argentina). Physis 45: 53-63.

Kaisin, F. J. 1989. Dinámica, producción y balance energético del zoobentos en un embalse norpatagónico. Doctoral Thesis, Fac. Cs. Exactas y Naturales, Universidad de Buenos Aires, 253 p.

Lindegaard, C. 1989. A review of secondary production of zoobenthos in freshwater ecosystems with special reference to Chironomidae (Diptera). Acta Biol. Debr. Oecol. Hung. 3: 231-240.

MacFarlane, M. B. and T. F. Waters. 1982. Annual production by caddisflies and mayflies in a western Minnesota plains stream. Can. J. Fish. Aquat. Sci. 39: 1628-1635.

Palmer, C., J. O'Keeffe; A. Palmer, T. Dunne and S. Radloff. 1993. Macroinvertebrate functional feeding groups in the middle and lower reaches of the Buffalo River, eastern Cape, South Africa. I. Dietary variability. Freshw. Biol. 29: 441-453.

Pritchard G. and J. Zloty. 1994. Life histories of two *Ameletus* mayflies (Ephemeroptera) in two mountain streams: the influence of temperature, body size, and parasitism. J. N. Am. Benthol. Soc. 13:557-568.

Rodgers, E. B. 1982. Production of *Caenis* sp (Ephemeroptera: Caenidae) in Elevated Water Temperatures. Freshw. Invert. Biol. 1: 2-16

Snyder, C. D.; L. D. Willis and A. C. Hendricks. 1991. Spatial and temporal variation in the growth and production of *Ephoron leukon* (Ephemeroptera: Polymitarcyidae). J. N. Amer. Benthol. Soc. 10: 57-67.

Sweeney, B. W. 1984. Factors influencing life history patterns of aquatic insects, pp. 56-100. In: The ecology of aquatic insects, V.H. Resh and D. Rosenberg (eds.), Praeger, New York.

Sweeney, B. W. and R. L. Vannote. 1986. Growth and production of a stream stonefly: influences of diet and temperature. Ecology 67: 1396-1410.

Sweeney, B. W., J. K. Jackson and D. H. Funk. 1995. Semivoltinism, seasonal emergence, and adult size variation in a tropical stream mayfly (*Euthyplocia hecuba*). J. N. Amer. Benthol. Soc. 14:131-146.

Vannote, R. L. and B. W. Sweeney. 1980. Geographic analysis of thermal equilibria: a conceptual model for evaluating the effect of natural and modified thermal regimes on aquatic insect communities. Amer. Nat. 115: 667-695.

Ward, J. V. 1992. Aquatic Insect Ecology. 1. Biology and Habitat. J. Wiley and Sons, New York, 438p.

Ward, J. V. and J. A. Standford. 1982. Thermal responses in the evolutionary ecology of aquatic insects. Ann. Rev. Ent. 27: 97-117.

Waters, T. F. 1977. Secondary production in inland waters. Adv. Ecol. Res. 10: 91-164.

Waters, T. F. 1979. Influence of benthos life history upon the estimation of secondary production. J. Fish Res. Bd Can. 36: 1425-1430.

Waters, T. F. and G. W. Crawford. 1973. Annual production of a stream mayfly population: a comparison of methods. Limnol. Oceanogr. 18: 286-296.

Zelinka, M. 1980. Differences in the production of mayfly larvae in partial habitats of a barbel stream. Arch. Hydrobiol. 90: 284-297.

EFFECTS OF ROTENONE TREATMENT ON MAYFLY DRIFT AND STANDING STOCKS IN TWO NORWEGIAN RIVERS

J. V. Arnekleiv, D. Dolmen, and L. Rønning

Norwegian University of Science and Technology
Museum of Natural History and Archaeology
N-7491 Trondheim, Norway

ABSTRACT

In Norway, the salmon parasite *Gyrodactylus salaris*, has spread to 40 rivers and caused a drastic reduction in the stocks of Atlantic salmon (*Salmo salar* L.). In order to eradicate the parasite, a total of 24 rivers has so far been treated with rotenone (0.5-1.0 ppm rotenone solution). Drift and standing stocks of Ephemeroptera from stretches treated with rotenone were compared with untreated stretches and with the faunal composition before rotenone treatment.

Rotenone treatment induced an immediate catastrophic drift and caused high mortalities. The different species varied with respect to the degree and timing of their response to rotenone. *Baetis* species were rapidly affected: A great number of dead individuals appeared in the drift samples at the start of the treatment, but then subsequently declined. The response was more slow in the genus *Heptagenia*, and a few larvae of *Ephemerella* occurred in the drift samples.

A significant reduction in standing stocks of most mayflies: *Ameletus inopinatus*, *Baetis rhodani*, *B. subalpinus*, *B. fuscatus/scambus*, *Metretopus borealis* and *Heptagenia* spp. was found in newly rotenone-treated stretches. However, *Ephemerella aurivillii* and *Cloeon simile* survived the treatment in high numbers. Thus, the species composition was altered during the first few months, but a fast recolonization of the mayfly fauna took place on the rotenone-treated stretches. Within one year, all the abundant species registered before the treatment occurred again in high numbers.

INTRODUCTION

The salmon parasite, *Gyrodactylus salaris* Malmberg 1957, was probably introduced to Norway with infested parr/smolt of Atlantic salmon (*Salmo salar* L.) in the early 1970s, and has since then spread mainly through the stocking of fish from infested salmon hatcheries. Populations of salmon parr have been severely reduced in infested rivers (Johnsen and Jensen 1986, 1991, 1992). Norwegian fishery authorities are attempting to prevent any further spread of *G. salaris* and to exterminate the parasite in as many infested rivers and hatch-

Trends in Research in Ephemeroptera and Plecoptera
Edited by E. Dominguez, Kluwer Academic/Plenum Publishers, 2001

eries as possible. So far the parasite has been found in 40 Norwegian rivers, and rotenone treatments have been accomplished in 24 rivers.

Knowledge of the toxic effect of rotenone on the benthos in Norwegian rivers, and on mayflies in particular, was lacking. The use and effect of rotenone in fishery management in North America and Scandinavia have been dealt with by Soleman (1950), Almquist (1959), Tobiasson (1979), Fox (1985) and Næss et al. (1991). Haley (1978) and Ugedal (1986) also provide literature reviews. Rotenone is highly toxic to fishes (Marking and Bills 1976) and some insect larvae (Fukami et al. 1969). However, its full effects on insects and other aquatic invertebrates is not well documented since most studies primarily have been concerned with fish. Among the more detailed field studies concerning toxic effects of rotenone on benthos are Chandler (1982), Koksvik and Aagaard (1984), Dudgeon (1990) and Dolmen et al. (1995). These field studies and also laboratory studies (Engstrom-Heg et al. 1978) have documented some taxon-specific responses to the poison.

Ephemeroptera are usually an abundant part of the benthic community in Norwegian salmon rivers. Among 45 species recorded in the country, *Baetis* spp. are especially abundant in running waters. The present study documents the influence of rotenone on drift and abundance of mayflies in two Norwegian rivers.

MATERIAL AND METHODS

The Study Area and the Rotenone Treatment

The rivers Ogna and Rauma are situated in Central Norway (Fig. 1). The river Ogna with a lowland catchment area of 571 km^2, has a length of 65 km, and an average annual discharge of 22 m^3/s. It flows into the fjord Trondheimsfjorden. The rotenone treatment was carried out on 4 July 1993. At Støafossen, 18 km from the river outlet, rotenone solution (Gullviks rotenone) was continuously emptied into the river, giving a rotenone concentration of 0.5-1.0 ppm for approximately five hours. Locally, the concentration of rotenone was probably higher, up to 2-5 ppm. Backwaters, small ponds and all tributaries where anadromous fish could ascend were also treated with rotenone.

The total catchment area of the river Rauma is 1240 km^2, and the river drains mountain areas. The river empties into the Romsdalsforden near the town Åndalsnes, and the annual discharge at the river mouth is 42 m^3/s. Atlantic salmon can ascend 42 km upstream, and the lower 45 km of the river were treated with rotenone on 26-28 September 1993 (PK-PW rotenone). Several ponds, two small lakes and three neighbouring streams were also treated with rotenone. The concentration of rotenone in water was calculated to be 0.5-1.0 ppm both in lakes and rivers, but locally it was probably higher. The river was exposed to rotenone for 7 hours, and the lowermost part was exposed on three successive days.

Sampling Methods

In the river Ogna, drift samples were taken at 1 h intervals in four periods before the rotenone treatment started, and at 0.5 h intervals continuously during the rotenone treatment period. The samples were taken from one location (st. 2) in a central part of the rotenone treated river stretch (Fig. 1). We used a 2.5 m long tube (opening 10 cm diam.) and a single net (70 cm, 0.25 mm mesh size). The opening of the drift sampler was raised about 5 cm above the substratum. All animals in the drift net were first examined for dead/living animals and then preserved in 70 % ethanol. All mayflies in a 10% subsample for each time interval were counted and identified.

Two benthic kick-samples (Frost et al. 1971), using a square framed net (opening 25x25 cm, mesh size 0.5 mm), were taken at four locations in the river Rauma and at three locations in the river Ogna. Each kick sample was carried out for one minute's duration. Additional qualitative benthic samples (e.g. z-sample, see Dolmen 1992) were taken from two small lakes or ponds

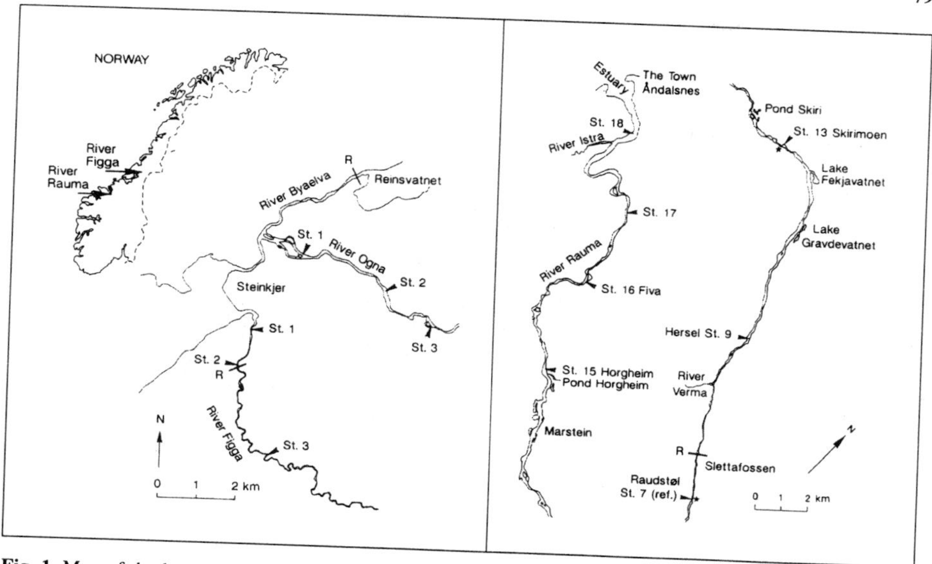

Fig. 1. Map of the lower parts of the rivers Ogna and Figga, showing the position of the sampling locations and the locations of rotenone application (R). The positions of the rivers Ogna and Rauma in Norway are shown at top, left.

near the river (Fig. 1). Samples were taken on 12 dates during the three years preceeding the rotenone treatment, and on 16 dates during the three years following the treatment. Mayflies were picked from the samples and preserved in 70 % ethanol. The animals were later sorted out in the laboratory and identified to species where possible.

RESULTS

Rotenone Effects on Mayfly Drift

A dramatic increase in drift densities was recorded following rotenone application, and the effect was still apparent during the five hours of treatment. Before rotenone was applied, the drift densities of mayflies were low (0.1-0.3 larvae per m^3). A slight indication of rotenone in the water was noticed at 14.00 p.m., resulting in an increase in drifting larvae from 0.3 to 4.7 individuals per m^3 . About one hour later, the full concentration of rotenone was reached (approximately 0.5- 1.0 ppm) as indicated by skimming water and the smell of rotenone. The total drift density of mayfly larvae peaked after 30-60 minutes, and increased from 4.7 to 97-125 individuals per m^3 followed by a decline in drift densities during the next few hours (Fig. 2). The total drift rates declined with time during the treatment and monitoring period of five hours ($r^2 = -0.805$, $n = 10$, $p < 0.001$).

The different species varied with respect to the degree and timing of their response to rotenone. Most *Baetis* species were rapidly affected: a great number of dead individuals appeared in the drift samples at the start of the treatment and then the number subsequently declined. The response was slower in *Baetis muticus* and in the genus *Heptagenia* (Fig. 3). For the species *B. rhodani, B. fuscatus/scambus* and *H. sulphurea* there was a significant decline in drift densities during the rotenone treatment period (*B. rhodani*: $r^2 = -0.648$, $n = 10$, $p < 0.001$, *B. fuscatus/scambus*: $r^2 = -0.506$, $n = 10$, $p < 0.01$, *H. sulphurea*: $r^2 = -0.681$,

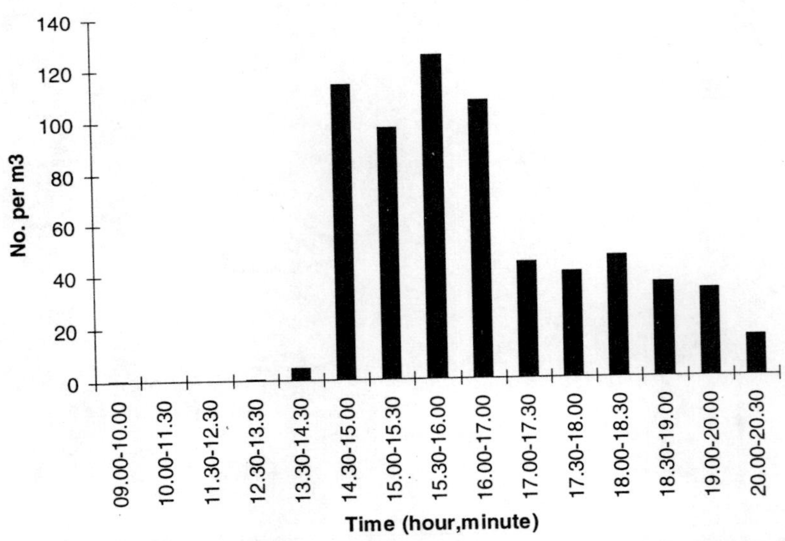

Fig. 2. Total drift densities (N/m³) before and during rotenone treatment of the river Ogna. The rotenone solution reached the station at approximatly 14.00

$n = 8$, $p < 0.01$). A few larvae of *Ephemerella mucronata*/sp. occurred in the drift samples through the whole treatment period, but their numbers did not decline with time ($r^2 = 0.037$, $n = 9$, $p = 0.310$).

About 95-99% of the drifting larvae were dead, especially *Baetis* spp. However, among living larvae *Ephemerella mucronata*/sp. was most common, although some living *Heptagenia sulphurea* were also found. The composition of rotenone-induced drift was markedly different from the ephemeropteran benthic community structure. *Heptagenia* sp./*sulphurea* comprised 60 % of the total mayfly drift, but only 19.3 % of mayfly benthic standing stock as indicated by kick samples. In the benthic samples baetid mayflies were clearly dominant (78 %), while in the drift samples they only made up 37.8 % (Table 1).

Rotenone Effects on Mayfly Standing Stocks

In the river Ogna, the mayfly fauna was almost eradicated from the treated stretches. One day after the application of rotenone, only a few specimens of *Heptagenia sulphurea* and *Ephemerella mucronata* was found alive in kick samples (Fig. 4). However, one and a half months after the treatment, the dominant species *Baetis rhodani* and *Ephemerella* sp. were common, and six species were detected compared with eigth species four days before the treatment.

In the river Rauma, 14 species of Ephemeroptera were found in the benthic samples in the investigated period. A significant reduction in standing stocks of most mayflies: *Ameletus inopinatus, Baetis rhodani, B. subalpinus, B. fuscatus/scambus, Metretopus borealis* and *Heptagenia* spp. was found on newly rotenone-treated stretches. The abundance of mayflies was reduced by 12-81% (between locations) during the month after the rotenone application, compared to the month before the rotenone treatment in late September 1993. In particular, baetid mayflies were strongly affected. *B. rhodani,* the most dominant species in the river, was absent from the samples in the autumn of 1993 and the spring of 1994 (Fig. 5). However, there was a

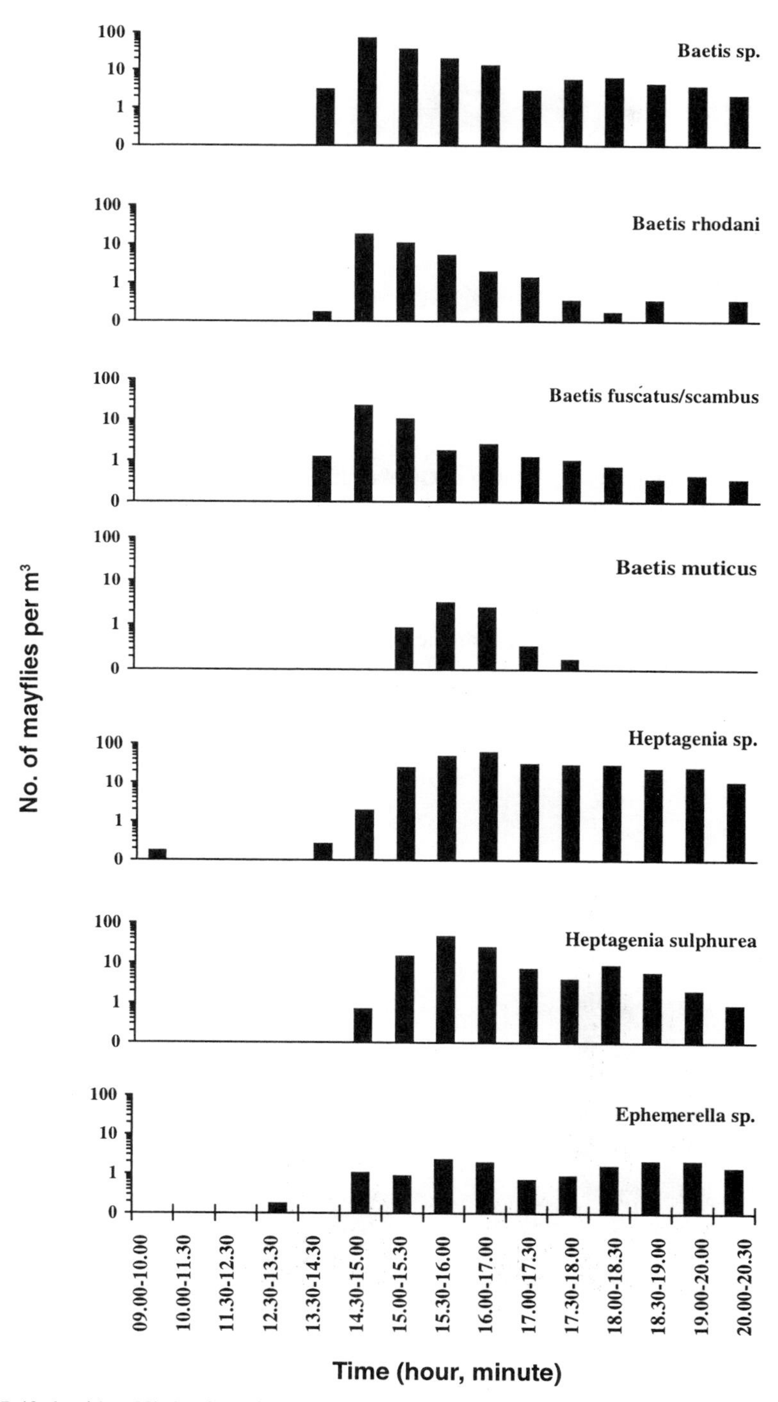

Fig. 3. Drift densities (N/m^3) of mayfly species before and during the rotenone treatment period in the river Ogna.

Table 1. The composition (%) of mayfly species in drift samples taken during the rotenone treatment, and the composition in benthic samples taken four days before the treatment.

Species	% in drift	% in benthos
Baetis spp.	24.7	7.4
Baetis rhodani	5.8	55.7
Baetis fuscatus/scambus	6.2	9.3
Baetis muticus	1.1	2.6
Baetis subalpinus	0.0	3
Heptagenia sp.	42.9	3.7
Heptagenia dalecarlica	0.0	1.5
Heptagenia sulphurea	17.1	14.1
Ephemerella mucronata/sp.	2.3	2.2
Caenis horaria	0.0	0.4

recolonization of numerous small larva in October 1994, and in 1995 and 1996 the species was as abundant as in the years preceeding the treatment. Its seasonal variation in abundance was also quite similar to that at the reference station (Fig. 5). Other baetid mayflies, as well as *A. inopinatus* died as a result of the rotenone application, and were absent from the river samples in the following month. However, they all recolonized the river stretches in 1994, and *A. inopinatus* was already numerous in May 1994 (Fig. 5). *E. aurivillii* was found alive in high numbers both during and after the rotenone treatment, and there was no significant reduction in larval standing stock (Fig. 5). The species occurred in especially high numbers during the year after the rotenone treatment. *Heptagenia sulphurea* also survived the treatment, as did the lentic species *Cloeon simile* in the pond Horgheimdammen (Fig. 6). These

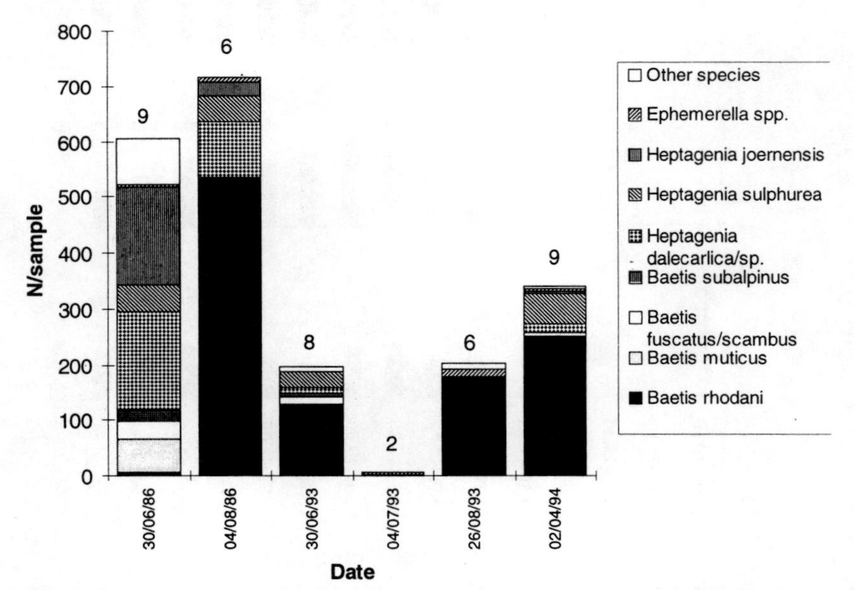

Fig. 4. The composition and abundance of mayfly species (N/sample) in the river Ogna before and after rotenone treatment. Pooled data from two locations. Number of species are given over each bar.

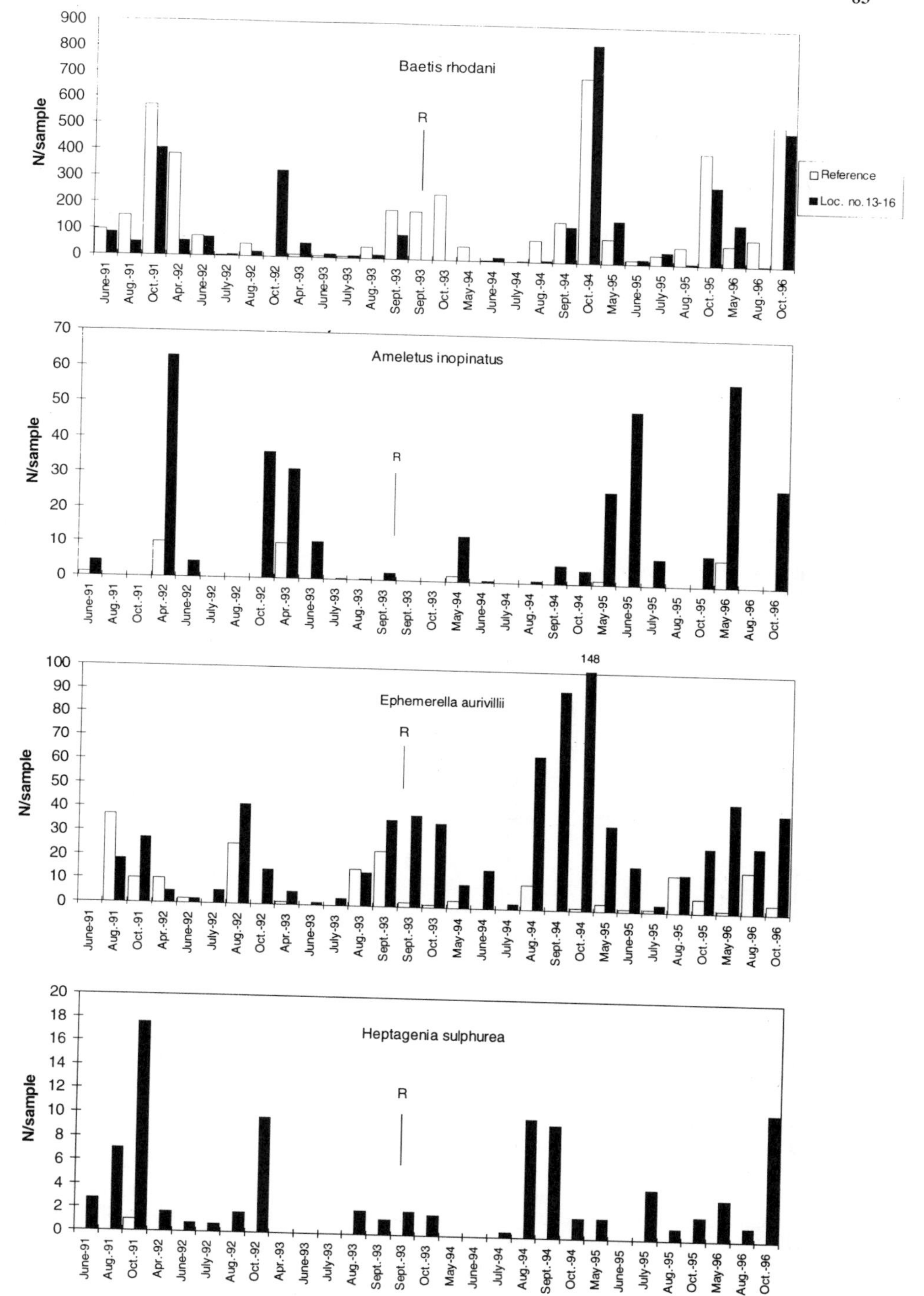

Fig. 5. Density of *Baetis rhodani, Ameletus inopinatus, Ephemerella aurivillii* and *Heptagenia sulphurea* in samples from the river Rauma before and after rotenone treatment (R).

results indicate that there is considerable variability in the effects of rotenone on different mayfly species.

At the localities exposed to rotenone, the overall mayfly fauna changed little compared to the three years preceeding the treatment (Table 2). *B. rhodani* was the dominant species in both periods (1991-1993 and 1994-1996). There was an increase in the abundance of *E. aurivillii* after the treatment, and a slight decrease in the abundance of *A. inopinatus* and *H. sulphurea*. Several species appeared sporadically in the samples both before and after the rotenone treatment: *Baetis subalpinus, B. fuscatus/scambus, Parameletus chelifer* and *Metretopus borealis*. A total of 12 species were observed in the samples during the three years before the rotenone treatment compared to 11 species during the three years after. Both the Shannon-Weaner diversity index and eveness were lower after the treatment. However, there was also a reduction in the mayfly diversity at the reference station (Table 2), so at least part of the reduction on the rotenone-treated stretches was probably not due to the poison. Among the more uncommon species, *H. dalecarlica* and *Paraleptophlebia werneri* were only detected in the period before the treatment, while *Leptophlebia vespertina* and *Cloeon simile* were observed only in the period following the treatment. Overall, total mayfly abundances did not change much in the long term as a result of the rotenone application.

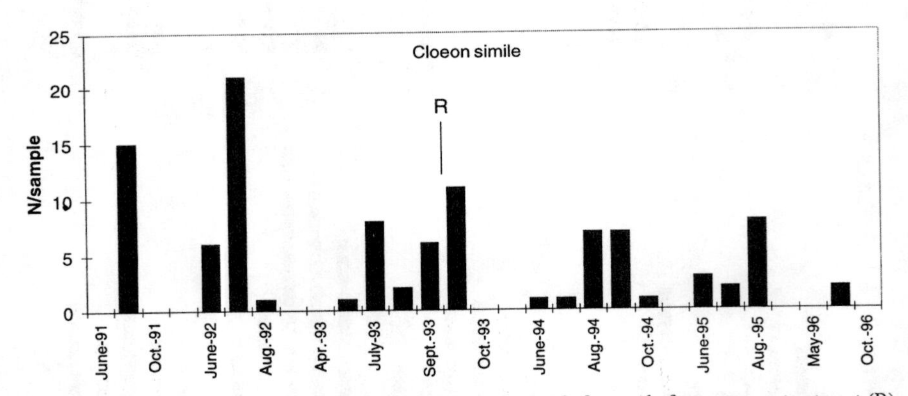

Fig. 6. Density of *Cloeon simile* in the pond Horgheimdammen before and after rotenone treatment (R).

DISCUSSION

The effects of rotenone on Scandinavian lotic mayflies is poorly known. This study shows that rotenone application induced an immediate catastrophic drift of mayflies, and caused large-scale mortality. Daytime drift densities before the rotenone application were low. The drift densities peaked 30-60 minutes after rotenone exposure, and were much higher than normal drift densities (cf. Brittain and Eikeland 1988). The different species differed in their tendency to enter the drift: most baetid mayfly species were particularly sensitive to rotenone. After a peak early in the treatment period, they decreased in densities. Others, like *Baetis muticus*, *Heptagenia* spp. and *Ephemerella* spp. were more slow in their response. Drift densities of *Ephemerella* did not decrease with time, and several specimens were found alive in drift samples. Different species of mayflies must therefore have widely different tolerances to rotenone, either for physiological reasons or as a result of ecological and/or behavioural differences between species. Koksvik and Aagaard (1984) describe taxon-specific effects of rotenone on lentic chironomids. In two Papua New Guinea streams, Dudgeon (1990) also found increasing drift densities after rotenone application. Baetid mayflies were particularly affected, and baetid densities decreased with time, as was also seen in the river Ogna. Leptophlebiid mayflies, by contrast, were found less sensitive to rotenone in Dudgeon's study (1990). Catastrophic drift in the Papua New Guinea streams comprised mainly living animals, in contrast to our findings in the river Ogna. However, Dudgeon does not give any data on the concentration of rotenone used and the time of exposure. Laboratory studies have indicated that tolerances vary greatly within single orders or families (Engstrom-Heg et al. 1978, Chandler and Marking 1982). Coleoptera, Hemiptera, Gastropoda and Bivalvia seem to have a high tolerance to rotenone (Engstrom-Heg et al. 1978). The pearl mussel (*Margaritifera margaritifera*) is also highly resistant to rotenone (Dolmen et al. 1995). Among ephemeropterans Engstrom-Heg et al. (1978) found Baetidae to be very sensitive and Heptageniidae and Ephemerella to be relatively unaffected. This agrees with our findings. Total mortality occurred after 6 ppm-hours of rotenone exposure in *Baetis* spp., in contrast to 72-144 ppm-hours in *Ephemerella* spp. (Engstrom-Heg et al. 1978). However, exposure to rotenone in laboratory studies does not necesserely mimic the way in which species respond to the poison in running water. As pointed out by Dudgeon (1990), the initial response to rotenone is behavioural, but we do not know whether the insects are drifting in response to detection of rotenone or as a result of its toxic effects. In the river Ogna, however, rotenone clearly had a toxic effect on most mayflies, causing high mortalities. The effects of rotenone at the species level may, moreover, depend on the microhabitat use of a species and also in the timing of the rotenone treatment versus life history events. The early instars of several mayfly species live in the hyporheic (Ward 1992, Williams 1984) and will not be exposed to rotenone as easily as later instars dwelling in the uppermost substrate. This may also be the case in later instars of *Baetis muticus* and *Heptagenia sulphurea,* which are more cryptic in their behaviour than *Baetis rhodani* and *Baetis subalpinus* for instance.

Reports on the short-time effects of rotenone treatment of rivers on standing-stocks of macroinvertebrates give highly different pictures. Some investigations have found a large temporary reduction in standing stocks of invertebrates (Binns 1967, Cook and Moore 1969, Arnekleiv 1997, Arnekleiv et al. 1997), while others report small or insignificant reductions (Dudgeon 1990, Morrison 1977). This depends probably both on the rotenone concentrations used and the design of investigations. In the river Ogna mayflies were almost eradicated from the rotenone treated stretches for some weeks. Dispite a higher total exposure to rotenone in the river Rauma (up to 10 ppm-hours in the river Ogna, versus 15-25 ppm-hours in the river Rauma) the mayfly fauna seemed less affected in the river Rauma. This can be due to several factors: The effects of rotenone are positively correlated by temperature (Meadows 1973, Gilderhus et al. 1986). The temperature was considerably higher during the rotenone treatment in Ogna in July (13.6 °C) compared to that in Rauma during the treatment in September (5.0-6.0 °C). Species composition and seasonal life history events were also different in July and September. In addition different types of rotenone were used in the two rivers.

In both rivers, however, we observed a rapid recolonization after the rotenone treatment. *Ameletus inopinatus* occured with large nymphs in May 1994 after the rotenone treatment in late September 1993. *Baetis rhodani* recolonized the treated stretches with small nymphs the following autumn. Within one year, all the abundant species registered before the treatment, again occurred in large numbers. Binns (1967) made similar observations when treating a river in Wyoming with emulsified rotenone: the invertebrate population was drastically reduced by the rotenone treatment, but within one month it had to a large extent been built up again. Drift was probably an important source in the recolonization of the rivers Ogna and Rauma, since both rivers have long stretches of different biotopes upstream of the rotenone treated reaches (cf. Brittain and Eikeland 1988, Mackay 1992). However, we do not have detailed information about the mechanisms of recolonization in these rivers. Engstrom-Heg et al. (1978) state that there is a strong tendency for the more sensitive insects to be highly mobile and/or to have short life cycles, and thus to repopulate depleted areas rapidly through drift, migration and oviposition. This is in parti-

Table 2. Relative abundance and diversity of mayfly species at the reference station and at three locations (pooled data) in the river Rauma before (12 dates) and after (16 dates) rotenone treatment in 1993.

1991-1993 (before r.)				1994-1996 (after r.)		
Loc. no. 7, reference	**Mean**	**± SE**	**% of total**	**Mean**	**± SE**	**% of total**
Baetis rhodani	131.46	50.88	90.63	172.50	51.29	96.02
Ephemerella aurivilli	10.21	3.57	7.04	3.88	1.35	2.16
Ameletus inopinatus	1.67	1.12	1.15	0.72	0.47	0.40
Baetis subalpinus	1.17	0.87	0.80	2.16	1.32	1.20
Parameletus chelifer	0.42	0.34	0.29	0.00	0.00	0.00
Heptagenia sp.	0.08	0.08	0.06	0.00	0.00	0.00
Baetis fuscatus/scambus	0.04	0.04	0.03	0.03	0.03	0.02
Siphlonurus sp.	0.00	0.00	0.00	0.38	0.34	0.21
Total	145.04	51.75	100.00	179.66	50.92	100.00
No. of species	7			6		
Shannon-Wener H'	0.169			0.092		
Eveness J'	0.200			0.118		

1991-1993, before rotenon				1994-1996, after rotenon		
Loc. no. 13, 15, 16	**Mean**	**± SE**	**% of total**	**Mean**	**± SE**	**% of total**
Baetis rhodani	113.86	34.04	72.7	147.14	41.87	71.94
Ephemerella aurivilli	14.32	2.97	9.14	38.89	10.32	19.01
Ameletus inopinatus	12.50	3.99	7.98	11.3	2.99	5.52
Siphlonuridae	9.53	6.26	6.08	0.01	0.01	0.01
Heptagenia sulphurea	2.14	0.93	1.37	1.81	0.77	0.88
Siphlonurus sp.	1.76	1.56	1.13	1.56	1.27	0.76
Heptagenia sp.	1.34	0.75	0.86	1.56	0.88	0.76
Paraleptophlebia werneri	0.58	0.43	0.37	0.00	0.00	0.00
Baetis subalpinus	0.24	0.13	0.15	0.41	0.22	0.20
Parameletus chelifer	0.08	0.06	0.05	0.11	0.11	0.06
Baetis sp.	0.08	0.06	0.05	0.57	0.56	0.28
Metretropus sp./*borealis*	0.08	0.08	0.05	0.01	0.01	0.01
Baetis fuscatus/scambus	0.07	0.04	0.04	0.01	0.05	0.01
Heptagenia dalecarlica	0.03	0.03	0.02	0.00	0.00	0.00
Metretropus borealis	0.01	0.01	0.01	0.00	0.00	0.00
Parameletus sp.	0.00	0.00	0.00	0.11	0.11	0.06
Siphlonurus lacustris	0.00	0.00	0.00	0.19	0.13	0.09
Cloeon sp.	0.0	0.00	0.00	0.01	0.01	0.01
Cloeon simile	0.00	0.00	0.00	0.01	0.01	0.01
Leptophlebia vespertina	0.00	0.00	0.00	0.01	0.01	0.01
Ephemerella sp.	0.00	0.00	0.00	0.76	0.49	0.37
Total	156.61	39.55	100.00	204.52	47.5	100.00
No. of species	12			11		
Shannon-Wiener H'	0.443			0.390		
Eveness J'	0.376			0.311		

cular the case for *Baetis rhodani* and *Ameletus inopinatus,* which are good swimmers and have an univoltine life cycle in Central Norway (Arnekleiv 1996). Experimental studies of recolonization of disturbed reaches by stream macroinvertebrates have given periods from a few days to approximately four months (Reice 1985, Malmquist et al. 1991, Tikkanen et al. 1994).

Heptagenia sulphurea, Ephemerella aurivillii and *Cloeon simile* all survived the rotenone treatment. In laboratory experiments the two former species were also the most rotenone-tolerant mayfly species (Engstrom-Heg et al. 1978). Many pond-dwelling insects, including Leptophlebiid mayflies, are known to tolerate oxygen deficiency (Brittain and Nagell 1981, see also Macan 1973). The reason why the lentic species *Cloeon simile* also survived the treatment is possibly that it can also tolerate low oxygen levels. However, the highest numbers of surviving individuals was found in *E. aurivillii*. This species even appeared more numerous the year following the rotenone treatment compared to the year before. The altered interspecific competition and lack of predators evidently gave *E. aurivillii* increased opportunity.

This study demonstrated a taxon-specific response to rotenone among mafly species. A significant reduction in standing stocks of most species was found in newly rotenone-treated streches. At least three species survived the treatment. A relatively fast recolonization of the mayfly fauna took place on the rotenone-treated stretches. Overall, total mayfly abundances did not change in the long term as a result of the rotenone treatment of the river Rauma. However, further investigations are needed to evaluate recolonization mechanisms and the long-term effects on the population biology and genetics of mayfly species.

ACKNOWLEDGMENT

The project was financed by The Directorate for Nature Management. Terje Dalen and Terje Bongard assisted in the species identification. We thank an anonymous referee for improving the English.

REFERENCES

Almqvist, E. 1959. Observations on the effect of rotenone emulsives on fish food organisms. Rep. Inst. Freshw. Res. Drottningholm 40: 146-160.

Arnekleiv, J. V. 1996. Life cycle strategies and seasonal distribution of mayflies (Ephemeroptera) in a small stream in central Norway. Fauna norv. Ser. B 43: 19-30.

Arnekleiv, J. V. 1997. Short-time effects of rotenone treatment on benthic macroinvertebrates in the rivers Ogna and Figga, Steinkjer municipality. Vitenskapsmuseet Rapp. Zool. Ser. 1997, 3: 1-28. (In Norwegian).

Arnekleiv, J. V., D. Dolmen, K. Aagaard., T. Bongard and O. Hanssen. 1997. Effect of rotenone treatment on the bottomfauna of the Rauma and Henselva watercourses, Møre and Romsdal. Part I: Qualitative investigations. Vitenskapsmuseet Rapp. Zool. Ser. 1997, 8: 1-48. (In Norwegian).

Binns, N. A. 1967. Effects of rotenone treatment on the fauna of the Green River, Wyoming. Fish. Res. Bull. 1: 1-114.

Brittain, J. E. and T. J. Eikeland. 1988. Invertebrate drift. A review. Hydrobiologia 166: 77-93.

Brittain, J. E. and B. Nagell. 1981. Overwintering at low oxugen concentraitons in the mayfly Leptophlebia vespertina. Oikos 36: 45-50.

Chandler, J. H. (Jr.) and L. L. Marking. 1982. Toxicity of rotenone to selected aquatic invertebrates and frog larvae. Prog. Fish. Cult. 44: 78-80.

Cook, S. F. (Jr.) and R. L. Moore. 1969. The effects of rotenone treatment on the insect fauna of a California stream. Trans. Amer. Fish. Soc. 98: 539-544.

Dolmen, D. 1992. Dammer i kulturlandskapet - makroinvertebrater, fisk og amfibier i 31 dammer i Østfold. NINA Forskningsrapport 20: 1-63. (In Norwegian).

Dolmen, D., J. V. Arnekleiv and T. Haukebø. 1995. Rotenone tolerance in the freshwater pearl mussel (*Margaritifera margaritifera*). Nordic J. Freshw. Res. 70: 21-30.

Dudgeon, D. 1990. Benthic community structure and the effect of rotenone piscicide on invertebrate drift and standing stocks in two Papua New Guinea streams. Arch. Hydrobiol. 119: 35-53.

Engstrom-Heg, R., R. T. Colesante and E. Silco. 1978. Rotenone tolerances of stream-bottom insects. N. Y. Game J. 25: 31-41.

Fox, R. C. 1985. Rotenone use for fisheries management. California Department of Fish and Game. 243 pp.

Frost, S., A. Huni and W. E. Kershaw. 1971. Evaluation of a kicking technique for sampling stream bottom fauna. Can. J. Zool. 49: 160-173.

Fukami, J. I., T. Shisido, K. Fukunaga and J. E. Casida. 1969. Oxidative metabolism of rotenone in mammals, fish, and insects, and its relation to selective toxicity. J. Agr. Food Chem. 17: 1217-1226.

Gilderhus, P. A., J. L. Allen and W. K. Dawson.1986. Persistence of rotenone in ponds at different temperatures. North Amer. J. Fish. Mgmt. 6: 129-130.

Haley, T. J. 1978. A review of the literature of rotenone. J. Environ. Pathol. Toxicol. 1: 315-337.

Johnsen, B. O. and A. J. Jensen. 1986. Infestations of Atlantic salmon (*Salmo salar*) by *Gyrodactylus salaris*, in Norwegian rivers. J. Fish Biol. 29: 233-241.

Johnsen, B. O. and A. J. Jensen. 1991. The *Gyrodactylus* story in Norway. Aquaculture 98: 289-302.

Johnsen, B. O. and A. J. Jensen. 1992. Infection of Atlantic salmon, *Salmo salar* L., by *Gyrodactylus salaris,* Malmberg 1957, in the River Lakselva, Misvær in northern Norway. J. Fish Biol. 40: 433-444.

Koksvik, J. I. and K. Aagaard. 1984. Effects of rotenone on the benthic fauna of a small eutrophic lake. Verh. Int. Ver. Limnol. 22: 658-665.

Macan, T. T. 1973. Ponds and lakes. London. George Allen and Unwin Ltd. 148 pp.

Mackay, R. J. 1992. Colonization by lotic macroinvertebrates: a review of processes and patterns. Can. J. Fish. aquat. Sci. 49: 617-628.

Malmquist, B., S. Rundle, C. Brönmark and Erlandson, A. 1991. Invertebrate colonization of a new, man-made stream in southern Sweden. Freshw. Biol. 26: 307-324.

Marking, L. L. and T. D. Bills. 1976. Toxicity of rotenone to fish in standardized laboratory tests. U.S. Fish Wildl. Serv., Invest. Fish Control 72: 1-11.

Meadows, B. S. 1973. Toxicity of rotenone to some species of coarse-fish and invertebrates. J. Fish Biol. 5: 155-163.

Morrison, B. R. S. 1977. The Effects of Rotenone on the Invertebrate Fauna of Three Hill Streams in Scotland. Fish. Mgmt. 8: 128-139.

Næss, T., K. E. Naas and O. B. Samuelsen. 1991. Toxicity of rotenone to some potential predators on marine fish larvae. An experimental study. Aquacult. Eng. 10: 149-159.

Reice, S. R., 1985. Experimental disturbance of species diversity in a stream community. Oecologia 67: 90-97.

Soleman, V. E. F. 1950. History and use of fish poisons in the United States. Can. Fish Cult. 8: 3-16.

Tikkanen, P., P. Laasonen, T. Muotka, A. Huhta, and K. Kuusela. 1994. Short-term recovery of benthos following disturbance from stream habitat rhabilitation. Hydrobiologia 273: 121-130.

Tobiasson, G. 1979. The use of rotenone in Sweden. Inf. Inst. Freshw. Res., Drottningholm 10: 1-33. (In Swedish).

Ugedal, O. 1986. A review of the litterature concerning the effects of rotenone in freswater ecosystems. DN-Reguleringsundersøkelsne 1986-14: 1-52. (In Norwegian).

Ward, J. V. 1992. Aquatic Insect Ecology. 1. Biology and habitat. John Wiley and Sons, Inc. New York. 438 s.

Williams, D. D. 1984. The hyporheic zone as a habitat for aquatic insects and associated arthropods, pp. 430-455 in V. H. Resh and D. R. Rosenberg (eds.), The Ecology of Aquatic Insects. Praeger, New York.

LONGITUDINAL DISTRIBUTION OF THE MAYFLY (EPHEMEROPTERA) COMMUNITIES AT THE CHOCANCHARAVA RIVER BASIN (CÓRDOBA, ARGENTINA)

M. del C. Corigliano, C. M. Gualdoni, A. M. Oberto, and G. B. Raffaini

Dpto. de Ciencias Naturales
Universidad Nacional de Río Cuarto
A.P. Nº 3, (X5804 BYA) Río Cuarto, Argentina

ABSTRACT

Distribution of Ephemeroptera communities in the longitudinal gradient of Chocancharava river basin was studied. Species and sample ordination was performed by means of DCA (Detrented Correspondence Analysis). Fourteen species of Ephemeroptera belonging to six families were collected. Sites with higher species richness were foothill stretches. There was an indication of distinct rhithron and potamic assemblages. There is a gradual change in abundance downstream and substitution of species along elevation gradients from mountain to lowland rivers, with a major discontinuity under conditions of environmental stress, as pollution and hydraulic shifts.

INTRODUCTION

Ephemeroptera larvae are major components of fluvial ecosystems in Córdoba Province (Corigliano *et al.,* 1996), and contribute significantly to zoobenthic biomass and are the main food resources for insectivore fishes in mountain streams (Corigliano y Malpassi, 1994).

In spite of their ecological importance there is a shortage of knowledge about basic issues such as distribution in longitudinal gradients and ecological requirements in stream and rivers of Córdoba Province. Besides, specific status of immature stages of the regional fauna has not been completely developed yet. There is a dubious species, *Paracloeodes* sp. not reported by Domínguez *et al.,* (1994). But meanwhile taxonomic research develops, it is necessary to make some ecological considerations about the taxa to understand their functional role in river ecology.

Questions about patterns of altitudinal distribution of invertebrates are of main interest in fluvial ecology. Whether species distribution is zonal (Illies and Botosaneanu, 1963) or continuous (Vannote *et al.,*1980) and what the importance of boundaries (Naiman *et al.,* 1988) and hydraulic stress (Statzner and Higler, 1986) are in the determination of patterns of benthic distribution, are hard-core questions in fluvial ecosystem theory.

Trends in Research in Ephemeroptera and Plecoptera
Edited by E. Dominguez, Kluwer Academic/Plenum Publishers, 2001

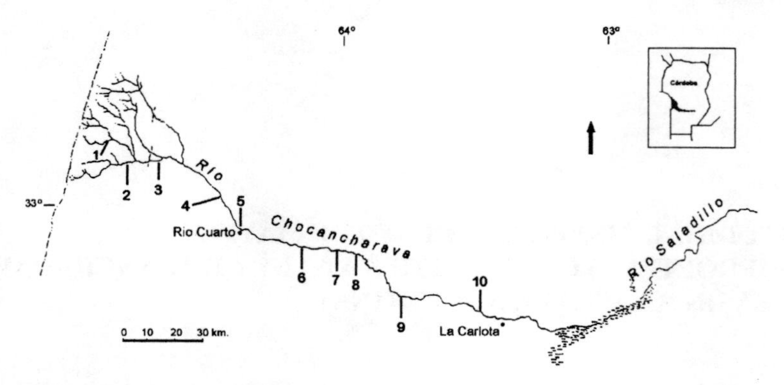

Fig. 1. Location of sampling stations in the Chocancharava river (Córdoba, Argentina).

Table 1. Physico-chemical and environmental characteristics of Chocancharava river. *Dominant sediment size: **3**: > 16 mm; **2**: < 16 > 0.25 mm; **1**: < 0.25 mm.

	Sampling stations									
	1	2	3	4	5	6	7	8	9	10
Altitude (m. asl)	700	618	548	468	433	325	290	260	200	165
Slope ($m.km^{-1}$)	80,6	13,7	7,7	3,3	2,9	3,7	1,8	1,4	1,5	0,9
Width (m)	7	15	17	50	80	50	90	70	60	45
Current velocity ($m.s^{-1}$)	1,26	0,44	0,47	0,55	0,53	0,45	0,45	0,45	0,45	0,35
Depth (cm)	15	30	20	30	20	35	30	35	40	40
Discharge ($m^3.seg^{-1}$)	1,36	2,99	3,56	3,56	3,50	3,50	3,80	3,80	3,80	5,86
Order number	2	5	6	7	7	7	7	7	7	7
Distance from the source (km)	9	35	46	70	82	111	130	151	191	229
Drainage area (km^2)	30	340	1450	1680	1770	1910	2050	2100	2265	2485
Sediment size (mm) *	3	3	2	2	2	2	2	2	1	1
Temperature (°C)	20,9	17,2	18,3	19,1	16,4	14,9	21,0	18,5	16,5	16,4
Termic amplitude (° C)	16,0	18,5	22,0	22,0	23,0	23,0	22,0	22,0	20,0	20,0
pH	8,1	8,4	8,1	8,3	8,3	8,3	8,3	8,3	8,3	8,4
Conductivity at 20 °C ($\mu S.cm^{-1}$)	105,3	155,9	168,5	237,6	236,1	303,8	308,2	348,7	421,4	526,9
Suspended solids (cm^3)	--	--	0,06	0,12	0,22	0,64	0,65	0,80	0,76	0,85
Dissolved oxygen ($mg.l^{-1}$)	7,40	9,50	8,90	8,40	8,50	9,90	8,40	7,90	8,50	8,80
Redox	20,00	19,52	19,00	19,73	19,50	15,00	18,00	19,80	19,60	19,60
COD ($mg.l^{-1}$)	0,20	0,61	0,70	0,77	1,88	0,78	1,33	1,57	1,99	1,99
Permanganate values ($mg.l^{-1}$)	2,00	3,98	3,00	2,13	2,29	3,16	1,34	2,50	3,00	3,00

The aim of this study has been the spatial characterization and distribution of Ephemeroptera communities in the Chocancharava basin of Córdoba Province as well as its relation to environmental variables.

MATERIALS AND METHODS

Ten sampling sites were selected along the longitudinal gradients of watercourses in Choncancharava river basin (32° 54' - 33° 21' S, 63° 28' - 64° 46' W) (Fig. 1).

Table 2. Density of Ephemeroptera taxa colected from sampling stations of Chocancharava river and number of total taxa by site.

	Sampling stations									
TAXA	1	2	3	4	5	6	7	8	9	10
Baetis sp. 1	147	8166	8071	4643	2331	338	121	179	274	100
Baetis sp. 2	0	350	2324	144	0	0	87	0	0	0
Baetodes sp.	1	1	13	0	0	0	0	0	0	0
Camelobaetidius penai	59	2476	1526	39	8	0	0	0	0	0
Paracloeodes sp. 1	0	0	903	3237	7499	734	829	123	5	5
Baetidae sp.	0	0	0	0	0	0	0	0	11	14
Caenis sp.	26	12	413	8	10	1	0	1	1	3
Leptohyphes sp. 1	907	12883	10346	280	6	0	0	0	0	0
Leptohyphes sp. 2	0	0	0	0	0	0	50	17	13	29
Tricorythodes sp.	32	1235	2340	171	10	0	1	0	0	1
Farrodes sp.	1	75	115	0	0	1	0	1	0	0
Traverella (Zonda) sp.	0	0	0	0	0	0	0	0	0	1
Homoeoneuria sp.	0	0	0	0	0	0	0	10	32	15
Polymitarcyidae sp.	0	0	0	0	0	0	0	0	7	0
Taxa/station	7	8	9	7	6	4	5	6	7	8

Headwaters are located in Sierras de los Comechingones at 2300 m. asl. Study sites were located from 700 to 165 m. asl, where there are second order streams to seventh order lowland collectors (Table 1). Periphyton assemblages are well developed year round and massive growth of *Cladophora glomerata* mats was observed during spring season in foothill stretch (Corigliano *et al.*, 1994). Native insectivore fishes belongs to Characidae and Loricariidae species, and rainbow trout *Onchorynchus mykiss* was introduced in mountain streams. Macroinvertebrates were collected in lotic habitats with a D frame handnet, with a mesh size of 300 m. The sampling period in each site lasted 10 minutes. The sampling program was developed for two years. Sites were visited twice each year during low flow and high flow conditions. A physicochemical water characterization was carried out at sampling sites. The organisms were counted under dissecting microscope and abundance was expressed in ind.10' which allows quantitative comparisons. From the

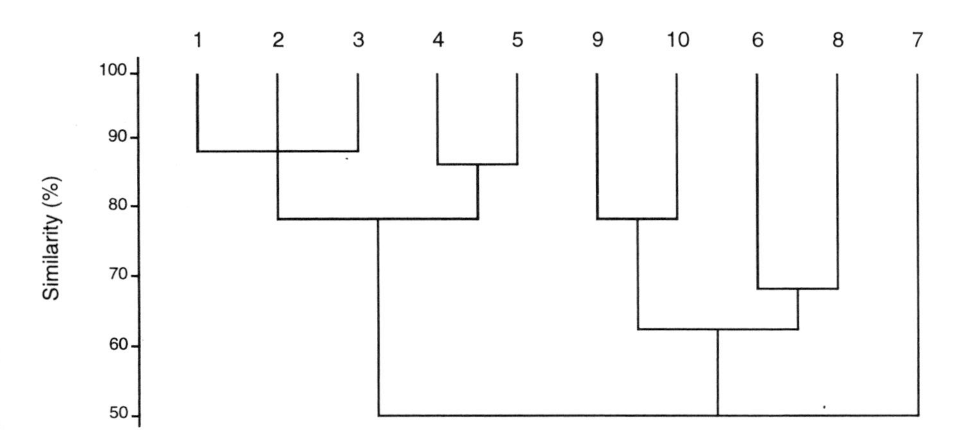

Fig. 2. Percentage similarity between sampling sites based on Ephemeroptera data.

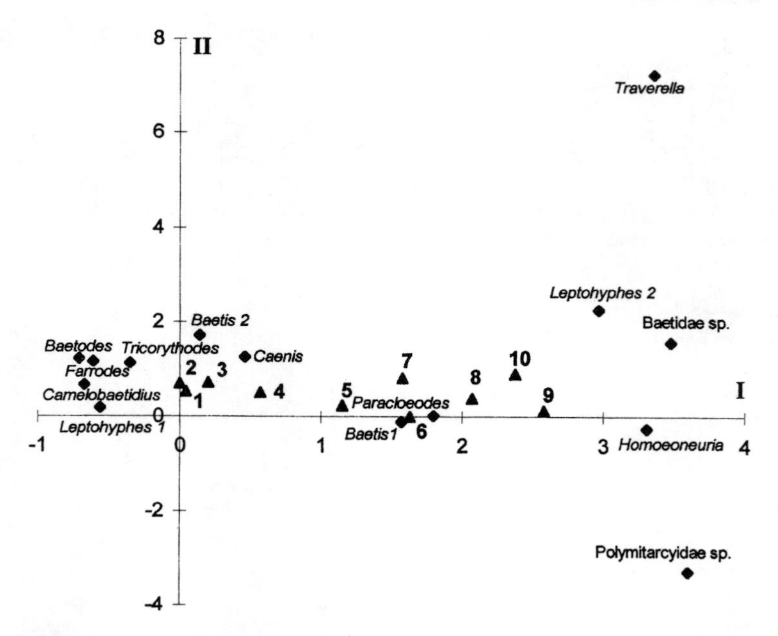

Fig. 3. DCA ordenation diagram (axis I and II) with sampling sites (▲) and Ephemeroptera taxa (◆).

abundance values of Ephemeroptera species the sample average was calculated for each site and with the arithmetic means a data matrix was created. Percentage similarity index was calculated by Jaccard coefficient. Species and sample ordination was performed by means of detrended correspondence analysis, DCA, by DECORANA (Hill, 1979). Pearson linear correlation between environmental parameters and sample scores of DCA axis and organism abundance (log x +1 transformed) were performed.

RESULTS

Fourteen species of Ephemeroptera belonging to six families (Baetidae, Caenidae, Leptohyphidae, Leptophlebiidae, Oligoneuriidae, Polymitarcyidae) were collected. (Table 2). Higher species richness was found at foothill stretches, and sites with lower richness were sampling stations 6, 7 and 8, where the river was recuperating from heavy sewage pollution. The percentage similarity matrix showed that the highest fauna similarity was between adjacent site pairs and there was also an indication of distinct rhithron and potamon assemblages (Fig. 2). Stations 6, 7 and 8 altered the longitudinal sequences of adjacent site pairs and were more dissimilar than the rest.

Arithmetic means of four samples for each site were calculated and DECORANA ordination analysis was performed in ten sample means. Samples and species were arranged in 4 axes whose eingenvalues were: 0.524, 0.049, 0.011 and 0.002. Figure 3 shows the ordination with respect to axes I and II.

The first DCA axis was positively correlated with depth, conductivity, COD, and negatively correlated with altitude and sediment size (Table 3). Higher scores for axis I belonged to Baetidae sp., *Leptohyphes* sp. 2, *Traverella* sp., *Homoeoneuria* sp. and Polymitarcyidae sp., which was the species assemblage from lowland stretches of the river.

Table 3. Linear correlations (r) of environmental parameters with DCA axis. ** = $p < 0.01$; *** = $p < 0.001$.

Parameters	AX1	AX2	AX3	AX4
Altitude (m. asl)	-0,964 ***	0,132	-0,360	-0,324
Slope ($m.km^{-1}$)	-0,543	0,080	-0,557	-0,543
Width (m)	0,699 **	0,714 **	0,148	0,090
Current velocity ($m.s^{-1}$)	-0,514	-0,210	-0,612 **	-0,614
Depth (cm)	0,757 ***	-0,145	0,580	0,577
Discharge ($m^3.seg^{-1}$)	0,693 **	-0,165	0,317	0,329
Order number	0,663 **	-0,191	0,439	0,409
Distance from the source (km)	0,943 ***	0,008	0,324	0,300
Drainage area (km^2)	0,865 ***	-0,147	0,310	0,269
Sediment size (mm)	-0,848 ***	0,116	-0,202	-0,161
Temperature (°C)	-0,329	0,316	-0,092	-0,100
Termic amplitude (° C)	0,347	-0,294	0,238	0,208
pH	0,467	0,029	0,337	0,366
Conductivity at 20 °C ($\mu S.cm^{-1}$)	0,920 ***	0,024	0,286	0,269
Suspended solids (cm^3)	0,942 ***	-0,103	0,357	0,326
Dissolved oxygen ($mg.l^{-1}$)	0,060	-0,250	0,181	0,219
Redox	-0,230	0,579	0,079	0,080
COD ($mg.l^{-1}$)	0,823 ***	-0,068	0,201	0,158
Permanganate values ($mg.l^{-1}$)	-0,139	0,131	0,355	0,401

Baetodes sp., *Camelobaetidius penai*, *Farrodes* sp., *Leptohyphes* sp. 1 and *Tricorythodes* sp. had the lowest score for axis I. This group of species characterized the mountain and foothill stretches. *Baetis* sp. 1, *Paracloeodes* ? sp. and *Caenis* sp. were euryzonic and ubiquitous.

Ephemeroptera community had different abundance distribution along the altitudinal gradient (Fig. 4). Major abundance was that of rhithron population, meanwhile *Paracloeodes* ? sp. peaked at potamon. Trends between abundance and environmental parameters were significant for altitude ($r = 0.71$, $p<0.01$), profundity ($r = -0.57$, $p<0.05$), headwater distance ($r = -0.76$, $p<0.01$), sediment size ($r = 0.56$, $p<0.05$), conductivity ($r = -0.74$, $p<0.01$) and COD ($r = -0.48$, $p<0.05$).

DISCUSSION

Community structure of Ephemeroptera was affected by environmental factors which depends on gradients of altitude. There is a gradual change in abundance downstream and a substitution of species along the elevation gradients from foothill to lowland river. A major discontinuity was observed under conditions of environmental stress, as pollution and hydraulic shifts.

Since the development of the concept of rhithron and potamon (Illies et Botosaneanu, 1963) and the River Continuum Concept (Vannote *et al.*, 1980) a lot of work has been carried out in order to answer the question of whether species distribution is clinal or zonal. Changes in abundance patterns of Ephemeroptera species along altitudinal gradients have been reported under different environmental lotic conditions (Gonzalez *et al.,* 1985; Dudgeon, 1990; Ward and Berner, 1980; Ward and Stanford, 1990; Domínguez and Ballesteros Valdez, 1992). Ward (1986), dealing with rivers whose lower basin are at high

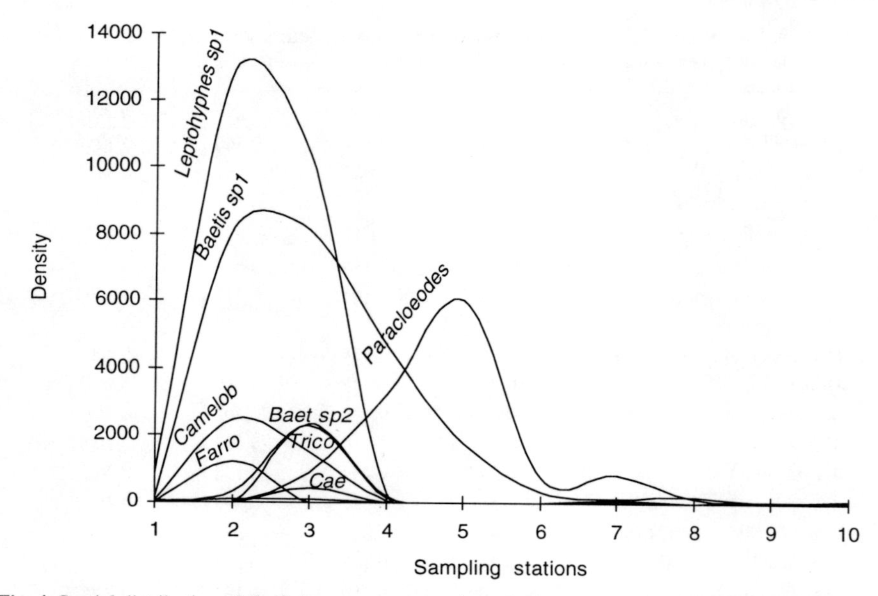

Fig. 4. Spatial distribution of the Ephemeroptera taxa along the longitudinal profile of Chocancharava river.

elevation plains, stated that the general pattern was the addition of species downstream without loss of those present in higher elevations. Gonzalez *et al.*, (1985) reported differences among the distribution pattern of five zonal groups overlapping from upper reaches to lower zones.

At Chocancharava river basin longitudinal profile is heterogeneous. Hierarchical structure is composed by short length streams in upper catchment area and a longer collector in lowland floodplain. Potamon reaches present different geomorphologic channel patterns: braided, anastomosing, meandering and straight. There is a loss of mountain species as one passes from rhithron to potamon conditions and there is a discontinuous addition of species downstream. Middle order streams have higher species richness in agreement with prediction of Vannote *et al.,* (1980). *Baetis* sp. 1 and *Caenis* sp. are euryzonal; *Baetis* sp. 2, *Camelobaetidius penai*, *Farrodes* sp. and *Baetodes* sp. are restricted to mountain streams. Some individuals of these species were found in lowland river but this fact is interpreted by drift phenomenon (Allan, 1995). Stenozonal lowland species were *Traverella* sp., *Homoeoneuria* sp. and Polymitarcyidae sp.

Although water temperature is reported as one of the most important factors affecting altitudinal zonation (Ward, 1992) we do not find correlation between temperature and DCA ordination axis, maybe because the study area is under more or less homogeneous climate conditions. Other physical-chemical variables indicative of water quality, such as COD and conductivity, were negatively correlated with Ephemeroptera abundance. Dudgeon (1990) found that pollution-related parameters were the major predictors of mayfly abundance in anthropic impacted running waters. In Chocancharava river basin good quality water conditions of rhithron allows the development of a mayfly community indicative of the ecological integrity of lotic systems. Other factors determining abundance distribution patterns such as predation, competition and disturbance must be interacting with the gradient of environmental stress (Mengue and Sutherland, 1987).

Hydraulic stress (Statzner and Higler, 1986), different combinations of geomorphological patterns and substrate size determinate habitat structure changes and longitudinal boundaries (Naiman *et al.*, 1988). Another environmental stress affecting the discontinuity of the distributions, is sewage effluent from Río Cuarto city (160,000 habitants). The hydraulic shift to potamon conditions is almost simultaneous with the entry of polluted effluents. The simultaneity of these two events affects synergically the composition of lowland communities. Some species do not pass through the hydrological barriers and others do not pass the pollution barriers.

ACKNOWLEDGMENTS

We thank to Dr. E. Domínguez for kindly giving advice and help to Ephemeroptera identification and Prof. M. E. Chiappello for correcting the english version of the manuscript.

This research has been supported by SeCyT (UNRC), CONICET and CONICOR.

REFERENCES

Allan, J. D. 1995. Stream ecology. Structure and function of running waters. Chapman and Hall, 388 pp., London.

Corigliano, M. del C., C. M. Gualdoni, A. M. Oberto and G. B. Raffaini. 1996. Macroinvertebrados Acuáticos de Córdoba. In: I. E. di Tada and E. Bucher (eds.). Biodiversidad de la Provincia de Córdoba. Vol. 1. Fauna. Ed. UNRC, Córdoba, Argentina.

Corigliano, M. del C. and R. Malpassi. 1994. La estructura trófica de un arroyo serrano. Tankay 1: 139-141.

Corigliano, M. del C., A. L. M. de Fabrizius, M. E. Luque and N. Gari. 1994. Patrones de distribución de variables fisicoquímicas y biológicas en el río Chocancharava (Cuarto) (Córdoba, Argentina). Rev. UNRC 14 (2): 177-194.

Domínguez, E., M. D. Hubbard and M. L. Pescador. 1994. Insecta Ephemeroptera. En Ageitos de Castellanos, Z. (dir.). Fauna de agua dulce de la República Argentina, Profadu, CONICET, La Plata, 33 (1): 1-142.

Domínguez, E. and J. M. Ballesteros Valdez. 1992. Altitudinal replacement of Ephemeroptera in a subtropical river. Hydrobiologia 246: 83-88.

Dudgeon, D. 1990. Determinants of the distribution and abundance of larval Ephemeroptera (Insecta) in Hong Kong running waters, pp. 221-232. In: I. C. Campbell (ed.). Mayflies and Stoneflies: Life Histories and Biology. Kluwer Academic, Dordrecht, The Netherlands.

González, G., X. Millet, N. Prat and M. A. Puig. 1985. Patterns of macroinvertebrates distribution in the Llobregat river basin (NE Spain). Verh. Int. Ver. Limnol. 22: 2081-2086.

Hill, M. O. 1979. DECORANA. A FORTRAN program for detrended correspondence analysis and reciprocal averaging. Ecology and Systematic. Cornell University. Ithaca, New York.

Illies, J. and L. Botosaneanu.1963. Problèmes et méthodes de la classification et de la zonation écologique des eaux courantes considerées surtout du point de vue faunistique. Mitt. Int. Ver. Limnol. 12: 1-57.

Mengue, G. and J. Sutherland. 1987. Community regulation: variation in disturbance, competition and predation in relation to environmental stress and recruitment. American Naturalist 130: 730-757.

Naiman, R. J., H. Decamps, J. Pastor and C. Johnston. 1988. The potential importance of boundaries to fluvial ecosystems. J. N. Amer. Benthol. Soc. 7 (4): 289-306.

Statzner B. and B. Higler. 1986. Stream hydraulics as a major determinant of benthic invertebrate zonation patterns. Freshw. Biol. 16: 126-139.

Vannote, R. L., G. W. Minshall, K. W. Cummins, J. R. Sedell and C. F. Cushing. 1980. The river continuum concept. Can. J. Fish. Aquat. Sci. 37: 130-137.

Ward, J. V. 1986. Altitudinal zonation in a Rocky Mountain stream. Arch. Hydrobiol. Suppl. 74: 133-199.

Ward, J. V. 1992. Aquatic Insect Ecology 1. Biology and habitat. John Wiley and Sons, New York.

Ward, J. V. and L. Berner. 1980. Abundance and altitudinal distribution of Ephemeroptera in a Rocky Mountain stream, pp. 169-177. In: J. F. Flanagan and K. E. Marshal (eds.). Advances in Ephemeroptera Biology. Plenum, New York.

Ward, J. V. and J. A. Stanford. 1990. Ephemeroptera of the Gunninson River, Colorado, U.S.A. In: I. C. Campbell (ed.). Mayflies and Stoneflies: Life Histories and Biology. Kluwer Academic. Dordrecht, The Netherlands.

EMERGENCE OF EPHEMEROPTERA FROM THE ASSINIBOINE RIVER, CANADA

J. F. Flannagan[1], J. Alba-Tercedor[2], R. G. Lowen[3], and D. G. Cobb[3]

[1] 456 Isabella Point Road, Saltspring, BC, Canada V8K 1V4
[2] Dept. de Zoologia, Universidad de Granada, Granada, Spain
[3] Freshwater Institute, 501 University Cres.,Winnipeg, MB, Canada R3T 2N6

ABSTRACT

Emergence traps set on sand and on cobble substrates in a riffle in the Assiniboine River, Manitoba, Canada over the open water seasons of 1990,'91,'93 and '94, collected mayflies over virtually all of the open water season. Individual traps produced 108 to 5,799 individuals, representing approximately 50 species, and average numbers/m^2/year varied from 500-3800 depending on the year. Number of individuals over sand and cobble were not significantly different. The diverse and abundant mayfly fauna is probably a result of the rich carbon sources and higher than expected temperature of the River. The emergence patterns were very different in each of the four years probably because summer thunder storms created large discharge increases which inhibited emergence. Flood control structures on the river appear to have increased low, under-ice flows and decreased peak spring and summer flows probably improving mayfly emergence success. A proposed new water withdrawal scheme may modify the effects of the control structures at least during the low water seasons.

INTRODUCTION

The Assiniboine River, at Headingley, Manitoba, drains an area of 153,000 km^2 including parts of southern Saskatchewan, northern North Dakota and southern and western Manitoba. In high water years it can also receive water from as far west as the Rocky Mountains via a series of flood control structures.

Within Manitoba, the main channel of the River is dammed at three places: A flood control storage structure near the mouth of the Shell River in western Manitoba, a hydro-electric structure at the city of Brandon, and a flood control diversion at the town of Portage la Prairie 100 km west of Winnipeg. This latter structure consists of a dam which backs water into an overflow channel draining flood water to Lake Manitoba. These various storage and flow control structures have resulted in changes in the peak spring meltwater flows and in winter, under-ice, flows.

A series of summer droughts and relatively low winter snowfalls in the 1980's resulted in the summer flows at Headingley being reduced to less than 10 m^3/s and the river bed being

Trends in Research in Ephemeroptera and Plecoptera
Edited by E. Dominguez, Kluwer Academic/Plenum Publishers, 2001

reduced to less than 25% of its normal summer width. Since, other than sporadic collections, (e.g. Ide, 1955) little or nothing was known of the aquatic insect fauna of the River, and since the low flows allowed collection of emerging insects with a box emergence trap (Flannagan, 1978), mayflies were sampled in 1990 and 1991. Subsequently, a proposal to divert water from the River to provide drinking water for a number of communities in southern Manitoba stimulated further study in 1993 and 1994.

METHODS

Over the open-water seasons of 1990, '91 '93 1 m^3 box emergence traps (Flannagan, 1978) were set on sand and on cobble substrates in a riffle approximately 5 km west of Headingley, at Lido Plage. In 1990, 4 traps were set over each substrate and in 1991 and 1993, 3 traps were set over each substrate. In 1994, because of continuous high water, 5 Dome Mesh emergence traps (Flannagan and Cobb, 1995) were used. These traps were used to quantify the mayfly emergence. In addition, a variety of other traps (Townsend trap, Dome trap, other box traps (Flannagan and Cobb, 1995)) were used, over the same period, at the River's edge, over deep water and over shifting sand substrate, to provide species lists for the area. All of these traps were emptied at least every second day. Water temperature, pH, conductivity and dissolved oxygen were measured on each sampling day. Discharge data were obtained from Environment Canada, Inland Waters Directorate publications (1983, 1985). Water chemistry analyses were carried out by the Freshwater Institute's chemistry laboratory.

Table 1. Mean numbers per square metre and ranges of Ephemeroptera emerging into emergence traps in the Assiniboine River in 1990, 1991, 1993 and 1994

Year	Substrate	Mean Number / m^2	Range
1990*	Cobble (N=4)	3812	1973-5799
	Sand (N=4)	2232	1689-3303
	Both Substrates	3022	
1991*	Cobble (N=3)	2596	1918-3683
	Sand (N=3)	2271	1287-3772
	Both Substrates	2434	
1993*	Cobble (N=3)	1432	976-1780
	Sand (N=3)	1179	1287-3772
	Both Substrates	1306	
1994**	Cobble(N=3)	538	367-787
	Sand (N=2)	527	108-945
	Both Substrates	533	

(*) Box Emergence Trap (Flannagan, 1978).
(**) New Mesh Trap (Flannagan and Cobb, 1994)

RESULTS

Quantitative Emergence

Density (numbers/m^2) data (Table 1) indicated that this riffle produced numbers of mayflies comparable to the highest ever recorded from flowing water. The mean density of both substrates together, was highest in 1990, a year in which the discharge gradually decreased over the summer; lower but similar in 1991, when there were low spring flow levels, but two high discharge periods in mid-summer resulting from summer storms; and much lower in 1993 when there was a very low meltwater flood, but many high discharge events resulting from summer storms (Fig.1). The emerging mayfly numbers for 1994 result from a different, somewhat less efficient, emergence trap (Flannagan and Cobb, 1995) and may not be directly comparable to the results of the other 3 years. However, the results should have been of the same order of magnitude, and the shape of the emergence curve should be comparable (Flannagan and Cobb, 1995).

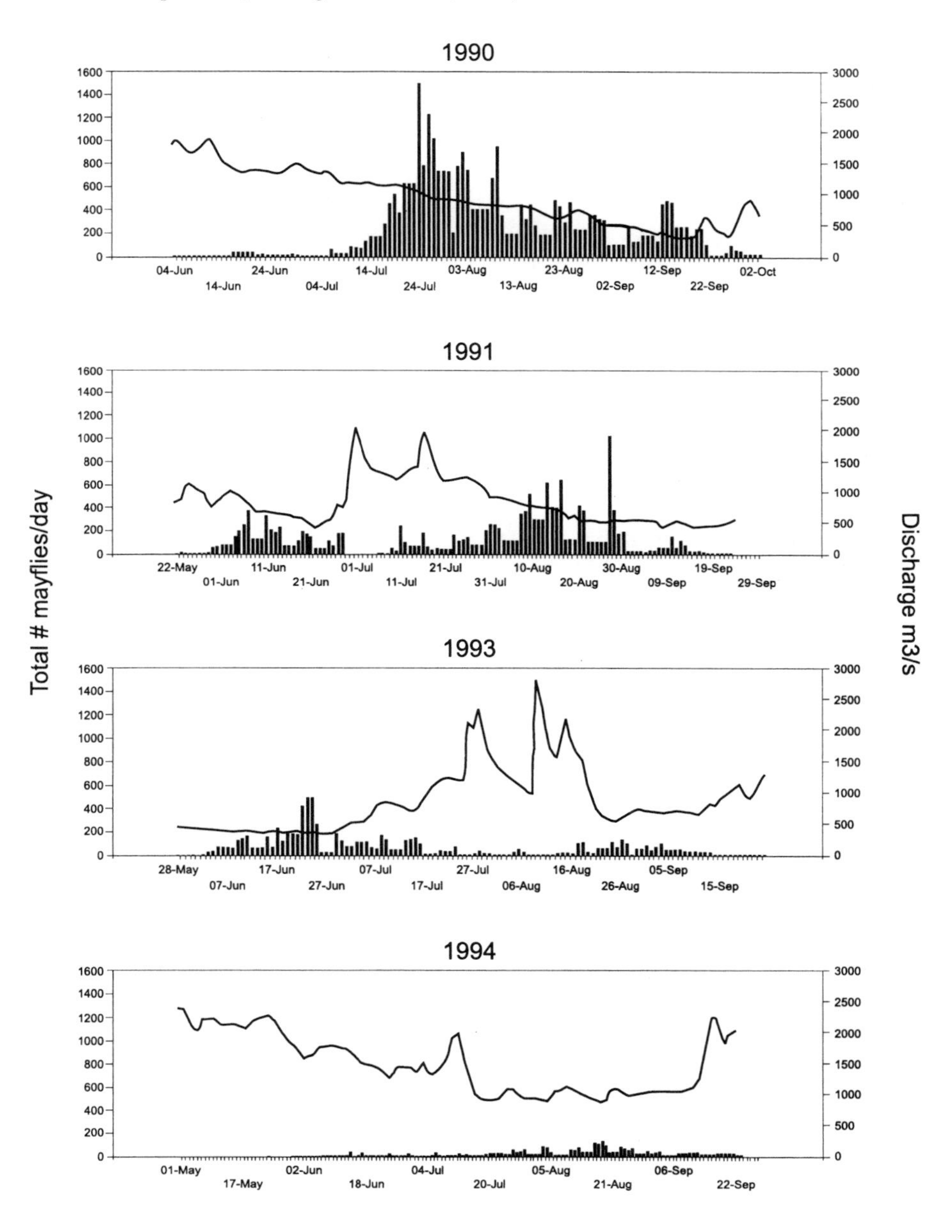

Fig. 1. Discharge and emergence patterns of mayflies of the Assiniboine River, Manitoba in 1990,'91,'93 and '94.

Table 2. Species composition and emergence times of mayflies emerging from the Assiniboine River in 1990, 1991, 1993 and 1994

Species	YR	MAY	JUNE	JULY	AUG	SEPT	OCT
Ametropus neavei [1]McD	93						
Apobaetis ? sp [1]	90						
	93						
Baetis flavistriga McD	91						
	93						
	94						
B. intercalaris McD	90						
	91						
	93						
	94						
Baetisca lacustris McD	91						
	93						
Brachycercus cf *prudens* (McD)	90						
	91						
	93						
	94						
Caenis amica Hagen	93						
C. hilaris (Say)	90						
	91						
	93						
	94						
C. latipennis Banks	93						
	94						
C. tardata McD	90						
	91						
	93						
	94						
Centroptilum bifurcatum McD	90						
	91						
	93						
	94						
C. terminatum	93						
C. walshi McD	90						
	91						
	93						
	94						
C. sp.	93						
Ephoron album (Say)	90						
	91						
	93						
	94						
Ephemera simulans Walker	91						
Heptagenia diabasia Burks	90						
	91						
	93						
	94						
H. flavescens (Walsh)	90						
	91						
Hexagenia limbata (Serville)	91						
	93						
	94						
Isonychia bicolor (Walker)	90						
	91						
	93						
I. rufa McD	90						
	91						
	93						
	94						
Leptophlebia cupida (Say)	94						

Table 2 (continued)

Species	YR	MAY	JUNE	JULY	AUG	SEPT	OCT
Leucrocuta maculipennis (Walsh)	90						
	91						
	93						
	94						
Macdunnoa persimplex [1] (McD)	90						
	91						
	94						
Paraleptophlebia sp.	93						
Pentagenia vittigera (Walsh)	90						
	91						
	93						
	94						
Plauditus dubius [1] (Walsh)	91						
	93						
	94						
P. ellioti [1](Daggy)	90						
	91						
	93						
	94						
Procloeon rufostrigatum (McD)	90						
	91						
	93						
	94						
Pseudocloeon dardanum (McD)	90						
	91						
	93						
	94						
P. ephippiatum [1](Traver)	93						
	94						
P. propinquum grp (Walsh)	90						
	91						
	93						
	94						
Raptoheptagenia cruentata McD	90						
Siphloplecton interlineatum (Walsh)	93						
Stenacron interpunctatum (Say)	90						
	91						
	93						
Stenonema mexicanum integrum (McD)	90						
	91						
	93						
	94						
S. luteum [1](Clemens)	90						
	91						
	93						
	94						
S. terminatum (Walsh)	91						
	93						
	94						
Tricorythodes cobbi [2] Alba-Tercedor&Flannagan	91						
	93						
	94						
T. mosegus Alba-Tercedor & Flannagan	93						
	94						
T. sp.	94						

Anthopotamanthus myops (Walsh)[3]
Hexagenia rigida McD[3]
Tortopus primus (McD)[3]

1. species new to Manitoba
2. species in this genus not separated in 1991
3. collected in area, not in traps

Qualitative Emergence

Emergence patterns (Fig. 1) showed little consistency among years.

Table 2 lists 42 of the species of mayflies collected, together with their emergence period. Three other known species were collected, but not in the traps, and are listed. At least four new species were collected and are not listed in this table, except for *Apobaetis ? sp. n.* which would be a new genus for Manitoba. All species collected are included in the emergence densities in Table 1. In addition to the new species, eight of the species collected are new to Manitoba. Fewer than half the species collected were collected in all four years and of these, many are represented by only one or two specimens in one or two years.

Physical and Chemical Results

As is expected from a river running over rich Prairie soils and through an intensively farmed area, the river is rich in nutrients (Table 3).

An example of the discrepancy between air temperatures and water temperatures on a daily (Fig. 2) and annual basis (Fig. 3) is given to demonstrate that this river is warmer than the surrounding air.

Based on cumulative water temperature (degree-days) for three of the four years emergence should be earliest in 1991 and latest in 1993 (Fig. 4)if cumulative temperature is involved in maturation and/or growth of mayflies.

The various impoundments and diversion of the river, described in the introduction, have led to an increase in winter flows and a decrease in spring melt-water floods (Fig. 5).

Table 3. Means and ranges of some pertinent physical and chemical attributes of the Assiniboine River

	Phosphorus µg / L		Nitrogen µg / L		Carbon		
	TSP	TDP	TSN	TDN	TSC	DIC	DOC
Mean	154	76	525	665	11431	5240	940
Minimum	28	15	131	300	990	5180	850
Maximum	251	134	722	1130	63260	5300	1030

	TSS mg / L	TDS mg / L	Turbidity NTU	Hardness µg / L	Conductivity µS / cm	pH
Mean	172	412	46	266	754	8.6
Minimum	106	324	12	209	490	7.7
Maximum	299	508	92	298	1201	9.0

DISCUSSION

The numbers of mayflies collected from the box traps (1990, '91,'93 - Table 1) are an order of magnitude higher than previously collected in Manitoba using the same traps (Flannagan *et al.*, 1990; Flannagan and Cobb, 1995). Clifford (1980) in a review of larval mayfly densities in the Holarctic region, indicated that the highest mean yearly abundance value was 1448 mayfly/m^2 and that the overall mean yearly value was 375/m^2. Harper and Harper (1984) recorded mayfly abundance from emergence traps ranging from 38 - 4992/m^2 in southern Ontario streams. Ide (1940) reported 399 - 6527 mayflies/m^2 in his studies. Thus the

Assiniboine River must be considered to be among the richest producers of mayflies in the country. The River, although very turbid during high water periods, due to eroded and resuspended fine sediments, has a relatively high suspended carbon load (Table 3). Once discharge drops below the level at which erosion and resuspension takes place, and the river clears up, (in most years by late June, early July) a very dense coating of filamentous and colonial algae covers most of the substrate in the riffle. Alba-Tercedor *et al.* (1995) discussed

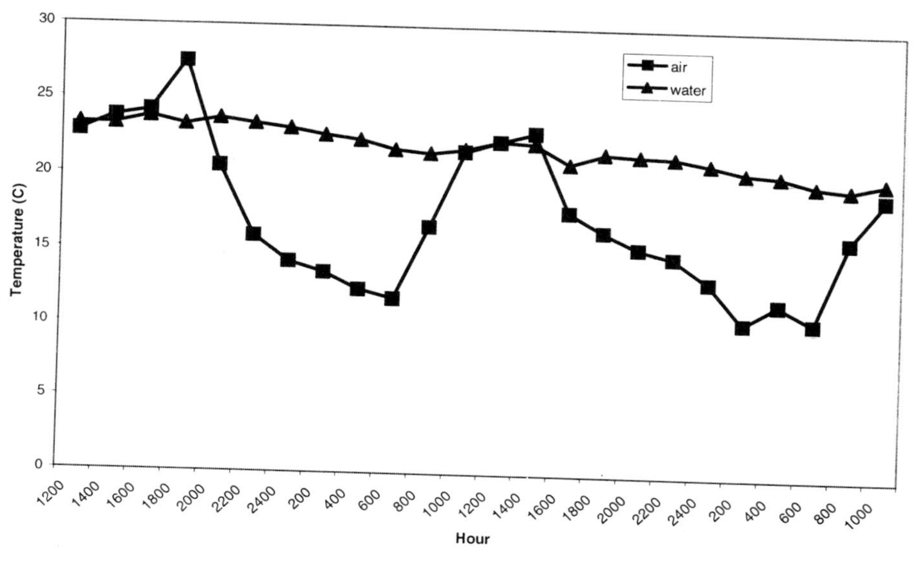

Fig. 2. River and air temperatures at the Assiniboine River sampling site. August 21-23, 1991.

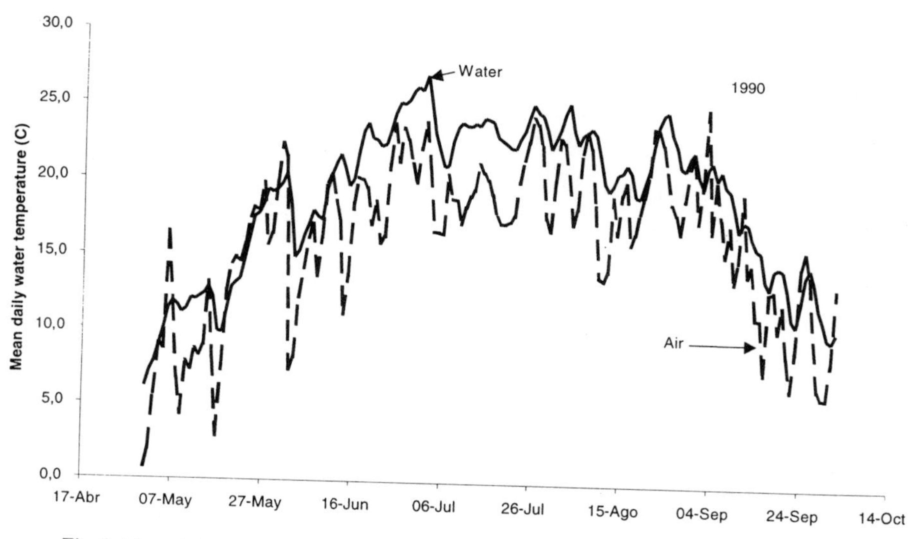

Fig. 3. Mean daily open-water air and water temperatures, Assiniboine River, Manitoba, 1990.

the sources of carbon and their incorporation into the mayflies of this River. The suspended and sessile carbon sources obviously provide a more than adequate food source for the mayflies. Ide (1940), Harper and Harper (1984) and Flannagan and Cobb (1991) in Canada and others elsewhere, have attributed high densities of mayflies to a number of habitat factors including substrate size and stability, water flow and water temperature. In this study, differences in mayfly abundance were not significant on cobble versus sand substrate within each year (Table 1). Water temperatures were not different among the stations (traps) and although quite different from year to year (Fig.4), the annual temperature differences were not correlated with either abundance of individuals or of species of mayflies.

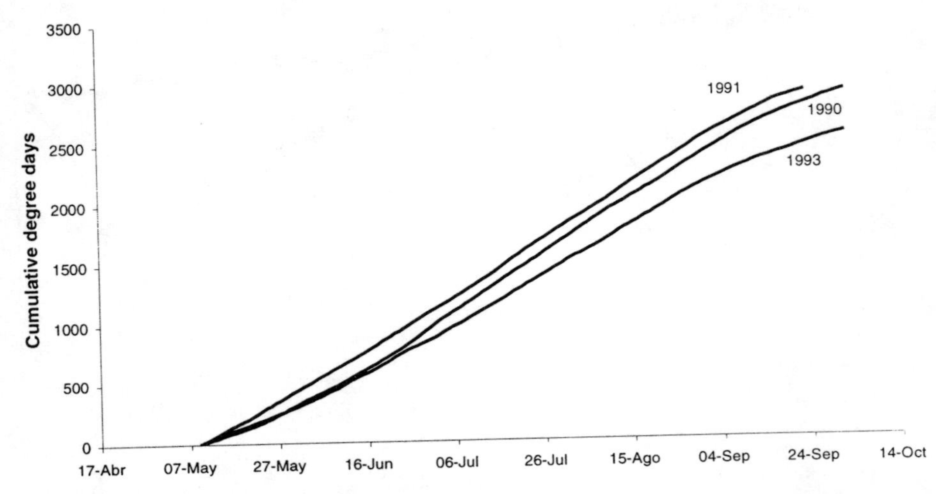

Fig. 4. Cumulative day-degrees, open-water season, Assiniboine River, 1990,'91 and '93.

A comparison of date of first emergence (Table 2) of those species which emerged in the three years for which temperature records are available (Fig.4) suggests that cumulative water temperature is not directly controlling emergence time of those species. In most cases, although 1991 (the year in which heat was accumulated fastest and to the highest level) emergence was earlier than the other two years, 1993 emergence often commenced earlier than 1990, the opposite to what might be expected from Fig.4. Water temperature may, however, be involved in the overall high abundance of mayflies in this river since the mean water temperature both on a daily basis (Fig.2) as well as on a seasonal basis (Fig.3) is unexpectedly higher than the air temperature and therefore higher than would normally be expected at this latitude. Examination of local clear streams (e.g.Roseau River (Flannagan 1978); South Duck River and Cowan Creek (Flannagan *et al.* 1990) have led us to the conclusion that the turbidity of the Assiniboine River allows it to absorb more heat from the sun during the day and lose less at night than do the clear streams.

Water flow is obviously involved in controlling the mayfly emergence in several ways. A comparison of the emergence patterns of mayflies (Fig.1), and of the mayfly densities (Table 1) with the discharge curves clearly shows that summer storms can severely interfere with emergence. The summer storms of June to mid July 1991 (Fig.1) at least delayed and perhaps reduced the total emergence of mayflies. The late summer storms of July, August 1993 appear to have removed the August peak emergence evident in 1990 and 1991, and the continuous high water of 1994 probably both accounted for the very low total abundance and the "flat" emergence patterns recorded in that year. The various impoundments and dams in existence have resulted in lower spring and summer discharges and higher winter, under-ice flows (Fig.5) and thus may have improved the success of mayfly emergence.

The proposed new diversion of water will remove water at a more or less constant rate throughout the year and thus may tend to reverse, to some extent, the effect of the existing structures.

Most species of mayflies collected from the Assiniboine River exhibited long emergence periods in most years (Table 2). The exceptions to this, *Baetisca lacustris*, *Leptophlebia cupida*, *Paraleptophlebia sp.*, *Pentagenia vittigera* and *Siphloplecton interlineatum* are all species with either very specialized niches or emergence areas. *L. cupida* migrates up small creeks and ditches to emerge; the *Paraleptophlebia sp.*, *S. interlineatum* and *B. lacustris* generally emerge very close to the river's edge; and *Pentagenia vittigera* buries into the side of underwater clay banks.

Previously published studies of Manitoba streams have recorded only 21-24 species in whole streams. Similarly, Harper and Harper (1984) recorded 25 species along its length of a southern Ontario stream, and Harper and Harper (1982) recorded 29 species in the middle section of a stream in Quebec. Thus the approximately 50 species collected here from one riffle indicates that the Assiniboine River has unusually diverse mayfly fauna.

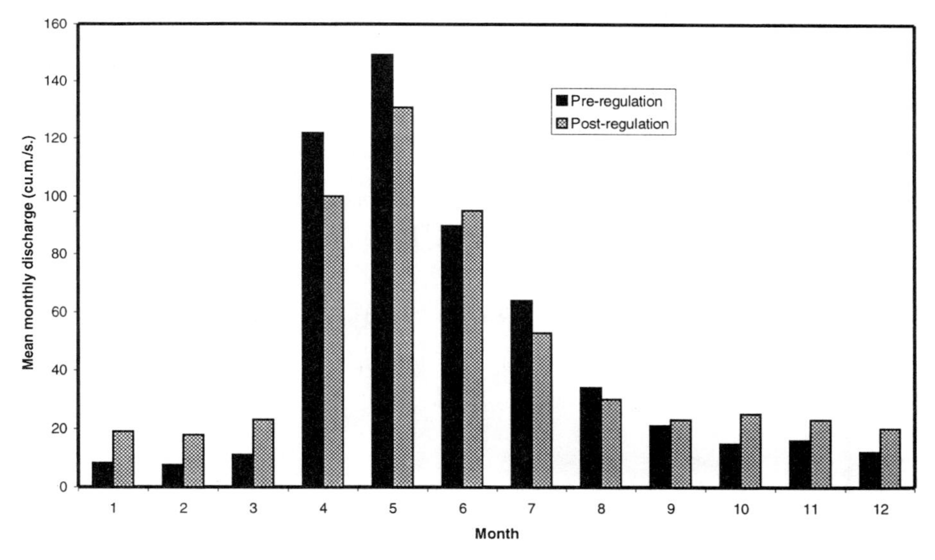

Fig. 5. Mean monthly discharges before and after flood control structures on the Assiniboine River, Manitoba.

CONCLUSIONS

1. The Assiniboine River supports a diverse and abundant mayfly fauna probably because of rich food (carbon) sources and higher than expected water temperatures.

2. Mayfly emergence appears to be lower in years of high discharge and is disrupted by summer floods.

3. The various impoundments and diversion along the river probably help the mayfly population by limiting the low winter and high spring/summer discharges. However, the proposed water withdrawal may modify this situation to some small extent, at least during the low water seasons.

ACKNOWLEDGMENTS

We are indebted to the many summer students employed by the Freshwater Institute who helped collect the samples; to Mr. Sandy Thompson for identifying some of the mayflies and to Patricia Flannagan for typing the manuscript and providing useful criticisms of it. The taxonomic nomenclature in Table 2 was updated from Dr. Pat McCafferty's internet website.

REFERENCES

Alba-Tercedor, J., J. F. Flannagan and R. Hesslein. 1995. Trophic characterization of mayfly and stonefly species of the Assiniboine River (Winnipeg, Canada). Abst. VÍIIth Internat. Confer. Ephemeroptera. Lausanne, Switzerland.

Clifford, H. E. 1980. Numerical abundance values of mayfly nymphs from the Holarctic regions, p. 503-509. In: J. F. Flannagan and K. E. Marshall (eds.). Advances in Ephemeroptera Biology. Plenum Press, New York.

Environment Canada. 1983. Historical steamflow summary, Manitoba - to 1982. Inland Waters Directorate, Water Resources Branch 247 p.

Environment Canada. 1985. Surface water data, Manitoba. 1984. Inland Waters Directorate, Water Resources Branch 160 p.

Flannagan, J. F. 1978. Emergence of caddisflies from the Roseau River, Manitoba. p. 183-197. In: M. I. Crichton (ed.). Proc. 2nd Internat. Symp. Trichoptera. 1997. W. Junk, The Hague.

Flannagan, J. F. and D. G. Cobb. 1991. Factors controlling the species diversity and density of mayflies (Ephemeroptera) emerging from an unstable river in Manitoba, Canada. p. 333-341. In: J. Alba-Tercedor and A. Sanchez-Ortega (eds.). Overview and strategies of Ephemeroptera and Plecoptera. Sandhill Crane Press. Gainesville, FL, USA.

Flannagan, J. F. and D. G. Cobb. 1995. Studies on some riverine insect emergence traps: Effects of sampling frequency and trap design. Can. Tech. Rep. Fish. Aquat.Sci. 1995: iv+10p.

Flannagan, J. F., D. G. Cobb and M. K. Friesen. 1990. The relationship between some physical factors and mayflies emerging from South Duck River and Cowan Creek, Manitoba, p. 233-242. In: I. Campbell (ed.). Mayflies and Stoneflies: Biology and life history. W. Junk, The Hague.

Harper, F. and P. P. Harper. 1982. Mayfly communities in a Laurentian watershed. Can. J. Zool. 60: 2828-2840.

Harper, F. and P. P. Harper. 1984. Phenology and distribution of mayflies in a southern Ontario lowland stream, p. 243-251. In: Proc. IVth Internat. Confer. Ephemeroptera. V. Landa, T. Soldan and M. Tonner (eds). CSAV.

Ide, F. P. 1940. Quantitative determination of the insect fauna of rapid water. Toronto Univ. Studies, Biol. Ser. 47, Pub. Ontario Fish. Lab. 59: 1-20.

Ide, F. P. 1955. Two species of mayflies representing southern groups occurring at Winnipeg, Manitoba (Ephemeroptera). Ann. ent. Soc. Amer. 48: 15-16.

MAYFLIES AS FOOD

Peter M. Grant

Department of Biological Sciences
Southwestern Oklahoma State University
Weatherford, Oklahoma 73096-3098, USA

ABSTRACT

Mayflies are ubiquitous in freshwater environments. As a result, they are a common and important component in the flow of energy through ecosystems, both aquatic and terrestrial. Many predators include mayflies on their menu of organisms consumed including invertebrates, vertebrates and at least one plant. This paper examines the diversity of organisms that consume mayflies. Some of the more interesting aspects of this predation are discussed. A list of 224 predators is included as a table.

INTRODUCTION

Each year I spend some time perusing the Ephemeroptera portion of the North American Benthological Society's (NABS) *Annual Bibliography on Benthic Biology*. Invariably, papers are encountered that discuss predators of mayflies. Some of these predators have rather intriguing names such as dipper, wagtail, bluegilled bully, shortjawed kokopu, gulf coast waterdog, edible nest swiftlet, or Pyrenean desman.

I have often wondered about how many different kinds of animals eat mayflies. Well, I have finally gotten around to compiling such a list. This is the first time that such a compilation has been made for mayflies.

The purposes of this paper are (1) to compile a list of the predators of mayflies and (2) to discuss some of the more interesting aspects of this predation. This is a rather unusual approach because, rather than studying the prey of a predator, I will be examining the predators of a prey.

METHODS AND MATERIALS

Initially, my goal was to compile an exhaustive list of all the predators of mayflies. It was soon apparent to me that this goal was not practical. Many of the papers I examined contained only small bits of information about mayflies as prey and thus could not be lo-

Trends in Research in Ephemeroptera and Plecoptera
Edited by E. Dominguez, Kluwer Academic/Plenum Publishers, 2001

cated using a keyword search of a database (e.g., BIOSIS). An exhaustive list could only be compiled by paging through all volumes of all appropriate journals.

A more efficient strategy that I adopted for this paper was to compile as many predators as I could in the time available. During this compilation, I tried to place more emphasis on the diversity of predators rather than the total number of predators.

I used the following sources to initially search for papers on predators: *NABS Annual Bibliography on Benthic Biology*, 1965-present (which, in later years, has been compiled by searching BIOSIS); *Eatonia*, 1954-1980; *The Mayfly Newsletter*, 1990-present; proceedings of previous mayfly conferences; books on mayflies, such Leonard and Leonard (1962); various books on aquatic entomology and ecology; and my reprint collection. The references cited at the end of papers examined were particularly valuable. These references led to many additional sources that could not have been located by keyword searches or would have been overlooked based on their title alone.

Predators, involved in manipulative studies of feeding behavior, are not included.

RESULTS AND DISCUSSION

I have compiled a total of 224 predators of mayflies. These predators, along with a published reference, are listed in Table 5 at the end of this paper. The scientific name of the predator is written exactly as it appeared in the reference. No attempt was made to update the taxonomy of these 224 species.

Unfortunately, space prohibits me from discussing all of these predators in this paper, so my emphasis will be on some of the more interesting aspects of predation on mayflies.

Were Mayflies Eaten?

The predators listed in Table 5 were identified as feeding on mayflies by gut or fecal analysis or by direct observation of feeding. The direct analysis of feces, though, may be misleading. Rabinowitz and Tuttle (1982) observed gray bats, *Myotis grisecens*, feeding on mayflies, but few of the bats' fecal pellets contained mayfly parts. They hand fed one bat a mixture of insects which included 60% mayflies. None of the fecal pellets from this animal collected after feeding contained any mayfly parts. Another bat was fed only mayflies (n = 274). Only 60% of this bat's fecal pellets contained parts of mayflies (wings). This species tends to remove the wings from mayflies before feeding, thereby eliminating the only body part that can pass through its digestive system. Thus, use of fecal pellets to analyze the importance of mayflies in the diet of the gray bat underestimates the actual amount consumed.

Turner (1983) encountered the same problem with the blue and white swallow, *Notiochelidon cyanoleuca*, in Venezuela. Mayflies were present in 53.7% of the food boluses of nestlings but only 0.6% of the feces.

Another difficulty with interpreting gut analyses is that the results can be reported three different ways: (1) percent occurrence - if half of the individuals consumed mayflies, the frequency of occurrence is 50%; (2) percent relative abundance – if 50 of the 100 food particles in the gut of a predator are mayflies, the percent relative abundance is 50%; and (3) percent volume – if mayflies make up half the volume of the gut, they contribute 50% of the volume. Knowing what the author means by "50% mayflies in the gut" is critical to a proper interpretation.

For example, the relative abundance of mayflies in the gut of the Pacific giant salamander, *Dicamptodon tenebrosus*, varies between 30 and 69%, yet mayflies occupy only 1-26% of the volume of the gut (Parker, 1994). Just the opposite occurs in the Oklahoma salamander, *Eurycea tynerensis* (Tumlison *et al.*, 1990): the relative abundance of mayflies in the gut is 19% but they occupy 46% of the volume. This is explained by examining the food items other than mayflies. The Pacific giant salamander eats a lot of mayflies (high relative frequency) but also consumes much larger prey items (such as megalopterans, fish and small salamanders) which greatly affect volume. The Oklahoma salamander consumes many more chironomids than mayflies, which results in a low relative frequency for mayflies. However, the mayflies consumed are much larger than the chironomids and thus occupy a larger volume of the gut.

The gut analyses of some predators require special techniques. For example, Reynoldson and Bellamy (1975) analyzed the guts of small turbellarians for the presence of mayfly proteins using serological techniques. Reilly and McCarthy (1990) used a similar immunological technique for corixids, while Giller (1986) used electrophoresis to determine the presence of mayfly proteins in the gut of notonectids.

Mayflies as Predators of Mayflies

Some mayflies are predators of other mayflies. *Siphlonisca aerodromia* nymphs begin their lives as detritivores but these agile, rapid swimmers become more carnivorous as they age, feeding on the likes of *Siphlonurus*, *Leptophlebia* and *Ephemerella* (Gibbs and Mingo, 1986). The sand-dwelling heptageniid, *Pseudiron centralis*, prefers chironomids as food, but 5% of its foregut contents were composed of early instars of *Baetis*, *Centroptilum* and *Ephoron* (Soluk and Clifford, 1984). Agnew (1962), in his study of *Centroptiloides bifasciata*, found remains of baetid nymphs in two of seven guts analyzed. Finally, Muttkowski and Smith (1929), while conducting a feeding analysis of *Ephemerella* sp., examined one nymph with 50% of its gut filled with *Heptagenia* nymphs.

Some Predators Prefer Mayflies

Fishes that feed predominantly on mayflies (based on percent relative abundance in the gut) are listed in Table 1. Young freshwater stingrays, *Potamotrygon magdalenae*, feed exclusively on mayflies (Caira and Orringer, 1995). These authors noted that 2 of the 15 juvenile stingrays examined were also parasitized by a cestode, which implicated its mayfly prey as being the intermediate host.

While the relative frequency of mayflies in the gut of the fish in Table 1 can be very high, some species exhibit quite a range of preference such as young smallmouth bass, *Micropterus dolomieu* (Easton *et al.*, 1996), bluegilled bully, *Gobiomorphus hubbsi*, and torrentfish, *Cheimarrichthys fosteri* (Scrimgeour and Winterbourn, 1987).

Downes (1978) has observed six species of ceratopogonids (Diptera) feeding on mayflies. Typically the female ceratopogonid enters the mayfly swarm, lands on the thorax

Table 1. Percent relative abundance of mayflies in the gut of eight species of fishes. Ranges are provided for some species. Number in parentheses indicates the mean value. Superscript after the name of the organisms identifies reference: [a]Caira and Orringer, 1995; [b]Easton *et al.*, 1996; [c]Scrimgeour and Winterbourn, 1987; [d]Glova *et al.*, 1992; [e]Vives, 1987; [f]Denoncourt and Stauffer, 1993.

Species	% Relative Abundance in Gut
Freshwater stingray (young)[a]	100
Smallmouth bass (young)[b]	19-99 (61)
Bluegilled bully[c]	13-91 (71)
Torrentfish[c]	1-82 (51)
Brown trout[d]	41-73
Slender madtom[e]	52-60 (58)
Common river galaxias[d]	27-60
American eel[f]	58

of a male, pierces his head with her mouthparts, and begins feeding. Quite often the mayfly prey are larger that the fly. Feeding may continue for up to 30 minutes in some species. The three genera listed in Table 5, *Bezzia, Probezzia* and *Palpomyia*, commonly feed on mayflies. In fact, the four *Palpomyia* species observed by Downes (1978) in Scotland fed only on mayflies during his observation period.

Over 50 species of *Podagritis* wasps (Hymenoptera: Sphecidae) are distributed throughout South America, Australia and New Zealand and they feed exclusively on Diptera (Harris, 1990). However, Harris (1990) discovered two species in New Zealand, *P. albipes* and *P. cora*, that are rather unusual – they provision their nests almost exclusively with *Deleatidium* mayflies. Female wasps wait on rocks in the stream until they encounter a *Deleatidium* nymph in the process of molting to the subimago. The wasps have been observed to actually pull the subimago from the nymphal exuviae. Females also collect subimagos as they float on the water surface. The mayflies are stung and are used to provision nests which the wasps make along the shores of the stream.

Two species of flies are also involved in this food web (Harris, 1990). A muscid fly, *Spilogona* sp., competes with the wasps for subimagos emerging on the rocks in the stream. Also, the larvae of *Anabarynchus* sp., a therevid fly, burrows through the soil along the edge of the stream. When this larva encounters a provisioned nest of *Podagritis*, it consumes the *Deleatidium*.

Variations in Feeding

Dudgeon (1989) has shown that not only does the preference for mayflies by the nymphs of the damselfly, *Euphaea decorata*, increases as they age, but also the number of genera consumed (Table 2). Twelve percent of the prey consumed by small nymphs was composed of two genera of mayflies, while large nymphs fed on five genera which represented 36% of the prey consumed.

Mayflies make up a larger portion of the food of young smallmouth bass, *Micropterus dolomieu*, and rock bass, *Ambloplites rupestris* (Rabeni, 1992), but make up a larger portion of the food of older white suckers, *Catostomus commersoni* (Chen and Harvey, 1995), and common river galaxias, *Galaxias vulgaris* (Cadwallader, 1975a). The change in the food preference of galaxias occurs when it moves from quiet water to riffles once it has attained a length of 40 mm (Cadwallader, 1975a). Riffles have a greater diversity of food items.

The diet of the Gulf coast waterdog, *Necturus beyeri*, changes with age also (Bart and Holzenthal, 1985). Table 3 shows how this salamander prefers *Leptophlebia* mayflies when younger and *Stenonema* mayflies when older.

The Eurasian dipper, *Cinclus cinclus*, consumes more mayflies during the breeding season (Ormerod and Tyler, 1991) as does the black duck, *Anas rubripes* (Reinecke and Owen, 1980).

Table 2. Relative abundance of mayflies in the gut of *Euphaea decorata* nymphs. Values for three age groups of *E. decorata* are shown along with the genera of mayflies consumed (Dudgeon, 1989).

Size of Nymphs		
Small 12%	Medium 26%	Large 36%
Choroterpes	*Baetis*	*Baetis*
Baetis	*Choroterpes*	*Compsoneuriella*
	Isca	*Ephemerellina*
	Compsoneuriella	*Choroterpes*
		Isca

Ormerod and Tyler (1991) also showed how the frequency of mayflies in the diet of dippers is affected by the type of stream in which it forages. For example, the frequency of mayflies in the diet is between 1% and 3% for adults and nestlings, respectively, when foraging in an acidic stream. The frequency greatly increases to 67% for adults and 38% for nestlings when foraging in a circumneutral stream. Mayflies were simply more abundant in the circumneutral stream.

Variety

The salamander, *Leurognathus marmorata* feeds on 10 species of mayflies (Martof and Scott, 1957). In addition to the genera listed in Table 2, *Euphaea decorata* also consumed six genera of mayflies (*Baetiella, Pseudocloeon, Epeorus, Habrophlebiodes, Serratella* and *Teloganodes*) in much smaller quantities for a total of 11 genera (Dudgeon, 1989). Teslenko (1997), however, may have discovered the premier mayfly predators – two species of stoneflies, *Skwala pusilla* and *Kamimuria exilis*, each of which consumes 18 species of mayflies.

Energetics

The production of a mayfly has been shown to directly influence the production of walleye, *Stizostedion vitreum*, in Ontario, Canada (Ritchie and Colby, 1988). During even years, when the production of *Hexagenia limbata* was high (7660 mg/m^2), the abundance of walleye was approximately three to five times higher than during odd years when *H. limbata* production was lower (1930 mg/m^2). Ritchie and Colby (1988) concluded that higher numbers of *Hexagenia* probably reduce carnivory and predation on young walleyes and increase fecundity of females.

The Pyrenean desman, *Galemys pryenaicus*, commonly feeds on mayflies, which occur in 96% of all the guts examined (Castién and Gosálbez, 1995). However, these mayflies only represent 16% of the food ingested by volume. While mayflies don't account for a large volume of the food consumed, Castién and Gosálbez (1995) believe the amount is critical to allow this homeotherm to thermoregulate. Mayflies are small but abundant in the stream and are easily captured and consumed.

Overall, mayflies represent 29% of the gross energy ingested for black ducks (Reinecke and Owen, 1980).

To show what some animals go through to acquire enough food, Ormerod and Tyler (1988) calculated that green sandpipers, *Tringa ochropus*, consume 9,500 to 11,000 *Baetis* nymphs per day!

Humans

Not to be outdone by other carnivores, some *Homo sapiens* have also acquired a taste for mayflies. Bodenheimer (1951) described how mayflies are prepared in North Vietnam (con-vo or phu-du), China and Japan for consumption. The people of Malawi make a paste out of mayflies (*Caenis kungu*) and mosquitoes called kungu (Fladung,

Table 3. Percent occurrence of two genera of mayflies in the gut of the Gulf Coast waterdog (*Necturus beyeri*) in different age classes (Bart and Holzenthal, 1985).

	% Occurrence in Gut					
Age Class (Years)	0-1	1-2	2-3	3-5	5-6	6-7
Leptophlebia	48	44	30	40	25	25
Stenonema	-	12	10	80	75	100

1924). The inhabitants near the shores of Lake Victoria collect emergent insects ("lakeflies"), which includes *Povilla* mayflies (Bergeron *et al.*, 1988). These insects are dried, ground into a flour, and formed into a dried cake for consumption. Swarms of mayflies (perhaps *Plethogenesia*) are collected in New Guinea, cooked and then eaten (Szent-Ivany and Ujházy, 1973). Even 17th century Incas ate raw nymphs (*Euthyplocia* or *Campylocia*?) or made them into a spicy sauce (Gillies, 1996).

Plants

I have not been able to locate any printed reference to carnivorous plants consuming mayflies. I do have, however, a photograph that I removed from an advertisement several years ago. It shows a small mayfly that has been captured by what appears to be a thread-leaf sundew, *Drosera filiformis*. I believe this photograph was part of an advertisement for a book on carnivorous plants, but, to date, I have not been able to locate this book or the source of this photograph.

Nutritional Analysis

Since a number of economically important animals, such as fish and ducks, consume mayflies, several individuals have analyzed their nutritional value. Table 4 represents a summary of the information found in these papers.

Table 4. List of papers that include nutritional analysis of mayflies. Abbreviations: ^{A}amino acids, ^{C}calories, ^{F}fiber, ^{H}carbohydrates, ^{L}lipids, ^{M}minerals, ^{P}proteins, ^{S}ash, ^{V}vitamins, ^{W}water. Analysis by Bergeron *et al.* (1988) is of an insect flour composed of *Povilla*, chaoborids and chironomids.

Source	A	L	V	M	C	H	P	S	W	F
Albrecht and Breitsprecher (1969)		•				•	•	•	•	
Albrecht and Wünsche (1972)	•									
Bell *et al.* (1994)		•								
Bergeron *et al.* (1988)	•	•	•	•	•		•	•	•	•
Block (1959)	•									
Cummins and Wuycheck (1971)					•					
Driver *et al.* (1974)	•				•		•	•	•	
Ghioni *et al.* (1996)		•								
Hanson *et al.* (1985)		•								
Okedi (1992)		•	•	•			•	•	•	•
Reinecke and Owen (1980)		•			•	•	•	•		•

Okedi (1992) observed that the traditional protein sources for some Africans (domestic animals, fish and wildlife) have declined over the years. He considered using the lakeflies mentioned previously as a source of protein. His analysis along with that of Bergeron *et al.* (1988) show that these insects are high in protein, minerals, B vitamins, and essential amino acids. They are also low in fat and moisture thus contributing to a long shelf life. The insect cakes have a high digestibility and the fact that *Povilla* occurs in huge swarms makes them relatively easy to capture. Suitable commercial harvesting techniques need to be designed before this potential source of food can be utilized.

Table 5. List of predators of mayflies. A reference that identifies each species as a predator is provided. Scientific names are exactly as they appear in the reference.

Predator	Reference
Platyhelminthes – Turbellaria	
Dugesia tigrina	Gee and Young (1993)
Polycelis nigra	Reynoldson and Bellamy (1975)
P. tenuis	"
Arthropoda – Crustacea	
Gammarus fossarum	Dumont and Verneaux (1976)
Orconectes propinquus	Capelli (1980)
Arthropoda – Chelicerata	
Tetragnatha elongata	Williams *et al.* (1995)
T. versicolor	"
Arthropoda – Uniramia	
Ephemeroptera	
Baetidae	
Centroptiloides bifasciata	Agnew (1962)
Ephemerellidae	
Ephemerella sp.	Muttkowski and Smith (1929)
Heptageniidae	
Pseudiron centralis	Soluk and Clifford (1984)
Siphlonuridae	
Siphlonisca aerodromia	Gibbs and Mingo (1986)
Odonata - Anisoptera	
Aeshnidae	
Anax imperator	Blois (1985)
A. junius	Folsom and Collins (1984)
Aeshna canadensis	Pritchard (1964)
A. eremita	"
A. interrupta lineata	"
A. cyanea	Blois (1985)
Corduliidae	
Cordulia shurtleffi	Pritchard (1964)
Gomphidae	
Lanthus vernalis	Wallace *et al.* (1987)
Ophiogomphus severus	Koslucher and Minshall (1973)
Libellulidae	
Leucorrhinia hudsonica	Pritchard (1964)
L. proxima	"
Libellula depressa	Blois (1985)
Odonata - Zygoptera	
Calopterygidae	
Hetaerina americana	McCafferty (1979)
Euphaeidae	
Euphaea decorata	Dudgeon (1989)
Coenagrionidae	
Enallagma anna	Koslucher and Minshall (1973)
Ischnura elegans	Thompson (1978)
Pyrrhosoma nymphula	Lawton (1970)
Plecoptera	
Perlidae	
Acroneuria abnormis	Johnson (1981b)

Table 5 (continued)

Predator	Reference
A. californica	Sheldon (1969)
A. carolinensis	Schmidt and Tarter (1985)
Agnetina capitata	Fuller and Hynes (1987)
Claassenia sabulosa	Allan (1982)
Dinocras cephalotes	Berthélemy and Lahoud (1981)
Hesperoperla pacifica	Fuller and Stewart (1977)
Kamimuria exilis	Teslenko (1997)
Neoperla clymene	Vaught and Stewart (1974)
Paragnetina mediá	Fuller and Hynes (1987)
P. immarginata	Johnson (1981b)
Perla bipunctata	Lucy *et al.* (1990)
P. marginata	Berthélemy and Lahoud (1981)
Perlesta placida	Snellen and Stewart (1979)
Phasganophora capitata	Johnson (1981b)
Perlodidae	
Arcynopteryx compacta	Berthélemy and Lahoud (1981)
Clioperla clio	Feminella and Stewart (1986)
Cultus aestivalis	Fuller and Stewart (1979)
Frisonia picticeps	Sheldon (1972)
Hydroperla crosbyi	Oberndorfer and Stewart (1977)
Isogenoides zionensis	Fuller and Stewart (1977)
Isoperla acicularis	Berthélemy and Lahoud (1981)
I. moselyi	"
I. difformis	Malmqvist *et al.* (1991)
I. grammatica	"
I. namata	Feminella and Stewart (1986)
I. fulva	Fuller and Stewart (1977)
I. viridinervis	Lavandier (1982)
Kogotus modestus	Allan (1982)
K. nonus	Walde and Davies (1987)
Megarcys ochracea	Teslenko (1997)
M. signata	Allan (1982)
Oroperla barbara	Sheldon (1972)
Perlinodes aurea	"
Perlodes microcephalus	Berthélemy and Lahoud (1981)
Skwala curvata	Sheldon (1972)
S. parallela	Fuller and Stewart (1977)
S. pucilla	Teslenko (1997)
Stavsolus sp.	Teslenko (1997)
Hemiptera	
Corixidae	
Cymatia bonsdorfi	Reilly and McCarthy (1990)
Cenocorixa bifida hungerfordi	Reynolds and Scudder (1987)
C. expleta	"
Notonectidae	
Notonecta hoffmanni	Fox (1975)
N. glauca	Giller (1986)
N. viridis	"
Megaloptera	
Sialidae	
Sialis fuliginosa	Dumont and Verneaux (1976)
Corydalidae	
Corydalus cornutus	Stewart *et al.* (1973)
Nigronia serricornis	Fuller and Hynes (1987)

Table 5 (continued)

Predator	Reference
Protohermes grandis	Yoshida *et al.* (1985)
Trichoptera	
Polycentropodidae	
Polycentropus variegatus	Dudgeon and Richardson (1988)
Hydropsychidae	
Arctopsyche irrorata	Wallace (1975)
Hydropsyche simulans	Rhame and Stewart (1976)
Parapsyche almota	Dudgeon and Richardson (1988)
P. elsis	"
Rhyacophilidae	
Rhyacophila acutiloba	Manuel and Folsom (1982)
R. carolina	"
R. minor	"
R. septentrionis	Dumont and Verneaux (1976)
R. vaccua	Thut (1969)
R. vaefes	"
R. vepulsa	"
Limnephilidae	
Drusus discolor	Bohle (1983)
Odontoceridae	
Odontocerum albicorne	Dumont and Verneaux (1976)
Hymenoptera	
Sphecidae	
Podagritus albipes	Harris (1990)
P. cora	"
Diptera	
Tipulidae	
Dicranota bimaculata	Dumont and Verneaux (1976)
Muscidae	
Spilogona sp.	Harris (1990)
Therevidae	
Anabarynchus sp.	"
Ceratopogonidae	
Bezzia varicolor	Downes (1978)
Palpomyia flavipes	"
P. nemorivaga	"
P. quadrispinosa	"
P. semifumosa	"
Probezzia venusta	"
Phylum Chordata	
Chondrichthyes	
Potamotrygon magdalenae	Caira and Orringer (1995)
Osteichthyes	
Anguillidae	
Anguilla anguilla	Sinha and Jones (1967)
A. australis schmidtii	Cadwallader (1975b)
A. dieffenbachii	"
A. japonica	Tzeng *et al.* (1995)
A. rostrata	Denoncourt and Stauffer (1993)
Balitoridae	
Noemacheilus barbatulus	Maitland (1965)

Table 5 (continued)

Predator	Reference
N. fasciolatus	Dudgeon (1987)
Catostomidae	
Catostomus commersoni	Chen and Harvey (1995)
Centrarchidae	
Ambloplites rupestris	Johnson and Dropkin (1993)
Lepomis auritus	"
L. gibbosus	"
L. megalotis peltastes	Laughlin and Werner (1980)
Micropterus dolomieu	Easton *et al.* (1996)
M. salmoides	Godinho and Ferreira (1994)
Cichlidae	
Crenicichla lipidota	Lobón-Cerviá *et al.* (1993)
Clariidae	
Clarias gariepinus	Adámek and Sukop (1995)
Cottidae	
Cottus cognatus	Petrosky and Waters (1975)
C. gobio	Dumont and Verneaux (1976)
Cyprinidae	
Barbus callensis	Kraiem (1996)
Gobio gobio	Przybylski and Banbura (1989)
Notropis atherinoides	Mendelson (1975)
N. dorsalis	"
N. hudsonius	Johnson and Dropkin (1993)
N. spilopterus	"
N. volucellus	"
Phoximus phoxinus	Maitland (1965)
Rhinichthys atratulus	Fuller and Hynes (1987)
Cyprinodontidae	
Fundulus catenatus	Fisher (1981)
F. diaphanus	Johnson and Dropkin (1993)
Eleotridae	
Gobiomorphus breviceps	Cadwallader (1975b)
G. hubbsi	Scrimgeour and Winterbourn (1987)
Esocidae	
Esox lucius	Ritchie and Colby (1988)
Galaxiidae	
Galaxias postvectis	McDowall *et al.* (1996)
G. vulgaris	Cadwallader (1975b)
Gasterosteidae	
Gasterosteus aculeatus	Maitland (1965)
Gobiidae	
Tukugobius wui	Dudgeon (1987)
Ictaluridae	
Noturus exilis	Vives (1987)
N. miurus	Burr and Mayden (1982)
Mormyridae	
Gnathonemus tamandua	Petr (1968)
G. cyprinoides	"
Hippopotamyrus pictus	Olatunde and Moneke (1985)
Hyperopisus bebe	Petr (1968)
Marcusenius senegalensis	Olatunde and Moneke (1985)
Momyrus rume	Petr (1968)
M. hasselquisti	"
M. macrophthalmus	"
M. deliciosus	"
Petrocephalus bane	"

Table 5 (continued)

Predator	Reference
P. bovei	Olatunde and Moneke (1985)
Percidae	
Etheostoma blennioides	Hlohowskyj and White (1983)
E. caeruleum	"
E. flabellare	Fuller and Hynes (1987)
E. lepidum	McClure and Stewart (1976)
E. rubrum	Knight and Ross (1994)
Gymnocephalus cernuus	Ogle *et al.* (1995)
Perca flavescens	Hayes *et al.* (1992)
Percina sçiera	McClure and Stewart (1976)
Stizostedion vitreum	Ritchie and Colby (1988)
Pinguipedidae	
Cheimarrichthys fosteri	Scrimgeour and Winterbourn (1987)
Poeciliidae	
Gambusia affinis	Miura *et al.* (1979)
Salmonidae	
Oncorhynchus kisutch	Johnson (1981a)
O. tshawytscha	"
O. mykiss	Dedual and Collier (1995)
Salmo salar	Levings *et al.* (1994)
S. trutta	Cadwallader (1975b)
Salvelinus fontinalis	Forrester *et al.* (1994)
Thymallus thymallus	Sempeski and Gaudin (1996)
Amphibia – Urodela	
Plethodontidae	
Desmognathus quadramaculaus	Martof and Scott (1957)
Eurycea tynerensis	Tumlison *et al.* (1990)
Leurognathus marmorata	Martof and Scott (1957)
Salamandridae	
Euproctus asper	Montori (1992)
Notophthalmus v. viridescens	Burton (1977)
Salamandra salamandra	Kuz'min (1992)
Triturus cristatus	Avery (1968)
T. helveticus	"
T. vulgaris	"
Dicamptodontidae	
Dicamptodon tenebrosus	Parker (1994)
Ambystomatidae	
Ambystoma texanum	Whitaker *et al.* (1980)
A. tigrinum nebulosum	Collins and Holomuzki (1984)
Proteidae	
Necturus beyeri	Bart and Holzenthal (1985)
N. punctatus	Meffe and Sheldon (1987)
Hynobiidae	
Onychodactylus fischeri	Kuz'min (1990)
Reptilia – Testudines	
Trionychidae	
Trionyx muticus	Williams and Christiansen (1981)
T. spiniferus	Cochran and McConville (1983)
Chelydridae	
Chelydra serpentina	Lagler (1943)
Kinosterniidae	
Sternotherus odoratus	"

Table 5 (continued)

Predator	Reference
Emydidae	
Chrysemys picta marginata	Lagler (1943)
Emys blandingii	"
Graptemys geographica	"
Aves	
Apodiformes	
Aerodramus fuciphagus	Langham (1980)
Chaetura pelagica	Leonard and Leonard (1962)
C. vauxi	Bull and Beckwith (1993)
Anseriformes	
Anas rubripes	Reinecke and Owen (1980)
Aythya valisineria	Bartonek and Hickey (1969)
Bucephala clangula	Eadie and Keast (1982)
Hymenolaimus malacorhynchos	Collier and Lyon (1991)
Charadriiformes	
Himantopus h. leucocephalus	Pierce (1986b)
H. novaezealandiae	"
Ibidorhyncha struthersii	Pierce (1986a)
Larus delawarensis	Welham (1987)
Tringa ochropus	Ormerod and Tyler (1988)
Sterna sandvicencis	Greenwood (1986)
Passeriformes	
Acrocephalus arundinaceus	Bibby and Green (1983)
A. schoenobaenus	"
A. scirpaceus	"
Bombycilla cedrorum	Leonard and Leonard (1962)
Cinclus cinclus	Ormerod and Tyler (1991)
Hirundo rustica	Loske (1992)
Locustella luscinioides	Bibby and Green (1983)
Motacilla cinerea	Bures and Král (1987)
Notiochelidon cyanoleuca	Turner (1983)
Tachycineta bicolor	Blancher and McNicol (1991)
Mammalia	
Rodentia	
Glaucomys sp.	Leonard and Leonard (1962)
Zapus princeps	Vaughan and Weil (1980)
Insectivora	
Galemys pyrenaicus	Castién and Gosálbez (1995)
Sorex palustris navigator	Conaway (1952)
Chiroptera	
Lasiurus borealis	Whitaker (1972)
Myotis grisescens	Rabinowitz and Tuttle (1982)
M. lucifugus	Anthony and Kunz (1977)
Pipistrellus pipistrellus	Swift *et al.* (1985)
Primates	
Homo sapiens	Fladung (1924)
Plantae – Tracheophyta	
Angiospermae	
Nepenthales	
Drosera filiformis?	See text for explanation

Behavior

Being so nutritionally rich, and being preyed upon by well over 200 species (see Table 5), mayflies have evolved defense mechanisms to reduce the chance of predation. Peckarsky (1996) has shown that behaviors such as drifting, swimming, scorpion posturing and timing of activity are all influenced by the presence of predators and are used to avoid predation. Edmunds and Edmunds (1980) have hypothesized that life history attributes of mayflies, such as short adult life, mass emergence, mating swarms, timing of emergence and remote nuptial flights all reduce the chance of predation.

ACKNOWLEDGMENTS

Funds to attend this meeting and prepare this paper were provided by the Department of Biological Sciences, Arts and Sciences Organized Research, and Sponsored Programs at Southwestern Oklahoma State University. Ms. Mary Roberson, Library Technician, worked diligently to obtain the many papers I required and I appreciate her efforts. I also thank my wife, Marci, whose technical and emotional support over the last 22 years has been unfailing.

REFERENCES

Adámek, Z. and I. Sukop. 1995. Summer outdoor culture of African catfish (*Clarias gariepinus*) and tilapias (*Oreochromis niloticus* and *O. aureus*). Aquat. Living Res. 8: 445-448.

Agnew, J. D. 1962. The distribution of *Centroptiloides bifasciata* (E.-P.) (Baëtidae: Ephem.) in Southern Africa, with ecological observations on the nymphs. Hydrobiologia 20: 367-372.

Albrecht, M.-L. and B. Breitsprecher. 1969. Untersuchungen über die chemische Zusammensetzung von Fischnährtieren und Fischfuttermitteln. Z. Fischerei NF 17: 143-163.

Albrecht, M.-L and J. Wünsche. 1972. The content of amino acids in the proteins of lower aquatic animals and its significance for fish nutrition. Arch. Tierernähr. 22: 423-430. (In German)

Allan, J. D. 1982. Feeding habits and prey consumption of three setipalpian stoneflies (Plecoptera) in a mountain stream. Ecology 63: 26-34.

Anthony, E. L. P. and T. H. Kunz. 1977. Feeding strategies of the little brown bat, *Myotis lucifugus*, in southern New Hampshire. Ecology 58: 775-786.

Avery, R. A. 1968. Food and feeding relations of three species of *Triturus* (Amphibia Urodela) during the aquatic phases. Oikos 19: 408-412.

Bart, H. L., Jr. and R. W. Holzenthal. 1985. Feeding ecology of *Necturus beyeri* in Louisiana. J. Herpetol. 19: 402-410.

Bartonek, J. C. and J. J. Hickey. 1969. Food habits of canvasbacks, redheads, and lesser scaup in Manitoba. Condor 71: 280-290.

Bell, J. G., C. Ghioni and J. R. Sargent. 1994. Fatty acid compositions of 10 freshwater invertebrates which are natural food organisms of Atlantic salmon parr (*Salmo salar*): a comparison with commercial diets. Aquaculture 128: 301-313.

Bergeron, D., R. J. Bushway, F. L. Roberts, I. Kornfield, J. Okedi and A. A. Bushway. 1988. The nutrient composition of an insect flour sample from Lake Victoria, Uganda. J. Food Comp. Anal. 1: 371-377.

Berthélemy, C. and M. Lahoud. 1981. Food and mouthparts of several pyrenean perlodids and perlids (Plecoptera). Ann. Limnol. 17: 1-24. (In French)

Bibby, C. J. and R. E. Green. 1983. Food and fattening of migrating warblers in some French marshlands. Ringing and Migration 4: 175-184.

Blancher, P. J. and D. K. McNicol. 1991. Tree swallow diet in relation to wetland activity. Can. J. Zool. 69: 2629-2637.

Block, R. J. 1959. The approximate amino acid composition of wild and hatchery trout (*Salvelinus fontinalis*) and some of their principal foods (*Gammarus* and *Hexagenia bilineata*). Contr. Boyce Thompson Inst. 20: 103-105.

Blois, C. 1985. The larval diet of three anisopteran (Odonata) species. Freshw. Biol. 15: 505-514.

Bodenheimer, F. S. 1951. Insects as human food. A chapter of the ecology of man. W. Junk, Publishers, The Hague.

Bohle, H. W. 1983. Drift-catching and feeding behaviour of the larvae of *Drusus discolor* (Trichoptera: Limnephilidae). Arch. Hydrobiol. 97: 455-470. (In German)

Bull, E. L. and R. C. Beckwith. 1993. Diet and foraging behavior of Vaux's swifts in northeastern Oregon. Condor 95:1016-1023.

Bures, S. and M. Král. 1987. Diet analysis and trophic ecology of the grey wagtail (*Motacilla cinerea* Tunst.) in Nízky Jeseník. Folia Zool. 36: 257-264.

Burr, B. M. and R. L. Mayden. 1982. Life history of the brindled madtom *Noturus miurus* in Mill Creek, Illinois (Pisces: Ictaluridae). Amer. Midl. Natural. 107: 25-41.

Burton, T. M. 1977. Population estimates, feeding habits and nutrient and energy relationships of *Notophthalmus v. viridescens*, in Mirror Lake, New Hampshire. Copeia 1977: 139-143.

Cadwallader, P. L. 1975a. The food of the New Zealand common river galaxias, *Galaxias vulgaris* Stokell (Pisces: Salmoniformes). Aust. J. Mar. Freshw. Res. 26: 15-30.

Cadwallader, P. L. 1975b. Feeding relationships of galaxiids, bullies, eels and trout in a New Zealand river. Aust. J. Mar. Freshw. Res. 26: 299-316.

Caira, J. N. and D. J. Orringer. 1995. Additional information on the morphology of *Potamotrygonocestus magdalensis* (Tetraphyllidea: Onchobothriidae) from the freshwater stingray *Potamotrygon magdalenae* in Colombia. J. Helminthol. Soc. 62: 22-26.

Capelli, G. M. 1980. Seasonal variation in the food habits of the crayfish *Orconectes propinquus* (Girard) in Trout Lake, Vilas County, Wisconsin, U.S.A. (Decapoda, Astacidea, Cambaridae). Crustaceana 38: 82-86.

Castién, E. and J. Gosálbez. 1995. Diet of *Galemys pyrenaicus* (Geoffroy, 1811) in the north of the Iberian Peninsula. Netherlands J. Zool. 45: 422-430.

Chen, Y. and H. H. Harvey. 1995. Growth, abundance, and food supply of white sucker. Trans. Amer. Fish. Soc. 124: 262-271.

Cochran, P. A. and D. R. McConville. 1983. Feeding by *Trionyx spiniferus* in backwaters of the Upper Mississippi River. J. Herpetol. 17: 82-86.

Collier, K. J. and G. L. Lyon. 1991. Trophic pathways and diet of blue duck (*Hymenolaimus malacorhynchos*) on Manganuiateao River: a stable carbon isotope study. N. Z. J. Mar. Freshw. Res. 25: 181-186.

Collins, J. P. and J. R. Holomuzki. 1984. Intraspecific variation in diet within and between trophic morphs in larval tiger salamanders (*Ambystoma tigrinum nebulosum*). Can. J. Zool. 62: 168-174.

Conaway, C. H. 1952. Life history of the water shrew (*Sorex palustris navigator*). Amer. Midl. Natural. 48: 219-248.

Cummins, K. W. and J. C. Wuycheck. 1971. Caloric equivalents for investigations in ecological energetics. Int. Ver. Theor. Angew. Limnol. 18: 1-158.

Dedual, M. and K. J. Collier. 1995. Aspects of juvenile rainbow trout (*Oncorhynchus mykiss*) diet in relation to food supply during summer in the lower Tongariro River, New Zealand. N. Z. J. Mar. Freshw. Res. 29: 381-391.

Denoncourt, C. E. and J. R. Stauffer, Jr. 1993. Feeding selectivity of the American eel *Anguilla rostrata* (LeSueur) in the Upper Delaware River. Amer. Midl. Natural. 129: 301-308.

Downes, J. A. 1978. Feeding and mating in the insectivorous Ceratopogoninae (Diptera). Mem. ent. Soc. Can. 104: 1-62.

Driver, E. A., L. G. Sugden and R. J. Kovach. 1974. Calorific, chemical and physical values of potential duck foods. Freshw. Biol. 4: 281-292.

Dudgeon, D. 1987. Niche specificities of four fish species (Homalopteridae, Cobitidae and Gobiidae) in a Hong Kong forest stream. Arch. Hydrobiol. 108: 349-364.

Dudgeon, D. 1989. Life cycle, production, microdistribution and diet of the damselfly *Euphaea decorata* (Odonata: Euphaeidae) in a Hong Kong forest stream. J. Zool. 217: 57-72.

Dudgeon, D. and J. S. Richardson. 1988. Dietary variations of predaceous caddisfly larvae (Trichoptera: Rhyacophilidae, Polycentropodidae and Arctopsychidae) from British Columbian streams. Hydrobiologia 160: 33-43.

Dumont, B. and J. Verneaux. 1976. Some tropic relationships in the upper reaches of a forest stream. Ann. Limnol. 12: 239-252. (In French)

Eadie, J. M. and A. Keast. 1982. Do goldeneye and perch compete for food? Oecologia 55: 225-230.

Easton, R. S., D. J. Orth and J. R. Voshell, Jr. 1996. Spatial and annual variation in the diets of juvenile smallmouth bass, *Micropterus dolomieu*. Environ. Biol. Fish. 46: 383-392.

Edmunds, G. F., Jr. and C. H. Edmunds. 1980. Predation, climate, and emergence and mating of mayflies. Pages 277-285. In, Flannagan, J. F. and K. E. Marshall (Eds.). Advances in Ephemeroptera Biology. Plenum, New York.

Feminella, J. W. and K. W. Stewart. 1986. Diet and predation by three leaf-associated stoneflies (Plecoptera) in an Arkansas mountain stream. Freshw. Biol. 16: 521-538.

Fisher, J. W. 1981. Ecology of *Fundulus catenatus* in three interconnected stream orders. Amer. Midl. Natural. 106: 372-378.

Fladung, E. B. 1924. Insects as food. Maryland Acad. Sci. Bull. 1924: 5-8.

Folsom, T. C. and N. C. Collins. 1984. The diet and foraging behavior of the larval dragonfly *Anax junius* (Aeshnidae), with an assessment of the role of refuges and prey activity. Oikos 42: 105-113.

Forrester, G. E. , J. G. Chace and W. McCarthy. 1994. Diel and density-related changes in food consumption and prey selection by brook charr in a New Hampshire stream. Environ. Biol. Fish. 39: 301-311.

Fox, L. R. 1975. Some demographic consequences of food shortage for the predator, *Notonecta hoffmanni*. Ecology 56: 868-880.

Fuller, R. L. and H. B. N. Hynes. 1987. Feeding ecology of three predacious aquatic insects and two fish in a riffle of the Speed River, Ontario. Hydrobiologia 150: 243-255.

Fuller, R. L. and K. W. Stewart. 1977. The food habits of stoneflies (Plecoptera) in the Upper Gunnison River, Colorado. Environ. Ent. 6: 293-302.

Fuller, R. L. and K. W. Stewart. 1979. Stonefly (Plecoptera) food habits and prey preference in the Dolores River, Colorado. Amer. Midl. Natural. 101: 170-181.

Gee, H. and J. O. Young. 1993. The food niches of the invasive *Dugesia tigrina* (Girard) and indigenous *Polycelis tenuis* Ijima and *P. nigra* (Müller) (Turbellaria; Tricladida) in a Welsh lake. Hydrobiologia 254: 99-106.

Ghioni, C., J. G. Bell and J. R. Sargent. 1996. Polyunsaturated fatty acids in neutral lipids and phospholipids of some freshwater insects. Comp. Biochem. Physiol. 114B: 161-170.

Gibbs, K. E. and T. M. Mingo. 1986. The life history, nymphal growth rates, and feeding habits of *Siphlonisca aerodromia* Needham (Ephemeroptera: Siphlonuridae) in Maine. Can. J. Zool. 64: 427-430.

Giller, P. S. 1986. The natural diet of the Notonectidae: field trials using electrophoresis. Ecol. Ent. 11: 163-172.

Gillies, M. T. 1996. Mayflies as food: a confused story from South America. The Mayfly Newsletter 6(2):1.

Glova, G. J., P. M. Sagar and I. Näslund. 1992. Interaction for food and space between populations of *Galaxias vulgaris* Stokell and juvenile *Salmo trutta* L. in a New Zealand stream. J. Fish. Biol. 41: 909-925.

Godinho, F. N. and M. T. Ferreira. 1994. Diet composition of largemouth black bass, *Micropterus salmoides* (Lacepède), in southern Portuguese reservoirs: its relation to habitat characteristics. Fish. Man. Ecol. 1: 129-137.

Greenwood, J. 1986. Sandwich terns feeding over fresh water. British Birds 69: 43.

Hanson, B. J., K. W. Cummins, A. S. Cargill and R. R. Lowry. 1985. Lipid content, fatty acid composition, and the effect of diet on fats of aquatic insects. Comp. Biochem. Physiol. 80B: 257-276.

Harris, A. C. 1990. *Podagritus cora* (Cameron) and *P. albipes* (F. Smith) (Hymenoptera: Specidae: Cabroninae) preying on Ephemeroptera and Trichoptera. Pan-Pac. Ent. 66: 55-61.

Hayes, D. B., W. W. Taylor and J. C. Schneider. 1992. Response of yellow perch and the benthic invertebrate community to a reduction in the abundance of white suckers. Trans. Amer. Fish. Soc. 121: 36-53.

Hlohowskyj, I. and A. M. White. 1983. Food resource partitioning and selectivity by the greenside, rainbow, and fantail darters (Pisces: Percidae). Ohio J. Sci. 83: 201-208.

Johnson, J. H. 1981a. Comparative food selection by coexisting subyearling coho salmon, chinook salmon and rainbow trout in a tributary of Lake Ontario. N. Y. Fish Game J. 28: 150-161.

Johnson, J. H. 1981b. Food habits and dietary overlap of perlid stoneflies (Plecoptera) in a tributary of Lake Ontario. Can. J. Zool. 59: 2030-2037.

Johnson, J. H. and D. S. Dropkin. 1993. Diel variation in diet composition of a riverine fish community. Hydrobiologia 271: 149-158.

Knight, J. G. and S. T. Ross. 1994. Feeding habits of the bayou darter. Trans. Amer. Fish. Soc. 123: 794-802.

Koslucher, D. G. and G. W. Minshall. 1973. Food habits of some benthic invertebrates in a northern cool-desert stream (Deep Creek, Curlew Valley, Idaho-Utah). Trans. Amer. Micros. Soc. 92: 441-452.

Kraiem, M. M. 1996. The diet of *Barbus callensis* (Cyprinidae) in northern Tunisia. Cybium 20: 75-85.

Kuz'min, S. L. 1990. Feeding of sympatric species Hynobiidae in the Primorye. Zool. Zh. 69: 71-75. (In Russian)

Kuz'min, S. L. 1992. Food of the salamander. Soviet J. Ecol. 22: 233-240. (Translated from Ékologiya, no. 4, pp. 34-42, 1991)

Lagler, K. F. 1943. Food habits and economic relations of the turtles of Michigan with special reference to fish management. Amer. Midl. Natural. 29: 257-312.

Langham, N. 1980. Breeding biology of the edible-nest swiftlet *Aerodramus fuciphagus*. Ibis 122: 447-461.

Laughlin, D. R. and E. E. Werner. 1980. Resource partitioning in two coexisting sunfish: pumpkinseed (*Lepomis gibbosus*) and northern longear sunfish (*Lepomis megalotis peltastes*). Can. J. Fish. Aquat. Sci. 37: 1411-1420.

Lavandier, P. 1982. Larval development, feeding and production of *Isoperla viridinervis* Pictet (Plecoptera, Perlodidae) in a cold river in the high mountains. Ann. Limnol. 18: 301-318. (In French)

Lawton, J. H. 1970. Feeding and food energy assimilation in larvae of the damselfly *Pyrrhosoma nymphula* (Sulz.) (Odonata: Zygoptera). J. Anim. Ecol. 39: 669-689.

Leonard, J. W. and F. A. Leonard. 1962. Mayflies of Michigan trout streams. Cranbrook Institute of Science, Bloomfield Hills.

Levings, C. D., N. A. Hvidsten and B. Ø. Johnsen. 1994. Feeding of Atlantic salmon (*Salmo salar* L.) postsmolts in a fjord in central Norway. Can. J. Zool. 72: 834-839.

Lobón-Cerviá, J., C. G. Utrilla, E. Querol and M. A. Puig. 1993. Population ecology of pike-cichlid, *Crenicichla lepidota*, in two streams of the Brazilian Pampa subject to a severe drought. J. Fish Biol. 43: 537-557.

Loske, K.-H. 1992. Nestling food of the swallow (*Hirundo rustica*) in Central Westphalia. Vogelwarte 36: 173-187. (In German)

Lucy, F., M. J. Costello and P. S. Giller. 1990. Diet of *Dinocras cephalotes* and *Perla bipunctata* (Plecoptera, Perlidae) in a South-West Irish stream. Aquat. Insects 12: 199-207.

Maitland, P. S. 1965. The distribution, life cycle, and predators of *Ephemerella ignita* (Poda) in the River Endrick, Scotland. Oikos 16: 48-57.

Malmqvist, B., P. Sjöström and K. Frick. 1991. The diet of two species of *Isoperla* (Plecoptera: Perlodidae) in relation to season, site, and sympatry. Hydrobiologia 213: 191-203.

Manuel, K. L. and T. C. Folsom. 1982. Instar sizes, life cycles, and food habits of five *Rhyacophila* (Trichoptera: Rhyacophilidae) species from the Appalachian Mountains of South Carolina, U.S.A. Hydrobiologia 97: 281-285.

Martof, B. S. and D. C. Scott. 1957. The food of the salamander *Leurognathus*. Ecology 38: 494-501.

McCafferty, W. P. 1979. Swarm-feeding by the damselfly *Hetaerina americana* (Odonata: Calopterygidae) on mayfly hatches. Aquat. Insects 3: 149-151.

McClure, R. G. and K. W. Stewart. 1976. Life cycle and production of the mayfly *Choroterpes* (*Neochoroterpes*) *mexicanus* Allen (Ephemeroptera: Leptophlebiidae). Ann. ent. Soc. Amer. 69: 134-144.

McDowall, R. M., M. R. Main, D. W. West and G. L. Lyon. 1996. Terrestrial and benthic foods in the diet of the shorjawed kokopu, *Galaxias postvectis* Clarke (Teleostei: Galaxiidae). N. Z. J. Mar. Freshw. Res. 30: 257-269.

Meffe, G. K. and A. L. Sheldon. 1987. Habitat use by dwarf waterdogs (*Necturus punctatus*) in South Carolina streams, with life history notes. Herpetologica 43: 490-496.

Mendelson, J. 1975. Feeding relationships among species of *Notropis* (Pisces: Cyprinidae) in a Wisconsin stream. Ecol. Monogr. 45: 199-230.

Miura, T., R. M. Takahashi and R. J. Stewart. 1979. Habitat and food selection by the mosquitofish, *Gambusia affinis*. Proc. Pap. California Mosq. Vector Contr. Assoc. 47: 46-50.

Montori, A. 1992. Alimentación de las larvas de tritón pirenaico, *Euproctus asper*, en el prepirineo de la Cerdaña, España. Amphibia-Reptilia 13: 157-167.

Muttkowski, R. A. and G. M. Smith. 1929. The food of trout stream insects in Yellowstone National Park. Roosevelt Wild Life Annals 2: 241-263.

Oberndorfer, R. Y. and K. W. Stewart. 1977. The life cycle of *Hydroperla crosbyi* (Plecoptera: Perlodidae). Gr. Basin Natural. 37: 260-273.

Ogle, D. H., J. H. Selgeby, R. M. Newman and M. G. Henry. 1995. Diet and feeding periodicity of ruffe in the St. Louis River estuary, Lake Superior. Trans. Amer. Fish. Soc. 124: 356-369.

Okedi, J. 1992. Chemical evaluation of Lake Victoria lakefly as nutrient source in animal feeds. Insect Sci. Applic. 13: 373-376.

Olatunde, A. A. and C. C. Moneke. 1985. The food habits of four mormyrid species in Zaria, Nigeria. Arch. Hydrobiol. 102: 503-517.

Ormerod, S. J. and S. J. Tyler. 1988. The diet of green sandpipers *Tringa ochropus* in contrasting areas of their winter range. Bird Study 35: 25-30.

Ormerod, S. J. and S. J. Tyler. 1991. Exploitation of prey by a river bird, the dipper *Cinclus cinclus* (L.), along acidic and circumneutral streams in upland Wales. Freshw. Biol. 25: 105-116.

Parker, M. S. 1994. Feeding ecology of stream-dwelling Pacific giant salamander larvae (*Dicamptodon tenebrosus*). Copeia 1994: 705-718.

Peckarsky, B. L. 1996. Alternative predator avoidance syndromes of stream-dwelling mayfly larvae. Ecology 77: 1888-1905.

Petr, T. 1968. Distribution, abundance and food of commercial fish in the Black Volta and the Volta man-made lake in Ghana during its first period of filling (1964-1966). I. Mormyridae. Hydrobiologia 32: 417-448.

Petrosky, C. E. and T. F. Waters. 1975. Annual production by the slimy sculpin population in a small Minnesota trout stream. Trans. Amer. Fish. Soc. 104: 237-244.

Pierce, R. J. 1986a. Observations on behaviour and foraging of the ibisbill *Ibidorhyncha struthersii* in Nepal. Ibis 128: 37-47.

Pierce, R. J. 1986b. Foraging responses of stilts (*Himantopus* spp.: Aves) to changes in behaviour and abundance of their riverbed prey. N. Z. J. Mar. Freshw. Res. 20: 17-28.

Pritchard, G. 1964. The prey of dragonfly larvae (Odonata; Anisoptera) in ponds in northern Alberta. Can. J. Zool. 42: 785-800.

Przybylski, M. and J. Banbura. 1989. Feeding relations between the gudgeon (*Gobio gobio* (L.)) and the stone loach (*Noemacheilus barbatulus* (L.)). Acta Hydrobiol. 31: 109-119.

Rabeni, C. F. 1992. Trophic linkage between stream centrarchids and their crayfish prey. Can. J. Fish. Aquat. Sci. 49: 1714-1721.

Rabinowitz, A. R. and M. D. Tuttle. 1982. A test of the validity of two currently used methods of determining bat prey preferences. Acta Theriol. 27: 283-293.

Reilly, P. and T. K. McCarthy. 1990. Observations on the natural diet of *Cymatia bonsdorfi* (C. Sahlb.) (Heteroptera: Corixidae): an immunological analysis. Hydrobiologia 196: 159-166.

Reinecke, K. J. and R. B. Owen, Jr. 1980. Food use and nutrition of black ducks nesting in Maine. J. Wildl. Manage. 44: 549-558.

Reynolds, J. D. and G. G. E. Scudder. 1987. Serological evidence of realized feeding niche in *Cenocorixa* species (Hemiptera: Corixidae) in sympatry and allopatry. Can. J. Zool. 65: 974-980.

Reynoldson, T. B. and P. Bellamy. 1975. Triclads (Turbellaria: Tricladida) as predators of lake-dwelling stonefly and mayfly nymphs. Freshw. Biol. 5: 305-312.

Rhame, R. E. and K. W. Stewart. 1976. Life cycles and food habits of three Hydropsychidae (Trichoptera) species in the Brazos River, Texas. Trans. Amer. ent. Soc. 102: 65-99.

Ritchie, B. J. and P. J. Colby. 1988. Even-odd year differences in walley year-class strength related to mayfly production. N. Amer. J. Fish. Manag. 8: 210-215.

Schmidt, D. A. and D. C. Tarter. 1985. Life history and ecology of *Acroneuria carolinensis* (Banks) in Panther Creek, Nicholas County, West Virginia (Plecoptera: Perlidae). Psyche 92: 393-406.

Scrimgeour, G. J. and M. J. Winterbourn. 1987. Diet, food resource partitioning and feeding periodicity of two riffle-dwelling fish species in a New Zealand river. J. Fish. Biol. 31: 309-324.

Sempeski, P. and P. Gaudin. 1996. Size-related shift in feeding strategy and prey-size selection in young grayling (*Thymallus thymallus*). Can. J. Zool. 74: 1597-1603.

Sheldon, A. L. 1969. Size relationships of *Acroneuria californica* (Perlidae, Plecoptera) and its prey. Hydrobiologia 34:85-94.

Sheldon, A. L. 1972. Comparative ecology of *Arcynopteryx* and *Diura* (Plecoptera) in a California stream. Arch. Hydrobiol. 69: 521-56.

Sinha, V. R. P. and J. W. Jones. 1967. On the food of the freshwater eels and their feeding relationship with the salmonids. J. Zool. 153: 119-137.

Snellen, R. K. and K. W. Stewart. 1979. The life cycle of *Perlesta placida* (Plecoptera: Perlidae) in an intermittent stream in northern Texas. Ann. ent. Soc. Amer. 72: 659-666.

Soluk, D. A. and H. F. Clifford. 1984. Life history and abundance of the predaceous psammophilous mayfly *Pseudiron centralis* McDunnough (Ephemeroptera: Heptageniidae). Can. J. Zool. 62: 1534-1539.

Stewart, K. W., G. P. Friday and R. E. Rhame. 1973. Food habits of hellgrammite larvae, *Corydalus cornutus* (Megaloptera: Corydalidae), in the Brazos River, Texas. Ann. ent. Soc. Amer. 66: 959-963.

Swift, S. M., P. A. Racey and M. I. Avery. 1985. Feeding ecology of *Pipistrellus pipistrellus* (Chiroptera: Vespertilionidae) during pregnancy and lactation. II. Diet. J. Anim. Ecol. 54: 217-225.

Szent-Ivany, J. J. H. and E. I. V. Ujházy. 1973. Ephemeroptera in the regimen of some New Guinea people and in Hungarian folksongs. Eatonia 0(17): 1-6.

Teslenko, V. A. 1997. Feeding habits of the predaceous stoneflies in a salmon stream of the Russian far east. Pages 73-78. IN, Landolt, P. and M. Sartori (Eds.). Ephemeroptera and Plecoptera: Biology-Ecology-Systematics. Mauron + Tinguely & Lachat SA, Fribourg.

Thompson, D. J. 1978. Prey size selection by larvae of the damselfly, *Ischnura elegans* (Odonata). J. Anim. Ecol. 47: 769-785.

Thut, R. N. 1969. Feeding habits of larvae of seven *Rhyacophila* (Trichoptera: Rhyacophilidae) species with notes on other life-history features. Ann. ent. Soc. Amer. 62: 894-898.

Tumlison, R., G. R. Cline and P. Zwank. 1990. Surface habitat associations of the Oklahoma salamander (*Eurycea tynerensis*). Herpetologica 46: 169-175.

Turner, A. K. 1983. Food selection and the timing of breeding of the blue-and-white swallow *Notiochelidon cyanoleuca* in Venezuela. Ibis 125: 450-462.

Tzeng, W.-N., J.-J. Hsiao, H.-P. Shen, Y.-T. Chern, Y.-T. Wang and J.-Y. Wu. 1995. Feeding habits of the Japanese eel, *Anguilla japonica*, in the streams of northern Taiwan. J. Fish. Soc. Taiwan 22: 297-302.

Vaughan, T. A. and W. P. Weil. 1980. The importance of arthropods in the diet of *Zapus princeps* in a subalpine habitat. J. Mammal. 61: 122-124.

Vaught, G. L. and K. W. Stewart. 1974. The life history and ecology of the stonefly *Neoperla clymene* (Newman) (Plecoptera: Perlidae). Ann. ent. Soc. Amer. 67: 167-178.

Vives, S. P. 1987. Aspects of the life history of the slender madtom *Noturus exilis* in northeastern Oklahoma (Pisces: Ictaluridae). Amer. Midl. Natural. 117: 167-176.

Walde, S. J. and R. W. Davies. 1987. Spatial and temporal variation in the diet of a predaceous stonefly (Plecoptera: Perlodidae). Freshwater Biol. 17: 109-115.

Wallace, J. B. 1975. The larval retreat and food of *Arctopsyche*; with phylogenetic notes on feeding adaptations in Hydropsychidae larvae (Trichoptera). Ann. ent. Soc. Amer. 68: 167-173.

Wallace, J. B., T. F. Cuffney, C. C. Lay and D. Vogel. 1987. The influence of an ecosystem-level manipulation on prey consumption by a lotic dragonfly. Can. J. Zool. 65: 35-40.

Welham, C. V. J. 1987. Diet and foraging behavior of ring-billed gulls breeding at Dog Lake, Manitoba. Wilson Bull. 99: 233-239.

Whitaker, J. O., Jr. 1972. Food habits of bats from Indiana. Can. J. Zool. 50: 877-883.

Whitaker, J. O., Jr., W. W. Cudmore and B. A. Brown. 1980. Foods of larval, subadult and adult smallmouth salamanders, *Ambystoma texanum*, from Vigo County, Indiana. Proc. Indiana Acad. Sci 90: 461-464.

Williams, D. D., L. G. Ambrose and L. N. Browning. 1995. Trophic dynamics of two sympatric species of riparian spider (Araneae: Tetragnathidae). Can. J. Zool. 73: 1545-1553.

Williams, T. A. and J. L. Christiansen. 1981. The niches of two sympatric softshell turtles, *Trionyx muticus* and *Trionyx spiniferus*, in Iowa. J. Herpetol. 15: 303-308.

Yoshida, T., K. Sugimoto and F. Hayashi. 1985. Notes on the life history of the dobsonfly, *Protohermes grandis* Thunberg (Megaloptera, Corydalidae). Kontyu 53: 734-742. (In Japanese.)

SEASONAL VARIATION OF EPHEMEROPTERA IN FOUR STREAMS OF GUATOPO NATIONAL PARK, VENEZUELA

V. Maldonado, B. Pérez, and C. Cressa

Universidad Central de Venezuela
Escuela de Biología - Instituto de Zoología Tropical
Apartado Postal 47058, Caracas 1041-A, Venezuela

ABSTRACT

The composition, seasonal variation and diversity of the mayfly fauna in four streams (two watersheds) of the Guatopo National Park were studied biweekly for one year (February 1996-February 1997). The density of the Ephemeroptera has a distinctive cycle with maximum abundance at the middle of the dry season and minimum during the rainy season. The abundance of this community is significantly different among rivers belonging to different watersheds. The Leptophlebiidae (*Thraulodes* sp.) was always the most abundant family, in all rivers, followed by the Leptohyphiidae. However, the latter showed a higher diversity than the former. Principal Component Analysis of the data indicated that the community structure of the mayfly fauna is more similar among rivers belonging to the same watershed regardless of differences in morphometric variables of the streams.

INTRODUCTION

Seasonal variations as well as distributional patterns in aquatic insects have been associated with different causes. Historically, since seasonal variation in tropical systems was smaller than in temperate ones, they were considered climatically stable (Klopfer, 1959). Thus, it follows that the fauna should also exhibit smaller fluctuations. However, as more research in tropical areas is being conducted, they indicate that aquatic insects exhibit seasonal fluctuations similar to temperate ones (Flecker and Feifarek, 1994; Flowers and Pringle, 1995; Cressa, 1998; Jacobsen and Encalada, 1998).

On the other hand, distributional patterns of insects were explained using three different approaches. First of all, the classification of rivers was based on empirical relationships between the riverine fauna and environmental factors. The environmental factors more frequently used were substrate composition (Wright *et al.*, 1984), hydrological conditions (Stanford and Ward, 1983; Resh *et al.*, 1988; Corkum, 1989; 1992). However, the prediction power was poor since models were area-specific (Hawkes, 1975; Corkum, 1989). Secondly, distributional patterns

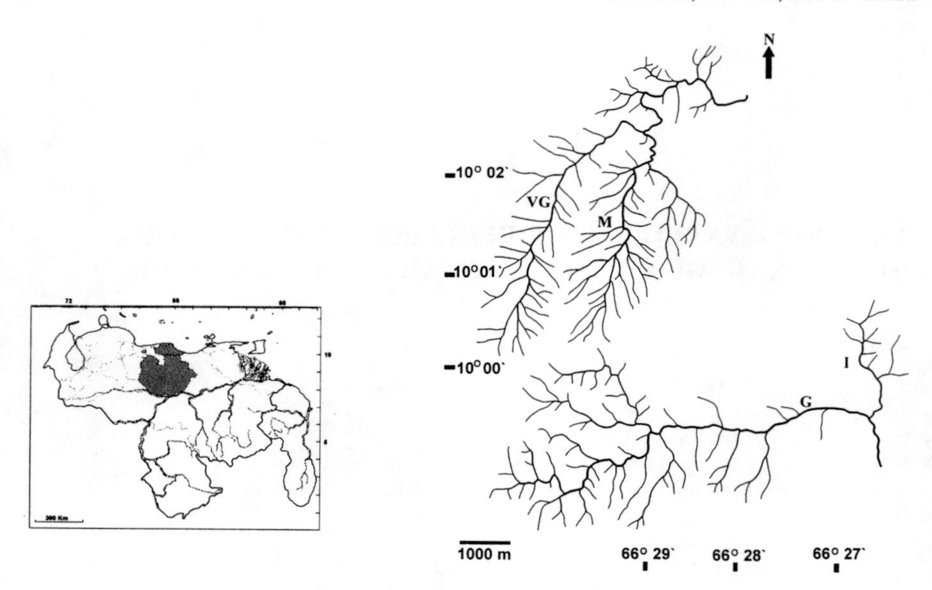

Fig. 1. Venezuela map including two watersheds studied.

were sought applying the continuum concept (Vannote *et al.,* 1980) which stated that longitudinal changes in physical variables along a river were associated with changes in biological process (production) and community composition of vertebrates (fishes) as well as invertebrates (feeding groups). In this case also several modifications were needed in order to apply the model to different biomes for the rest of the world (Wiggins and Mackay, 1978; Minshall *et al.*, 1985; Winterbourn *et al.,* 1981; Cressa, 1994). Finally, the biome dependency hypothesis originally expressed by Ross (1963) indicated that caddisflies were associated with the biome through which the river flowed. Corkum (1989; 1990) expanded on this idea and suggested that similar assemblages of macroinvertebrates were likely to be found within drainage basins.

In this study, we present data from one year's sampling of the mayfly community of four streams located in Guatopo National Park, Venezuela. We studied the seasonal fluctuations of mayflies as well as their distributional patterns in these two watersheds and evaluated the importance of the hydrological variables on their assemblages.

Study Area

This study was conducted on four streams of Guatopo National Park, Venezuela (9° 57' – 10° 5' N and 66° 24' – 66° 30' W, Fig. 1) where the rivers are unperturbed and in pristine state. The National Park lies on the Cordillera del Interior of Cenozoic origin (Tertiary period); its geological characterization is metamorphic-sedimentary (shale) and metamorphic-igneous rocks. The region has narrow valleys, abrupt and pronounced slopes, and narrow areas for deposition of material and terraces formed as by-products of climatic changes during the Quaternary. Due to the extension of the Park (122,464 Ha) there is a difference in altitude (200-1,430 m) easily recognized in the vegetation that changes from humid to a dry tropical forest as the altitude dismishes.

The rivers studied belong to two different watersheds (Fig. 1), Quebrada Martinera (M) and Quebrada Vuelta Grande (VG) are the headwaters of the river Taguaza which flows into the Caribbean Sea. The other streams, Quebrada Ingenio (I) and Quebrada Guatopo (G), are part of the Orituco basin of the Orinoco river system.

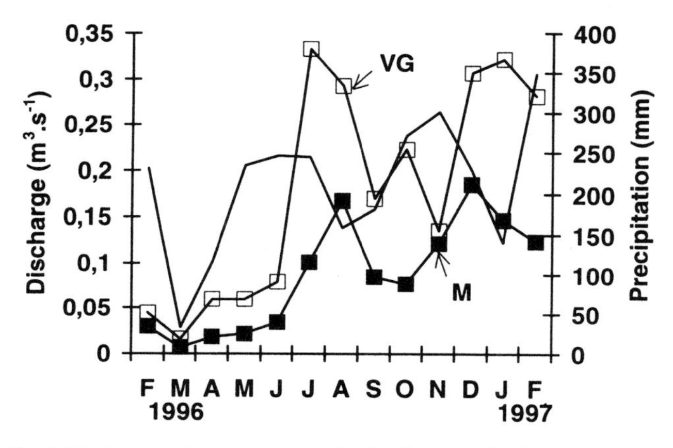

Fig. 2. Discharge and precipitation in Vuelta Grande – Martinera watershed.

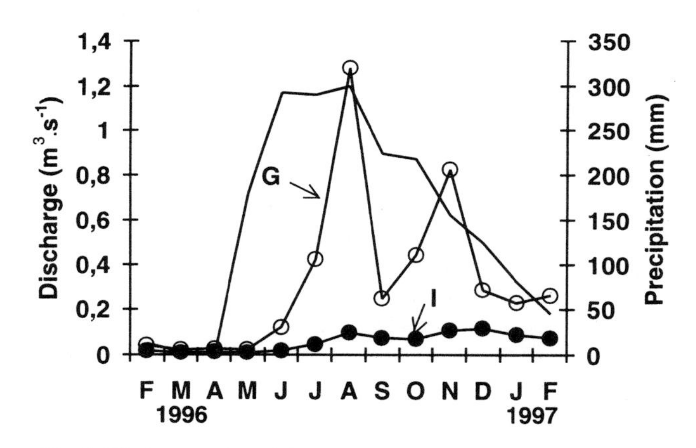

Fig. 3. Discharge and precipitation in Guatopo – Ingenio watershed.

The weather data were obtained from the Ministerio del Ambiente y los Recursos Naturales Renovables from stations located near the sampling sites (Agua Blanca for VG-M and Guatopo for G-I). Mean annual rainfall based on 37 records (1960-1997) is 2611.4 mm and 2166.4 mm for VG-M and G-I respectively. However, during the sampling period VG-M had more rainfall that this average (2737.9 mm) while for G-I it was lower (1925.5). In general, precipitation is bimodal; the first rainy season begins in April and continues until July. The second period starts in August and lasts until December. VG-M follows this pattern more closely (Fig. 2) since the bimodality was present during the sampling period.

Discharge annual variation resembles the rainfall pattern (Figs. 2 and 3) as previously recorded by Cressa and Senior (1987) for the G-I watershed. The highest discharge is for G (Table 1) followed by VG. Interestingly, even though VG and I are streams of the same order (3rd) the latter presented the lower discharge (Table 1).

Table 1. Morphometric characteristics of four streams in Guatopo National Park

River	Vuelta Grande	Martinera	Guatopo	Ingenio
Order	3rd	4th	5th	3rd
Altitude (masl)	396	300	620	620
Length (km)	6	5	7.75	3
Watershed area (km^2)	6.08	6.24	11.83	1.786
Mean Depth (m)				
Annual	0.17	0.12	0.15	0.12
Dry	0.15	0.10	0.11	0.10
Wet	0.19	0.15	0.21	0.14
Mean Width (m)				
Annual	3.02	3.14	3.89	1.87
Dry	3.14	2.90	2.93	1.22
Wet	2.86	7.85	5.25	2.22
Bank Full (m)	9.2	10.3	8.4	6.4
Mean Velocity (m/s)				
Annual	0.53	0.35	0.57	0.38
Dry	0.42	0.31	0.40	0.33
Wet	0.70	0.41	0.82	0.44
Mean Discharge (m^3s)				
Annual	0.16	0.08	0.30	0.05
Dry	0.11	0.05	0.10	0.03
Wet	0.24	0.11	0.6	0.08

The river channel in Qda. Ingenio and Martinera is completely covered by the riparian vegetation with canopies interlocking over the stream while in Vuelta Grande and Guatopo canopies do not cover the entire river bed. The streams substrates were mainly cobble (5-10 cm) and gravel in both watersheds. In Martinera and Ingenio the rock bottom was clearly visible in several stretches of the river. The riparian vegetation was by far dominated by the Leguminosae which were the most abundant and diverse component, followed by the Euphorbiaceae, Rubiaceae and Moraceae. The main representatives of the riparian vegetation in Martinera were *Bauhinia multinervia, Acacia glomerosa, Guazuma tomentosa, Ficus* sp., *Inga* sp. *and Erythrina poeppigiana*. In Guatopo *Hura crepitans, Brownea leucantha, Simira longifolia, Zanthoxylum monophyllum*, and *Jacaranda* sp. were the more representative species.

MATERIALS AND METHODS

The macroinvertebrates were collected biweekly from February 1996 to February 1997, with a Surber net (0.1296 m^2, 0.286 mesh size) using a stratified design to cover all the habitats of the highly heterogeneous substrate as well as to obtain a good representation of the early instars. Organisms were separated from the sediment in the field and preserved with ethanol (70%) for counting and identification at the laboratory. The identification of the specimens were determined, as far as possible to morphospecies, using available literature (Roldan, 1988; Domínguez *et al.*, 1992; Merritt and Cummins, 1996). Some mayflies were directly identified by Dr. E. Domínguez, Universidad de Tucumán, Argentina.

At the same time that macroinvertebrates were collected, the following variables were recorded: velocity (Ott meter), depth, width, temperature, pH and conductivity as well as water samples for chemical analysis. These data will be presented in another paper.

Table 2. Mean density (no.m^{-2}), number of morphospecies, Fisher's alpha diversity of the log series (μ) and Shannon (H') index in four streams in Guatopo National Park. All variables are giving as and annual mean as well as for the dry and wet season.

River	Vuelta Grande	Martinera	Guatopo	Ingenio
Mean Density (no/m^2)	12.814	14.618	13.048	16.590
Annual	13.382	17.048	19.435	22.787
Dry	12.056	11.083	4.533	8.317
Wet				
Number of morphospecies				
Annual	11	13	11	10
Dry	11	11	11	10
Wet	11	10	10	10
a diversity index				
Annual	1.791	2.167	1.691	1.506
Dry	1.944	1.824	1.784	1.634
Wet	2.119	2.006	2.511	2.017
Shannon Index (H')				
Annual	0.624	0.548	0.588	0.614
Dry	0.583	0.532	0.542	0.574
Wet	0.661	0.541	0.774	0.687

RESULTS

The maximum number of morphospecies was 13 (Table 2) which could be considered low when compared with the thirty-seven found by Flowers and Pringle (1995) in the Río Sábalo-Esquina in Costa Rica. The highest number of morphospecies as well as the highest density was found in the Leptophlebiidae followed by the Leptohyphidae. The same family was also the most important in R. Sábalo (Flowers and Pringle, 1995).

The mean annual density of mayflies was similar between watersheds (Table 2). However, a statistically significant difference (Two-way Anova, $p < 0.05$) was found between seasons, with maximum and minimum abundance during the dry (March, Fig. 4) and wet season (August), respectively. The watersheds did not behave similarly since in G-I this difference was an order of magnitude while for VG-M the variation was much smaller (Table 2). Similar results in the magnitude of the seasonal variation on density was reported by Flecker and Feifarek (1994) for Río las Marías, Venezuela.

This seasonal variation in density, indicates changes in mayfly composition since some morphospecies appear only during one of the seasons (dry or wet). Thus, in Martinera, *Leptohyphes* 3, *Camelobaetidius* sp. and *Hydrosmilodon* sp. were present only during the dry season as was *Camelobaetidius* sp. in Guatopo.

On the other hand, *Thraulodes* sp. was the most abundant morphospecies. Its highest density was in Qda. Martinera during the dry season (66 %) while the lowest was found in Qda. Ingenio during the wet season (21 %).

Diversity is also shown in Table 2, calculated as Fisher's alpha of the log series and Shannon's index. The alpha values give a better representation of the morphospecies variation than Shannon's. This was expected since Fisher's alpha of the log series is independent of the sample size and does not give excessive weight to the most common species in a sample (Wolda, 1981; 1983a; 1983b). However, peaks in diversity, for both indexes, relatively coincide with peaks in number of species and less so with density (Fig. 4). Similar results were found by Wolda (1983b) for tropical cockroaches which do not have abrupt changes in density, similar to the results for aquatic insects presented here.

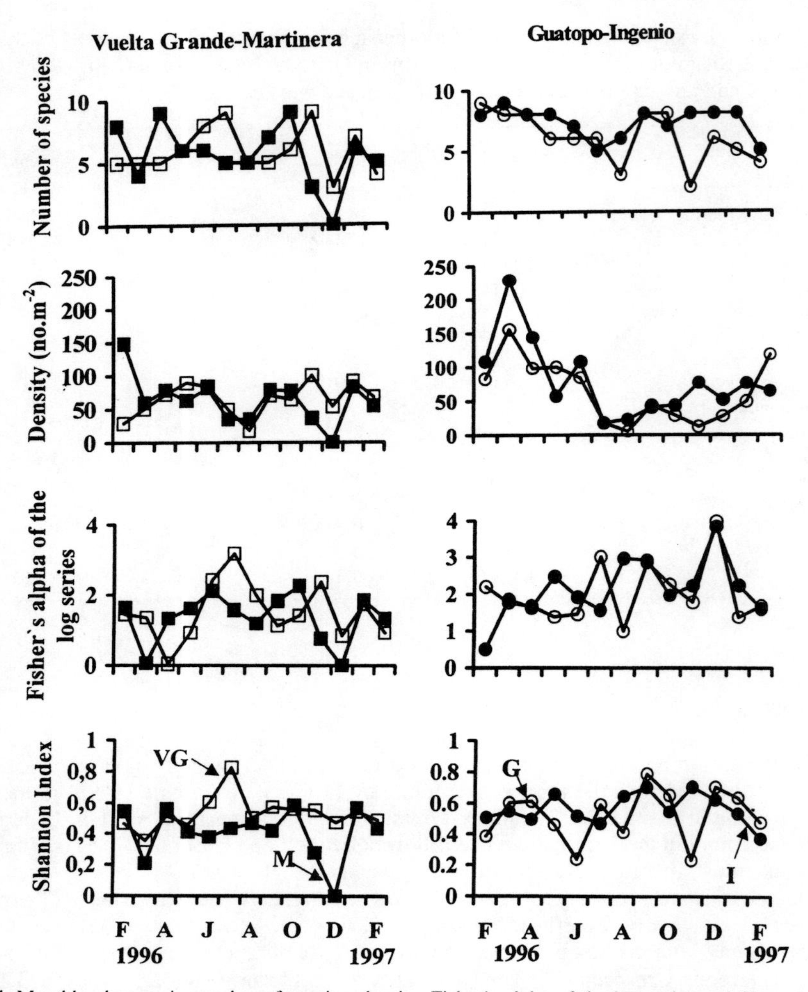

Fig. 4. Monthly changes in number of species, density, Fisher's alpha of the log series and Shannon index for two watersheds.

Since the seasonal variation in mayflies abundance was so pronounced, the data were analyzed using Principal Component Analysis (PCA) in order to associate changes in the community with seasons and rivers. The first axis of the PCA was able to separate the streams, and the species more representative were *Terpides* sp. (94%), *Euthyplocia* sp. (86%) and *Haplohyphes* sp. (63%). The second axis of the PCA was able to separate the dry and wet season for all streams (Fig. 5) and the species with greater weight were *Thraulodes* sp. (69%), *Leptohyphes* 1 (56%) and *Baetodes* sp. (55%). The PCA also indicates that the difference during seasons in VG was not as great as in the other rivers.

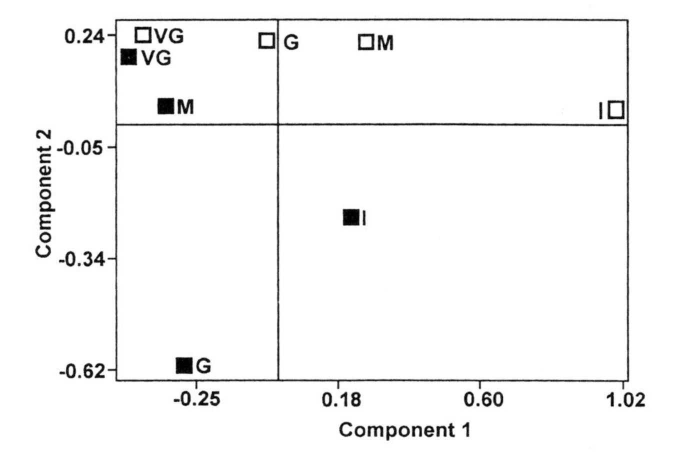

Fig. 5. Principal Component Analysis. White square: dry season and black square: wet season (VG: Vuelta Grande, M: Martinera, G: Guatopo and I: Ingenio).

Furthermore, during the wet season sites belonging to the same watershed were more similar because they shared the same faunal composition. During the dry season the association between streams belonging to the same watershed was not as clear as during the wet season because Guatopo and Martinera are closer. This similarity could be explained by the physical characteristics of both streams during this season, since their width, depth, velocity and discharge were similar (Table 1).

During the dry season the changes on community composition and density were greater for Ingenio (Table 2, Fig. 5) than for the other rivers. Ingenio had mainly the less abundant species (*Terpides* sp. and *Euthyplocia* sp.) as is indicated by axis 1. On the other hand, Guatopo was characterized by the more common species (*Thraulodes* sp. and *Leptohyphes* 1) as is clearly indicated by the axis 2 (Fig. 5). Therefore, these patterns are more similar among rivers draining similar watersheds than between rivers from different watersheds, as expected according to the biome theory of Ross (1963) and Corkum (1989).

DISCUSSION

This study indicates the importance of seasons in determining the mayfly community composition in the four streams studied. The seasonal variation can produce alterations in environmental variables which in turn can regulate the mayfly community composition (discharge, substrate, mean width), as was demonstrated by Dudgeon (1993) for a river in Hong Kong, Flecker and Feirarek (1994) for Andean Venezuelan rivers and Jacobsen and Encalada (1998) for Andean Ecuatorian rivers.

The low number of morphospecies found in these streams is surprising, and even though there are few studies on tropical streams from which to draw meaningful comparisons, the data that exist indicate higher numbers (Stout and Vandermeer, 1975; Pearson *et al.,* 1986; Flowers and Pringle, 1995). Only the recent paper by Jacobsen and Encalada (1998) reported lower numbers (7) for rivers of Ecuador. In our case, there are three reasons that could explain this difference. First, that sample size was not adequate; however, even though there is always the possibility that a greater sample size could capture the rarer species, we feel that sample size was appropriate since it follows the suggestion of Vinson and Hawkins (1996). Second, that fluctuations of the composition of the mayfly fauna are greater than a year sampling and thus were not observed. This also could not be disregarded completely, particularly in view of the data presented by Flowers and Pringle (1995) where only thirteen taxa were found only one of the three years that lasted the study. However, in our case this is also unlikely since sampling in one of the rivers (Qda. Guatopo) has been going on since 1983. Even though this sampling has not been continuous, we feel that our collections should indicate the actual morphospecies present in the stream. Finally, we feel that taxonomic separation could be our major problem due to the high number of undescribed nymphal stages. However, care was taken to separate species as morphospecies, although the small size of organisms could easily have caused some species to be overlooked and thus the diversity could be underestimated. Accounting for this factor even if we double our numbers of morphospecies, they are still low and could really mean that this is characteristic of the area.

In any case, these results emphasize both the need for more studies in tropical systems in order to characterize these systems appropriately and the importance that long-term studies have in evaluating diversity in lotic systems (Flowers and Pringle, 1995). This is even more important if a great variation in composition and abundance is present, as was in this case.

The seasonal patterns shown by the mayflies fauna are similar to the studies of Hynes (1975) in Ghana, Dudgeon (1993) in Hong Kong, Flecker and Feifarek (1994) in Venezuela and Jacobsen and Escalada (1998) in Ecuador. In this study as well as in those mentioned above, the density of the aquatic fauna is dependent on the hydrological conditions in the environment. In our case the hydrological regime is different in the four rivers studied. Martinera and Ingenio being smaller in size are more subject to changes in morphometry during high floods and the variation in density should reflect this higher washing out of the fauna during the wet season. Flecker and Feifarek (1994) reported similar results for two rivers of Venezuela, where they found that magnitude of changes of the density of the macroinvertebrates was lower in the stream with the less pronounced changes in the hydrological regime. Flooding them must be one of the major causes of mortality in tropical streams and given that rain is also a pseudo-random event it should not be surprising that short life cycles and a fast recolonization rate are common events in tropical systems.

ACKNOWLEDGMENTS

We sincerely thank Habil Veroes, parkguard of the Station La Colonia in Guatopo National Park, Corabel Barrios for field assistance, Manuel Guillermo Rada for his invaluable support during our field work and to the Institute of National Parks for allowing use of its facilities. This study was supported partially by the European Commission (Grant No. CI1*CT94-0100) and the Consejo de Desarrollo Cientifico y Humanístico (Grant No. 03-31.3653/95).

REFERENCES

Corkum L. A. 1989. Patterns of benthic invertebrate assemblages in rivers of northwestern North America. Freshw. Biol. 21: 191-205.

Corkum L. D. 1990. Intrabiome distributional patterns of lotic macroinvertebrate assemblages. Can. J. Fish. Aquat. Sci. 47: 2147-2157.

Corkum L. D. 1992. Spatial distributional patterns of macroinvertebrates along rivers within and among biomes. Hydrobiologia. 239: 101-114.

Cressa C. 1994. Structural changes of the macroinvertebrate community in a tropical river. Verh. Int. Ver. Limnol. 25:1853-1855.

Cressa C. 1998. Community composition and structure of macroinvertebrates of the river Camurí Grande, Venezuela. Verh. Int. Ver. Limnol. 26: 1008-1011.

Cressa C. and C. Senior. 1987. Aspects of the chemistry and hydrology of the Orituco river, Venezuela. Acta Cientif. Venez. 38: 99-105.

Domínguez, E., M. D. Hubbard and W. L. Peters. 1992. Clave para ninfas y adultos de las familias y generos de Ephemeroptera (Insecta) Sudamericanos. Biología Acuática. (16): 1-24.

Dudgeon, D. 1993. The effects of spate-induced, predation and enviromental complexity on macroinvertebrates in a tropical stream. Freshw. Biol. 30: 189-197.

Flecker A. S. and B. Feifarek. 1994. Disturbance and the temporal variability of invertebrate assemblages in two Andean streams. Freshw. Biol. 31: 131-142.

Flowers, R. W. and Pringle C. M. 1995. Yearly fluctuations in the mayfly community of a tropical stream draining lowland pasture in Costa Rica, pp. 131-150. In: L. D. Corkum and J. J. H. Ciborowski (eds.). Current Directions in Research on Ephemeroptera. Canadian Scholar's Press Inc.

Hawkes, H. A. 1975. River zonation and classiffication. In: B. A. Whitton (ed.). River Ecology. Blackwell Scientific Press, Oxford.

Hynes, J. D. 1975. Annual cycles of macro-invertebrates in a river in southern Ghana. Freshw. Biol. 5: 71-83.

Jacobsen, D. and A. Encalada. 1998. The macroinvertebrate fauna of Ecuadorian highland streams in the wet and dry season. Arch. Hydrobiol. 142: 53-70.

Klopfer, P. H. 1959. Environmental determinants of faunal diversity. Amer. Natural. 93: 337-342.

Merritt, R. W. and C. W. Cummins. 1996. An introduction to the aquatic insects of North America. Third Edition. Kendall/Hurt Publishing. USA.

Minshall, G. W., K. W. Cummins, R. C. Petersen, C. E. Cushing, D. A. Bruns, J. R. Sedell and R. L. Vannote. 1985. Developments in stream ecosystem theory. Can. J. Fish. Aquat. Sci. 42: 1045-1055.

Pearson, R. G., L. J. Benson and R. E. W. Smith. 1986. Diversity and abundance of the fauna in Yuccabine Creek, a tropical rainforest stream. In: P. Deckker and W. D. Williams (eds.). Limnology in Australia. 329-342. Melbourne: CSIRO and Dordrecht: Dr. W. Junk Publishers.

Resh, V. H., A. V. Brown, A. P. Covich, M. E. Gurtz, H. W. Li, G. W. Minshall, S. R. Reice, A. L. Sheldon, J. B. Wallace and R. C. Wissmar. 1988. The role of disturbance in stream ecology. J. N. Amer. Benthol. Soc. 7: 433-455.

Roldán, G. 1988. Guía para el estudio de macroinvertebrados acuáticos del Departamento de Antioquia. Fondo para la protección del ambiente. Universidad de Antioquia. Colombia.

Ross H. H. 1963. Stream communities and terrestrial biomes. Arch. Hydrobiol. 59: 235-242.

Stanford, J. A. and J. V. Ward. 1983. Insect species divertsity as a function of enviromental variability and disturbance in stream systems, p. 265-78. In: J. R. Barnes and G. W. Minshall (eds.): Stream Ecology: Application and Testing of General Ecology Theory: 265-278. Plenum Press, New York.

Stout, J. and J. Vandermeer. 1975. Comparison of species richness for stream-inhabiting insects in tropical and mid-latitude streams. Amer. Natural. 109: 263-280.

Vannote, R. L., G. W. Minshall, K. W. Cummins, J. R. Sedell. and C. E. Cushing. 1980. The rivers continuum concept. Can. J. Fish. Aqua. Sci. 37: 130-137.

Vinson, M. R. and C. P. Hawkins. 1996. Effects of sampling area and subsampling procedure on comparisons of taxa richness among streams. J. N. Amer. Benthol. Soc. 15: 392-399.

Wiggins, G. B. and R. J. Mackay. 1978. Some relationships between systematics and trophic ecology in nearctic aquatic insects, with special reference to Trichoptera. University of Toronto Press. Toronto.

Winterbourn, M. J., J. S. Rounick and B. Cowie. 1981. Are New Zeland stream ecosystems really different? New Zeland J. of Marine Freshw. Res. 16: 271-281.

Wolda, H. 1981. Similarity Indices, Sample Size and Diversity. Oecologia. 50: 296-302.

Wolda, H. 1983a. 'Long term' stability of tropical insect populations. Res. Pop. Ecol. Suppl. 3: 112-126.

Wolda, H. 1983b. Diversity, diversity indices and tropical cockroaches. Oecologia. 58: 290-298.

Wright, J. F., P. D. Hiley, D. A. Cooling and A. C. Cameron. 1984. The invertebrate fauna of a small chalk stream in Berkshire, England, and the effect of intermittent flow. Arch. Hydrobiol. 99: 176-199.

ABUNDANCE AND ALTITUDINAL DISTRIBUTION OF EPHEMEROPTERA IN AN ANDEAN-PATAGONEAN RIVER SYSTEM (ARGENTINA)

M. Laura Miserendino and Lino A. Pizzolón

Laboratorio de Ecología Acuática
Sarmiento 849
Universidad Nacional de la Patagonia, Sede Esquel
(9200) Esquel, Chubut, Argentina

ABSTRACT

The distribution patterns of the Ephemeroptera found in a river system of the Patagonian Cordillera (Chubut, Argentina) were analyzed. The system extends along a 1000 m altitudinal gradient and exhibits a marked perturbation in its middle reach owing to the organic sewage discharged by the city of Esquel. Eight ephemeropteran species were identified. Species distributions were determined by altitude, stream order, water temperature, conductivity, total alkalinity, and biochemical oxygen demand. *Metamonius* sp and *Meridialaris chiloeensis* proved to be stenothermic, *M. laminata* characterized the reaches of higher stream order, while *Baetis* sp. and *M. chiloeensis* were the species most affected by organic sewage.

INTRODUCTION

The nymphs of ephemeropterans are one of the most important insect groups within the macrobenthic communities of lotic environments. Most mayfly nymphs consume epilithic algae and fine particulate organic matter (Merritt and Cummins, 1978; Ward, 1992), and also form part of the diet of fishes of temperate zones (Allan, 1995). Thus they are considered one of the chief links within the riverine food web. Ephemeropterans are also among the dominant organisms involved in drift and because of their marked sensitivity to organic pollution, they are widely used as water quality indicators (Rosenberg and Resh, 1993; Allan, 1995).

Investigations of the distribution patterns of ephemeropterans along altitudinal gradients are scarce in Argentina, and have only been carried out in the subtropical zone (Domínguez and Ballesteros Valdez, 1992). Most of the studies on the riverine species of the group in Argentina refer to taxonomic and biogeographical aspects. A few qualitative surveys have dealt

Trends in Research in Ephemeroptera and Plecoptera
Edited by E. Dominguez, Kluwer Academic/Plenum Publishers, 2001

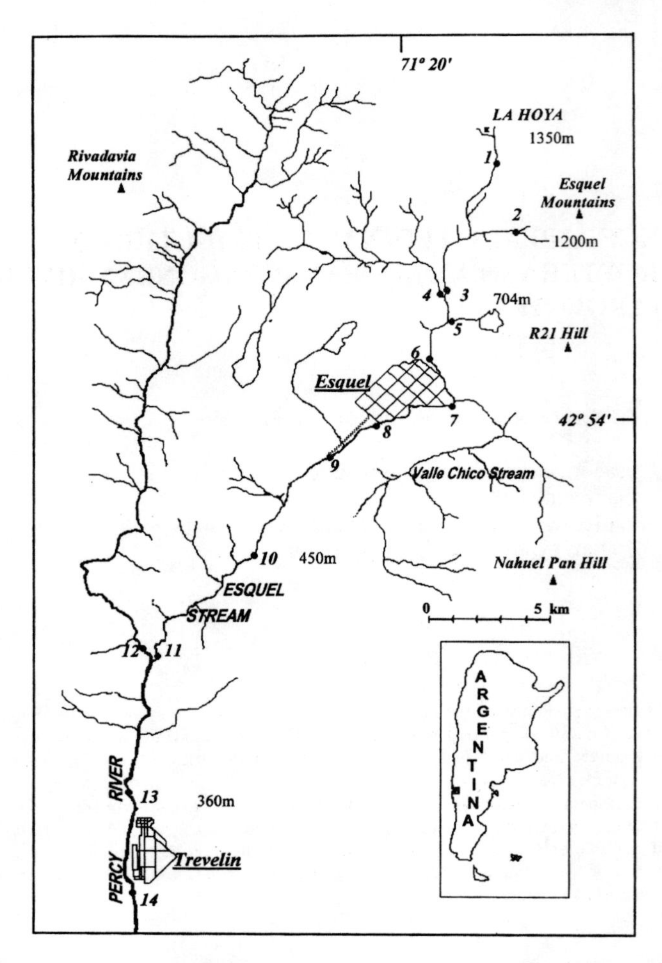

Fig. 1. Location of Esquel-Percy system and study sites. Representative heights are given.

with mountain environments of the Argentinian Patagonia (Wais, 1987, 1990; Wais and Campos, 1984; de Cabo and Wais, 1991). This situation contrasts with the numerous studies on insect zonation along altitudinal gradients carried out in northern hemisphere countries (e.g. Ward, 1986; Ward and Standford, 1991).

The present contribution is aimed at determining the distribution patterns of ephemeropterans along an altitudinal gradient in a lotic system of the Patagonian Cordillera, and exploring the influence of different physical and chemical variables on the distributions of the different species.

STUDY AREA

The system studied is formed by Esquel Stream and Percy River. It is located in the northwestern region of the Province of Chubut, in the Patagonian Cordillera (42°54'S-71°20'W). The system belongs to the catchment area of the rivers Futaleufú-Yelcho, which drains to the Pacific Ocean through the Yelcho River in Chile. The drainage basin of Esquel Stream extends over 349 km^2, whereas the Percy River basin occupies 1093 km^2. There are two periods of maximum water flow, which coincide with winter precipitation and the spring thaw. The annual mean water flow of Esquel Stream ranges from 1 to 2 $m^3 s^{-1}$ and that of Percy River is about 15 $m^3 s^{-1}$ (EVARSA, 1994).

A remarkable west-east pluviometric gradient characterizes this region, located in the eastern fringe of the Andean cordillera, where the strong westerly winds prevail (Jobbágy *et al.*, 1995). The climate is clearly continental, with an annual mean temperature of 8.6°C, a coldest month mean of 2.98°C, and a warmest month mean of 15.6°C. Some stretches of the system are ice-covered during winter, particularly those in the upper reaches.

The native forest bordering the upper stream course are dominated by *Nothofagus pumilio, Austrocedrus chilensis*, and *Maitenus boaria*. In the lower stretches Salicaceae become more abundant, *Salix nigra* being the predominant species.

The upper course of Esquel Stream receives the very high conductivity waters from the shallow lake Willimanco. The stream then crosses the city of Esquel (25,000 inhabitants) and later joins Percy River, to cross the city of Trevelin (5,000 inhabitants) (Fig. 1). Untreated domestic sewage of the city of Esquel is discharged in the middle section of the stream, producing biological consequences that have been discussed elsewhere (Pizzolón *et al.*, 1992; Miserendino and Pizzolón, 1992; Miserendino, 1995).

Table 1. Physical and chemical features of the sampling sites. K20 (conductivity at 20°C), TA (total alkalinity), DO (dissolved oxygen), %DO (percentage oxygen saturation), BOD (biochemical oxygen demand) (Annual mean values n = 8).

Sampling site	Stream order	Water Depth. max-min cm	Veloc. max-min cm s^{-1}	Water temp °C	pH	K20 µS cm^{-1}	TA meq l^{-1}	DO mg l^{-1}	%DO	BOD_5 mg l^{-1}
1	2	40-15	120-67	3.8	7.5	28	0.45	12.91	·116	1.31
2	2	40-11.5	120-43	4.8	7.7	97	1.31	12.94	116	1.35
3	4	40-15	125-68	8.9	7.5	60	0.87	12.08	114	1.11
4	3	10-10	125-36	7.8	7.8	89	1.23	12.74	116	1.84
5	3	10-16	110-50	9.7	8.0	545	2.41	12.10	115	1.79
6	4	10-10	125-60	9.3	7.9	184	1.40	12.73	118	2.11
7	5	60-0	120-60	7.7	7.7	265	1.84	11.09	98	2.79
8	5	40-13	110-60	8	8.1	304	1.89	14.67	136	3.71
9	5	40-15	111-60	8.6	7.9	342	2.49	9.6	78	31.83
10	5	50-35	116-52	10.4	8.1	306	2.23	12.50	116	5.92
11	5	60-26	115-81	8.8	8.0	283	2.18	12.25	110	2.89
12	5	60-39	133-53	7.7	7.8	58	0.85	13.94	121	3.00
13	6	60-39	111-50	6.9	7.8	83	0.97	13.71	116	2.75
14	6	60-40	111-50	7.2	7.9	81	0.96	14.07	120	2.86

METHODS

The study was carried out between November 1990 and October 1991. Samples were taken on a monthly basis during the spring-summer period, and bimonthly during the autumm-winter season. Fourteen monitoring stations, situated between 1350 and 300 m a.s.l., were established along an extension of 51 km of the river system (Fig. 1). Data on water velocity and depth were recorded during low and high water periods. Substratum characterization as described in Ward (1992). Conductivity, pH, dissolved oxygen, biochemical oxygen demand, and total alkalinity were recorded monthly, following APHA recommendations (1978). Air and water temperature were recorded with a mercury thermometer (-10°C–60°C) (Table 1). The organisms were collected with a modified Surber sampler (Winget and Mangum, 1979), integrating eight subsamples. Samples were fixed *in situ* with 4% formaldehyde, and the specimens subsequently preserved in 70° ethanol. Ephemeropterans were identified according to the keys of Domínguez *et al.* (1994). Density and wet weight biomass per square metre, as well as the relative contributions by each species were calculated for every site and date (Kownacki, 1971). Abundance matrices by species, site, and date were then generated. A non-parametric analysis using Spearman's rank correlation coefficients was performed between species and abiotic variables.

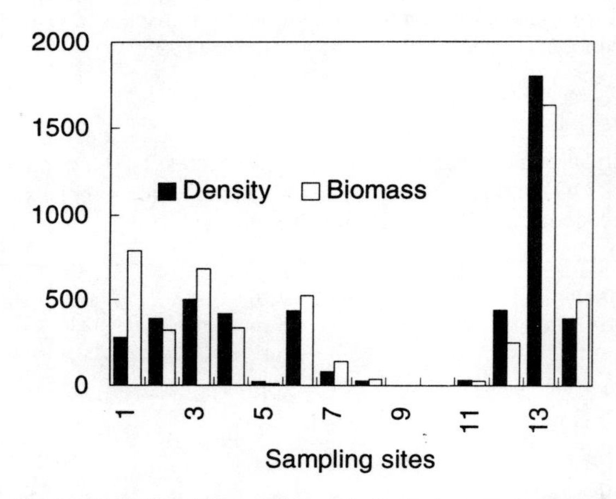

Fig. 2. Annual mean density (ind.m^{-2}) and biomass (mg.m^{-2}) of ephemeropterans in the Esquel-Percy system.

RESULTS

Ephemeropterans were present along the whole system, except for its middle reach. There was a strong decline at station 5, although species richness was high. After an increase in density at station 6, number again decreased from station 7 onwards. No Ephemeroptera were recorded from station 9 and 10 (Table 2, Fig. 2). Annual mean abundance ranged from 0 to 1798 ind.m^{-2} and annual mean wet weight biomass ranged from 0 to 1,627 mg.m^{-2}. Density and biomass both increased towards the lowest zones of the river system. Mayfly abundance was three times greater in station 13 in the Percy river than in the headwater. Station 13 showed two annual maximums of 10,002 and 1,885 ind.m^{-2} and in station 14 one peak of 2,086 ind.m^{-2} was recorded (Fig. 2). Annual mean biomass in station 1 reached 1,000 mg.m^{-2} due to Leptophlebbidae and Siphlonuriidae. In the other sampling sites Leptophlebiidae comprised most biomass values.

Table 2. Abundance ranges of ephemeropterans along the Esquel-Percy system according to Kownacki's Index (Kownacki, 1971). A: dominant (10-100), B: subdominant (1-9.99), C: non-dominant a (0.1-0.99), and D: non-dominant b (0-0.099).

Elevation →	1350	1200		704		600		560		450		400		340
Species: ↓ Sampling Sites →	1	2	3	4	5	6	7	8	9	10	11	12	13	14
Fam. Baetidae														
Baetis sp	A	A	A	A	B	B	C	C			C	B	B	B
Fam. Caenidae														
Caenis sp					C		C							
Fam. Leptophlebiidae														
Meridialaris laminata					D	D	B	D			D	B	A	A
Meridialaris diguilina			B	B	D	A	D	C			D	A	D	C
Meridialaris chiloeensis	A	A	A	A	D	B	C							
Penaphlebia chilensis			D			D		D			B		D	
Fam. Ameletopsiidae														
Chiloporter eatoni		D				D								
Fam. Siphlonuriidae														
Metamonius sp	B	D												
Specific richness	3	4	4	3	5	6	5	4	0	0	4	3	4	3

Eight species belonging to five families were identified. Only six of them were dominant or subdominant throughout the year (Table 2). In spite of being the most widely distributed taxon, *Baetis* sp was not the most abundant mayfly, *Meridialaris laminata* and *M. diguilina* were more numerous than *Baetis* sp. especially in the lower part of the system. *Meridialaris chiloeensis* was recorded only at the upper six sampling stations, whereas *M. diguilina* appeared in almost all the monitoring stations, except in the headwaters. *M. laminata* had very high densities in the Percy River. *Metamonius* sp, *Penaphlebia chilensis, Caenis* sp, and *Chiloporter eatoni* showed low densities and a restricted distribution.

The distribution of *Baetis* sp, *Meridialaris chiloeensis*, and *Metamonius* sp was positively correlated with altitude and negatively correlated with stream order, total alkalinity, and conductivity. The first two species were also negatively correlated with BOD (Table 3). The distribution of *M. laminata* was negatively correlated with altitude and positively correlated with stream order. All correlations were highly significant, but the distribution of *Caenis* sp was significantly, although weakly correlated with conductivity and total alkalinity.

Concerning seasonal factors, the distribution of *M. chiloeensis* and *Metamonius* sp was negatively correlated with water temperature.

DISCUSSION

Recent studies carried out in Patagonian rithral environments showed that ephemeropterans are among the insect groups with highest densities in the macrobenthic community (Miserendino, 1994; Miserendino and Pizzolón, 1996; Miserendino, 1997). Although well represented in the Esquel-Percy system, mayflies were absent both in polluted reaches and in highly conductive waters. The highest specific richness was recorded in the upper stations, although not in the headwaters. In spite of the high diversity recorded in station 5, densities were very low, probably owing to the high conductivity of the waters flowing in from Willi-

Table 3. Spearman's rank correlation coefficients between ephemeropteran species and the abiotic parameters: K20 (conductivity at 20°C), TA (total alkalinity), DO (dissolved oxygen), %DO (percentage oxygen saturation), BOD (biochemical oxygen demand).

	Abiotic parameters								
Species	Altitude	Stream order	Water temp.	pH	K20	TA	DO	%DO	BOD
Baetis sp.	0.42^{E}	-0.42^{E}	ns	ns	0.53^{E}	-0.52^{E}	ns	0.27^{B}	-0.64^{E}
Caenis sp.	ns	ns	ns	ns	0.27^{B}	0.18^{A}	-0.21^{A}	ns	ns
Meridialaris laminata	-0.57^{E}	0.56^{E}	ns	ns	-0.24^{A}	-0.26^{B}	ns	ns	ns
Meridialaris diguilina	ns	ns	ns	ns	ns	ns	ns	0.28^{B}	ns
Meridialaris chiloeensis	0.74^{E}	-0.69^{E}	-0.33^{C}	ns	-0.42^{E}	-0.36^{C}	ns	ns	-0.40^{D}
Metamonius sp.	0.74^{E}	-0.46^{E}	-0.31^{C}	-0.21^{A}	-0.43^{E}	-0.45^{E}	ns	ns	-0.24^{A}

$^{A} = p < 0.05$ $^{B} = p < 0.005$ $^{C} = p < 0.0005$ $^{D} = p < 0.00005$ $^{E} = p < 0.000005$

manco Lake. Mayflies composition presented similarities with different studies in other basins in Patagonia (Wais, 1990), however a major number of species was recorded in the present work. A previous study has mentioned the record of Heptageniidae in Patagonia (Wais and Campos, 1984); nevertheless this family was not included in the list of Ephemeroptera in Argentina (Domínguez *et al.*, 1994) and was not recorded in the Esquel-Percy system.

There has been a great deal of work on the effects of organic pollution on macroinvertebrates communities and mayflies are a frequently used group to asses organic enrichment (Mason, 1991). Studies carried out in New Zealand showed that certain ephemeropterans are even more sensitive as indicator species than plecopterans, mayflies appear to be useful in the detection of organic pollution as they are one of the first groups to disappear and the last to reappear downstream of an effluent outfall (Winterbourn, 1981). However, several mayflies species have been used as indicators of acid stress, rather than organic pollution in northern Europe, and some studies showed that Ephemeroptera was the most susceptible order of Insecta to acidification (Fjellheim and Raddum, 1990).

The abundance of the ephemeropterans found in this Patagonian system were similar to those reported from comparable environments in the northern hemisphere. Another similarity observed was the increase in density values in rivers of higher stream order (Ward, 1986).

The distribution of the different species along the altitudinal gradient was quite marked in this river system. The restricted presence of the siphlonurid, *Metamonius* sp, in the headwaters was expected, since the species belonging to this genus have been cited as dwellers of cold and well-oxygenated waters (Domínguez *et al.*, 1994). This mayfly nymph found in this system, but never recorded in any other basin of the region, is probably *M. fueguiensis*, the only species of this genus recorded for Argentina. It should be taken into account however, that previous studies did not cover altitudinal ranges of the magnitude herein considered (Miserendino, 1997). On the other hand, the only species that could characterize the lower parts of the system would be *Meridialaris laminata*, which was positively correlated with higher stream order. Succession of species belonging to the same functional group is a strategy that tends to minimize interspecific competition (Vannote *et al.*, 1980). The existence of three congeneric species, *Meridialaris laminata*, *M. chiloeensis*, and *M. diguilina*, showing partial spacial overlap in the different stretches of the system, is probably due to their different ecological requirements. All three species probably belong to the same trophic guild (collectors-scrapers), and as such they all exploit fine particulate organic matter, but in different stretches of the system. A comparative study of Chilean and Argentine basins has associated the density of *Meridialaris* with the relative abundance of detritus; furthermore this was the most represented genus all along the continuum (Wais and

Campos, 1984). The increase of *M. laminata* in Percy River may be attributed to the availability of fine particulate organic matter from the Esquel Stream. Sewage may contribute to develop dense communities of epilithic algae, as has been observed on several occasions in the field throughout the year. These three species of Leptophlebiidae are also most common mayflies in other basins of the Patagonian Cordillera (Miserendino, 1997).

It has been suggested that temperature is one of the main factors determining the distribution patterns of ephemeropterans (Ward and Standford, 1982). In this study, only two species showed clear correlation with water temperature: *Metamonius* sp and *M. chiloeensis*. Correlation analysis suggests that the factors determining ephemeropteran distribution in the Esquel-Percy system were both physical and chemical. Altitude and stream order seem to be important for the abundance of *Baetis* sp, *Meridialaris laminata*, and *M. chiloeensis*. However, all three species were negatively correlated with conductivity. *Baetis* sp, which had the widest distribution among the species recorded, showed a strong negative correlation with biochemical oxygen demand. Although baetid species tolerant of organic enrichment have been recorded in Europe and USA (Hynes, 1974; Rosenberg and Resh, 1993), the representatives of this family in the Esquel-Percy system did not tolerate high or even moderate levels of organic enrichment. Considering their diversity and biological importance, this family urgently require systematic revision in Argentina (Domínguez *et al.*, 1994).

Studies on altitudinal zonation in subtropical rivers (Tucumán, Argentina) show that besides ecological factors, zoogeographic factors may also affect species distribution. Thus, the absence of a taxa in a certain mountain system may allow the expansion of another taxa beyond is expected limits (Domínguez and Ballesteros Valez, 1992).

Previous studies corroborate the absence of ephemeropterans in other rivers of the Patagonian Cordillera subject to organic pollution (Pizzolón *et al.*, 1997). Previous observations together with the present results demostrate mayflies value as a water quality indicator, and can be to recommend for the assessment of the effects of organic sewage in other lotic systems of this region.

ACKNOWLEDGMENTS

We wish to thank E. Domínguez and an anonimous reviewer for helpful suggestions on the manuscript. This study was financed by: "Contaminación por efluentes cloacales en el arroyo Esquel y en el río Percy" Dir. Lino Pizzolón. PI N° 118 Consejo de Investigación de la Universidad Nacional de la Patagonia Res. "CS" 143/90.

REFERENCES

Allan, D. 1995. *Stream Ecology*. Structure and function of running waters. Chapman and Hall, London, 388 pp.

American Public Health Association. 1989. Standard Methods for the examination of water and waste water. 17th edn. (APHA) Washington DC, 1550 pp.

de Cabo L and I. Wais. 1991. Macrozoobenthos Prospection in Central Neuquen Streams, Patagonia, Argentina. Verh. Int. Ver. Limnol. Stuttgart. 24: 2091-2094.

Domínguez, E. and J. M. Ballesteros Valdez. 1992. Altitudinal replacement of Ephemeroptera in a subtropical river. Hydrobiologia 246: 83-88.

Domínguez, E., M. D. Hubbard and M. L. Pescador. 1994. Los Ephemeroptera de Argentina. Fauna de Agua Dulce de la República Argentina. 33.(l):1-142

E.V.A.R.S.A. 1994. Estadística hidrológica . Secretaría de Energía. Ministerio de Economía Obras y Servicios públicos de la Nación. Tomo I y II. 651 pp.

Fjellheim, A and G. G. Raddum. 1990. Acid precipitation: biological monitoring of stream and lakes. Science of the total environment 96: 57-66.

Hynes, H. B. 1974. The Biology of the Polluted Waters. University of Toronto Press. 200 pp.

Jobbágy E. G., J. M. Paurelo and R. León. 1995. Estimación del régimen de precipitación a partir de la distancia a la cordillera en el noroeste de la Patagonia. Ecologia Austral. 5 (1): 49-53.
Kownacki, A., 1971. Taxocens of Chironomidae in streams of the Polish High Tatra Mountains. (Str.) Acta Hydrobiol. 13: 439-464.
Mason, C. F. 1991. Biology of Freshwater Pollution. Longman Scientific and Technical. New York. pp 250.
Merrit, R. W. and K. W. Cummins. 1978. An Introduction to the Aquatic Insects of North America. Dubuque, Kendall-Hunt. 441 pp.
Miserendino, M. L. 1994. Estructura de la comunidad de macroinvertebrados del arroyo Esquel. Tankay. 1: 167-169.
Miserendino, M. L. 1995. Composición y distribución del macrozoobentos en un arroyo andino. Ecología Austral. 5 (2). 133-142.
Miserendino M. L. 1997. Composición y estructura de la comunidad de macroinvertebrados de los ríos Azul y Quemqueutreu (Río Negro-Chubut). Resúmenes del "II Congreso Argentino de Limnología". Septiembre de 1997. Buenos Aires. pp 114.
Miserendino, M. L and L. A. Pizzolón. 1992. Un índice biótico de calidad de aguas corrientes para la región andino-patagónica. Resúmenes del Segundo congreso Latinoamericano de Ecología . Caxambú, Minais Gerais, Brasil. pp: 39-40.
Miserendino, M. L. and L. A. Pizzolón. 1996. Bentos y características físicas y químicas de los ambientes lóticos de la cuenca del Futaleufú. Resúmenes de las "III Jornadas Patagónicas de Medio Ambiente". Sede Esquel. Pp 63.
Pizzolón, L., M. L. Miserendino, L. B. Arias and R. Benedetti. 1992. Patrones de perturbación por efluentes cloacales en un sistema lótico andino-patagónico (Chubut: Argentina) Resúmenes del Segundo Congreso Latinoamericano de Ecología. Caxambú, Minais Gerais, Brasil. pp 36-37.
Pizzolón, L., M. L. Miserendino, L. Arias and R. Bennedetti. 1997. Impacto de las descargas cloacales de Cholila sobre el arroyo Las Minas. Ing. Sanitaria y Ambiental. 31: 56-58.
Rosenberg D. M. and V. Resh. 1993. Freshwater Biomonitoring and Benthic Macroinvertebrates. Chapman and Hall. New York-London. 487 pp.
Vannote R. L., Minshall, G. W., Cummins, K. W., Sedell, J. R. and Cushing, C. E., 1980. The river continuum concept. Can. J. Fish. Aquat. Sci. 37: 130-137.
Wais, I. 1987. Macrozoobenthos of Negro River Basin, Argentine, Patagonia. Stud. Neotrop. Fauna Environ. 22: 73-91.
Wais, I. R., 1990. A Checklist of the benthic macroinvertebrates of the Negro River Basin, Patagonia, Argentina, including an approach to their functional feeding groups. Acta Limnol. Brasil. Vol. III. 829-845.
Wais I. R. and Campos, H. H. 1984. The ephemeroptera of creeks and rivers of two South American Basins and its relative presence along the river continuum. Proc. 4th international Conference on Ephemeroptera, Czechosl. Acad. Sci: 229-230.
Ward, J. V. 1986. Altitudinal zonation in a Rocky Mountain stream. Arch. Hydrobiol. Suppl. 74. 2: 133-199. Stuttgart.
Ward, J. V. 1992. Aquatic Insect Ecology. John Wiley and Sons, Inc. 438 pp.
Ward, J. V. and J. A. Standford. 1982. Thermal responses in the evolutionary ecology of aquatic insects. Ann. Rev. Ent. 27: 97-117.
Ward, J. V. and J. A. Standford. 1991. Benthic faunal patterns along the longitudinal gradient of a Rocky Mountain river system. Verh. Int. Ver. Limnol. 24: 3087-3094.
Winget, R. N. and Mangum, F. A., 1979. Aquatic ecosystem inventory. Macroinvertebrates analisys. Biotic condition index: integrated biological, phisical and chemical stream parameters for management. US. Dept. Agriculture - Intermountain Region, Spec. Forest. Serv. Rep., 51 pp.
Winterbourn, M. J. 1981. The use of aquatic invertebrates in studies of stream water quality. Standing Biological Working Party of the Water Resources Council. N. Z. Water Soil Pub. 22: 5-16.

EPHEMEROPTERA: ABUNDANCE AND DISTRIBUTION IN REGULATED STREAMS (SAN LUIS, ARGENTINA)

Ana I. Medina and E. Adriana Vallania

Area de Zoología. Fac. de Qca. Bioqca. y Fcia. U.N.S.L.
Chacabuco y Pedernera
(5700) San Luis, Argentina

ABSTRACT

The faunistic composition of the mayflies community was analized in Potrero y Chorrillo streams at 7 sampling sites. Six taxa which belong to three families were collected: *Tricorythodes popayanicus* Dominguez, *Leptohyphes* sp., *Baetis* sp., *Baetodes* sp., *Dactylobaetis* sp. and *Caenis* sp.. The cluster analysis among the study sites showed two well defined groups, the first one grouped the sites above dams and the second below dams. In the below dam sites a decrease of the species richness was observed. *Baetis* sp., was the only species that was present in the whole fluvial channel studied and it dominated downstream of impoundments. *Tricorythodes popayanicus* was dominant in the rest of the locations, except in one tributary, in which *Caenis* sp. predominated. *Dactylobaetis* sp. was completely eliminated downstream of the impoundment. Other species were also affected to varying degrees. The importance of the Ephemeroptera species varied along the longitudinal profile of the river system, mainly related to the effects of dams.

INTRODUCTION

The fluvial systems of the arid and semi-arid basins, like San Luis province's systems, are subject to natural variation (droughts and floods), causing an irregular flow regime. The water flow depends both on the pattern of rains, the precipitation/evaporation balance and groundwater level. The basins of the mountains of San Luis have uniform geological and hydrological features; they are endorrheic, have steep slopes, low flows and a short course, although some channels are permanent while others are temporary. Furthermore, some of the streams have been regulated by one or more dams.

River regulation invariably modifies both flow patterns and discharge, altering the structural and functional attributes of biotic communities (Ward and Stanford, 1987). Mayflies (Ephemeroptera) are a major order of benthic insects in running waters. They form an important link between primary production and the secondary consumers such as fish and

Trends in Research in Ephemeroptera and Plecoptera
Edited by E. Dominguez, Kluwer Academic/Plenum Publishers, 2001

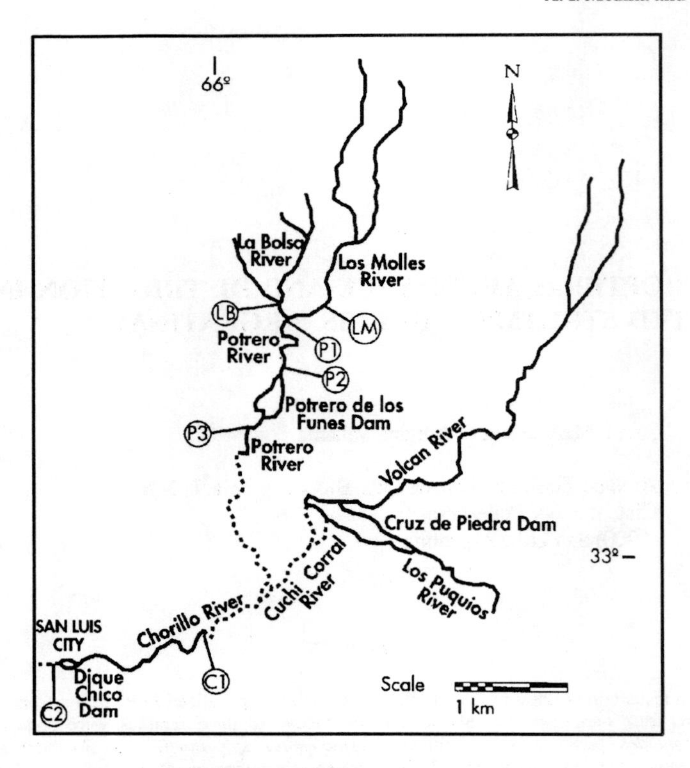

Fig. 1. Sampling sites along the Chorrillo and Potrero Rivers.

their potential as indicator of water quality and river regulation has attracted increasing attention (Brittain and Saltveit, 1989).

The aims of the present study were: a) to determine the distribution of the mayflies in Potrero and Chorrillo streams and b) to determine the effect of impoundment on mayfly composition and abundance.

Study Site

The Potrero river is a third order stream formed by the confluence of two second order streams; Los Molles and La Bolsa (Fig.1). Downstream of the confluence, the river is dammed. Dam releases are scarce and the river flow depends on groundwater infiltration, severely reducing flows. Chorrillo River, is formed by the confluence of the Potrero and Cuchi-Corral Rivers (Gez, 1938). This latter river comes from Cruz de Piedra dam and it is temporary, since it only contains water when the dam overflows. The Chorrillo channel which is temporary and fed by groundwater, runs a short distance above ground before running into Dique Chico Reservoir, used for irrigation. Below Dique Chico the stream flow depends on rainfall and the aquifer level.

MATERIALS AND METHODS

Samplings was carried out during the high and low water periods in 1993, at 7 stations: Los Molles (LM), La Bolsa (LB), Potrero 1(P1), Potrero 2 (P2), Potrero 3 (P3), Chorrillo 1 (C1) and Chorrillo (C2) (Fig. 1). A Surber sampler with mesh of 300 mm and 0,09 m^2 in area was

Table 1. Hydrological and physical - chemical characteristics of the sampling sites in the Chorrillo and Potrero Rivers

Sampling sites	LM	LB	P1	P2	P3	C1	C2
Wetted mean width (m)	6	2.8	9.63	8.5	2.66	7.41	9.76
Mean depth (m)	0.32	0.3	0.23	0.25	0.17	0.31	0.27
Distance from source (km)	8.3	4.4	9.3	10	13	23.1	25.8
Stream order	2	2	3	3	3	3	3
Altitude (m a.s.l.)	1000	1000	980	940	920	810	780
Dominant Substrate *	1	1	1	2	3	2	4
Temperature (°C)	18.5	19	18.5	17	16.5	18.5	14.5
pH	7.8	7.8	8.3	9	7.7	7.8	7.7

Dominant Substrate *: 1- Bedrock, coarse and median sand 2- Gravel, coarse and median sand 3- Bedrock, gravel, median sand and mud 4- Median and fine sand and mud.

used. Eight replicates were taken and integrated into single unit. A physicochemical water characterization was made at the sampling sites.

The organisms were identified using Domínguez et al. (1995). Each species was assigned to functional feeding group according to Merrit and Cummins (1984).

A cluster analysis was made using the Pearson correlation coefficient (median linkage method), in order to rank the similarity among the sampling sites (Fig. 2). The species distribution for each location was represented in a rank/abundance diagram. Species whose numbers were less than 5% of the most abundant were considered low density species. Those between 5-50% were considered intermediate, while those > 50% were considered high density species.

RESULTS

The hydrological, physical and chemical features are summarized in Table 1. The pH values always remain in alkaline range.

Mayflies were represented by six taxa: *Tricorythodes popayanicus* Domínguez, *Leptohyphes* sp., *Baetis* sp., *Baetodes* sp., *Dactylobaetis* sp. and *Caenis* sp. *T. popayanicus* with 8386 ind. m^2 at site P1 had the highest mean density while *Baetodes* sp. at C2 had the lowest with 24 ind. m^2. A decrease in species richness was observed at sites P3 and C2 (Table 2).

The cluster analysis (Fig. 2) showed a group by the sites LM, P1 and P2, to which C1 was associated with a lesser degree of similarity; and a second group formed by the sites P3 and C2. Station LB was clearly different from the others.

The rank/abundance diagram indicated that the mayfly community varied along the river, although the proportions of abundant, intermediate and rare species was different. *Baetis* sp., *T. popayanicus* and *Leptohyphes* sp. were present at all sites, but *Baetis* sp. dominated at the two stations below the dams (P3, C2) while *T. popayanicus* and *Leptohyphes* sp. were absent at P3. Codominance was observed among the following species: *Caenis* sp. (LB), *Baetodes* sp. (LB), *T. popayanicus* (LB, C1) and *Baetis* sp. (C1). *T. popayanicus* dominated at the other locations (LM, P1, P2) (Fig. 3).

Table 2. Mean density (ind.m^2), richness and percentage ocurrence. FG: functional group. CG: collector-gatherers and SCR: scrapers at sites in the Potrero and Chorrillo Rivers.

	Site						
Taxa/ FG	LM	LB	P1	P2	P3	C1	C2
T. popayanicus CG	6625	1313	8326	5711	0	3116	1430
Leptohyphes sp. CG	139	178	755	78	0	830	880
Baetis sp. CG/SCR	1055	35	157	35	104	1915	4460
Baetodes sp. SCR	542	836	781	0	0	24	0
Dactylobaetis sp. CG	94	0	521	69	0	0	0
Caenis sp. CG/SCR	209	1547	78	36	0	1551	0
Taxa Richness	6	5	6	5	1	5	3
Abundance %	1: >50% 2: 5-50% 3: < 5%	3: >50% 1: 5-50% 1: < 5%	1: >50% 3: 5-50% 2: < 5%	1: >50% 4: < 5%	1: >50%	2: >50% 2: 5-50% 1: < 5%	1: >50% 2: 5-50%

DISCUSSION

The non-impacted sites had higher species richness and grouped together. The sites below the dams have lower species richness. Site C1 could be considered as the beginning of a new more varied community, with an increase in species richness, not only among mayflies but also other macroinvertebrates (Medina et al, 1997).

The characteristic responses of the zoobenthos in regulated streams is a reduction in species diversity, alterations of community composition and feeding guilds. There is also often an increase (constant flow) or decrease (fluctuating flow) in abundance. Many species of benthic invertebrates are eliminated or reduced in numbers in regulated streams, but a few taxa flourish under the altered environmental conditions. These are sometimes present in lower numbers prior to impoundment (Ward and Stanford, 1987). According to Corigliano (1994), the studied mayfly communities showed a decrease in the species richness below dams. While *Dactylobaetis* sp. was eliminated, *Baetis* sp. showed an increase in their abundance. There were no changes in the trophic structure since collectors gatherers were dominant throughout the whole river system (Table 2).

Ward (1976) recognized four major types of flow regime in regulated rivers: reduced flow, seasonal flow constancy, increased flow and short-term flow fluctuations. Discharges from the present study reservoirs do not fit into with any of these categories, since they are interrupted and the downstream flow depends solely on groundwaters infiltration. However, our results are similar to the conclusions of Brittain and Saltveit (1989), for the stabilised flow conditions.

The two species of Tricorythidae persist after second impoundment. The occurrence of *Tricorythodes*, often a dominant mayfly genus, below North American dams has been related

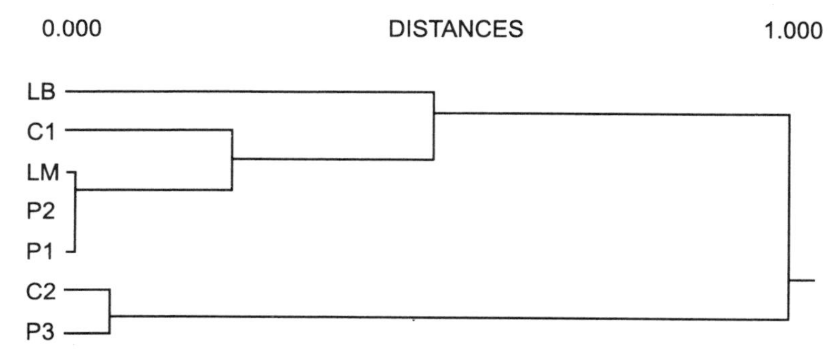

Fig. 2. Similarity cluster among sampling sites in the Potrero and Chorrillo Rivers.

to the distribution of well–developed submerged macrophytes (Ward, 1976; Ward and Short, 1978). Neverttheless, below a Córdoba (Argentine) impoundment, Corigliano (1994) found that *Tricorythodes* was eliminated. Brittain and Saltveit (1989), also mention other facets of this genus that are advantageous in impacted situations, like life cycle flexibility. Vallania and Corigliano (1990) reported that *T. popayanicus* was flexible in its voltinism, with two or more generations per year in the Chorrillo River.

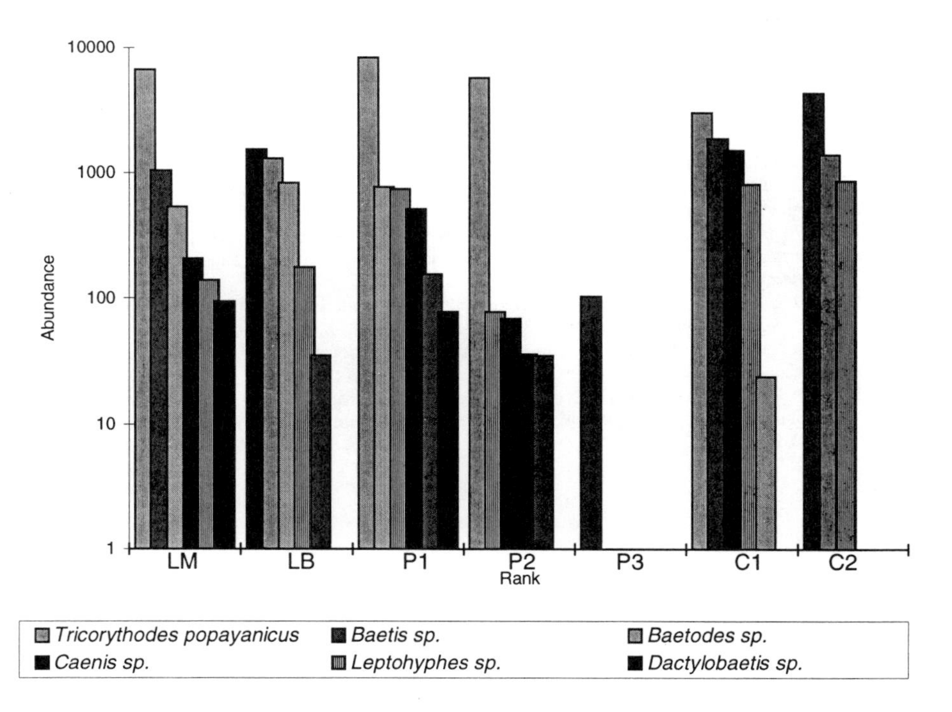

Fig. 3. Abundance of myflies in Potrero and Chorrillo Rivers for each location.

In a regulated river, an altered ecosystem, the species which are unable to adapt to the new environment conditions are eliminated, but generalists and opportunists species tend to dominate (Ward, 1976; Ward and Stanford, 1982).

The mayfly community may be greatly changed in terms of species composition and diversity as a result of river regulation (Brittain and Saltveit ,1989) Interrupted discharges for very long periods, low flows and the development of filamentous algae and macrophytes, contribute to the sucess of *Baetis* sp. in regulated rivers in semiarid zones. However, *T. popayanicus* is also able to adapt to new environmental conditions, which can probably be explained by the characteristics of its life cycle. Nevertheless, others factors are also important in determining a species response to man-made environmental disturbance and change (Brittain,1991).

ACKNOWLEDGMENTS

We thank J. Lasko for drawing the map. This study has been financed partially by the Science and Technique Secretary, Fac. de Qca. Bioqca. and Fcia. UNSL.

REFERENCES

Brittain, J. E. 1991. Life history characteristics as a determinant of the response of mayflies an stoneflies to man-made environmental disturbance (Ephemeroptera an Plecoptera), pp. 539-545. In: J. Alba-Tercedor and A. Sanchez-Ortega (eds.). Overview an Strategies of Ephemeroptera an Plecoptera. Ed. Sandhill Crane Press, Gainesville, Fla. USA.

Brittain, J. E. and S. J. Saltveit 1989. A review of the effect of river regulation on mayflies (Ephemeroptera). Regulated Rivers: Res. Manag., 3: 191-204.

Corigliano, M. del C. 1994. El efecto de los embalses sobre la fauna plantónica y bentónica del río Ctalamochita (Tercero), (Córdoba, Argentina). Rev. UNRC, 14 (1): 23-38.

Domínguez, E., M. D. Hubbard and W. L. Peters 1995. Insecta Ephemeroptera, pp. 1069-1075. In: E. C. Lopretto and G. Tell (eds.). Ecosistemas de aguas continentales. Ed. Ediciones Sur, La Plata, Rep. Argentina.

Gez, J. W. 1938. Geografia de la provincia de San Luis. Ed. Peuser, Buenos Aires. 487 pp.

Medina, A. I., E. A. Vallania, E. S. Tripole and P. A. Garelis 1997. Estructura y composición del zoobentos de ríos serranos (San Luis). Ecología Austral, 7: 28-34.

Merrit, R. W. and K. W. Cummins 1984. An Introduction to the Aquatic Insects of North America. 2nd Ed. Kendall-Hunt Publishing Company, Dubuque, Iowa. 722 pp.

Vallania, E. A. and M. Corigliano 1990. La historia de vida de *Tricorythodes popayanicus* Domínguez (Ephemeroptera) en el río Chorrillo (San Luis, Arg.). Rev. UNRC, 9 (2): 125-133.

Ward, J. V. 1976. Effects of flow patterns below large dams on stream benthos: A review, pp. 235-253. In: J. F. Orsborn and C. H. Allman (eds.). Instream flows needs symposium, Amer. Fish Soc. , Bethesda, Maryland. Vol. II.

Ward, J. V. and R. A. Short 1978. Macroinvertebrate community structure of four lotic habitats in Colorado, USA. Verh. Int. Ver. Limnol., 20: 1382-1387.

Ward, J. V. and J. A. Stanford 1982. Thermal responses in the evolutionary ecology of aquatic insects. Ann. Rev. Ent., 27: 97-117.

Ward, J. V. and J. A. Stanford 1987. The ecology the regulated streams: Past accomplishments and directions for future research, pp. 391-409. In: J. F. Craig and J. B. Kemper (eds.). Regulated Streams. Plenum, New York.

EPIRHITHRAL COMMUNITIES OF MAYFLIES (EPHEMEROPTERA) OF THE ODRA RIVER BASIN (CZECH REPUBLIC)

A. Mergl

Department of Zoology and Ecology, Faculty of Science
Masaryk University, Kotlářská 2
611 37 Brno, Czech Republic

ABSTRACT

Eleven localities with minimal anthropic impact were sampled in the Odra River basin in 1995-1996. The main aim was (i) to define taxocoenoses of mayflies within the epirhithral benthic communities (ii) to compare data obtained recently with those from 1955-1960 in order to define long-term changes in species composition and biodiversity. The programmes and methods of TWINSPAN (Two Way INdicator SPecies ANalysis) and DECORANA (DEtrended CORrespondence ANAlysis) were used to define mayfly taxocoenoses. *Rhithrogena iridina, Baetis alpinus* and *Epeorus sylvicola* were found to be the most important indicators of mayfly communities in this area. Long-term environmental changes are discussed, no pronounced tendencies to deterioration within the area studied are apparent.

INTRODUCTION

Several authors (Zelinka, .1953; Losos and Marvan, 1957; Obrdlík, 1979, 1981; Simanov and Kantorek, 1987; Landa and Soldán, 1989) dealt with surface water quality in the Odra River basin and several faunistic studies of localities of this area (Kolenati, 1859; Tomaszewski, 1932; Zelinka, 1950, 1951, 1953; Tuša, 1974a, 1974b; Zelinka, 1977, 1979) were published. During 1950-1965 the Institute of Entomology of the Academy of Sciences of the Czech Republic together with the Dept. of Zoology and Ecology of Masaryk University organized a large-scale faunistic programme of research of aquatic insect distribution (e.g. so-called Research Project No. 210). Three aquatic insect orders important from the biomonitoring point of view (i.e. Ephemeroptera, Plecoptera and Trichoptera) were used as models. Landa and Soldán (1989), Soldán et al. (1998) defined long-term changes of mayfly taxocoenoses of epirhithral localities situated in this area on the basis of samples taken in 1950-1965, 1970-1985 and 1994-1996.

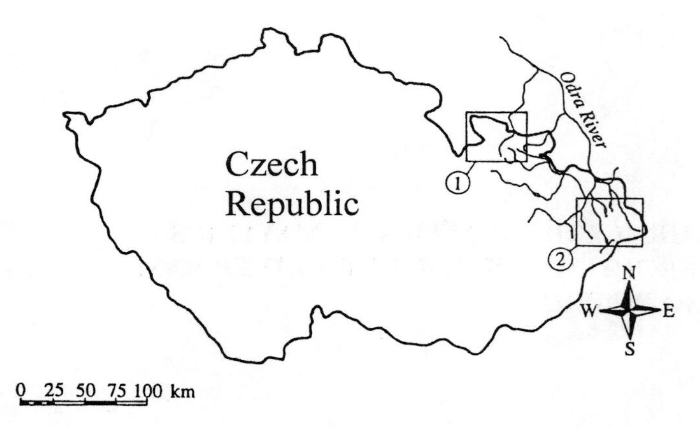

Fig. 1. Location of study areas – the Jeseníky (1) and Beskydy Mts. (2)

A new classification method of rivers was established in Great Britain (Wright et al., 1989). This method – RIVPACS (River InVvertebrate Prediction And Classification System) – is using the concept of natural reference localities, which are predicting for specific places on the base of recent levels of natural abiotic factors. Actually present macroinvertebrate community is compared to community predicted.

The main aim of this study is to define the taxocoenoses of mayflies in epirhithral benthic communities and to compare the data obtained with those from 1955 – 1960 in order to define long-term changes in species composition and biodiversity.

STUDY AREA AND METHODS

The samples were taken at 11 localities of the Odra River basin (the Baltic Sea drainage area) in four seasons in 1995 (July, October) to 1996 (March, June). Only undisturbed localities were selected. Study areas were situated in two geographically different parts of the Odra River basin: the Jeseníky and Beskydy Mts. (Fig. 1). The localities were situated in the upper reaches (epirhithral). The Jeseníky Mts. belong to the Hercynian subprovince and are mainly formed from mica schist, gneisses, phyllites – acide rocks, poor in nutriments. Feeding of rivers of this area with cool water during the year has considerable influence on cool stenotherm benthic fauna composition. Average annual temperature is 7.1 °C and annual total of precipitation is 846 mm (Mt. Jeseník). Six localities (L1,.., L6) at altitudes of 350 – 770 m above sea level were studied.

The Beskydy Mts. belong to the Carpathian subprovince. Flysh sandstone and claystone represent prevailing types of substratum in this area. Erosion of these rocks results in the abundant occurrence of flat stones. Air temperature conditions are similar to the Jeseníky Mts. Average annual temperature is 7.4 °C (Frenštát) but average precipitations are higher than in the Jeseníky Mts. Annual total precipitations exceed 1000 mm in the whole area (Culek, 1996). Five localities (L7,..., L11) at the altitudes of 540 – 720 m above sea level were studied here. Localities were sampled semiquantitively. Sampling was limited to 5 minutes at each locality in order to compare species composition and numbers of speciemens more correctly at all localities. All samples were taken with a hydrobiological net using the kicking-technique (Lillehammer, 1974). Samples were taken from all types of habitats. Species representatives were collected from each macrozoobenthos sample and the rest were counted. The samples were immediately fixed with 4 % formaldehyde. Imagines were fixed with 70 % alcohol. Characteristics of bottom and banks, stream width and depth, and current speed were

Table 1. List of mayfly taxa collected in the Jeseníky and Beskydy Mts.

Taxon	Studied sites										
	The Jeseníky Mts.						The Beskydy Mts.				
	L1	L2	L3	L4	L5	L6	L7	L8	L9	L10	L11
Rallidentidae											
Ameletus inopinatus	•										•
Baetidae											
Alainites muticus	•	•		•	•	•	•	•	•	•	•
Baetis alpinus	•	•			•	•	•	•	•	•	•
Baetis buceratus			•								
Baetis lutheri						•					
Baetis melanonyx						•	•	•		•	
Baetis rhodani	•	•	•	•	•	•	•	•	•	•	•
Baetis vernus	•		•					•			
Baetis sp. juveniles	•	•	•	•				•	•		
Centroptilum luteolum			•								
Heptageniidae											
Epeorus sylvicola	•	•		•	•	•	•	•	•	•	•
Rhithrogena carpatoalpina	•	•		•	•	•		•	•	•	•
Rhithrogena iridina	•		•	•		•	•	•	•	•	•
Rhithrogena semicolorata	•	•				•		•	•		
Rhithrogena sp. juveniles	•							•	•	•	
Ecdyonurus aurantiacus			•								
Ecdyonurus starmachi									•		
Ecdyonurus subalpinus	•	•					•	•		•	
Ecdyonurus submontanus	•						•				
Ecdyonurus venosus	•		•	•	•	•	•	•	•	•	•
Ecdyonurus sp. juveniles	•						•	•	•		
Electrogena quadrilineata	•						•	•	•		•
Leptophlebiidae											
Habroleptoides confusa	•	•	•	•	•	•	•	•	•	•	•
Habrophlebia lauta				•				•	•		
Ephemeridae											
Ephemera danica			•								
Ephemerellidae											
Ephemerella ignita		•	•	•		•		•	•		
Ephemerella mucronata	•	•	•	•	•	•		•		•	•
Torleya major						•					
No. of species	15	10	11	10	8	14	11	16	13	11	11

measured at each locality investigated. Temperature, pH, DO (dissolved oxygen), and conductivity were measured by HORIBA U-10 multimeter. General classification of surface sediment were estimated according to current scales (boulders, cobbles, coarse gravel, gravel, sand and silt). Determination of mayflies and analysis of water chemistry were conducted in the laboratory. Samples of water were taken three times (point sample) and total organic carbon, nitrates, nitrites, total phosphorus, and total alcalinity were measured.

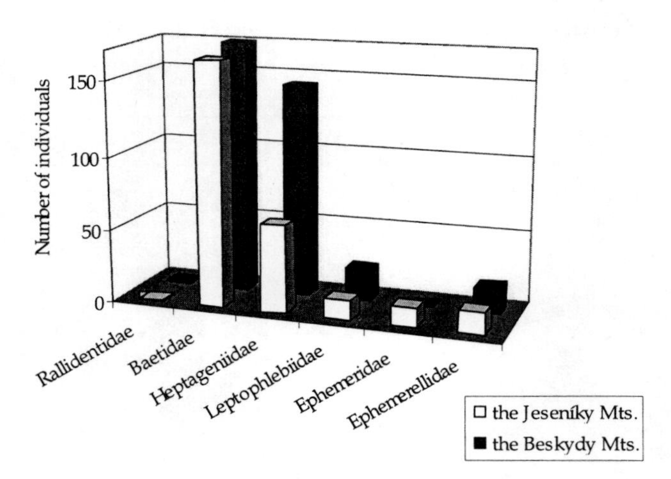

Fig. 2. Average abundance per locality.

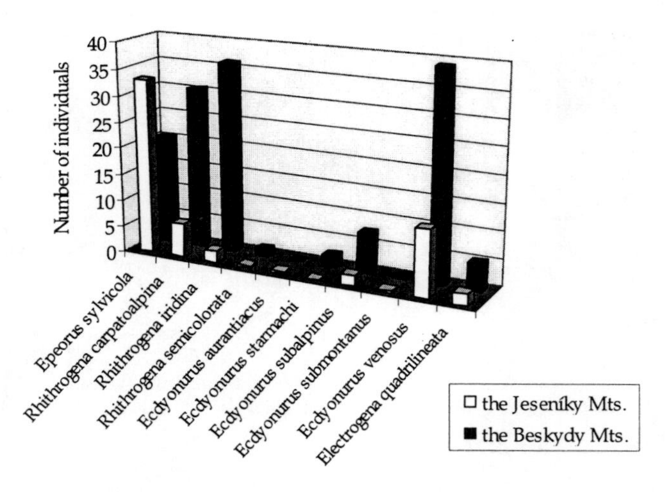

Fig. 3. Average abundance of family Heptageniidae per locality.

Estimation dominancy of individual species and diversity were estimated according to current diversity indices (Margalef, Shannon–Weaver) (cf. Washington, 1984). The saprobial indices were calculated according to Czech State Norm No. 83 05 32 (Sládeček et al., 1981). The programmes and methods of TWINSPAN (Hill, 1979a) and DECORANA (Hill, 1979b) were used to define mayfly taxocoenoses and indicator species.

RESULTS AND DISCUSSION

At the localities studied, 25 mayflies species were found, most of them belonging to the Heptageniidae (10 species) (Tab. 1). The highest species richness was found at the site

Table 2. The parameters of analysed localities which were used in DCA analysis. (The result of Monte Carlo permuted test)

No.	Locality		Altitude	Distance from source (km)	Max. measured temperature of water (°C)	Average conductivity (μS.cm^{-1})
L1	Ramzovský brook	Studied sites	590	5.1	12.2	87
L2	Stříbrný brook		750	6.8	11.6	110
L3	Černý brook		350	1.7	14.9	198
L4	Červený brook		400	2.5	14.3	89
L5	Střední Opava		770	2.9	13.8	60
L6	Černá Opava		610	9.9	12.8	83
L7	Trojanovický brook		600	1.5	10.9	80
L8	Mohelnice		540	1.1	10.2	84
L9	Morávka		707	2.0	11.7	406
L10	Ropičanka		670	2.8	10.3	76
L11	Tyrka		720	1.5	10.2	90
12	Svratka	Added sites	490	13.6	14.8	148
13	Sitka		320	19.0	16.0	197
14	Branná		440	17.1	12.3	150
15	Jezerní brook		630	1.5	14.0	162
16	Trnava		570	2.1	13.0	518
17	Litava		320	3.0	15.0	507
18	Fryšávka		690	5.4	13.0	133
19	Kyjovka		425	2.5	13.0	384
20	Opava		350	38.1	15.9	152
21	Jihlava		185	79.6	23.1	300

Mohelnice - L8 (16 species) in the Beskydy Mts. and at the locality Ramzovský brook - L1 (15 species) in the Jeseníky Mts. *Baetis* species were the most abundant, mainly *Baetis alpinus* and *B. rhodani. B. rhodani* was eudominant at 9 sites and *B. alpinus* at 8 sites. *Alainites muticus* showed the highest dominancy during the second sampling period (up to 87 %). The highest values of dominancy were recorded in the case of *Ephemera danica* at Černý brook – L3 (up to 85 %) where a lot of patches of sand substrate occurred. Species of the genera *Baetis* and *Rhithrogena* predominated between eudominant species in the Beskydy Mts. (Figs 2, 3). A similar situation was observed in the Jeseníky Mts. although no species of the genus *Rhithrogena* belonged to eudominant species here. In the Beskydy Mts., the species of the genus *Rhithrogena* were frequently found due to different substrate roughness (below flat stones). This particular microhabitat is typical for larvae of this genus. Species composition seems to be affected also by other factors, namely glaciation and presence of Carpathian elements. Effects of the last continental glaciation reaching to the Moravian Gate is recognizable here (Soldán, pers. comm.) while the Beskydy Mts. were not influenced. We can see this phenomenon on the example of *Ecdyonurus starmachi*, which was found in the Beskydy Mts. but not in the Jeseníky Mts. It represents a typical Carpathian species.

Diversity index was determined for each sample. Decrease of diversity was detected during autumn (sample 2) and in spring (sample 4) in some cases. It is caused by the end of emergence of some species in the first case and, on other hand, by occurrence of a large quantity of juvenile stages – especially *Baetis* spp. and *Ecdyonurus* spp. accumulated into a single taxon (*Baetis* sp. juveniles or *Ecdyonurus* sp. juveniles, respectively). In the second case, the diversity decrease was influenced by spring snow melting and drift loses following increase of discharge.

Saprobial indices show very good water quality. Xeno- or oligosaprobity was determined at all types of sites, only Černý brook had ß-mezosaprobity character. This site generally differed from the others.

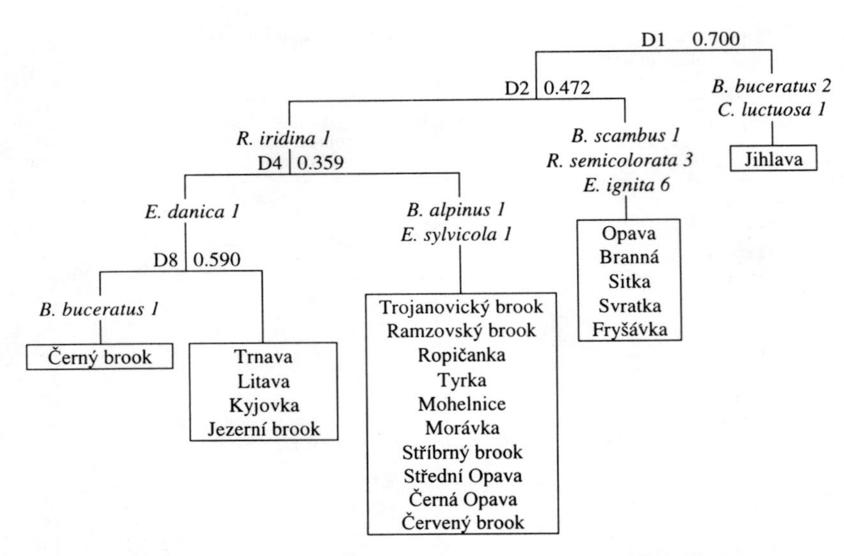

Fig. 4. Divisive hierarchic classification (TWINSPAN) with depicted indicator species. Separated groups of localities are marked out in five boxes.
[1]No. of division, [2]eigenvalue, [3]pseudospecies level.

Comparison of these results with data from 1955 – 1960, when the large-scale faunistic research of aquatic insect was organised, shows long-term environmental changes not very pronounced at studied sites. There were a lot of taxonomical shifts during the last 40 years, especially into the genera *Rhithrogena, Baetis* and *Ecdyonurus.* Several new species were recently defined. Consequently, more species than ascertained by the Research Project No. 210 are now recorded from the area studied. Only *Rhithrogena hybrida* was not found, but this species evidently occur in other parts of the Jeseníky Mts. (Landa, 1989).

For precision of method TWINSPAN analogical data from the Odra and Morava Rivers basins were added (Tab. 2). Analysis supported considerable similarity of the 10 localities investigated. The first division (D1) – Jihlava is an epipothamal lowland river with *Baetis buceratus* and *Caenis luctuosa* as indicator species. During the second division (D2) the group of large rivers from the Morava, Odra and Dyje Rivers basins was divided according to *Baetis scambus, Rhithrogena semicolorata* and *Ephemerella ignita* as indicator species. The fourth division (D4) divided the group of sandy and slowly flowing rivers (including site Černý brook - L3) according to *Ephemera danica* as indicator species from group of epirhithral localities. Indicator species for the group of epirhithral localities were *Rhithrogena iridina, Baetis alpinus* and *Epeorus sylvicola.* This is the result of divisive hierarchic classification (Fig. 4). *Baetis alpinus, Baetis vernus* and *Alainites muticus* were determined to be eudominant and euconstant species of this community. *Baetis lutheri, Ecdyonurus starmachi* and *Torleya major* were found out to be specific species, collected only at one locality.

Detrended correspondence analysis was used for considering the influence of environmental characteristics on mayfly communities. Significance of environmental variables was tested by Monte Carlo permuted test (Ter Braak, 1991). Altitude (alt), conductivity (cond), distance from source (kmsource) and the maximum annual water temperature (tmax) were found to be important (Fig. 5). The most expressive positive correlation was found between the maximum annual temperature of water and distance from source. On other hand, the maximum negative correlation was found between altitude and the maximum annual water tem-

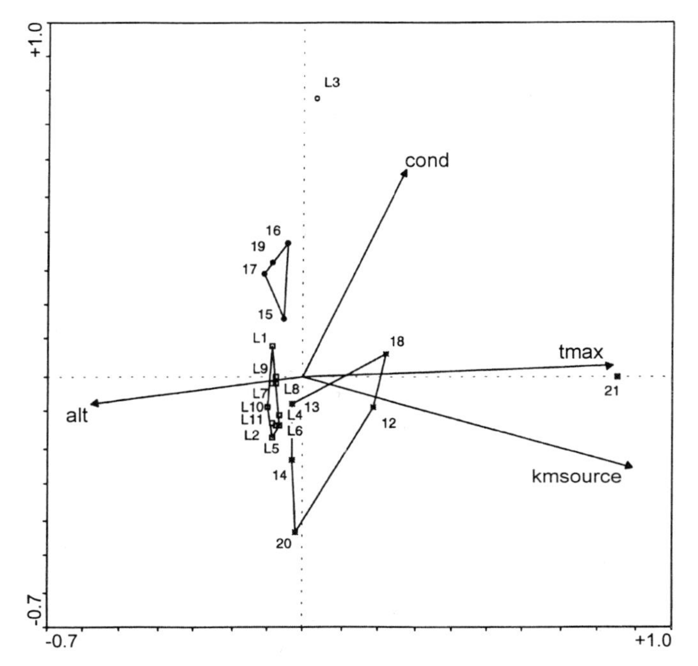

Fig. 5. Detrended correspondence analysis (DCA), ordination plot of localities and environmental variables on the first two DCA axes.

perature (Fig. 5). Low correlation was found between conductivity and other environmental factors. The most expressive correlation was found between altitude and occurrence of species of epirhithral community. The results of this study represent basic data for definition of characteristic communities of macroinvertebrates at anthropically undisturbed river segments and will be included into recently constructed predicting models analogical to the British RIVPAC system which are planned to be used at numerous localities within the whole Czech Republic.

ACKNOWLEDGMENTS

My sincere thank are due to Dr. S. Zahrádková for valuable help during determination of larvae and, statistical processing of data, encouragement and comments and suggestions. I am grateful to Dr. T. Soldán for comments on the manuscript.

REFERENCES

Culek, M. (ed.). 1996. Biogeographical classification of the Czech Republic. Enigma, Praha, 347 pp. (in Czech, English summary).

Hill, M. O. 1979a. TWINSPAN - A FORTRAN program for arranging multivariate data in an ordered two-way table by classification of the individuals and attributes. Cornell University Press, Ithaca, N. York, 99 pp.

Hill, M. O. 1979b. DECORANA - A FORTRAN program for detrended correspondence analysis and reciprocal overaging. Cornell University Press, Ithaca, N. York, 52 pp.

Kolenati, F. A. 1859. Fauna des Altvaters (hohen Gesenkes der Sudeten). Jahersheft der naturwissenschaftlichen Section der k.k.mähr. schles. Gessellschaft für Ackerbau, Natur- und Landeskunde, 83 pp.

Landa, V. and T. Soldán . 1989. Distribution of mayflies (Ephemeroptera) in Czechoslovakia and its changes in connection with water quality changes in the Elbe basin. Studie CSAV, 17, Academia, Praha, 172 pp. (in Czech, English summary).

Lillehammer, A. 1974. Norwegian stoneflies. Norsk. ent. Tidss. 21: 195-250.

Losos, B. and P. Marvan. 1957. Hydrological ćonditions of the Moravice River and its tributaries the Podolský and Černý stream. Sbor. Vys. šk. zeměd. a lesn. fak. v Brně, A 4: 41-69. (in Czech).

Obrdlík, P. 1979. Rheobenthos and water quality of the Stříbrný stream in the Rychlebské Mts. Čas. Slez. Muz. Opava, A 28: 69-75. (in Czech).

Obrdlík, P. 1981. To the knowledge of hydrobiology of the Borový and Šumný stream from the quality of water poin of view. Čas. Slez. Muz. Opava, A 30: 89-95. (in Czech).

Research Project No. 210. (1955-1956). Katedra zoologie a antropologie Univerzity Brno, unpublished. (in Czech).

Simanov, L. and J. Kantorek. 1987. The biological quality of the water of the Czechoslovak part of the Odra River. Dept. of Biology, Pedagogical Faculty Ostrava, Acta hydrochim. hydrobiol. (15) 3: 263-274.

Sládeček, V., M. Zelinka, J. Rothchein, and V. Moravcová. 1981. Biological analysis of surface water. Czech State Norm. Office for Normalisation and Measurement. Prague, 186 pp. (in Czech).

Soldán, T., S. Zahrádková, J. Helešic, L. Dušek and V. Landa. 1998. Distributional and quantitative patterns of Ephemeroptera and Plecoptera in the Czech Republic: A possibility of long - term environmental changes of aquatic biotopes. Folia Fac. Sci. Nat. Univ. Masarykianae Brunensis, Biologia 98: 1-305.

Ter Braak, C. J. F. 1991. CANOCO version 3.12. Agricultural mathematics group, Staringebow, P. O. Box 100, 6700 AC Wageningen, The Netherlands, 93 pp.

Tomaszewski, W. 1932. Beitrag zur Kenntnis der Tierwelt Schlesischer Bergbäche. Abh. naturf. Ges. Görlitz 31: 1-80.

Tuša, I. 1974a. Mayfly larvae (Ephemeroptera) in current habitat of Bělá creek (the northwestern part of Moravia, Czechoslovakia). Acta Hydrobiol., Kraków 15: 311-320.

Tuša, I. 1974b. Mayfly larvae (Ephemeroptera) in current habitats of three trout streems with stony bottom (the northwestern part of Moravia, Czechoslovakia). Acta Hydrobiol., Kraków 16: 417-429.

Washington, H. G. 1984. Diversity, biotic and similarity indices. A review with special relevance to aquatic ecosystems. Water Res. 18: 653-694.

Wright, J. F., P. D. Armitage, M. T. Furse, and D. Moss. 1989. Prediction of invertebrate communities using stream measurements. Regulated rivers, Res. Manag. 4: 147-155.

Zelinka, M. 1950. To the knowledge of fauna of streams of the Slezské Beskydy Mts. Zvl. příl. Přírodov. sbor. ostrav. kraje 11: 3-28. (in Czech).

Zelinka, M. 1951. A contribution to knowledge of fauna of the Bílá Opava River. Sbor. Klubu přírodov. Brno 29: 201-205. (in Czech).

Zelinka, M. 1953. Larvae of mayflies (Ephemeroptera) of the Moravice River and their relationships to water quality. Práce Moravskoslez. akad. přír. věd. 25: 181-200. (in Czech).

Zelinka, M. 1977. The production of Ephemeroptera in running waters. Hydrobiologia 56: 121-125.

Zelinka, M. 1979. Differences in the production of mayfly larvae in partial habitats of a barbel stream. Arch. Hydrobiol. 90: 284-297.

ATOPOPHLEBIA FORTUNENSIS FLOWERS (EPHEMEROPTERA: LEPTOPHLEBIIDAE) AND *CAENIS CHAMIE*, ALBA-TERCEDOR AND MOSQUERA (EPHEMEROPTERA: CAENIDAE). NOTES ON THEIR BIOLOGY AND ECOLOGY

S. Mosquera[1], M. del C. Zúñiga[1], and J. Alba-Tercedor[2]

[1] Universidad del Valle
Departamentos de Biología y Procesos Químicos y Biológicos
Apartado Aéreo 25360, Cali, Colombia
[2] Universidad de Granada
Departamento de Ecología Animal
18071, Granada, España

ABSTRACT

Caenis chamie and *Atopophlebia fortunensis* were collected in a water supply channel of the Pavas aqueduct (Departamento del Valle del Cauca- Cordillera Occidental, Colombia), that comes from natural sources with a run along a stretch with abundant riparian vegetation. *A. fortunensis* is the first record of a species co-ocurring in Central and South America.

The activity of each species was clearly related with timing and intensity of rain periods. Emergence and molting occurs during the day at different hours. Most winged individuals of *A. fortunensis* were observed at the end of rain periods. Observations in the laboratory have shown that subimagos molted to imagos within 24 hours after they had emerged. Time elapsed between emergence and ovoposition of *C. chamie* was very short (5-6 minutes), and the highest emergence peaks took place during cloudy and rainy days. Observations in the laboratory have shown that males emerged first. Females did not undergo an imaginal molt and remained with the submaginal skin.

INTRODUCTION

Atopophlebia fortunensis was described from nymphs and adults collected in western Panama and Costa Rica by Flowers (1987). Zúñiga et al. (1997) reported this species in Colombia, but its biology and ecology there were unknown. *Atopophlebia* is a strictly Neotropical genus with a close morphological relation to the *Thraulodes* complex (Flowers, 1987). Nymphs of *Atopophlebia* have been collected from good quality waters (Mosquera, 1996). According to the Pan-American Ephemeroptera fauna evaluated by McCafferty (1998), with respect to the interchange of species between the Americas, *A. fortunensis* is the first record of a species co-occurring in Central and South America.

Caenis chamie was recently described by Alba-Tercedor and Mosquera (1999). This genus belongs to the family *Caenidae* and was first reported in Colombia in water supply channels to Pavas aqueduct (Zúñiga et al. 1997). The family is widespread, but not well-known in South America. *Caenis* is a cosmopolitan genus of considerable antiquity, possibly of Pangaean origin, and no species are known to be common to Central and South America (McCafferty,1998).

This paper contains some notes about the biology and ecology of *A. fortunensis* and *C. chamie* found in the southwestern Colombia.

METHODOLOGY

Area of Study

Samples of individuals were taken from September 1994 to March 1995 in water supply channels to Pavas aqueduct, which consisted of a lake and two streams: one with a sandy bed, and the other one with a muddy bed and abundant vegetable detritus. These correspond to high quality water bodies that come from natural sources protected by a small forest. This study was conducted in an area located in the Department of Valle del Cauca, Cordillera Occidental (3° 40' 48" northern latitude, 76° 33' western longitude) at 1,275 meters above sea level, in a life zone classified Subtropical Humid Forest (SHF), according to Holdridge (Espinal,1968), with an annual rainfall range of 1,000 – 2,000 mm and an average temperature of 22° C (Figure 1).

Collection Methods and Field and Laboratory Observations

Manual search and aquatic nets were used for collecting live mature nymphs. Animals were then taken to the laboratory and placed under simulated temperature and oxygenation conditions, in two emergence chambers. Malaise and light traps, aerial nets, and searches of surrounding vegetation, were the methods used for collecting subimagos and imagos in the field. Breeding in the laboratory is an unambiguous way of associating nymphs with adults.

RESULTS

Caenis chamie, Alba-Tercedor and Mosquera

This species, recently described from Colombia, has been just collected in all water supply channels to Pavas aqueduct. Nymphs were most frequently found in the muddy bottom stream with abundant vegetable detritus. The average values for some parameters of water quality in this stream were: Temperature 19,50°C; pH 7,52 units and 64 % oxygen saturation.

Field observations: The time that elapsed from emergence to ovoposition was very short. Individuals were seen to emerge sporadically in the field at the beginning of dawn, for approximately one hour, between 6:00 and 7:00 hours. Individuals flew in a slow and irregular pattern. The highest emergence peaks occurred during rainy periods, in cloudy, rainy and shadowed areas. It is believed that this species makes no real swarms. Males do not fly together, but as they emerge, both males and females rapidly complete their growth and mate. Males die shortly after they mate, and females do so shortly after laying their eggs.

In the laboratory, males were the first to emerge, molting to the imago phase within 5 or 6 minutes. Females did not undergo a final molt from subimagos to imagos. This was confirmed by evaluating their emergence in the laboratory, where it was found that only males left a subimaginal skin. Their emergence occurred between 6:30 and 7:30 hours. Edmunds

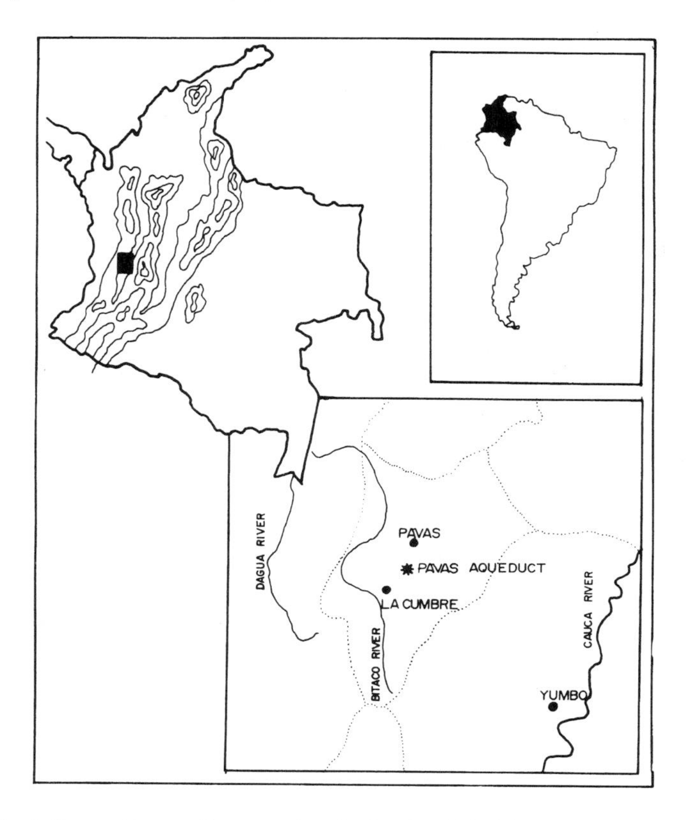

Fig. 1. Geographic area. Pavas aqueduct. La Cumbre. Valle del Cauca. Colombia.

and McCafferty (1988) state that females of some specialized species do not molt from subimagos to adults, and that maturation of their eggs is completed in the final nymphal phase.

Atopophlebia fortunensis, Flowers

This species in Colombia has only been recorded in the sandy-bed stream of Pavas aqueduct, in shadowed places with abundant fallen leaves. The average values for some parameters of water quality to this stream were: Temperature 19,40 ° C; pH 7,70 units and 85 % of oxygen saturation.

While nymphs in younger stages are a grayish brown color, which turns into a brownish orange color as they approach emergence, adults are a lighter orange color. Males differ from females because they have divided compound eyes. This characteristic is visible in the late nymphal stage, and can be useful for both collecting individuals to be raised in the laboratory, and identifying them at a later stage.

Male and female subimagos emerged from 20:00 to 21:00 hours in the laboratory, males emerged first followed by the females; they both molted to imagos 24 hours later. In the field, however, only one subimago was observed as it emerged at 20:15 hours on a clear night. The largest number of subimagos and imagos was observed at the end of a rain period in clear and sunny days, lying under leaves or tree trunks at a height between 1.50 m. and 2.30 m. No swarms or oviposition was observed for this species.

ACKNOWLEDGMENTS

We thank Dr. R. W. Flowers of Florida Agricultural and Mechanical University for the identification of *A. fortunensis*; Martha Rojas of Universidad del Valle for her valuable suggestions and advise. This work is based upon Bioindicators Water Quality Program of the Universidad del Valle and was supported in part by its Research Committee.

REFERENCES

Alba-Tercedor, J. and S. Mosquera. 1999. *Caenis chamie,* a new species from Colombia (Ephemeroptera: Caenidae). Pan-Pac. Ent.75 (2): 61-67.

Edmunds, G. F. (Jr.) and W. P. McCafferty. 1988. The mayfly subimago. Ann. Rev. Ent. 33: 509-529.

Espinal, L. S. 1968. Visión Ecológica del Departamento del Valle del Cauca. Imprenta Universidad del Valle. Cali. 103 p.

Flowers, R. W. 1987. New Species and Life Stages of *Atopophlebia* (Ephemeroptera: Leptophlebiidae: Atalophlebiinae). Aquat. Insects. 9 (4): 203-209.

McCafferty, W. P. 1998. Ephemeroptera and the great American interchange. J. N. Amer. Benthol. Soc. 17 (1): 1-20.

Mosquera, S. 1996. Emergencia, formación de enjambres y distribución de algunas especies de Ephemeroptera del suroccidente colombiano. Trabajo de grado. Universidad del Valle. Departamento de Biología. Cali, Colombia.

Zúñiga de C., M. del C., A. M. Rojas de H. and S. Mosquera de A. 1997. Biological aspects of Ephemeroptera in rivers of southwestern Colombia (South America), pp. 261-268. In: P.Landolt and M. Sartorius (eds). Ephemeroptera and Plecoptera: Biology- Ecology-Systematics. Fribourg, Switzerland.

DISTRIBUTION OF EPHEMEROPTERA IN THE ANDEAN PART OF THE RIO BENI DRAINAGE BASIN (BOLIVIA): REGIONAL PATTERN OR CONTROL AT THE LOCAL SCALE?

Giovanna Rocabado[1], Jean-Gabriel Wasson[2], and Faviany Lino[1]

[1] Universidad de La Paz (U.M.S.A.)
Instituto de Ecología, Unidad de Limnología
P.O. box 10077, La Paz, Bolivia
[2] IRD (Institut de Recherche pour le Développement)
BIOBAB project IRD-UMSA
P.O. box 9214, La Paz, Bolivia

ABSTRACT

A regionalization of the Rio Beni Andean basin based on geomorphology and climate led to hydro-ecoregions (HER). Ephemeroptera were sampled in 13 streams representative of 6 HER. Abiotic factors included slope, granulometry, and chemical parameters. Fauna is dominated by 6 genera (*Baetodes, Camelobaetidius, Baetis, Leptohyphes, Tricorythodes, Thraulodes*). Ephemeropteran communities does not present a strong hydro-ecoregional pattern, but major trends were evidenced. At the stream scale, slope governs the densities. HER clearly discriminate the streams on their physical and chemical parameters, and constitute a valid framework to explain the densities of Ephemeroptera.

INTRODUCTION

Aquatic ecosystems of the Bolivian Amazonian drainage basin are still very poorly known, especially for the invertebrate fauna. A better knowledge of the ecology and distribution of Ephemeroptera of this region will be of great interest to better understand the functioning of these ecosystems, as mayflies generally play an important paper in the faunistic structure. Predicting the distribution pattern of Ephemeroptera in natural streams according to their regional characteristics will be also of critical importance to further develop bioindication tools. In the northern countries (North America and Europe), the taxonomy, ecology and distribution of mayflies are well known. In the tropical Andes, the taxonomy and distribution of Ephemeroptera is documented by works carried out mainly in Colombia and Argentina (Needham and Needham, 1978; Roldán, 1980, 1985, 1988; Domínguez, *et al.*, 1992,1995; Flowers and Domínguez, 1992; Rojas *et al.*, 1993; Zúñiga and Rojas, 1995). In

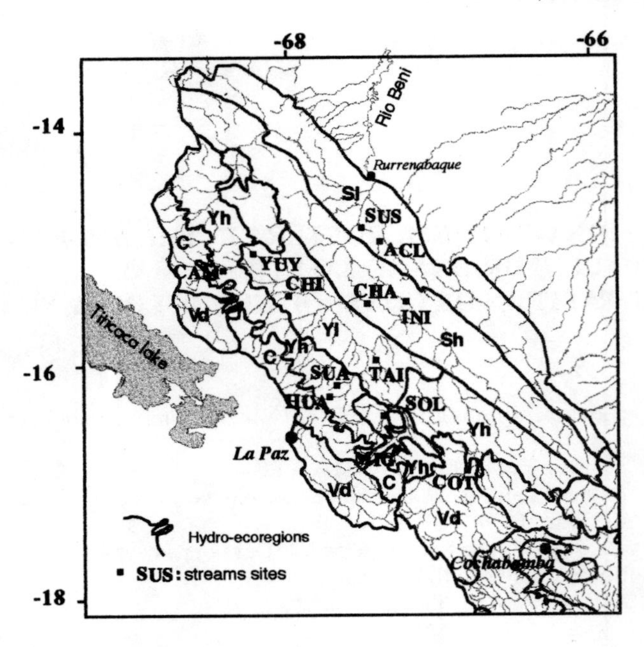

Fig. 1. Localization of the stream sites in the Hydro-ecoregions of the Rio Beni Andean basin.

Bolivia, the ephemeropteran fauna of the tropical Andean region is almost unknown. Thus, the aims of this paper are: 1) to get a preliminary list of the mayflies genera living in natural streams of the Andean Rio Beni drainage basin, 2) to give a quantitative distribution pattern of ephemeropteran fauna during low flow, and 3) to investigate at the regional and local scales the environmental factors that may explain this pattern.

STUDY AREA

Regionalization of the Basin

The Rio Beni is one of the main tributaries of the Rio Madeira in the Amazonian basin. The studied part of its basin (fig 1) encompasses all the geomorphologic structures of the eastern slope of the Bolivian Andes. Altitude ranges from 6400 m in the Cordillera Real to 200 m at the base of the Andes. The climate is mainly tropical humid, but some sheltered valleys are semi-arid. The anthropogenic alteration of the whole basin is very low, except in some areas of recent colonization, and extended areas are still covered with primary vegetation.

A preliminary regionalization of the basin (Gourdin, 1997), based on geology, geomorphology, climate and natural vegetation, led to the delimitation of hydro-ecoregions (HER *sensu* Wasson, 1996). These HER provide *a priori* a geographical frame for the regionalization of running waters ecosystems, with the hypothesis that physical and ecological structures, as well as the longitudinal gradient, will significantly differ from one region to another. The figure 1, derived from the definitive map of hydro-ecoregions of the Bolivian Andes (Wasson and Barrère 1999), presents the localization of the study sites in the HER of the basin. Main characteristics of the 6 studied regions are summarized in table 1.

Table 1. Characteristics of the six Hydro-ecoregions studied. Stream sites endogenous of a single region are in the left column; sites in the right column have a significant part of their watershed in the region Cordillera (code Y+C). See table 2 for stream sites codes.

Hydro-ecoregions	code	Altitude	Relief	Geology	Rainfall	Stream sites	
Dry Valleys	Vd	1000 -4000	V shaped valley	sedimentary	500-1000	COT	
Cordillera	C	3000 -6000	Steep valleys	metamorphic	600-900		Y+C:
High Yungas	Yh	1000 - 3000	V shaped valley	metamorphic	1500-3000	SUA	HUA MIG SOL CAM
Low Yungas	Yl	500 - 2500	open valleys	both	1500-2500	TAI CHI YUY	
High Sub-Andean	Sh	400 - 2000	mountain ridges	sedimentary	1300-1700	INI CHA	
Low Sub-Andean	Sl	200 - 1500	depression	sedimentary	2000-2500	SUS ACL	

Studied Streams

A total of 13 stream sites of similar size (about 15 - 20 meters wide in low flow) were selected on three criteria: representative of a region, natural state or low anthropogenic alteration of the watershed, and accessibility of the site. Representativeness is based on the watershed extension : most of the studied streams are endogenous of single HER, but the watersheds of four streams sites situated in the region "High Yungas" encompasses also significantly the region "Cordillera" (see table 1). Due to the difficulty of access, two regions (High Yungas and Dry Valleys) have only one representative site sampled at the present time. To emphasize on the regional and local control factors, the well known effects of the altitudinal and thermal gradients were limited by selecting stream sites in a restricted altitudinal range, between 1300 and 240 m. The slopes of the studied sites lie in the range of median values for streams of similar size in each HER (Gourdin, 1997) (table 2).

METHODS

All streams were sampled once during the dry season (May-October 1997). Stream sites were selected to represent the morphodynamic features of the reach, each site including two major riffle/pool sequences. The total length of a site represents about 12 times the bankfull width. The physical habitat of the stream sites was characterized by slope and granulometry. The slope of the water level was measured using topographic instruments, and calculated for the whole site and each major morphodynamic units. By the same way was evaluated the percentage of lentic units. Surface granulometry was evaluated by measuring the B-axis of 50 randomly selected elements, in both rapid and flat morphodynamic units (100 elements measured in total). Instability of the substrate was evaluated from its structure and the shape of the stones. Chemical data included temperature measured over a 24 h period, pH and conductivity measured in the field with WTW portable equipment, alkalinity and suspended solids analyzed in the laboratory following standard methods (Laboratorio de Calidad Ambiental, Instituto de Ecologia, UMSA La Paz). Benthic fauna was sampled using a Surber net (area 0.1 m^2, mesh size 0.250 mm). Six samples were taken to represent the main morphodynamic features of the site. Samples were conserved in 10% formalin for the transfer to the laboratory, then Ephemeroptera were sorted and identified to the genus level using the keys of Roldán (1988), Edmondson (1959), and Domínguez *et al.* (1992, 1995). Data were analyzed using the software ADE-4 (Chessel and Doledec, 1996) for multivariate analysis and SYSTAT 5 for other statistical analyses.

Table 2. Characteristics and localization of the stream sites.

Stream	Code	Slope %	Altitude (m)	Latitude (S)	Longitude (W)
				decimal	degrees
Huarinilla	HUA	1.8	1300	16.2167	67.8333
Miguillas	MIG	1.33	1272	16.5926	67.3087
Solacama	SOL	2.01	1269	16.3982	67.4705
Camata	CAM	1.28	1100	15.2174	68.6464
Suapi (Yungas)	SUA	2.86	1271	16.1152	67.7864
Yuyo	YUY	0.73	720	15.0424	68.4593
Chimate	CHI	0.61	560	15.4058	68.1536
Taipiplaya	TAI	0.55	804	15.9136	67.5023
Cotacajes	COT	0.93	1100	16.7474	66.7399
Inicua	INI	0.76	531	15.5065	67.1678
Chamaleo	CHA	0.27	450	15.4142	67.5786
Suapi (subandino)	SUS	0.25	240	14.8346	67.6220
Agua Clara	ACL	0.29	240	14.9217	67.4292

RESULTS

Physical and Chemical Data

The 13 stream sites were classified according to 14 physical and chemical parameters using a normalized Principal Component Analysis (PCAn) (fig. 2). The first two axes summarize 67% of the total inertia (fig. 2C). Axis 1 is structured by the physical parameters: altitude, slope and granulometry on the positive side, opposite to temperature, substrate instability and percentage of lentic units on the negative side. Axis 2 correspond to chemical parameters, suspended and dissolved solids being strongly correlated (fig 2A), together with pH and alkalinity. When plotted on these first two axes, the stream sites grouped by HER spread out along the F1 according to a gradient of their physical characteristics (fig. 2B). Stream sites of the High Yungas and Cordillera (Yh, Y+C) present higher slopes and granulometry, while sub-Andean sites (Sh, Sl) are characterized by lower values for these parameters, substrate instability, a high proportion of lentic units, and higher temperature. The second axis differentiates the sites of two HER: the Cotacajes in the Dry Valleys (Vd) due to high dissolved and suspended solids concentrations, and the sites of the Low Yungas (Yl) characterized by low pH and alkalinity. As a whole, the different hydro-ecoregions are well separated by the physical and chemical parameters of their representative stream sites.

Ephemeropteran Fauna

Faunistic Distribution. A total of 5 families and 16 genera were identified, with total densities varying from 15 to 1073 ind./m^2 (table 3).

The table was analyzed by the mean of a centered PCA after logarithmic transformation of the density values (log x + 1) (fig. 3). The first two axes account for 68% of the total inertia (fig. 3A). The faunistic map separates three groups of Ephemeroptera (fig. 3B) : group I includes three genera of Baetidae (*Baetodes, Camelobaetidius* and *Baetis*) ; group II includes three genera of Leptohyphidae and Leptophlebiidae (*Leptohyphes, Tricorythodes* and *Thraulodes*); group III includes all remaining taxa. The six genera in groups I and II are situated on the negative side of F1 and are the most abundant. The genera of the group III, located near the origin of the axis, are less abundant and more evenly distributed. Among them, the genus *Euthyplocia* was found only in two sites of the High sub-Andean region (Sh).

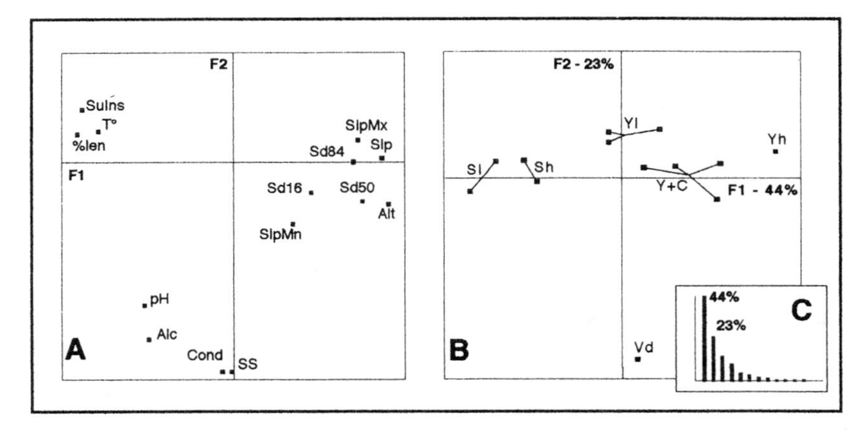

Fig. 2. Normalized PCA of the stream sites according to their physical and chemical characteristics.
A) Factor map of the 14 physical and chemical parameters. Alt: altitude (m); Slp: mean slope (%); SlpMx: maximum slope of the site; SlpMn: minimum slope of the site; Sd50: median granulometry (D50 of the granulometric curve); Sd84: coarse granulometry (D84); Sd16: fine granulometry (D16); SuIns: substrate instability, (qualitative evaluation, see text); T°: temperature; %len: percentage (in length) of lentic morphodynamic units. pH; Alc: Alkalinity; Cond: conductivity; SS: suspended solids ($mg.l^{-1}$).
B) Stream sites map. The 13 stream sites are grouped according to the hydro-ecoregions of their watershed. See table 1 for region codes.
C) Eigenvalues of the PCA axis.

The site map separates also three clusters of stream sites (fig. 3C). The cluster located on the positive side of the F1 includes the sites with a poor Ephemeropteran fauna, i.e. with 1 to 4 genera and less than 100 ind./m^2. The second cluster, negative on F1 and positive on F2, corresponds to stream sites dominated by the family Baetidae, with high ephemeropteran densities (500 to 1000 ind./m^2), and 5 or 6 genera. The third cluster, negative on F2, is dominated by genera of the families Leptohyphidae and Leptophlebiidae, have intermediate densities (200 to 800 ind./m^2), but the highest number of genera (8 to 11). One stream site (SUS) lies in an intermediate position.

When grouping the stream sites by hydro-ecoregions (fig. 3D), the streams sampled in the High Yungas but coming from the Cordillera (Y+C), appears with a dominance of Baetidae (group I) and relatively isolated from the others. Comparatively, the stream (SUA) endogenous from the High Yungas (Yh), although also supporting a high density of Baetidae (group I), presents a dominance of Leptohyphidae and Leptophlebiidae (group II). Another stream, the Cotacajes (COT) flowing in the Dry Valleys region (Vd), lies apart due to its very poor fauna dominated by *Baetodes*. The three other regions are not separated by their Ephemeropteran fauna, although a slight gradient might appear from the Low Yungas to the Low sub-Andean with increasing abundance and diversity, and increasing dominance of Leptohyphidae and Leptophlebiidae (group II).

Quantitative analysis. At the site level, control factors of the Ephemeropteran fauna quantitative structure were investigated using regression models on raw data. The entries of the models were the number of genera, total density, densities of each family, and densities of the groups I and II as dependent variables, and as independent variables the 14 physical and chemical parameters. The only significant regressions ($p< 0.05$) were: total density vs. mean slope (fig 4A), densities of Baetidae vs. mean slope ($r^2 = 0.601$, $p = 0.002$) and altitude ($r^2 = 0.527$, $p = 0.005$), density of group I (Baetide) vs. mean slope ($r^2 = 0.610$, $p = 0.002$) and altitude ($r^2 = 0.542$, $p = 0.004$), and densities of Euthyplociidae vs. fine

Table 3. Densities of Ephemeroptera per stream site (number of individuals / m^2). See table 2 for sites codes.

Taxa	Code	Stream sites												
		HUA	MIG	SOL	CAM	YUY	SUA	TAI	COT	INI	CHA	SUS	ACL	CHI
Euthyplociidae														
Euthyplocia	Eut	0	0	0	0	0	0	0	0	13	1,7	0	0	0
Leptohyphidae														
Leptohyphes	Lep	463	168	78	8.3	0	137	68	0	1.7	160	32	58	0
Tricorythodes	Tri	0	0	0	0	0	78	1.7	0	0	225	6.7	3.3	0
Leptophlebidae														
Thraulodes	Tra	230	35	3.3	57	25	233	53	3.3	1.7	28	63	100	17
Traverella	Trv	0	0	0	0	0	0	0	0	0	1.7	0	0	0
Terpides	Ter	0	0	0	0	0	10	0	0	0	0	0	0	0
Ulmeritoides	Ulm	0	1.7	0	0	0	0	0	0	0	0	0	0	0
Farrodes	Far	0	0	0	0	0	0	3.3	0	0	0	0	0	0
Meridialaris	Mer	12	0	0	0	0	0	17	0	0	3.3	1.7	47	0
Baetidae														
Baetis	Bae	90	8.3	23	0	0	6.7	12	0	0	40	0	1.7	8.3
Baetodes	Bao	263	185	248	17	0	248	5	12	0	10	27	0	0
Moribaetis	Mor	0	0	3.3	0	0	0	0	0	0	0	0	0	0
Cloeodes	Clo	0	0	0	0	0	0	1,7	0	0	0	0	6.7	0
Camelobaetidius	Cam	15	77	5	0	0	0	0	0	0	5	1.7	0	5
Paracloedoes	Par	0	0	0	0	0	5	0	0	0	5	0	5	0
Caenidae														0
Brachycercus	Bra	0	0	0	1.7	0	1.7	0	0	0	1.7	0	3.3	0
Total density		1073	475	362	83	25	720	162	15	17	482	132	225	30
Nb of genera		6	6	6	4	1	8	8	2	3	11	6	8	3

granulometry ($r^2 = 0.881$, $p = 0.000$) and percentage of lentic units ($r^2 = 0.330$, $p = 0.040$). However, the regressions for Euthyplociidae are only driven by one station (INI). For the family Baetidae and group I, regressions are significant with both altitude and mean slope but these two parameters are highly correlated ($r = 0.84$). When using a multiple regression analysis which eliminates the effect of partial correlation, only the mean slope enters in the regression model to explain the densities of both Baetidae and group I. The two models are very similar, as the 3 genera of the group I (*Baetodes, Camelobaetidius, Baetis*) make up more than 90% of the Baetidae in all sites except one (ACL); so only the regression for Baetidae is given in fig 4B.

The pattern of mayflies densities was also investigated at the region scale. As the slope appears as a main factor in discriminating the stream sites of the different hydro-ecoregions, mean densities were plotted against HER ranked following their position along the F1 of the PCAn on physical and chemical parameters (fig. 5). The global pattern evidences at the region scale the positive relationship between the total density of Ephemeroptera and the position of HER along the F1 axis, that integrates the physical characteristics of the streams sites in these regions. However, the two HER previously discriminated in the PCAn by their chemical particularities (Low Yungas and Dry Valleys) appears also with densities much lower than expected only by their position along the F1.

The differences in Ephemeropteran densities between HER were tested with an ANOVA. We used the logarithm of number of individuals per Surber sample, in order to downsize the problem of the small number of stream sites by HER. Differences were tested with a Mann-Whitney's test (0.05 significance level) for total densities of Ephemeroptera (Table 4), and for the densities of the two dominant groups I and II identified in the faunistic

PCAc. Despite the few number of samples, the results indicate that the stream sites of both Low Yungas and Dry Valleys differ significantly from those of the other Yungean regions (High Yungas and Yungas + Cordillera) in almost every case (11/12). The differences between the High Yungean sites (Yh and Y+C) and the Sub-Andean ones (Sh and Sl) are significant for 2/3 of the tested cases.

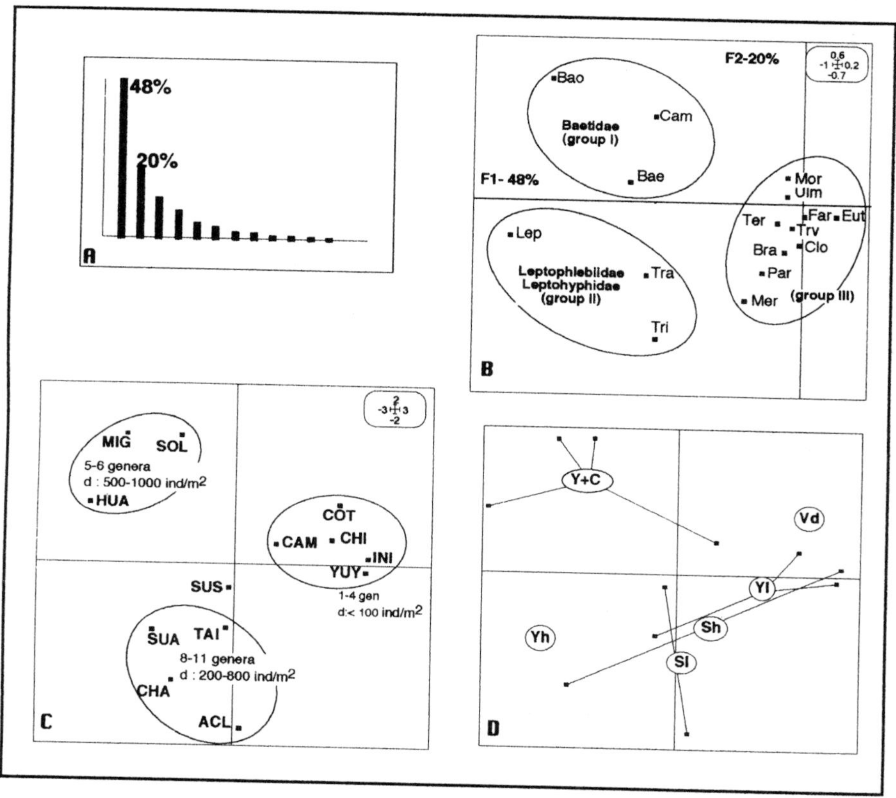

Fig. 3. Centered PCA of the stream sites according to their Ephemeropteran fauna.
A) Eigenvalues of the PCA axis.
B) Factor map of the 16 Ephemeropteran genera. See table 3 for genera codes and text for explanations.
C) Map of the 13 stream sites. See table 2 for sites codes.
D) Map of the 13 stream sites grouped according to the hydro-ecoregions of their watershed. See table 1 for region codes.

DISCUSSION

Validity of the Regionalization

Hydro-ecoregions were delimited on the basis of macro-scale determinants: geology, geomorphology, climate, and natural vegetation, available on maps or by remote sensing. We assume that these highest control factors will determine the basic physical and chemical characteristics of the streams, and the gradient of their longitudinal evolution. Thus, the hypothesis is that streams of similar size will present different physical and chemical

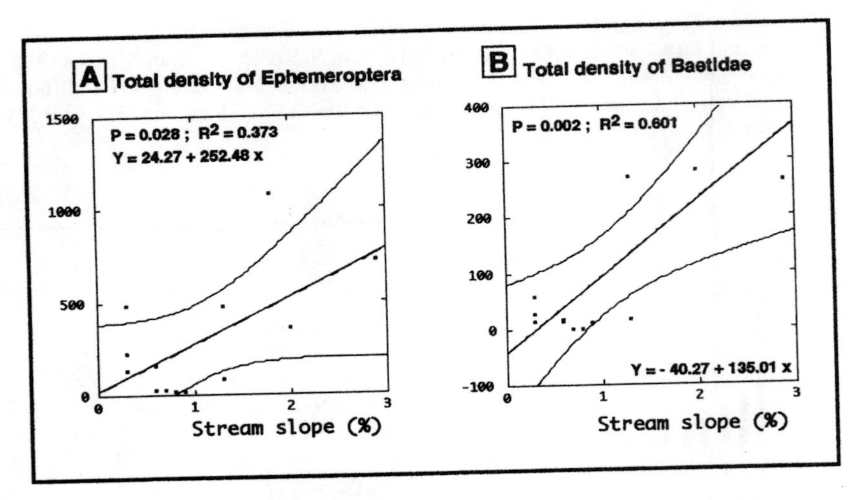

Fig. 4. Linear regressions models between A) total density of Ephemeroptera (Ind./m^2) and B) density of Baetidae (Ind./m^2) vs. mean stream slope (%).

characteristics according to the HER they flow through. Despite the small number of stream sites studied at the present time, the results obtained with physical and chemical parameters measured at the local scale support this hypothesis. The discrimination of HER by these parameters is good as evidenced by the PCAn (fig. 2B). This discrimination is in accordance with the global characteristics of HER. The geomorphologic sequence of HER along the eastern slope of the Bolivian Andes determines the gradient of physical characteristics of the stream sites (evidenced by the F1 axis of the PCAn). Noticeably, this gradient is not equivalent to a longitudinal or altitudinal zonation, as the streams of similar size were selected in different watersheds, without longitudinal relations between them. Another important point is the following: the hydro-ecoregions encompassed by the watersheds must be taken into account to explain the physical characteristics at the stream scale. In the High Yungas region, the endogenous stream (Yh) presents different characteristics (greater slope and granulometry) than the other streams flowing down from the Cordillera (Y+C). The climatic features, regulating the vegetal cover, explain some important variations in the chemical characteristics, such as the high dissolved and suspended solid content in the Dry Valleys region (Vd) where erosion is very high. Similarly, the low pH could be expected from the soil and vegetation characteristics; in the Yungean region, the humid tropical forest cover generates the accumulation of organic matter upon superficial uncarbonated rocks, thus leading to acid soils (Ribera *et al.* 1996). However, the lowest pH registered in the Low Yungas region (Yl) are probably related to the presence of pyrite veins in some basins (Guyot 1993), hardly predictable on the basis of the macro-scale determinants we used. Despite that, we assume that the hydro-ecoregions provide a valid frame for the regionalization of the physical and chemical characteristics of the streams. Such a correspondence has yet been observed in the Loire river basin (France) (Wasson, 1996). Now, the question is to what extent the Ephemeropteran fauna is controlled by this abiotic frame?

Distribution of Ephemeropteran Communities

All the genera identified in this study are encountered either in northern Argentina, or in Colombia. Thus we did not observed a particularity of the Bolivian fauna at the genus level. Globally speaking, the most salient feature of the mayflies fauna is the predominance of two

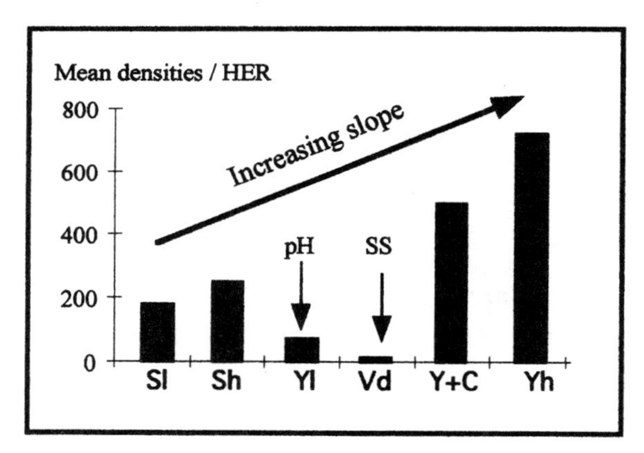

Fig. 5. Histogram of the mean total densities of Ephemeroptera by hydro-ecoregion. HER are ranked following their position along the axis F1 of the PCAn on physical and chemical parameters (fig. 2B). See table 1 for region codes.

groups of taxa (fig. 3B). Group I is composed of three genera of Baetidae (*Baetodes, Camelobaetidius* and *Baetis*) and group II by three genera of Leptophlebiidae and Leptohyphidae (*Leptohyphes, Tricorythodes* and *Thraulodes,*). These two groups account for most of the faunistic structure. When looking at the stream scale, three groups of sites can be distinguished (fig. 3C). The group "poor" as regarding Ephemeroptera includes sites where the fauna is impoverished for various reasons: low pH (CHI, YUY), high suspended solids content (COT), extended sandy substrate (INI). The second group of sites, whose ephemeropteran fauna is dominated by Baetidae, is formed by three streams flowing down from the Cordillera. The third group of sites, with high abundance of Leptophlebiidae and Leptohyphidae, includes only streams endogenous from the Yungean or Sub-Andean regions.

When searching for regional differentiation, the distribution of Ephemeropteran communities does not match very tightly the hydro-ecoregional structure. Three HER might appear with a relatively particular faunistic structure. The clearest one is that of the High Yungean streams flowing down from the Cordillera (Y+C), with a fauna dominated by the group I (Baetidae), and 4 to 6 genera. The endogenous stream of the High Yungas (Yh) differs from the former in sheltering both group I (Baetidae) and group II (Leptophlebiidae and Leptohyphidae) with high densities and 8 genera. The Dry Valleys (Vd) lies apart of the other regions due to its very poor fauna (2 genera and the lowest density) dominated by the genus *Baetodes*. The three other regions present only a slight gradient of differentiation, with high internal variability. Thus, these results did not evidence a strong differentiation of the Ephemeropteran communities on the basis of the Hydro-ecoregions, at least at the generic level. A high variability of the faunistic composition occurs at the local scale.

However, a general trend appears with the dominance of the Baetidae (group I) in the streams coming from the Cordillera, and of Leptophlebiidae and Leptohyphidae (group II) in the others. This could be related to a thermal effect, as streams flowing down from the Cordillera arrive in the Yungas valleys with mean water temperature much lower than the temperature of the air, and probably 2°C lower than the endogenous streams of this region (Wasson *et al.* 1989). Thus, the distribution of the family Leptophlebiidae, typically known as polystenotherm (Dominguez *et al.* 1995), could be limited in these streams by a colder mean temperature.

More sites and more precise data would be necessary to evidence a stronger hydro-ecoregional pattern of the ephemeropteran fauna, if it does exist. An interesting question is if the identification of species would lead to a better regional differentiation. The six genera that make up most of the faunistic structure are relatively common and widespread in the neotropical area, and presumably some species will tolerate a narrower range of abiotic conditions than the genus they belong to. Unfortunately, this question will remain unanswered for a while due to taxonomic limitations.

Table 4. Non parametric ANOVA (Mann-Whitney) for total density of Ephemeroptera between hydro-ecoregions. Probabilities for non-significant differences based on log transformed densities by Surber samples.

	Y + C	Yh	Yl	Vd	Sh	Sl
Y + C	1	0.565	0.001	0.002	0.141	0.149
Yh		1	0.001	0.006	0.008	0.004
Yl			1	0.095	0.442	0.442
Vd				1	0.047	0.003
Sh					1	0.73
Sl						1

Quantitative Structures of the Ephemeropteran Fauna

This high variability at the local scale oriented the investigation in search of quantitative models that could at least explicate the general parameters of the Ephemeropteran communities: total densities and number of genera. As a result, once eliminated the effects of partial or auto-correlation between abiotic parameters, the only significant relationships were found between stream slope vs. total mayflies densities, and slope vs. densities of the family Baetidae. However, even if these relationships are not doubtful, plotting the data (fig. 4) evidences the limited predictive capacity of these models at the local scale. Despite of that, the relationship between mayfly densities and stream slope is interesting because its offers a causal factor explaining the quantitative pattern of Ephemeroptera at the regional scale, as evidenced in the figure 5. Among the various parameters that contribute to the ranking of the six HER according to their physical characteristics, stream slope rises up as the only one significant in explaining the correlative increase of mayfly densities. The altitudinal effect seems less important as it disappears behind the slope effect in the multiple regression models, and we did not observe a significant difference between the two lowest HER (Sh and Sl) that could be related to the transition observed at 300 m by Flowers (1991) in the Ephemeropteran fauna of Panama.

But the stream slope alone does not predict the significantly lower densities in the two regions Dry Valleys (Vd) with high suspended solids, and Low Yungas (Yl) with low pH. Although pH and suspended solids were included in the entry parameters of the regression models, they did not contribute to the explanation of densities, perhaps because although they are inversely correlated, they have the same negative biological effect. The inverse relationship between suspended sediment and benthic fauna density has been clearly evidenced in the Yungean region following the impact of road construction on a clear water river (Salinas *et al.* 1999). Although not surprising, the effect of the low pH provides an interesting insight in the functioning of aquatic ecosystems in the Andean zone, and urges to pay more attention to the pH factor even in fast flowing streams. In fact, if the effects of a low pH are well documented in the lower Amazonian basin, we are not aware of any case study in natural streams of the Andean part of this basin.

Thus, the explanation of the density pattern of Ephemeroptera in the six HER appears relatively evident: there is a general trend of increasing densities with increasing stream slope, but limiting chemical factors such as pH and suspended solid can alter this pattern in lowering significantly the densities. Therefore, as these three parameters, stream slope, pH and suspended solids are fairly well predicted by HER, we assume that the hydro-ecoregions make up a valid frame to predict a range of abundance of Ephemeroptera at the stream scale.

CONCLUSION

The Andean part of Rio Beni drainage basin, with its largely pristine state over a wide range of geomorphologic and climatic features, constitute a choice laboratory for ecological studies on stream ecosystems. Our data present the first insight on Ephemeropteran fauna, identified at the genus level, in natural streams of this basin. This faunistic list including 16 genera may serve as a reference for future bioindication studies.

The aim of this work was to identify at the regional and local scales the control factors that may explain the distribution and quantitative structure of Ephemeropteran communities. A previous regionalization of the basin led to the delimitation of hydro-ecoregions on the basis of geological, geomorphologic and climatic features. Ours results demonstrate that hydro-ecoregions make up a valid frame to regionalize the basic physical and chemical characteristics of the streams. The sites studied in the different hydro-ecoregions are well discriminated on the basis of abiotic features measured at the stream scale, including slope, granulometry, temperature, dissolved and suspended solids, and pH. The question was to what extent the Ephemeropteran fauna do respond to this abiotic frame ?

On the basis of the faunistic distribution, we did not observed a strong regional pattern. However, some trends are relatively clear, such as the dominance of Baetidae in streams flowing down from the Cordillera, and Leptohyphidae and Leptophlebiidae in the other regions. The High Yungas region and the Dry Valleys appear also different from the others, but here our dataset is presently too weak to lead to definitive conclusions. In many regions, the internal variability of the Ephemeropteran communities is very high, due to the influence of local factors or limited sampling.

When looking at the quantitative structure of theses communities, two HER (Dry Valleys and Low Yungas) appear with densities much lower than the others. In regression models, the mean slope rise up as the only control factor explaining the abundance of Ephemeroptera at the stream scale; but if this fàctor can explain the difference between Yungean (Yh, Y+C) and Sub-Andean (Sh, Sl) regions, it cannot explain the low densities in the Dry Valleys and Low Yungas. In fact, both physical and chemical parameters governs the abundance of Ephemeroptera. Although not doubtful, the general trend of increasing abundance with increasing stream slope is a poor predictor of the quantitative structure, which may be significantly altered by adverse chemical parameters such as suspended solids or low pH.

As a whole, when looking both qualitative and quantitative characteristics, five of the six studied HER are discriminated on the basis of their Ephemeropteran fauna; only the two Sub-Andean regions (Sh, Sl) appear with undifferentiated communities. Thus, hydro-ecoregions allow a rather good regionalization of the Ephemeropteran communities, and to some extent the prediction at the stream scale of abiotic factors that explain important differences in their quantitative structure. We expect similar or even stronger results when taking into account the whole invertebrate fauna. These preliminary results support, at least partially, the hypothesis that hydro-ecoregions provide a valid frame for the regionalization of stream ecosystems. In such a large and almost unknown basin, this result will be of critical importance to devèlop bioindication tools, and to recommend management practices adapted to the regional ecosystems characteristics.

REFERENCES

Chessel, D. and S. Doledec. 1996. Programmatèque ADE: analyses multivariées et représentations graphiques de données écologiques, v4.0. Université de Lyon I, France.

Domínguez, E., M. Hubbard and W. Peters. 1992. Clave para Ninfas y Adultos de las Famílias y Géneros de Ephemeroptera (Insecta) Sudamericanos. Biología Acuática, ILPLA, .16: 41 p.

Domínguez, E., M. Hubbard and M. Pescador. 1995. Los Ephemeroptera en Argentina. Coll. Fauna de agua dulce de la República de Argentina , Vol 33 (1): 142 p.

Edmunds, G. F. 1959. Ephemeroptera, pp. 908-916. In: W. Edmondson, W. Freshwater Biology. Second Edition, John Wiley & Sons, Inc., New York.

Edmunds, G. F. 1984. Ephemeroptera, 72-84 p. In: R. Merrit and K. Cummins. An Introduction to the Aquatic Insects of North America. Second Edition. Kendall/Hunt publ.

Flowers, R. W., 1991. Diversity of stream-living insects in Northwestern Panama. J. N. Amer. Benthol. Soc. 10 (3): 322-334.

Flowers R. W. and E. Domínguez. 1992. New Genus of Leptophlebiidae (Ephemeroptera) from Central and South America. Ann. Ent. Soc. Amer., 85 (6): 655-661.

Gourdin, F. 1997. Regionalisation des déterminants géomorphologiques des hydrosystèmes dans le bassin amazonien Bolivien - Cas de la zone andine. Rapport de stage ENGEES, Strasbourg et ORSTOM, La Paz - Bolivia, 109 p.

Guyot, J. L., 1993. Hydrogéochimie des fleuves de l'Amazonie Bolivienne. Coll. Etudes et Thèses, ORSTOM, Paris, 261 p.

Needham, J. G. and P. R. Needham. 1978. Guía para el estudio de los seres vivos de las aguas dulces. Reverté publ., 131 p.

Ribera, M., M. Liberman, E. Beck and M. Moraes. 1996. Vegetación de Bolivia, pp. 171-221. In: Mihotek, Comunidades, territorios indígenas y biodiversidad en Bolivia. Univ. Gabriel René Moreno - CIMAR, Santa Cruz.

Rojas, M., M. Baena, C. Serrato, G. Caicedo and M. C. Zúñiga. 1993. Clave para las Familias y Géneros de Ninfas de Ephemeroptera del Departamento del Valle del Cauca, Colombia. Bol. Mus. Ent. Univ. Valle, 1 (2): 33-46.

Roldán, G. 1980. Estudio limnológico de cuatro ecosistemas neotropicales diferentes con especial referencia a su fauna de ephemeropteros. Actualidades Biológicas 9 (34): 103-116.

Roldán, G. 1985. Contribución al conocimiento de las Ninfas de los Efemerópteros (Clase: Insecta, Orden: Ephemeroptera) en el departamento de Antioquía, Colombia. Actualidades Biológicas 14 (51): 3-13.

Roldán, G. 1988. Guía para el Estudio de los Macroinvertebrados Acuáticos del Departamento de Antioquía. Universidad de Antioquía. 38p.

Salinas, G., R. Marín R and J. G. Wasson. 1999. Efecto de la materia en suspención sobre las comunidades bénticas en los ríos de aguas claras en los Yungas de Bolivia. Actas del Congreso Boliviano de Limnología y Recursos Acuáticos. Cochabamba, Rev. Bol. Ecol. y Cons. Amb. In press.

Wasson, J. G., J. L. Guyot., C. Dejoux and A. Roche. 1989. Régimen térmico de los ríos de Bolivia. Publicación ORSTOM - PHICAB. La Paz, 35 p.

Wasson, J. G., 1996. Structures régionales du bassin de la Loire. La Houille Blanche, 338 (6/7): 25-31.

Wasson, J. G. and B. Barrère. 1999. Regionalización de la cuenca Amazónica boliviana: Las hidro-ecorregiones de la zona Andina. Rev. Bol. Ecol. y Cons. Amb.: 111-119.

Zúñiga, M. and A. Rojas. 1995. Contribución al conocimiento del orden Ephemeroptera en Colombia y su utilización en estudios ambientales. Actas del Seminario "Invertebrados acuáticos y su utilización en estudios ambientales". Universidad Nacional de Colombia, Santafé de Bogotá - Colombia. p 121-146.

ABUNDANCE AND DIVERSITY OF A MAYFLY TAXOCENE IN A SOUTH AMERICAN SUBTROPICAL MOUNTAIN STREAM

F. Romero[1] and H. R. Fernández[2]

[1] Fundación Miguel Lillo
[2] Facultad de Ciencias Naturales e Instituto Miguel Lillo, UNT - CONICET
Miguel Lillo 251, (4000) San Miguel de Tucumán, Argentina

ABSTRACT

The structure of the Ephemeroptera taxocene was studied under adverse summer conditions. Our study examined eigth stations in a South American mountain stream from moist temperate cloud forest to premontane subtropical forest including an altitudinal range of 950 m. To make them more comparable, the two samplings were carried out in the rainy season (1989 and 1990). We collected nymphs of nine mayfly species from four different families. The mayfly density ranged from 4 ind/m^2 to 1,214 ind/m^2. Our data suggest some degree of persistence of the taxocene in the five lower stations, probably induced by the dominant *Baetodes* sp. This is attributed to invariable physical conditions in the channel. We did not observe any pattern in other species.

INTRODUCTION

The subtropical mountain forest is a conspicuous biome in this part of South America. It is very important in the water production aspects and its regulation. The tree richness of this mountain forest and its variations along altitudinal levels has been documented (Morales et al., 1995).

The altitudinal gradient offers appropriate opportunities to investigate factors which control the diversity, composition and abundance of stream organisms (Ward and Berner, 1980) although little previous work on stream zonation have been conducted in Northwestern argentine (Argañaraz, 1984, Domínguez and Ballesteros Valdez, 1992).

The Ephemeroptera taxocene were investigated within a major study on the macroinvertebrate benthic fauna of mountain streams.

MATERIAL AND METHODS

The De los Reales River belongs to the Salí River drainage basin (Tucumán, Argentina). The stream begins at approximately 2,500 m.a.s.l (Fig. 1) and drains into Pueblo Viejo River (Tucumán, Argentina) at 800 m.a.s.l.

Trends in Research in Ephemeroptera and Plecoptera
Edited by E. Domínguez, Kluwer Academic/Plenum Publishers, 2001

Table 1. Mayfly composition at the stations investigated and abundance (ind/0,27 m^2) in two years.

	Station															
	1989								1990							
Species	a1	b1	c1	d1	e1	f1	g1	h1	a2	b2	c2	d2	e2	f2	g2	h2
Baetis sp. 1	14	0	1	13	8	28	60	5	0	0	0	2	24	4	1	0
Baetis sp. 2	0	5	1	1	10	7	6	1	0	0	0	0	13	0	2	0
Baetodes sp.	3	0	0	54	69	82	328	163	0	0	2	19	55	67	93	101
Camelobaetidius sp.	27	1	158	22	11	0	35	61	5	6	12	13	12	8	4	9
Thraulodes sp.	0	1	4	0	2	6	13	25	0	0	0	0	0	0	0	0
Meridialaris sp.	1	0	0	0	0	0	0	0	1	4	1	0	1	1	3	0
Leptohyphes sp.	3	16	16	12	15	15	28	6	18	0	8	12	4	15	23	0
Lachlania sp.	0	0	0	0	0	0	8	1	0	0	0	0	0	0	0	0
Tricorythodes sp.	0	0	0	0	0	0	0	0	0	0	0	0	0	0	1	0
Total Ephemeroptera	48	23	180	102	115	138	47	261	24	10	23	46	109	95	127	110
Ephem. Richness	4	4	5	5	6	4	7	6	3	2	4	4	6	5	7	2
Total other taxa	67	407	143	170	236	379	282	197	528	60	102	48	63	171	279	177
Mean taxa/sample	5	6	6.6	6.3	7.6	6	7.6	8	6.6	4.6	5.3	6	7.3	7	8.3	6

The study area is situated in the «Yungas» phytogeographical province, typical of the mountainous subtropical region. We had two collections, in summer (January) 1989 and 1990. Samples were taken every 110 m change in elevation from 2,070 to 1,120 m.a.s.l. of elevation. Eight stations (Fig. 1) were sampled with three subsamples of Surber sampler (300 µm mesh) from cobble riffles at each site.

The structure of the mayfly community was studied under summer conditions (rainy season with 80-100% of the annual total) to differentiate from of the other authors (Argañaraz, 1984, Dominguez and Ballesteros Valdez, 1992, Fernández et al., 1995). On average, the summer monthly precipitation is near 120 mm (Villalba et al., 1992). In this season the faunal assemblage is variable, caused by synchronous transformation of nymphs into adults (Domínguez and Ballesteros Valdez, 1992) and disturbs caused by torrential rains. Floods with discharges ranging from 400 to 800 L/s (Hunzinger, 1997) are frequent.

Analysis of Data

Heterogeneity from mayfly taxocene was calculated for each year using Shannon-Wiener index:

$$H' = -\sum_{i=1}^{k} p_i \log p_i \quad \text{(Zar,1984).}$$

The Spearman rank correlation coefficient (Zar, 1984) was used in order to compare the taxocene between years. For this it was used the total number of individuals from 9 taxa in each year. A Correspondence Analysis (CA) was performed using CANOCO (Ter Braak, 1987), without transforming the data and including rare species

RESULTS

Nine mayfly species from four different families were found (Table 1), and the density ranged from 4 ind/m^2 to 1,214 ind/m^2.

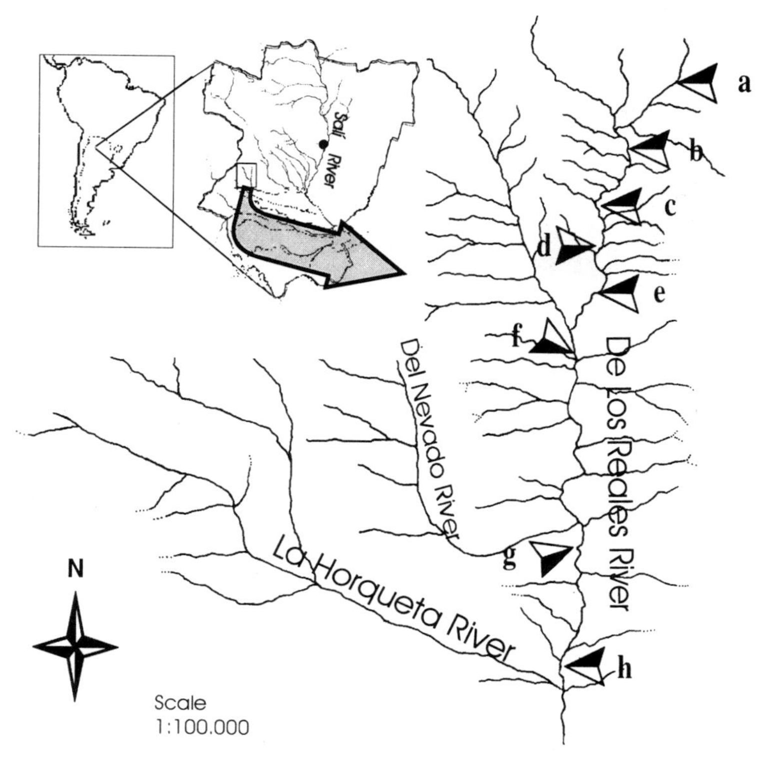

Fig. 1. Course of De los Reales River (Tucumán, Argentina) with sampling stations locations.

Among aquatic macroinvertebrates, the mayflies were important in 1989, with a percentage of total catches of 41%; however in 1990 the mayflies represented only 27.5%. Diversity measure (H') was 1.96 and 1.72 in each year respectively.

Baetidae, collectively constituted the majority of the mayfly fauna at sampled stations and *Baetodes* sp. is the species most important numerically, over the 50% in both years (Table 1).

Altitudinal distribution and abundance of the mayfly species along the profile of the De los Reales River is presented in both years (Figs. 2a and b). Baetodes sp. becomes more important downstream, while other species do not present a clear pattern. The abundance and richness exhibited a general increase downstream. There is a significative correlation between both years ($r_s = 0.738$, $p < 0.05$).

In the ordering (Fig. 3) can be noticed that stations d_1-h_1 from 1989 and d_2-h_2 from 1990 are grouped in one extreme of the axis II gradient ($\lambda = 0.175$). This axis is inversely correlated with depht ($r=-0.578$, $p<0.05$), these stations are depht ≥30 cm. Moreover both axis are significantly correlated ($p<0.05$) with the altitude ($r=0.50$ and $r=0.53$, respectively).

DISCUSSION

The results obtained under the summer conditions were interesting, especially because of adverse conditions for the benthic community and the coincidence with the emergence period for most of aquatic taxa.

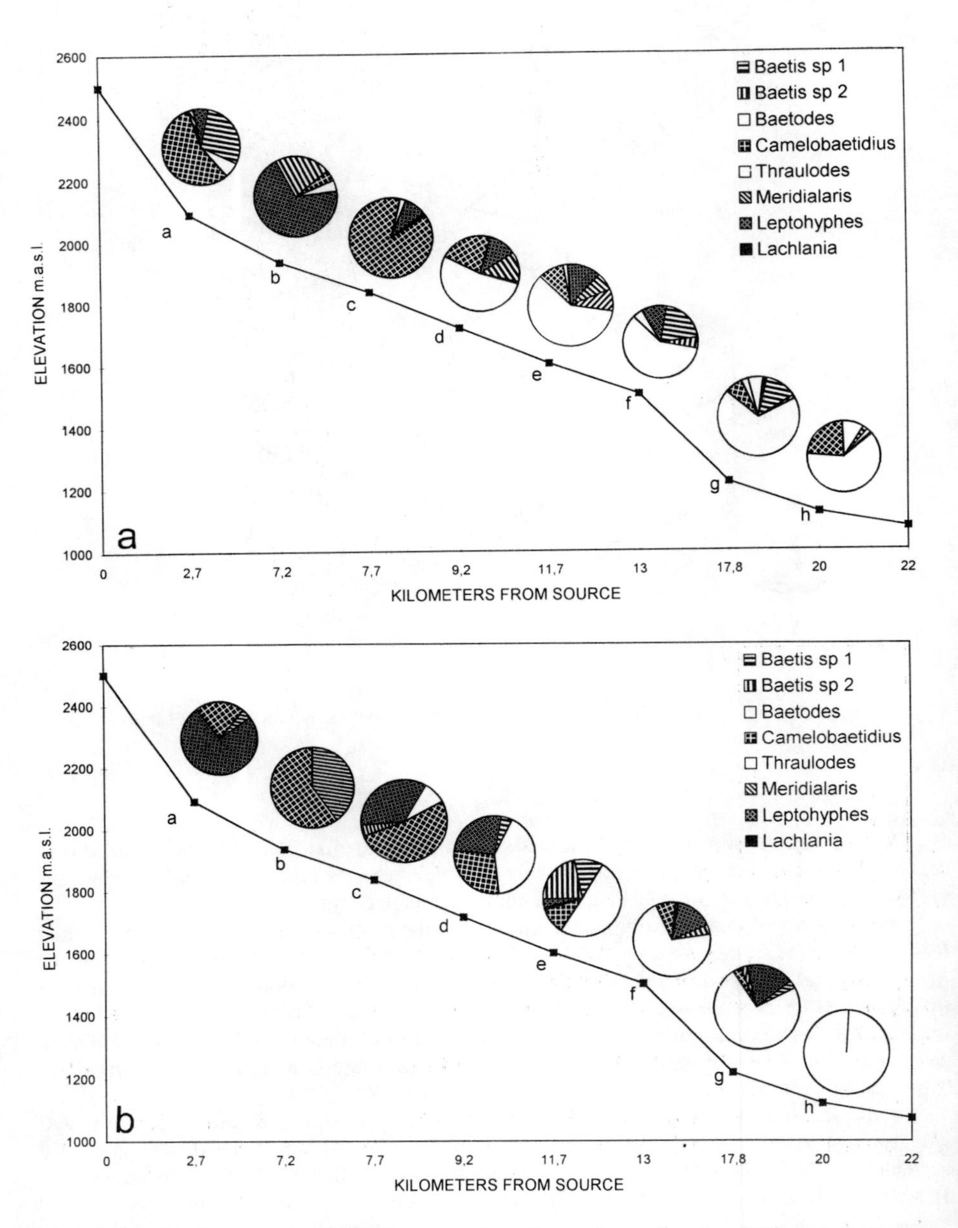

Figs. 2a-b. Numerical percentages of different genera of mayfly nymphs as a function of altitude in De los Reales River in summer of each year. Sites locations are indicated by letters on the longitudinal profile (a:1989, b:1990).

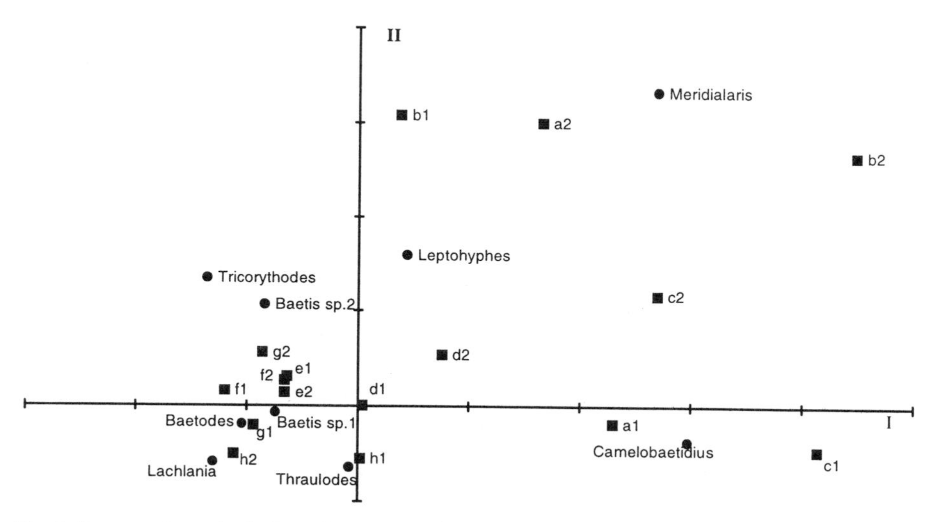

Fig. 3. Correspondence Analysis on the presence of the mayfly in De los Reales river. Distribution of the samples in the plane defined by the I and II components.

The diversity index was smaller than in some prealpine streams (Switzerland) from 1,030 m to 1,670 m.a.s.l. (Breitenmoser-Würsten and Sartori, 1995), which is in agreement with generalized concepts of a greatest richness and diversity in the benthic fauna in temperate regions.

The strong correlation between both years show some degree of persitence, in Townsend et al. (1987) sense, of the taxocene in the stations.

The ordering of stations *d-h* (Fig. 3) is probably induced by *Baetodes* sp., mostly because is more abundant downstream. These five stations are very similar especially in depth, displaying less variable conditions from year to year. The five upper stations are situated in the temperate cloud forest, a monoespecific forest of *Alnus acuminata* (Betulaceae). Within of these stations, the three upper ones (*a-c*) are situated in a very unstable area.

The station b_2 is important on the first axis due to low richness and abundance. This localization in the ordenation is also influenced by the presence of four individuals of *Meridialaris* sp., a negligible species. On the other hand, it is a rare species for the analysis (Ter Braak, 1987) in the previous year.

Domínguez and Ballesteros Valdez (1992) collected mayflies over the similar altitudinal range in a drainage system immediately to the north (24 Km) of our study area. They listed 10 species of nymphs in winter and they found a faunal break at an altitude of 770-670 m, however this range is not included in our study. Noticeably, the overall richness and the specific constitution were similar (60%). The altitudinal (1,700 - 1,100 m.a.s.l.) comparable stations showed lower richness (generally one species in each) Domínguez and Ballesteros Valdez (1992) with respect to data from 1989.

In this same river at lower elevations (450 m.a.s.l.) in winter, Fernández et al. (1995) collected six species of mayflies, with similar constitution and representing 31.2% of the benthic community; *Baetodes* sp. was also the dominant mayfly species.

Within the general pattern of altitudinal diversity, station *b* is very particular from the abundance and richness point of view, atributable to the conditions of this station, which as influencing CA.

The importance of altitude in the taxocenotic analysis, was also observed in the forest trees (Morales et al., 1995) leading other causes to a second place.

ACKNOWLEDGMENTS

We thanks E. Domínguez for improving the language and assistance in the field and two anonymous reviewers for suggestion and constructive criticism.

REFERENCES

Argañaraz, M. S. 1984. Estudio de la composición y variación del bentos del río de Medina, desde la cabecera hasta su confluencia con el río Seco, Provincia de Tucumán. Seminario Univ. Nac. de Tucumán, 117 pp.

Breitenmoser-Würsten, C. and M. Sartori. 1995. Distribution, diversity, life cycle and growth of a mayfly community in a prealpine stream system (Insecta, Ephemeroptera). Hydrobiologia 308: 85-100.

Domínguez, E. and J. M. Ballesteros Valdez. 1992. Altitudinal replacement of Ephemeroptera in a subtropical river. Hydrobiologia 246: 83-88.

Fernández, H. R., F. Romero, L. Grosso, M. L. de Grosso, M. Peralta and M. C. Rueda. 1995. La diversidad del zoobentos en ríos de montaña del NOA, I: el río Zerda, Provincia de Tucumán, República Argentina. Acta zool. lilloana, 43 (1): 215-219.

Hunzinger, H. 1997. Hydrology of montane forest in the Sierra de San Javier, Tucumán, Argentina. Mount. Res. Devel. 17 (4): 299-308.

Morales, J. M., M. Sirombra y A. D. Brown. 1995. Riqueza de árboles en las Yungas argentinas, pp. 163-174. In: A. D. Brown and H. R. Grau (eds.). Investigación, Conservación y Desarrollo en Selvas Subtropicales de Montaña. Proyecto de Desarrollo Agroforestal / L.I.E.Y.

Ter Braak, C. J. F. 1987. Ordination, pp. 90-173. In: R. H. G. Jongman, C. J. F. Ter Braak, O. F. R. Van Tongeren (eds.). Data Analysis in Community and Landscape Ecology. Pudoc, Wageningen.

Townsend, C. R., A. G. Hildrew and K. Schofield. 1987. Persistence of stream invertebrate communities in relation to environmental variability. J. Anim. Ecol. 56: 597-613.

Villalba, R., R. L. Holmes and J. A. Bonninsegna. 1992. Spatial patterns of climate and tree growth variations in subtropical northwestern Argentina. J. Biogeogr. 19: 631-649.

Ward, J. V. and L. Berner. 1980. Abundance and altitudinal distribution of Ephemeroptera in a Rocky Mountain stream, pp. 169-177. In: J. F. Flannagan and K. E. Marshall (eds.). Advances in Ephemeroptera Biology. Plenum Press, New York.

Zar, J. H. 1984. Biostatistical analysis. Prentice-Hall, New Jersey.

A PREDICTION MODEL OF RUNNING WATERS ECOSYSTEM IN THE CZECH REPUBLIC BASED ON MAYFLY TAXOCENES OF UNDISTURBED RHITHRAL STREAMS

Světlana Zahrádková[1], Tomáš Soldán[2], and Jiří Kokeš[3]

[1] Department of Zoology and Ecology, Faculty of Science
Masaryk University, Kotlárská 2, 611 37 Brno, Czech Republic
[2] Institute of Entomology, Academy of Sciences of the Czech Republic
Branišovská 31, 370 05 Ceské Budejovice, Czech Republic
[3] Water Research Institute T.G.M. Prague, Dept. Brno
Drevarská 12, 657 57 Brno, Czech Republic

ABSTRACT

Samples of mayfly larvae and main environmental variables from 80 undisturbed localities evenly distributed in the Czech Republic were taken and measured during 1994-1997. Data were examined by TWINSPAN and by canonical correspondence analysis (CCA) to define mayfly taxocenes characteristic of different areas and to prepare a mayfly database for a prediction model of effects of environmental changes (HOBENT). To test the general applicability of the model, mayfly species composition at 13 localities was compared to a hypothetical, desired taxocenes. Although restricted only to a part of aquatic macroinvertebrate communities (mayflies), this procedure permitted evaluation of environmental degradation of other 13 localities. Despite some restrictions, the planned total number of 400 background localities seems to be sufficient for successful application of this prediction model.

INTRODUCTION

Aquatic macroinvertebrates studies have been widely used to estimate water quality based on saprobial (e. g. Zelinka and Marvan, 1961) and diversity or species richness indices. In many places, organic pollution is being gradually decreased and recovery of aquatic biotopes (restoration) represents one of the most important research problems. Consequently, determination of desired environmental characteristics of watercourses and respective target communities is urgently needed. As much as possible, the natural (or original) conditions are most desirable. To meet this need, the system PERLA, a new approach to evaluate the environmental quality of running waters based on macro-invertebrate communities analyses, is being gradually developed in the Czech Republic. The

Trends in Research in Ephemeroptera and Plecoptera
Edited by E. Dominguez, Kluwer Academic/Plenum Publishers, 2001

most important tool of this system is the recently developed program HOBENT (Kokeš, 1997) constructed in a similar way to the British RIVPACS (for details see e.g. Armitage et al. 1983, Wright 1995, Wright et al. 1989, 1993). This system is based on a comparison of the actual macroinvertebrate diversity at particular place with a respective target or desired community. Target community (or taxocoene) is defined as an unaffected community of the (reference) localities with no or minimal human-mediated environmental pressure. The comparison of targets with actual community composition enables evaluation of the extent of disturbance and prediction of the probability of successful recovery. Careful selection of localities, though to be reasonably pristine, analysis of their biodiversity and abiotic variables, definition of interrelations and computer processing of data represent the first step in a development of a reference database. The objectives of this paper are: (i) to develop a mayfly database for the prediction model; (ii) to define characteristic mayfly taxocenes of a modeled area (rhithral of the Czech Republic); and, (iii) to test the prediction model.

Table 1. List of reference localities with basic characteristics and associated TWINSPAN groups

No. of locality	Name of watercourse	Sampling site	River basin	Distance from source	Altitude [m]	Orographic unit	Biogeographic subprovince of CR	Coordinates*	TWINSPAN group
1	ŘÍČKA	Říčky	L	7.7	625	IV	Hercynicum	57-64	A
2	ORLIČKA	Orlička	L	0.5	473	IV	Hercynicum	59-66	G
3	BROOK	Horní Lipka	L	8.9	605	IV	Hercynicum	58-66	G
4	JIZERA	Splzov	L	93.7	279	IV	Hercynicum	53-57	J
5	MUKAŘOVSKÝ BROOK	Splzov	L	2.0	327	IV	Hercynicum	53-57	G
6	VLTAVA	Pěkná	L	58.7	725	I	Hercynicum	71-49	J
7	KRASETÍNSKÝ BROOK	Krasetín	L	1.8	585	I	Hercynicum	71-51	G
8	ČERNÁ	Benešov	L	16.5	654	I	Hercynicum	73-54	G
9	LUŽNICE	Stará Hlína	L	98.5	430	II	Hercynicum	70-54	H
10	NOVÁ ŘEKA	Mláka	L	2.2	432	II	Hercynicum	69-55	H
11	ŽIDOVA STROUHA	Nuzice	L	2.8	375	II	Hercynicum	67-52	H
12	FILIPOHUŤSKÝ BROOK	Filipova Huť	L	2.8	1020	I	Hercynicum	69-47	A
13	ZHŮŘSKÝ BROOK	Turnerova chata	L	0.5	945	I	Hercynicum	69-47	A
14	PRÁŠILSKÝ BROOK	Prášily	L	4.8	905	I	Hercynicum	68-46	A
15	OTAVA	Sušice	L	94.3	490	I	Hercynicum	67-47	A
16	OSTRUŽNÁ	Sušice	L	3.3	475	I	Hercynicum	67-46	J
17	BLANICE	Blažejovice	L	78.6	748	I	Hercynicum	70-49	J
18	ZLATÝ BROOK	Záhoří	L	24.2	659	I	Hercynicum	70-50	G
19	ZÁVIŠÍNSKÝ BROOK*	Bezdědovice	L	3.6	435	II	Hercynicum	65-49	I
20	STRŽSKÝ BROOK	Cikháj	L	10.0	650	II	Hercynicum	63-61	I
21	SÁZAVA	Ledeč	L	93.4	351	II	Hercynicum	62-57	L
22	TRNÁVKA	Hrádek	L	30.5	475	II	Hercynicum	65-56	I
23	BLANICE	Světlá	L	30.5	365	II	Hercynicum	63-55	I
24	ZAHOŘANSKÝ BROOK	Davle - Libřice	L	2.5	225	V	Hercynicum	61-52	I
25	SENNÝ BROOK	Drmoul	L	4.5	554	I	Hercynicum	60-41	G
26	MŽE	Milíkov	L	53.0	380	V	Hercynicum	62-43	J
27	KLABAVA	Kamenné	L	40.0	601	V	Hercynicum	62-47	G
28	STŘELA	Ondřejov	L	24.6	348	III	Hercynicum	60-45	L
29	KRALOVICKÝ BROOK	Dolní Hradiště	L	13.5	320	V	Hercynicum	60-47	I
30	VELKÁ LIBAVA	Arnoltov	L	11.0	568	III	Hercynicum	58-41	G
31	TEPLÁ	Teplička	L	18.0	453	III	Hercynicum	58-43	J
32	LOMNICE	Kyselka	L	160.0	360	III	Hercynicum	57-44	G
33	OHŘE	Kadaň	L	125.0	285	III	Hercynicum	56-43	L
34	BROOK	Prackovice	L	0.9	225	VI	Hercynicum	54-50	C
35	TELNICKÝ BROOK	Adolfov	L	6.5	645	III	Hercynicum	52-49	C
36	PLOUČNICE	Mimoň	L	76.0	283	VI	Hercynicum	53-54	H
37	LESNÍ BROOK	Šluknov	L	1.8	349	IV	Hercynicum	50-52	G
38	TROJANOVICKÝ BROOK	Trojanovice	O	1.5	720	IX	Hercynicum	64-75	C
39	ČERNÁ OPAVA	Mnichov	O	9.9	610	IV	Hercynicum	58-70	C
40	STŘEDNÍ OPAVA	Vidly	O	2.9	770	IV	Hercynicum	58-69	C

* Coordinates according to European uniform grid system.

Table 1 (cont.)

No. of locality	Name of watercourse	Sampling site	River basin	Distance from source	Altitude [m]	Orographic unit	Biogeographic subprovince of CR	Coordinates*	TWINSPAN group
41	MORAVICE	Karlov	O	6.2	620	IV	Hercynicum	59-69	D
42	MORÁVKA	Strongy	O	2.0	707	IX	Carpathicum	67-77	C
43	MOHELNICE	Zlatník	O	3.8	625	IX	Carpathicum	64-77	C
44	LOMNÁ	Dolní Lomná	O	7.5	495	IX	Carpathicum	64-78	E
45	OLŠE	Třinec	O	20.8	309	IX	Carpathicum	63-78	K
46	TYRKA	Tyra	O	1.5	600	IX	Carpathicum	63-77	C
47	ROPIČANKA	Řeka	O	1.1	540	IX	Carpathicum	63-77	C
48	STŘÍBRNÝ BROOK	Nýznerov	O	6.8	750	IV	Hercynicum	57-68	C
49	ČERVENÝ BROOK	Červená Voda	O	2.5	400	IV	Polonicum	56-68	F
50	RAMZOVSKÝ BROOK	Ramzová	O	5.1	590	IV	Hercynicum	57-68	C
51	MORAVA	Dolní Morava	M	9.8	610	IV	Hercynicum	58-66	D
52	KRUPÁ	Seninka	M	2.0	700	IV	Hercynicum	57-67	C
53	BRANNÁ	Branná	M	3.0	750	IV	Hercynicum	58-68	C
54	BRANNÁ	Jindřichov	M	15.0	510	IV	Hercynicum	58-68	D
55	MERTA	Vernířovice	M	4.3	610	IV	Hercynicum	59-68	C
56	MORAVSKÁ SÁZAVA	Albrechtice	M	9.5	430	IV	Hercynicum	60-65	F
57	HYNČINSKÝ BROOK	Hynčina	M	3.0	420	IV	Hercynicum	61-67	C
58	MORAVSKÁ SÁZAVA	Lupěné	M	45.5	255	IV	Hercynicum	61-66	K
59	MÍROVKA	Mírov	M	10.0	350	IV	Hercynicum	62-67	F
60	SITKA	Šternberk	M	20.6	320	IV	Hercynicum	62-69	F
61	BYSTŘICE	Hluboč ky	M	37.4	335	IV	Hercynicum	63-70	E
62	VSETÍNSKÁ BEČVA	Velké Karlovice	M	8.1	620	IX	Carpathicum	66-76	E
63	JEZERNÍ BROOK	Velké Karlovice	M	2.1	570	IX	Carpathicum	66-75	F
64	SENICE	Leskovec	M	27.2	370	IX	Carpathicum	67-74	K
65	ROŽNOVSKÁ BEČVA	above dam	M	1.5	510	IX	Carpathicum	65-76	D
66	JUHYNĚ	Troják	M	3.8	500	IX	Carpathicum	66-72	F
67	RÁZTOKA	Rusava	M	2.5	400	IX	Carpathicum	66-72	F
68	TRNÁVKA	Hrobice - Neubuz	M	9.0	290	IX	Carpathicum	67-72	F
69	OLŠAVA	Pitín	M	1.7	375	IX	Carpathicum	69-73	B
70	VELIČKA	Vápenky	M	3.9	470	IX	Carpathicum	71-71	C
71	BOLIKOVSKÝ BROOK	Lipnice	M	6.7	550	II	Hercynicum	68-57	I
72	DYJE	Podhradí	M	96.6	355	II	Hercynicum	71-59	L
73	TRHONICKÝ BROOK	Jimramov	M	5.0	530	II	Hercynicum	63-63	F
74	SVRATKA	Unčín	M	45.4	515	II	Hercynicum	63-63	K
75	NEDVĚDIČKA	Pernštejn	M	27.6	400	II	Hercynicum	66-64	E
76	LITAVA	Zástřizly	M	3.3	320	IX	Carpathicum	68-69	B
77	JIHLAVA	Horní Ves	M	3.1	615	II	Hercynicum	67-57	F
78	JIŘÍNSKÝ BROOK	Hlávkov	M	8.3	560	II	Hercynicum	65-58	I
79	JIHLAVA	Ivář	M	179.6	185	VIII	Pannonicum	70-65	L
80	KYJOVKA	Staré Hutě	M	2.2	425	IX	Carpathicum	68-69	B

STUDY SITES AND METHODS

There are three main river basins – the Labe (Elbe), the Odra (Oder) and the Morava in the Czech Republic situated in four biogeographic subprovinces (Culek, 1996) and ten orographic subprovinces (Demek, 1987). Altogether, 80 localities of reasonably pristine streams (Tab. 1, Fig. 1) are evenly distributed in all biogeographic subprovinces (Fig. 2) and eight orographic subprovinces (Fig. 3). Reference localities were those sampled during the 1950's (so called Project No. 210) and which in the 1990's did not exhibit any pronounced differences in either species composition or in abiotic variables (for details see Soldán et al., 1998).

Kick samples of larvae were obtained at early spring, spring, summer and autumn during 1994-1997. Equal attention was paid to all types of substratum and pure collection time was restricted to 10-15 minutes at each locality in each season. Main environmental variables (altitude, slope, distance from source, mean stream width and depth, mean current velocity, instantaneous discharge, mean substrate roughness, water temperature, dissolved oxygen, biochemical oxygen demand, pH, conductivity, concentration of total phosphorus, ammonium and nitrate and data on submerged and riparian vegetation were measured. Index of species richness (Margalef, 1958) and index of diversity (Shannon and Weaver, 1963) were calculated.

Table 2. List of species found and abbreviations of species names used in other tables and figures

Species		Species name abbrev.	Species		Species name abbrev.
Ameletus inopinatus	Eaton, 1887	Amel inop	*Heptagenia flava*	Rostock, 1877	Hept flav
Siphlonurus aestivalis	(Eaton, 1903)	Siph aest	*Heptagenia fuscogrisea*	(Retzius, 1783)	Hept fusc
Siphlonurus alternatus	(Say, 1824)	Siph alte	*Heptagenia longicauda*	(Stephens, 1836)	Hept long
Siphlonurus armatus	(Eaton, 1870)	Siph arma	*Heptagenia sulphurea*	(Müller, 1776)	Hept sulp
Siphlonurus lacustris	(Eaton, 1870)	Siph lacu	*Rhithrogena beskidensis*	Alba et Sowa, 1987	Rhit besk
Alainites muticus	(Linné, 1758)	Alai muti	*Rhithrogena carpatoalpina*	Klonowska et al., 1985	Rhit carp
Baetis alpinus	(Pictet, 1843 - 1845)	Baet alpi	*Rhithrogena germanica*	Eaton, 1885	Rhit germ
Baetis buceratus	Eaton, 1870	Baet buce	*Rhithrogena hercynia*	Landa, 1970	Rhit herc
Baetis calcaratus	Keffermüller, 1972	Baet calc	*Rhithrogena iridina*	(Kolenati, 1859)	Rhit irid
Baetis fuscatus	(Linné, 1761)	Baet fusc	*Rhithrogena landai*	Sowa et Soldán, 1984	Rhit land
Baetis lutheri	Müller-Liebenau, 1967	Baet luth	*Rhithrogena loyolaea*	Navás, 1922	Rhit loyo
Baetis melanonyx	Pictet, 1843 - 1845	Baet mela	*Rhithrogena savoiensis*	Alba et Sowa, 1987	Rhit savo
Baetis rhodani	Pictet, 1843 -1845	Baet rhod	*Rhithrogena semicolorata*	(Curtis, 1834)	Rhit semi
Baetis scambus	Eaton, 1870	Baet scam	*Choroterpes picteti*	(Eaton, 1871)	Chor pict
Baetis vernus	Curtis, 1834	Baet vern	*Habroleptoides confusa*	Sartori et Jacob, 1986	Habr conf
Nigrobaetis niger	(Linné, 1761)	Nigr nige	*Habrophlebia fusca*	(Curtis, 1834)	Habr fusc
Centroptilum luteolum	(Müller, 1776)	Cent lute	*Habrophlebia lauta*	Eaton, 1884	Habr laut
Centroptilum pennulatum	Eaton, 1870	Cent penn	*Leptophlebia marginata*	(Linné, 1767)	Lept marg
Cloeon dipterum	(Linné, 1761)	Cloe dipt	*Leptophlebia vespertina*	(Linné, 1758)	Lept vesp
Procloeon bifidum	(Bengtsson, 1912)	Proc bifi	*Paraleptophlebia cincta*	(Retzius, 1783)	Para cinc
Oligoneuriella rhenana	(Imhoff, 1852)	Olig rhen	*Paraleptophlebia submarginata*	(Stephens, 1835)	Para subm
Ecdyonurus aurantiacus	(Burmeister, 1839)	Ecdy aura	*Potamanthus luteus*	(Linné, 1767)	Pota lute
Ecdyonurus austriacus	Kimmins, 1958	Ecdy aust	*Ephoron virgo*	(Olivier, 1791)	Epho virg
Ecdyonurus dispar	(Curtis, 1843)	Ecdy disp	*Ephemera danica*	Müller, 1764	Ephe dani
Ecdyonurus forcipula	Pictet, 1843 - 1845	Ecdy forc	*Ephemera lineata*	Eaton, 1870	Ephe line
Ecdyonurus insignis	(Eaton, 1870)	Ecdy insi	*Ephemera vulgata*	Linné, 1758	Ephe vulg
Ecdyonurus starmachi	Sowa, 1971	Ecdy star	*Ephemerella ignita*	(Poda, 1761)	Ephe igni
Ecdyonurus subalpinus	Klapálek, 1905	Ecdy suba	*Ephemerella mucronata*	(Bengtsson, 1909)	Ephe mucr
Ecdyonurus submontanus	Landa, 1970	Ecdy subm	*Ephemerella notata*	Eaton, 1887	Ephe nota
Ecdyonurus torrentis	Kimmins, 1942	Ecdy torr	*Torleya major*	(Klapálek, 1905)	Torl majo
Ecdyonurus venosus	(Fabricius, 1775)	Ecdy veno	*Brachycercus harrisella*	Curtis, 1834	Brac harr
Electrogena affinis	(Eaton, 1885)	Elec affi	*Caenis luctuosa*	(Burmeister, 1839)	Caen luct
Electrogena lateralis	(Curtis, 1834)	Elec late	*Caenis macrura*	Stephens, 1835	Caen macr
Electrogena quadrilineata	(Landa, 1970)	Elec quad	*Caenis pseudorivulorum*	Keffermüller, 1960	Caen pseu
Epeorus sylvicola	(Pictet, 1865)	Epeo sylv	*Caenis rivulorum*	Eaton, 1884	Caen rivu
Heptagenia coerulans	Rostock, 1877	Hept coer			

TWINSPAN (Hill, 1979) was used to classify the data obtained. No samples were deleted or weighted, no species were deleted, but ubiquitous species (such as *Baetis rhodani* and *Ephemerella ignita*) were downweighted. Pseudo-species (the way of substituting a quantitative variable - abundance - by several quantitative variables used in program TWINSPAN) were defined at 5 cut levels.

Canonical correspondence analysis - (CCA, Ter Braak, 1988) was selected for spatial evaluation of species and environmental data. As basic input for the analysis, the complete data set comprising 80 samples, 71 species and 15 environmental variables was used. No samples were deleted or weighed. No species were deleted, but ubiquitous species were downweighted in the same way as in TWINSPAN. Environmental variables were individually tested by the Monte Carlo permutation test and only significant (p=0.01, 999 unconstrained permutations) outcomes were used for CCA. During the forward selection the

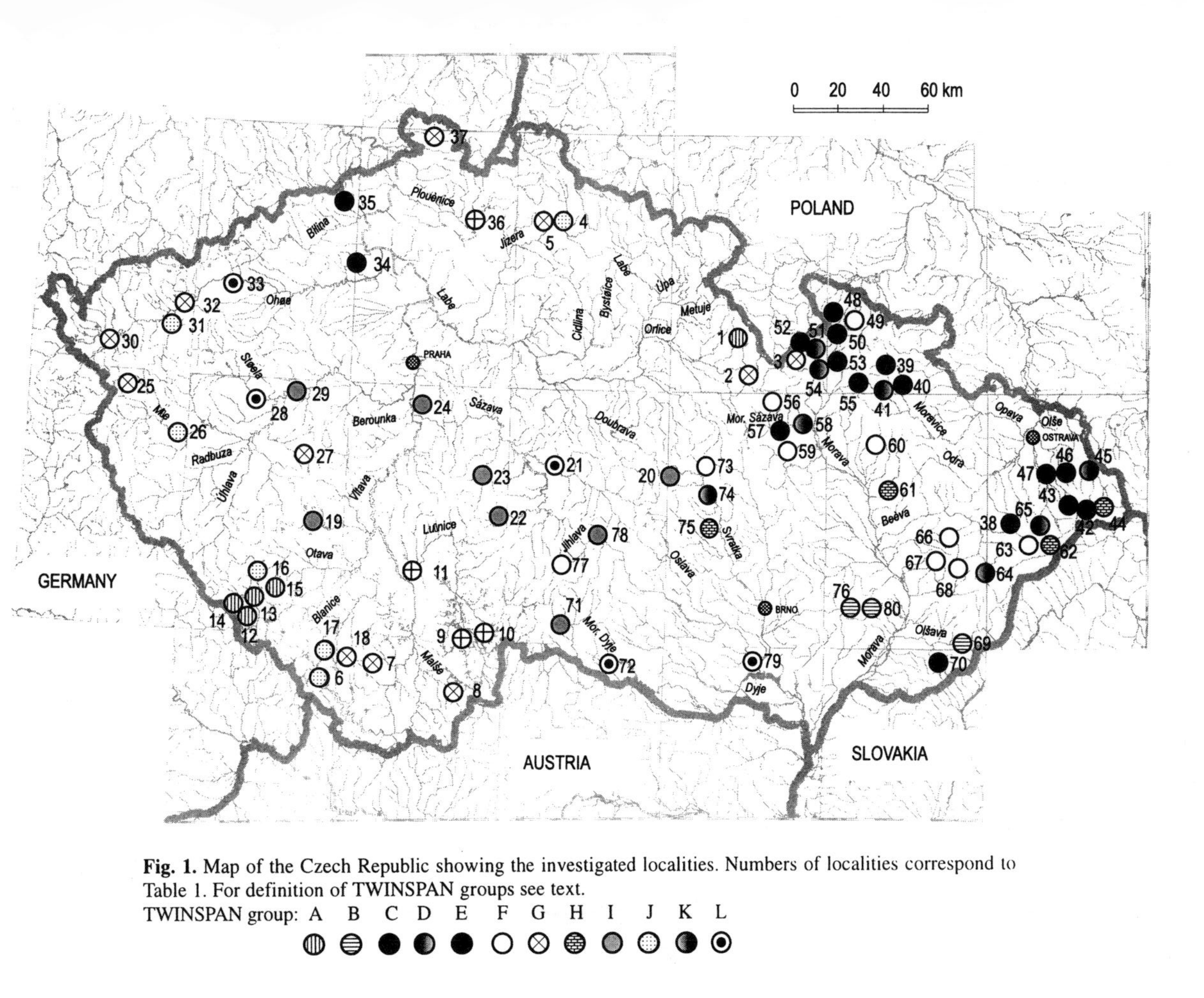

Fig. 1. Map of the Czech Republic showing the investigated localities. Numbers of localities correspond to Table 1. For definition of TWINSPAN groups see text.
TWINSPAN group: A B C D E F G H I J K L

variables were used in order to follow their fit. Significance of the first canonical axis was tested by the Monte Carlo permutation test.

The program HOBENT (Kokeš, 1997) used to predict and compare the species composition is based on the procedures as follows: (i) localities of the reference database are classified into individual final TWINSPAN groups, (ii) these are characterised by the set of the most significant environmental variables using Multiple Discriminant Analysis (MDA) (Klecka, 1975), (iii) the locality in question is then classified with respect to its particular environmental variables to belong, with defined probability, into one or more respective group(s) of reference database, (iv) based on these probabilities a list of species expected at this particular locality is defined, (v) this list and real species occurrence are compared to define the coefficient B expressing similarity of real and predicted taxocoene (B = 0: taxocenes completely different, B =1: taxocenes identical). The B values thus enable to evaluate the ecological quality of the locality.

To test the applicability of the model, mayfly species composition at 13 localities (see Tab. 7) was compared to a hypothetical, desired community.

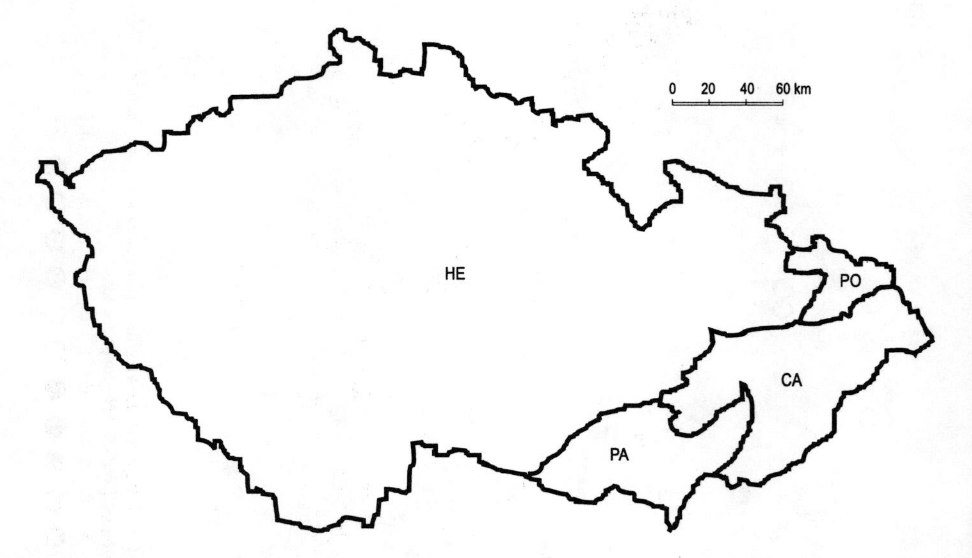

Fig. 2. Map of the Czech Republic showing main biogeographic subprovinces according to Culek (1996). HE - Hercynicum, PO - Polonicum, CA -Carpathicum, PA - Pannonicum. For position of localities studied in individual subprovinces see Table 1.

RESULTS

TWINSPAN Classification

In the first step (D1, Fig. 4), the TWINSPAN principally separated the data into class *0 (i.e. localities of typical rhithral and transitional zone between crenal and rhithral) and class *1 on the other hand (localities of both typical potamal zone and transitional zone between rhithral and potamal). The second division (D2) separated headwater localities (class *00) from other rhithral localities (class *01). Class *00 was divided by the fourth division (D4) into classes *000 and *001. The former class represented final group A, indicator species *Baetis alpinus* (for full scientific names of species and their abbreviations used in Fig. 4 see

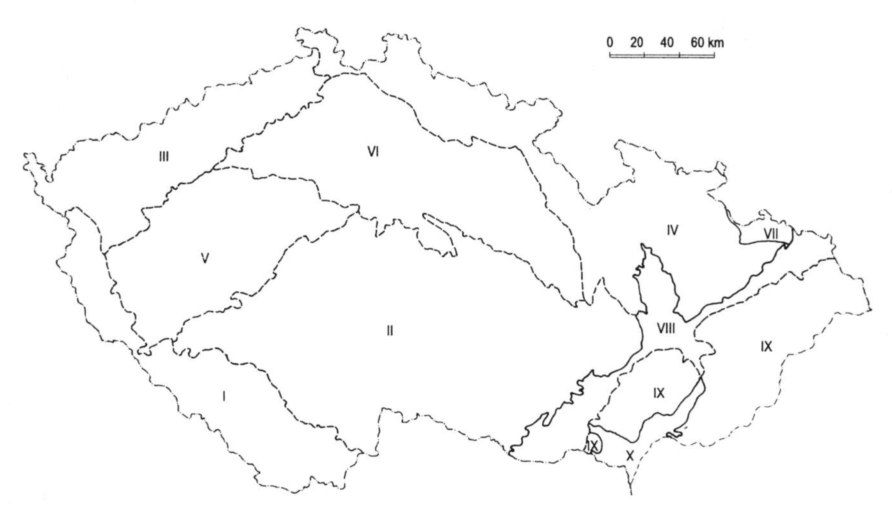

Fig. 3. Map of the Czech Republic showing main orographic subprovinces according to DEMEK (1987). For position of localities studied in individual biogeographic subprovinces see Table 1.
I - the Šumava Mts. system, II - the Czech Moravian Highland System, III - the Krušné Hory Mts. system, IV - the Krkonoše - Jeseníky Mts. system (East Sudete system), V - the Berounka Highland system, VI - the Czech Plateau, VII - the Central Poland lowlands, VIII - depressions of the Outer Carpathians, IX - the Outer West Carpathians, X - the Vienna Basin. Note that no localities in question are situated in the orographic unit No. VII and X.

Tab. 2), largely montane, epirhithral stretches with naturally low pH values (peat-bog water) and lower water temperature, the Labe basin. The latter class (*001) gave rise to final group B (D9 - *0010, no specific indicator, lower elevation brooks of the Carpathians, the Morava basin) and class *0011 with localities of typical epirhithral stretches. Next division of this class (D19) resulted in the class *00110 (final group C, indicators *Habroleptoides confusa, Ecdyonurus subalpinus, Ephemerella mucronata* and *Baetis alpinus*, mostly localities of epirhithral of predominantly higher coline zone, with low distance from source, high slope and lower water temperature) and class *00111 (final group D, without a specific indicator, epirhithral river stretches of middle elevations, more distant from source). Water temperatures were lower and discharges were higher than those at the localities of group C. The fifth division (D5) separated final group G (the class *011, indicators *Ephemera danica, Baetis vernus, Ephemerella ignita, Baetis alpinus*, mainly localities of rhithral of the Labe basin) and followed by division of D10 separated class *010 into two final groups: E (*0100, indicator *Torleya major,* metarhithral stretches of the Morava and Odra basins with higher discharge rates) and F (*0101, indicator *Ephemerella mucronata*, epirhithral stretches predominantly in the Morava basin).

Class *1 (right side of dichotomy, Fig. 4) was divided into class *10 and class *11 (final group L with indicator *Potamanthus luteus*, potamal localities of the Labe and Morava basins).

The division D6 separated classes *100 and *101. Class *100 gave rise to classes *1000 and *1001. The former represents the final group H (indicators *Heptagenia flava, Procloeon bifidum* and *Ephemerella ignita*, localities of specific, nearly potamal character at lower elevations, with low slope), the latter class *1001 included rhithral localities predominantly of lower elevations, situated mostly in the Labe basin. The division D25 separated the final groups I (*10010, no specific indicator, brooks with sandy or gravel bottom,

Table 3. Hierarchic classification (TWINSPAN) of reference localities – two-way ordered table. For number of localities see Table 1, for full species names see Table 2, for characterisation of individual groups (A-L) see in the text.

No. of locality / Species	13	12	14	1	15	69	76	80	38	43	53	50	39	40	46	47	48	52	55	57	70	42	35	34	65	51	41	54	44	62	61	75	60	67	73	49	59	56	66	63	68	77	Class of species
Ecdy suba	-	-	-	-	-	3	2	4	2	2	2	2	-	-	-	2	1	-	1	2	2	-	-	2	-	-	-	-	-	-	-	-	-	-	-	-	1	-	-	-	-	-	0 0 0
Rhit irid	2	-	-	-	2	3	-	4	2	3	-	2	1	-	2	3	-	2	1	-	2	3	3	2	1	2	-	2	-	3	-	-	-	-	-	1	-	-	-	3	-	-	0 0 0
Amel inop	-	3	2	-	-	-	-	-	-	-	-	1	-	-	2	-	-	-	-	-	-	-	-	-	-	-	-	-	-	-	-	-	-	-	-	-	-	-	-	-	-	-	0 0 0
Rhit loyo	-	-	-	-	-	-	-	-	-	-	-	-	-	-	-	-	-	-	-	-	-	-	-	-	-	1	-	-	-	-	-	-	-	-	-	-	-	-	-	-	-	-	0 0 0
Rhit land	-	-	-	-	-	-	-	-	-	-	-	-	-	-	-	-	-	-	-	-	-	-	-	-	-	-	1	-	-	-	-	-	-	-	-	-	-	-	-	-	-	-	0 0 0
Ecdy veno	-	2	-	-	2	-	-	1	2	3	3	2	2	2	3	2	-	1	-	-	-	2	2	-	1	2	2	-	-	2	-	-	-	-	-	2	-	-	-	-	-	-	0 0 0
Bact mela	-	-	-	-	-	-	-	-	2	2	3	-	1	-	-	3	-	2	2	2	-	-	-	-	-	-	-	-	-	-	-	-	-	-	-	-	2	2	-	-	-	-	0 0 0
Ecdy forc	2	-	-	-	1	-	-	-	-	-	-	-	-	-	-	-	-	-	-	-	-	-	-	-	-	-	-	-	-	-	-	-	-	-	-	-	-	-	-	-	-	-	0 0 0
Ecdy aust	3	-	-	-	-	-	-	-	-	-	-	-	-	-	-	-	-	-	-	-	-	-	-	-	-	-	-	-	-	-	-	-	-	-	-	-	-	-	-	-	-	-	0 0 0
Epeo sylv	-	-	-	-	3	-	-	-	2	2	2	2	3	3	2	2	2	2	3	2	2	3	2	-	2	3	2	2	3	3	2	3	1	-	3	2	3	3	1	2	-	-	0 0 0
Rhit carp	-	-	-	-	-	-	2	-	-	3	3	2	2	2	1	3	2	3	-	2	3	2	-	-	2	2	-	1	2	3	-	-	-	2	2	2	3	3	4	3	1	-	0 0 0
Ecdy star	-	-	-	-	-	-	-	-	-	-	-	-	-	-	-	-	-	-	-	-	1	2	-	-	-	-	-	-	-	-	-	-	2	2	-	-	-	-	-	3	3	-	0 0 0
Elec quad	-	-	-	-	-	2	2	3	2	2	1	2	-	-	1	-	-	1	-	-	-	2	-	-	-	-	-	-	-	-	-	-	-	-	-	-	-	-	1	2	2	2	0 0 1
Bact alpi	4	4	4	3	5	1	-	-	2	2	2	2	3	3	4	4	3	3	3	-	2	3	4	3	2	2	2	2	3	1	-	-	-	-	-	-	-	-	-	-	-	-	0 0 1
Ephe mucr	-	-	-	2	3	-	-	1	-	2	-	1	2	1	3	2	2	2	-	3	2	-	-	-	-	-	-	-	1	-	-	-	2	2	2	1	3	2	2	2	-	-	0 0 1
Habr conf	2	-	-	-	2	3	-	2	1	2	1	2	2	2	1	2	3	-	2	2	2	3	3	3	2	-	-	-	-	3	3	2	4	1	3	2	4	3	3	5	2	-	0 0 1
Ecdy subm	-	-	-	-	-	-	-	-	2	-	2	1	-	-	-	-	-	-	-	-	-	-	-	-	-	-	-	-	-	2	-	-	-	-	-	-	-	-	-	-	2	-	0 1 0
Rhit semi	2	-	-	-	-	-	-	-	-	2	-	1	1	-	-	-	1	3	2	-	2	2	4	3	2	2	2	-	-	-	1	2	2	-	3	-	3	-	-	-	-	2	0 1 0
Bact rhod	5	4	3	3	3	3	4	4	3	4	2	3	4	3	3	2	3	2	2	4	3	3	4	3	3	3	2	3	3	3	4	5	4	3	5	4	5	5	4	4	4	3	0 1 0
Alai muti	-	-	-	-	2	2	-	2	2	3	-	2	2	3	5	2	2	2	1	3	2	1	-	2	-	2	2	2	2	2	3	2	2	-	3	2	3	-	-	2	-	-	0 1 0
Siph lacu	-	-	2	2	2	-	-	-	-	-	-	-	-	-	-	-	-	-	-	-	-	-	-	-	-	-	-	-	-	-	-	-	-	-	-	-	-	-	-	-	-	-	0 1 0
Elec late	-	-	-	-	-	-	-	-	-	-	-	-	-	-	-	-	-	-	-	-	-	-	-	-	-	-	-	-	-	-	-	-	-	-	-	-	-	-	-	-	-	-	0 1 1
Elc affi	-	-	-	-	-	-	-	-	-	-	-	-	-	-	-	-	-	-	-	-	-	-	-	-	-	-	-	-	-	-	-	-	-	-	-	-	-	-	-	-	-	-	0 1 1
Caen pseu	-	-	-	-	-	-	-	-	-	-	-	-	-	-	-	-	-	-	-	-	-	-	-	-	-	-	-	-	-	2	1	2	2	-	-	-	-	-	-	-	-	-	0 1 1
Habr laut	-	-	-	-	2	-	-	-	-	2	-	-	-	-	-	-	-	-	-	-	2	3	-	-	1	-	-	-	-	2	2	1	2	1	2	2	-	1	1	3	1	2	0 1 1
Bact vern	-	4	3	2	4	-	-	-	-	2	2	2	-	-	-	-	-	2	1	-	-	-	4	-	-	1	-	-	-	-	-	-	-	-	-	-	2	-	-	-	2	-	1 0 0
Ephe dani	-	-	-	-	-	2	-	2	-	-	-	-	-	-	-	-	-	-	-	-	2	-	-	-	-	-	-	-	-	-	-	-	-	-	-	-	-	-	1	2	1	2	1 0 0
Ephe igni	-	-	-	-	4	-	-	-	-	2	-	-	3	-	-	-	1	-	-	-	1	2	-	-	-	2	-	2	2	2	-	-	3	2	2	2	2	2	1	2	2	1	1 0 0
Torl majo	-	-	-	-	-	-	-	-	-	-	-	-	2	-	-	-	-	-	-	-	-	-	-	-	2	-	-	1	2	2	2	2	-	-	-	-	-	-	-	2	-	-	1 0 0
Ecdy torr	-	-	-	-	-	-	-	-	-	-	-	-	-	-	-	-	-	-	-	-	-	-	2	-	-	-	-	-	3	1	2	3	2	-	3	-	2	-	-	1	1	2	1 0 0
Ecdy aura	-	-	-	-	-	-	-	-	-	-	-	-	-	-	-	-	-	-	-	-	-	-	-	-	-	-	-	-	-	-	-	2	-	-	-	-	-	-	-	-	-	-	1 0 1
Bact luth	-	-	-	-	-	-	-	-	-	-	-	-	1	-	-	-	-	-	-	-	-	-	-	-	-	-	-	-	-	-	2	-	-	-	-	-	-	2	-	2	-	-	1 0 1
Habr fusc	-	-	-	-	-	-	-	-	-	-	-	-	-	-	-	-	-	-	-	-	-	-	-	-	-	-	-	-	-	-	-	-	-	1	-	-	-	-	-	-	-	-	1 1 0
Baet bucc	-	-	-	-	-	-	-	-	-	-	-	-	-	-	-	-	-	-	-	-	-	-	-	-	-	-	-	-	-	-	1	3	-	-	-	-	-	-	-	-	-	-	1 1 0
Bact scam	-	-	-	-	-	-	-	-	-	-	-	-	-	-	-	-	-	-	-	-	-	-	-	-	-	2	-	2	3	-	-	-	-	-	-	-	-	-	-	-	-	-	1 1 0
Rhit herc	-	-	-	-	-	-	-	-	-	-	-	-	-	-	-	-	-	-	-	-	-	-	-	-	-	1	-	-	-	-	-	-	-	-	-	-	-	-	-	-	-	-	1 1 0
Rhit besk	-	-	-	-	-	-	-	-	-	-	-	-	-	-	-	-	-	-	-	-	-	-	-	-	-	1	-	1	-	-	-	-	-	-	-	-	-	-	-	-	-	-	1 1 0
Lept vesp	-	-	-	-	-	-	-	-	-	-	-	-	-	-	-	-	-	-	-	-	-	-	-	-	-	-	-	-	-	-	-	-	-	-	-	-	-	-	-	-	-	-	1 1 0
Rhit germ	-	-	-	-	-	-	-	-	-	-	-	-	-	-	-	-	-	-	-	-	-	-	-	-	-	-	-	-	1	-	-	-	-	-	-	-	-	-	-	-	-	-	1 1 0
Cloe dipt	-	-	-	-	-	-	-	-	-	-	-	-	-	-	-	-	-	-	-	-	-	-	-	-	-	-	-	-	-	-	-	-	-	-	-	-	-	-	-	-	-	-	1 1 0
Para subm	-	-	-	-	-	-	-	-	-	-	-	-	-	-	-	-	-	-	-	-	-	-	2	2	-	-	-	-	-	-	-	-	-	-	-	-	-	2	-	-	1	-	1 1 0
Lept marg	-	-	-	-	-	-	-	-	-	-	-	-	-	-	-	-	-	-	-	-	-	-	-	-	-	-	-	-	-	-	-	-	-	-	-	-	-	-	-	-	-	-	1 1 0
Nigr nige	-	-	-	-	2	-	-	-	-	-	-	-	-	-	-	-	-	-	-	-	-	-	-	-	-	-	-	-	-	-	-	-	-	-	-	-	-	-	-	-	-	-	1 1 0
Siph aest	-	-	-	-	-	-	-	-	-	-	-	-	-	-	-	-	-	-	-	-	-	-	-	-	-	-	-	-	-	-	-	-	-	-	-	-	-	-	-	-	-	-	1 1 0
Ecdy disp	-	-	-	-	-	-	-	-	-	-	-	-	-	-	-	-	-	-	-	-	-	-	-	-	-	-	-	-	-	-	-	-	-	-	-	-	-	-	-	-	-	-	1 1 0
Rhit savo	-	-	-	-	2	-	-	-	-	-	-	-	-	-	-	-	-	-	-	-	-	-	-	-	-	-	-	-	-	-	-	-	-	-	-	-	-	-	-	-	-	-	1 1 0
Siph arma	-	-	-	-	-	-	-	-	-	-	-	-	-	-	-	-	-	-	-	-	-	-	-	-	-	-	-	-	-	-	-	-	-	-	-	-	-	-	-	-	-	-	1 1 0
Cent lute	-	-	-	-	-	-	-	-	-	-	-	-	-	-	-	-	-	-	-	-	-	-	-	-	-	-	-	-	-	-	-	-	-	-	-	-	-	-	-	-	-	-	1 1 0
Ephe vulg	-	-	-	-	-	-	-	-	-	-	-	-	-	-	-	-	-	-	-	-	-	-	-	-	-	-	-	-	-	-	-	-	-	-	-	-	-	-	-	-	-	-	1 1 0
Cent penn	-	-	-	-	-	-	-	-	-	-	-	-	-	-	-	-	-	-	-	-	-	-	-	-	-	-	-	-	-	-	-	-	-	-	-	-	-	-	-	-	-	-	1 1 0
Baet calc	-	-	-	-	-	-	-	-	-	-	-	-	-	-	-	-	-	-	-	-	-	-	-	-	-	-	-	-	-	-	-	-	-	-	-	-	-	-	-	-	-	-	1 1 0
Brac hari	-	-	-	-	-	-	-	-	-	-	-	-	-	-	-	-	-	-	-	-	-	-	-	-	-	-	-	-	-	-	-	-	-	-	-	-	-	-	-	-	-	-	1 1 0
Hept long	-	-	-	-	-	-	-	-	-	-	-	-	-	-	-	-	-	-	-	-	-	-	-	-	-	-	-	-	-	-	-	-	-	-	-	-	-	-	-	-	-	-	1 1 0
Para cinc	-	-	-	-	-	-	-	-	-	-	-	-	-	-	-	-	-	-	-	-	-	-	-	-	-	-	-	-	-	-	-	-	-	-	-	-	-	-	-	-	-	-	1 1 0
Ecdy insi	-	-	-	-	-	-	-	-	-	-	-	-	-	-	-	-	-	-	-	-	-	-	-	-	-	-	-	-	-	-	-	-	-	-	-	-	-	-	-	-	-	-	1 1 0
Hept fusc	-	-	-	-	-	-	-	-	-	-	-	-	-	-	-	-	-	-	-	-	-	-	-	-	-	-	-	-	-	-	-	-	-	-	-	-	-	-	-	-	-	-	1 1 0
Chor pict	-	-	-	-	-	-	-	-	-	-	-	-	-	-	-	-	-	-	-	-	-	-	-	-	-	-	-	-	-	-	-	-	-	-	-	-	-	-	-	-	-	-	1 1 0
Siph alte	-	-	-	-	-	-	-	-	-	-	-	-	-	-	-	-	-	-	-	-	-	-	-	-	-	-	-	-	-	-	-	-	-	-	-	-	-	-	-	-	-	-	1 1 0
Baet fusc	-	-	-	-	3	-	-	-	-	-	-	-	-	-	-	-	-	-	-	-	-	-	-	-	-	-	-	-	-	-	2	-	-	-	-	-	-	-	-	-	-	-	1 1 1
Proc bifi	-	-	-	-	-	-	-	-	-	-	-	-	-	-	-	-	-	-	-	-	-	-	-	-	-	-	-	-	-	-	-	-	-	-	-	-	-	-	-	-	-	-	1 1 1
Ephe nota	-	-	-	-	-	-	-	-	-	-	-	-	-	-	-	-	-	-	-	-	-	-	-	-	-	-	-	-	-	-	-	-	-	-	-	-	-	-	-	-	-	-	1 1 1
Ephe line	-	-	-	-	-	-	-	-	-	-	-	-	-	-	-	-	-	-	-	-	-	-	-	-	-	-	-	-	-	-	-	-	-	-	-	-	-	-	-	-	-	-	1 1 1
Hept sulp	-	-	-	-	-	-	-	-	-	-	-	-	-	-	-	-	-	-	-	-	-	-	-	-	-	-	-	-	-	-	-	-	-	-	-	-	-	-	-	-	-	-	1 1 1
Caen rivu	-	-	-	-	-	-	-	-	-	-	-	-	-	-	-	-	-	-	-	-	-	-	-	-	-	-	-	-	-	-	-	-	-	-	-	-	-	-	-	-	-	-	1 1 1
Olig rhen	-	-	-	-	-	-	-	-	-	-	-	-	-	-	-	-	-	-	-	-	-	-	-	-	-	-	-	-	-	-	-	-	-	-	-	-	-	-	-	-	-	-	1 1 1
Pota lute	-	-	-	-	-	-	-	-	-	-	-	-	-	-	-	-	-	-	-	-	-	-	-	-	-	-	-	-	-	-	-	-	-	-	-	-	-	-	-	-	-	-	1 1 1
Caen luct	-	-	-	-	-	-	-	-	-	-	-	-	-	-	-	-	-	-	-	-	-	-	-	-	-	-	-	-	-	-	-	-	-	-	-	-	-	-	-	-	-	-	1 1 1
Epho virg	-	-	-	-	-	-	-	-	-	-	-	-	-	-	-	-	-	-	-	-	-	-	-	-	-	-	-	-	-	-	-	-	-	-	-	-	-	-	-	-	-	-	1 1 1
Hept cocr	-	-	-	-	-	-	-	-	-	-	-	-	-	-	-	-	-	-	-	-	-	-	-	-	-	-	-	-	-	-	-	-	-	-	-	-	-	-	-	-	-	-	1 1 1
Hept flav	-	-	-	-	-	-	-	-	-	-	-	-	-	-	-	-	-	-	-	-	-	-	-	-	-	-	-	-	-	-	-	-	-	-	-	-	-	-	-	-	-	-	1 1 1
Caen macr	-	-	-	-	3	-	-	-	-	-	-	-	-	-	-	-	-	-	-	-	-	-	-	-	-	-	-	-	-	-	-	-	-	-	-	-	-	-	-	-	-	-	1 1 1
Group of loc.	A	A	A	A	A	B	B	B	C	C	C	C	C	C	C	C	C	C	C	C	C	C	C	C	D	D	D	D	E	E	E	E	F	F	F	F	F	F	F	F	F	F	

Table 3 (cont.)

No. of		3	1	2				2	3	3			1	1	3	1	2	2	2	2	2	7	7		1	1	3		2	6	7	4	5	3	2	2	7	7	Class
locality	3	0	8	5	5	2	7	7	7	2	8	9	1	0	6	9	0	4	2	3	9	1	8	4	6	7	1	6	6	4	4	5	8	3	1	8	2	9	of
Species																																							
Ecdy suba	-	2	-	-	-	-	-	2	-	-	-	-	-	-	-	-	-	4	-	-	-	-	-	-	-	-	-	-	-	-	-	-	-	-	-	-	-	-	0 0 0
Rhit irid	3	2	-	-	-	-	-	-	-	-	-	-	-	-	-	-	-	-	-	-	-	-	-	-	-	-	-	2	-	-	-	-	-	-	-	-	-	-	0 0 0
Amel inop	-	-	-	-	-	-	-	-	-	-	-	-	-	-	-	-	-	-	-	-	-	-	-	-	-	-	-	-	-	-	-	-	-	-	-	-	-	-	0 0 0
Rhit lovo	-	-	-	-	-	-	-	-	-	-	-	-	-	-	-	-	-	-	-	-	-	-	-	-	-	-	-	-	-	-	-	-	-	-	-	-	-	-	0 0 0
Rhit land	-	-	-	-	-	-	-	-	-	-	-	-	-	-	-	-	-	-	-	-	-	-	-	-	-	-	-	-	-	-	-	-	-	-	-	-	-	-	0 0 0
Ecdy veno	-	-	2	-	-	-	-	-	-	-	2	-	-	-	-	-	-	-	-	-	-	-	-	-	-	2	-	-	-	-	-	-	-	-	-	-	-	-	0 0 0
Baet mela	-	-	-	-	-	-	-	-	-	-	1	-	-	-	-	-	-	-	-	-	-	-	-	-	-	-	-	-	-	-	-	-	-	-	-	-	-	-	0 0 0
Ecdy forc	-	-	-	-	-	-	-	-	-	-	1	-	-	-	-	-	-	-	-	-	-	-	-	-	-	-	-	-	-	-	-	-	-	-	-	-	-	-	0 0 0
Ecdy aust	-	-	-	-	-	-	-	-	-	-	2	-	-	-	-	-	-	-	-	-	-	-	-	-	-	-	-	-	-	-	-	-	-	-	-	-	-	-	0 0 0
Epeo sylv	-	2	2	-	-	-	-	-	-	-	2	-	-	-	-	-	-	-	-	-	-	-	-	3	2	-	-	-	-	-	-	2	-	-	-	-	-	-	0 0 0
Rhit carp	2	-	-	3	2	2	2	2	-	2	-	-	-	-	1	2	-	-	-	-	-	-	2	-	-	-	-	-	-	3	-	-	1	-	-	-	-	-	0 0 0
Ecdy star	-	-	-	-	-	-	-	-	-	-	-	-	-	-	-	-	-	-	-	-	-	-	-	-	-	-	-	-	-	1	-	-	-	-	-	-	-	-	0 0 0
Elec quad	-	-	-	-	-	-	-	-	-	-	-	-	-	-	-	-	-	-	-	-	2	-	2	-	-	-	-	-	-	-	-	-	-	-	-	-	-	-	0 0 1
Baet alpi	4	3	3	3	3	-	-	-	-	3	2	-	-	-	-	-	-	-	-	-	-	-	-	3	4	4	4	4	-	-	-	-	-	-	-	-	-	-	0 0 1
Ephe mucr	-	2	3	2	2	2	-	-	-	-	2	-	-	-	-	-	-	-	-	-	-	-	-	2	-	2	2	2	2	2	1	-	-	-	-	-	-	-	0 0 1
Habr conf	-	3	3	2	2	3	3	2	3	3	2	-	-	-	3	3	3	-	3	3	3	3	-	2	2	2	3	2	-	3	1	3	-	2	-	-	-	-	0 0 1
Ecdy subm	-	-	-	-	-	-	-	1	-	2	2	-	-	-	-	2	-	-	2	-	-	-	-	-	-	-	2	-	-	-	2	-	-	-	-	-	-	-	0 1 0
Rhit semi	-	3	3	2	3	2	3	-	2	-	-	-	-	2	2	4	2	3	-	-	2	3	-	2	3	2	3	2	2	-	3	-	1	-	-	-	-	-	0 1 0
Baet rhod	3	5	4	4	4	4	5	3	3	4	5	-	4	4	3	5	4	5	4	4	4	4	5	5	4	-	4	5	4	3	4	2	3	3	-	4	2	2	0 1 0
Alai muti	2	-	2	-	2	2	-	2	-	2	2	-	3	-	2	3	2	4	2	3	2	-	-	-	3	2	3	-	2	-	-	-	-	2	-	2	-	-	0 1 0
Siph lacu	-	-	-	-	-	-	-	-	-	-	-	-	-	-	-	-	-	-	-	-	-	-	-	2	-	-	-	2	-	-	-	-	-	-	-	-	-	-	0 1 0
Elec late	-	-	-	-	-	2	3	2	-	-	2	-	-	-	-	-	-	-	-	-	2	-	-	-	-	-	-	-	2	-	-	-	-	1	-	-	-	-	0 1 1
Ele affi	-	-	-	-	2	-	-	2	-	2	2	-	-	-	-	-	-	-	-	-	-	-	-	-	3	-	2	-	-	-	-	-	-	-	-	-	-	-	0 1 1
Caen pseu	-	-	-	-	-	-	-	-	-	-	-	-	-	-	-	-	-	-	-	-	-	-	-	2	-	-	-	-	-	1	-	-	-	2	2	-	-	-	0 1 1
Habr laut	-	-	-	2	2	2	2	2	-	-	3	-	3	-	-	3	-	2	3	2	2	2	2	2	2	2	2	2	-	-	-	-	-	-	-	-	-	-	0 1 1
Baet vern	3	3	4	3	-	3	4	3	3	4	-	3	4	4	4	-	4	4	3	4	3	1	-	4	-	-	3	4	-	1	2	-	-	4	2	-	-	-	1 0 0
Ephe dani	4	4	3	3	-	3	3	3	3	4	5	-	4	4	4	4	3	4	3	3	3	2	2	3	4	3	-	3	3	2	-	-	-	-	-	3	-	-	1 0 0
Ephe igni	3	4	-	3	3	3	2	3	2	3	3	4	5	4	4	-	3	5	3	4	3	3	4	4	4	3	3	3	4	2	2	2	2	3	4	4	2	1	1 0 0
Torl majo	-	-	-	-	-	-	-	-	-	2	-	-	-	-	-	-	-	3	-	-	1	-	-	2	-	-	2	-	2	1	2	-	2	-	-	-	-	-	1 0 0
Ecdy torr	2	-	-	2	1	-	-	-	-	-	-	-	3	-	-	3	3	4	1	2	2	-	2	2	2	1	3	-	2	3	2	3	-	-	-	-	-	-	1 0 0
Ecdy aura	-	-	-	-	-	-	-	-	-	-	-	-	-	-	-	-	-	-	-	-	-	-	-	-	-	-	-	-	-	-	-	-	2	-	-	-	-	1	1 0 1
Baet luth	-	-	-	-	-	-	-	-	-	-	-	-	-	-	-	-	-	-	-	1	-	-	-	-	-	-	-	-	2	1	-	2	2	-	2	2	2	1	1 0 1
Habr fusc	-	-	-	-	-	-	-	2	-	-	-	-	2	-	2	-	-	-	-	-	-	-	-	-	-	-	-	-	-	-	-	-	-	2	-	-	-	-	1 1 0
Baet buce	-	-	-	-	-	-	-	-	-	-	-	-	-	-	-	-	-	-	-	-	-	-	-	-	-	-	-	-	2	2	2	2	2	-	-	-	-	2	1 1 0
Baet scam	-	-	-	-	-	-	-	-	-	-	-	-	-	-	-	-	-	-	-	-	2	-	-	-	2	-	-	-	-	2	3	3	2	-	-	-	-	-	1 1 0
Rhit herc	-	-	-	-	-	-	-	-	-	-	-	-	-	-	-	-	-	-	-	-	-	-	-	-	-	2	-	-	-	-	-	-	-	-	-	-	-	-	1 1 0
Rhit besk	-	-	-	-	-	-	-	-	-	-	2	-	-	-	-	2	-	-	-	-	-	-	-	2	-	-	-	-	2	-	-	-	-	-	-	-	-	-	1 1 0
Lept vesp	-	-	-	-	-	-	-	-	-	-	2	-	-	-	-	-	-	-	-	-	-	-	-	-	-	-	-	3	-	-	-	-	-	-	-	-	-	-	1 1 0
Rhit germ	-	-	-	-	-	-	-	-	-	-	-	-	-	-	-	-	-	-	-	-	-	-	-	-	-	-	-	-	2	-	-	-	-	-	-	-	-	-	1 1 0
Cloe dipt	-	-	-	-	-	-	-	-	-	-	3	3	-	-	-	-	-	3	-	-	-	-	-	-	-	-	-	2	-	-	-	-	-	-	-	-	-	-	1 1 0
Para subm	2	3	-	2	-	2	2	2	-	-	-	2	4	2	-	3	3	3	3	3	2	2	2	2	2	2	-	3	-	-	-	2	1	-	-	-	-	-	1 1 0
Lept marg	-	-	2	2	-	2	-	-	-	-	-	2	-	-	2	4	-	3	-	-	-	-	-	-	2	-	2	2	2	-	-	-	-	-	-	-	-	-	1 1 0
Nigr nige	2	2	2	2	-	-	-	-	-	-	-	4	-	3	-	2	1	2	-	2	-	3	-	-	3	3	3	3	3	-	-	-	-	-	-	-	-	-	1 1 0
Siph aest	-	-	2	2	-	-	-	-	-	-	-	2	-	2	-	3	3	-	-	-	-	-	-	-	-	3	-	4	2	-	-	-	-	-	-	-	-	-	1 1 0
Ecdy disp	2	2	-	-	-	-	-	2	-	3	-	-	-	-	2	3	3	3	-	2	2	2	2	2	3	2	3	-	2	2	-	2	2	-	-	-	-	-	1 1 0
Rhit savo	-	-	-	-	-	-	-	-	-	-	-	-	-	-	-	-	-	-	-	-	-	-	-	2	2	2	-	2	-	-	-	-	-	-	-	-	-	-	1 1 0
Siph arma	-	-	-	-	-	-	-	-	-	-	-	-	-	-	-	2	-	-	-	-	-	-	-	-	-	-	-	-	-	-	-	-	-	-	-	-	-	-	1 1 0
Cent lute	-	-	-	-	-	-	-	-	-	-	3	2	4	2	-	-	3	4	2	3	1	1	-	-	3	-	-	3	2	-	-	-	-	-	2	-	-	-	1 1 0
Ephe vulg	-	-	-	1	-	-	-	-	-	-	-	2	3	-	-	2	-	-	-	2	-	-	-	-	-	-	-	-	-	-	-	-	-	-	-	-	-	-	1 1 0
Cent penn	-	-	-	-	-	-	-	-	-	-	-	-	2	-	-	-	-	-	-	-	-	-	-	-	-	-	-	-	-	-	-	-	-	-	-	-	-	-	1 1 0
Baet calc	-	-	-	-	-	-	-	-	-	-	-	3	-	-	-	-	-	-	-	-	-	-	-	-	-	-	-	-	-	-	-	-	-	-	-	-	-	-	1 1 0
Brac hari	-	-	-	-	-	-	-	-	-	-	-	1	-	2	-	-	-	-	-	-	-	-	-	-	-	-	-	-	-	-	-	-	-	-	-	-	-	-	1 1 0
Hept long	-	-	-	-	-	-	-	-	-	-	-	-	-	2	-	-	-	-	-	-	-	-	-	-	-	-	-	-	-	-	-	-	-	-	-	-	-	-	1 1 0
Para cinc	-	-	-	-	-	-	-	-	-	-	-	-	-	-	-	-	-	-	2	-	-	-	-	-	3	-	-	-	-	-	-	-	-	-	-	-	-	-	1 1 0
Ecdy insi	-	-	-	-	-	-	-	-	-	-	-	-	-	-	-	-	-	-	-	-	-	-	-	-	3	-	-	-	3	-	-	-	-	-	-	-	-	-	1 1 0
Hept fusc	-	-	-	-	-	-	-	-	-	-	-	2	-	-	-	-	-	-	-	-	-	-	-	-	2	2	-	2	2	-	-	-	-	-	-	-	-	-	1 1 0
Chor pict	-	-	-	-	-	-	-	-	-	-	-	-	-	2	-	-	-	-	-	-	-	-	-	-	-	-	-	-	2	-	-	-	-	-	-	-	-	-	1 1 0
Siph alte	-	-	-	-	-	-	-	-	-	-	-	-	-	2	-	-	-	-	-	-	-	-	-	-	-	-	-	2	-	-	-	-	-	-	-	-	-	-	1 1 0
Baet fusc	-	2	-	2	-	-	-	-	-	-	-	3	3	-	2	2	-	-	-	3	3	-	-	4	3	3	3	-	4	4	2	-	-	4	3	3	3	3	1 1 1
Proc bifi	-	-	-	-	-	-	-	-	-	-	2	2	2	2	-	-	-	-	-	-	2	-	-	-	-	-	-	2	-	-	-	-	-	2	2	-	-	-	1 1 1
Ephe nota	-	-	-	-	-	-	-	-	-	-	-	-	-	1	-	-	-	-	-	-	-	-	-	-	-	-	-	-	1	-	-	-	-	-	-	-	-	1	1 1 1
Ephe line	-	-	-	-	-	-	-	-	-	-	-	-	-	2	-	-	-	-	-	-	-	-	-	-	-	-	-	-	-	-	-	-	-	-	-	1	-	-	1 1 1
Hept sulp	-	-	-	-	-	-	-	-	-	-	-	-	-	3	2	-	-	-	-	2	-	-	2	-	-	-	3	2	2	-	2	-	-	3	3	3	2	3	1 1 1
Caen rivu	-	-	-	-	-	-	-	-	-	-	-	-	-	-	-	-	-	-	-	-	-	-	-	-	-	-	-	-	-	-	-	-	1	-	-	-	-	-	1 1 1
Olig rhen	-	-	-	-	-	-	-	-	-	-	-	-	-	-	-	-	-	-	-	-	-	-	-	-	-	-	-	-	-	-	-	-	-	2	2	-	-	2	1 1 1
Pota lute	-	-	-	-	-	-	-	-	-	-	-	-	-	-	-	-	-	-	-	-	-	-	-	-	-	-	-	-	3	-	-	-	-	2	3	3	4	4	1 1 1
Caen luct	-	-	-	-	-	-	-	-	-	-	-	-	-	-	-	-	-	-	-	-	-	-	-	-	2	-	-	-	2	-	-	-	-	-	2	-	2	2	1 1 1
Epho virg	-	-	-	-	-	-	-	-	-	-	-	-	-	2	-	-	-	-	-	-	-	-	-	-	-	-	-	-	-	-	-	-	-	-	1	-	-	2	1 1 1
Hept coer	-	-	-	-	-	-	-	-	-	-	-	-	-	2	-	-	-	-	-	-	-	-	-	-	-	-	-	-	-	-	-	-	-	-	2	-	-	-	1 1 1
Hept flav	-	-	-	-	-	-	-	-	-	-	-	3	-	2	2	-	-	-	-	-	-	-	-	-	-	-	-	-	-	-	-	-	-	3	3	3	-	2	1 1 1
Caen macr	-	-	-	-	-	-	-	-	-	-	-	-	4	-	-	3	2	3	-	3	-	-	-	-	-	-	-	-	3	-	-	2	-	3	3	3	3	2	1 1 1
Group of loc.	G	G	G	G	G	G	G	G	G	G	G	H	H	H	H	I	I	I	I	I	I	I	I	J	J	J	J	J	J	K	K	K	K	L	L	L	L	L	

mainly of the Labe basin and those of the Morava basin situated close to the watershed) and J (*10011, indicators *B. alpinus* and *E. mucronata*, localities of meta- and hyporhithral of the Labe basin). These two groups differed in discharge rates. The final group K (*101, indicator *B. buceratus*, comprised hyporhithral/epipotamal stretches of the Morava and Odra basins). For detail species composition of individual taxocenes see Tab. 3, for summary of important environmental variables and parameters evaluated with respect to TWINSPAN groups see Tabs. 4 and 5.

Table 4. Summary of important variables evaluated with respect to TWINSPAN groups – environmental variables. Order of values: minimum, maximum, (median), average (if present). L – Labe River basin, M – Morava River basin, O – Odra River basin.

TWIN-SPAN group	No. of localities	River basin	Distance from source [km]	Altitude [m]	Slope [m.km-1]	Mean width [m]	Maximal water temperature [oC]	pH minimal value
A	5	L	2.4, 18.7, (6.8)	490, 1020, (905)	15, 80, (30)	0.9, 7.5, (1.7)	11.8, 15.2, (13.5)	4.7, 6.2, (5.9)
B	3	M	1.7, 3.3, (2.2)	320, 425, (375)	30, 50, (50)	0.7, 1.2, (1.2)	12.9, 15.4, (14.9)	7.4, 7.9, (7.7)
C	16	L,O,M	0.5, 9.9, (3)	225, 770, (618)	20, 90, (55)	0.6, 8.0, (1.9)	10.2, 14.8, (12.2)	6.3, 7.7, (7.2)
D	4	O,M	1.5, 12.5, (8)	510, 620, (560)	20, 40, (30)	3.5, 9.0, (6.5)	11.5, 15.5, (12.1)	6.9, 7.5, (7.1)
E	4	O,M	7.5, 37.4, (17.9)	335, 620, (448)	10, 20, (18)	3.5, 8.0, (6.8)	17.0, 18.2, (17.9)	7.0, 8.1, (7.3)
F	10	O,M	2.1, 20.6, (4.4)	290, 615, (415)	10, 50, (20)	1.8, 5.0, (2.5)	13.0, 16.0, (15.0)	6.8, 8.0, (7.3)
G	11	L	0.5, 11.3, (8.9)	327, 659, (568)	5, 40, (10)	1.0, 5.5, (1.6)	13.5, 22.9, (19.1)	6.1, 6.9, (6.8)
H	4	L	17.7, 55.0, (42.4)	283, 432, (403)	5, 10, (5)	2.5, 6.0, (4.0)	16.2, 20.5, (19.1)	6.7, 7.2, (6.9)
I	8	L,M	1.4, 30.5, (9.9)	225, 650, (455)	5, 40, (10)	1.0, 10.0, (2.0)	16.0, 23.1, (17.2)	6.6, 7.0, (6.9)
J	6	L	14.7, 93.7, (49.9)	279, 748, (464)	5, 20, (8)	3.0, 15.0, (6.0)	15.6, 24.5, (18.9)	5.8, 7.2, (6.7)
K	4	L,O,M	20.8, 45.5, (36.3)	255, 515, (340)	5, 5, (5)	10.0, 20.0, (11.6)	14.8, 23.6, (19.5)	7.2, 8.4, (8.2)
L	5	L,M	72.8, 179.6, (96.6)	185, 355, (348)	5, 10, (5)	9.0, 27.0, (20.0)	18.9, 23.1, (21.6)	6.7, 7.0, (6.9)

Canonical Correspondence Analysis (CCA)

Five statistically significant environmental variables, namely distance from source (correlated with the first ordination axis); minimum pH measured (correlated with the second ordination axis), altitude, slope and water temperature (summer maximum) were found. For general statistical characteristics (eigenvalues and percentage of variance) see Tab. 6.

Table 5. Summary of important parameters evaluated with respect to TWINSPAN groups – biodiversity measures. Order of values: minimum, maximum, (median), average (if present). L – Labe River basin, M – Morava River basin, O – Odra River basin.

TWIN-SPAN Group	No. of localities	River basin	No. of species N	Species richness D	Species diversity H'
A	5	L	5, 17 (5) 7.8	1.36, 5.24, (1.81)	1.63, 3.16, (1.96)
B	3	M	4, 9 (8) 7.0	1.21, 2.89, (2.71)	0.88, 2.43, (2.04)
C	16	L,O,M	8, 16 (11) 11.4	2.61, 5.65, (3.29)	1.68, 3.00, (2.46)
D	4	O,M	7, 14 (10) 10.3	3.22, 5.70, (4.10)	2.26, 2.93, (2.55)
E	4	O,M	11, 14 (11) 12.0	3.63, 4.79, (4.07)	1.79, 2.94, (2.64)
F	10	O,M	7, 15 (10) 10.4	2.49, 4.75, (3.41)	1.65, 2.75, (2.20)
G	11	L	6, 21 (13) 13.2	3.13, 6.60, (4.12)	2.62, 3.51, (2.81)
H	4	L	14, 20 (15) 16.0	4.27, 6.20, (4.66)	2.77, 3.38, (3.16)
I	8	L,M	10, 19 (15.5) 14.8	2.99, 6.03, (4.55)	1.16, 3.62, (3.19)
J	6	L	18, 27 (21) 21.3	5.72, 8.47, (6.41)	3.26, 3.82, (3.72)
K	4	L,O,M	11, 16 (12.5) 13.0	4.17, 5.58, (4.96)	2.67, 3.12, (2.94)
L	5	L,M	8, 15 (14) 12.6	2.70, 5.15, (4.71)	1.99, 3.15, (2.79)

Depicting of respective envelopes of the TWINSPAN classification groups of localities (A-L in Fig. 5) shows evident differences in their nature. Rhithral localities (e.g. groups A or C) are situated at the left side of the diagram, potamal one (group L) at the right side. Localities of the rhithral/potamal transition zone (e.g. group K) occupy the central part. Upper part of the diagram (Fig. 6) comprises mainly the Labe basin localities, lower part mainly the Odra and the Morava basins localities. This distribution is due mainly to rather low values of pH in the Labe basin caused by geological conditions and effects of peat-bog waters in comparison with the Morava and Odra basins. Moreover, the localities of TWINSPAN group A (high elevations of the Labe basin, lower pH) are situated in the left upper quadrant. The localities of the groups B (small streams of the Morava basin Carpathians), D (epirhithral of the Morava and Odra basins, higher discharges), F (epirhithral of the Morava basin), and most of localities of the group E (metarhithral of the Morava and Odra basins) are situated in the left lower quadrant. The group C (cold epirhithral brooks of higher coline zone of all basins, but predominantly of the Odra basin) is located on the left side of diagram but just in intermediary position. The group G (consisting of rather not clearly defined rhithral localities of the Labe basin) occurs just in the central part of the diagram. The group H (specific localities of lowland rivers) is situated at the right lower part of the diagram, very closely to groups G and I. The group K (localities of hyporhithral/epipotamal of the Morava and Odra basins) is placed in the same quadrant although more distally owing to its transitional character. Similar localities of the Labe basin (group J) are situated mainly in the right upper quadrant. The group L (typical potamal sites) is located also in this quadrant but shifted positively along the first ordination axis.

Table 6. Eigenvalues and percent variance explained for the first four ordination axes of CCA

Axis	Eigenvalue	% variance
1	0.494	8.4
2	0.352	14.4
3	0.245	18.6
4	0.098	20.3

Sum of all canonical eigenvalues: 1.245
Monte Carlo permutation test, 999 perm.

Differences between the Labe basin on one hand and the Odra+Morava basins on the other hand are very apparent from the ordination the whole set of localities (Fig. 6). The envelope comprising the Labe basin localities is conspicuously shifted to the upper part of the diagram, only partially superimposing those of the Odra+Morava basin localities. Moreover, the position of the Odra basin localities indicated close relationships to the Morava basin.

Ordination of species is apparent from Figs. 7 and 8. The positions of individual species clearly depend on the distribution of environmental variables. For instance, *Electrogena quadrilineata* or *Ecdyonurus subalpinus* are situated at the left lower quadrant preferring places in short distance from source, middle elevations and slope, lower water temperature and higher pH. On the other hand, the potamal species *Potamanthus luteus* is located distally at the right side of diagram. This position means preferences of places at long distance from source, localities at low elevations with minimal slope higher pH and high summer water temperatures.

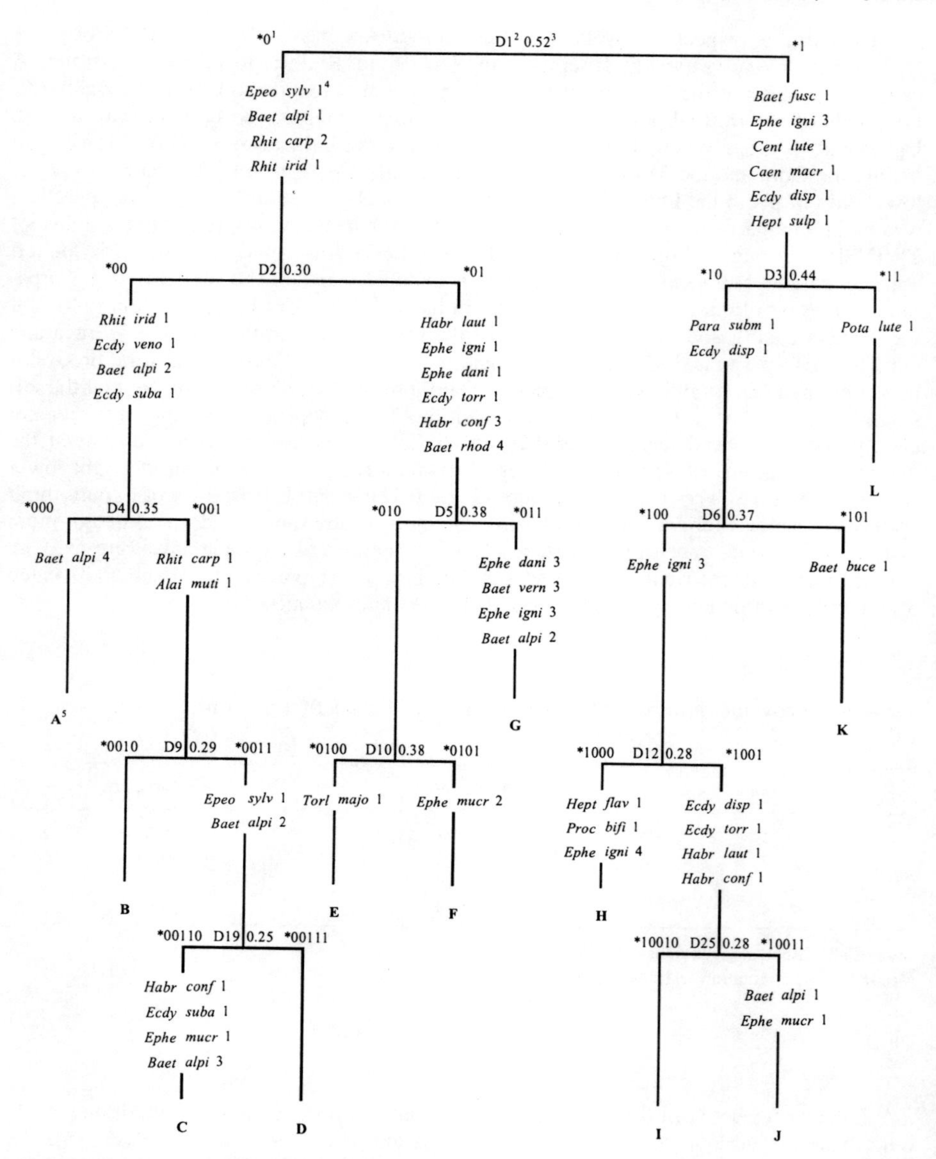

Fig. 4. Hierarchic classification (TWINSPAN) of reference localities and associated indicator species. For abbreviations of indicator species names see Table 2, for definition of individual groups (A -L) see the text.
[1] TWINSPAN class, [2] No. of division, [3] eigenvalue, [4] pseudospecies cut level

HOBENT

The complete data set (80 localities, 71 species) formed the reference database. HOBENT was used to verify that mayfly communities could be predicted from the reference database. Further thirteen rhithral samples taken at the localities different from the above reference data set were tested (No. C1-C13, see Tab. 7). A wide spectrum of information on

Table 7. List of control localities with basic characteristics and results of evaluation of control localities by HOBENT. L – Labe River basin, M – Morava River basin, O – Odra River basin. For definition of individual groups (A - L) see in the text.

No. of locality	Name of watercourse	Sampling site	Basin	Distance from source [km]	Altitude [m]	Slope [m.km^{-1}]	Mean width [m]	Max. water temperature [°C]	pH minimal	Conductivity [µS.cm^{-1}]	No. of species expected	No. of species found	Coefficient B	TWINSPAN Group	Probability %
C1	Mumlava	Muml. Bouda	L	5.5	705	40	7.5	10.1	5.5	150	3.4	2	0.59	A	100.0
C2*	Chomutovka River	Chomutov	L	12.7	382	15	3.0	16.2	12.7	286	10.2			I	81.4
														G	14.1
C3	Bystřice River	Komárov	L	21.7	235	10	4.0	25.6	21.7	600	8.7	1	0.11	H	91.2
C4	Vltava River	Rožmberk	L	120.6	528	5	8.0	18.9	120.6	225	8.6	1	0.12	L	99.8
C5	brook	Pernštejn	M	1.5	400	30	1.4	14.4	1.5	390	5.6	5	0.89	B	98.0
C6	Bystřice River	Tesák	M	1.7	560	50	1.5	14.9	1.7	186	7.1	9	1.27	F	94.6
C7	Oskava River	Bedřichov	M	4.2	550	50	1.5	12.5	4.2	75	8.6	8	0.93	C	95.0
C8	Habřina Brook	Zastávka	M	6.3	350	20	2.6	14.5	6.3	560	5.7	3	0.53	B	99.9
C9	Gránický Brook	Znojmo	M	7.9	285	15	2.5	16.0	7.9	950	5.7	3	0.53	B	100.0
C10	Luha Brook	Sloup	M	8.6	540	20	3.0	18.0	8.6	250	7.9	5	0.63	E	87.6
C11	Velička River	Louka	M	20.0	260	10	6.0	22.4	20.0	500	8.6	4	0.47	E	93.4
C12	Křetínka River	Letovice	M	29.0	361	10	4.2	16.0	29.0	450	8.7	2	0.23	H	98.8
C13	Oslava River	Náměšť	M	65.1	360	10	16.0	18.0	65.1	410	8.2	6	0.73	K	99.9

* included in 2 TWINSPAN groups

environmental variables (e.g. physico-chemical analyses, discharge rates etc., for the most important variables see Tab. 7) of these control localities was gathered, so that the degree of their denaturalisation could be evaluated independently of species composition. Taxocenes of the two control localities (No. C5 and C7, see Tab. 7) were classified as very similar to the predicted ones (B > 0.8). Favourable results agree with real status - these localities might in fact represent further (not yet used) background localities. Control locality C13 was classified as less similar to predicted one (B = 0.73). The influence of water reservoir and moderate organic pollution influenced species composition of this locality. Control localities No. C3 and C12 represent strongly disturbed river stretches (B < 0.4), e.g. the C12 is influenced by combined effects of extreme discharge rates (situated below dam) and municipal wastewater. Locality C10 represents moderately influenced biotope, localities C8 and C11 heavily influenced ones. Localities C4 and C9 are examples of a certain weakness of the reference database – respective types of these localities are missing from the reference database. For example, the locality No. C9 represents only slightly disturbed lowland brook, a very rare biotope in the Morava basin not yet included into the reference database and thus the value of B coefficient (0.53) does not express the real status. On the other hand, loc. No. C6 is a Carpathian brook of very high species richness even higher than those of localities used. The locality was selected as an example of situation, when value of

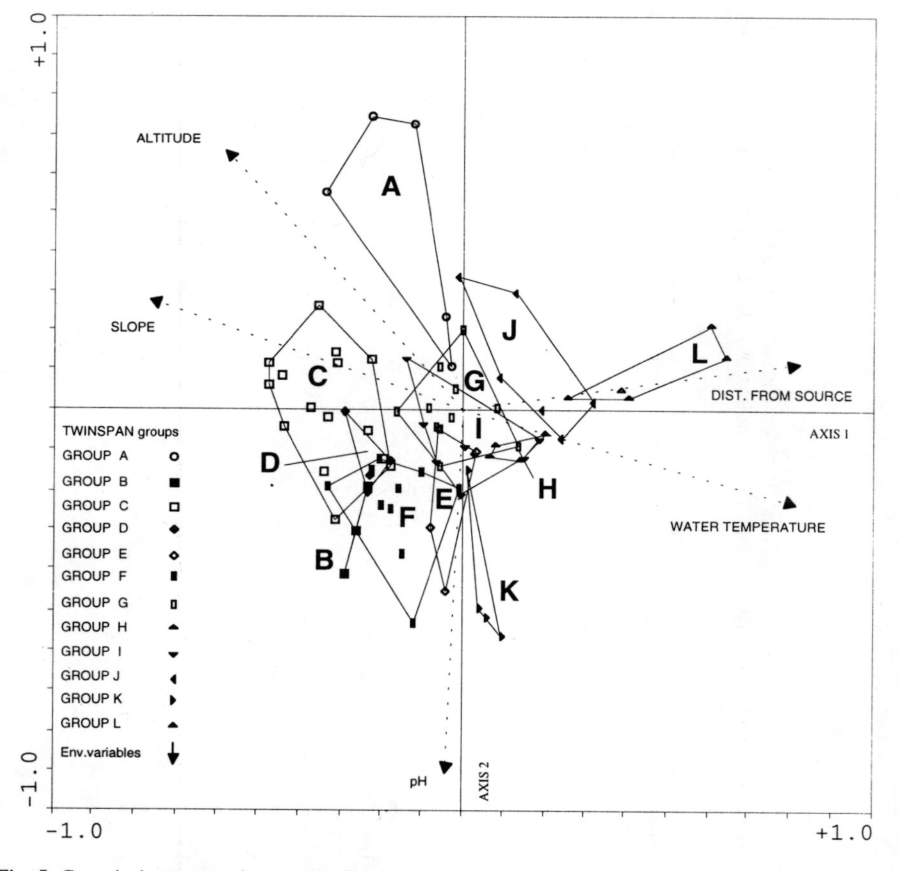

Fig. 5. Canonical correspondence analysis (CCA), ordination plot of localities, species and main environmental variables on the first two CCA axes with envelopes showing TWINSPAN groupings.

coefficient B > 1 can be reached. The localities C1 and C2 are examples of deterioration by acidification. These extremes underscore the necessity of large-scale reference localities, comprising all types of aquatic biotopes actually present in the interest area.

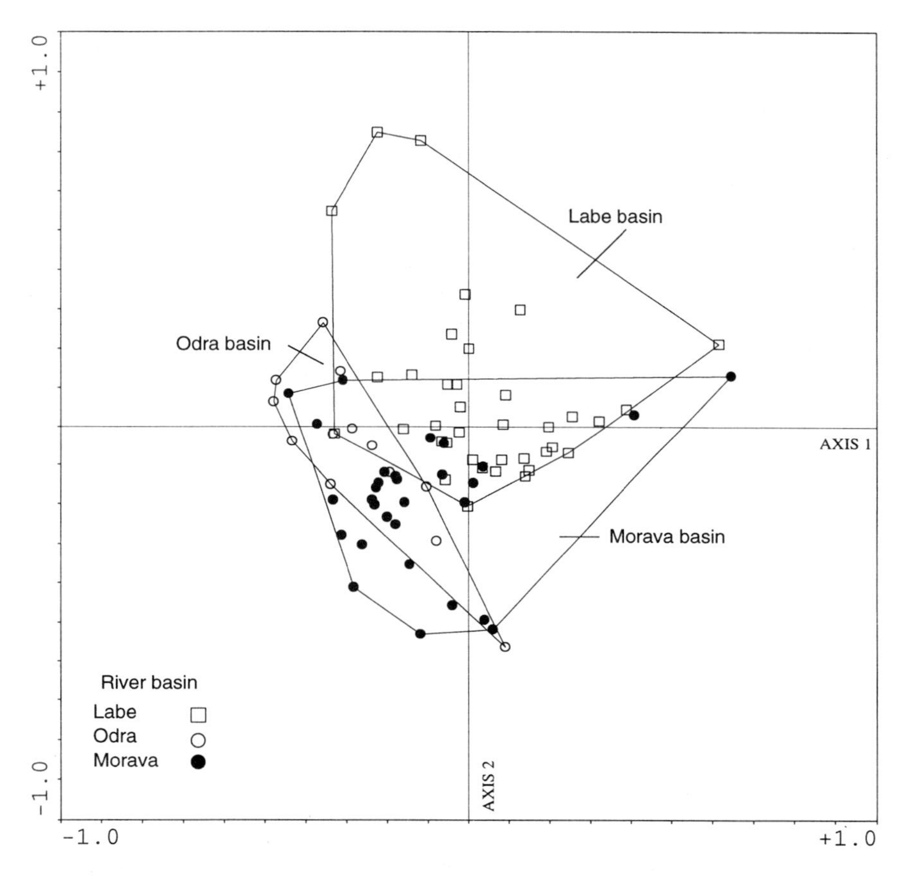

Fig. 6. Canonical correspondence analysis (CCA), ordination plot of localities on the first two CCA axes with envelopes grouping localities of the Labe River, the Morava River and the Odra River basins.

DISCUSSION AND CONCLUSIONS

Several attempts to classify the mayfly taxocenes in the area studied have been made. For instance, Zelinka (1953) distinguished 6 running waters zones according to the occurrence of characteristic genera: the *Ameletus* zone (hypocrenal), *Rhithrogena* zone (epirhithral), *Ecdyonurus* zone (meta- and hyporhithral), *Oligoneuriella* zone (epipotamal) and *Ephoron* zone (metapotamal). This classification, although roughly corresponding to that by Illies and Botosaneau (1963), seems to be based mostly on empirical estimations of dominance, and moreover, only at the generic level. Since then a lot of information on mayfly species distribution, life cycles, habitat preferences, interspecific interactions and origin and chorology has been summarised by Landa (1969, 1984), Landa and Soldán (1981, 1982, 1985, 1989, 1995), Landa et al. (1997), Soldán (1992), Zelinka (1953), Zelinka and

Skalníková (1959), Krpal and Zelinka (1990) and an attempt to predict the long-term development of taxocenes has been made (Lepš et al., 1989, 1990) in the Czech Republic. Krno and Šporka (1991), Krno et al., (1994) and Deván and Mucina (1986) studied similar problems in the Danube basin in Slovakia. However, even this large amount of information allows little more than intuitive assessment of taxocoenoses expected at a particular locality. In contrast, we believe the combination of TWINSPAN and CCA classification better defines the differences in taxocenes among river basins. By taking into account principal environmental variables, these differences are clearly seen e.g. from shifts of respective localities envelopes. Apart from several exceptions, TWINSPAN groups presenting rhithral consisted either of the Labe basin localities or of the Morava+Odra basin localities where differences are not so pronounced (cf. Mergl, 1997, Soldán et al., 1998, see also Tab. 3, and Fig. 4). For instance, *Siphlonurus alternatus, Baetis calcaratus* and *Leptophlebia vespertina* are known from the Labe basin only. Quantitative presentation of e.g. *Rhithrogena carpatoalpina, Alainites muticus* and *Habrophlebiá lauta* seems to be higher in the Morava basin, that of e.g. *Paraleptophlebia submarginata* in the Labe basin. Due to a lot of Hercynian mountain system elements in the Czech part and the Carpathian and/or Pannonian elements in the Danube basin, these differences become even more apparent (cf. Krno and Šporka, 1991, Krno et al., 1994, Deván and Mucina, 1986). The variations of the species composition of the Czech rivers were found depending on basin (cf. Tab. 4) and even, within the same basin, on different biogeographic subprovinces (Soldán et al., 1998).

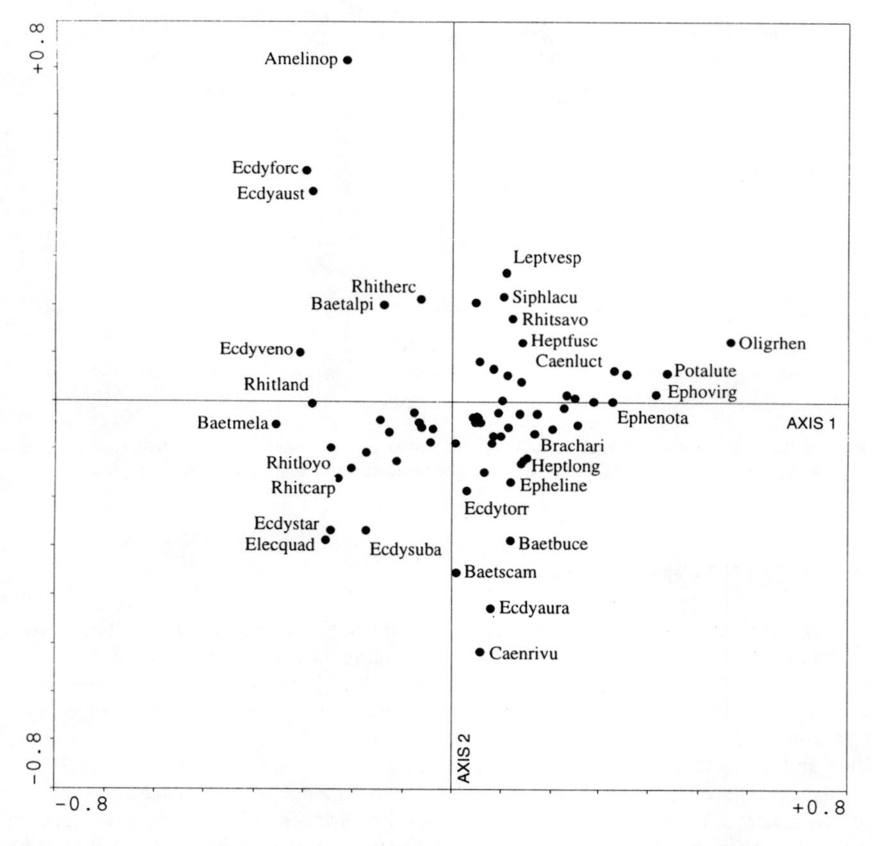

Fig. 7. Canonical correspondence analysis (CCA), ordination plot of species on the first two CCA axes. For full species names see Table 2.

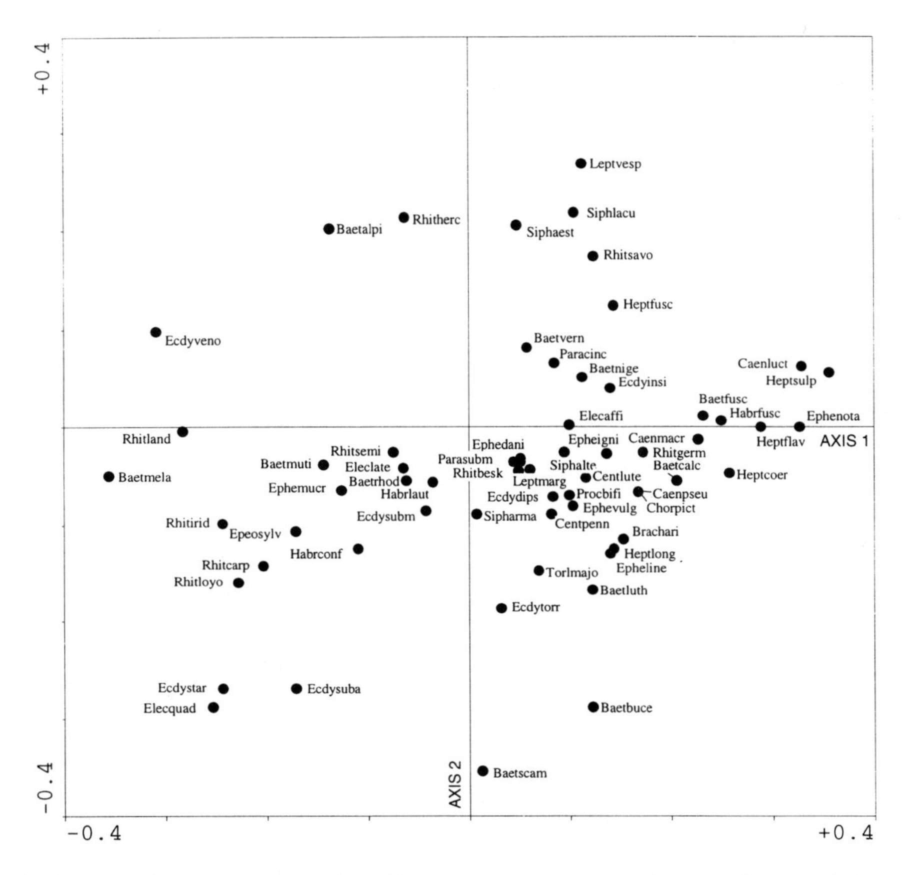

Fig. 8. Canonical correspondence analysis (CCA), ordination plot of species on the first two CCA axes - central part of ordination plot. For full species names see Table 2.

Consequently, an unambiguous ecoregional approach seems to be required for the proposed model because even in such relatively restricted area, as former Czechoslovakia presents. Moreover, lack of some type of biotope may lead to misclassification and the establishment of database of reference localities covering the whole area of investigation and all biotopes types seems to be necessary.

In the Czech Republic, the occurrence of 97 mayfly species has been documented so far (Soldán et al., 1998); however, at least 3 species are considered extinct. Of the remaining 94 species, 71 were found and are contained in this data set. We recommend determination at the species level when differentiating spatial distribution of taxocenes. This enables higher accuracy of predictions, allows more precise definition of locality status and it is also advantageous for environmental conservation management.

This study illustrates that intensity of river deterioration (and/or restoration) can be evaluated by the prediction model HOBENT, i.e. by comparison of target and actual status, even by using only mayflies rather than the entire macroinvertebrate community. Naturally, like in other macroinvertebrate groups, mayflies cannot be used to evaluate spring areas (crenal), peat bogs water, and acidified and heavily polluted stretches because these biotopes usually lack mayflies. Hence, we expect that complete macroinvertebrate community composition will be preferred for practical application. The necessity of a large-scale database of localities comprising all main types of biotopes and covering the entire interest area was demonstrated. Despite some reservation, we believe the total number of 400 background localities (planned to be evaluated during the first and second phase of the project in 1997 - 2001) will be sufficient for successful application of this prediction model in the Czech Republic.

ACKNOWLEDGMENTS

This research was supported by the grant of the Council of the Government of the Czech Republic for Research and Development No. 510/2/96 and partly by the Research Project No. S5007015 of the Grant Agency of the Academy of Science of the Czech Republic. We would like to sincerely thank Prof. J. Stanford, Flathead Lake Biological Station, The University of Montana, for critical review of the manuscript. Our thanks are due to Prof. F. Kubícek and Prof. R. Rozkošný for encouragement and consultations and Dr. E. Domínguez who carefully revised final version of manuscript. We wish to thank also Dr. A. Mergl for data from the Odra river basin, and Dr. J. Schenková for technical assistance.

REFERENCES

Armitage, P. D., D. Moss, J. F. Wright and M. T. Furse. 1983. The performance of a new biological water quality score system based on macroinvertebrates over a wide range of unpolluted running-water sites. Water Res. 17 (3): 333-347.

Culek, M. 1996. (Ed.) Biogeographical classification of the Czech Republic. Enigma, Praha, 347 pp. (in Czech, English summary)

Demek, J. 1987. Geographical lexicon of Czech Republic. Mountains and lowlands. Academia, Praha, 584 pp. (in Czech)

Deván, P. and L. Mucina. 1986. Structure, zonation and species diversity of the mayfly communities of the Belá basin, Slovakia. Hydrobiology, 135: 155-165.

Hill, M. O. 1979. TWINSPAN - A FORTRAN program for arranging multivariate in an ordered two-way table by classification of the individuals and attributes. Cornell University Press, Ithaca, N. York, 99 pp.

Illies, J. and L. Botosaneanu. 1963. Problèmes et méthodes de la classification et de la zonation écologique des eaux courantes, considerées surtout du point de vue faunistique. Mitt. Int. Ver. Limnol. 12: 1-57.

Klecka, W. R. 1975. Discriminant analysis, pp- 434-467. In: N. H. Nie, C. H. Hull, J. G. Jenkins, Steinbrenner and D. H. Bent (eds.). SPSS Statistical Package for Social Sciences. McGraw-Hill, New York.

Kokeš, J. 1997. Evaluation of the influence of anthropogenic factors on chosen components of biocoenoses of running waters. Report, Water Research Institute T.G.M. Prague, Dept. Brno, 32 pp. (in Czech)

Krno, I. and F. Šporka. 1991. Some notes on hydrobiological investigations in the Turiec river basin. Acta Fac. Rerum nat. Univ. Comenianae - Zoologia. 35: 71-76.

Krno, I., D. Illéšová and J. Hagaloš J. 1994. Temporal fauna of the Gidra Brook (Little Carpathians, Slovakia). Acta Zool. Univ. Comenianae. 38: 35-46.

Krpal, J. and M. Zelinka. 1990. Statistical evaluation of some effects on distribution of macrozoobenthos of running water. Scripta Fac. Sci. nat. Univ. Purk. Brun., Biologia, 20 (9-10): 451-460.

Landa, V. 1969. Jepice - Ephemeroptera. Fauna of Czechoslovakia, Vol. 18, Academia Praha, 352 pp. (in Czech, German Summary)

Landa, V. 1984. Studies on aquatic insects in Czechoslovakia with regard to changes in the quality of water in the last 20-30 years, pp. 317-321. In: V. Landa, T. Soldán and M. Tonner (eds.). Proc. IVth Int. Conf. Ephemeroptera Bechyne, Czechoslovak Academy of Sciences, Ceské Budejovice.

Landa, V. and T. Soldán. 1981. Some faunistic and biogeographic aspects of the mayfly fauna of the Hercynian and Carpathian mountain systems in Czechoslovakia (Ephemeroptera). Acta Mus. Reginaehradecensis, Suppl. 1980: 58-60.

Landa, V. and T. Soldán. 1982. Changes in distribution of mayflies (Ephemeroptera) in Bohemia with regards to the South Bohemian region during the past 20-30 years. Sbor. Jihoces. Muz. C. Budejovice., Prír. vedy, 22: 21-28 (in Czech).

Landa, V. and T. Soldán. 1985. Distributional patterns, chorology and origin of the Czechoslovak fauna of mayflies (Ephemeroptera). Acta ent. Bohemoslov., 82: 241-268.

Landa, V. and T. Soldán. 1989. Distribution of mayflies (Ephemeroptera) in Czechoslovakia and its changes in connection with water quality changes in the Elbe basin. Studie CSAV, 17, Academia, Praha, 172 pp. (in Czech, English summary).

Landa, V. and T. Soldán. 1995. Mayflies as bioindicators of water quality and environmental changes on a regional and global scale, pp. 21-30. In: Corkum L.D. and Ciborowski J.J.H. (Eds): Current Directions in Research on Ephemeroptera. Canadian Scholar's Press Inc., Toronto.

Landa, V., S. Zahrádková, T. Soldán and J. Helešic. 1997. The Morava and Elbe river basins, Czech Republic: a comparison of long-term changes in mayfly (Ephemeroptera) biodiversity, pp. 219-226. In: P. Landolt and M. Sartori (eds.). Ephemeroptera and Plecoptera: Biology-Ecology-Systematics. Mauron, Tinguely and Lachat, CH-Fribourg.

Lepš, J., T. Soldán and V. Landa. 1989. Multivariate analysis of compositional changes in communities of Ephemeroptera (Insecta) in the Labe basin, Czechoslovakia - a comparison of methods. Coenoses, 4: 39-37.

Lepš, J., T. Soldán and V. Landa. 1990. Prediction of changes in ephemeropteran communities - a transition matrix approach, pp. 281-287. In: I. C. Campbell (ed.). Mayflies and Stoneflies. Life Histories and Biology. Kluwer Academic Publ., Dordrecht.

Margalef, R. 1958. Information theory in ecology. Gen. Syst. 3: 36-71.

Mergl, A. 1997. Characteristic mayfly taxocoenes of epirhithral of the Odra river basin. Diploma thesis, Dept. of Zool. and Ecol., Masaryk Univ. Brno, 81 pp. (in Czech).

Shannon, C. E. and W. Weaver . 1963 Mathematical Theory of Communication. University of Illinois Press, Urbana, 182 pp. Sládecek V. (1973) System of water quality from the biological point of view. Arch. Hydrobiol./Ergebn. Limnol., 7: 1-218.

Soldán, T. 1992. Mayflies - Ephemeroptera. In: Red Book of Endangered and Rare Species and Plants of Czechoslovakia. Príroda, Bratislava, pp. 60-62. (in Czech).

Soldán, T., S. Zahrádková, J. Helešic, L. Dušek and V. Landa. 1998. Distributional and quantitative patterns of Ephemeroptera and Plecoptera in the Czech Republic: a possibility of detection of long-term environmental changes in aquatic biotopes. Folia Fac. Sci. nat. Univ. Masaryk. Brunensis, Biologia 98, 305 pp.

Ter Braak, C. J. F. 1988. CANOCO an extension of DECORANA to analyse species environment relationships. Vegetatio 75: 159-160.

Wright, J. F. 1995. Development and use of a system for predicting the macroinvertebrate fauna in flowing waters. Aust. J. Ecol. 20: 181-197.

Wright, J. F., P. D. Armitage and M. T. Furse. 1989. Prediction of invertebrate communities using stream measurements. Reg. Rivers: Res. Manag. 4: 147-155.

Wright, J. F., M. T. Furse and P. D. Armitage. 1993. RIVPACS - a technique for evaluating the biological quality of rivers in the UK. European Water Poll. Cont. 3 (4): 15-25.

Zelinka, M. and P. Marvan. 1961. Zur Präzisierung der biologischen Klasifikation der Reinheit fliessender Gewässer. Arch. Hydrobiol., 57: 389-407.

Zelinka, M. and J. Skalníková J. 1959. To the knowledge of mayflies (Ephemeroptera) of the Morava river basin. Spisy prír. fak. Univ. Brno, 401: 89-96. (in Czech)

Zelinka, M. 1953. Larvae of mayflies (Ephemeroptera) of the Moravice River and their relationships to water quality. Práce Moravskoslez. akad. prír. ved., 25: 181-200. (in Czech).

Zelinka, M. and P. Marvan. 1961. Zur Präzisierung der biologischen Klasifikation der Reinheit fliessender Gewässer. Arch. Hydrobiol. 57: 389-407.

GEOGRAPHICAL AND SEASONAL OCCURRENCE OF WINTER STONEFLIES (PLECOPTERA: CAPNIIDAE) OF THE SOUTHERN SIERRA NEVADA, CALIFORNIA, USA

C. Riley Nelson[1,3] and Derham Giuliani[2]

[1] Division of Biological Sciences
and Brackenridge Field Laboratory
University of Texas, Austin, Texas 78712, USA
[2] P. O. Box 265, Big Pine, California 93513, USA
[3] Current address and to whom correspondence
may be directed: Department of Zoology
Brigham Young University, Provo, Utah 84602, USA

ABSTRACT

During the winter of 1988-1989, 5 streams in the southern Sierra Nevada of California were visited once a week and adult stoneflies collected. Additional collecting visits from 1986-1992 were made to a larger number of streams draining the southeastern slopes of the Sierra Nevada. Capniidae dominated the collections. We collected a total of 5726 specimens in 10 species of Capniidae. Two species accounted for 89% of the individuals. *Capnia inyo* (18.6%) and *C. utahensis* (70.5%) were the numerically dominant species. In 3 of the 5 streams surveyed which had stretches with both *C. inyo* and *C. utahensis* present, *C. utahensis* appeared sooner and disappeared later than *C. inyo*. Winter emergence of the Capniidae is again substantiated but there is little evidence of a relationship between timing of emergence and altitude. We found no evidence of protandry but evidence of postgyny.

INTRODUCTION

The Sierra Nevada of California and Nevada in the United States of America are the tallest mountains in temperate North America and provide an important natural laboratory for the study of patterns of capniid zoogeography. The stonefly fauna in the creeks and rivers of the Sierra Nevada is rich. Jewett (1959) compiled a synthesis of the known distributions of Californian stoneflies in which he found 101 of the 350 stonefly species then known to occur in North America. Of these 101 species 21 were capniids. Much collecting and systematic work has been done in California since Jewett's report such that now over 150 species are known from California (Stark, Szczytko, and Baumann 1986; Nelson, C. R.

Trends in Research in Ephemeroptera and Plecoptera
Edited by E. Dominguez, Kluwer Academic/Plenum Publishers, 2001

Table 1. Capniidae of the southeastern Sierra Nevada, California, USA

Name	Number of Individuals	Percentage of Total	Total Distribution
Capnia utahensis	4039	70.54	Southern Great Basin and Sierra Nevada
Capnia inyo	1066	18.62	Southeastern Sierra Nevada
Capnia mono	309	5.40	Southeastern Sierra Nevada
Capnia shepardi	234	4.09	Southern Sierra Nevada
Capnia scobina	43	0.75	Central Sierra Nevada
Eucapnopsis brevicauda	15	0.26	Widespread in western North America
Capnia mariposa	12	0.21	Southern Sierra Nevada
Mesocapnia lapwae	6	0.10	Western California and Northern Idaho
Utacapnia sierra	1	0.02	Central Sierra Nevada
Capnia giulianii	1	0.02	Southeastern Sierra Nevada
Total	5726	100.00	Western North America

unpublished notes). Of these 150, 44 are capniids (Nelson, C. R. unpublished data) with 27 species known from the Sierra Nevada. This is the richest fauna known for a single mountain range in the world. Knowledge of how these species interact with abiotic factors as well as among themselves would help us understand adult Plecoptera communities. We thus chose to sample capniids from a series of parallel streams in the southern Sierra Nevada.

Stoneflies are limited in time and space by narrow ecological tolerances (Hynes 1976; Hitchcock 1974) both as terrestrial adults and aquatic nymphs. Stonefly nymphs are almost all restricted to cold, running waters (Baumann 1979; Stewart and Stark 1988). These restrictions are especially apparent in the winter stonefly family Capniidae. In contrast with slightly more tolerant stoneflies such as some Perlidae (Baumann 1979) which can have rather extended emergence seasons, capniids begin emerging as adults during the seemingly coldest and harshest parts of winter and continue emergence only into springtime. No capniids are known to emerge during the summer except at very high elevations (greater than 3000 m, Hitchcock 1969). The interaction of temperature and altitude in stonefly community structuring has been noted in numerous studies (as reviewed by Ward and Stanford 1982; Ward 1984; Ward 1986; Radford and Hartland-Rowe 1971) but general attempts to note zoogeographic trends related to emergence of winter stonefly adults in streams in close proximity to each other are lacking (but see Nebeker 1967; Baumann 1967).

Anecdotal records from the northern hemisphere indicate that emergences of individual species start at low latitudes and move north as the season progresses (Nebeker and Gaufin 1967). These same authors report that within streams the emergence begins at low elevation and proceeds to higher, but indicate that more extensive data than what they had would be necessary to fully substantiate this claim.

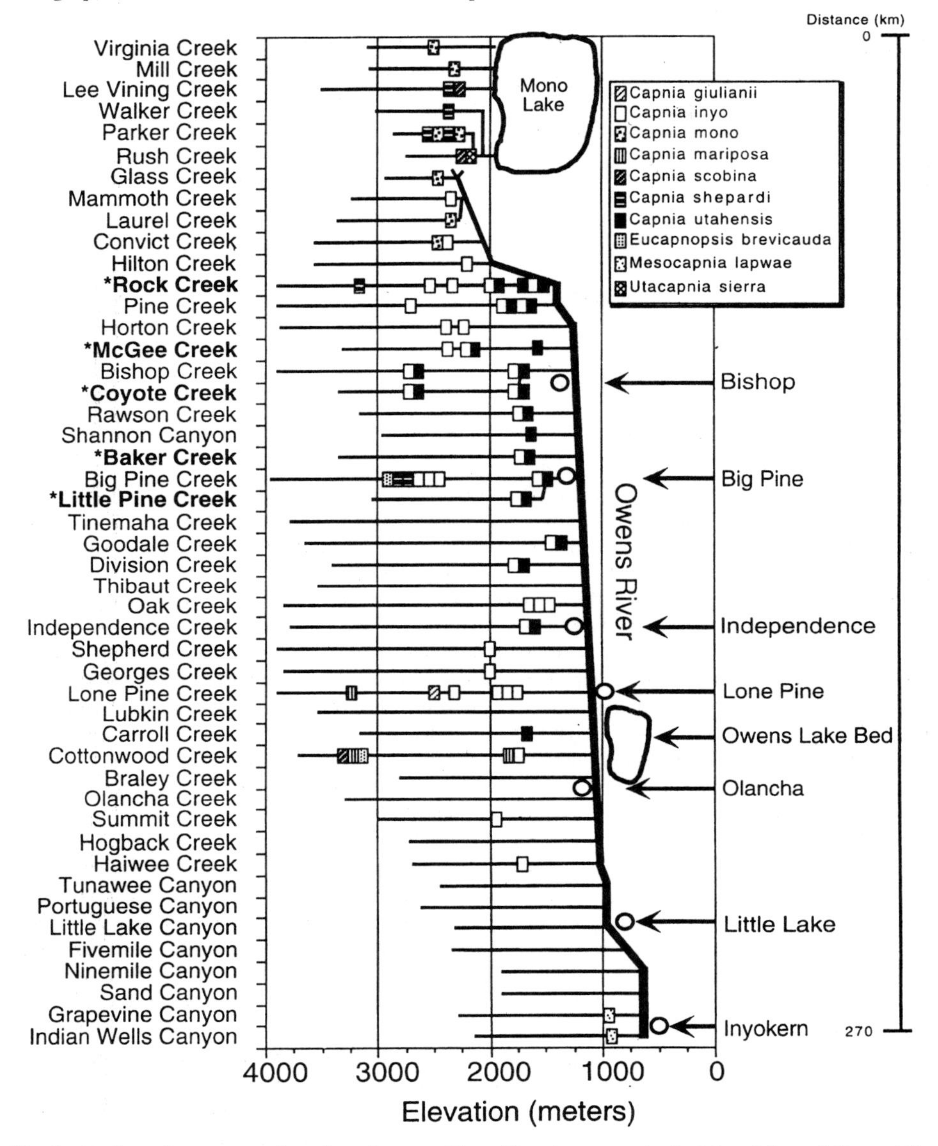

Fig. 1. A schematic representation of southeastern Sierra Nevada streams in the Mono lake and Owens River drainages, with distributions of 10 species of Capniidae. Distance from Inyokern to Bishop is 200 km (125 mi). Creeks marked **bold** and with * were studied in greater detail.

In addition it is thought that males emerge from the streams ahead of the females (protandry), both within a day and within the season (Illies 1952; Brinck 1949) and that females remain after the males have disappeared for the season (postgyny) as noted by Sheldon and Jewett (1967), Illies (1952), and Brinck (1949). Our study attempts to document these four notions of adult emergence in stoneflies as well as broader geographic patterns of emergence for these species. We report herein an intensive study of the capniid communities of several streams from the southern Sierra Nevada in California.

Table 2. Sex of first specimens of the season of *Capnia inyo* and *C. utahensis* to be collected from the banks of five streams of the east slope of the central Sierra Nevada of California, USA, winter 1988-1989. n = number of creeks (creeks by species) sampled.

Species	n	Males	Females	Both
Capnia inyo	5	1	2	2
Capnia utahensis	5	3	1	1
Total	(10)	4	3	3

MATERIALS AND METHODS

During the winter of 1988-1989, 5 streams (from north to south, Rock Creek, McGee Creek, Coyote Creek, Baker Creek, and Little Pine Creek) were visited approximately once a week. Monitoring sites on each stream were chosen to coincide with locations where previous collecting had yielded adults of two or more species.

Each visit to a site consisted of a fixed sampling procedure: walking a loop over a prescribed route encompassing both sides of the stream at each location. The five loops ranged in length from 35 to 135 m with length dependent on available and accessible habitat. All stoneflies seen within easy reach were collected along the route and a beating stick was used to propel adult specimens from the shrubs and trees on to the snow or a beating sheet. This fixed procedure was adhered to at each site during each of the 17 or 18 visits. Sampling intensity, as measured by area sampled, at each site was constant but sampling intensity varied from site to site because the habitat available on each stream's loop limited the length of the loop . Thus caution is advised in comparing exact numbers of specimens from site to site.

Additional collecting visits were made to a larger number of streams draining the southeastern slopes of the Sierra Nevada. These streams were sporadically sampled at wide ranges of elevations. Sampling at these sites consisted of searching for adult specimens in a wider variety of microhabitats in a qualitative fashion. Specimen distribution data for these streams is presented simply as presence. These data help show the broader distribution, both latitudinally and altitudinally, for the study species. Historical records from the literature and from recent efforts of many collectors are used to show presence of capniid species in particular streams.

All collected specimens were preserved in 70% ethanol and labeled in the field with information regarding stream, locality, elevation, and date. All identifications were made by the first author using a Wild M5 stereomicroscope equipped with an ocular micrometer and following the keys in Nelson and Baumann (1989). Elevations are given as meters above sea-level (masl).

We approached the issues of seasonal protandry and postgyny by noting the sexes of the individuals of the two dominant species for the first and last detection dates at each creek site. Each species was detected at each of the 5 creek sites so the total number of detection events (species by creek) was 10 for the protandry and postgyny studies. These data were subjected to chi squared tests for independence (X^2) with the null hypothesis for protandry being that males and females occurred first with equal frequency.

We approached the postgyny issue in a similar fashion by noting the sexes of the individuals caught during the last sampling event at each creek, for each of the two species. Our null hypothesis was that the last date on which males occurred was not different from the last date on which females occurred. Thus our null hypotheses for all studies of protandry and postgyny were that the sexes were present on the first and last dates in equal frequencies.

Table 3. Sex of last specimens of the season of *Capnia inyo* and *C. utahensis* to be collected from the banks of five streams of the east slope of the central Sierra Nevada of California, USA, winter 1988-1989. n = number of creeks (creeks by species) sampled.

Species	n	Males	Females	Both
Capnia inyo	5	0	4	1
Capnia utahensis	5	0	4	1
Total	(10)	0	8	2

RESULTS AND DISCUSSION

Winter emerging stoneflies consisted almost entirely of Capniidae. We collected and identified a total of 5726 specimens of winter-emerging stoneflies in 10 species of Capniidae from streams of the southeastern Sierra Nevada (Table 1, Fig. 1). The only other stonefly adults collected during our winter and early spring sampling were a very few specimens (less than 100) of *Zapada cinctipes* (Banks) and *Paraleuctra occidentalis* (Banks) both of whose emergence patterns extend beyond the winter season (Sheldon and Jewett 1967). Seven of these 10 species of capniids are endemic to the Sierra Nevada, one species' range also includes the Great Basin, and the remaining two are much more widespread.

Two species *Capnia inyo* and *C. utahensis* comprised 89% of the total specimens. The next two species accounted for another 9.5% of the total so that the four most abundant species accounted for 98.5% of the total. The last six species divided the remaining 1.5%.

Space, time, and their apparent interaction were noted in this study. On the spatial scale, latitude and altitude both played roles in structuring the emergence patterns of the winter stonefly communities of the streams. Considering first the latitudinal scale: 4 species, *C. mono*, *C. scobina*, *C. shepardi*, and *Utacapnia sierra* were found in the northern streams (Fig. 1, Virginia to Rush Creeks); 8 species were found in the central streams (Fig. 1, Glass to Haiwee Creeks), *C. giulianii*, *C. inyo*, *C. mariposa*, *C. mono*, *C. scobina*, *C. shepardi*, *C. utahensis*, and *Eucapnopsis brevicauda*, and the southern streams (Fig. 1, Tunawee to Indian Wells Canyon) had only a single species, *Mesocapnia lapwae*. Note the overlap in distribution of *C. mono*, *C. scobina*, and *C. shepardi* in the northern and central areas. There is no overlap between the capniids of these areas and that of the southernmost streams.

Second, the communities could be divided into three altitudinal series: high (>3000 m), middle (1000 - 2999 m), and low (<1000 m). High altitude species included *C. mariposa*, *C. scobina*, *C. shepardi*, and *E. brevicauda*. Middle elevation species were *C. giulianii*, *C. inyo*, *C. mariposa*, *C. mono*, *C. scobina*, *C. shepardi*, *C. utahensis*, *E. brevicauda*, and *U. sierra*. The low elevation species was *M. lapwae*, which was caught only in the southern streams. Once again, elevational overlap was apparent only between the high and middle reaches with *C. mariposa*, *C. scobina*, *C. shepardi*, and *E. brevicauda* participating.

Species sharing the north-central and high-middle series included *C. scobina* and *C. shepardi*, but the widespread species *E. brevicauda* will probably share this pattern when even more intensive collecting has been done. An interaction between latitude and altitude in three species: *C. scobina*, *C. shepardi*, and *E. brevicauda;* is apparent. In these species southern representatives were collected at higher elevations than those in the north. This pattern was noticeably absent in *C. inyo*, *C. mariposa*, and *C. utahensis*.

Three scales of time of decreasing period could possibly be discerned using the detailed 5 creeks studies: winter emergence, length of species presence, and protandry or postgyny (both within a day and throughout the season). Once again capniids were shown to be winter emergers. This detailed study of 5 central area creeks allowed patterns of emergence, or at least presence to be determined (Fig. 2) using the two dominant winter stone-

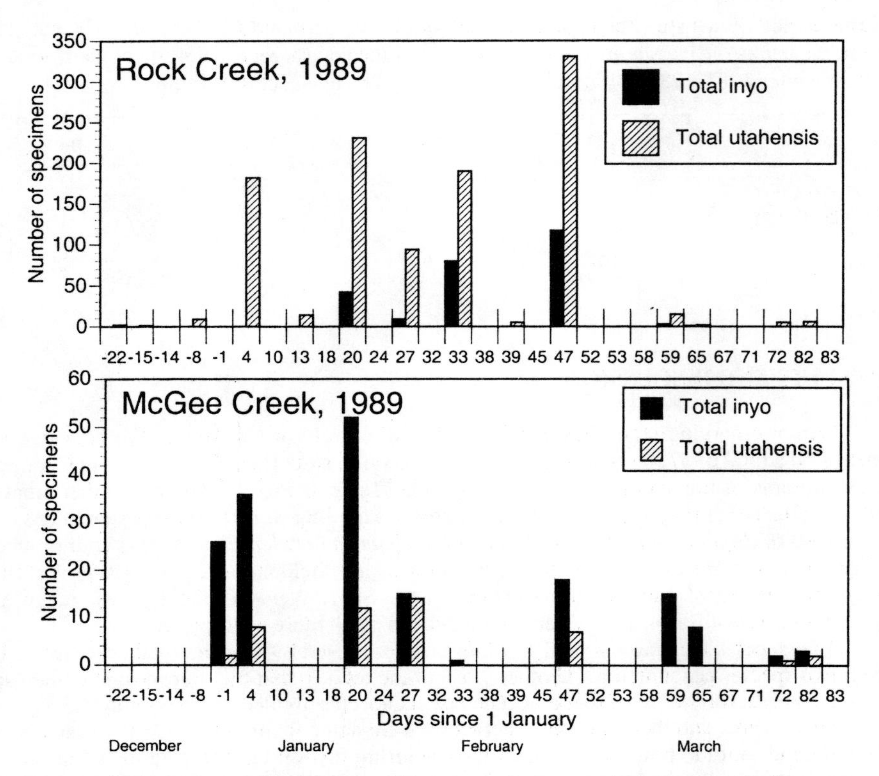

Fig. 2. Seasonal distribution of *Capnia inyo* and *C. utahensis* in 5 creeks of the southeastern Sierra Nevada of California. Sampling began December, 1988.

flies. In 3 of the 5 creeks *C. inyo* was collected over a much shorter period than *C. utahensis*. In McGee Creek and Coyote Creeks, both species were collected over similar time periods (Fig. 2) but the small number of *C. inyo* taken in Coyote, Baker and Little Pine Creeks make any committment to pattern suspect.

We found little evidence for a linear relationship between timing of presence and altitude in the two dominant species (Figs. 3-5). Regressions for altitude and first collections were not significant (Fig. 3), nor were those for last collections (Fig. 5). However *C. inyo* (Fig. 4) did show a significant relationship ($p = 0.0001$) for peak of presence (the mode). The same analysis for *C. utahensis* (Fig. 5) was not significant ($p = 0.71$). It is tempting to interpret this significance for the peak in *C. inyo* as being related to its short emergence period. A shorter emergence season at a given elevation would allow maximum contact between the sexes while limiting exposure to streamside predation. Is this species reading environmental cues more closely than the other capniids to better synchronize emergence? This is worthy of further investigation.

Our analyses of seasonal protandry and postgyny in the two dominant species are summarized in Tables 2 and 3. In the protandry study, only males were in the first detected lot of specimens during four events and only females at three events. Both males and females were in the first detection sample during three events. No significant differences were noted for the X^2 tests of the protandry hypothesis ($X^2 = 0.14$; df = 1; $p = 0.705$). Thus there is no evidence of seasonal protandry in these species in these streams and our sampling protocol could not detect protandry (or postgyny) on a daily scale. As a result of this study it is clear that another study involving the daily emergence patterns of winter stoneflies would be a logical next step in our understanding of protandry in these insects.

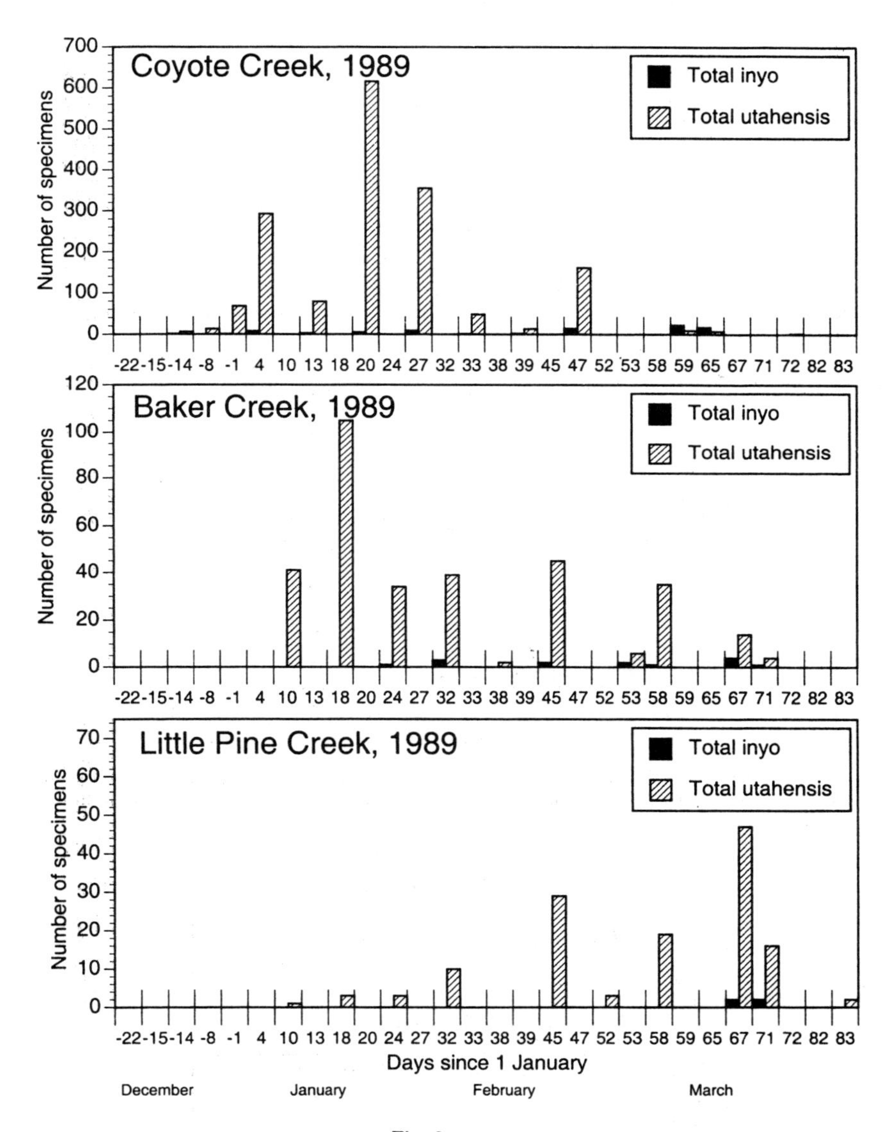

Coyote Creek, 1989
Baker Creek, 1989
Little Pine Creek, 1989
Total inyo
Total utahensis
Number of specimens
Days since 1 January
-22 -15 -14 -8 -1 4 10 13 18 20 24 27 32 33 38 39 45 47 52 53 58 59 65 67 71 72 82 83
December
January
February
March

Fig. 2 (cont.)

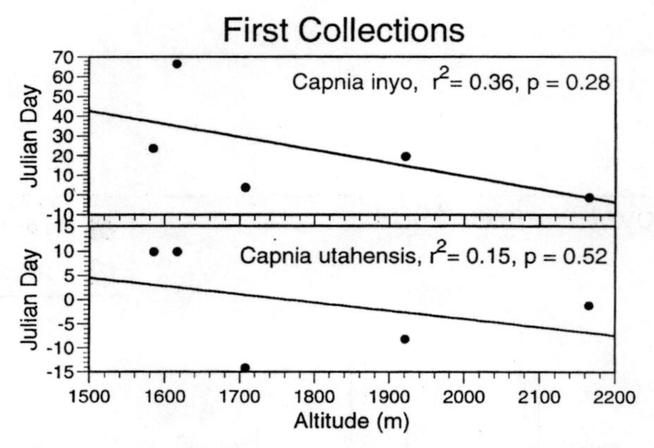

Fig. 3. First collections of the season of *Capnia inyo* and *C. utahensis* at various altitudes in 5 creeks of the southeastern Sierra Nevada of California. Regression of altitude against Julian day not significant in either species.

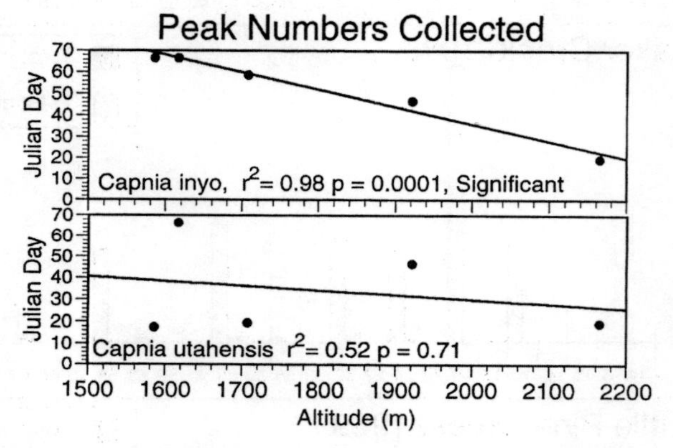

Fig. 4. Peak of numbers in collections of *Capnia inyo* and *C. utahensis* at various altitudes in 5 creeks of the southeastern Sierra Nevada of California. Regression of altitude against Julian day significant ($p = 0.05$) in *Capnia inyo*, not in *C. utahensis*.

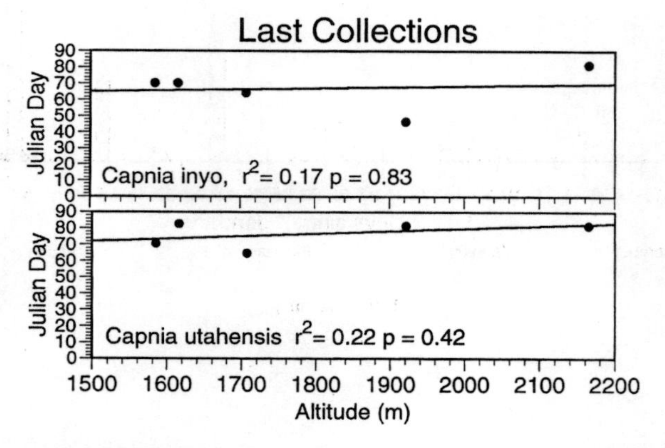

Fig. 5. End of season collections of *Capnia inyo* and *C. utahensis* at various altitudes in 5 creeks of the southeastern Sierra Nevada of California. Regression of altitude against Julian day not significant in either species.

In the postgyny study, during eight of the last detection events only females were found and two last detection events yielded both males and females. We did detect significant differences in the frequencies of males and females in the postgyny hypothesis ($X^2 = 6.80$; df = 1; p = 0.009). The equal frequency hypotheses are rejected at $p < 0.05$: females disappeared from the sites later than did males. Thus we did find evidence of seasonal postgyny (where females persist later in the season than males). Males alone were never collected on a species' last detection date for each of the creeks. Females alone were collected on these last dates during 8 of the 10 events. This supports the generally accepted idea that females persist longer than males streamside (Sheldon and Jewett 1967; Illies 1952; and Brinck 1949). A next step in the study of postgyny should include a mark and recapture study of individuals of both sexes to monitor longevity in the field.

CONCLUSIONS

1. Capniidae dominate winter-emerging stonefly communities.
2. Seven of the 10 capniid species collected were limited in distribution to the Sierra Nevada.
3. Rare species (<10% of total numbers caught) outnumbered common ones 8 to 2.
4. 89% of the total number of specimens belonged to these 2 common species, wtih one, *C. utahensis* being the dominant (70.5% of specimens caught).
5. Latitude and altitude both played roles in structuring the winter stonefly communities of the streams.
6. There was overlap in distribution of several species in the northern and central areas, but none with the south.
7. We propose three altitudinal series: >3000 m, 1000-3000 m, and <1000 m.
8. Elevational overlap was apparent only between the high and middle reaches and southern representatives of the same species were collected at higher elevations than those in the north in 3 species.
9. We found little evidence for a linear relationship between timing of collection and altitude in the two species most frequently encountered.
10. We found no evidence for protandry.
11. Postgyny was apparent.

ACKNOWLEDGMENTS

We thank the members of the California Academy of Sciences for logistic support for both authors. The Tilton Fellowship afforded C. R. Nelson the opportunity to pursue these research interests during his stay at the California Academy of Sciences. We also thank Dr. L. E. Gilbert of the Brackenridge Field Laboratory of the University of Texas for support through the years.

REFERENCES

Baumann, R. W. 1967. A study of the stoneflies (Plecoptera) of the Wasatch Front, Utah. Unpublished Master's thesis, University of Utah, Salt Lake City, Utah.

Baumann, R. W. 1979. Nearctic stonefly genera as indicators of ecological parameters (Plecoptera: Insecta). Gr. Basin Natural. 39: 241-244.

Brinck, P. 1949. Studies on swedish stoneflies (Plecoptera). Opus. Ent., suppl. 11. 250 pp.

Hitchcock, S. W. 1969. Plecoptera from high altitudes and a new species of *Leuctra* (Leuctridae). Ent. News 80: 311-316.

Hitchcock, S. W. 1974. Guide to the insects of Connecticut. Part VII. The Plecoptera or stoneflies of Connecticut. Bull. Conn. Geol. Nat. Hist. Surv. 107: 1-262.

Hynes, H. B. N. 1976. The biology of Plecoptera. Ann. Rev. Ent. 21: 135-153.

Illies, J. 1952. Die Plecopteren und das Monardsche prinzip. Ber. Limnol. Flusst. Freudenthal 3: 53-69.

Jewett, S. W. 1960. The stoneflies (Plecoptera) of California. Bull. California Insect Survey 6: 122-177.

Nebeker, A. V. 1966. The taxonomy and ecology of the family Capniidae (Plecoptera) of the western United States. Unpublished Ph. D. dissertation, University of Utah, Salt Lake City, Utah.

Nelson, C. R. and R. W. Baumann. 1989. Systematics and zoogeography of the winter stonefly genus *Capnia* (Plecoptera: Capniidae) in North America. Gr. Basin Natural. 49: 289-363.

Radford, D. S. and R. Hartland-Rowe. 1971. Emergence patterns of some Plecoptera in two mountain streams in Alberta. Can. J. Zool. 49: 657-662.

Sheldon, A. L. aand S. G. Jewett, Jr. 1967. Stonefly emergenc in a Sierra Nevada stream. Pan-Pac. Ent. 43: 1-8.

Stark, B. P., S. W. Szczytko and R. W. Baumann. 1986. North American stoneflies (Plecoptera): systematics, distribution, and taxonomic references. Gr. Basin Natural. 46: 383-397.

Stewart, K. W. and B. P. Stark. 1988. Nymphs of North American stonefly genera (Plecoptera). Thomas Say Foundation. Ent. Soc. Amer., no. 12.

Ward, J. V. 1984. Diversity patterns exhibited by the Plecoptera of a Colorado mountain stream. Ann. Limnol. 20: 123-128.

Ward, J. V. 1986. Altitudinal zonation in a Rocky Mountain stream. Arch. Hydrobiol. 2: 133-199.

Ward, J. V. and J. A. Stanford. 1982. Thermal responses in the evolutionary ecology of aquatic insects. Ann. Rev. Ent. 27: 97-117.

EGG DEVELOPMENT IN *DINOCRAS CEPHALOTES* (PLECOPTERA, PERLIDAE) AT ITS ALTITUDINAL LIMIT IN NORWAY

Ketil Sand and John E. Brittain

Zoological Museum, University of Oslo
Sars gate 1, 0562 Oslo, Norway

ABSTRACT

Egg development of *Dinocras cephalotes* (Curtis) was studied in a population from the Jotunheimen Mountains of central southern Norway. At 16, 20 and 24 °C hatching success was >50%, while at 12 °C only a few eggs hatched. The number of degree days required for hatching was less, but the lower threshold temperature for development was higher than found in earlier studies. The mean length of first instar nymphs increased significantly ($P<0.001$) with increasing egg incubation temperature. Eggs probably hatch throughout the ice free period when temperatures exceed 10 °C and *D. cephalotes* probably has a 5-6 year life cycle including egg development.

INTRODUCTION

Dinocras cephalotes (Curtis) is a warm stenothermal plecopteran, occurring throughout most of Europe (Lillehammer, 1988). Its life cycle, especially egg development, has been investigated by several workers in recent years (Lillehammer, 1987a; Huru, 1987, Sanchez-Ortega and Alba-Tercedor, 1991; Elliott, 1995; Frutiger, 1996; Zwick, 1996 a, b). The eggs of *D. cephalotes* require a relatively high incubation temperature to develop, the threshold temperature increasing at higher altitudes and latitudes. The minimum threshold temperature at 46°N is 6°C (Zwick, 1996a), but is 12 -14°C at 61°N (Lillehammer, 1987a; Zwick, 1996a). Three years are required to complete its life cycle through most of Europe (Frutiger, 1987; Sanchez-Ortega and Alba Tercedor, 1991). However, in northern Scandinavia it takes at least 4 -5 years to complete its life cycle (Huru, 1987; Ulfstrand, 1968). In the present study *D. cephalotes* has been studied in a mountain population from one of the highest known Norwegian localities. The aim of this investigation was thus to study a population at its altitudinal limit, with emphasis on egg development.

STUDY AREA

The investigation was carried out in the river, Hinøgla, at 1080 m a.s.l. in the valley, Øvre Heimdalen, on the eastern slopes of the Jotunheimen Mountains in central southern Norway (61°25'N, 8°52'E). The area is about 50 km north of the small town of Fagernes (379 m a.s.l.) where *D. cephalotes* was collected for previous egg development studies (Lillehammer, 1987a; Zwick, 1996a). The river where the species was collected in the present study is about 3 km long, 5 - 10 m wide and situated between the two lakes Øvre and Nedre Heimdalsvatn in the subalpine birch forest. Birch and willow, interspersed with areas of bog border the river. The climate is harsh with long winters and short summers. Ice cover on the river extends from October/November until May. The mean temperature in the river was 11.6°C (range 2.7 - 16.4°C) and 11.4°C (range 3.0 - 15.1°C) from 15 June - 15 September during 1995 and 1996, respectively (Fig. 1). *D. cephalotes* was found in an area with large stones and rocks about 500 m downstream of Øvre Heimdalsvatn. The population is relatively isolated in the stream. The nearest possible locations are in the river flowing out of the 10 km long Nedre Heimdalsvatn and on the other side of a 1800 m high mountain in the rivers Sjoa or Sikkelsdalselva, about 10 km away. The valley and the lake, Øvre Heimdalsvatn, have been extensively studied and are well described in Vik (1978).

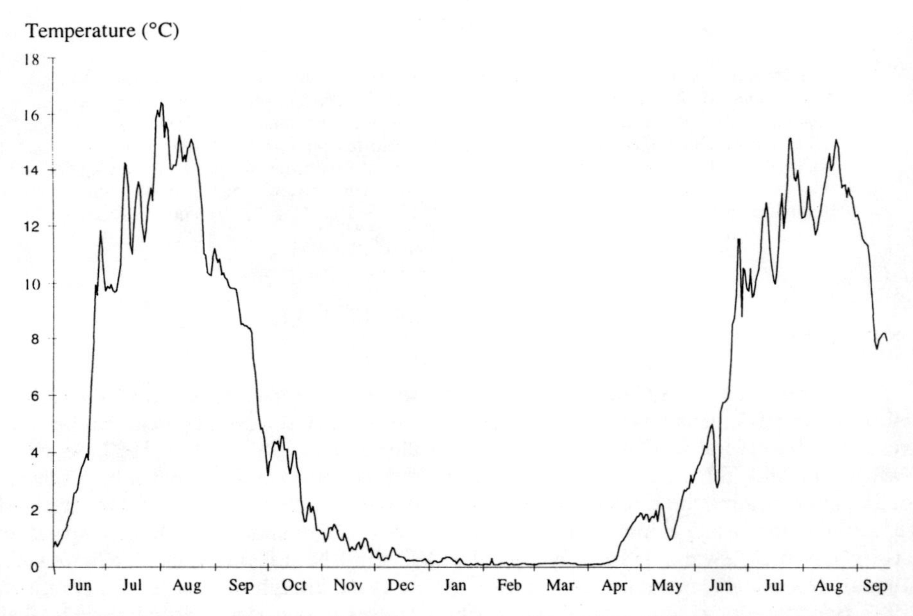

Fig. 1. Water temperatures in the river, Hinøgla, Norway, from 1 June 1995 to 19 September 1996.

METHODS

Adults of *D. cephalotes* were collected under stones along the river bank. Males and females were held together in plastic containers until they had copulated, which took about 24 hours. After copulation each female was placed in a separate container in the laboratory until the egg masses were extruded. The eggs were then removed and each egg mass allocated to six different small Petri dishes filled with water. All water used was taken from the sampling site. The eggs were then incubated at different constant temperature regimes (4, 8

12, 16, 20 and 24°C) in total darkness (Lillehammer 1987a). The eggs were generally inspected daily.

After 107 days incubation, the eggs that did not hatch at 4°C and 8°C were divided into two samples, one kept at the same temperature as earlier and one moved to 12°C. Hatched nymphs were counted and removed by a pipette. The length of first instar nymphs was measured from those hatching at the mean incubation time for the three temperatures: 16°C, 20°C and 24°C. Nymphal length was measured from the tip of the head to the base of the cerci using a stereo microscope with 50x magnification.

The stream was sampled on three different occasions using the standard kick sampling technique (net with mesh size 350 mm and opening 30 x 30 cm). All samples were preserved in 70 % alcohol and sorted in the laboratory using a stereo microscope with 6x magnifica-

Table 1. Hatching time and number of degree days for 10, 50 and 90 % of the eggs that eventually hatched. Hatching success is the % of eggs that hatched and duration of hatching is the number of days between 10 and 90 % hatching.

Temperature (°C)	Female no.	No. of eggs	10 % °days/days	50 % °days/days	90 % °days/days	Hatching success (%)	Duration of hatching (days)
4	1	111	0	0	0	0	0
4	2	186	0	0	0	0	0
4	3	163	0	0	0	0	0
8	1	197	0	0	0	0	0
8	2	168	0	0	0	0	0
8	3	119	0	0	0	0	0
12	1	149	1824/152	1824/152	1824/152	0.7	1
12	2	103	0	0	0	0	0
12	3	171	1764/147	1764/147	2076/173	1.2	26
16	1	161	656/41	688/43	688/43	91.3	3
16	2	117	688/43	720/45	752/47	76.9	5
16	3	96	656/41	688/43	68/43	95.8	3
16	total/mean	374	656/41	688/43	720/45	88.0	5
20	1	114	600/30	600/30	640/32	99.1	3
20	2	237	580/29	620/31	780/39	97.9	11
20	3	111	580/29	600/30	660/33	96.5	5
20	total/mean	462	580/29	620/31	720/36	97.8	7
24	1	135	624/26	648/27	696/29	53.3	4
24	2	197	0	0	0	0	0
24	3	128	624/26	624/26	648/27	48.4	2
24	total/mean*	263	624/26	648/27	672/28	51.0	3
4 - 12**	1	76	0	0	0	0	0
4 - 12**	2	107	1356/113	1356/113	1692/141	1.9	28
4 - 12**	3	98	1584/132	1764/147	1956/163	3.1	31
8 - 12**	1	103	1428/119	1428/119	1848/154	3.9	35
8 - 12**	2	67	1536/128	1536/128	2028/169	3.0	41
8 - 12**	3	79	1956/163	1956/163	1956/163	1.3	1

* Only the two batches that hatched are used in the calculations.

** Eggs initially incubated at 4 and 8°C and subsequently transferred to 12°C.

tion. Nymphal length was measured from the tip of the head to the base of the cerci using 6 - 40 x magnification, depending on size.

Temperature in the river was recorded every 6 hours, from which daily mean temperatures have been calculated. The temperature was taken at a site about 400 m upstream, but there is unlikely to be significant differences between the two sites.

RESULTS AND DISCUSSION

Egg Development

At 16 and 20°C, three out of five batches hatched successfully while at 24°C only two batches hatched (Tab. 1). Highest hatching success was at 20°C, where 97.8 % of the eggs hatched. Hatching success decreased to 50.8 % at 24°C and 88.0 % at 16°C. At 12°C only three eggs hatched, the first one after 147 days. No hatching occurred at 4°C and 8°C. At 16°C - 24°C eggs hatched sporadically over a long period. The last egg hatched after 39 days at 24°C while the last egg hatched after 78 days at 20°C and 140 days at 16°C. Remaining eggs were removed from the incubators and counted after 249 days. However, the main hatching period (10 - 90 % hatching) was short at all temperatures, varying between 3 and 7 days, being shortest at 24°C and longest at 20°C. Elliott (1995) stated that *D. cephalotes* has a very short hatching period. However, Zwick (1996 a, b) showed that the hatching period could be prolonged (200 - 300 days) when incubation was interrupted by cold periods, below the lower threshold temperature. In Frutiger (1996) hatching period (5 - 95 % hatch) was long for several batches with maximum length of 233 days. Hatching was obtained at

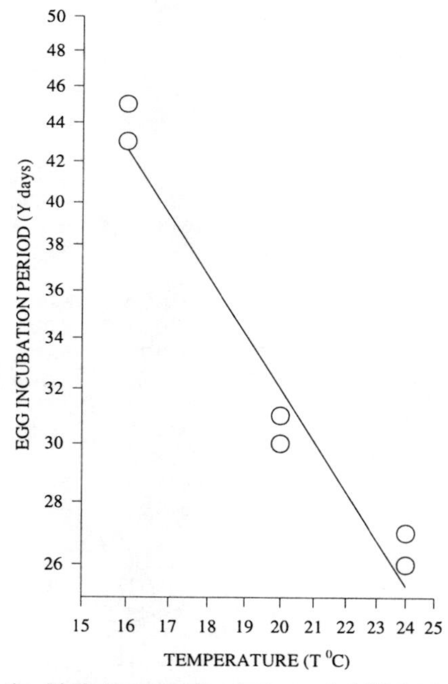

Fig. 2. The relationship between egg incubation period (Y days) and temperature (T°C) for *Dinocras cephalotes* from the Norwegian mountains plotted on logarithmic scales. The relationship is expressed by the following equation: $Y = 1489T^{-1.28}$ ($r^2 = 0.943$, $p < 0.0001$).

temperatures as low as 6.3°C with 5 % hatching after 385 days. In the present study incubation was terminated after 249 days, and prolonged incubation may perhaps have increased the hatching success at 12°C.

Dinocras cephalotes shows high variability in the minimum incubation temperature needed for egg development, ranging from 6.3°C to 14°C, with increasing temperature at higher latitudes and altitudes (Lillehammer, 1987a; Elliott, 1989; Frutiger, 1996; Zwick, 1996a, b). In a Norwegian population, Lillehammer (1987a) showed hatching success ranging from 30 to 85 % at 12°C. Zwick (1996a) obtained very low hatching success (two out of three batches hatched with less than 10 % hatching success) at 12°C in his Norwegian material. At 14°C hatching success was in the range 50 - 90 %. Samples were taken from the same locality in these two studies, in the outlet of the lake, Sæbufjorden (375 m a.s.l.), just north of Fagernes (61°00'N, 9°15'E). In the present study samples were taken from almost the same latitude, but at a higher altitude. A lower hatching success at 12°C in this study indicates a higher threshold temperature than found by Lillehammer (1987a) and Zwick (1996a). Lillehammer (1987a) found hatching success in the range of 89.3 - 97.3 % for his three batches at 16°C. At 20°C hatching success was similar in both studies and Lillehammer (1987a) observed a somewhat higher hatching success at 24°C than in the present study.

Eggs initially incubated at 4°C and 8°C and subsequently transferred to 12°C had a slightly better hatching success than those incubated at a constant temperature of 12°C, but no distinct hatching period was observed. Only 12 eggs hatched, over a period of two months, with the first after 113 days.

Whereas the threshold temperature for egg development was higher at higher altitude the number of degree days required was lower. The number of degree days needed for 50% of the eggs to hatch was lowest at 20°C, 620 degree days, increasing to 648 degree days at 24°C and 688 degree days at 16°C. This is lower than found in other studies: Lillehammer (1987a) found the mean degree days requirement to be 700 at 20°C, while Zwick (1996a) found 830 degree days at 22°C.

In the present study the relationship between incubation time (Y days) and water temperature (T°C) was well described by the linear form of the power law equation:

$Y = a - T^{-b}$,

where the constant a was 1489 and the constant b was 1.282 ± 0.119. The relationship was highly significant ($r^2 = 0.943$, $p < 0.0001$; fig. 3).

In Lillehammer (1987a) and Elliott (1989) the equivalent regression constants were: a = 2383 and b = 1.402 and a = 2209 and b = 1.354, respectively. Although not significantly different, they suggest a lower thermal demand (Brittain 1990) in the Øvre Heimdalen population compared to the other populations. Incubation time was also less temperature dependent (Brittain 1990).

Plecopteran egg development studies under fluctuating temperature regimes have been undertaken in natural temperature regimes (Elliott, 1989; Frutiger, 1996), fixed fluctuating regimes with daily amplitudes of 4°C at temperatures higher than the threshold temperature (Frutiger, 1996) and using differences in temperatures much higher and lower than the threshold temperature to study egg dormancy (Zwick, 1996 a, b). However, fluctuating temperatures around the threshold temperature, which is the case under field conditions in Heimdalen, have not been used.

The minimum temperature for 50 % hatch lies in the range 12 to 16°C. In the present study, the mean incubation time was 43 days at 16°C, while in Lillehammer (1987a) incubation time at 16°C varied between 46 and 48 days. Zwick (1996a) showed a hatching success of 50 - 90 % at 14°C and in the present study the threshold temperature is probably in the same range, around 13 - 14°C. Such temperatures were reached by 11-12 July in 1995 and 21-25 July in 1996 (Fig. 1). In 1995 the highest daily mean temperature was 16.4°C, and 28 days had a higher mean temperature than 14°C, 42 days higher that 12°C and 67 days higher than 10°C. For 1996 the highest mean temperature was 15.1°C and only 9 days exceeded 14°C, 47 days exceeded 12°C and 70 days exceeded 10°C.

To be able to calculate the egg development time in Heimdalen it is necessary to know at which temperature egg development stops when it has first started. Lillehammer et al. (1991) found that eggs of *Leuctra fusca* (L.) hatched at very low temperature (0.5°C) as long as initial egg development had been initiated at relatively high temperatures (8 - 12°C). Eggs held at a constant temperature of 2°C did not hatch and hatching success of eggs reared initially at 8 and 12°C and subsequently transferred to 0.5°C was higher than for those kept at a constant temperature of 4°C. The same situation may also be valid for *D. cephalotes*. Eggs may continue to develop at temperatures below threshold temperature when development has started at a higher temperature. Lillehammer (1987b) found that eggs of *D. cephalotes* developing at 16°C stopped developing when they were cooled to 8°C.

In the summer of 1995 adults were caught during the first half of August when temperatures varied between 14 and 15°C. If development stops at 10°C, eggs laid on 10 August would have 29 days during the first summer with temperatures above 10°C (mean 12.6°C) and 30 days in 1996. If egg development did not start before temperatures reached 14°C the next year it would start on 25 July. Then there would be 47 days with temperatures above 10°C (mean 13.0°C) in both 1995 and 1996. If development stops at 12°C only 15 days (mean temperature 14.4°C) would be available for development after oviposition during 1995 and 21 days (mean temperature 13.3°C) in 1996. During both 1995 and 1996, 41 days had a higher mean temperature than 12°C.

By using the power law equation (Fig. 2), 56 days would be required at 13°C and 51 days at 14.0°C. A decrease in development time (about 10 %) has been recorded for development under fixed/natural daily fluctuations in temperature (Frutiger, 1996). However, no changes in development time have been found for natural light regimes compared to total darkness (Frutiger, 1996; Zwick, 1996a). Incubation time increased when development was interrupted by cold (4°C) periods (Zwick, 1996b). Eggs needed two weeks at the higher (16°C) temperature before development continued. Egg development time calculated on the basis of this should be about 60 - 65 days at mean temperatures of 13 -14°C. Thus, in warm summers, eggs can probably hatch during the autumn of the year after they are laid, while in cooler summers development time will be extended to two years. However, if development continues down to 10°C, eggs can hatch during the following year.

Table 2. Lengths of first instar nymphs measured at the mean hatching time for each temperature

Temperature (°C)	Incubation period (days)	Degree days above 0 °C	Number measured	Mean length ± 95% CL
24	27	648	52	1.20 ± 0.02
20	31	620	46	1.14 ± 0.04
16	43	688	54	1.08 ± 0.01

First Instar Nymphs

The mean length of the first instar nymph increased significantly ($p < 0.001$) with increasing incubation temperature (Table 2). Individual lengths varied between 0.9 and 1.4 mm. Brittain et al. (1984) also found differences in body length of the first instar nymphs related to incubation temperature for the stonefly, *Capnia atra* Morton. Eggs incubated at the optimum temperature (8°C) gave rise to significantly larger nymphs than at other temperatures, while 4, 12 and 16°C gave nymphs of intermediate size and 20 and 24°C gave the smallest sizes. Sweeney and Vannote (1978) in their thermal equilibrium hypothesis suggest that body size and fecundity decrease at both sides of the optimum temperature.

Brittain et. al. (1984) suggest that this is also already reflected in the egg stage. *C. atra* is a cold adapted species, most abundant in the upland areas of northern Europe and therefore has the largest nymphs at the lower incubation temperatures. *D. cephalotes* is a warm adapted species and therefore the higher temperatures are more optimal and body size of first instar nymphs increases with increasing temperature.

Life Cycle

The nymphal development of *D. cephalotes* has not been as intensively studied as egg development. In published studies the length of the life cycle has been suggested to vary from three years including egg development in Spain (Sanchez-Ortega and Alba-Tercedor, 1991) to four to five years, excluding egg development, in northern Scandinavia (Huru, 1987; Ulfstrand, 1968).

It is often difficult to interpret the life cycle of *D. cephalotes* due to long periods of hatching, frequent low densities and the large size difference between male and female nymphs. Twelve of the nymphs collected on 25 July 1995 were in the range 1.6 to 2.0 mm (fig. 4). The first instar nymphs are 0.9 - 1.4 mm in length so the smallest nymphs collected on 25 July could be second and third instars. The first instar of *D. cephalotes* does not eat (Harper, 1978) and its duration is relatively short and at 16°C it took less than one week in the present study. The smallest nymphs collected in the field probably hatched during the previous autumn. The next size group consisted of seven nymphs in the range 2.6 to 5.5 mm. These nymphs probably belong to a single year class. Growth is slow during the final instars due to development of wingpads and gonads (Frutiger, 1996). An increase in body length from 5.5 mm to 19.9 mm (mean of the largest nymphs in the sample), giving a growth rate of 1.5 %/day, is too high and one generation is probably missing in the sample. All the larger nymphs belong to the same generation and were ready to emerge. The large size difference is due to sexual differences. The imagos (29 males and 5 females) collected during the first three weeks of August showed large differences between the two sexes. Adult males varied between 13.0 and 17.3 mm with a mean length of 15.0 ± 0.4 mm (± 95 % CL), while the females varied between 19.7 and 21.9 mm with a mean length 20.7 ± 1.2 mm (± 95 % CL).

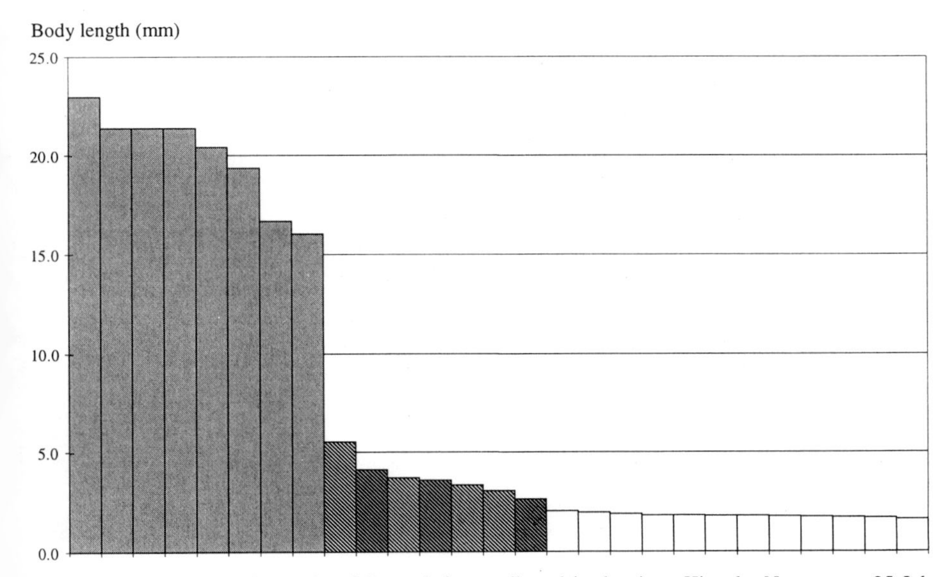

Fig. 3. Size distribution of nymphs of *D. cephalotes* collected in the river, Hinøgla, Norway, on 25 July 1995 (each bar represents a single nymph). The different generations are indicated by shading.

Dinocras cephalotes has the potential to use summer warmth during successive years by extending its embryogenesis over more than one year. Low temperatures resulting in a temporary stop in development, increase the incubation time and also lengthen the hatching period (Zwick, 1996b). Eggs hatched in pulses may reflect earlier cold periods (Zwick, 1996b). Eggs collected in the stream and incubated at suboptimal temperatures (10°C over 350 days) had a very different hatching pattern, 9 out of 11 batches had a hatching period of 200 - 300 days (Zwick, 1996b). In Øvre Heimdalen incubation is interrupted during the winter and the mean summer temperature is suboptimal for the species. This leads to an extended hatching period, and hatching probably occurs throughout the whole ice-free period, and the generations are therefore difficult to separate. A four year nymphal development is supported by the four nymphs found in the 14 June 1996 sample, they seemed to belong to three different generations; 3, 9 and 15 mm in length. Thus, the length of the life cycle is probably five years in Heimdalen including egg development, but this may extend to six years in cooler summers.

REFERENCES

Brittain, J. E. 1990. Life history strategies in Ephemeroptera and Plecoptera, pp. 1-12. In: I. C. Campbell (ed.). Mayflies and Stoneflies: Life Histories and Biology. Kluwer, Dordrecht, The Netherlands.

Brittain, J. E., A. Lillehammer and S. J. Saltveit. 1984. The effect of temperature on intraspesific variation in egg biology and nymphal size in the stonefly, *Capnia atra* (Plecoptera). J. Anim. Ecol., 53: 161-169.

Elliott J. M. 1989. The effect of temperature on egg hatching for three populations of *Dinocras cephalotes* (Plecoptera: Perlidae). Ent. Gaz. 40: 153-158.

Elliott, J. M. 1995. Egg hatching and ecological partitioning in carnivorous stoneflies (Plecoptera). Compt. Rend. Acad. Sci., Paris, Biol. Path. Anim. 318: 237-243.

Frutiger, A. 1987. Investigations on the life-history of the stonefly *Dinocras cephalotes* Curt. (Plecoptera: Perlidae). Aquat. Insects 9: 51-63.

Frutiger, A. 1996. Embryogenesis of *Dinocras cephalotes*, *Perla grandis* and *P. marginata* (Plecoptera: Perlidae) in different temperature regimes. Freshw. Biol., 36: 497-508.

Harper, P. P. 1978. Observations on the early instars of stoneflies (Plecoptera). Gewäss. Abwäss., 64: 18-28.

Huru. H. 1987. Occurrence and life cycle of *Dinocras cephalotes* (Curtis, 1827) (Plec. Perlidae) in North Norway. Fauna Norv. Ser. B, 34: 14-18.

Lillehammer, A. 1987a. Egg development of the stoneflies *Siphonoperla burmeisteri* (Chloroperlidae) and *Dinocras cephalotes* (Perlidae). Freshw. Biol. 17: 35-39.

Lillehammer, A. 1987b. Diapause and quiescence in eggs of Systellognatha stonefly species (Plecoptera) occurring in alpine areas of Norway. Ann. Limnol. 23: 179-184.

Lillehammer, A. 1988. Stoneflies (Plecoptera) of Fennoscandia and Denmark. Fauna Ent. scand. 21: 1-165.

Lillehammer, A., S. J. Saltveit and M. Brusven. 1991 The influence of variable temperatures on the incubation period of stonefly eggs (Plecoptera), pp. 377-385. In: J. Alba-Tercedor and A. Sanchez-Ortega (eds.). Overview and Strategies of Ephemeroptera and Plecoptera, Sandhill Crane Press, Gainesville.

Sanchez-Ortega, A. and J. Alba-Tercedor. 1991. The life cycle of *Perla marginata* and *Dinocras cephalotes* in Sierra Nevada (Grenada, Spain) (Plecoptera: Perlidae), pp. 387-401. In: J. Alba-Tercedor and A. Sanchez-Ortega (eds.). Overview and Strategies of Ephemeroptera and Plecoptera, Sandhill Crane Press, Gainesville.

Sweeney, B. W. and R. L. Vannote. 1978. Size variation and the distribution of hemimetabolous aquatic insects: Two thermal equilibrium hypotheses. Science 200: 444-446.

Ulfstrand, S. 1968. Life cycles of benthic insects in Lapland streams (Ephemeroptera, Plecoptera, Trichoptera, Diptera Simuliidae). Oikos 19: 167-190.

Vik, R. 1978. (ed.) The lake, Øvre Heimdalsvatn, a subalpine freshwater ecosystem. Holarct. Ecol. 1: 81-320.

Zwick, P. 1996a. Variable egg development of *Dinocras* spp. (Plecoptera, Perlidae) and the stonefly seed bank theory. Freshw. Biol. 35: 81-100.

Zwick, P. 1996b. Capacity of discontinuous egg development and its importance for the geographic distribution of the warm water stenotherm. *Dinocras cephalotes* (Insecta: Plecoptera: Perlidae). Ann. Limnol. 32: 147-160.

VIBRATIONAL COMMUNICATION (DRUMMING) AND MATE-SEARCHING BEHAVIOR OF STONEFLIES (PLECOPTERA); EVOLUTIONARY CONSIDERATIONS

Kenneth W. Stewart

Department of Biological Sciences
University of North Texas
Denton, Texas, USA 76203-5220

ABSTRACT

A long recognized but little explored mode of intersexual communication in insects is use of low-frequency substrate-borne vibrational signals. Representatives of 10 insect orders are known to have adopted this mode; range of communication, informational content and receiver integration of signals and energy costs are discussed. Arctoperlarian stoneflies represent the epitome of evolution of vibrational communication. Their ancestral signals were monophasic volleys of evenly spaced drumbeats. Derived signals to achieve species-specificity and possibly to enable sexual selection or some measure of reproductive fitness has involved modification of the ancestral form toward complex signals through: (1) changes in the rhythmic patterning of calls, (2) patterns of ♂-♀ duetting, and/or changes in the method of signal production such as rubbing or tremulation. Proposed paradigms for the evolution of vibrational communication and evolution of signal patterns are presented, with examples of the signals of several arctoperlarian species. The entire mating system of Arctoperlaria is discussed, and searching behavior in relation to vibrational communication is presented for *Pteronarcella badia, Claassenia sabulosa, Perlinella drymo* and *Suwallia sp.*

INTRODUCTION

Mine and my students investigations of the vibrational signals of 140 North American and New Zealand stonefly species, and the work on European species by Rupprecht (1968, 1969, 1981, 1982), Membiela (1990) and Tierno de Figueroa and Sanchez-Ortega (1998) have revealed that the stonefly suborder Arctoperlaria has developed the most diverse and complex system of vibrational communication known in insects. Only the leafhoppers and delphacid planthoppers, recently studied by Heady (Heady et al., 1986, Heady and Denno, 1991, Heady and Nault, 1991) possibly have evolved signals as complex as those of stoneflies.

Table 1. Insect orders known to use vibrational communication[1]

Order	Method(s) of vibration production
1. Orthoptera; Longhorned Grasshoppers, Katydids, Crickets	Percussion, Stridulation
2. Blattodea; Cockroaches	Percussion, Stridulation
3. Plecoptera; Stoneflies (representatives of all nine Arctoperlarian families)	Percussion, Stridulation, Tremulation
4. Psocoptera; Booklice	Percussion
5. Heteroptera; True Bugs (representatives of 16 families)	Stridulation, Tremulation, Tymbal Clicking
6. Homoptera; Leafhoppers, Planthoppers (representatives of three families)	Tymbal Clicking
7. Neuroptera; Green Lacewings, Alderflies, Spongillaflies (representatives of three families)	Tremulation
8. Coleoptera; Death Watch Beetles, Darkling Beetles	Percussion
9. Diptera; Chloropid Flies	Tremulation
10. Trichoptera; Caddisflies (representatives of seven families)	Percussion, Tremulation

[1] A more detailed analysis of taxa, sexes or castes producing signals, signal functions and key references for each order were presented by Stewart (1997).

Communication using low-frequency, substrate-borne vibrational signals is a mode of intersexual communication long recognized, but little explored in insects, and only in the past few decades has there been much effort to distinguish this mode from airborne, acoustical communication (Morris, 1980, Gogala, 1985) to determine how widespread it is in insects, and to explore its quantitative aspects and evolution in major groups such as Plecoptera that have adopted and refined it. Representatives of 10 insect orders have adopted vibrational communication (Table 1), primarily for locating mates. The signals produced by most species consist of simple volleys of evenly spaced vibrations, produced variously as indicated by percussion (drumming, tokking [a term derived from an Afrikaans word meaning "to knock"]), tremulation (abdominal or body jerking), tymbal clicking, stridulation (rubbing of a body scraper across a body file or body scraper across substrate) or combinations of these methods. Percussion is considered to be the ancestral method, and the various forms of stridulation and tremulation are behaviors derived from it.

Vibrational signaling is a viable evolutionary strategy, particularly for small insects that have limited body space necessary to accommodate the elaborate song-producing structures that can produce air-transmitted waves at effective frequencies for acoustical communication. The biophysical aspects of low-frequency, substrate-borne vibrations in solid or liquid substrates are complex, but we do know that the various wave forms offer good characteris

Table 2. Pre - 1968 Observations of drumming of stoneflies

Newport (1851)	*Acroneuria abnormis* Pictet
Hagen (1877)	*Pteronarcys regalis* Newman
Briggs (1897)	5 European species
Schwermer (1914)	2 European species
McNamara (1926)	ALarge Stonefly@ - Ontario
Gaunitz (1935)	*Taeniopteryx nebulosa* L.
Miller (1939)	*Pteronarcys proteus* Newman
Brinck (1949)	*Capnia bifrons* Newman
Zwick (1965)	*Isoperla goertzi* Illies
Schwarz (1968)	*Diura bicaudata* L.
	15 species

tics for the location of a caller by a receiver, and for supplying recognition characters in the caller's signal. In ways little understood, males or females may be able to measure a series of vibrational outputs against some selectively arrived-at neuronal template (Morris and Fullard, 1983), encoded to respond only to a conspecific mate, and possibly even to a more fit conspecific mate. Pairing experiments and playback experiments with several stonefly species, using computer models and computer-modified male calls have proven that male calls do contain conspecific recognition characters (Stewart and Maketon, 1990). Females of the stonefly *Pteronarcella badia* (Hagen) can discern the fitness of particular males that are duetting with her by measuring the time required by them to find her. This results in a duetting female remaining stationary only for a time, selectively allocated, for a fit male to find her. If this time requirement is not met, she moves to another position to potentially respond to the vibrational calls of a fit male (Abbott and Stewart, 1993). This reduces her vulnerability to vibration-detecting predators such as spiders.

The range of vibrational communication is somewhat restricted by the distortion and dampening of waves in natural substrates, but this poses no major problem, since transmission through dead plant stems has been demonstrated in the medium-sized stonefly *Perlinella drymo* (Newman) for up to 8 meters (Stewart and Zeigler, 1984), that translates to 800 to 1,000 times the length of the insects body. Effective transmission distances can probably also be achieved in live plants, connected leaf mats and other stream side substrates.

Although there is little research to support it, directional location of a vibrational signal producer is probably determined by the differential time delays in reception of vibrational waves by the sensors on a receivers planted tarsi. Scorpions can determine direction of a call signal by integrating time delays as small as 0.2 milliseconds, received by subgenual organs of the different legs (Brownell and Farley, 1979). The planted tarsi of insects would similarly establish positions like points on a compass, and the source of a directional vibrational wave signal could be read from the time delays of the signal in reaching these positions. Distance to a caller is probably read by intensity of wave signals received above some threshold.

Little is known of the energy cost of vibrational signaling as a component of reproductive effort, but generally it is considered less than for producing high-frequency airborne sounds. Lighton (1987) found that percussion assisted mate-finding in the tenebrionid beetle *Psammodes sp.* was less than 10% of the cost of random searching over its communicable range. This low cost has probably been a major factor in the adoption of the vibrational mode by many small insects.

Thirty years ago, we knew very little about stonefly drumming, and even less overall about vibrational communication in insects. Ten scientists had observed the drumming of about 15 species (Table 2), but the behavior had not been quantified beyond timing the duration of signals, and its function was thought to be the calling of mates. Rupprecht (1968,

Table 3. The typical mate-finding system of Arctoperlaria

1. Encounter site aggregation of sexes.
2. Species-specific calling by males during ranging search.
3. Duet establishment with receptive females within communicable range.
4. Localized search for now-stationary females (with continued duetting).
5. Location, contact with female by male, and immediate mating.

1969) was the first to quantify the behavior of eight European species in two classic papers. On reading Rupprects' papers, I was immediately spellbound by his exciting findings and wondered about the extent of the behavior in Plecoptera and what the informational content of the signal language might be. Not quite 10 years later my student D.D. Zeigler and I (Zeigler and Stewart, 1977) reported the first quantified descriptions of drumming in 11 North American species. By this time we knew that: (1) signals were produced by percussion with either the unmodified or specialized distal, ventral portion of the abdomen, (2) duets were either "2-way" (male call-female answer sequences) or "3-way" (male call-female answer-male reply sequences), (3) male calls were more complex than female answers or male replies, and the signals of both sexes and duet pattern were species-specific and therefore probably fixed action behaviors, and (4) during duetting, males searched for stationary females, further suggesting that the behavior was part of the mate-finding system. Since 1973, my students D.D. Zeigler, S.W. Szczytko, R.K. Snellen, M. Maketon, J.C. Abbott, S.R. Moulton and K.D. Alexander and I have studied the language of 140 species, a substantial sampling representing about 22% of the North American fauna, and the related searching behavior of four species. Our experimental methodology has been set forth in the 33 papers published on this continuing research.

From all of this work, I have been able to formulate some reasonably sound theoretical considerations and conclusions about the mating system of Arctoperlaria, and a working hypothesis of the evolution of vibrational communication in this suborder. The high degree of species-specificity in this fixed action, genetically programmed behavior in Arctoperlaria make it a valuable line of evidence for delineating morphologically cryptic species and for resolving phylogenetic relationships (Stewart and Zeigler, 1984).

THE ARCTOPERLARIAN MATING SYSTEM

The entire mating system of stoneflies involves not only their intersexual communication, but also the associated aggregation and movement behaviors of both sexes (Stewart, 1994). The typical system in Arctoperlaria involves the sequence of behaviors indicated in Table 3 by males and virgin females. Males are polygamous and continue calling and searching during their short reproductive lives. Typically, mated and unguarded females reject subsequent male advances by raising and curving their abdomens. At the International Stonefly Symposium in Treehaven, Wisconsin, and in Stewart (1994), I proposed the probable encounter site conventions of stoneflies, based on my own experience and observations of fellow workers. I adapted descriptors for these such as bankscrambler, bushtopper and treetopper from the concept of hilltopping in butterflies described by Shields (1967). These encounter site behaviors have undoubtedly been selectively reinforced by reproductive success, and represent the initial aggregation of sexes and mate encounter that get them close enough together for vibrational communication to effectively come into play.

In Stewart (1997), I presented numerous examples of call signal and duetting patterns that male and female stoneflies use to get together, and that represent possible evolutionary steps that I will present later in this paper. In these examples it is evident that male calls are much more complex than female answers, since they must convey to females the critical in-

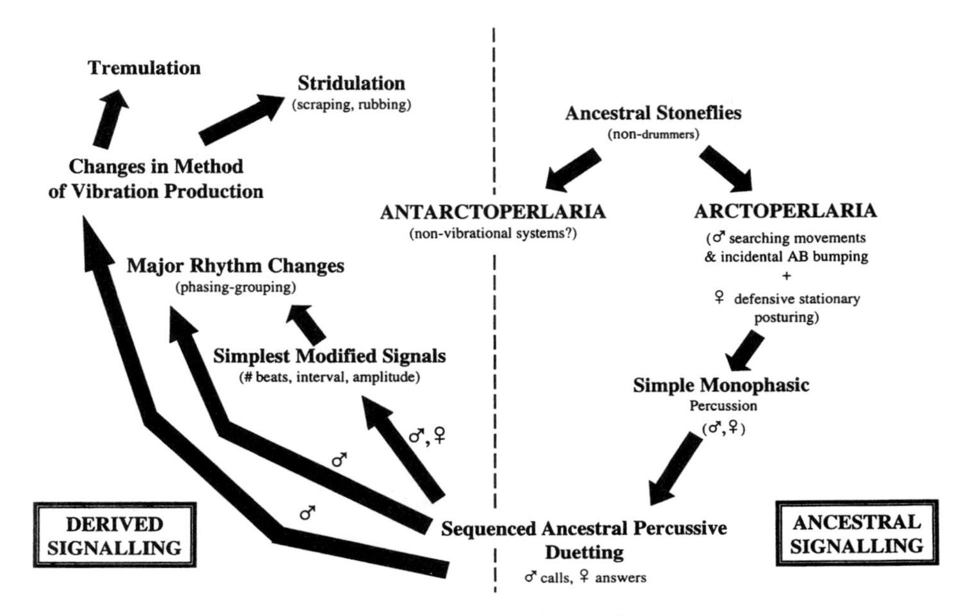

Fig. 1. Drumming evolution paradigm.

formation of species-specificity and possibly reproductive fitness to stimulate female receptivity. Assuming that females can discern this information in the call, her answers need only acknowledge his calls and repeatedly report her stationary location, neither of which require complex signals.

THE EVOLUTION OF VIBRATIONAL COMMUNICATION IN ARCTOPERLARIA

There is obviously no fossil record for vibrational communication in stoneflies, so formulation of an evolutionary scenario of the behavior must rely solely on the patterns revealed from extant species. My investigations of several species of the Southern Hemisphere suborder Antarctoperlaria, and observations by several of my colleagues in Australia and New Zealand over several years, indicate that members of this suborder are non-drummers, and have never adopted vibrational communication. Their mating systems are unknown, but I believe that they may have evolved highly specific encounter site aggregation behaviors, possibly supplemented by as yet unrevealed intersexual communication modes, that bring the sexes sufficiently close together to accommodate mate-finding.

Figure 1 illustrates my proposed paradigm of evolution of vibrational communication in the Arctoperlaria. Out-group comparisons with the Grylloblattodea, other orthopteroid groups and the suborder Antarctoperlaria (Maketon and Stewart, 1998), indicate that ancestral stoneflies were non-drummers, and that only the Arctoperlaria have adopted and refined the vibrational mode of communication as part of their mating system. This is further reinforced by the fact that special ventral abdominal structures have not appeared in males of the Antarctoperlaria. The first "drumming" in arctoperlarian stoneflies probably resulted from the accidental bumping of an unmodified abdomen of males, while searching for females,

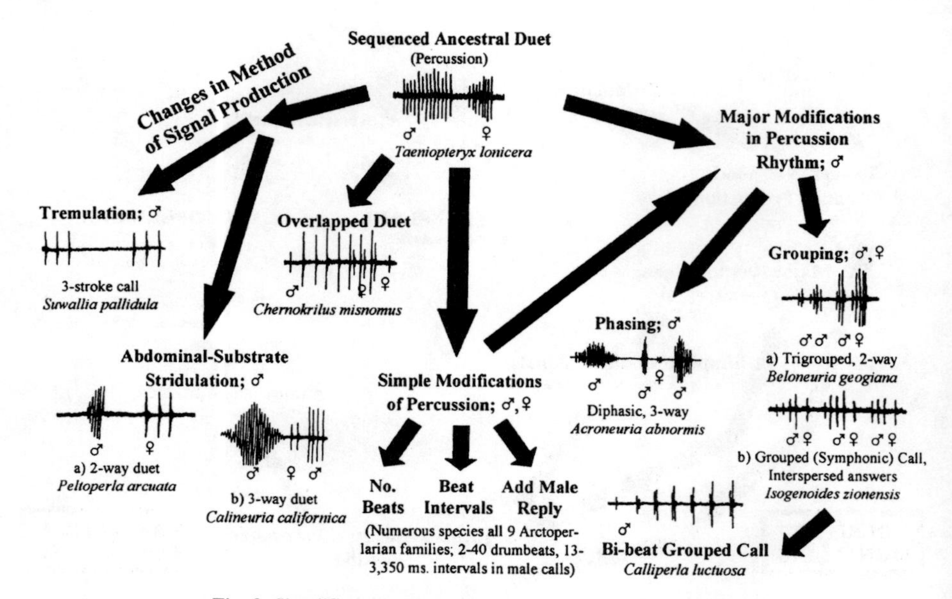

Fig. 2. Simplified diagram of evolution of vibrational signals.

and the possible defensive response of females to the vibrations by becoming motionless and stationary.

Selection progressively reinforced the male bumping into a behavioral action, and a similar sequence in females followed, until the relatively simple, sequenced duetting (♂ calls, ♀ answers) became an ancestral communication. These ancestral calls and answers were produced by percussion, and were monophasic volleys of even spacing and little amplitude modulation. This, along with active male searching and stationary female response became the ancestral mating system.

Out-group comparisons among arctoperlarian families (Maketon and Stewart, 1988) indicate that species-specificity and behavioral isolation were then derived from this ancestral system, and are presented in extant arctoperlarian patterns variously by: (1) retention of the ancestral pattern, with simplest modifications of signal characteristics (no. of beats, overlapping of ♂-♀ duet, changing of beat interval rhythm, amplitude modulation) in both male and female signals, (2) major rhythm changes, particularly in male calls, by phasing or grouping of signal beats, and (3) changes in the method of signal vibration production from ancestral percussion to abdominal-substrate stridulation, or in rare cases, tremulation, with associated changes in rhythm (Fig. 1). The result, that we have discovered in the many extant species studied, is the exciting, highly derived and complex system of vibrational communication in the Arctoperlaria suggestive that they represent the epitome of evolution of the vibrational mode of communication in insects.

Figure 2 more specifically uses examples of the signals of selected species to illustrate how currently known patterns of arctoperlarian duets correspond with the general evolutionary paradigm of Figure 1. The ancestral pattern is illustrated by the percussion-produced monophasic duet of *Taeniopteryx lonicera* Frison; both male and female signals are monophasic, of evenly spaced beats, and little amplitude modulation.

The simplest ways to derive specificity from the ancestral model would have been to: (1) shorten or lengthen the monophasic signals by changing the number of beats or the even time intervals between beats, (2) add a male reply, or (3) start chorusing the female answer into the male call as shown by the duet of the perlodid *Chernokrilus misnomus* (Claassen) (Fig. 2). Numerous species of all nine Northern Hemisphere stonefly families have derived such simply modified specific signals ranging from 2-40 drumbeats having distinct intervals of 13-3350 milliseconds (Stewart and Maketon, 1991).

The most complex evolved signals in Arctoperlaria appear to have resulted from major modifications in percussion rhythm, either directly from the ancestral pattern or after the simpler modifications, or from modified methods of producing the vibrations and changes in rhythm. The major modifications in percussion rhythm have resulted in phased calls of some Peltoperlidae, Perlodidae, Chloroperlidae and Perlidae species having two or more phases of different numbers of beats and often very different interbeat spacing, as illustrated in Fig. 2 by the 3-way, diphasic call of *Acroneuria abnormis* (Newman), or by grouped calls having two or more distinct groups of drumbeats as illustrated by the tri-grouped call with sequenced female answer of the perlid *Beloneuria geogiana* (Banks) and the grouped call with the interspersed female answers (symphonic signaling) of the perlodid *Isogenoides zionensis* Hanson (Fig.2). In *zionensis*, the signal is very complex in that females, after a short symphony, start to sequence their answers to the 3-4-beat male call, and the male then acknowledges the female answer with a 2-or 3-beat signal. The perlodids *Kogotus modestus* (Banks) and *Calliperla luctuosa* (Banks) carry grouped calling a step further by producing rapid, bi-beat call groups. We have determined with computer-simulated variations of the *K.modestus* calls that the bi-beat character and spacing within and between the bi-beat groups are critical language variables for female recognition.

The most specialized calls of Arctoperlaria that involve abdominal contact are made by males rubbing the substrate with a ventral abdominal knob or hammer (Stewart, 1991). Each rub is a prolonged percussion stroke with the textured ventral surface of the knob or hammer, that may be ridged or papillous, producing on resonant substrates a squeaking sound. I have described this unique, modified vibration-producing behavior, presumably derived from percussion, as an abdominal-to-substrate stridulation. Calls of various species we have studied produce 1-7 rubs, always answered by females with percussion. *Acroneuria abnormis* may represent an evolutionary link in this percussion-to-rubbing method of vibration production, since some males produce both the first and second phases of their diphasic call by percussion, but other males produce phase 1 by rubbing and phase 2 by percussion. Abdominal-substrate stridulation with single male rub calls is illustrated in Figure 2 by the 2-way duet of *Peltoperla arcuata* (Needham) and 3-way duet of *Calineuria californica* (Banks).

Three stonefly species in the subfamily Chloroperlinae have derived the very specialized tremulation method for producing vibrational calls. *Siphonoperla montana* Pictet and *Siphonoperla torrentium* Pictet produce signals on plants by rapid non-contact vertical movements of the abdomen (Rupprecht 1981), and *Suwallia pallidula* (Banks) males produce a 3-stroke tremulation signal (Fig. 2) by a forward-backward rocking motion of the whole body (Alexander and Stewart, 1997). I consider tremulation in Plecoptera to be behavior derived from drumming because it is known to occur only in this one chloroperlid subfamily; species in the other subfamily Paraperlinae produce vibrations by percussion.

So, we know quite a lot about the encounter sites and vibrational calling and duetting components of Arctoperlarian mate-finding systems, but very little about the critically important follow up step of the localized search, whereby a male utilizes the vibrational information from a female to actually find her. We began approaching this question of searching behavior about six years ago, with experiments and observations of four species selected from different encounter site conventions. This approach was based on the hypothesis that ground scramblers would have different searching behavior and patterns on contoured, streambank surfaces, than say bushtoppers or treetoppers on a branched, plant substrate, and that the currently arrived-at system of any species should combine effective vibrational signaling with an efficient search modality.

Our first experiments with *Pteronarcella badia* (Abbott and Stewart, 1993) showed that in an experimental arena the average find time for pairs engaging in strong, continuous duetting was significantly shorter than those for non-duetting or anomalously duetting pairs, and that their local search patterns involved a male triangulation, aided by the vibrational cues from answering females. Alexander and Stewart (1996a) videotaped a ground scrambling population of *Claassenia sabulosa* (Banks) under red light at night and discovered that males drummed and searched near the shoreline on isolated stones that protruded above the shallow water surface. They ran across the shallow water surface and circled and scrambled over high areas and angular surfaces of individual rock encounter sites; these corresponded with female emergence sites. Alexander and Stewart (1996b) videotaped the treetopper *Perlinella drymo* (Newman) on experimental branches in the laboratory and found that duetting increased male searching activity, influenced their directional movements and decreased the time required to find females, in contrast to non-duetting pairs. Males, interestingly, never called at branch bifurcations; when a male reached a fork, he randomly traveled up one side or the other, drumming at some point beyond the fork. If he first traveled up the side not leading to the female, then her answer signal came from behind him, causing him to turn and travel back to the fork and take the other, correct side. This process was repeated at each fork until he achieved the stem leading directly to the answering female. Females were found by males in most trials within 10 minutes, the actual time depending on distance traveled, number of forks tested, and possibly fitness of the male selected for a given trial. These experiments proved that female answers provided information on her stationary location that males were able to integrate to correct his turns and directional movements toward her.

Our most recent work (Alexander and Stewart, 1997) videotaping the chloroperlid *Suwallia pallidula* in natural streamside habitats has shown that this bushtopper-tremulator uses a "fly-tremulate-search" pattern similar to the fly-and-call pattern of delphacid leafhoppers (Heady and Nault, 1991). Males flew and landed on riparian shrubs in late afternoon on sunny days, and began calling on leaves with its typical 3-stroke tremulation signal. Females answered with a 1-stroke signal, and then moved to the petiole-leaf junction and became stationary. This typical positioning insures that the male does not have to search entire leaf surfaces. The male searched the petioles and bases of leaves as they were encountered, while intermittently duetting with the female, and repeating until she was found. Copulation occurred on the upper surface of leaves, and alder was the preferred primary encounter site despite its lower density than willow at streamside.

These studies indicate that the finding systems of these four stonefly species all conform to the hypothetical systems proposed by me (Stewart, 1994). They all utilize an encounter site for initial aggregation of sexes, and their males engage in a ranging search at the encounter site while intermittently calling for females with species-specific vibrational signals. Except for *Claassenia sabulosa*, duets are established that allow the male to begin a more localized search for the female whose answers give him repeated information about her location.

As expected, search patterns relate to the type of substrate at the encounter site. *Suwallia pallidula* had a "fly-tremulate-search" pattern; *Claassenia sabulosa* had a "rock-to-rock" pattern at the shallow-water shoreline and *Perlinella drymo* had a "fly-run-drum-search" pattern on branches. Further research in this area is expected to reveal more exciting and diverse patterns of behavior.

REFERENCES

Abbott, J. C. and K. W. Stewart. 1993. Male searching behavior of the stonefly, *Pteronarcella badia* (Hagen) (Plecoptera: Pteronarcyidae) in relation to drumming. J. Insect Behav. 6: 467-481.

Alexander, K. D. and K. W. Stewart. 1996a. Description and theoretical considerations of mate finding and other adult behaviors in a Colorado population of *Claassenia sabulosa* (Banks) (Plecoptera: Perlidae). Ann. ent. Soc. Amer. 89: 290-296.

Alexander, K. D. and K. W. Stewart. 1996b. The mate searching behavior of *Perlinella drymo* (Newman)(Plecoptera: Perlidae) in relation to drumming on a branched system. Bull. Swiss Ent. Soc. 69: 121-126.

Alexander, K. D. and K. W. Stewart. 1997. Further considerations of mate searching behavior and communication in adult stoneflies (Plecoptera); first report of tremulation in *Suwallia* (Chloroperlidae), pp. 107-112. In: P. Landolt and M. Sartori (eds.). Ephemeroptera and Plecoptera Biology-Ecology-Systematics, Fribourg: Mauron + Tinguely and Lachat SA.

Brownell, P. and R. D. Farley. 1979. Orientation to vibrations in sand by the nocturnal scorpion *Paruroctonus mesaensis*: mechanism of target localization. J. Comp. Physiol. 131: 31-38.

Gogala, M. 1985. Vibrational communication in insects (biophysical and behavioral aspects), pp. 117-126. In: K. Kalmring and N. Elsner (eds.). Acoustic and vibrational communication in insects. Proceedings XVII International Congress of Entomology. Verlag Paul Parey, Berlin.

Heady, S. E. and R. F. Denno. 1991. Reproductive isolation in *Prokelisia* plant hoppers (Homoptera: Delphacidae): Acoustic differentiation and hybridization failure. J. Insect Behav. 4: 367-390.

Heady, S. E. and L. R. Nault. 1991. Acoustic signals of *Graminella nigrifrons*. Gr. Lakes Ent. 24: 9-16.

Heady, S. E., L. R. Nault, G. F. Shambaugh and L. Fairchild. 1986. Acoustic and mating behavior of *Dalbulus* leafhoppers (Homoptera: Cicadellidae). Ann. ent. Soc. Amer. 79: 727-736.

Lighton, J. R. B. 1987. Cost of tokking: the energetics of substrate communications in the tok-tok beetle *Psammodes striatus*. J. Comp. Physiol. B 157: 11-20.

Maketon, M. and K. W. Stewart. 1988. Patterns and evolution of drumming behavior in the stonefly families Perlidae and Peltoperlidae. Aquat. Insects 10: 77-98.

Membiela, P. 1990. The mating calls of *Perla madritensis* Rambur, 1842 (Plecoptera, Perlidae). Aquat. Insects 12 (4): 223-226.

Morris, G. K. 1980. Calling display and mating behavior of *Copiphora rhinoceros* Pictet (Orthoptera: Tettigoniidae). Anim. Behav. 28: 42-51.

Morris, G. K. and J. H. Fullard. 1983. Random noise and congeneric discrimination in *Conocephalus (Orthoptera: Tettigoniidae)*, pp. 73-96. In: D. T. Gwynne and C. K. Morris (eds.). Orthopteran mating systems, sexual competition in a diverse group of insects. Westview, Boulder, CO.

Rupprecht, R. 1968. Das Trommeln der Plecopteren. Z. Vergl. Physiol. 59: 38-71

Rupprecht, R. 1969. Zur artspezifizität der trommelsignale der Plecopteren (Insecta). Oikos 20: 26-33.

Rupprecht, R. 1981. A new system of communication within Plecoptera and a signal with a new significance. Proc. VII International Symposium Plecoptera. Biol. of Inland Waters 2: 30-35.

Rupprecht, R. 1982. Drumming signals of Danish Plecoptera. Aquat. Insects 4: 93-103.

Shields, O. 1967. Hilltopping. J. Res. Lepid. 6: 69-178.

Stewart, K. W. 1994. Theoretical considerations of mate finding and other adult behaviors of Plecoptera. Aquat. Insects 16: 95-104.

Stewart, K. W. 1997. Vibrational communication in insects: epitome in the language of stoneflies? Amer. Entomol. 43: 81-91.

Stewart, K. W. and M. Maketon. 1990. Intraspecific variation and information content of drumming in three Plecoptera species, pp. 259-268. In: I. Campbell (ed.), Mayflies and stoneflies: life histories and biology. Kluwer Academic, The Netherlands.

Stewart, K. W. and M. Maketon. 1991. Structures used by Nearctic stoneflies (Plecoptera) for drumming, and their relationship to behavioral pattern diversity. Aquat. Insects 13: 33-53.

Stewart, K. W. and D. D. Zeigler. 1984. The use of larval morphology and drumming in Plecoptera systematics, and further studies of drumming behavior. Ann. Limnol. 20: 105-114.

Tierno de Figueroa, J. M. and A. Sanchez-Ortega. 1999. The male drumming call of *Isoperla nevada* Aubert (Plecoptera: Perlodidae). Aquat. Insects 21: 33-38

Zeigler, D. D. and K. W. Stewart. 1977. Drumming Behavior of Eleven Neartic Stonefly (Plecoptera) Species. Ann. ent. Soc. Amer. 70: 495-505.

PARTHENOGENETIC AND BISEXUAL POPULATIONS OF *EPHEMERELLA NOTATA* EAT. IN POLAND

Adam Głazaczow

Department of Systematic Zoology
Adam Mickiewicz University
Fredry 10, 61-701 Poznań, Poland

ABSTRACT

The sex ratio (males: females) of final instar nymphs of *Ephemerella notata* varies from 0.12: 1.00 to 0.92: 1.00 in isolated populations of the Pomeranian Lake District in Poland. These data support the view that facultative parthenogenesis occurs in this species. The incidence of parthenogenesis was inversely related to population density. Parthenogenetic populations were less variable in size (determined by a principal component analysis of a total body-size index) and more synchronous in their development.

INTRODUCTION

Parthenogenesis is found in most animal groups (Suomalainen et al, 1987), including many mayfly species (Degrange, 1960). Mayfly parthenogenesis is usually facultative and automictic. After chromosome reduction during meiosis the nuclear phase in the egg becomes azygoid (haploid). The zygoid phase is then restored by the fusion of two azygoid nuclei of different mayflies in fertilization or by the fusion of two nuclei of a single individual (automixis). Generally, only a small per cent of unfertilized eggs hatch and their development is slower and not well synchronized, leading Brittain (1982) to argue that parthenogenesis has no importance in population dynamics. However, Landolt et al. (1997) observed that half of the unfertilized eggs of *Palingenia longicauda* develop and so parthenogenesis could be very important.

Parthenogenesis is usually detected by the lack of males in particular populations. Unisexual (female) populations of *Ephemerella notata* were first recorded by Landa (1969) in Czechia, and Jażdżewska (1976) made similar observations in central Poland. *E. notata* males appear only in northern Poland (Pomeranian Lake District) but the sex ratio is variable (Głazaczow, 1994); there is a small predominance of males near the Baltic Coast (the ratio of males to females - 1.00: 0.90) but a very large predominance of females in the southern portion of this region (the ratio of males to females - 0.05: 1.00).

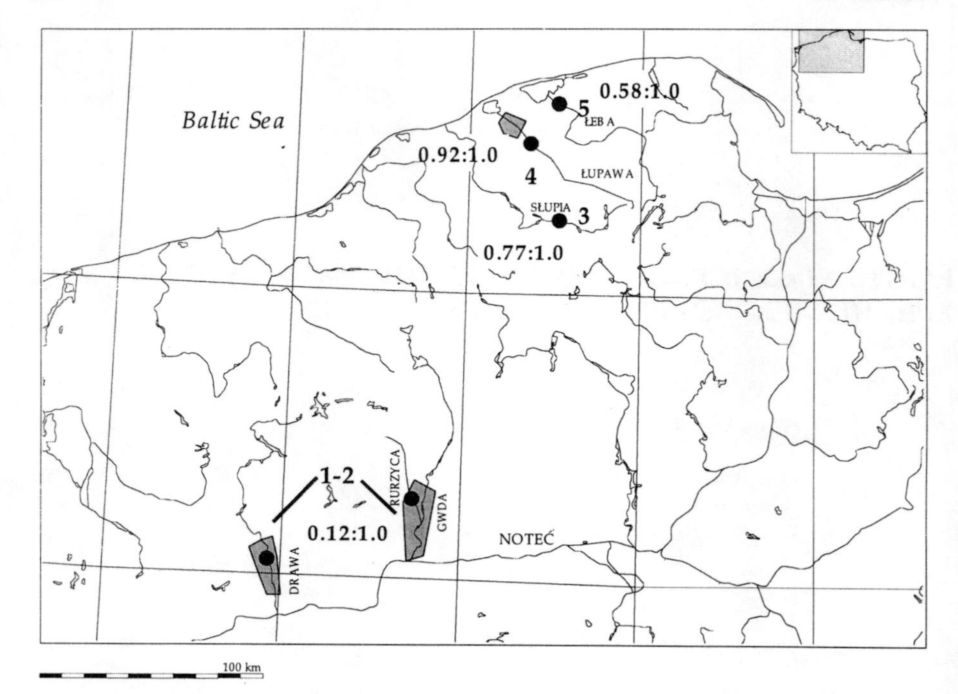

Fig. 1. Distribution of *E. notata* populations in the Pomeranian Lake District.

E. notata is widely distributed in Europe, occurring in England and France in the west, through the Central Europe to the Balkans in south and West Russia in the east. It was assigned by Sowa (1975) to the south-central European group. In spite of the species wide distribution, males are known from the northern part of its range only, such as in England (Kimmins and Frost, 1943), Denmark (Jensen, 1961), and northern Poland. In Poland they decrease and disappear over a distance of 150 km and the river Noteć seems to be a southern limit of their distribution.

The purpose of this study was to compare populations over this transition zone in Poland.

METHODS

Final instar nymphs of *E. notata* were collected in 1998 from five sites in the Pomeranian lake district of Poland (Fig. 1). A bottom scraper was used in which substratum from about 0.05 m^2 was gathered. The sex of nymphs was determined by differences in the structure of their eyes and eighth sternite. Measurements of body length and width of the head, thorax and abdomen were recorded. To analyse the differences in the nymphal sizes between the populations, a total body-size index, derived from principal component analysis of these four

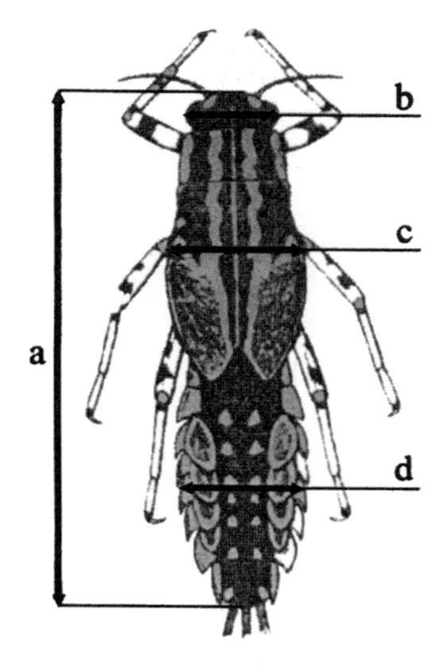

Variable	Loading on PC1	
	Females	Males
Body length (a)	0.90	0.90
Head width (b)	0.96	0.91
Thorax width (c)	0.91	0.84
Abdomen width (d)	0.94	0.94
Eigenvalue	3.45	3.23
% variance explained	86.3	80.7

Fig. 2. Total body-size index in *E. notata* populations derived as first component from principal component analysis.
Loading on PC1 - only 1 principal component is analysed; it shows in which degree various measurements influence (load) on the principal component e.g. new measure of body size (1.00 is the highest degree).
Eigenvalue - self-value of the correlation matrix; it should be > 1 but it can not exceed the value of the number of the analysed variables (in this case 4); if it is higher it explain the grater part of the variance.
% variance explained - informs which % of variation of the all variables (4) is explained by 1 principal component.

measurements, was used. All morphometric variables had positive loadings on their first component (being positively correlated) and similar magnitudes (0.84-0.96) (Fig. 2).

Considering that populations 1 and 2 had low numbers and were not significantly different (HSD Tuckey's test), they were combined. Nymphs that were preserved with the substratum were significantly different in their total body-size from the ones preserved immediately in alcohol, and were not used.

RESULTS

Pomeranian rivers are divided by a watershed line so that they flow either to the north or the south. *E. notata* does not occupy all water courses, and in the south was found in the lower sections only. These conditions lead to the isolation of populations. *E. notata* is a rheophilic mayfly, occurring mainly in patches of *Elodea canadensis* and *Fontinalis antipyretica* and roots of riparian plants submerged in the current. Their abundance is very variable and in some habitats (roots of alder) difficult to measure.

Density was highest in the river Lupawa (Fig.1, population 4), where it was the dominant mayfly species and its abundance reached $30x0.05m^{-2}$. The lowest numbers were in the rivers Drawa and Rurzyca (population 1-2), where densities never exceeded $3x0.05m^{-2}$. Intermediate densities were observed in the other stations.

Males occurred in the whole area of investigation but the sex ratios in the particular populations were very different and females predominated at all sites. The sex ratio most approached equality - 0.92:1.00 (166 males, 181 females) - in the most abundant population (4). Very low representation of males was typical for southern populations where they were

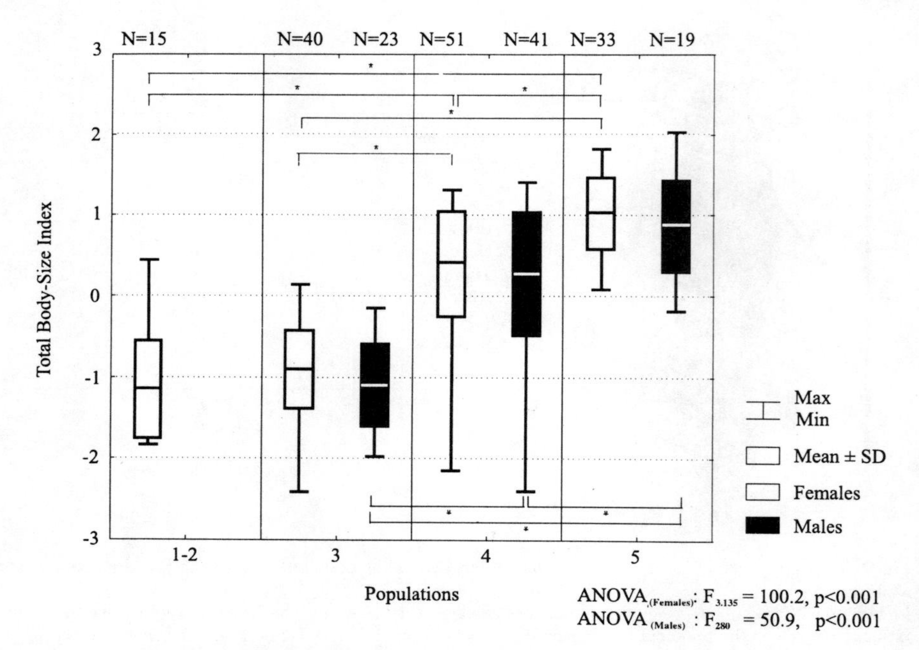

Fig. 3. Changes in total body-size index of *E. notata* in various populations.
* - statistically significant differences; p<0.05

found only sporadically in ratios of 0.12:1.00 (2 males, 17 females). These populations must have, in part, reproduced parthenogenetically. Changes in the sex ratio were proportional to the changes in the number. When populations were more abundant, the sex ratio was more or less equal, but when their abundances were low, males became scarce.

Because all measurements were correlated, they were calculated as a total body-size index. Populations 1-2 and 3 were geographically distant and differed in the degree of parthenogenesis but their body sizes were not significantly different. Body size was significantly different between populations 3,4 and 5 (Fig. 3). Though their distance of separation was not as great, they were located on separate rivers. The observed changes in the size increase to the North, but the degree of parthenogenesis varies in the other direction. The size of the populations in which males were numerous (3,4) are much more variable than in those where parthenogenetic females predominated (1-2, 5). Somewhat younger and slightly smaller nymphs were lacking in the southern populations, but also in the northernmost population (5).

DISCUSSION

As expected populations of *E. notata* which develop in the most productive environments, reach the greatest abundance, although it is sometimes difficult to measure accurately. So, considering the patchy distribution of nymphs, maximum density does not reflect the real numbers in the whole river-bed. However, it appears that the highest relative densities were reached in populations 4 and 3 from the rivers Łupawa and Słupia. In these populations both the greatest percentage of males, and the greatest variation in body size were observed. Populations, in which the lower numbers of males were found, so at least partly parthenogenetic, were less variable in body size, although the difference between the populations was statistically significant. Also, development was more synchronous in the parthenogenetic populations. This study agrees with general suggestion that facultative parthenogenesis is an adaptation to unfavourable conditions, but does not support the view that parthenogenetic populations show a decay in reproductive synchrony, which was also observed by Sweeney and Vannote (1982).

ACKNOWLEDGMENTS

I would like to thank Dr. G. Pritchard and B. A. Young for editing my text. I also thank R. Bajaczyk for help with creating the figures.

REFERENCES

Brittain, J. E. 1982. Biology of mayflies. Ann. Rev. Ent. 27: 119-147.

Degrange, C. H. 1960. Recherches sur la reproduction des Ephéméroptères. Trav. Lab. Hydrobiol. Piscic. Grenoble, 51: 7-193.

Głazaczow, A. 1994. Mayflies (Ephemeroptera) from the rivers Gwda and Drawa (in Pomeranian Lake District of North West Poland) and from some waters of their river basins. Pol. Pismo ent., 63: 213-257.

Jażdżewska, T. 1976. *Ephemerella mucronata* (Bengtsson) i *Ephemerella notata* Eaton (Ephemeroptera) w dorzeczach Pilicy i Warty. Acta Univ. lodziensis, 3: 95-109.

Jensen, C. F. 1961. *Ephemerella notata* Etn., *Caenis undosa* Ts. og *Heptagenia longicauda* (Steph.) nye for Danmark (Ephemeroptera - døgnfluer). Særtryk af 'Flora og Fauna', 67: 97-104.

Kimmins D. E. and W. E. Frost,. 1943. Observations on the nymph and adult of *Ephemerella notata* Eaton (Ephemeroptera). Proc. R. ent. Soc. Lond., 18: 43-49.

Landa V. 1969. Jepice - Ephemeroptera. Fauna ČSSR, 18: 1-347.

Landolt P., M. Sartori and D. Studemann, 1997. *Palingenia longicauda* (Ephemeroptera, Palingeniidae): from mating to the larvulae stage. Ephemeroptera and Plecoptera: Biology-Ecology-Systematics. In: P. Landolt and M. Sartori (eds.). MTL, Fribourg, 15-20.

Sowa R. 1975. Ecology and biogeography of mayflies (Ephemeroptera) of running waters in the Polish part of the Carpathians. 1. Distribution and quantitative analysis. Acta Hydrobiol. 17: 223-297.

Suomalainen E., Saura A. and J. Lokki, 1987. Cytology and evolution in parthenogenesis. CRC Press, Boca Raton, Florida, 216 pp.

Sweeney, B. W. and R. L. Vannote, 1982. Population synchrony in mayflies (*Dolania americana*). Evolution, 35: 810-821.

GENETIC DIVERGENCE IN POPULATIONS OF *DINOCRAS CEPHALOTES* (CURTIS, 1827) FROM THREE DIFFERENT CATCHMENTS IN CENTRAL ITALY (PLECOPTERA, PERLIDAE)

R. Fochetti[1], V. Iannilli[1], V. Ketmaier[2], and E. De Matthaeis[2]

[1] Dipartimento di Scienze Ambientali, Università della Tuscia
Via S. Camillo De Lellis, I-01100 Viterbo, Italy
[2] Dipartimento di Biologia Animale e dell'Uomo
Università "La Sapienza", Viale dell'Università
32 -I-00185 Roma, Italy

ABSTRACT

We studied the genetic structure of *Dinocras cephalotes* (Curtis, 1827) in Central Italy, by means of starch-gel electrophoresis. Eleven populations from three different rivers (Aniene, Nera, Velino) were analysed. 21 enzymes were screened, yelding data for 27 presumptive loci. Genetic interpopulation distances (D; Nei, 1978) were relatively low, ranging from D=0.000 to D=0.052. These results show little differentiation among the populations and agree to some extent with the data coming from other genetic studies of Plecoptera. Nevertheless genetic distances among some contiguous subpopulations of each river seem fairly high for taxa spatially contiguous (D max=0.052).

INTRODUCTION

The analysis of enzyme polymorphism by means of electrophoresis has been a very effective technique in population genetics particularly for investigating the genetic structure of populations representing different species. Aquatic insects, despite their importance in the biocenoses of riverine ecosystems, have not been studied exhaustively. Studies on genetic structure of the Plecoptera have been carried out only recently (Funk and Sweeney, 1990; Wright and White, 1992; Fochetti et al., 1997). The limited data indicate that they are a rather homogeneous order genetically, with limited genetic variability, uniform allele frequencies, and low inter and intraspecific genetic distances. These aspects probably correlate with the stenoecy of stoneflies (Lees and Ward, 1987; Robinson et al., 1992). In this paper, we study the genetic structure of a species, *Dinocras cephalotes*, that shows considerable interpopulation variability with regard to features, such as the length of the life cycle and the length of embryonic development (Elliot, 1989; Lillehammer, 1987; Zwick, 1996; pers. com.). Our intent was to examine how genetic structure of *D. cephalotes* varied geographically among natural populations from three river systems.

Table 1. List of enzymes, electrophoretic techniques and references for the present study (for the specific references see Harris and Hopkinson, 1978; Richardson et al. 1986)

Codes	Enzyme	E.C.N°	Buffer system	Staining reference	Loci scored
ACPH	Acid phosphatase	3.1.3.2	A	Tracey et al., 1975	*Acph1, Acph2*
ADA	Adenosine deaminase	3.5.4.4	E	Ward & Beardmore, 1977	*Ada1, Ada2*
ALDO	Aldolase	4.1.2.13	B1	Ayala et al., 1972	*Aldo*
AO	Aldehyde oxidase	1.2.3.1	B1	Ayala et al., 1974	*Ao*
APH	Alkaline phosphatase	3.1.3.1	A	Ayala et al., 1972	*Aph2*
CA	Carbonic anydrase	4.2.1.1	G	Brewer e Sing, 1970	*Ca1, Ca2, Ca3*
EST	Esterase	3.1.1.1	A	Ayala et al., 1972	*Est1, Est2, Est3*
GOT	Glutammic oxalacetic transaminase	2.6.1.1	B1	Ayala et al., 1975	*Got1, Got2*
G6PD	Glucose-6-phosphate dehydrogenase	1.1.1.49	B1	Ayala e Powell, 1974	*G6pd1, G6pd2*
HK	Hexokinase	2.7.1.1	C	Ayala et al., 1974	*Hk*
IDH	Isocitrate Dehydrogenase	1.1.1.42	E	Brewer e Sing, 1970	*Idh1, Idh2*
LDH	Lactate dehydrogenase	1.1.1.27	A	Selander et al., 1971	*Ldh*
MDH	Malate dehydrogenase	1.1.1.37	A	Ayala et al., 1972	*Mdh*
MPI	Mannose phosphate isomerase	5.3.1.8	F	Harris et al., 1977	*Mpi*
PEP	Peptidase	3.4.11	A	Ward & Beardmore, 1977	*Pep*
PGM	Phosphoglucomutase	2.7.5.1	D	Brewer e Sing, 1970	*Pgm*
PHI	Phosphoglucose isomerase	5.3.1.9	C	Brewer e Sing, 1970	*Phi*
TO	Tetrazolium oxidase	1.15.1.1	B1	Ayala et al., 1972	*To*

MATERIAL AND METHODS

We analysed 182 specimens of *Dinocras cephalotes* belonging to eleven populations collected from three different rivers of Central Italy: Aniene, Nera and Velino. Only eighteen of these specimens were adults, with all adults belonging to a single population (Trevi: Aniene River, see below). Larvae were sampled using the standard kick sampling method, while the adults were collected with an aerial net. The individuals collected were either returned to the laboratory in liquid nitrogen or were brought to the laboratory alive in refrigerated boxes filled with water. Afterwards all the samples were stored at –80°C prior electrophoresis. Samples were collected from May 1996 to June 1997. The sites and dates of sampling (the geographic codes for these sites are listed in parentheses) are listed below:

R. Aniene: Trevi (TRE), m 821, 05/30/1996, 09/20/1996, 06/12/1997; T.Simbrivio (SIM), m 550, 09/20/1996; Subiaco (SUB), m 408, 05/30/1996, 09/20/1996, 05/13/1997; Madonna della Pace (MPA), m 390, 05/13/1997; Vicovaro (VIC), m 308, 05/30/1996.

R. Velino: Posta (POS), m 715, 05/23/1997

R. Nera: Castelsantangelo (CAS), m 780, 05/27/1997; Ponte Ceselli (NE1), m 279, 10/31/1996, 26/11/1997; Terria (NE2), m 270, 10/31/1996, 11/26/1997; Castellonalto (NE3), m 390, 10/31/1996; Arrone (NE4), m 230, 10/31/1996, 11/26/1997.

Horizontal starch-gel electrophoresis was conducted on the crude homogenate of the whole body. Samples were screened for several enzymes (Tab. 1); alleles at each locus were ranked according to electrophoretic mobility. The divergence between populations was estimated using the genetic distance index, D (Nei, 1978), calculated with the Biosys-1 computer program (Swofford and Selander, 1981). A UPGMA cluster analysis (Sneath and Sokal, 1973) was performed and relative dendrogram drawn for the distance matrix.

RESULTS

We surveyed banding patterns interpreted as the products of 27 gene loci. No differences at study loci were recorded between larval and adult specimens. On the basis of genotype frequen-

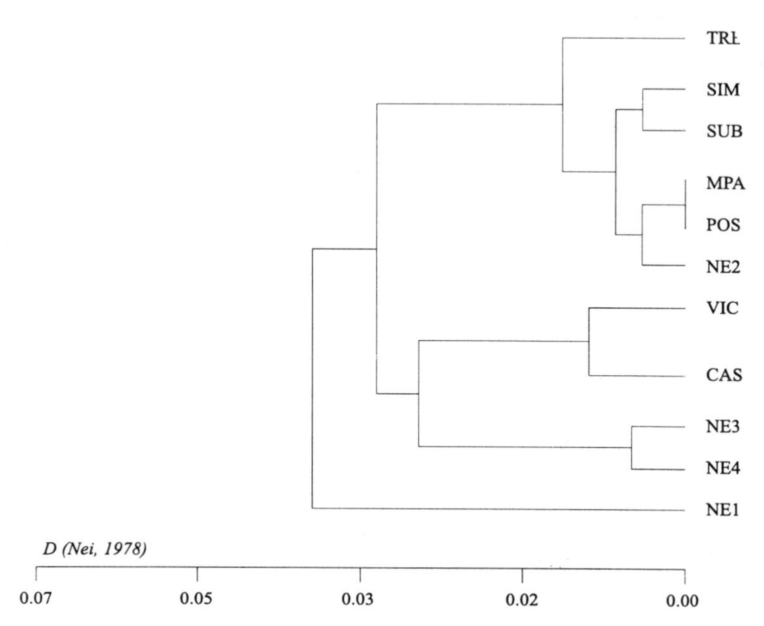

Fig. 1. Genetic relationships among studied populations as portrayed by a dendrogram elaborated after UPGMA Cluster Analysis.

cy data, 13 loci were monomorphic and fixed for the same allele in all populations. 14 loci (*Acph2, Ada1, Ada2, Aldo, Ao, Aph2, Ca3, Est2, G6pd2, Hk, Ldh, Pep, Pgm, Phi*) were polymorphic in at least one population. Allele frequencies are reported in Table 2. Nei's genetic distance values are shown in matrix form in Table 3. The highest value found was D=0.052 between the populations of NE1 and NE4 (R. Nera) and between populations NE1 (R.Nera) and VIC (R. Aniene). The lowest distance was D=0.000 between POS (R. Velino) and MPA (R. Aniene). Genetic distances among the populations within the Aniene River (Tab. 3) were relatively high, ranging from D=0.002 (between SIM and SUB) to D= 0.044 (between TRE and VIC), as well as genetic distances among the populations belonging to Nera River (Tab. 3), that ranged from 0.002 (between NE3 and NE4) to 0.052 (between NE4 and NE1). A UPGMA dendrogram summarizing the relationships among all populations shows (Fig. 1) a cluster including most of the Aniene River populations and the only population studied from the Velino River. A second cluster includes three populations from the Nera River and one from the Aniene River (VIC). The last one includes only one population from the Nera River (NE1). However, genetic distances used to construct the UPGMA dendrogram were rather low. Therefore, a clear separation among populations could not be observed.

DISCUSSION

Absolute genetic distance values obtained among the 11 populations of *D. cephalotes* analyzed were relatively low (Tab. 3). Data presented here should be viewed with relative caution, due to the small sample size of study populations, they are in general agreement with what has been reported elsewhere for stoneflies (see for instance Fochetti et al., 1994; Funk and Sweeney, 1990; White, 1989). Indeed, many published genetic distances among conspecific populations in Plecoptera are around 0. The low values found in the present study sup-

Table 2. Allele frequencies observed for 14 polymorphic loci studied at 11 sites on 3 rivers for *D. cephalotes*

			Populations										
			R. Aniene					R. Velino	R. Nera				
Locus			TRE	SIM	SUB	MPA	VIC	POS	CAS	NE1	NE2	NE3	NE4
N tot			54	7	26	8	2	27	11	20	9	5	13
Acph2	(N)		20	6	8	5	1	15	6	4	9	1	9
		A	0.500	0.000	0.187	0.000	0.000	0.067	0.000	0.750	0.167	0.000	0.056
		B	0.500	1.000	0.812	1.000	1.000	0.933	1.000	0.250	0.833	1.000	0.944
Ada1	(N)		11	1	9	1	2	10	8	6	2	2	5
		A	0.091	0.000	0.000	0.000	0.000	0.050	0.000	0.167	0.000	0.000	0.000
		B	0.682	1.000	0.833	1.000	0.750	0.850	0.937	0.833	1.000	1.000	1.000
		C	0.227	0.000	0.167	0.000	0.250	0.100	0.062	0.000	0.000	0.000	0.000
Ada2	(N)		33	3	20	4	2	13	8	6	3	3	10
		A	0.985	1.000	0.950	1.000	0.500	0.962	1.000	1.000	1.000	1.000	1.000
		B	0.015	0.000	0.050	0.000	0.500	0.038	0.000	0.000	0.000	0.000	0.000
Aldo	(N)		45	7	17	8	2	27	11	19	9	5	13
		A	0.089	0.000	0.059	0.187	0.000	0.111	0.091	0.000	0.000	0.000	0.192
		B	0.911	1.000	0.941	0.812	1.000	0.889	0.909	1.000	1.000	1.000	0.808
Ao	(N)		47	7	21	8	2	20	11	16	6	5	10
		A	0.032	0.214	0.048	0.000	0.000	0.000	0.045	0.094	0.167	0.000	0.150
		B	0.968	0.786	0.952	1.000	1.000	1.000	0.955	0.906	0.833	1.000	0.850
Aph2	(N)		34	5	15	4	2	24	6	19	8	4	6
		A	0.176	0.300	0.133	0.000	0.000	0.042	0.250	0.579	0.125	0.000	0.167
		B	0.824	0.700	0.867	1.000	1.000	0.958	0.750	0.421	0.875	1.000	0.833
Ca3	(N)		23	7	13	4	2	13	2	6	6	2	2
		A	0.978	0.929	0.923	0.875	1.000	0.923	1.000	1.000	0.917	1.000	1.000
		B	0.022	0.071	0.077	0.125	0.000	0.077	0.000	0.000	0.083	0.000	0.000
Est2	(N)		36	3	16	4	2	6	11	13	2	5	8
		A	0.000	0.000	0.000	0.000	0.000	0.000	0.091	0.000	0.000	0.000	0.125
		B	1.000	1.000	1.000	1.000	1.000	1.000	0.909	1.000	1.000	1.000	0.875
Gopd2	(N)		19	4	8	8	2	19	5	4	9	2	13
		A	0.368	0.000	0.000	0.000	0.000	0.000	0.000	0.000	0.000	0.000	0.000
		B	0.632	1.000	1.000	1.000	1.000	1.000	1.000	1.000	1.000	1.000	1.000
Hk	(N)		32	3	13	1	2	13	9	16	1	5	5
		A	0.062	0.000	0.000	0.000	0.000	0.115	0.111	0.000	0.000	0.60	0.800
		B	0.828	1.000	1.000	1.000	1.000	0.885	0.889	1.000	1.000	0.400	0.200
		C	0.078	0.000	0.000	0.000	0.000	0.000	0.000	0.000	0.000	0.000	0.000
		D	0.031	0.000	0.000	0.000	0.000	0.000	0.000	0.000	0.000	0.000	0.000
Ldh	(N)		43	6	20	8	2	23	9	18	8	5	11
		A	0.849	0.833	0.825	1.000	1.000	0.957	0.778	0.917	1.000	1.000	0.955
		B	0.151	0.167	0.175	0.000	0.000	0.043	0.222	0.083	0.000	0.000	0.045
Pep	(N)		52	7	25	8	2	25	11	19	9	5	13
		A	0.231	0.429	0.400	0.062	1.000	0.100	0.818	0.474	0.167	0.700	0.500
		B	0.769	0.571	0.600	0.937	0.000	0.900	0.182	0.526	0.833	0.300	0.500
Pgm	(N)		54	7	26	8	2	27	11	20	9	5	13
		A	1.000	1.000	1.000	0.937	1.000	0.981	1.000	1.000	1.000	0.900	1.000
		B	0.000	0.000	0.000	0.062	0.000	0.019	0.000	0.000	0.000	0.100	0.000
Phi	(N)		45	7	19	8	2	21	11	14	8	5	13
		A	0.111	0.000	0.105	0.062	0.250	0.048	0.000	0.036	0.125	0.100	0.077
		B	0.856	1.000	0.763	0.875	0.750	0.952	1.000	0.964	0.875	0.900	0.923
		C	0.033	0.000	0.132	0.062	0.000	0.000	0.000	0.000	0.000	0.000	0.000

Table 3. Genetic distance (lower left) and genetic similarity (upper right) among all study populations of *D. cephalotes*

Populations	TRE	SIM	SUB	MPA	VIC	POS	CAS	NE1	NE2	NE3	NE4
1 TRE	*****	0.980	0.990	0.980	0.957	0.986	0.970	0.983	0.988	0.964	0.963
2 SIM	0.020	*****	0.998	0.991	0.977	0.993	0.996	0.977	0.998	0.981	0.976
3 SUB	0.010	0.002	*****	0.993	0.983	0.996	0.992	0.980	0.998	0.981	0.973
4 MPA	0.020	0.009	0.007	*****	0.960	1.000	0.974	0.958	0.999	0.972	0.968
5 VIC	0.044	0.024	0.017	0.041	*****	0.965	0.990	0.950	0.967	0.980	0.958
6 POS	0.014	0.008	0.004	0.000	0.036	*****	0.979	0.966	0.999	0.978	0.975
7 CAS	0.030	0.004	0.008	0.026	0.010	0.022	*****	0.970	0.981	0.989	0.979
8 NE1	0.017	0.023	0.020	0.043	0.052	0.035	0.030	*****	0.977	0.951	0.949
9 NE2	0.013	0.002	0.002	0.001	0.034	0.001	0.019	0.023	*****	0.976	0.972
10 NE3	0.036	0.019	0.020	0.028	0.020	0.022	0.011	0.050	0.024	*****	0.998
11 NE4	0.038	0.025	0.028	0.032	0.043	0.025	0.021	0.052	0.029	0.002	*****

port the notion that our study populations are conspecific. In particular, data for populations from different rivers such a TRE and POS, showed a level of genetic differentiation comparable to those found by Ayala (1983) among conspecific local populations. The genetic divergence between TRE and POS is mainly due to some differences in allele frequencies at *Acph2* and *G6pd2* loci, probably related to the limited among-rivers dispersal ability of stoneflies. There was some indication of genetic differentiation among subpopulations living in the same river (Tab. 3), especially river Nera. For example, Nei's distance between NE1 and NE4 was 0.052 and the two populations differed greatly in allele frequency at both the *Acph2* and *Hk* loci. These data, which are based on reasonable sample size, suggest little or no recent gene flow between the two locations.

Values of genetic distances among conspecific populations can be found in literature with regards to stoneflies. Among the lowest values found in the literature for conspecific populations are those for the genus *Taeniopteryx*. Populations of European species (collected from Central Italy) showed null distances (Fochetti et al., 1994) while populations of American species had D values that varied from a minimum of 0.0004 to a maximum of 0.002 (Funk and Sweeney, 1990). In *Isoperla insularis,* distance values ranged from D=0.000 to D=0.010 (Fochetti, 1993), while in *Pteronarcys proteus* the distances varied from a minimum of D=0.004 to a maximum of D=0.010 (White, 1989). Conspecific populations of the genus *Protonemura* seemed to be relatively more isolated, with distance values ranging from 0.001 to 0.096 (Fochetti, 1994). Also, in *Hesperoperla pacifica* the distance between the two populations studied was D=0.054 (Robinson et al., 1992). It must be stressed that all these papers deal with conspecific populations sampled in different rivers. The genetic divergence in these cases could be partially explained by the reduced dispersal ability of the stoneflies.

Genetic distance data obtained in the present study for populations sampled in the same river seem to be elevated as compared to values for species representing other orders of aquatic insects. A comparison of our data with mayflies may be meaningful because they are also quite primitive insects and are similar to the Plecoptera from the ecological point of view. From a research carried out on 40 populations of fifteen species of Ephemerellidae, the mean genetic distance (among conspecific populations) turned out to be D=0.006, with a minimum of 0.002 and a maximum of 0.023 (Sweeney et al., 1987). The authors point out that these are among the lowest values ever found in insects. In the species *Dolania americana,* interpopulation genetic distances ranged from D=0.000 to D=0.029 (Sweeney and Funk, 1991). Similarly for dragonflies, conspecific populations of species of the genus *Ischnura* (Zygoptera: Coenagrionidae) showed distances from a minimum of 0.000 to a maximum of 0.060 (Carchini et al., 1994).

Field observations suggest that the natural populations investigated in the present study have small effective population size. In fact, it was difficult to collect more than a small number of specimens for some of the sampled sites. Thus, stochastic factors, such as genetic drift, may play a pivotal role in shaping the observed pattern of genetic variation.

Also, there may be barriers preventing genetic exchange of individuals among populations (e.g. dams), which can isolate them from each other. This is a good possibility for the Aniene River, where dams are numerous over its short length.

The genetic data obtained in this study generally corroborate the results of other researches on Plecoptera. We have found a low genetic divergence, similar to that described for other orders of aquatic insects. All the populations studied belong to the same species and are genetically homogeneous. Notwithstanding subpopulations coming from the same river are clearly separated from one another. The values of genetic distance clearly show that the populations analyzed are not completely panmictic. This fact is caused probably by both the autecology of *D. cephalotes*, and by the environmental disturbance that man has exerted on its habitat. Further researches on a higher number of individuals and populations are still needed in order to provide a careful description of the genetic structuring of this species.

ACKNOWLEDGMENTS

We wish to thank Rick Jacobsen (Florida, USA) for smoothing the English text and for providing useful suggestions. This research was supported by grants from the Ministero della Ricerca Scientifica e Tecnologica, MURST (40%) and Consiglio Nazionale delle Ricerche, CNR (Comitato Ambiente). We wish to thank anonymous reviewers for useful criticism on a first draft of this paper.

REFERENCES

Ayala, F. J. 1983. Enzymes as a taxonomic characters, pp. 3-26. In: G. S. Oxford and D. Rollinson (eds.). Protein polymorphism: adaptive and taxonomic significance. Academic Press. London.

Carchini, G., M. Cobolli, E. De Matthaeis and C. Utzeri. 1994. A study on genetic differentiation in the Mediterranean *Ischnura charpentier* (Zygoptera: Coenagrionidae). Adv. Odonatol. 6: 11-20.

Elliot, J. M. 1989. The effect of temperature on egg hatching for three populations of *Dinocras cephalotes* (Plecoptera: Perlidae). Ent. Gaz. 40: 153-158.

Fochetti, R. 1993. Il genere *Isoperla* nel sistema sardo-corso: dati elettroforetici. Fragm. Ent. 25: 11-19.

Fochetti, R. 1994. Biochemical systematics and biogeographical patterns of the Italian and Corsican species of the *Protonemura corsicana* species group. Aquat. Insects 16: 1-15.

Fochetti, R., M. Cobolli, E. De Matthaeis and M. Oliverio. 1994. Tassonomia biochimica del genere *Taeniopteryx*. Atti XVII Congr. Naz. Ital. Entomol.: 83-86.

Fochetti, R., M. Cobolli, E. De Matthaeis and M. Oliverio. 1997. Allozyme variation in the genus *Isoperla* (Plecoptera; Perlodidae) from Mediterranean islands, with remarks on genetic data on stoneflies, pp. 476-483. In: P. Landolt and M. Sartori (eds.). Ephemeroptera and Plecoptera: Biology-Ecology-Systematics. MTL, Friburg.

Funk, D. H. and B. W. Sweeney. 1990. Electrophoretic analysis of species boundaries and phylogenetic relationships in some taeniopterygid stoneflies (Plecoptera). Trans. Amer. ent. Soc. 116: 727-751.

Harris, H. and D. A. Hopkinson. 1978. Handbook of enzyme electrophoresis in human genetics. North Holland Publishing Co. Amsterdam NL.

Lees, J. and R. D. Ward. 1987. Genetic variation and biochemical systematics of British Nemouridae. Biochem. Syst. Ecol. 57: 117-125.

Lillehammer, A. 1987. Egg development of the stoneflies *Siphonoperla burmeisteri* (Chloroperlidae) and *Dinocras cephalotes* (Perlidae). Freshw. Biol. 17: 35-39

Nei, M. 1978. Estimation of average heterozygosity and genetic distance from a small numbers of individuals. Genetics 89: 583-590.

Richardson, B. J., P. R. Baverstock and N. Adams. 1986. Allozyme electrophoresis. Academic Press, Australia.

Robinson, C. T., L. M. Reed and G. W. Minshall. 1992. Influence of flow regime on life history, production and genetic structure of *Baetis tricaudatus* and *Hesperoperla pacifica*. J. N. Amer. Benthol. Soc. 11: 278-289.

Sneath, P. H. A. and R. R. Sokal. 1973. Numerical taxonomy. Freeman, San Francisco.

Sweeney, B. W. and D. H. Funk. 1991. Population genetics of the burrowing mayfly *Dolania americana*: geographic variation and the presence of a cryptic species. Aquat. Insects, 13: 17-27.

Sweeney, B. W., D. H. Funk and R. L. Vannote. 1987. Genetic variation in stream mayfly (Insecta: Ephemeroptera) populations of eastern North America. Ann. ent. Soc. Amer. 80: 600-612.

Swofford, D. L. and R. B. Selander. 1981. BIOSYS-1: a FORTRAN program for the comprehensive analysis of electrophoretic data in population genetics and systematics. J. Hered. 72: 281-283.

White, M. M. 1989. Age class and population genic differentiation in *Pteronarcys proteus*. Amer. Midl. Nat. 122: 242-248.

Wright, M. and M. M. White. 1992. Biochemical Systematics of the North American *Pteronarcys* (Pteronarcydae: Plecoptera). Biochem. Syst. Ecol. 20: 515-521.

Zwick, P. 1996. Capacity of discontinuous development and its importance for the geographic distribution of the warm water stenotherm, *Dinocras cephalotes* (Insecta: Plecoptera: Perlidae). Ann. Limnol. 32: 147-160.

CONSTRUCTION AND EVALUATION OF A NEW LABORATORY SYSTEM FOR REARING MAYFLIES

Kyla J. Finlay

Dept of Biological Sciences, Monash University
Co-operative Research Centre for Freshwater Ecology
Land and Water Resources Research and Development Corporation
Clayton, VIC 3168, Australia

ABSTRACT

A new laboratory system was developed for rearing adults of Australian Leptophlebiidae (Ephemeroptera). The space-efficient chambers, made from readily available materials, are an inexpensive alternative to conventional rearing systems and worked just as well, and in many cases better, than a comparable larger system. Genera differed significantly in rearing success. No genera were significantly affected by the rearing conditions except *Austrophlebioides* Campbell and Suter (1988), which responded strongly to all external factors. This suggests that a more restricted environmental regime may be required to maximise emergence success for this genus. Sex ratios were biased towards females for all genera, indicating the possible occurrence of parthenogenesis in Australian Leptophlebiidae.

INTRODUCTION

Positive identification of species requires examination of all life stages for most aquatic insects. However, in most cases, identification of species has been based on the nymphs or adults only with no association being made between the two (Hynes, 1970; Smock, 1996; Merritt *et al.*, 1996). Field collecting of nymphs and adults in one location is an accepted method of identifying all insect life stages but has the inherent problem with discriminating between different species, especially if one has to rely on immature nymphs for initial identification. An insect reared from an immature stage to an adult, with the subsequent larval skin moult kept for comparison, provides the definitive association.

Many authors have made suggestions for rearing aquatic insects to adults (see review by Merritt *et al.*, 1996). The two main approaches are field and laboratory rearing. Most field rearing techniques involve a mechanism for containing the nymphs within the existing water body and providing room for the animal to emerge while safe from drowning (Speith, 1938; Fremling,1967; Day, 1968; Schnieder, 1967; Edmunds *et al.*,1976). An alternative is the use of emergence traps (Hynes, 1941; Southwood, 1978; Merritt *et al.*, 1996). Despite the relative

Fig. 1. The new rearing chamber. The aquarium tubing is connected to an air supply.

simplicity of design, the disadvantage of field rearing is that it may require the researcher to be away from the workplace for extended periods. Laboratory methods range from simple to extremely complex as authors have tried to address the problem of recreating stream flow conditions. Covered aquariums are used in conjunction with various methods of inducing a current such as a magnetic stirrer (Mason and Lewis, 1970) or directed air (Craig, 1966). Gravity flow systems were first used by Hynes (1941) and improved upon by Mason and Lewis (1970). Later came the development of large flow tanks powered by propellers (Vogel and LaBarbera, 1978) and complicated systems designed more specifically for the purposes of toxicity testing (Buikema and Voshell, 1993). All have the disadvantage of being suitable only for mass rearing and often requiring large inputs of electrical power. None address the issues of cost-effectiveness and simplicity in a field where rearing is likely to be of secondary concern.

The rearing of mayflies can be especially difficult because of the presence of a fragil subimago stage which has characteristics different to those of the adult. With all these factor in mind I have designed and tested a new laboratory rearing system for mayflies. Each cham ber houses one individual and allows the animal to pass through the stages of nymph an subimago without disturbance. The chambers are made from readily obtainable material an are easy to construct. They are space-efficient and inexpensive, costing less than US$2.0 each when an air supply is available. Such systems have been described before in the litera ture (Merritt *et al.*, 1996) but I have yet to find a published account of their use.

An analysis of the success of the new rearing system by imago emergence success rate in relation to genus, sex, photoperiod, year of collection, temperature, and for an alternativ chamber type has been conducted. The influences of altitude of collection and refrigerate storage of nymphs, which was often necessary after a prolonged collecting trip, were als investigated. These data were collected as a consequence of rearing mayfly adults for tax nomic review rather than experimental purposes. The data collected provides indications how these genera may respond to factors influencing emergence in the field, however, the l boratory outcomes are, at present, analysed at genus level only. Within each genus there m well be species with very different responses to these environmental factors. Further, on one controlled temperature room was used for this work, therefore all individuals from ea

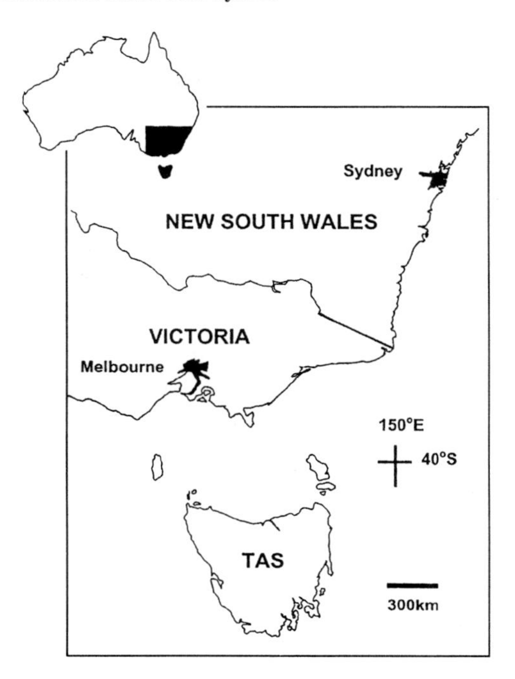

Fig. 2. South-east Australia: area from which mayflies were collected.

separate field collection were placed in the same temperature and photoperiod regimes. As a result the species compositions of the groups receiving each set of conditions may well have been different. In these circumstances it is difficult to be confident that the results obtained mirror responses in the field. Nevertheless, these observations may be of interest to workers concerned with field responses of mayfly nymphs to environmental conditions.

METHODS

Chamber Design

The new rearing chamber (Figure 1) was constructed from a 1.25 litre plastic soft drink bottle. This was cut in two at about two-thirds its length, at the point where the sides start to converge towards the lid. The open container was lined with nylon mesh, which can either be glued in place or simply wetted. Two hooks were attached facing outwards and opposite each other on the outside of the bottle using electrical tape, so that a rubber band could be stretched between them across the open end. A small hole (diameter of 6 -7 mm) was drilled in the plastic bottle lid. The top third of the bottle with the lid was then inverted to sit in the chamber, lid downwards, and secured by the rubber band. The chamber was then half filled with water. Compressed air was supplied to the chamber by means of PVC aquarium tubing (interior diameter 4 mm) attached to a pump or laboratory air supply. Up to ten chambers can be aerated from one s. .all 240V air pump linked through aquarium tubing, although each chamber requires a two-way controller so flow can be balanced. It is also advisable to attach a plastic micropipette tip to the end of the tubing to restrict the flow of bubbles to a small stream. Glass pipettes proved too fragile and, being heavier, were prone to blockage by resting on the chamber bottom.

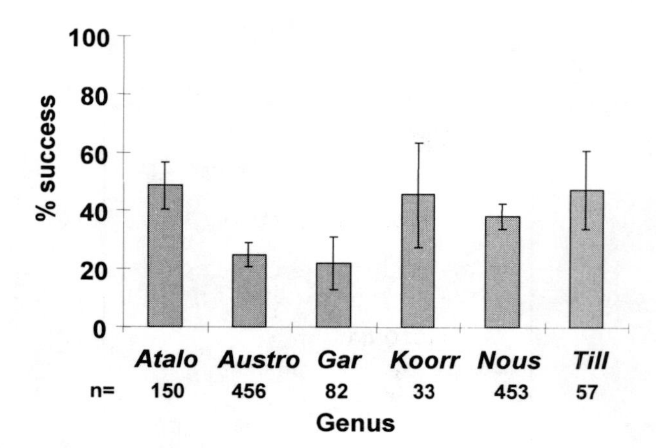

Fig. 3. Imago emergence success by genus.
Abbreviations for genera are as follows: Atalo=*Atalophlebia*, Austro=*Austrophlebioides*, Gar=*Garinjuga*, Koorr=*Koornonga*, Nous=*Nousia*, Till=*Tillyardophlebia*. Bars represent 95% confidence intervals.

Data Collection

Mayfly nymphs from the family Leptophlebiidae were collected over two extended summer periods; October 1996 to April 1997 (year 1) and October 1997 to April 1998 (year 2) from 123 sites throughout Victoria, Tasmania and New South Wales (Figure 2). Animals were collected from altitudes on the shoreline (<10 m) to near the summit of Australia's highest point, Mt Kosciusko (1650m), and therefore represent species from a wide range of climatically diverse regions. Two closely related Leptophlebiids, *Nousia* Navás (1918) and *Koorrnonga* Campbell and Suter (1988), were targeted for collection as part of a larger taxonomic study being undertaken, although all Leptophlebiidae were collected if found. *Nousia* and *Koorrnonga* are relatively common in stream riffles in association with logs and organic matter (Peters and Campbell, 1991), so the sampling regime favoured these areas.

Late instar nymphs were carefully removed from the substrate with a paintbrush and placed in a bottle of the stream's water. The bottle was sealed and placed on ice for transportation to the laboratory. During the day the water was adequately aerated through the motion of the vehicle but at night a battery-operated pump was employed to aerate each chamber.

Laboratory Rearing

Each rearing chamber was half filled with water from a particular site; and one late instar individual from that site added. Twigs collected from the site were added to each chamber for the insect to use as a food source and as a platform for emergence. Each chamber was attached to an air supply and placed in controlled temperature room environments at 16°C, 18°C, 20°C or 22°C. Photoperiods of 12 hours daylight and darkness (12:12) or 14 hours daylight, 10 hours darkness (14:10) were used. Some nymphs were placed in much larger chambers designed by Campbell (1983). These were made from a cube-shaped frame of wood to which fly-screen or mesh is stapled on all sides. A container, which can hold volumes up to 500 ml, was placed inside the frame and connected to an air supply by aquarium tubing through a small hole. Surplus nymphs were stored in an aerated container refrigerated to 7°C. In the controlled environment the nymphs were checked every second day and the life cycle stage of the individual noted. Once emergence (or death) occurred the animals were removed and genus and sex determined by observation using a stereomicroscope. Empty chambers were thoroughly washed and nymphs replaced from refrigerated stock. These new nymphs were acclimatised to the controlled temperatures for a period of 20 to 30 minutes.

RESULTS

The full data set included 1251 individuals of which *Austrophlebioides* and *Nousia* predominated, comprising 36.5% and 36.2% respectively. Next came *Atalohplebia* Eaton (1881) at 12%, then *Garinjuga* (Campbell and Suter (1988)) at 6.6%, *Tillyardophlebia* Dean (1997) at 4.6%, *Koornonga* (Campbell and Suter (1988)) (2.6%) and *Ulmerophlebia* Demoulin (1955) at 0.9%. A few individuals of other genera, such as *Atalomicra, Jappa, Kirrara* and an undescribed one, were also collected and represented the remaining 0.6%. Of the ten named genera in south-east Australia, all were sampled except *Neboissophlebia* Dean (1988), despite this genus having been found previously in many of the sites where I collected (Dean, 1988). Data analysis will focus on the six most prevalent genera.

Emergence Success

Imago emergence success rate for the full data set was 34.1% with 10.0% reaching the subimago stage before dying and 55.9% dying as nymphs. Individual genera, however, differed significantly in emergence success (χ^2=46.071, df=5, p<<0.001, Figure 3). *Atalophlebia* was reared most successfully with 48.7% becoming imagos. Other highly successful genera were *Tillyardophlebia* (47.4%) and *Koorrnonga* (45.5%). The genus with the lowest success rate was *Garinjuga* (22.0%) followed by *Austrophlebioides* (25.0%).

The new rearing chamber was compared with that designed by Campbell (1983). The new chamber produced higher imago emergence rates compared with the 'old' one for all the genera examined (except *Nousia*) (Table 1). A significant difference was found only for *Austrophlebioides* (χ^2= 5.993, df=1, p=0.014), where the emergence success rate more than doubled in the new chambers (27.1% versus 12.7%).

Emergence success did not differ between the sexes for any genus excep *Tillyardophlebia* (Table 1) where a much greater proportion of females (58.8%) than male (30.4%) emerged successfully (χ^2= 4.435, df=1, p=0.035). The effect of varying th photoperiod could be analysed only for year 1 as there were no individuals reared unde 12:12 conditions during year 2. Within the restricted data set a significant difference due to photoperiod was found only for *Austrophlebioides* (χ^2= 17.810, df=1, p<<0.001, Table 1) yet this went against the trend for all other genera where success rates, although not significantly different, were higher under a 12:12 cycle. Similarly, the effect of year of collection could only be examined in relation to photoperiod 14:10. Again there was a significant difference between year 1 and 2 only for *Austrophlebioides* (χ^2= 56.936, df=1, p<<0.001, Table 1). There was no apparent trend for the other genera.

For most genera the proportion successfully emerging was highest at 18°C (for *Nousia* this occurred at 16°C but the difference in success rate from 18°C was very slight; 0.1%). The temperature which produced the lowest proportion of successful emergence was 22°C. The effect of temperature was significant for both *Austrophlebioides* (χ^2= 47.838, df=3, p<<0.001) and *Nousia* (χ^2= 10.517, df=3, p=0.015, Table 1). The effect of temperature was also considered in relation to time spent in the rearing system for a restricted number of genera (Figure 4). Time taken to emerge successfully was greatest at 18°C followed by 16°C, 20°C then 22°C. Although patterns of response to temperature were similar for each genus, the time taken to reach outcome varied considerably. For example, at 18°C mean time to emerge varied from 7.7 days for *Nousia* to 12.9 days for *Atalophlebia.* Data were log transformed to meet the assumption of normality and an ANOVA run to test for a significant effect of temperature. All genera had significant temperature effects (*Nousia,* F-ratio = 0.635,

Table 1. Percentages of successful emergences as affected by collection and rearing conditions

	Atalo-phlebia	*Austro-phlebioides*	*Garinjuga*	*Koorrnonga*	*Nousia*	*Tillyard-phlebia*
Chamber Type						
p value	0.628	**0.014**	**	**	0.926	**
n	150	*454*	*82*	33	*453*	57
% success old cage	44.4	12.6	15.4	37.5	38.7	50.0
% success new cage	49.6	27.1	23.2	48.0	38.1	47.0
Sex						
p value	0.795	0.909	0.542	0.435	0.234	**0.035**
n	150	456	*82*	*33*	*453*	57
% success female	47.8	24.8	20.0	40.0	40.0	58.8
% success male	50.0	25.3	25.9	53.8	34.2	30.4
Photoperiod (year 1) *						
p value	**	**0.000**	**	**	0.293	**
n	*62*	*164*	82	*3*	166	*14*
% success 12:12	62.3	24.0	28.2	100.0	46.8	53.8
% success 14:10	33.3	56.5	18.0	100.0	38.4	0.0
Year of Coll (photoperiod 14:10)						
p value	**	**0.000**	**	**	0.687	**
n	97	377	50	31	*391*	*44*
% success – year 1	33.3	56.5	13.0	100.00	38.5	0.0
% success – year 2	42.0	16.1	22.2	40.00	36.2	46.5
Temperature						
p value	**	**0.000**	**	**	**0.015**	**
n	*150*	*456*	82	*33*	*452*	*57*
% success 16°C	43.4	23.6	20.0	42.1	46.4	51.4
% success 18°C	59.0	45.4	25.6	100.0	46.3	53.8
% success 20°C	50.0	24.3	42.8	0	38.9	25.0
% success 22°C	38.2	10.8	9.5	40.0	29.3	0
Altitude						
p value	0.126	**0.000**	**	**	0.118	**
n	*150*	*456*	*82*	*33*	*453*	*57*
% success <400m	46.9	33.1	24.5	12.5	39.4	50.0
% success 400-800m	44.4	22.5	-	54.5	34.5	16.7
% success 801-1200m	73.3	9.6	0.0	-	15.4	100.0
% success >1200m	-	6.1	19.0	66.7	56.3	-
Storage						
p-value	**	**0.000**	**	**	0.856	**
n	*150*	*456*	*82*	*33*	*453*	*57*
% success 0-2 days	43.8	32.3	20.9	50.0	38.5	62.5
% success 3-5 days	75.0	19.8	33.3	33.3	35.0	28.6
% success > 5 days	62.5	12.0	16.7	50.0	39.1	27.8

Effect of each factor analysed by χ^2 Significant p values (at < 0.05) shown in bold.
* analysed on restricted data set as there were no mayflies reared under the 12:12 regime in year 2.
** one or more categories with small sample size: χ^2 test not reliable.

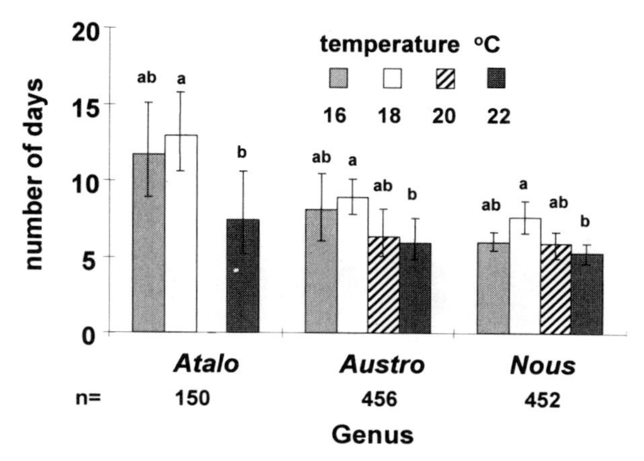

Fig. 4. Effect of temperature on time taken to successful emergence by genus.
Abbreviations for genera as in Figure 3. Data back transformed from logged values. Bars represent back transformed standard error. Within each genus temperatures without a letter in common are significantly different (p-value less than 0.05) by Tukeys HSD tests.

df=3, p=0.010; *Austrophlebioides,* F-ratio=4.269, df=3, p=0.007; *Atalophlebia,* 20°C degrees removed, F-ratio = 4.354, df=2, p=0.017) although multiple r^2 showed that very little variability (around 10% or less) in rearing time was explained by temperature differences. Tukey's HSD tests showed significant differences between the temperatures 18°C and 22°C only for all genera (Figure 4).

The effect of altitude of collection on rearing success by genus was examined by chi-squared for 4 altitude categories. (1. <400 m, 2. 400-800 m, 3. 801-1200 m, 4. > 1200 m) and was found to be significant only for *Austrophlebioides* (χ^2= 29.826, df=3, p<<0.001, Table 1) where the highest success rates were found at progressively lower altitudes.

Division of storage time into three categories (1. 0-2 days, 2. 3-5 days, 3. >5 days) and subsequent analysis by chi-squared also showed a significant effect only for the genus *Austrophlebioides* (χ^2=18.297,df=2, p<<0.001, Table 1) with a higher success rate for progressively less days in storage. However, altitude of collection and storage time were positively correlated as it took longer to return to the laboratory from high altitude collection sites. A logistic regression of both factors against emergence success showed altitude probably was more influential than storage.

Sex-Ratios

Ratios of males to females showed a female bias for all individual genera in the range of 1: 1.5 for *Tillyardophlebia* to 1: 2.1 for *Nousia* (Figure 5). Chi-squared analysis was used to determine departure from the expected 1:1 sex ratio and was found to be significant for all genera except *Koorrnonga* and *Tillyardophlebia.*

DISCUSSION

The overall imago emergence success rate was 34%. Clearly, however, success rates depend on the genus in question. Taxa preferring slow waters would be expected to emerge more successfully in aerated tanks with low flow regimes (Edmunds *et al.*,1976). Therefore,

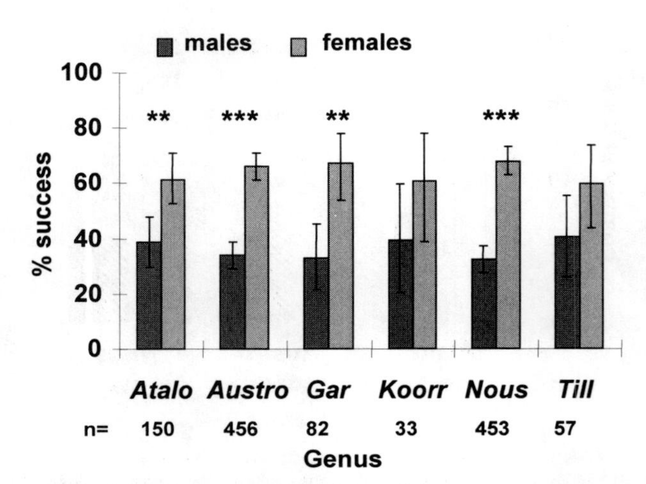

Fig. 5. Mayfly sex-ratios of six genera captured over 2 extended summer periods. Abbreviations for genera as in Figure 3. Bars represent 95% confidence intervals. Asterisks indicate significant departures (at less than 5% significance level) from 1:1 ratio by chi-squared.

it is not surprising that *Atalophlebia,* predominantly found in standing or slowly flowing waters (Peters and Campbell, 1991), had the greatest proportion of imagos that emerged. The relatively low success rates of *Garinjuga* may be due to the presence of a possibly new high altitude species for which the emergence success was very low, thereby reducing the average for the whole genus.

Overall, the new rearing system was a success in that imago emergence rates were generally higher in the new chamber compared with the older one for all genera. In particular, the success rate for *Austrophlebioides* was much greater in the new chamber. The considerable advantages of the new chambers in terms of time saved in construction and set-up, space efficiency and low cost indicate that the system could be widely used even if just for routine species identification.

In general, few factors had an effect on overall emergence success for any genera other than *Austrophlebioides*. Sex appeared to affect *Tillyardophlebia* but, because this was a relatively small data set, may not prove to be biologically significant. There are also some temperature effects for *Nousia* and *Atalophlebia* which must be considered. (Table 1, Figure 4). Most striking, however, were the results for the genus *Austrophlebioides* which appeared to be affected strongly by all factors except sex. This is almost certainly related to habitat requirements. For example, the greater rearing success in smaller chambers with much less water could be indicative of an inherent need for highly oxygenated water in this genus. Neither chamber adequately simulates current flow, but the new chamber may provide more oxygen per volume than the 'old' chamber. A high rate of water movement may be necessary for the development of this genus. Similarly, the highly positive response to a 14 hour photoperiod, which is typical of an Australian summer light regime, may be required as a cue for *Austrophlebioides* to emerge. If this is so, this is the first record of an enhanced emergence response to photoperiod for Australian Leptophlebiidae as photoperiod has been shown previously to have no effect on mayfly egg hatching or emergence (Brittain, 1982; Suter and Bishop, 1990; Newbold *et al.,* 1994) despite the suggestion that it is important for aquatic insects in general by Hynes (1970).

Austrophlebioides fared better in year 1 possibly because they were stored for much less time (0.1 mean days in year 1 compared with 4.9 mean days in year 2). Progressively shorter storage times produced significantly higher success rates. Another contributing factor may be that altitude range for each year was considerably skewed, with animals being collected no higher than 500 m in year 1, yet up to 1560 m in year 2; progressively lower altitudes produced greater emergence success rates.

It is acknowledged that animals do not respond to altitude *per se* but rather environmental variables associated with altitude (see reviews by Minshall, 1988; Power *et al.*, 1988) such as temperature, substrate, dissolved oxygen and hydraulic variation. One can speculate that there is greater temperature differential between higher altitude sites and the laboratory compared with lower altitude sites, possibly making the physiological stress on the animals greater.

Temperature not only determines abundance, distribution and diversity of stream insects (Hynes, 1970; Ward and Stanford, 1982; Zamora-Muñoz *et al.*, 1993) but is considered to be one of the most important influencing factors affecting insect development (Corkum, 1978; Elliott, 1978; Brittain, 1982; Wallace and Anderson, 1996). Indeed, in this study, temperature was the only external factor significantly affecting emergence success of a genus other than *Austrophlebioides*. Success rates were much higher at 18°C than at 22°C, yet summer water temperatures would be within the range 16-22°C for all but the highest altitudes; so it is puzzling to find such a restricted temperature preference for emergence. This restricted preference also applied to the length of time taken to emerge. It is possible that the nymphs have a narrow temperature requirement for development which is in line with the theory of Sweeney and Vannote (1978) that an optimal thermal regime exists for a given species. Adult size and fecundity and, presumably emergence rates, may diminish outside the bounds of the optimal regime for the species.

Trends in the data are not significant with the clear exception of *Austrophlebioides*. Species level data within this genus would therefore be especially valuable.

Sex-Ratios

Insect sex-ratios in nature are generally expected to be 1:1 although skewed ratios due to inbreeding occur and will be biased towards females (Thornhill and Alcock, 1983). Female biased sex ratios in Ephemeroptera have been recorded only for parthenogenetic taxa of which 50 species are known worldwide (Brittain, 1982) and only in 3 or 4 families (McCafferty and Huff, 1974). It appears obligatory in only a few species (Peters and Campbell, 1991). In general, parthenogenetic eggs develop more slowly, causing a delayed female bias in the sex ratio of the nymphs which is perpetuated through the life cycle. For example Harker (1997) found sex ratios of *Cloeon similae* (Baetidae) increased from 1:1 in the summer to 2:1 in Spring and early Winter over 13 consecutive years due to a longer development time for the unfertilised eggs and subsequent late appearance of parthenogenetic progeny (females). Data for the present study were collected from a wide range of sites over two years, so it is possible that the observed sex ratio actually reflects what occurs in nature. As far as I am aware there are no previous records of parthenogenesis occurring in Australian Leptophlebiidae.

REFERENCES

Brittain, J. E. 1982. Biology of mayflies. Ann. Rev. Ent. 27: 119-147.

Buikema, A. L. (Jr.) and J. R. Voshell (Jr.). 1993. Toxicity studies using freshwater benthic macroinvertebrates, pp. 344-398 In: D. M. Rosenberg and V. H. Resh (eds.). Freshwater biomonitoring and Benthic macroinvertebrates. Chapman and Hall, New York and London.

Campbell, I. C. 1983. Studies on the Taxonomy and Ecology of the Australian Siplonuridae and Oligoneuriidae (Insecta: Ephemeroptera) PhD Thesis, Monash University, Mebourne, Australia.

Campbell, I. C. and P. J. Suter. 1988. Three new genera, a new subgenus and a new species of Leptophlebiidae (Ephemeroptera) from Australia. J. Aust. ent. Soc. 27: 259-273.

Corkum, L. D. 1978. The nymphal development of *Paraleptophlebia adoptiva* (McDunnough) and *Paraleptophlebia mollis* (Eaton) (Ephemeroptera: Leptophlebiidae) and the possible influence of temperature. Can J. Zool. 56: 1842-1846.

Craig, D. A. 1966. Techniques for rearing stream dwelling organisms in the laboratory. Tuatara 14: 65-92.
Day, W. C. 1968. Ephemeroptera, pp. 79-105. In: R. L. Usinger (ed.) 3rd edn. Aquatic Insects of California. University of California Press, Berkeley and Los Angeles.
Dean, J. C. 1988. Description of a new genus of Leptophlebiid mayfly from Australia (Ephemeroptera: Leptophlebiidae: Atalophlebiinae). Proc. roy. Soc. Vict. 100: 39-45.
Dean, J.C. 1997. Descriptions of a new genus of Leptophlebiidae (Insecta: Ephemeroptera) from Australia. I. *Tillyardophlebia* gen. nov. Mem. Mus. Vic. 56(1): 83-89.
Demoulin, G. 1955. Note sur deux nouveaux genres de Leptophlebiidae d'Australie. Bull. Ann. Soc. r. Ent. Belg. 91: 227-229.
Eaton, A. E. 1881. An Announcement of new genera of the Ephemeridae. Entomol. Mon. Mag. 17: 191-197.
Edmunds, G. F. (Jr.), S. L. Jensen and L. Berner. 1976. The Mayflies of North and Central America. University of Minnesota Press, Minneapolis.
Elliott, J. M. 1978. Effect of temperature on the hatching time of eggs of *Ephemerella ignita* (Poda) (Ephemeroptera: Ephemerellidae). Freshw. Biol. 8: 51-58.
Fremling, C. R. 1967. Methods for mass-rearing *Hexagenia* mayflies (Ephemeroptera: Ephemeridae) Trans. Amer. Fish. Soc. 96: 407-410.
Harker, J. E. 1954. The Ephemeroptera of eastern Australia. Trans. r. ent. Soc. Lond. 105: 241-268.
Harker, J. E. 1997. The role of parthenogenesis in the biology of two species of mayfly (Ephemeroptera) Freshw. Biol. 37: 287-297.
Hynes H. B. N. 1941. The taxonomy and ecology of the nymphs of British Plecoptera with notes on the adults and eggs. Trans. R. ent. Soc. Lond. 91(10): 459-557.
Hynes, H. B. N. 1970. The Ecology of Stream Insects. Ann. Rev. ent. 15: 25-42.
Mason, W. T. and Lewis, P. A. 1970. Rearing devices for stream insect larvae. Prog. Fish-Cult. 32(1): 61-62.
McCafferty, W. P. and Huff, B. L. 1974. Parthenogenesis in the mayfly *Stenonema fermoratum* (Say) Ephemeroptera: Heptageniidae. Ent. News 85: 76-80.
Merritt, R. W, K. W. Cummins and V. H. Resh. 1996. Design of Aquatic Insect Studies: collecting, sampling and rearing procedures, pp. 12-28. In: R. W. Merritt and K. W. Cummins (eds.). An Introduction to the Aquatic Insects of North America. 3rd edn. Kendall Hunt Publishing Co, Iowa.
Minshall, G. W. 1988. Stream ecosystem theory: a global perspective. J. N. Amer. Benthol. Soc. 7(4): 263-288.
Navás, L. 1918. Insectos chilenos. Bol. Soc. Arag. Cienc. Nat. 17: 212-230.
Newbold, J. D, B. W. Sweeney and R. L. Vannote. 1994. A model for seasonal synchrony in stream mayflies. J. N. Amer. Benthol. Soc. 13 (1): 3-18.
Peters, W. L. and I. C. Campbell. 1991. Ephemeroptera, pp. 279-293. In: The Insects of Australia. 2nd edn. Melbourne University, Press, Melbourne.
Power, M. E., R. J. Stout, C. E. Cushing, P. P. Harper, F. R. Hauer, W. J. Matthews, P. B. Moyle, B. Statzner and I. R. Wais de Badgen. 1988. Biotic and abiotic controls in river and stream communities. J. N. Amer. Benthol. Soc. 7(4): 456-479.
Schneider, R. F. 1967. An Aquatic Rearing Apparatus for Insects. Turtox News 44: 90.
Smock, L. A. 1996. Macroinvertebrate movements: drift, colonisation, and emergence, pp. 371-390. In: F. R. Hauer and G. A. Lamberti (eds.). Methods in Stream Ecology. Academic Press, San Diego.
Southwood, T. R. E. 1978. Ecological Methods - with particular reference to the study of Insect Populations. Chapman and Hall, London.
Speith H. T. 1938. A method of rearing *Hexagenia* nymphs (Ephemeridae) Ent. News 49 (2): 29-32.
Suter. P. J. and J. E. Bishop. 1990. Post-oviposition development of eggs of South Australian mayflies, pp. 85-94. In: I. C. Campbell (ed.). Mayflies and Stoneflies: Life History and Biology. Kluwer Academic Publishers, Dordrecht.
Sweeney, B. W. and R. L. Vannote. 1978. Size variation and the distribution of hemimetabolous aquatic insects: two thermal equilibrium hypotheses. Science 200: 444-446.
Thornhill, R. and J. Alcock. 1983. The Evolution of Insect mating Systems. Harvard University Press, Cambridge.
Vogel, S. and M. LaBarbera. 1978. Simple flow tanks for research and teaching. BioScience 28(10): 638-643.
Wallace, J. B. and N. H. Anderson. 1996. Habitat, Life History and Behavioural adaptations of Aquatic Insects, pp. 41-73. In: R. W. Merritt and K. W. Cummins (eds.). An Introduction to the Aquatic Insects of North America. 3rd edn. Kendall Hunt Publishing Co, Iowa.
Ward, J. V. and J. A. Stanford. 1982. Thermal responses in the evolutionary ecology of aquatic insects. Ann. Rev. ent. 27: 97-117.
Zamora-Muñoz, C., A. Sánchez-Ortega and J. Alba-Tercedor. 1993. Physio-chemical factors that determine the distribution of mayflies and stoneflies in a high mountain stream in southern Europe (Sierra Nevada, Southern Spain). Aquat. Insects 15(1): 11-20.

A COMPARISON OF METHODS FOR ANALYSIS OF A LONG AQUATIC INSECT LIFE HISTORY: *PTERONARCYS CALIFORNICA* (PLECOPTERA) IN THE CROWSNEST RIVER, ALBERTA

G. D. Townsend and G. Pritchard

Department of Biological Sciences
The University of Calgary, 2500 University Drive N.W.
Calgary, Alberta, Canada T2N 1N4

ABSTRACT

Larval development time for *Pteronarcys californica* was estimated by 1) Multifan™, a software programme designed to provide an objective interpretation of size-frequency data, and 2) a growth simulation model based on larval growth rates in the laboratory and temperature data from the field. The results were compared with the interpretation obtained by visual inspection of size-frequency distributions (Townsend and Pritchard, 1998). The three methods gave different results. Multifan™ did not incorporate extended larval recruitment, cohort splitting, diapause, step-wise growth due to moulting, and different male and female growth patterns. The growth simulation method did not incorporate larval diapause, and laboratory rearing probably underestimated field growth rates.

INTRODUCTION

The length of aquatic insect life cycles and the number of generations completed each year (voltinism) are often determined by visual inspection of serial size-frequency distributions, a method that includes a degree of subjectivity, and the results are often not clear-cut. Sample size, sampling interval, variation in growth rates between individuals and between sexes, as well as taxonomic problems are some factors that can complicate the analysis (Butler, 1984). Only synchronous, univoltine life histories, with a short annual period of adult emergence and oviposition, lend themselves to unequivocal interpretation.

We attempted an objective estimate of the length of larval life of the large, slow-growing stonefly, *Pteronarcys californica* Newport, in the Crowsnest River, Alberta, Canada, by using two other methods, and we compared the results with an interpretation based on visual inspection of serial size-frequency distributions (Townsend and Pritchard, 1998). Multifan™ (Fournier *et al.*, 1990) is a software programme designed to interpret serial size-frequency data from fish populations. Multifan™ was developed to avoid the use of costly procedures (such as tagging or measuring scales or otoliths) to estimate age of fish, and to make objective corrections for samples in which one or more age-classes are absent or poorly represented. Multifan™ has been used to analyse multiple sets of length-frequency data from fish populations (Fournier *et al.,* 1990, Terceiro *et al.,* 1992, Labelle *et al.*, 1993), a shrimp population (Fournier *et al.*, 1991), and a squid population (Welch and Morris, 1992). Our second alternative approach was rearing of larvae at constant temperatures in the laboratory, and we then used observed growth rates to simulate growth at field temperatures.

STUDY SITE

Field data were collected from a 1 km stretch of the third-order Crowsnest River in the Rocky Mountain foothills of southwestern Alberta (49°35'N, 114°11'W). A Ryan Model-J thermograph was anchored to the substrate for the duration of the study. We read temperatures from the graph paper to the nearest 0.25°C at 2h intervals. Monthly mean temperatures and monthly and annual day-degrees (d°) above 0°C were higher in 1992 than in 1993 (d° accumulated May-November 1992 = 2112, May-November 1993 = 1814).Townsend and Pritchard (1998) give more details.

METHODS

Larval Collection

Larvae were collected monthly during the ice-free periods (April to November) of 1992 and 1993. We used a kick net with an inner 1 mm mesh bag and an outer 0.2 mm mesh bag. Larvae were sorted under a dissecting microscope, and head-capsule widths (across the eyes) and mesothoracic wing-pad lengths were measured to the nearest 0.05 mm with the ocular scale.

Visual Inspection of Size-Frequency Distributions

Frequency distributions for each sample date were plotted from head-capsule widths in 0.15 mm intervals and wing-pad lengths in 0.50 intervals. The two sexes were plotted separately. We then followed the life history by visually tracing the movement of age-classes in the size-frequency plots through time. This procedure was aided by fixing the number of generations in Multifan™ plots. Townsend and Pritchard (1998) should be consulted for further details.

Analysis of the Size-Frequency Distributions by Multifan™

The Multifan™ model provides a relatively objective means of evaluating size-frequency distributions. Fournier *et al.* (1990) give a detailed description. Multifan™'s basic assumptions are that sizes in each age-class are normally distributed in field data, that the standard deviation is identical in all cohorts, and that mean sizes for each age lie on a von Bertalanffy Type I growth curve (von Bertalanffy, 1957). However, three basic modifications are usually incorporated (Labelle *et al.*, 1993): a correction for sampling bias for the first cohort in which individuals may be so small that they are not properly represented; an age-dependent standard deviation that allows the width of a mode to increase or decrease with age; seasonal variation in growth rates. Also, the procedure must be modified for aquatic insects to allow final instar larvae to leave the water as adults, because the default in Multifan™ is for survivors to accumulate in the largest size-class.

As age-classes are defined by modes in the frequency distributions, it is important that the modes accurately represent age-classes. Thus, the size-class interval must be carefully chosen. An interval that is too small will exaggerate the number of age-classes, and one that is too large will combine age-classes. The recommended class-width is <1.0 standard deviation around the mean of an evident cohort determined from the raw data (D.A. Fournier, Otter Software, P.O. Box 265, Nanaimo, B.C., pers. comm.). The standard deviation of the recruitment cohort in July 1992 was 0.2, and so we used 0.15 mm as the class-width, because our measuring interval was 0.05 mm. Given that Multifan™ is designed to save time, we did not separate the sexes in this analysis, although we did remove final instar females, which stood out as a clearly observable mode.

Multifan™ is not completely objective. The user must at least define a sample in which a single pulse of recruitment occurs, the modal size of that recruitment class, and place quite wide bounds on that modal size to allow for deviations of actual modal sizes from an exact fit to the von Bertalanffy growth curve. We used the sample from July 1992 as the recruitment sample, and removed individuals in the smallest size-class from the three previous months. This represents minimal user input. Other user inputs may be made and all may be applied to any age-class, but clearly the more made the less objective the procedure becomes. We systematically fitted all possible combinations of the three basic modifications to the field data and allowed Multifan™ to estimate the best fit. We ran the analysis with the recruitment class defined (level 1) and with the proportion in the recruitment class estimated (level 2). Then we defined a modal length for final instar males in May 1992, set bounds on this, and defined the proportion of the total sample in this class (level 3). We then repeated this procedure and added a restriction on the largest-sized individuals in June 1992 after the previous cohort had emerged, defining a modal length, and establishing bounds and proportion in the class (level 4). Multifan™ selects the best fit to the data and may ignore these additional user inputs if they lead to poorer fits.

Growth Rates of Larvae at Constant Temperatures and Simulation at River Temperatures

To determine the response of larval growth rate to temperature, we reared larvae with head widths ranging from 0.42 mm (first instar) to 5.00 mm (final instar females) at nominal constant temperatures of 5.0, 7.5, 10.0, 12.5, 15.0, 17.5 or 20.0°C, and a photoperiod of 16L:8D. We kept larvae in 300 ml jars containing stream water, and an alga-encrusted stone provided a substrate. The water was aerated in jars above 10°C. Larvae were fed conditioned *Alnus* and *Populus* leaves. We replenished larval food and changed the water twice weekly for temperatures >10.0°C and once weekly for temperatures ≤10.0°C. Food was available in excess in all treatments. Every day we recorded the temperature and checked larvae for moulting. We measured head-capsule width and wing-pad length on each larva one day after it moulted, and recorded the period between moults for each larva until it died or emerged. The experiment was continued until some larvae of most sizes had moulted in each treatment. Although each larva was kept at the same nominal temperature throughout the experiment, different larvae spent different periods of time in that temperature. And because temperatures varied a little from day to day in each chamber, each larva experienced a slightly different temperature regime. However, no overall daily mean temperature differed by more than 0.3°C from the nominal temperature, and most were considerably closer. The temperature variation about the means was also small, standard errors being mainly <0.5°C.

For each treatment, mean specific growth rates (G) were calculated for each instar with the function:

$$(1) \qquad G_{(n)} = \frac{L_{(n+1)} - L_{(n)}}{T_{(n)}} \cdot \frac{2}{L_{(n+1)} + L_{(n)}}$$

where, $G_{(n)}$ = percent growth rate of instar n (mm mm^{-1} d^{-1}),
$L_{(n)}$ = mean head capsule width of instar n (mm),
$L_{(n+1)}$ = mean head capsule width of instar $n+1$ (mm),
$T_{(n)}$ = mean developmental period of instar n (d).

A Mixed Model ANCOVA was used to determine the effect of temperature on larval growth rate. The analysis also differentiated between growth rates of larvae that moulted only twice (and so yielded only one data point) and those that moulted three or more times (yielding two or more non-independent data points). Larval size was the covariate.

At constant temperatures an animal's growth rate curve reflects enzyme dynamics, rising logistically to a maximum then dropping steeply (Cossins and Bowler, 1987). We used a two-stage growth function to simplify this curve. The first stage is described by a linear equation of the form:

$$G = a + bTemp \quad (2)$$

where, G = mean specific growth rate (mm mm^{-1} d^{-1}),
a = a constant,
b = the regression coefficient for temperature,
$Temp$ = temperature (°C).

The second stage is a constant growth rate for all river temperatures that exceed the temperature at which growth rate is maximum. The *Temp* term in Equation (2) is fixed at the thermal maximum irrespective of actual temperature. This assumes that river temperatures above the thermal maximum occur for only very short periods and do not slow larval growth rates. Thus, our model differs from that of Newbold *et al.* (1994), which assumes quiescence at temperatures above the thermal maximum.

Size must be incorporated into the equation, because insect larvae of different sizes grow at different rates at the same temperature:

$$G = a + bTemp + cSize \quad (3)$$

where, b = partial regression coefficient for temperature,
c = partial regression coefficient for size,
$Size$ = head capsule width (mm).

To predict the time required to complete larval growth in the river, we substituted river temperature sequences at two-hour intervals in Equation (3). A simulation was started in each ice-free month, April through November, with first instar larvae (0.42 mm head capsule width). Because larvae with head capsule widths >4.00 mm did not moult in the laboratory, the simulation was run until a head capsule width of 4.00 mm was reached. We ran the simulation with warm years (1992 temperatures), cool years (1993 temperatures), and alternating warm and cool years, starting with a warm year then with a cool year, until growth was complete.

RESULTS

Visual Inspection of Larval Size-Structure Data

Townsend and Pritchard (1998), in a detailed study using changes in head capsule width and wing pad length, and timing of larval recruitment and adult emergence, determined that *Pteronarcys californica* spends four years in the larval stage in the Crowsnest River. Female larvae grow at a faster rate than males and so achieve a larger size at emergence. Extended larval recruitment leads to variation in the sizes of larvae of the same year-class and probably to cohort splitting. Adult emergence is highly synchronous in May, perhaps as a result of a summer diapause in antepenultimate instar larvae. Eggs hatched over periods of 130-322 days at different temperatures in the laboratory (Townsend and Pritchard, 2000), and over an 11-

month period in the field, with the pulse of larval recruitment in the field population occurring between April and August, 11 to 15 months after oviposition (Townsend and Pritchard, 1998).

Multifan™ Analysis of Larval Size-Structure Data

Multifan™ found 10 year-classes in the field data when only the recruitment mode was defined (levels 1 and 2). Definition of the emergence cohort (level 3) reduced the estimated number of year-classes to eight (Fig. 1). Attempts to further constrain the size variation in older larvae (level 4) led to an increase in the estimated number of age-classes back to 10.

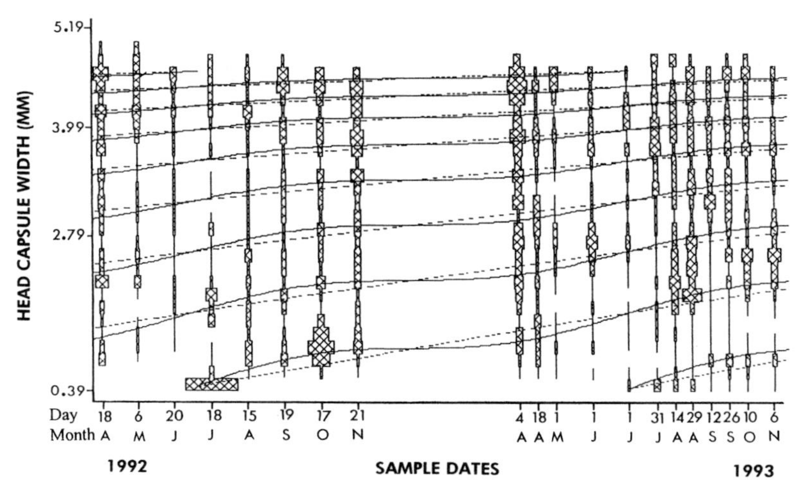

Fig. 1. Frequency distributions of larval sizes in the Crowsnest River *P. californica* population (final instar females removed) with the Multifan™ best fit estimate of year-classes. Dashed lines define estimated size-class modes when no seasonality in growth rate is incorporated; continuous lines incorporate slower growth in winter.

Growth Rates of Larvae at Constant Temperatures

Larval growth rate changed non-linearly with size at each temperature (Fig. 2), and so we used the natural logarithm of size in the analyses. The residual plots for the effects on growth rate of *lnSize, Temp*, and use of several data points from the same individual, showed that the data were normally distributed. Therefore, no further data transformation was necessary. In the initial run of the Mixed Model ANCOVA, the *lnSize* x *Temp* interaction was non-significant ($F_{1,17} = 2.891$; $P = 0.11$). Therefore, we removed the interaction term and ran the ANCOVA again. For this test there was a significant *lnSize* effect and a significant *Temp* effect. However, there was no significant difference between growth rates of individuals that yielded one data point and individuals from which more than one data point was obtained (Table 1).

Growth rates decreased above 15°C, except in first instar larvae (Fig. 2). Therefore, we used 15°C as the temperature separating the two stages of the growth model. The first stage was then described by the linear equation ($r^2 = 0.65$):

$$G = 0.000096 + 0.00027Temp - 0.00164lnSize \quad (4)$$

Because the linear equation for the growth rate simulation includes a size component, developmental zero varies with size. Larvae with a head capsule width of 0.42 mm have a predicted developmental zero of -5.65°C, and larvae with a head capsule width of 4.00 mm have a predicted developmental zero of 8.11°C.

Table 1. ANCOVA of the effects of temperature (Temp), natural logarithm of size (lnSize), and collection of several data points from the same individual (Ind), on *P. californica* larval growth rate, with lnSize as the co-variate.

Source of variation	DF	MS	F	P > F
Temp	6	3.456×10^{-6}	5.932	0.0001
lnSize	1	4.493×10^{-6}	8.925	0.0079
Ind	62	6.047×10^{-7}	1.201	0.3436

Growth Simulation

We first simulated growth rates under different temperature regimes with eggs hatching in April, the first of four consecutive months of the pulse of larval recruitment in the field (Townsend and Pritchard 1998). The shortest period predicted for development from hatching to a head width size of 4 mm was 5.2 yr, and this occurred with 1992 temperatures (a warm regime). The longest period of 6.4 yr occurred with 1993 temperatures (a cool regime). Starting a simulation of alternating yearly river temperature sequences with either a warm or

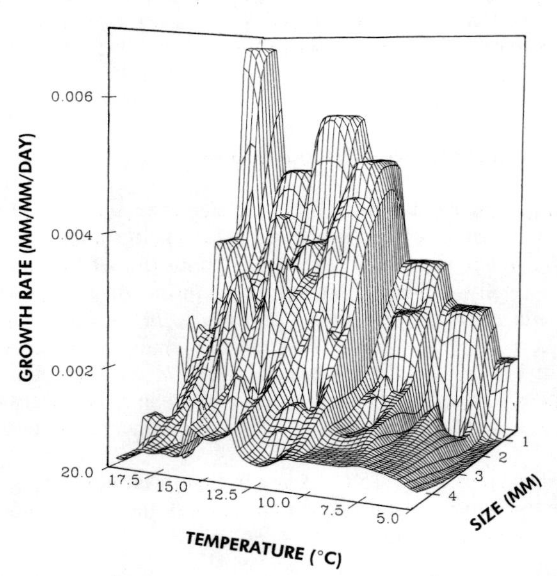

Fig. 2. Relationships between *P. californica* larval growth rate, temperature, and size, in laboratory experiments at constant temperatures.

cool year resulted in only a two-month difference in the time required to attain a 4 mm head width (5.33 yr compared with 5.50 yr). However, there was a 7 mo difference between larvae that entered a system of alternating cool and warm years in April or May and those that entered in June-November (Fig. 3). This was because the June-November hatch required an extra winter when no growth occurred. The model predicted that larvae recruited in April and May would reach 4 mm head width in August and September five years later. With addition of the remaining development, adults would not emerge until May six years after egg hatch. Larvae recruited in June-November reached 4 mm head width in May-October six years later, and would not emerge until May at least seven years after egg hatch.

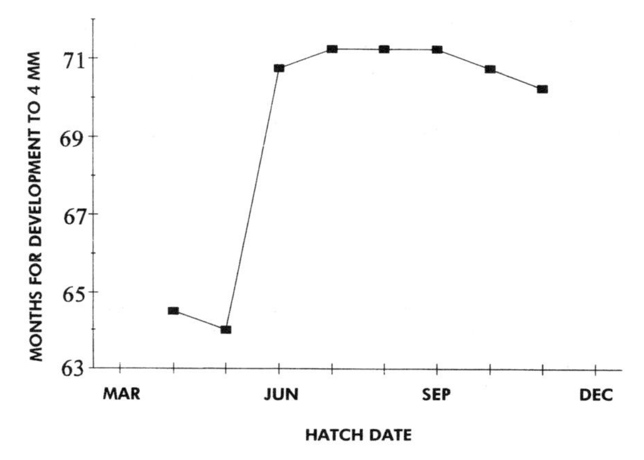

Fig. 3. Growth simulation model estimates of time for *P. californica* larval development from hatching to a head-capsule width of 4 mm. Eggs hatch on the first day of each ice-free month. A temperature regime of alternating cool (1993 temperatures) and warm (1992 temperatures) years was followed.

DISCUSSION

The principal problem with Multifan™ for analysis of insect life cycles is that individual larvae cannot be precisely aged and so the model's predictions cannot be objectively assessed. Furthermore, Multifan™'s inflexibility once the parameters have been entered limits its application to insect populations. With the complexity of the life history of *Pteronarcys californica*, involving extended egg hatch, diapause, variation in individual growth rates, cohort splitting, and sexual dimorphism, total objectivity is impossible. One reason for developing Multifan™ was to save effort in the analysis of fish populations. An analogy with measurement of otoliths is determination of sex of insect larvae. This is rarely possible in holometabolous insects, and then only with considerable effort (e.g. Pritchard, 1976). It is easier in hemimetabolous insects such as *P. californica*, but is still time consuming. Therefore, except for separating the sexes in the final larval instar, we ran Multifan™ with the sexes combined. However, sexual size differences are clearly picked up by Multifan™. A further problem is that fish growth is continuous, whereas growth of external dimensions in insects is step-wise. This will average out in the long run but will lead to difficulties in fitting a growth model in the short term (see Townsend and Pritchard, 1998).

One indication that Multifan™ is not correctly tracking cohorts is that the modes become closer together as the population ages (see Fig 2). This is in part because differences between the sexes increase with increasing size, and Multifan™ is partitioning this variation. More important is the inapplicability of Type I von Bertalanffy growth to insects. This type of growth, which applies well to fish, plots increase in linear dimensions as a monotonic rise to a steady state. Thus, the model searches for ever-decreasing size differences over unit time intervals. However, increase in linear dimensions of insects is sigmoid over time, with a long inflection, and so size differences increase or remain steady over time if temperature remains constant and diapause is not present. Of course, these latter two constraints are rarely realized in insect populations, and a simple growth curve can be fitted only over very short portions of an insect's life cycle in the field.

In theory the growth rate simulation model should give the best estimate of growth rates in the field. However, although the diet used for laboratory rearing was sufficient for larval growth, it probably did not compare favourably with food in the river and a larval life of 6-7 years is certainly an over-estimate. *In situ* growth experiments in the field would perhaps alleviate some of the problems of laboratory studies, but the logistical problems of keeping track of individual larvae in a fast-flowing river that is frozen for five months of the year are prohibitive.

We conclude that these alternative methods, while giving some insight into complex aquatic insect life histories, are not substitutes for careful visual analysis of size-frequency data.

ACKNOWLEDGMENTS

We thank the Natural Sciences and Engineering Research Council of Canada, the Alberta Recreation, Parks and Wildlife Foundation, and the Faculty of Graduate Studies, University of Calgary, for research grant support.

REFERENCES

Bertalanffy, L. von. 1957. Quantitative laws in metabolism and growth. Quart. Rev. Biol. 32: 217-231.

Butler, M. G. 1984. Life histories of aquatic insects, pp. 24-55. In: V. H. Resh and D. M. Rosenberg (eds.). The Ecology of Aquatic Insects. Praeger Publishers, New York.

Cossins, A. R. and K. Bowler. 1987. Temperature Biology of Animals. Chapman and Hall, London.

Fournier, D. A., J. R. Sibert, J. Majkowski and J. Hampton. 1990. Multifan a likelihood-based method for estimating growth parameters and age composition from multiple length frequency data sets illustrated using data for southern bluefin tuna (*Thunnus maccoyii*). Can. J. Fish. Aquat. Sci. 47: 301-317.

Fournier, D. A., J. R. Sibert and M. Terceiro. 1991. Analysis of length frequency samples with relative abundance data for the Gulf of Maine northern shrimp (*Pandalus borealis*) by the Multifan method. Can. J. Fish. Aquat. Sci. 48: 591-598.

Labelle, M., J. Hampton, K. Bailey, T. Murray, D. A. Fournier and J. R. Sibert. 1993. Determination of age and growth of South Pacific albacore (*Thunnus alalunga*) using three methodologies. Fish. Bull. 91: 649-663.

Newbold, J. D., B. W. Sweeney and R. L. Vannote. 1994. A model for seasonal synchrony in stream mayflies. J. N. Amer. Benthol. Soc. 13: 3-18.

Pritchard, G. 1976. Growth and development of larvae and adults of *Tipula sacra* Alexander (Insecta: Diptera) in a series of abandoned beaver ponds. Can. J. Zool. 54: 266-284.

Terceiro, M., D. A. Fournier and J. R. Sibert. 1992. Comparative performance of Multifan and Shepherd's Length Composition Analysis (SRLCA) on simulated length-frequency distributions. Trans. Amer. Fish. Soc. 121: 667-677.

Townsend, G. D. and G. Pritchard. 1998. Larval growth and development of the stonefly *Pteronarcys californica* (Insecta: Plecoptera) in the Crowsnest River, Alberta. Can. J. Zool. 76: 2274-2280.

Townsend, G. D. and G. Pritchard. 2000. Egg development in the stonefly *Pteronarcys californica* Newport (Plecoptera: Pteronarcyidae). Aquat. Insects 22: 19-26.

Welch, D. W. and J. F. T. Morris. 1992. Age and growth of flying squid (*Ommastrephes bartrami*). N. Pac. Com. Bull. 53: 183-190.

DISTRIBUTION IN POLAND OF SPECIES OF THE *BAETIS* GROUP (EPHEMEROPTERA, BAETIDAE)

Teresa Jażdżewska

Department of Invertebrate Zoology and Hydrobiology
University of Lódz
12/16 Banacha st., 90-237 Lódz, Poland

ABSTRACT

In the present paper the regions of Poland where investigations on Ephemeroptera were carried out are shown and, against this background, the distribution of 19 species of the *Baetis* group is presented. *B. vernus*, *A. muticus*, *B. rhodani* and representatives of *B. fuscatus* group are very common, widely distributed and usually abundant species, whereas *B. pentaphlebodes*, *N. digitatus* and *B. tracheatus* are rare and never abundant. Distributional maps are based on critically examined literature and the author's unpublished data; a Universal Transverse Mercator (UTM) grid was used for presentation of species distribution.

INTRODUCTION

Studies on the diversity of living organisms in the last decade are gaining more and more attention among students of biology (Ricklefs and Schluter, 1993; Hawksworth, 1996) and even among politicians, receiving official support during major international meetings (see Rio - "The Earth Summit"; Systematics Agenda 2000, Charting the Biosphere, 1994).

Studies on species distributions and their possibly universal presentation are a base for biodiversity monitoring and regional and global biogeographic considerations (Ruffo, 1996).

Depending on the level of knowledge of national faunas, the results of biodiversity studies are published as ample faunistic lists like those edited in Italy by Minelli et al. (1993-95) or in Poland by Razowski (1990-1997), or as distribution maps published by "Centre suisse de cartographie de la faune" (CSCF) in Switzerland (Lubini et al., 1996). In Poland, inventory activity is successfully realized by the Museum and Institute of Zoology, Polish Academy of Sciences (MIZ, Pol. Acad. Sci.) in Warsaw as successive volumes of *Katalog fauny Polski* (*Catalogus faunae Poloniae*).

The first studies on Ephemeroptera in Poland were undertaken some 150 years ago, but only in the last four decades have these studies become intensive, covering most of the country. Sowa (1990) has published a comprehensive list of Polish mayflies that includes 120 species; out of these species 21 were placed in the genus *Baetis* Leach. However during the last decade new related genera were established in several revisions (i. e. Novikova and Kluge 1984, 1987: Waltz et al., 1994; McCafferty and Waltz, 1995). In the present paper the distribution in Poland of 19 mayfly species is presented; until recently these species were treated in Polish literature as *Baetis* Leach. Now some of them are placed in another genera *Acentrella* Bengtsson, *Alainites* Waltz and McCafferty and *Nigrobaetis* Novikova and Kluge.

Baetis larvae are very common and usually abundant freshwater animals; in Carpathian mayfly collection of Sowa (1975a) they constituted ca 54% of the material, in Mazurian collections of Lewandowski (1989) ca 77% and in Pomeranian collection of Glazaczow (1994) ca 69%. However, well known difficulties in species identification have led to the situation that many old data are not credible. The milestone in the knowledge of European *Baetis* larvae was a monograph by Müller-Liebenau (1969); only after its publication it was possible to verify part of the old information and, as a result, the knowledge of Polish *Baetis* species distinctly increased.

There are still, however, difficulties in the identification of larval stages in some species and many old data are doubtful. Therefore in this paper the distribution of very common larvae of the *Baetis fuscatus* group is presented jointly, without separation into *B. fuscatus* and *B. scambus*. Excluded are also two species of *Labiobaetis* Novikova et Kluge: *L. tricolor* (Tshern.) and *L. calcaratus* Keffermüller, because the materials available to the author indicate the necessity of further taxonomic studies on these species.

In this paper, the distribution of baetid species in Poland is presented using a Universal Transverse Mercator (UTM) grid according to the system propagated by the Faunistic Documentation Centre of MIZ, Pol. Acad. Sci. in Warsaw and widely used in Europe (Van Goethem and Grootaert, 1992). Critically verified literature information, as well as the author's unpublished data, were used.

Fig. 1 shows the distribution of the mayfly research effort in Poland against the background of basic physiographical units of this country according to Dylikowa (1973). These regions are approximately latitudinal belts running more or less parallel to the Baltic Sea coast in the north and to the Carpathian Mts in the south; altitude generally increases from the north to the south. The regions are: Pomerania and Mazuria, belt of Central Polish Lowlands (Great Poland, Mazovia and Silesia lowlands), belt of old mountains and uplands (Sudety Mts, Silesian Upland, Little Poland Upland with Swietokrzyskie Mts, Lublin Upland with Roztocze), belt of submontane lowlands (Oswiecim basin, Sandomierz basin) and the Carpathian Mts. In Figs. 2-11 the distribution of successive species is presented.

RESULTS AND DISCUSSION

Despite rather intensive searching for Ephemeroptera in Poland some regions remain understudied, for instance Mazovia and Silesia lowlands (Fig. 1).

The most common species, occurring in the whole country were *B. vernus*, *B. rhodani*, *A. muticus* and *B. fuscatus* group (Fig. 2, 3, 4 and 5). *B. vernus* (in some papers as *B. tenax*) in many studies appeared to be the most abundant *Baetis* species. The densiest network of *B. rhodani* and *A. muticus* localities occurred in southern Poland. *B. rhodani* was often the most abundant baetid species there. On the other hand, its comparatively rare occurrence in the otherwise well studied western part of the Central Lowlands, is striking. Larvae of *Baetis rhodani* and *B. vernus* were also collected outside running waters, namely in springs (Dratnal, 1976).

Baetis fuscatus group is formally represented in Polish fauna by three species: *B. fuscatus*, *B. scambus* and *B. beskidensis*. Larvae of two first species were noted in many regions; they occured often in common or in neighbouring localities. Considering the difficulties in the discrimination between these species many authors have treated them combined as *B. fuscatus* and *B. scambus*. If we might trust all published information *B. fuscatus* is more common and abundant than *B. scambus* in the regions north of the sub-

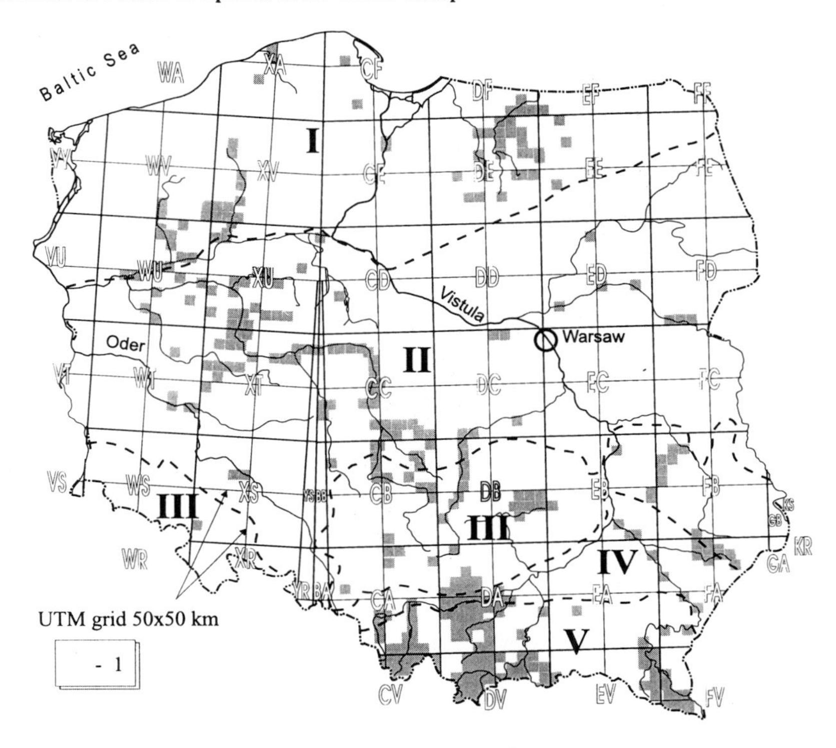

Fig. 1. Distribution of collecting locations for Ephemeroptera in Poland against the background of the main physiographic regions of this country [1-75].
I) Pomerania and Mazuria. II) Central Polish Lowlands. III) Old Mountains and Uplands. IV) Submontane Lowlands. V) Carpathians. 1) Collecting location.

montane lowlands. Winged forms were comparatively rarely reported. Imagines of *B. fuscatus* were collected in the Carpathians at the Raba river and Poprad river as well as in nothern Poland at the Vistula river and in the basins of the Warta and Gwda rivers (Mikulski 1929; Keffermüller 1960; Wójcik 1963; Sowa 1975a, 1975b; Glazaczow 1964). Winged forms of *B. scambus* were caught in the Carpathians by Sowa (1975a) and in western Pomerania by Glazaczow (1994).

The third species - *B. beskidensis* Sowa (Fig. 5) was described from Polish Carpathians by Sowa (1973) on the basis of larval material only. Besides the different colour pattern, other morphological features differentiating these larvae from akin species are not clear. All Polish localities of *B. beskidensis*, were reported by the author of this species or by his collaborators and are located in Polish Carpathians. The species is much less common and abundant than other species of *B. fuscatus* group. Further studies are needed to clarify the proper systematic position of this form.

Localities of *B. buceratus* and *N. niger* are scattered over the whole country; however these species are distinctly less common than other baetids and their abundances are usually low. *B. buceratus* was recorded mainly in large rivers: San, Vistula, Pilica and Warta.

NOTE:
Captions to illustrations: Numbers in brackets indicate papers used in preparation of particular maps; see "References".

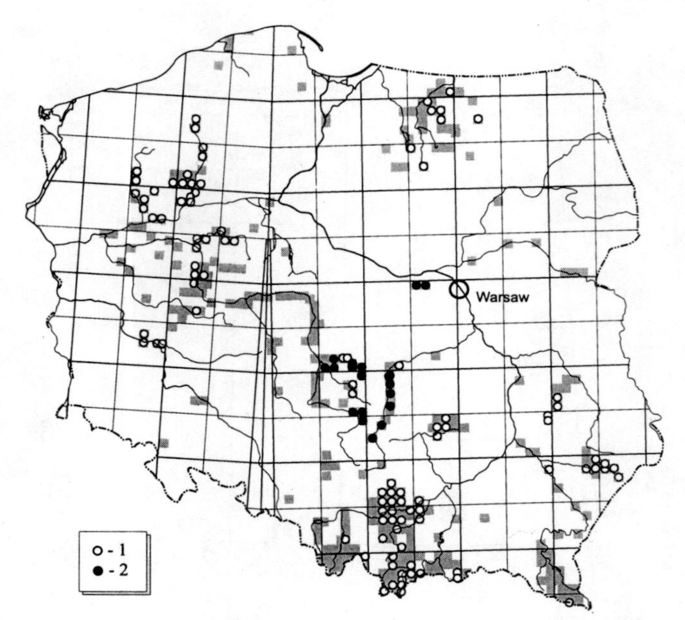

Fig. 2. Distribution of *Baetis vernus* Curtis [1, 2, 4, 5, 7, 9, 11, 15, 16, 17, 18, 19, 23, 33, 35, 37, 44, 45, 46, 48, 52, 53, 56, 62, 66, 70].
1) Literature data. 2) New author's data.

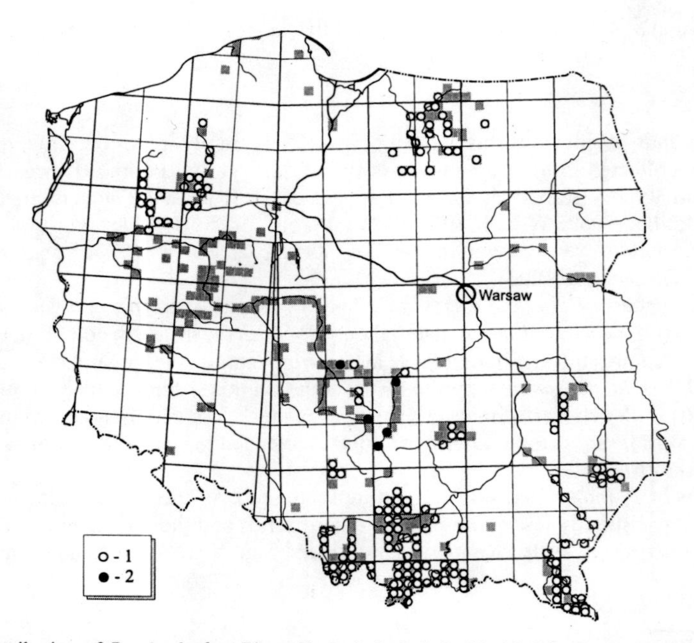

Fig. 3. Distribution of *Baetis rhodani* Pictet [1, 3, 4, 5, 6, 7, 9, 11, 15, 16, 17, 18, 30, 32, 33, 34, 35, 37, 38, 39, 40, 41, 42, 43, 44, 48, 51, 52, 56, 57, 60, 62, 66, 69, 70, 71, 73, 74, 75].
1) Literature data. 2) New author's data.

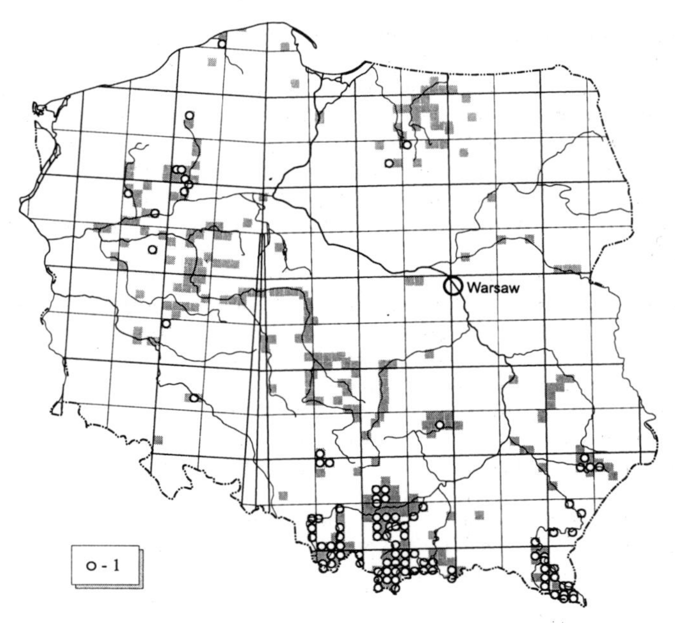

Fig. 4. Distribution of *Alainites muticus* (L.) [1, 4, 6, 7, 9, 15, 18, 29, 33-35, 37, 38, 40, 42-44, 51, 52, 56, 57, 60, 62, 66, 70, 74, 75].
1) Literature data.

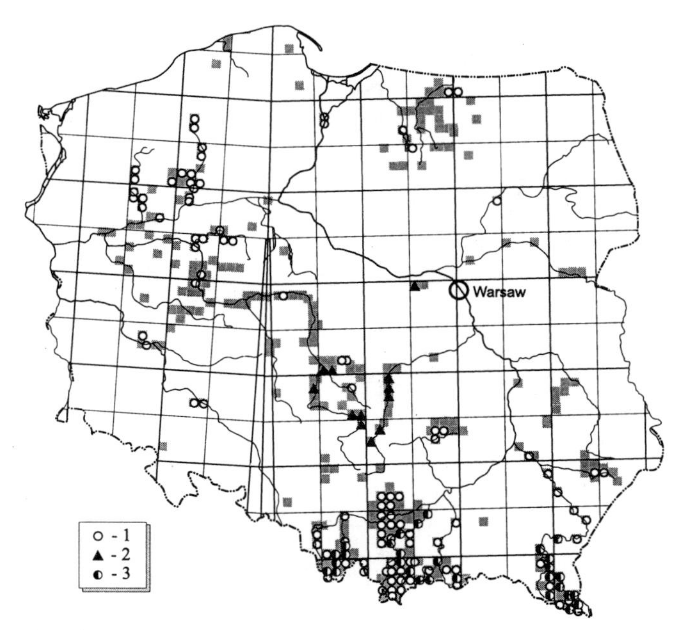

Fig. 5. Distribution of *Baetis fuscatus* and/or *Baetis scambus* [1, 4, 6-8, 11, 15-18, 20, 22, 33-35, 38, 43-45, 48, 52-54, 62, 66, 70, 72, 73] and *Baetis beskidensis* [4, 33, 65, 66, 70].
1) Distribution of *Baetis fuscatus* (L.) and/or *Baetis scambus* Eaton, literature data. 2) Distribution of *Baetis fuscatus* (L.) and/or *Baetis scambus* Eaton, new author's data. 3) Co-occurrence of *Baetis fuscatus* (L.) and/or *Baetis scambus* Eaton and *Baetis beskidensis* Sowa.

Separate group includes species clearly associated with montane regions. These are *B. beskidensis*, *B. alpinus*, *B. lutheri*, *B. vardarensis*, *B. melanonyx*, *N. gracilis* and *A. sinaica* (Figs. 5, 7, 8, 9, 10 and 11). The most common of them, *B. alpinus* (sometimes cited as *B. carpathica*), was also collected in lakes in the highest Carpathians (Tatra Mts). Larvae of *B. lutheri* (before 1968 known as *B. venustulus*), *B. vardarensis* and *B. melanonyx* were often very abundant. *A. sinaica*, *B. beskidensis* and *N. gracilis* were usually rare and not numerous. Localities of *A. gracilis* are concentrated in the eastern part of the Polish Carpathians, in the San river basin; this distribution is related to the south-east European range of this species.

A much smaller group of species is restricted to the lowlands. These are *B. pentaphlebodes*, *B. liebenauae*, *B. tracheatus*, and *N. digitatus* (Figs. 6, 7, 9 and 10). Rather recent descriptions of these species may be the reason for their still poorly recognized distribution. Identification of *N. digitatus* and *B. pentaphlebodes* is rather difficult and their reported localities in Poland are not numerous.

The richest area in *Baetis* species appears to be the Carpathian montane running waters; rather rich in species were also swift Pomeranian streams in north-western Poland running down the morainic hills. Such a tendency is in accordance with the well known altitudinal and habitat preferences of this mayfly group.

The present survey is a rather preliminary step; however this summarized state of knowledge allows us to indicate rare and possibly endangered species as well as the regions where studies on mayflies should be undertaken.

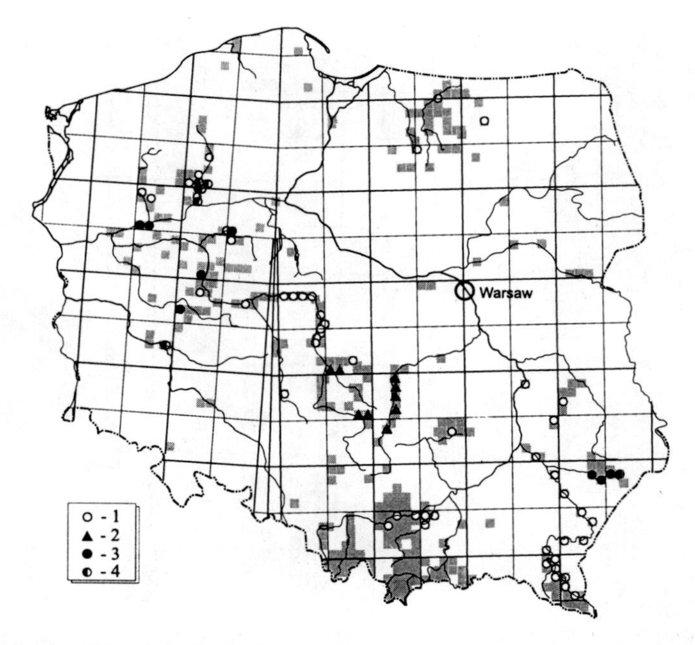

Fig. 6. Distribution of *Baetis buceratus* Eaton [5, 7, 9, 15, 17, 25, 44, 45, 53, 56, 61, 66] and *Baetis pentaphleboues* Ujhelyi [7, 18, 27, 45].
1) Distribution of *Baetis buceratus* Eaton, literature data. 2) Distribution of *Baetis buceratus* Eaton, new author's data. 3) Distribution of *Baetis pentaphlebodes*, literature data. 4) Co-occurrence of both species.

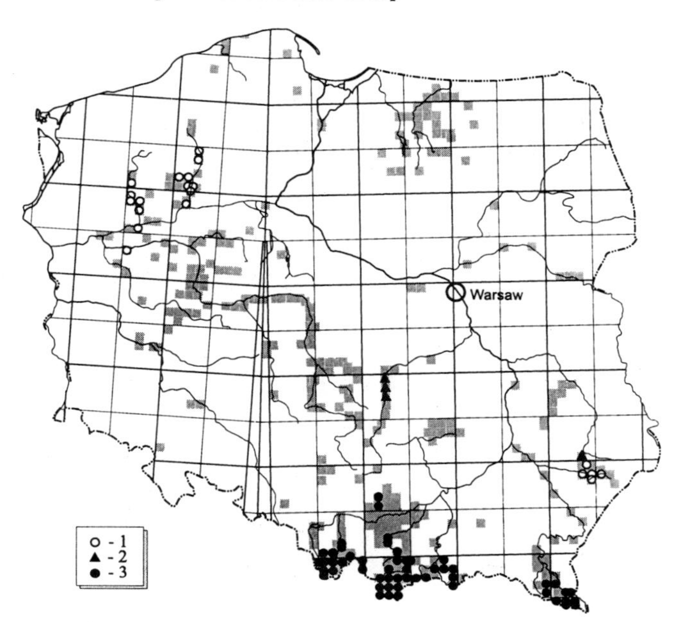

Fig. 7. Distribution of *Baetis liebenauae* Keffermüller [7, 18, 28] and *Baetis alpinus* Pictet [4, 9, 20, 22, 34-38, 43, 48, 50-52, 56, 62, 66, 69, 70, 73].
1) Distribution of *Baetis liebenauae* Keffermüller, literature data. 2) Distribution of *Baetis liebenauae* Keffermüller, new author's data. 3) Distribution of *Baetis alpinus*, literature data.

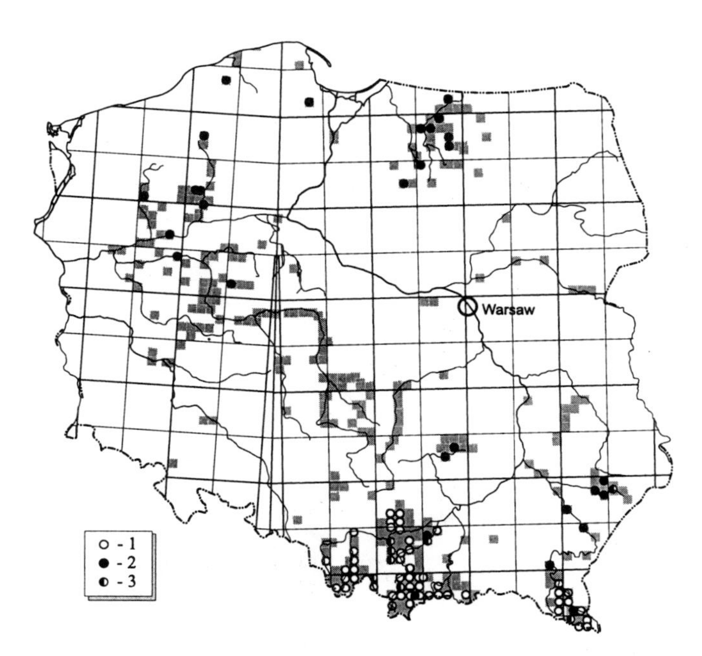

Fig. 8. Distribution of *Baetis lutheri* Müller-Liebenau [1, 4, 6, 9, 18, 33, 35, 38, 40, 43, 52, 56, 57, 62, 66, 70] and *Nigrobaetis niger* (Linnaeus) [7, 9, 15, 18, 19, 24, 44, 56, 61, 66].
1) Distribution of *Baetis lutheri* Müller-Liebenau, literature data. 2) Distribution of *Nigrobaetis niger* (Linnaeus), literature data. 3) Co-occurrence of both species.

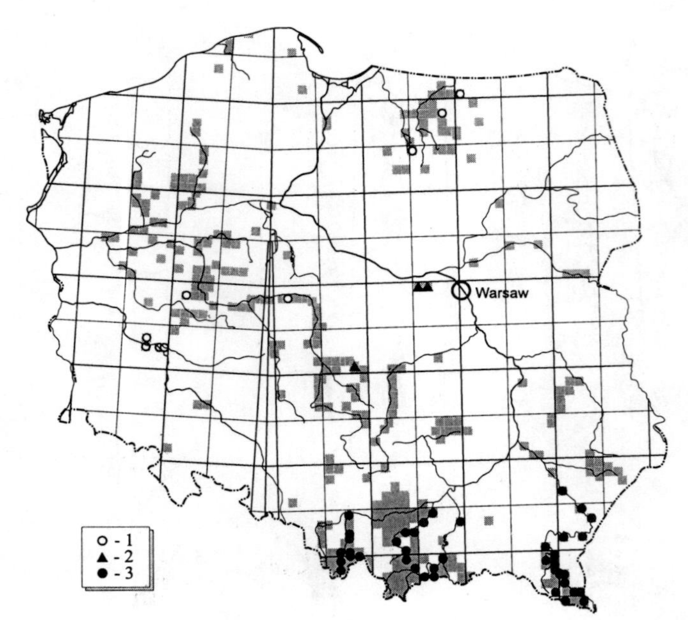

Fig. 9. Distribution of *Baetis tracheatus* Keffermüller et Machel [30, 44, 45] and *Baetis vardarensis* Ikonomov [4, 6, 32, 56, 66].
1) Distribution of *Baetis tracheatus* Keffermüller et Machel, literature data. 2) Distribution of *Baetis tracheatus* Keffermüller et Machel, new author's data. 3) Distribution of *Baetis vardarensis* Ikonomov, literature data.

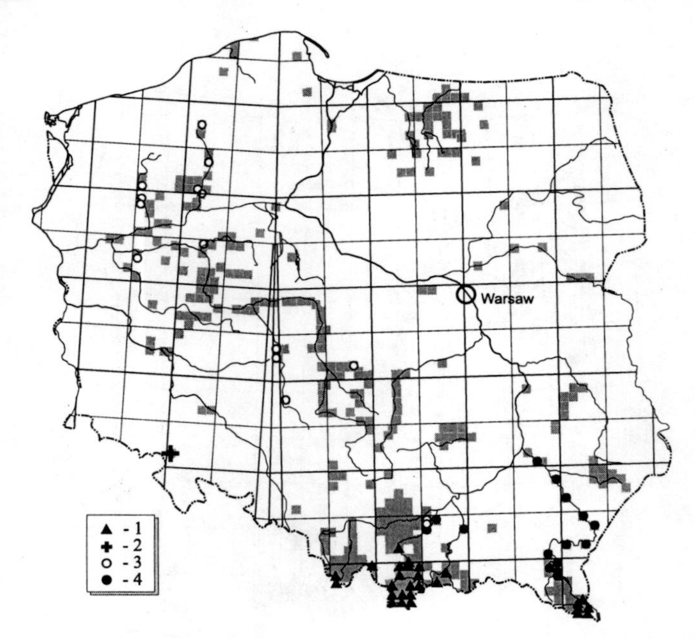

Fig. 10. Distribution of *Baetis melanonyx* (Pictet) [4, 20, 33, 34, 43, 52, 66], *Nigrobaetis digitatus* (Bengtsson) [7, 17, 19, 24, 29, 66,] and *Nigrobaetis gracilis* (Bogoescu et Tabacaru) [61, 66].
1) Distribution of *Baetis melanonyx* (Pictet), literature data. 2) Distribution of *Baetis melanonyx* (Pictet), new author's data. 3) Distribution of *Nigrobaetis digitatus* (Bengtsson), literature data. 4) Distribution of *Nigrobaetis gracilis* (Bogoescu et Tabacaru), literature data.

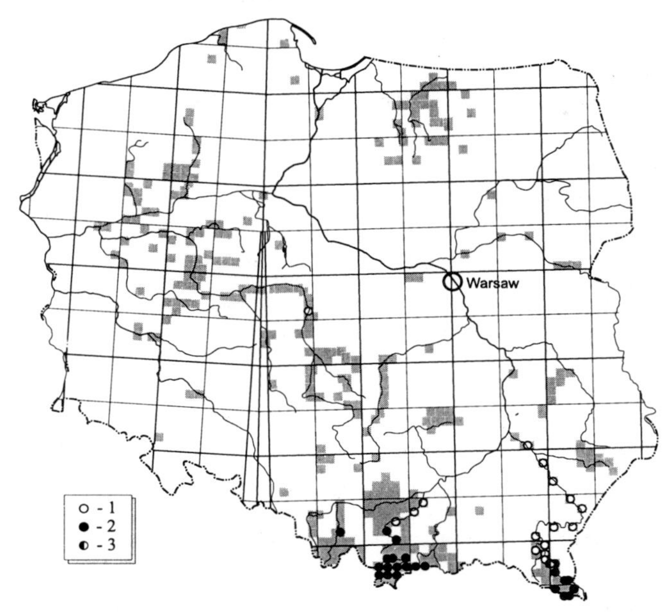

Fig. 11. Distribution of *Baetis inexpectatus* (Tshernova) [24, 61, 66] and *Acentrella sinaica* Bogoescu [4, 33-35, 43, 52, 61, 66].
1) Distribution of *Baetis inexpectatus* (Tshernova), literature data. 2) Distribution of *Acentrella sinaica* Bogoescu, literature data. 3) Co-occurrence of both species.

ACKNOWLEDGMENTS

My husband, Prof. K. Jazdzewski, is acknowledged for his valuable comments. Thanks are also due to the referee, Dr. Michel Sartori, for his constructive and helpful review.

REFERENCES

Numbers in brackets indicate papers used in preparation of particular maps; see Figs. 1-11.

Dratnal, E. 1976. The benthic fauna of the Pradnik stream below an inlet of dairy waste affluents. Arch. Ochr. Srod., 2: 235-270. [1]

Dratnal, E. and E. Dumnicka. 1982. Composition and zonation of benthic invertebrate communities in some chemically stressed aquatic habitats of Niepołomice Forest (South Poland). Acta Hydrobiol., 24: [2]

Dratnal, E. and K. Kasprzak. 1980. The response of the invertebrate fauna to organic pollution in a well oxygenated karst stream exemplified by the Pradnik Stream (South Poland). Acta Hydrobiol., 22: 263-278. [3]

Dratnal, E., R. Sowa, B. Szczęsny. 1979. Zgrupowanie bezkręgowców bentosowych Dunajca na odcinku Harklowa - Sromowce Nizne. Ochr. Przyr., 42: 183-215. [4]

Dylikowa, A. 1973. Geografia Polski. Krainy geograficzne. PZWS, Warszawa, 816 pp.

Fall, J. 1976. Materialy do znajomosci fauny jetek (Ephemeroptera) rzeki Bystrzycy lubelskiej. Ann. UMCS, Lublin, Sect. C, 31: 211-220. [5]

Fleituch, T. (Jr.). 1985. Macroinvertebrate drift in the middle course of the River Dunajec (Southern Poland). Acta Hydrobiol., 27: 49-61. [6]

Glazaczow, A. 1994. Mayflies (Ephemeroptera) from the rivers Gwda and Drawa (in the Pomeranian Lake District of North West Poland) and from some waters of their river basins. Pol. Pismo Ent., 63: 213-257. [7]

Glazaczow, A. 1997. Observations on the psammophilous mayfly species *Procloeon nanum* in the northeast Poland, pp. 83-87. In: P. Landolt and M. Sartori (eds.). Ephemeroptera and Plecoptera: Biology-Ecology-Systematics. [8]

Glowacinski, Z. 1968. Badania nad fauną jetęk (Ephemeroptera) okolic Krakowa.. Acta Hydrobiol., 10: 103-130. [9]

Hawksworth, D. L. (ed.) 1996. Biodiversity. Measurement and estimation. The Royal Society, Champan and Hall, London, Weinheim, New York, Tokyo, Melbourne, Madras, 140 pp.

Jażdżewska, T. 1971. Jętki (Ephemeroptera) rzeki Grabi. Pol. Pismo Ent., 41: 244-304. [10]

Jażdżewska, T. 1972. Fauna Niebieskich Zródel. Jętki (Ephemeroptera) na terenie rezerwatu. Zesz. Nauk. Uniw. Łódz., ser. 2, mat.-przyr., 46: 35-39. [11]

Jażdżewska, T. 1973. Notes on the biology and ecology of the mayfly *Ametropus eatoni* Brodskij (Ephemeroptera). Pol. Pismo Ent., 43: 469-477. [12]

Jażdżewska, T. 1976. *Ephemerella mucronata* (Bengtsson) i *Ephemerella notata Eato*n (Ephemeroptera) w dorzeczach Pilicy i Warty. Zesz. Nauk. Uniw. Łódz., ser. 2, mat.-przyr., 3: 95-109. [13]

Jażdżewska, T. 1979. Premiers résultats des recherches sur la faune des Ephéméroptères de la rivière Pilica, pp. 133-137. In: K. Pasternak and R. Sowa (eds.). Proc. 2-nd Int. Conf. Ephemeroptera, Kraków. [14]

Jażdżewska, T. 1984. Les Ephéméroptères de la rivière Lubrzanka (Montagnes Świętokrzyskie, Pologne Centrale), pp. 231-242. In: V. Landa *et al.* (eds.). Proc. IVth Intern. Confer. Ephemeroptera. CSAV. [15]

Jażdżewska, T. 1985. Influence des aménagements de la rivière Widawka sur la faune des Ephéméroptères. Verh. Int. Ver. Limnol., 2: 2063-2068. [16]

Jażdżewska, T. 1997. Mayflies (Ephemeroptera) of the sandy bottom of the River Grabia (Central Poland), pp. 157-166. In: P. Landolt and M. Sartori (eds.). Ephemeroptera and Plecoptera: Biology-Ecology- Systematics. [17]

Jażdżewska, T. and Górczynski, A. 1991. Les Ephéméroptères des rivières qui franchissent la zone marginale du Roztocze Central, pp. 263-270. In: J. Alba-Tercedor (ed.). Overview and Strategies of Ephemeroptera and Plecoptera. The Sandhill Crane Press. [18]

Jop, K. 1981. Ecology of the forest stream Lane Bloto in the Niepolomice Forest. Acta Hydrobiol., 23: 107-123. [19]

Kamler, E. 1960. Notes on the Ephemeroptera Fauna of Tatra streams. Pol. Arch. Hydrobiol., 8: 107-127. [20]

Kamler, E. 1965. Thermal condition in mountain water and their influence on the distribution of Plecoptera and Ephemeroptera larvae. Ekol. Polska, Ser. A., 13, 20: 377-414. [21]

Kamler, E. and W. Riedel. 1960. The Effect of Drought on the Fauna Ephemeroptera, Plecoptera and Trichoptera of a Mountain Stream. Pol. Arch. Hydrobiol., 8: 87-94. [22]

Keffermüller, M. 1960. Badania nad fauną jętek (Ephemeroptera) Wielkopolski. Pr Kom. Biol. Pozn. TPN, 19: 1-57. [23]

Keffermüller, M. 1964. Uzupelnienie badań nad fauna jetek (Ephemeroptera) Wielkopolski. Bad. Fizjogr. Pol. Zach., 14: 69-86. [24]

Keffermüller, M. 1967. Badania nad fauną jętek (Ephemeroptera) Wielkopolski. III. Bad. Fizjogr. Pol. Zach., 20: 15-28. [25]

Keffermüller, M. 1972a. Badania nad fauną jętek (Ephemeroptera) Wielkopolski. IV. Analiza zmienności *Baetis tricolor* Tsher. wraz z opisem *B. calcaratus* sp.n. Pr. Kom. Biol. Pozn. TPN, 35: 1-30. [26]

Keffermüller, M. 1972b. Badania nad fauną jętek (Ephemeroptera) Wielkopolski V. Pol. Pismo Ent., 42:527-533. [27]

Keffermuller, M. 1974. A new Species of the Genus *Baetis* Leach (Ephemeroptera) from Western Poland. Bull. Acad. Pol. Cl. II., 22: 183-185. [28]

Keffermüller, M. 1978. Badania nad fauną jętek (Ephemeroptera) Wielkopolski. VI. Bad. Fizjogr. Pol. Zach., 31: 95-103. [29]

Keffermüller, M. and Machel M., 1967. *Baetis tracheatus*, sp. n. (Ephemeroptera) (Baetidae). Bad. Fizjogr. Pol. Zach., 20: 7-14. [30]

Keffermüller, M. and R. Sowa, 1975. Les espèces du groupe *Centroptilum pulchrum* Eaton (Ephemeroptera, Baetidae) en Pologne. Pol. Pismo Ent., 45: 479-486. [31]

Klonowska M., 1986. The food of some mayfly (Ephemeroptera) nymphs from the streams of the Kraków - Częstochowa Jura. Acta Hydrobiol., 28: 181-197. [32]

Klonowska-Olejnik, M. 1995. Jętki (Ephemeroptera), pp. 207-224. In: B. Szczęsny (ed.). Degradacja fauny bezkręgowców bentosowych Dunajca w rejonie Pienińskiego Parku Narodowego. Ochr. Przyr., 52.

Klonowska-Olejnik, M. 1997. Ephemeroptera of the river Dunajec near Czorsztyn dam (Southern Poland), pp. 83-87. In: P. Landolt and M. Sartori (eds.). Ephemeroptera and Plecoptera: Biology-Ecology-Systematics. [33]

Kownacka, M. 1971. Fauna denna potoku Sucha Woda (Tatry Wysokie) w cyklu rocznym. Acta Hydrobiol., 13: 415-438. [34]

Kownacka, M. and A. Kownacki. 1965. The bottom fauna of the river Bialka and of its Tatra tributaries, the Rybi Potok and Potok Roztoka. Limnol. Invest. Tatra Mts and Dunajec River Basin. Kom. Zagosp. Ziem Górskich PAN 11: 129-151. [35]
Kownacka, M. and A. Kownacki. 1968. Wplyw pokrywy lodowej na faunę denną potoków tatrzańskich. Acta Hydrobiol., 10: 95-102. [36]
Kownacki, A. 1977. Biocenoza potoku wysokogórskiego pozostającego pod wpływem turystyki. 4. Fauna denna Rybiego Potoku. Acta Hydrobiol., 11: 293-312. [37]
Kownacki, A. 1982. Stream ecosystems in mountain grassland (West Carpathians). 8. Benthic invertebrates. Acta Hydrobiol., 24: [38]
Kownacki, A., E. Dumnicka, E. Grabacka, B. Kawecka and A. Starzecka. 1985. Stream ecosystems in mountain grassland (West Carpathians). 14. The use of the experimental stream method in evaluating the effect of agricultural pollution. Acta Hydrobiol., 27: 381-400. [39]
Krzanowski, W., E. Fiedor, T. Kuflikowski. 1965. Fauna denna kamienisto-prądowych siedlisk dolnych odcinków Białego Dunajca, Rogoznika i Lepietnicy. Zesz. Nauk. U.J., 103, Prace Zool., 9: 43-60. [40]
Krzyzanek, E. 1971. Bottom fauna in the Tresna dam reservoir in 1966. Acta Hydrobiol., 13: 335-342. [41]
Kuflikowski, T. 1974. Fauna fitofilna zbiornika zaporowego w Goczałkowicach. Acta Hydrobiol., 16:189-207. [42]
Kukula, K. 1991. Mayflies (Ephemeroptera) of the Wołosatka stream and its main tributaries (The Bieszczady National Park, southern Poland). Acta Hydrobiol., 33: 31-45. [43]
Lewandowski, K. 1989. Mayflies (Ephemeroptera) of running water units in the Olsztyn province. Pol. Pismo Ent., 59: 387-392. [44]
Lubini, V., Knispel, S., Landolt, P. and Sartori, M. 1996. Geographical distribution of mayflies and stoneflies (Insecta: Ephemeroptera, Plecoptera) in Switzerland - preliminary results. Mitt. schweiz. ent. Ges., 69: 127-133.
Machel, M. 1969. Fauna jętek (Ephemeroptera) okolic Głogowa. Bad. Fizjogr. Pol. Zach., 22: 7-26. [45]
McCafferty, W.P. and Waltz, R.D. 1995. *Labiobaetis* (Ephemeroptera: Baetidae): new status, new north american species, and related new genus. Ent. News, 106: 19-28.
Mielewczyk, S. 1981. Evaluation of occurrence, density and biomass of Ephemeroptera in the drainage channel near the village of Turew (Region of Poznan). Acta Hydrobiol., 23: 363-373. [46]
Mielewczyk, S. 1983. Density and biomass of Ephemeroptera larvae in Lake Zbęchy (the Poznán region). Acta Hydrobiol., 24: 253-165. [47]
Mikulski, J.St. 1929. Przyczynek do znajomości fauny doliny Popradu w okolicy Muszyny: Ephemeroptera i Neuroptera. Spr. Kom. Fizjogr., Kraków, 65: 81-92. [48]
Mikulski, J.St. 1937. Materialy do poznania fauny jętek (Ephemeroptera) Beskidu Wyspowego i Gorców. Fragm. Faun. Mus. Zool. Pol., 3: 47-56. [49]
Mikulski, J.St. 1950. Fauna jętek (Ephemeroptera) zródłowych potoków Wisły. Pr. Biol. Wyd. Śląskie PAU, 2: 143-162. [50]
Minelli, A., Ruffo, S. and La Posta, S. (eds.) 1993-1995. Checklist delle specie della fauna italiana. Calderini, Bologna.
Musial, L., Turoboyski, L., Chobot, M. and Łabuz, W. 1958. Badania nad zanieczyszczeniem rzeki Soły i jej zdolnościa samooczyszczania. Pol. Arch. Hydrobiol., 4: 221-250. [51]
Müller-Liebenau, I. 1969. Revision der europäisches Arten der Gattung *Baetis* Leach, 1815 (Insecta, Ephemeroptera). Gewäss. Abwäss., 48/49: 1-214.
Novikova E.A. and Kluge N. 1987. Systematics of the genus *Baetis* (Ephemeroptera, Baetidae) with description of a species from Middle Asia. Vest. Zool. 1987(4): 8-19.
Novikova, E.A. and Kluge, N.Y. 1994. Mayflies of the subgenus *Nigrobaetis* (Ephemeroptera, Baetidae, *Baetis* Leach, 1815). Ent. Obozr., 73: 624-644.
Olechowska, M. 1982. Zonation of mayflies (Ephemeroptera) in several streams of Tatra Mts and the Podhale region. Acta Hydrobiol., 24: 62-71. [52]
Razowski, J. 1990-1997. Wykaz zwierząt Polski (Checklist of Animals of Poland). Vol. I, II (1990-1991). Zaklad Narodowy im. Ossolińskich. Wydawnictwo Polskiej Akademii Nauk, Wroclaw, Warszawa, Kraków; Vol. III (1991) Krakowskie Wydawnictwo Zoologiczne, Kraków; Vol. IV-V (1997). Wydawnictwa Instytutu Systematyki i Ewolucji Zwierząt PAN, Kraków.
Ricklefs, R.E. and Schluter, D. (eds.) 1993. Species Diversity in Ecological Communities. The University of Chicago Press, Chicago and London, 414 pp.
Ratajczak, E. 1976. Jętki (Ephemeroptera) rzeki Wełny. Pol. Pismo Ent. 46: 749-756. [53]
Ruffo, S. 1996. Il Progetto Checklist della specie della fauna italiana. Museol. Sci. 13, suppl. Atti 10° Congresso A.N.M.S. Bologna 1994: 165-169.
Schneider, W.G., 1885. Verzeichniss der Neuropteren Schlesiens. Z. Ent., 7: 17-32. [54]
Siemińska, J. 1954. Nowy gatunek jętki w faunie Polski - *Eurycaenis harrisella* (Curtis). Pol. Arch. Hydrobiol., 2: 185-188. [55]

Sowa, R. 1959. Przyczynek do poznania fauny jętek (Ephemeroptera) okolic Krakowa. Acta zool. Cracov., 4: 655-697. [56]

Sowa, R. 1961a. Fauna denna rzeki Bajerki. Acta Hydrobiol., 3: 1-32. [57]

Sowa, R. 1961b. Nowe stanowisko jętki *Ephemerella karelica* (Tiensuu) (= *Eurylophella karelica* Tiensuu). Acta Hydrobiol., 3: 59-62. [58]

Sowa, R. 1961c. *Oligoneuriella mikulskii* n.sp. (Ephemeroptera). Acta Hydrobiol., 3: 287-294. [59]

Sowa, R. 1961d. Nowe i rzadkie w faunie Polski gatunki widelnic (Plecoptera). New and rare species of stoneflies (Plecoptera) in the fauna of Poland. Acta Hydrobiol., 3: 295-302. [60]

Sowa, R. 1962. Materiały do poznania Ephemeroptera i Plecoptera w Polsce. Acta Hydrobiol., 4: 205-224. [61]

Sowa, R. 1965. Ecological characteristics of the bottom fauna of the Wielka Puszcza stream. Acta Hydrobiol., 7, suppl. 1: 61-92. [62]

Sowa, R. 1971. *Ecdyonurus starmachi* sp.n. et *E. submontanus* Landa des Carpates polonais (Ephemeroptera, Heptageniidae). Bull. Acad. Polon. Sci., Ser. sci. biol., 19: 407-412. [63]

Sowa, R. 1972. *Baetis beskidensis* n.sp. des Carpates polonais (Ephemeroptera: Baetidae). Bull. Acad. Polon. Ser. Sci. biol., Cl. II, 20: 711-712. [64]

Sowa, R. 1973. Contribution à l'étude des *Oligoneuriella* Ulm. européenes (Ephemeroptera, Oligoneuriidae). Bull. Acad. Polon. Sci., Ser. sci. biol., Cl. II, 21: 657-665. [65]

Sowa, R. 1975a. Ecology and biogeography of mayflies (Ephemeroptera) of running waters in Polish part of the Carpathians. 1. Distribution and quantitative analysis. Acta Hydrobiol., 17: 223-297. [66]

Sowa, R. 1975b. Ecology and biogeography of mayflies (Ephemeroptera) of running waters in Polish part of the Carpathians. 2. Life cycles. Acta Hydrobiol., 17: 319-353. [67]

Sowa, R. 1990. Ephemeroptera - Jętki. In: J. Razowski (Ed.), Checklist of Animals of Poland. Vol. 1. Zaklad Narodowy im. Ossolińskich. Wydawnictwo PAN, Wrocław, Warszawa, Kraków; 33-38.

Sowa, R. and Degrange, Ch. 1987. Sur quelques espèces europeéennes de *Rhithrogena* du groupe *semicolorata* (Ephemeroptera, Heptageniidae). Acta Hydrobiol., 29: 523-534. [68]

Szczęsny, B.1968. Fauna denna potoku Saspówka na terenie Ojcowskiego Parku Narodowego. Ochrona Przyrody, 33: 215-233. [69]

Szczęsny, B. 1974. Wpływ ścieków z miasta Krynicy na zbiorowiska bezkręgowych dna potoku Kryniczanka. Acta Hydrobiol., 16: 1-29. [70]

Szczęsny, B. 1990. Benthic macroinvertebrates in acidified streams of the Świętokrzyski National Park (Central Poland). Acta Hydrobiol., 32: 155-169. [71]

Systematics Agenda 2000. Charting the Biosphere 1994. Technical Report. Amer. Mus. Nat. Hist. and New York Botanic Gardens, New York.

Van Goethem, J.L. and Grootaert, P. (eds.), Faunal inventories of sites for cartography and nature conservation. Proc. 8th Int. Coll. European Invertebrate Survey, Brussels, 9-10 September 1991. Inst. roy. Sci. nat. Belgique, Bruxelles, 247 pp.

Waltz R.D., W.P. McCafferty, and A. Thomas, 1994. Systematics of *Alainites* n. gen., *Diphetor*, *Indobaetis*, *Nigrobaetis* n. stat., and *Takobia* n. stat. (Ephemeroptera, Baetidae). Bull. Soc. Hist. nat., Toulouse, 130: 33-36.

Wójcik, S. 1963. Fauna jetek (Ephemeroptera) Wisly pod Tczewem. Zesz. Nauk. Uniw. Poznań., Biol., 4: 102-120. [72]

Zaćwilichowska, K. 1968. Fauna denna Kamienicy Nawojowskiej. Acta Hydrobiol., 10: 319-341. [73]

Zięba, J. 1985. Ecology of some waters in the forest-agricultural basin of the River Brynica near the Upper Silesian Industrian Region. 10. Bottom insects with special regard to Chironomidae. Acta Hydrobiol., 27: 547-560. [74]

Zięba, J. 1986. Bottom insects, chiefly Chironomidae, in the River Mala Panew and its tributaries polluted with industrial wastes (Southern Poland). Acta Hydrobiol., 28: 429-441. [75]

BIOMASS OF EPHEMEROPTERA AND PLECOPTERA IN THREE SWISS RIVERS

Peter Landolt, Judith Thüer, and Denise Studemann

Institute of Zoology, Entomology
University of Fribourg
Pérolles, CH-1700 Fribourg, Switzerland

ABSTRACT

The fauna composition and the biomass of Ephemeroptera and Plecoptera were observed in a meandering lowland river, a prealpine stream and a regulated river. The faunal diversity in the regulated river was poor compared to the other two rivers due to the limnic character of the water, specially the suspended particles floating in the water. Growth of *Serratella ignita* depended mainly on the water temperature. The largest biomass of Ephemeroptera was observed in the lowland river, whereas that of Plecoptera was observed in the prealpine stream. The dam-regulated river produced the most biomass, composed principally of Gammaridae. Their inorganic compounds were three times higher than those of Ephemeroptera and Plecoptera.

INTRODUCTION

The biocenosis of a stream is determined by abiotic factors such as the slope (Ambühl, 1959; Statzner and Higler, 1986), the sediment (Hefti et al., 1985; Reynoldson, 1987; Maier, 1994; Palmer et al., 1995), the altitude (Margreiter-Kownacka, 1990; Ward, 1994), the temperature regime (Humpesch, 1980a, b, 1981; Ward and Stanford, 1982), the oxygen concentration (Jacob and Walther, 1981; Jacob et al., 1984), the chemical compounds in the water (e.g. Buikema et al., 1982; Smock, 1983; Townsend et al., 1983; Dewey, 1986; Winterbourn and Collier, 1987) and also by biotic interactions of the organisms, specially on the trophic level (Hawkins, 1986). The three investigated rivers, a prealpine stream, a meandering lowland river and a dam-regulated river, have different characteristics which influence the diversity of the species, the growth of the organisms and also the production of the biomass.

The aim of the present study is to show the influence of physical and chemical parameters on Ephemeroptera and Plecoptera, specially in the regulated river.

Trends in Research in Ephemeroptera and Plecoptera
Edited by E. Dominguez, Kluwer Academic/Plenum Publishers, 2001

MATERIAL AND METHODS

Geographical Situation

The three rivers Nessleraa, Neirigue and Petite Sarine are located in the Canton of Fribourg (Fig. 1). The water of the Sarine river is used several times for the production of hydroelectricity. A seventy-meter high dam near Rossens forms the Lac de la Gruyère, water from a catchment area of about 640 km^2. Some 20 m^3 water per second pass through the power plant turbines 12 km downstream from the dam and the remaining water (rest water) at 1 m^3 per second is released in the river bed between the dam and the hydroelectric station and forms the Petite Sarine. One of the stations is situated near the dam (400 m downstream), whereas the second station, Hauterive, is located 12 km downstream between a waste-water cleaning plant and the power plant releasing the water into the Petite Sarine. The Nessleraa and the Neirigue have a catchment area of 30 and 57 km^2, respectively. Their normal water quantity amounts to about 0.5 m^3 per second. However, after thunder-storms in summer, the water level rises quickly up to 80 cm above the normal water level. The Neirigue, in particular, experiences extremely high water occurences with 180 cm higher water levels. The water of these two rivers joins the Sarine near Fribourg and via the Aare and the Rhine finally empties in the North Sea.

Physical Characteristics and Chemical Properties

The values of pH, temperature, conductivity and oxygen content were measured directly in the field. The amounts of nitrogen and phosphorus were determined by standard procedures in the Laboratory for Environmental Protection of the Canton of Fribourg.

The composition of sediment in the rivers were determined by analyzing a volume of 50 dm^3 at a depth of 0-30 cm from the river bottom.

Sampling of the Organisms and Their Analysis

From March 1997 to March 1998 samples were taken from four sites (Fig. 1). Sampling was carried out by a surber sampler of 41 cm side length and a mesh size of 0.7 mm. Three lentic and three lotic probes of 0.168 m^2 each were taken. The sampling was carried out up to a depth of 20 cm in the sediment. The organisms were floated in a solution of 350 g/l $MgSO_4$ in a flat basin, gathered and conserved in 80% alcohol for further investigations.

The organisms captured from April to November were determined principally by the keys of Aubert (1959), Brittain and Salveit (1996), Studemann et al. (1992), Tachet et al. (1984) and Warninger and Graf (1997).

The measurements of body length and head capsule width of *Serratella ignita* were carried out with a binocular microscope. The measured individuals (if available at least 40 specimens) were grouped by size classes of 0.35 mm.

The biomass of Ephemeroptera and Plecoptera were determined by the dry weight (organic and inorganic compounds) after drying the probes at 100°C over four days. The animals were incinerated at 900°C for one hour, and the amount of inorganic compounds was estimated from the ash. The determination of the biomass was carried out ten times from April 1997 until March 1998.

RESULTS

Physical and Chemical Characteristics

The temperature regimes in the Nessleraa and in the Neirigue are comparable as well as their oxygen contents (Table 1). The temperature in the Petite Sarine shows a different pattern

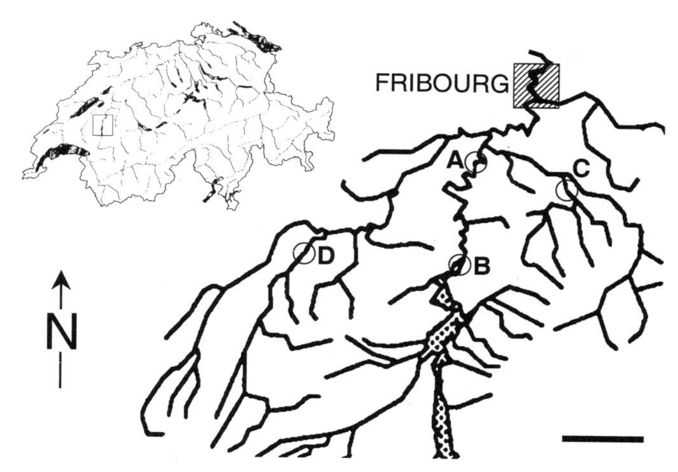

Fig. 1. General map of the main Swiss rivers and detailed map of the four investigated localities near Fribourg. A: Petite Sarine/Hauterive; B Petite Sarine/dam; C: Nessleraa and D: Neirigue. The scale represents 5 km.

Table 1. Physical and chemical characteristics (mean ± SD) of the stations investigated (*measured 5 cm above ground)

	Nessleraa	Neirigue	Petite Sarine / dam	Petite Sarine / Hauterive
Altitude [m a.s.l.]	666	652	610	577
River slope [‰]	22	11	3	3
Water current* [cm/s]	0-95	0 - 120	0 - 60	0 - 115
Conductivity [µS/cm]	464 ± 74	526 ± 51	415 ± 46	442 ± 66
O_2 content [%] in Summer	90	96	81	84
Annual mean temperature [°C]	10.2 ± 2.1	10.2 ± 2.7	8.1 ± 2.6	11.6 ± 3.0
Daily amplitude [°C]				
- Summer	12.4 - 15.1	14.0 - 17.5	10.2 - 11.7	14.0 ± 25.6
- Winter	1.8 - 4.9	2.3 - 4.3	1.3 - 3.5	-1.7 - 4.4
pH	8.0 ± 0.4	7.7 ± 0.6	8.0 ± 0.9	7.7 ± 0.5

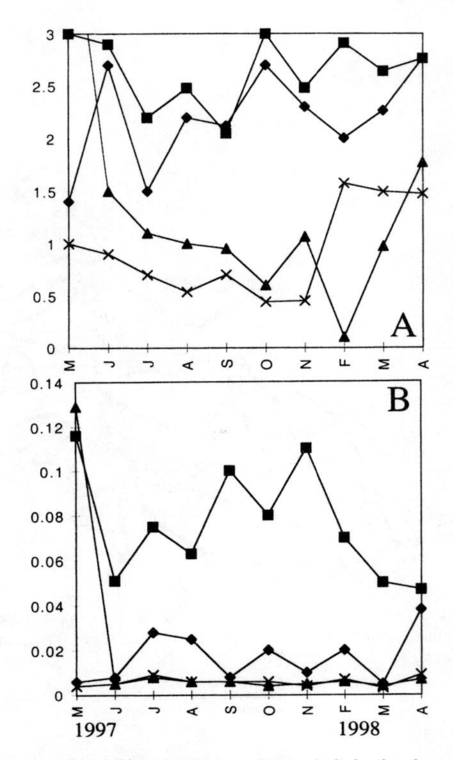

Fig. 2. Concentration of chemical compounds (mg/ml) in the three rivers. A: nitrate; B: orthophosphate; ◆ Nessleraa, ■ Neirigue, ▲ Petite Sarine/Hauterive and × Petite Sarine/dam.

between the dam and in Hauterive, 12 km downstream; the station upstream is under the influence of the 60 m deep lake water and the temperature at the dam reflects the lake temperature at 60 m depth (a low mean temperature and low temperature amplitudes within a year). At the Hauterive station, after the meandering of the river in the 200 to 500 m broad canyon, the environmental temperature influences are shown; the water temperature resembles the air temperature (low winter and high summer temperatures with high daily amplitudes).

The pH, buffered by the carbonate substrate, remained slightly alcaline and stayed stable throughout the observation period.

The mean oxygen levels in Neirigue and Nessleraa are higher than in the regulated Petite Sarine. In winter, the oxygenation levels rise about 5% in all the rivers. The conductivity amounts to about 500 µS/cm for the Neirigue and Nessleraa and nearly 430 µS/cm in the Petite Sarine (Table 1).

The amount of nitrate exceeded the Federal legal limit level of 3 mg/ml (Federal edict 1975) once in the Petite Sarine/Hauterive (12 mg/ml in April) near a waste-water cleaning plant (Fig. 2). Otherwise the amounts of the nitrate fluctuated between 0.5 mg/ml and 1 mg/ml in the regulated river. They were two times smaller than the amounts in the Nessleraa and Neirigue (Figure 2A). The amounts of nitrite were similar to those of nitrate, and the values for ammonium never exceeded the legal limits (0.1 mg/ml) (data not shown). The amounts of orthophosphate were generally low; in the Petite Sarine lower (less than 0.01 mg/ml) than in

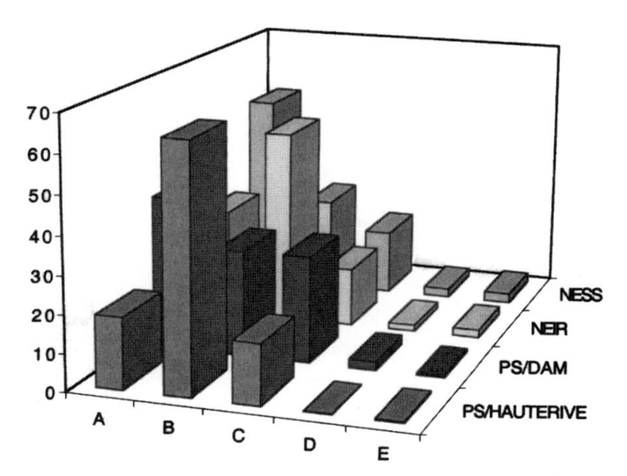

Fig. 3. Proportion (%) of different substrate classes: A: boulders (> 100 mm in diameter), B: large gravel (20-100 mm), C: fine gravel (2-20 mm), D: sand (1-2 mm) and silt (< 1 mm) at the sampling sites NESS (Nessleraa), NEIR (Neirigue), PS/DAM (Petite Sarine/dam) and PS/HAUTERIVE (Petite Sarine/Hauterive).

the two other rivers. The concentration in the Nessleraa was normally less than 0.03 mg/ml, but reached 0.13 mg/ml in May. In the Neirigue however, the mean value of orthophosphate was constantly much higher (mean value 0.08 mg/ml, Fig. 2B). The progress of the curves for the total phosphorus were comparable to those of the orthophosphate (data not shown).

The composition of sediment in the rivers is presented in Fig. 3. Boulders represented the major proportion in the Nessleraa and in the Petite Sarine near the dam, whereas large gravel was the largest fraction in the Neirigue and Petite Sarine/Hauterive. Sand (0.1 to 2.3 %) and silt (0.4% to 2.6 %) were minor constituents in the investigated rivers.

Faunal Composition

The presence of insects, crustaceans, nemathelminthes and annelides were recorded and determined from March to November 1997. *Erpobdella octoculata* and *Helobdella stagnalis* (Hirundinea) occurred all the time in high numbers in the Petite Sarine/Hauterive, whereas in the other stations these two species were found once. Crustacea (Gammaridae and Asselidae) were collected all over the year in the Petite Sarine, at Hauterive. *Gammarus pulex* lived also in huge amounts at the station near the dam. *Gammarus* represents the big mass of the macroinvertebrates of the Petite Sarine (see section biomass), these animals were present only sparsely in the Neirigue and Nessleraa.

The captured insects were representatives of Ephemeroptera and Plecoptera (Table 2), Trichoptera (species from 4 to 8 families, depending on the station), Diptera (5-7 families), Coleoptera (2-4 families) and some sparse individuals of Heteroptera (*Plea* sp. and *Corixa* sp.) and Megaloptera (*Sialis* sp.).

Table 2 gives the occurrence of the representatives from March to November 1997. The most common and abundant Ephemeroptera species was *Baetis rhodani*. *Alainites muticus*, *Baetis vernus* and *B. alpinus* were also observed. *Serratella ignita* was present in all the investigated rivers (Table 2), but the other Ephemerellidae species (*Ephemerella mucronata*) was observed only once in the Petite Sarine near the dam. A large amount of young *Ecdyonurus* larvae (especially *E. venosus*) were collected in spring, *Rhithrogena*

Table 2. Genera of Ephemeroptera en Plecoptera recorded in the stations investigated from March to November 1997 (● = presence).

Nessleraa	10.3	26.3	10.4	29.4	20.5	3.6	24.6	8.7	21.7	1.8	21.8	1.9	16.9	29.9	5.11
EPHEMEROPTERA															
Baetis	●	●	●	●	●	●	●	●	●	●	●	●	●	●	●
Epeorus	●				●			●					●		●
Rhithrogena	●	●	●	●	●	●	●	●							
Ecdyonurus	●			●		●		●	●	●	●	●	●	●	●
Heptagenia				●	●										
Serratella						●	●	●	●	●	●	●	●	●	●
Caenis												●			
Paraleptophlebia	●				●							●		●	●
Habroleptoides			●									●	●	●	●
Habrophlebia							●								
Ephemera									●						
PLECOPTERA															
Brachyptera	●	●	●		●										
Protonemoura					●										
Amphinemoura				●											
Nemoura				●			●								
Leuctra	●	●	●	●	●	●	●	●	●	●	●	●	●	●	●
Perlodes				●										●	●
Isoperla	●	●													

Neirigue	10.3	26.3	10.4	29.4	20.5	3.6	24.6	8.7	21.7	1.8	21.8	1.9	16.9	29.9	5.11
EPHEMEROPTERA															
Baetis	●	●	●	●	●	●	●	●	●	●	●	●	●	●	●
Epeorus	●			●		●	●	●	●	●				●	●
Rhithrogena	●	●	●	●	●	●			●	●				●	●
Ecdyonurus	●	●	●	●	●	●	●	●	●	●	●	●	●	●	●
Heptagenia						●	●								
Serratella						●	●	●	●	●	●	●	●	●	●
Caenis												●		●	
Paraleptophlebia	●			●	●				●						
Habroleptoides		●	●												
Habrophlebia				●	●	●	●		●	●					
PLECOPTERA															
Brachyptera	●	●	●												
Amphinemoura	●			●		●									
Nemoura	●			●		●									
Leuctra	●	●	●	●	●	●	●	●	●	●	●	●	●	●	●
Isoperla	●	●		●											
Perla							●					●			

Petite Sarine / dam	10.3	26.3	10.4	29.4	20.5	3.6	24.6	8.7	21.7	1.8	21.8	1.9	16.9	29.9	5.11
EPHEMEROPTERA															
Baetis	●	●	●	●	●	●	●	●	●	●	●	●	●	●	●
Epeorus	●														
Rhithrogena							●								
Ecdyonurus		●	●	●				●	●	●		●			●
Serratella							●	●	●	●	●	●		●	
Ephemerella					●										
PLECOPTERA															
Brachyptera	●	●													
Nemoura						●									
Leuctra	●														

Petite Sarine/Hauterive	10.3	26.3	10.4	29.4	20.5	3.6	24.6	8.7	21.7	1.8	21.8	1.9	16.9	29.9	5.11
EPHEMEROPTERA															
Baetis	●	●	●	●	●	●	●	●	●	●	●	●	●	●	●
Epeorus	●														
Rhithrogena	●	●	●	●		●									
Ecdyonurus		●	●	●	●			●	●	●				●	●
Heptagenia			●	●	●	●									
Serratella						●	●	●	●	●	●	●		●	
Paraleptophlebia	●		●	●	●						●				
Habroleptoides			●												
PLECOPTERA															
Nemoura			●												

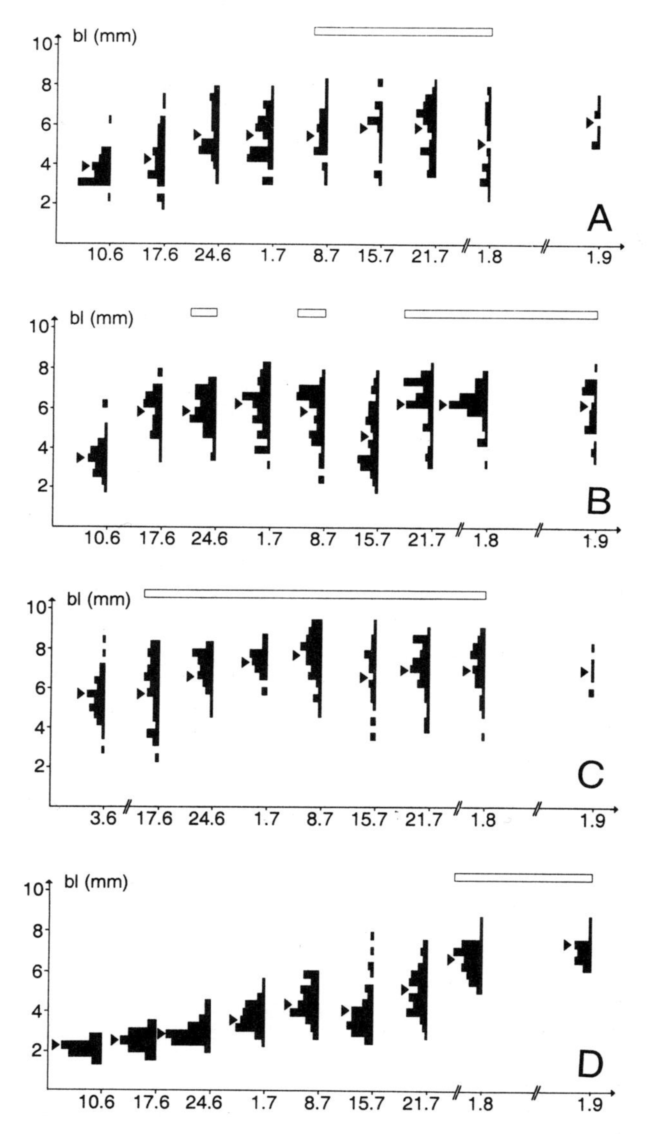

Fig. 4. Size distribution (bl=body length in mm) of *Serratella ignita* larvae for each sample from June to September 1997 at the stations A Nessleraa, B Neirigue, C Petite Sarine/Hauterive and D Petite Sarine/dam. The arrows represent the mean length of the larvae, and the white bars represent the presence of mature larvae.

semicolorata was abundant, too. The *Heptagenia* representative was the common *H. sulfurea*. Both *Epeorus* species (*E. alpicola* and *E. sylvicola*) were observed in the Nessleraa; in the other rivers only *E. sylvicola* occurred. Leptophlebiidae (*Paraleptophlebia submarginata*, *Habroleptoides confusa* and *Habrophlebia lauta*) were found occasionally.

The diversity of Plecoptera genera was relatively high in the Nessleraa and Neirigue with seven and six genera respectively. In the Petite Sarine only three genera occurred near the dam and some individuals of *Nemoura* sp. were caught in April in Hauterive.

Table 3. Dimensions of *Serratella ignita* nymphs (at least 15 specimens tested, mean ± SD).

	Nessleraa	Neirigue	Petite Sarine / dam	Petite Sarine / Hauterive
First appearance	July 8	June 24	August 1	June 17
Mean head width (hw) [mm]	1.30	1.27	1.24	1.37
Mean body length (bl) [mm]	6.31	5.96	6.32	6.72
bl/hw (male)	4.82 ± 0.55	4.63 ± 0.27	5.15 ± 0.52	4.70 ± 0.48
bl/hw (female)	4.90 ± 0.15	5.14 ± 0.50	5.20 ± 0.28	5.00 ± 0.25

Growth of *Serratella ignita*

The larval growth of *Serratella ignita* is shown in Fig. 4. The animals in the Petite Sarine near the dam presented a synchronous growth. Their growth was slower than those of the populations in the other stations. The maximal mean length was achieved only in September, the first larvae of the last instar (larvae with black coloured wing pads, also called nymphs), appeared late at the beginning of August. The station Petite Sarine/Hauterive presented the other extreme situation: the earliest appearance of nymphs were observed already in the middle of June but not later than the beginning of August. The most variable length of the larvae per sample and the maximum mean length of the larvae (7.4 mm in July) also were found in Petite Sarine/Hauterive. The maximum mean length of the animals in the Neirigue was achieved at the beginning of July (6.24 mm), whereas in the Nessleraa specimen with their maximum length were detected two weeks later (5.7 mm). The nymphs were observed in the Neirigue from June to the beginning of September, in the Nessleraa concentrated on two weeks in July.

The ratio body length to head width (bl/hw) for nymphs was always larger for females than for males due to the smaller head width of the females (Table 3).

Biomass of Ephemeroptera and Plecoptera Larvae

Figure 5 shows the biomass of Ephemeroptera per square meter in the rivers (lentic and lotic zones together). The Neirigue is the most productive river for Ephemeroptera with the biomass maxima in spring (1.64 g/m^2) . After the hatching of the spring species, the weight dropped during the summer and autumn comparable to the production of other rivers. In the winter season, an increasing number of first small and then larger individuals of Baetids and Heptagenids was observed. The maximum amounts per square meter was 0.83 g in the Nesslera, 0.28 g in the Petite Sarine/Hauterive and 0.39 g at the dam station. There was about three times less biomass of Ephemeroptera collected during the ten quantitative collection trips in the Nessleraa (2.3 g) than in the Neirigue (6.8 g). The station in the Petite Sarine near the dam was more productive (1.8 g) than in Hauterive (1.2 g). During the summer months, the main portion of the Ephemeroptera in the Petite Sarine were Baetidae. The general distribution during the year is comparable between the Nesslera and the Neirigue, but the peaks were different.

The biomass of the Plecoptera is presented in Figure 6. No prominent amounts of biomass were registered in the Petite Sarine: only three samples contained Plecoptera weighting 20 mg near the dam and none at Hauterive. The highest amount of 0.33 g/m^2 was captured March 1998 in the Nessleraa and 0.15g/m^2 in the Neirigue. The whole sampled biomass was

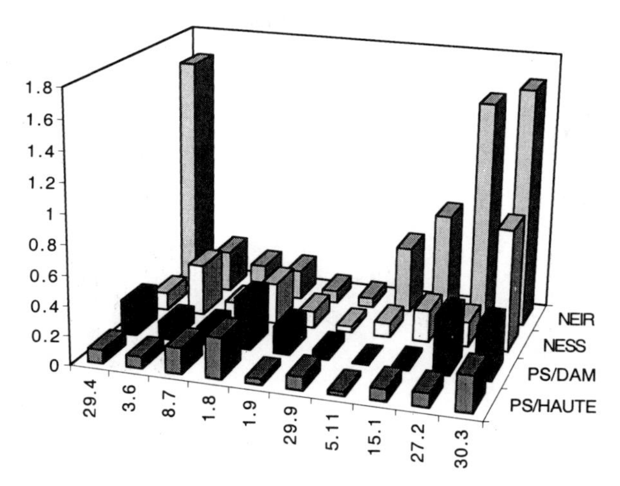

Fig. 5. Biomass of Ephemeroptera in gramme per square metre (g/m^2). Samplings were taken from April 1997 till March 1998 at the stations NEIR (Neirigue), NESS (Nessleraa), PS/DAM (Petite Sarine/dam) and PS/HAUTE (Petite Sarine/Hauterive).

0.52 g in the Neirigue and 0.71 g in the Nessleraa. In both rivers, the main proportion resulted from captures in January, February and March.

The production of invertebrate biomass in the Petite Sarine was dominated by the crustacea, mainly gammarids. Up to 19 g/m^2 dry weight was measured in August in Hauterive and 21 g/m^2 were collected near the dam. The productivity in the Neirigue and Nessleraa was much lower, due to another more diversified fauna composition. During summer and autumn, Diptera (chironomids and simulids) and Trichoptera were mainly present; the Diptera fauna reached 1.2 g/m^2 in the Neirigue and 0.2 g/m^2 in the Nessleraa, the caddisflies 4 g/m^2 in the Neirigue and near 1 g/m^2 in the Nessleraa.

For Ephemeroptera and Plecoptera, the inorganic matter, equivalent to the ash weight, reached comparable amounts of about 10% of the dry weight (Table 4). The inorganic proportion was slightly higher in January than 10 weeks later, when the larvae had matured. The proportion of ash in Trichoptera (5.3%) amounted to about half of that measured for Ephemeroptera. For gammarids, the high content of ash (31.6%) determined in the Petite Sarine in July reflected well the amounts measured in other probes (other stations and periods, data not shown). It was always about three times higher than those for Ephemeroptera and Plecoptera. The comparison between the dry weight of single specimen of gammarids with those of Ephemeroptera (e.g. Baetidae) revealed values of about five times higher the weight of gammarids (about 50 mg for a *Gammarus pulex* and about 10 mg for a *Baetis* specimen). The proportion of the organic material was calculated finally to 34 mg for a gammarid and 9 mg for a *Baetis* individual.

DISCUSSION

The permanent chemical charges of nitrogenous and phosphorous compounds, issuing principally from agricultural sources, are comparable to other Swiss rivers like the Glatt, Urnäsch, Sitter, Thur, Necker, Aare, Kander or Simme (reports from the Environmental Offices of various cantons). The values in the Petite Sarine and the Nessleraa reach the average values measured in Switzerland, whereas the Neirigue reaches the upper limit. This

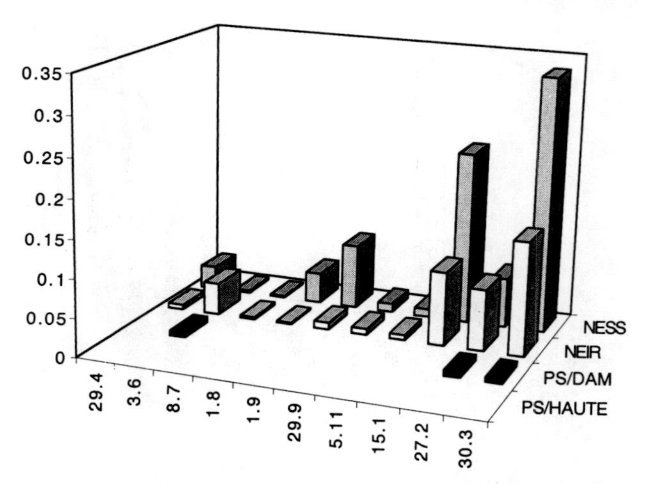

Fig. 6. Biomass of Plecoptera in g/m². Ten samplings were taken from April 1997 till March 1998 at the stations NESS (Nessleraa), NEIR (Neirigue), PS/DAM (Petite Sarine/dam) and PS/HAUTE (Petite Sarine/ Hauterive).

river runs in a valley surrounded by fields highly improved by agriculture. The waste-water purification plant just upstream from the station Hauterive in the Petite Sarine has no measurable influence on the river when the new plant works well. The dilution of the released cleaned water in the river amounts to about 1:50. The exceptional high amounts of nitrate and orthophosphate on 29.4.1997 could originate from a malfunctioning of the waste-water cleaning plant or another source such as agriculture. The chemical parameters measured for this study are similar to those from Noël and Fasel (1985). After the system of Klee (1993), all four stations are classified as slightly organic charged. The constant monitoring of the rivers in the canton of Fribourg by Noël reveals a stable biotic index after the system of Verneaux and Tuffery (1967) since 1990 (Noël's personal communication), indicating the stability of the situation.

The presence of species depends more on the physical properties of the biotope, such as substrate or slope (Higler, 1987; Verdonschot, 1990), than on dissolved chemical substances. Marten and Reusch (1992) suggest that other factors, such as current velocity or substrate, modify or superpose the effects of waste water.

The water flow determines temperatures and oxygenation, as well as the composition and changing events of the sediment. The low and constant velocity in the regulated Petite Sarine differs from the characteristic of the Neirigue and Nessleraa. These rivers react with high water levels after rainfall, specially after regular thunder storms from May to September. These events change the river course in the Neirigue-bed frequently, or at least wash out primary producers like bacteria. In the Petite Sarine however, a black layer on the lower surface of the stones shows the presence of reducing anaerobic bacteria, which indicates the low amount of oxygen in this zone. This factor influences the embryonic development of deposited eggs from macroinvertebrates and fish (Dedual, 1990). The upper surface of the stones is covered by a centimeter-thick coat of stromatolite-like structures. These formations are produced by different organisms, such as coccoid bacteria, cyanophytes, eucaryotic algaes and bryophytes (Freytet and Plet, 1996). The principal composition of blocs in the Nessleraa varies only slightly, even in high water levels, but the «cleaning-effect» happens in contrast to the Petite Sarine. The comparable composition of the sediment in Neirigue and in the Petite Sarine/Hauterive on the one hand, and Nesslera and Petite Sarine/dam on the other, indicates only a modest influence of the substrate on the faunal composition in these three rivers.

Table 4. Proportion (%) of inorganic substances corresponding to the ash content in Eph (Ephemeroptera), Plec (Plecoptera), Trich (Trichoptera) and Gammarus.

	ash [%]		
	January 15	March 30	July 7
Eph: Neirigue	12.7	9.4	
Eph: Nessleraa	12.5	10.3	
Plec: Ness + Neir	12.3	10.6	
Trich: Neirigue			5.3
Gammarus: Petite Sarine / dam			31.6

The sediment composition, the water velocity and the pH values are similar at Petite Sarine/Hauterive compared to the other two rivers. The poor fauna diversity in this station results therefore from other factors. This is also true for the station near the dam.

The majority of Ephemeroptera and Plecoptera species are spring species in the rivers on this altitude. In summer *Baetis rhodani*, *Alainites muticus*, *Serratella ignita* and *Ecdyonurus venosus* have been observed. Imhof et al. (1988) showed the existence of two populations of *E. venosus* with shifted developmental times in the Petite Sarine and another river near Fribourg. The very similar genetic pattern of the populations assumes that interbreeding between spring and autumn populations happens; that means that these populations are partially bivoltine and larvae can be found throughout the year (Imhof et al., 1988).

Probably, the lower oxygen content diminishes the presence of Plecoptera in the Petite Sarine compared to the Nessleraa and the Neirigue. In the last two rivers, only *Leuctra* species were captured all year. This observation agrees with those of Gonser and Schwörbel (1985) in the Gutach-Wutach in southern Germany.

The Gammaridae have a broad spectrum of distribution but prefer to live in organic charged waters (Engelhardt, 1974). The organic material from washed lake-sediments improves their diet.

One important factor in the development of Ephemeroptera and Plecoptera is the water temperature. The two stations in the Petite Sarine differ principally in this characteristic. A time shift of development is generally observed for the captured macroinvertebrates as well as for the composition of the representatives of three Plecoptera genera in the cool water near the dam. The only single observation of Plecoptera 12 km downstream in the summer heated water can be explained by the effect of the temperature.

The growth of *Serratella ignita* clearly demonstrates the influence of temperature which was shown even in the embryonic development by Jazdzewska (1980) and the growth of larvae by various authors (e.g. Hefti and Tomka, 1990; Riano et al., 1997; Tiunova, 1997). Larvae in the Hauterive station are caught early in the season and they mature early too. The larval growth near the dam starts later, but runs more synchronously, and nymphs are formed only late in the autumn. The ratio body length/head width of male and female Serratella nymphs is clearly larger in the cool water environment than in the other stations. The growth of *S. ignita* is comparable to that of the individuals in the Sense, a prealpine Fribourg river (Hefti and Tomka, 1990).

The productivity of different individual Ephemeroptera species has been investigated by various authors (e.g. Welton et al., 1982; Zelinka, 1984, Hefti and Tomka, 1990). The present study presents the total biomass of all Ephemeroptera and Plecoptera species, respectively. In the Neirigue, their biomass is dominant in winter and spring months with a maximum of 1.6 g/m^2 in March. The amount in the Nessleraa at the same time is about half (0.8 g/m^2). The total dry weight of the captured Ephemeroptera amounts to 6.8 g in the Neirigue, 2.3 g in the Nessleraa, 1.8 g in the Petite Sarine near the dam and 1.2 g in Hauterive. The Plecoptera are more abundant in the Nessleraa (0.71 g in ten samples) than in the Neirigue (0.52 g). In the Petite Sarine, the quantitative captures were successful only three times with 0.02 g at the dam station.

Elliott et al. (1988) calculated the production of the Ephemeroptera to be 25 % of the total zoobenthos. The stoneflies do not predominate either in the Nessleraa or in the Neirigue in agreement with the findings of Teslenko (1992). Kocharina et al. (1988) estimated the annual biomass of the stonefly larvae in Far-East rivers to 12.6 % of the entire benthos biomass.

The negligible findings of Plecoptera and the only small proportion of Ephemeroptera in the Petite Sarine are presumbably due to the influence of the dam. The particles suspended in the water, issuing from the bottom of the sediments of the storage lake behind the dam, are released constantly into the Petite Sarine and they affect considerably the fauna elements. This influence, combined with the small and constant water volume over the year and the gently flowing limnic water in the broad river-bed, determine principally the poor fauna diversity with the dominant presence of the *Gammarus* which represent 81 % to 99 % of the biomass (depending on station and period). The drop in mayfly species by a dam is reported by Klonowska-Olejnik (1997) too. Usseglio-Polatera (1997) observed that dams lead to a simplified redistribution of ecological niches through the growing uniformity of the substrate and possible sources of food.

Bargos et al. (1990), Landa and Soldan (1991), Buffagni (1997) and other authors considered the mayflies as good biological indicators. The observation of long-term changes in mayfly and stonefly fauna are necessary to determine trends on a large scale (Landa et al., 1997a,b). Further investigations in collaboration with the producers of hydroelectricity will clear the impact of the interference of the particles on the faunal diversity as well as the water regime.

ACKNOWLEDGMENTS

We wish to thank the Environmental Offices of the Cantons of Appenzell, Berne, Fribourg, St.Gallen and Zürich for their river reports with chemical and biological results. M. F. Noël and D. Folly of the Environmental Office of the Canton of Fribourg are thanked for the chemical measurements of the water and their river reports. The Gravière at Tuffière carried out the granulometric measurements. Mrs. Knispel determined Plecoptera. We are indebted to M. H. Gachoud and Mrs S. Watson for their helpful technical assistance. Finally, we thank an anonymous reviewer for his valuable suggestions.

REFERENCES

Ambühl, H. 1959. Die Bedeutung der Strömung als ökologischer Faktor. Schweiz. Zeitsch. Hydrol. 21: 133-264.

Aubert, J. 1959. Plecoptera. Insecta Helvetica 1. Ed. Soc. Ent. Suisse. 1-140.

Bargos, T., J. M. Mesanza, A. Basaguren and E. Orive. 1990. Assessing river quality by means of multifactorial methods using macroinvertebrates. A comparing study of main courses of Biscay. Water Res. 24: 1-10.

Brittain, J. E. and S. J. Salveit. 1996. Plecoptera, Stoneflies. In: Nilsson, A.N.(Ed.). Aquatic Insects of North Europe. Vol. 1. pp. 55-75. Appolo Books, Stenstrup, Denmark.

Buffagni, A. 1997. Mayfly community composition and biological quality of streams, pp. 235-246. In: P. Landolt and M. Sartori (eds.). Ephemeroptera and Plecoptera: Biology-Ecology-Systematics. MTL, Fribourg.

Buikema, A. L., E. F. Benfield and B. R. Niederlehner. 1982. Effects of pollution on freshwater invertebrates. Journal WPCP 54 (6): 862-868.

Dedual, M. 1990. Biologie et problèmes de dynamique de population de nas (*Chondrostoma nasus nasus*) dans la Petite Sarine. PhD Thesis Université de Fribourg: 1-169.
Dewey, S. L. 1986. Effects of the herbicide Atrazine on aquatic insect community structure and emergence. Ecology 67 (1): 148-162.
Elliott, J. M., U. H. Humpesch and T. Macan. 1988. Larvae of the British Ephemeroptera. A key with ecological notes. Freshwater Biological Association. Sientific Publication 49: 1-145.
Engelhardt, W. 1974. Was lebt in Tümpel, Bach und Weiher. Verlag Franckh-Kosmos, Stuttgart: 1-257.
Freytet, P. and A. Plet. 1996. Modern Freshwater Microbial Carbonates: The Phormidium Stromatolites (Tufa-Travertine) of Southeastern Burgundy (Paris Basin, France). Facies 34: 219-238.
Gonser, T. and J. Schwörbel. 1985. Chemische und biologische Untersuchung des Gutach-Wutach Flusssystems zwischen Neustadt und Weizener Steg. Beih. Veröff. Naturschutz Landschaftspflege Baden-Württemberg 44: 9-112.
Hawkins, C. P. 1986. Variation in individual growth rates and population densities of Ephemerellid mayflies. Ecology 67 (5): 1384-1395.
Hefti, D. and I. Tomka. 1990. Abundance, growth and production of three mayfly species (Ephemeroptera, Insecta) from the Swiss Prealps. Arch. Hydrobiol. 120 (2): 211-228.
Hefti, D., I. Tomka and A. Zurwerra. 1985. Recherche autécologique sur les Heptageniidae (Ephemeroptera, Insecta). Bull. Soc. ent. Suisse 58: 87-111.
Higler, L. W. G. 1987. Geschiedenis van de biologische waterbeoordeling. In: Werkgroep Biologische Waterbeoordeling Leersum: 15-22.
Humpesch, U. H. 1980a. Effect of temperature on the hatching time of eggs of five *Ecdyonurus* spp. (Ephemeroptera) from Austrian streams and English streams, rivers and lakes. J. Anim. Ecol. 49: 317-333.
Humpesch, U. H. 1980b. Effect of temperature on the hatching time of parthenogenetic eggs of five *Ecdyonurus* spp. and two *Rhithrogena* spp. (Ephemeroptera) from Austrian streams and English rivers and lakes. J. Anim. Ecol. 49: 927-937.
Humpesch, U. H. 1981. Effect of temperature on larval growth of *Ecdyonurus dispar* (Ephemeroptera: Heptageniidae) from two English lakes. Freshw. Biol. 11: 441-457.
Imhof, A., I. Tomka and G. Lampel, G. 1988. Autökologische und enzymelektrophoretische Untersuchungen an zwei *Ecdyonurus venosus*-Populationen (Ephemeroptera, Heptageniidae). Bull. Soc. Frib. Sc. Nat. 77 (1/2) 55-129.
Jacob, U. and H. Walther. 1981. Aquatic insect larvae as indicators of limiting minimal contents of dissoved oxygen. Aquat. Insects 4: 219-224.
Jacob, U., H. Walther and R. Klenke, R. 1984. Aquatic insect larvae as indicators of limiting minimal contents of dissoved oxygen- Part 2. Aquat. Insects 3: 185-190.
Jazdzewska, T. 1980. Structure et fontionnement des écosystèmes du Haut-Rhône français. Bull. Ecol. 11 (1): 33-43.
Klee, O. 1993. Wasser untersuchen: einfache Analysenmethoden und Beurteilungskriterien. Ed. Quelle und Meyer, Wiesbaden: 1-245.
Klonowska-Olejnik, M. 1997. Ephemeroptera of the River Dunajec near Czorsztyn dam (Southern Poland), pp. 282-287. In: P. Landolt and M. Sartori (eds.). Ephemeroptera and Plecoptera: Biology-Ecology-Systematics. MTL, Fribourg.
Kocharina, S., M. Makarchenko, E. Markachenko, E. Nikolayeva, T. Tiunova and V. Teslenko. 1988. Bottom invertebrates in the ecosystems of the salmon streams in the South of the Far East of the USSR, pp. 86-108. In: Fauna, systematics and biology of freshwater invertebrates. FEB AS USSR, Vladivostok.
Landa, V. and T. Soldan. 1991. The possibility of mayfly faunistics to indicate environmental changes of large areas, pp. 559-565. In: J. Alba Tercedor and J. Sanchez-Ortega (eds.). Overview and Strategies of Ephemeroptera and Plecoptera, Sandhill-Crane Press.
Landa, V., S. Zahradkova, T. Soldan and J. Helesic. 1997a. The Morava and Elbe river basins, Czech Republic: a comparison of long-term changes in mayfly (Ephemeroptera) biodiversity, pp. 219-226. In: P. Landolt and M. Sartori (eds.). Ephemeroptera and Plecoptera: Biology-Ecology-Systematics. MTL, Fribourg.
Landa, V., J. Helesic, T. Soldan and S. Zahradkova. 1997b. Stoneflies (Plecoptera) of the river Vltava, Czech Republic: a century of extinction, pp. 288-295. In: P. Landolt and M. Sartori (eds.). Ephemeroptera and Plecoptera: Biology-Ecology-Systematics. MTL, Fribourg.
Maier, K. J. 1994. Effects of spates on the benthic macroinvertebrate community of a prealpine river (First results). Verh. Int. Ver. Limnol. 25: 1605-1608.
Margreiter-Kownacka, M. 1990. Einfluss der Gletscherbachfassung auf die Biozönosen der unmittelbar anschliessenden Entnahmestrecke.Österreichische Wasserwirtschaft 42 (3/4): 94-94.
Marten, M. and H. Reusch. 1992. Anmerkungen zur DIN "Saprobienindex" (38410 Teil 2) und Forderung alternativer Verfahren. In: Kohlhammer, W. Natur Landschaft, Stuttgart 67: 544-547.

Noël, F. and D. Fasel. 1985. Etude de l'état sanitaire des cours d'eau du canton de Fribourg. Bull. Soc. Frib. Sc. Nat. 74(1/2/3): 1-332.

Palmer, M. A., P. Arensburger, P. S. Botts, C. C. Hakenkamp and J. W. Reid. 1995. Disturbance and the community structure of stram invertebrates: patch-specific effects and the role of refugia. Freshw. Biol. 34: 343-356.

Reynoldson, T. B. 1987. Interactions between sediment contaminants and benthic organisms. Hydrobiologia 149: 53-66.

Riano, P., A. Basaguren and J. Pozo. 1997. Diet variations of *Ephemerella ignita* (Poda) (Ephemeroptera, Ephemerellidae) in relation to the development stage, pp. 79-82. In: P. Landolt and M. Sartori (eds.). Ephemeroptera and Plecoptera: Biology-Ecology-Systematics. MTL, Fribourg.

Smock, L. A. 1983. Relationship between metal concentrations and organism size in aquatic insects. Freshw. Biol. 13: 313-321.

Statzner, B. and B. Higler. 1986. Stream hydraulics as a major determinant of benthic invertebrate zonation patterns. Freshw. Biol. 16: 127-139.

Studemann, D., P. Landolt, M. Sartori, D. Hefti and I. Tomka. 1992. Ephemeroptera. Fauna 9, Insecta Helvetica. Ed. Soc. Ent. Suisse: 1-175.

Tachet, H., M. Bournaud and Ph. Richoux. 1984. Introduction à l'étude des macroinvertébré des eaux douces. Assoc. française Limnol. 2: 1-155.

Teslenko, V. A. 1992. Role of stonefly nymphs in the freshwater invertebrate communities in the small Kedrovaya River. Diploma thesis, Sankt-Petersburg university: 1-22

Tiunova, T. M. 1997. Growth of rheophilic mayfly larvae (Ephemeroptera), pp. 65-72. In: P. Landolt and M. Sartori (eds). Ephemeroptera and Plecoptera: Biology-Ecology-Systematics. MTL, Fribourg.

Townsend, C. R., A. G. Hildrew and J. Francis. 1983. Community structure in some southern English streams: the influence of physicochemical factors. Freshw. Biol. 13: 521-544.

Usseglio-Polatera, Ph. 1997. Long-term changes in the Ephemeroptera of the River Rhône at Lyon, France, assessed using fuzzy coding aproach, pp. 227-234. In: P. Landolt and M. Sartori (eds.). Ephemeroptera and Plecoptera: Biology-Ecology-Systematics. MTL, Fribourg.

Verdonschot, P. M. 1990. Ecological characterization of surface waters in the province of Overijssel. PhD Thesis Landbouw University Wageningen: 1-255.

Verdonschot, P. M. 1992. Typifying macrofaunal communities of larger disturbed waters in the Netherlands. Aquatic conservation: marine and frehwater ecosystems 2: 223-242.

Verneaux, J. and G. Tuffery. 1967. Une méthode zoologique pratique de détermination de la qualité biologique des eaux courantes. Ann. Sc. Univ. Besançon 3: 79-91.

Ward, J. V. and J. A. Standford. 1982. Thermal responses in the evolutionary ecology of aquatic insects. Ann. Rev. Ent. 27: 97-117.

Ward, J. V. 1994. Ecology of alpine streams. Freshw. Biol. 32: 277-294.

Warninger, J. and W. Graf. 1997. Atlas der österreichischen Köcherfliegenlarven (unter Einschluss der angrenzenden Gebiete). Facultas-Universitäts-Verlag, Wien 1: 1-286.

Welton, J. S., M. Ladle and J. A. B. Bass. 1982. Growth and production of five species of Ephemeroptera larvae from an experimental recirculation stream. Freshw. Biol. 12: 103-122.

Winterbourn, M. J. and K. J. Collier. 1987. Distribution of benthic invertebrates in acid, brown water streams in the South Island of New Zealand. Hydrobiologia 153: 277-286.

Zelinka, M. 1984. Production of several species of Mayfly larvae. Limnologica 15 (1): 21-41.

EPHEMEROPTERA IN SWITZERLAND

Peter Landolt[1] and Michel Sartori[2]

[1] Institute of Zoology, Pérolles
CH-1700 Fribourg, Switzerland
[2] Museum of Zoology, C.P. 448
CH-1000 Lausanne, Switzerland

ABSTRACT

The Swiss mayfly fauna, to date 85 species, has been documented for the first time (Sartori and Landolt, 1999). Most of the 10,700 records were collected during the last twenty years. Over 1800 locations were investigated in streams, rivers, ponds and lakes situated on the northern and southern slopes of the Alps, in prealpine regions, on the Swiss Plateau and the Jura Mountains. Faunal distributions are attributed to geographical factors such as latitude, East-West extension and altitude, geological properties, physical characteristics such as current velocity and stream order.

INTRODUCTION

A distributional atlas, indicating the spatio-temporal spread of populations within a certain area, necessitates a solid systematic base. The work on Ephemeroptera in Swiss Fauna (Studemann et al., 1992) covers this need.

We have established that for some years now an important number of Ephemeroptera populations have been declining and some are even on the point of extinction. It is therefore urgent, to have a document that gives an objective appreciation of the current situation and that also serves as a measure for any future analysis of the evolution and distribution of these insects.

BACKGROUND

1. Geography of Switzerland

Switzerland lies in the heart of Europe and is influenced by all the major climatic regions of the continent —Atlantic, Mediterranean, Continental and Boreal. Its relief was moulded by alpine orogenesis and glaciations. A tough a small country (41,293 km^2) these elements have resulted in an extraordinarily rich diversity in fauna and flora.

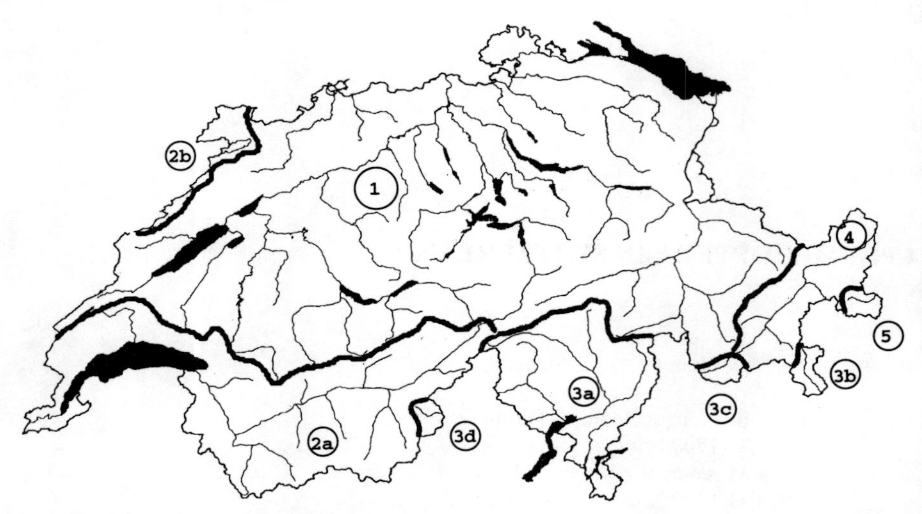

Fig. 1. The five catchment areas of Switzerland. 1 Rine; 2 Rhone (2a) and Doubs (2b); 3 Tichino (3a), Posciavino (3b) and Mera (3c); 4 Inn and 5 Rombach.

Switzerland is traditionally divided into four major geographical regions: the Jura (a low limestone mountain range up to 1500 m a.s.l.), the Swiss Plateau (from Tertiary and Quartenary deposits), the Pre-alps (formed by the folding of secondary sedimentary rocks) and the Central Alps (from crystalline rocks reaching over 4000 m a.s.l.) and the Southern Alps (with crystalline rocks forming the foothills of the great Italian plains). The lowest point in Switzerland lies on the shores of Lake Maggiore (190 m a.s.l.), the highest is the Dufour peak (Monte Rosa Massive) at 4634 m a.s.l..

2. Hydrology

Switzerland is often described as the water tower of Europe, for the great rivers, Rhine and Rhone, have their sources there. Glaciers cover 4 % of the country and those are 2000 lakes and approx. 50,000 km of running waters (Anonymus, 1992).

The waters are divided into five catchment areas of varied size and importance (Fig. 1). The Rhine catchment (1) covers more than 50% of the total area of Switzerland; the Rhone (with the Doubs) about 20% (2); the Po (Ticino, Posciavino, Mera, Diviera) about 5% (3); the Danube (Inn) approx. 3.5% (4); and the Etsch (Rombach) less than 1% (5). Table 1 presents the ten longest Swiss rivers with their annual mean water discharge.

3. Database

The distribution atlas for the 85 species, currently or formerly found in Switzerland is based on 1814 investigated locations, 10,610 data sets and 87,020 identified individuals (13,175 imagos or subimagos and 73,845 larvae). In general, the area was well covered, although with a few exceptions such as the Upper Rhine, the Aargauer Mittelland and some valleys in Upper Wallis. Figure 2 illustrates the distribution of *Baetis rhodani* and Figure 3 that of *Ecdyonurus helveticus*.

The species analysis in Sartori and Landolt (1999) describes seven themes for every species:

Table 1. The ten longest Swiss rivers, their catchment area and annual mean water discharge measured at the confluence with other rivers or at the Swiss border, respectively

River	Length (km)	Catchment area (km^2)	Water discharge (m^3/sec)
Rhine	375	35925	1040
Aare	295	11750	306
Rhone	264	10299	338
Reuss	158	3382	141
Limmat	140	2176	99
Sarine	128	1269	42
Thur	125	1696	45
Inn	104	1945	59
Ticino	91	1515	71
Broye	86	392	8

altitudinal distribution (steps of 200 m which we considered sufficiently precise), flight periods (in months), typology (stream order, see Table 2), life cycle (according to Clifford, 1982), ecology, distribution (including details about presence in other European countries, Table 3) and rarity status.

We determined stream order, using the maps with the scale of 1: 200,000 in the Swiss Hydrological Atlas. We assumed that the smallest watercourses automatically belonged to class 1 or 2. Consequently any eventual codification errors are the same for all locations. Only those locations with larvae were considered. Table 2 sums up the six selected typology classes.

This zonation seems to correspond well to the circumstances in Switzerland. No separate category was created for source zones (crenal). The Ephemeroptera colonising the springs as well as their outflows do not differ from those situated downstream (epirhithral), although this is not the case for other aquatic insects (Zollhöfer, 1997).

The "ecology" section has a number of details on each species observed; in particular phenology, altitudinal limits, preferred habitat, and, if known, any special ecological requirements.

The "distribution" section gives information on the known distribution of each species in Switzerland, and in a few cases elsewhere. Table 3 gives bibliographic references consulted to establish the presence or absence of a species in other European countries. A map of all sites of capture completes the details about the Swiss distribution.

With regards to status, the following classification was used:
if a species is present in less than 10 localities it is considered as "very rare", between 11 and 25 as "rare", between 26 and 50 as "uncommon", between 51 and 100 as "common" and beyond 100 as "widespread" (Table 4). It should be noted that these categories do not necessarily relate to the dangers threatening the species (Sartori et al., 1994). A species regarded as "uncommon" may be in greater danger of extinction than one regarded as "rare" (see also Discussion).

RESULTS AND DISCUSSION

1. Catchment Area

Table 5 illustrates the special status of the Rhine within Switzerland. All Swiss species of mayfly are found in its catchment area . This species diversity is not surprising as the Rhine is the largest catchment area in Switzerland and all hydrological, geomorphological and altitudinal units are covered.

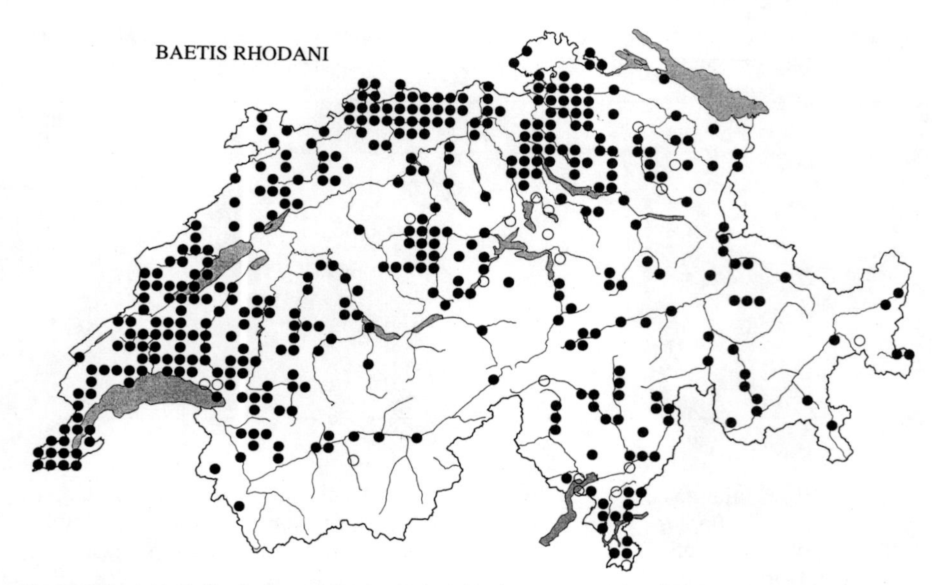

Fig. 2. Geographical distribution of *Baetis rhodani*. Territory covered for the atlas based on a grid of 5x5 km. O - stations with no data after 1970; ● - stations with data before and after 1970.

A second characteristic is that the smaller the catchment area, the fewer the number of species. There is a logarithmic relationship between the size of the catchment area (x) and the number of mayfly species (y): $y= 17.6 \ln x - 98.3$; $r^2=0.94$; $p<0.001$).

Ten species, typical of higher regions, are found in all catchment areas. This is not surprising as all the river systems have an alpine origin. The matrix of similarity of the different regions, calculated according to Jaccard's index by dividing the number of species common and exclusive to each by the number of the species common to two regions shows that the greatest affinity exists between the Rhone and the Rhine (0.71), the Inn and the Etsch (0.68), and the Rhone and the Ticino (0.50).

2. Altitudinal Distribution

Species diversity in Switzerland is inversely proportional to the altitude i.e. with increasing altitude the number of species decreases (Fig. 4). Reasons for this are to be found in the brevity of the ice free period, the limited food supply and the extreme physiographical conditions at higher altitudes. The hill region (200-800 m) is the richest with an average of 61 species, then the mountain region (800-1400 m) with 44, the subalpine (1400-2000 m) with 19, and finally the alpine region with only 6. These numbers do not take into consideration any human influence and confirm what is often found in European specialist literature that the mayfly is greatest at lower and middle altitudes. This characteristic is diametrically opposed to that of the Plecoptera which prefer much higher situations (Brittain, 1983).

The altitudinal range of the individual species is interesting. Twenty-eight species are only found within the hill region, two species only in the mountain region and none are restricted soley to the sub-alpine or alpine regions.

A single species (*Baetis alpinus*) inhabits all altitudes. This pioneer species is adapted to a wide range of environmental conditions and, quickly becomes dominant when there is little competition.

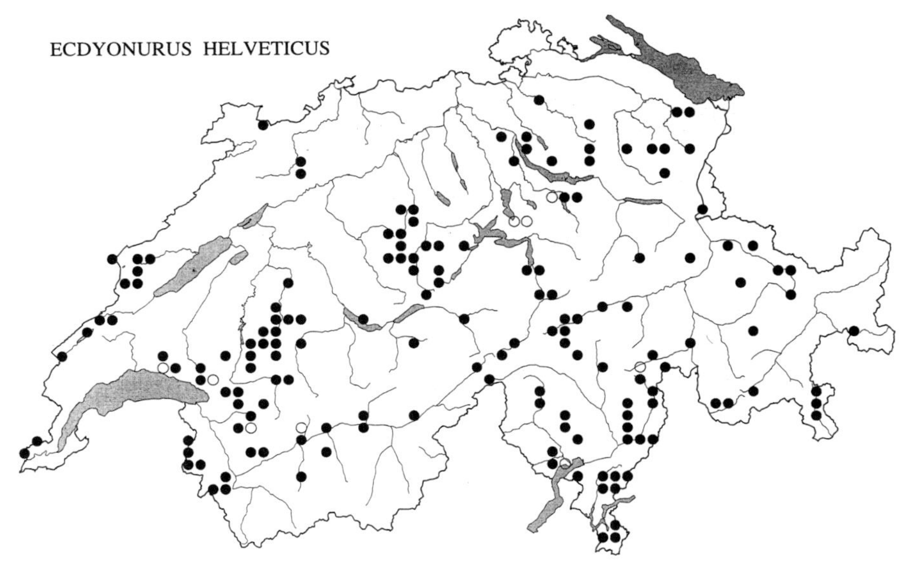

Fig. 3. Geographical distribution of *Ecdyonurus helveticus*. The same specifications as in Fig. 2.

3. Phenology

The flight periods of adults in Switzerland are mainly during spring or summer (Fig. 5). A single species (*Baetis rhodani*) has been observed throughout the whole year. Two other species limit their flight phase mostly to the end of the winter and the beginning of spring: *Rhithrogena germanica* and *Rh. gratianopolitana*. These two species can emerge already in February. The former is not found after April, whereas the latter may still be flying at the beginning of May depending on the altitude. At least a dozen species are found mostly in summer and autumn. This group includes as well as lowland species like *Caenis pusilla*, *Ecdyonurus insignis*, *Baetis liebenauae*, species from higher situations such as *Rhithrogena loyolaea*, *Rh. nivata* and *Ecdyonurus parahelveticus*.

The flight times of mayflies also differ clearly from those of stoneflies which have their greatest emergence activity in early spring or autumn. The considerable duration of the flight period of some species has two main reasons. One is the presence of several generations in a year (polyvoltinism), a second is the time-lag because of the altitude. Figure 6 illustrates this with the example of *Baetis alpinus*. Development in lower regions shows two emergence peaks, one in spring, the other in autumn, typical for a bivoltine species. The higher the altitude, the later the peaks appear in the year. At about 1400 m development is reduced to a single emergence period in summer that takes place later at higher altitudes; over 2000 m (September/October).

4. Development Cycles

a) Voltinism. The species found in Switzerland are mainly univoltine. *Ephemera* species are all partivoltine, probably semivoltine with a generation every two years. High altitude populations of the Heptageniidae (e.g. *Rhithrogena loyolaea*, *Rh. nivata*, *Ecdyonurus alpinus*) may also belong to this group, exact autoecological data on the species in extreme habitats is lacking.

Table 2. Typology of rare Swiss mayfly species

Categorie	Stream order	Description	Number of stations	% of stations	Zonation
A	1-2	chiefly large and small streams	688	43%	epirhithral
B	3-4	small and medium rivers	545	34%	metarhithral
C	5-6	large rivers at foothills	175	11%	hyporhithral
D	7-8	large rivers in lowlands	63	4%	epipotamal
E		canals, drainage channels	19	1%	
F		lakes, ponds	116	7%	

The majority of the Baetidae and Caenidae are bivoltine (two generations a year), but for the Baetidae at least, this is not valid at higher altitudes where they are univoltine (e.g. *Baetis alpinus,* Fig. 6).

b) Embryonic development. One can separate the Swiss species into two large groups; one that presents a continuous embryonic development and one with an obligatory embryonic diapause. This interruption in the development very often allows the species to survive in unfavourable conditions (too high or too low temperatures). The conditions needed to restart the development are complex (see Ruffieux, 1997 for a view).

In contrast to voltinism there is no strategy common to one genus or family, as both types of development are found within the same genus (e.g. *Baetis*, *Rhithrogena*, *Ecdyonurus*).

5. Typology

Mayfly distribution in various habitats has been analysed. It is striking that most of the species are present in the three subunits of the rhithral: 63 in the epirhithral, 67 in the metarhithral, and 57 in the hyporhithral. The other three habitats are clearly poorer in species: 38 in the epipotamal, 30 in stagnant waters and 27 in canals. The drop in the diversity is caused by the loss of suitable habitats.

The majority of species have more or less clear habitat affinities. The following 13 species were found in all the habitats, were widespread in Switzerland and had a broad ecological valency: *Baetis alpinus*, *B. lutheri*, *B. rhodani*, *B. vernus*, *Cloeon dipterum*, *Centroptilum luteolum*, *Ecdyonurus venosus*, *Heptagenia sulphurea*, *Serratella ignita*, *Caenis luctuosa*, *C. macrura*, *Paraleptophlebia submarginata* and *Ephemera danica*.

Few species colonised only one type of habitat and also no species limited their presence to the hyporhithral or the artificial waterways. The following list indicates the species exclusive to the different habitats.

In the epirhithral *Ecdyonurus parahelveticus, E. zelleri* and *Rhithrogena colmarsensis* were present; in the epipotamal *Heptagenia coerulans*, *H. longicauda* and *Ephoron virgo*; in lentic waters *Caenis lactea*, *Leptophlebia marginata*, *L. vespertina*, *Choroterpes picteti* and *Ephemera glaucops*.

Table 3. References consulted for the faunal analyses of European mayfly species

Austria	Bauernfeind, 1990 a, b, c; Moog, 1995; Weichselbaumer, 1997
Belgium	Müller-Liebenau, 1980; Mol, 1987
Bulgarien	Russev, 1993
Czech Republic	Landa and Soldán, 1985
Denmark	Jensen, 1974; Engblom, 1996
Finland	Saaristo and Savolainen, 1980; Engblom, 1996
France	Jacquemin and Coppa, 1996; Thomas and Masselot, 1996
Germany	Schönemund, 1930; Zimmermann, 1986
Great Britain	Elliot *et al.*, 1988
Greece	Puthz, 1978
Hungary	Ujhelyi, 1966
Ireland	Puthz, 1978
Italy	Gaino *et al.*, 1982; Belfiore, 1983 a, b, 1994; Belfiore and D'Antonio, 1991; Buffagni, 1992
Netherland	Mol, 1983, 1985 a, b, c
North Africa	Kraïem, 1986; Boumaiza and Thomas, 1986, 1995; Gagneur and Thomas, 1988
Norway	Engblom, 1996
Poland	Sowa, 1962, 1975 a, b, 1992
Portugal	Alba-Tercedor, 1981
Roumania	Bogoescu, 1958
Slovakia	Landa and Soldán, 1985
Slovenia	Zabric and Sartori, 1997
Spain	Alba-Tercedor, 1981
Sweden	Engblom, 1996
ex-Yugoslavia	Ikonomov, 1960

If species restricted to one type of habitat (Fig. 7) the number of occasional and accidental species is always greater. Habitats with the most restricted fauna are the epi- and metarhithral as well as lentic habitats (in ≥ 40% of the sites in a habitat). The hyporhithral (80%) and the epipotamal (85%) have many occasional and accidental species. This can be explained by the downstream position of the habitats, thus receiving an important part of the fauna established upstream that drifts down. Artificial waterways appear of little interest to mayflies. A single species (*Caenis robusta*) is occasionally found in this type of habitat; any others are there by accident.

This spread is a compromise determined by factors such as altitude, temperature, food supply and substrate.

6. Status of the Species

The status of the Swiss mayflies is shown in Table 4. About 40% (common and widespread) of the species are well represented in Switzerland. A further third (rare and uncommon) consists of species that are found at only a few sites and the remaining third of very rare species.

In his work on the concept of the rarity of living organisms Gaston (1994) suggests a simple definition for rarity "... I suggest that a useful cut-off point is the first quadrile of the frequency distribution of species abundances or range sizes (i.e. a cut-off of 25%)". If this rule is applied to Swisss mayflies and is used as a measure of frequency for the number of sites colonised by a species, then the pattern in Figure 8 is obtained. According to Gaston (1994) those species to the right of the quadrile should be considered as rare. The species considered by Gaston as rare correspond well to those estimated by us as very rare (25% to 28%), validating our category. Based on current knowledge, classification in the category of very rare species in Switzerland is presence in less than ten locations.

Table 4. Status of the species according to their present occurence

Number of localities	Status	Number of species	Percentage
< 10	very rare	24	28%
11-25	rare	15	18%
26-50	uncommon	13	15%
51-100	common	14	16%
> 100	widespread	19	23%

7. The Future of Swiss Mayfly Fauna

The mayflies in Switzerland arerelatively well known and are represented today by 85 species. A comparison of the diversity with that of other central and southern European countries shows that the size of the territory and the number of species corresponds well (see Sartori, this volume, p. 47). The rise in the number of species shows a positive correlation with the increase in the number of habitats. For the Nordic countries the relationship is clearly different (Engblom, 1996). Their northern latitude and the effects of the last glaciations are the main reason for this difference. In the case of the British Isles their insular nature has limited the number of species.

Certain species have been described for the first time from populations in Switzerland. These are *Baetis nubecularis* Eaton, 1898, *Habroleptoides auberti* (Biancheri, 1954), *Habroleptoides confusa* Sartori and Jacob, 1986, *Rhithrogena enedenensis* Metzler, Tomka and Zurwerra, 1985, *Rhithrogena germanica* Eaton, 1885, *Rhithrogena grischuna* Sartori and Oswald, 1988, *Rhithrogena nivata* (Eaton, 1871), *Ecdyonurus alpinus* Hefti, Tomka and Zurwerra, 1987, *Ecdyonurus helveticus* Eaton, 1883 and *Ecdyonurus parahelveticus* Hefti, Tomka and Zurwerra, 1986.

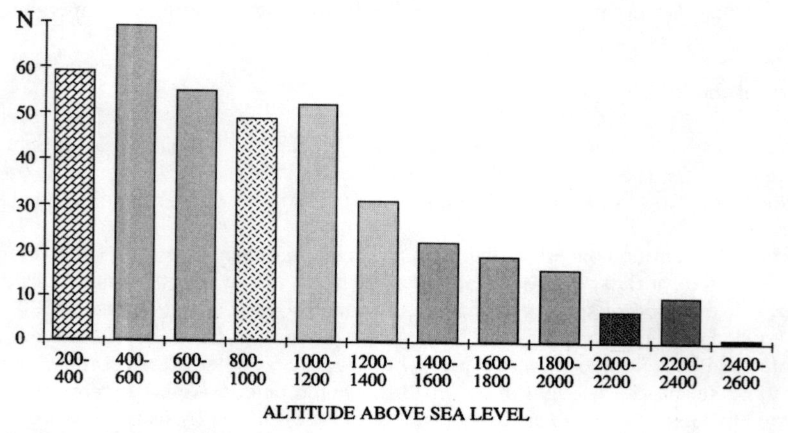

Fig. 4. Distribution of the species according to altitude. The categories correspond to the different altitudinal zones given in the text. N = number of species .

a) Physical changes threatening the mayfly fauna. The increase in erosion in the alpine region because of, among other things, changes in the precipitation (rain instead of snow) increases sediment loads. This change light conditions, hinder the respiration of macroinvertebrates, reduce substrate interstitial spaces and hampers embryonic and larval development of aquatic organisms, including mayflies. Moreover, organic pollution increases the danger of eutrophication of surface waters.

Human activites cause physical changes at various levels. In the mountainous regions the increase in hydroelectric power development is noticed, even when legal minimal flows are respected. The regulation of the flow produces a dramatic change in the natural composition of the fauna and flora.

Table 5. Distribution of species in the five Swiss catchment areas. Grey: species exclusive to Rhine; black: species found in all catchment areas.

Catchment area	Rhine	Rhone	Ticino	Inn	Adige
Siphlonuridae					
Siphlonurus aestivalis	+	+	-	-	-
S. lacustris	+	-	-	+	+
Ameletidae					
Ameletus inopinatus	+	-	-	-	-
Baetidae					
Baetis alpinus	+	+	+	+	+
B. buceratus	+	-	+	-	-
B. fuscatus	+	+	-	-	-
B. liebenauae	+	-	+	-	-
B. lutheri	+	+	-	-	-
B. melanonyx	+	+	+	-	+
B. nubecularis	+	-	-	-	-
B. rhodani	+	+	+	+	+
B. scambus	+	+	-	-	-
B. vardarensis	+	-	-	-	-
B. vernus	+	+	+	+	-
Alainites muticus	+	+	+	-	+
Nigrobaetis niger	+	+	-	-	-
Acentrella sinaica	+	-	-	-	-
Centroptilum luteolum	+	+	+	-	-
Procloeon bifidum	+	-	-	-	-
P. pennulatum	+	+	-	-	-
Cloeon dipterum	+	+	+	+	-
C. simile	+	+	-	-	-
Oligoneuriidae					
Oligoneuriella rhenana	+	-	-	-	-
Heptageniidae					
Epeorus alpicola	+	+	+	+	+
E. sylvicola	+	+	+	-	-
Rhithrogena allobrogica	+	-	-	+	-
Rh. alpestris	+	+	+	+	+
Rh. beskidensis	+	+	-	-	-
Rh. carpatoalpina	+	+	+	-	-
Rh. colmarsensis	+	-	-	-	-
Rh. degrangei	+	+	+	+	+
Rh. dorieri	+	-	-	-	-
Rh. endenensis	+	-	+	+	+
Rh. germanica	+	-	-	-	-
Rh. gratianopolitana	+	+	+	+	-
Rh. grischuna	+	-	+	+	+
Rh. hybrida	+	+	+	-	-
Rh. iridina	+	+	+	-	-
Rh. landai	+	-	-	-	-
Rh. loyolaea	+	+	+	+	+
Rh. nivata	+	+	+	+	+
Rh. puthzi	+	-	-	-	-
Rh. puytoraci	+	+	+	-	-
Rh. savoiensis	+	+	-	-	-
Rh. semicolorata	+	+	+	-	-

Catchment area	Rhine	Rhone	Ticino	Inn	Adige
Ecdyonurus alpinus	+	+	+	+	+
E. dispar	+	+	-	-	-
E. helveticus	+	+	+	+	+
E. insignis	+	+	-	-	-
E. parahelveticus	+	+	-	-	-
E. picteti	+	+	+	+	+
E. torrentis	+	+	-	-	-
E. venosus		+	+	+	-
E. zelleri	+	+	-	-	-
Electrogena lateralis	+	+	+	-	-
E. ujhelyii	+	+	-	-	-
Heptagenia coerulans	+	-	-	-	-
H. longicauda	+	+	-	-	-
H. sulphurea	+	+	-	-	-
Ephemerellidae					
Serratella ignita	+	+	+	-	-
Ephemerella mucronata	+	+	-	-	-
E. notata	+	-	-	-	-
Torleya major	+	+	-	-	-
Caenidae					
Caenis beskidensis	+	+	+	-	-
C. horaria	+	+	+	-	-
C. lactea	+	+	-	-	-
C. luctuosa	+	+	-	-	-
C. macrura	+	+	-	-	-
C. pusilla	+	-	-	-	-
C. rivulorum	+	+	-	-	-
C. robusta	+	+	-	-	-
Leptophlebiidae					
Leptophlebia marginata	+	-	-	-	-
L. vespertina	+	-	-	-	-
Habroleptoides auberti	+	+	-	-	-
H. confusa	+	+	+	-	-
Habrophlebia lauta	+	+	+	-	-
Paraleptophlebia cincta	+	-	-	-	-
P. submarginata	+	+	-	-	-
Choroterpes picteti	+	+	+	-	-
Ephemeridae					
Ephemera danica	+	+	+	-	-
E. glaucops	+	+	-	-	-
E. lineata	+	+	+	-	-
E. vulgata	+	+	-	-	-
Potamanthidae					
Potamanthus luteus	+	-	-	-	-
Polymitarcyidae					
Ephoron virgo	+	-	-	-	-
TOTAL	85	60	36	18	15

In the lowlands the water extraction (e.g. for irrigation) diminishes the volume of water and in summer leads to higher water temperatures that accelerate the growth of the population and may lead to the emergence of smaller individuals with reduced fertility, in some cases causing the disappearance of the population (Sweeney and Vannote, 1978). The damming of rivers, the reinforcement of the banks and the straightening of the rivers standardise the drainage and the sediment composition and leads, therefore, to a uniformity of habitat. Especially in the Swiss Plateau this has resulted in the disappearance of certain mayfly species.

The restoration of certain reaches is often the most promising method of creating new habitats. These "islands" encourage re-colonisation of other sectors.

b) Chemical changes threatening the mayfly fauna. Chemical changes can have important consequences for aquatic organisms. Even in low concentrations the combined effect of certain substances can insidiously cause sub-lethal effects, such as impaired respiration or a decrease in fertility. Increased concentration of nitrogen and phosphorus derivatives, mainly from farming, pesticides and heavy metals may endanger mayfly populations.

Pollution may be of short duration and damages only the individuals exposed. However, if the harmful substances are bond to the substratum they can remain there for a long time and endanger the organisms developing there. Floods can liberate these substances and their toxicity can endanger populations elsewhere. These dangers can appear in all habitats, although the greatest danger potential lies in the lower reaches of running waters and in lakes. An example is the eutrophication of the Lake of Geneva at the end of the 1950s. In this period *Choroterpes picteti*, *Ephemera glaucops* and *E. lineata* became extinct. In the last fifty years the lower reaches of the Rhine have shown an decrease in diversity with the disappearance of, among others, *Oligoneuriella rhenana*, *Heptagenia coerulans, H. longicauda* and *Ephoron virgo*.

Most chemical changes, are however reversible. The macroinvertebrates and especially the mayflies show a remarkable capacity to re-colonise. A good example is the current situation of *Rhithrogena germanica*. In the 1970s it was thought to be extinct, but today as a result of a clear improvement in its original habitats, it is present in several locations.

The re-establishment of a varied fauna in the Rhine after the Schweizerhalle (Sandoz) accident, near Basel, demonstrates how important affluents are as reservoirs for the re-colonisation of the original habitats (Schröder and Rey, 1991). In lakes recovery is much more difficult and is linked to the presence of nearby faunal reservoirs which often do not exist.

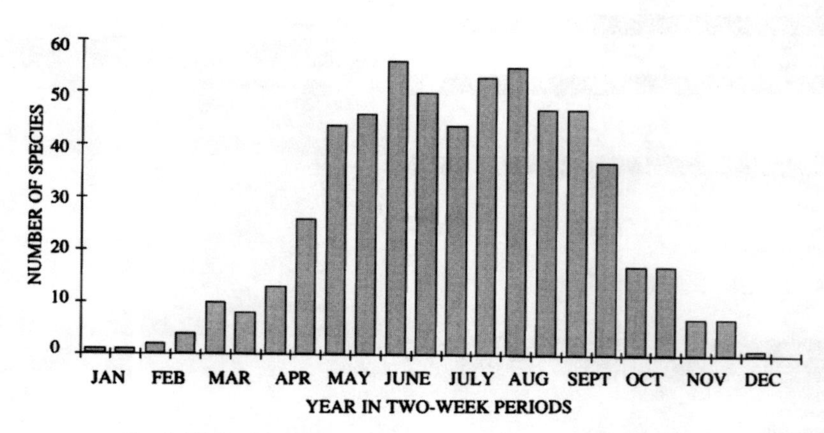

Fig. 5. Flight periods of mayfly species in Switzerland at all altitudes.

c) Typology of Rare Species. The rarity of a species is not only linked to its status in Switzerland, certain other factors must be taken into consideration. The whole distributional area plays a role; it can be vast and cover the greater part of Europe, or be small and limited to a certain region. The habitat requirements can be narrow or broad. Finally the size of the population in the benthic community can be dominant or rare (Rabinowitz, 1981). If one uses this classification system (Table 6), all species fit into one of eight categories. It is interesting that of the 24 species regarded as rare in Switzerland, the majority (75%) occur over a wide distributional area. These species are found in most regions of Europe. One would have expected first to find endemic and alpine species. According to the definition of Gaston (1994), the majority of alpine species do not comply with the criteria necessary to be regarded as rare. The majority of the above species with a wide distributional area inhabit either still waters or the epipotamal. Our conclusions concerning rare Swiss Ephemeroptera may also be valid for other countries. It is therefore desirable that comparable data on a European level be made available.

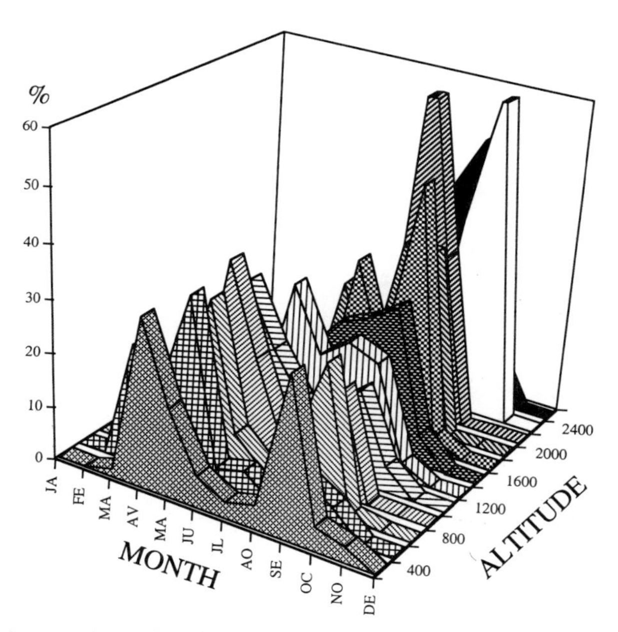

Fig. 6. Emergence of *Baetis alpinus* throughout a year, in relation to altitude. The scale used is the percentage of the numbers of specimens caught per 200 m section.

CONCLUSION

The present state of Ephemeroptera populations in Switzerland is still far from optimal. However, the current change of attitude towards our waters encourages hope for the future. The regeneration of several waterways, above all in German Switzerland, and the use of new technologies in water management point to a change of mentality and an increased sensitivity to the problems linked to the preservation of habitats. The greater the variety that a river offers within its course, the richer the flora and fauna.

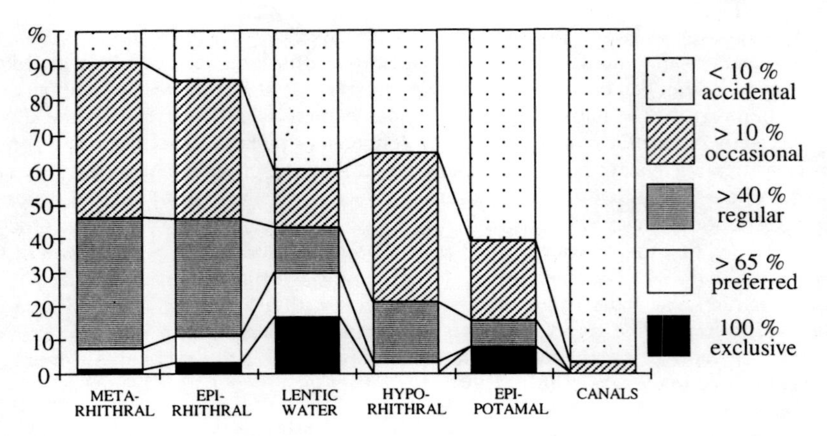

Fig. 7. Distribution by habitat of species (% of the species in the habitats).

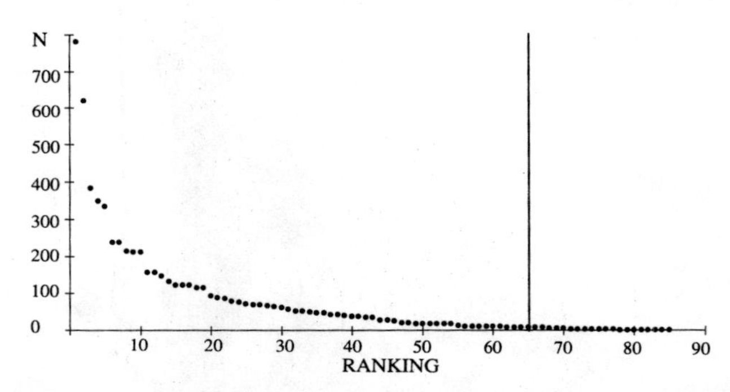

Fig. 8. Occurence of the species according to the number of sites where they were found. The vertical line marks the limit of the 25% quadrile which delimits those species classified as rare.

ACKNOWLEDGMENTS

Without the collaboration of many colleagues and institutions the work of the authors would not have had the same extensive scope. Our sincerest thanks go to all those who enriched our data bank with their numerous captures; to those in charge of institutions (museums, university departments) that put their collections at our disposal and likewise other public and private offices;

Special thanks for their collaboration in taxonomic matters over many years to:

Richard Sowa † (Poland), Alain Thomas (France), Volker Puthz (Germany), Peter Malzacher (Germany), Carlo Belfiore (Italy), Nikita Y. Kluge (Russia), Andreas Zurwerra and Daniel Hefti (Switzerland).

Sheila Watson and Hubert Gachoud are thanked for their technical help. The authors would like to warmly thank Denise Studemann and Laurence Ruffieux for their many comments and criticisms throughout this work.

Table 6. Attempt at a typology of mayfly species in Switzerland with respect to rarity, modified according to Rabinowitz (1981).

Distribution	Habitat specificity	Size and dominance of population	Examples
wide	wide	large, often dominant	*common species*
wide	wide	small, seldom dominant	*Ephemera vulgata* *Ecdyonurus insignis*
wide	narrow	large, often dominant	*Siphlonurus aestivalis* *Ephemera glaucops* *Ephemera lineata* *Leptophlebia marginata* *Choroterpes picteti* *Ephoron virgo*
wide	narrow	small, seldom dominant	*Ameletus inopinatus* *Nigrobaetis niger* *Acentrella sinaica* *Procoleon bifidum* *Baetis vardarensis* *Heptagenia longicauda* *Heptagenia coerulans* *Leptophlebia vespertina* *Paraletophlebia cincta*
restricted	wide	large, often dominant	*Baetis libenauae*
restricted	wide	small, seldom dominant	*Rhithrogena puthzi*
restricted	narrow	large, often dominant	*Baetis nubecularis*
restricted	narrow	small, seldom dominant	*Rhithrogena allobrogica* *Rhithrogena colmarsensis* *Rhithrogena landai* *Ecdyonurus zelleri*

The investigation to establish the atlas of the Swiss Ephemeroptera was financelly supported by grants from diverse institutions, including the Swiss Agency for the Environment, Forests and Landscape (Berne), the Swiss National Foundation of Scientific Research (Berne) and the Centre Suisse de Cartographie de la Faune (CSCF) in Neuchâtel. The Swiss Academy of Science (Berne) and the "Fondation du Fonds de la Recherche de l'Université de Fribourg" allow one of us (P.L.) to present this study at the congress in Argentina. Finally, this publication was made possible through the courtesy of the CSCF allowing us to re-use part of our data.

An anonymus reviewer is thanked for his most valuable comments.

REFERENCES

Alba-Tercedor, J. 1981. Recopilacion de citas de efemeropteros en la Peninsula iberica e islas Baleares.Trab. Monogr. Dep. Zool. Univ. Granada (N.S.) 4 (2): 41-81.

Anonymus, 1992. Atlas Hydrologique de la Suisse. Service hydrologique et geologique national, Berne.

Bauernfeind, E. 1990a. Der derzeitige Stand der Eintagsfliegen-Faunistik in Österreich (Insecta, Ephemeroptera). Verh. Zool.-Bot. Ges. Österreich 127: 61-82.

Bauernfeind, E. 1990b. Einige für Österreich neue oder wenig bekannte Eintagsfliegen (Insecta, Ephemeroptera). Linzer Biol. Beitr. 22: 341-347.

Bauernfeind, E. 1990c. Eintagsfliegen. Nachweise aus Oberösterreich (Insecta, Ephemeroptera), die Sammlung Adlmannseder am O.Ö. Landesmuseum Linz. Linzer Biol. Beitr. 22: 349-356.

Belfiore, C. 1983a. Efemerotteri. Guide per il ricinoscimento delle specie animale delle acque interne italiane. AQ/1/2011-113.

Belfiore, C. 1983b. Note su alcune specie del genere *Habroleptoides* Schönemund, con signalazione per l'Italia di *H. auberti* Biancheri, 1954. Boll. Soc. ent. ital. 115(1-3): 5-6.

Belfiore, C. 1994. Gli Efemerotteri dell'Appenino marchigiano (Insecta, Ephemeroptera). Biogeographica 17:173-181 (1993).

Belfiore, C. and D'Antonio C. 1991. Faunistic, taxonomic and biogeographical studies of Ephemeroptera from Southern Italy. In: Alba-Tercedor, J. and Sanchez-Ortega, A. (Eds). Overview and strategies in Ephemeroptera and Plecoptera. Sandhill Crane, Gainsvile: 253-262.

Bogoescu, C. 1958. Ephemeroptera. Fauna Rep. Pop. Rom. Vll (3):1-187.

Bouzmaia, M. and A. G. B. Thomas. 1986. Repartition et ecologie des Ephemeropteres en Tunisie:lère partie (lnsecta,Ephemeroptera). Arch. lnst. Pasteur Tunis 63 (4): 567-600.

Bouzmaia, M. and A. G. B. Thomas. 1995. Distribution and ecological limits of Baetidae vs the other mayfly families in Tunisia: a first evaluation (Insecta, Ephemeroptera). Bull. Soc. Hist. Nat.Toulouse 131: 27-33.

Brittain, J. E. 1983. The influence of temperature on nymphal growth rates in mountain stoneflies (Plecoptera). Ecology 64: 440-446.

Buffagni, A. 1992. *Baetis liebenauae* Keffermüller, 1974 (Ephemeroptera, Baetidae) in Pianura Padana. Boll. Mus. reg. Sci. nat.Torino 10: 333-340.

Clifford, H. C. 1982. Life cycle of mayflies, with special reference to voltinism. Quest. Ent. 18 (1-4): 15-90.

Elliott, J. M., U. H. Humpesch and T. T. Macan. 1988. Larvae of the British Ephemeroptera: a key with ecological notes. Sci. Publs Freshw. Biol. Assoc. 49: 1-145.

Engblom, E. 1996. Ephemeroptera, Mayflies. In: Nilsson, A. (ed.). Aquatic Insects of North Europe - A taxonomic handbook. Apollo Books, Stenstrup: 13-53.

Gagneur, J. and A. G. B. Thomas. 1988. Contribution à la connaissance des Ephéméroptères d'Algerie. I. Répartition et écologie (lère partie) (Insecta, Ephemeroptera). Bull. Soc. Hist. Nat.Toulouse 124: 213-223.

Gaino, E., C. Belfiore and S. Spano. 1982. Gli Efemerotteri delle Alpi Liguri. Lav. Soc. ital. Biogeogr. 9: 1-19.

Gaston, K. J. 1994. Rarity. Population and community biology series 13. Chapman and Hall, London: 1-205.

Ikonomov, P. 1960. Die Verbreitung der Ephemeroptera in Macedonien. Acta Mus. Mac. Sc. Nat. Skopje 7: 59-74.

Jacquemin, G. and G. Coppa. 1996. Inventaire des Ephémères de Lorraine et de Champagne-Ardenne (N-E France): premiers résultats (Ephemeroptera). Mitt. schweiz. ent. Ges. 69: 141-155.

Jensen, C. F. 1974. Dagsländor (Ephemeroptera) i Kaltisjokkomrädet. Norrb. Natur smäsk 1.

Kraïem, M. 1986. Contribution a l'étude hydrobiologique de trois cours d'eau du nordouest de la Tunisie. Bull. mens. Soc. Lin. Lyon 55 (3): 96-104.

Landa, V. and Soldan, T. 1985. Distributional patterns, chronology and origin of the czechoslovak fauna of mayflies (Ephemeroptera). Acta ent. bohemoslov. 82 (4): 241-268.

Mol, A. W. M. 1983. *Caenis lactea* (Burm.) in the Netherlands. Ent. Bericht. (Deel) 43: 119-123.

Mol, A. W. M. 1985a. Een overzicht van de Nederlandse haften (Ephemeroptera). 1. Siphlonuridae, Baetidae en Heptageniidae. Ent. Ber. 45: 105-111.

Mol, A. W. M. 1985b. Een overzicht van de Nederlandse haften (Ephemeroptera). 2. Overige families. Ent. Ber. 45: 128-135.

Mol, A. W. M. 1985c. Enkele interessante en nieuwe Nederlandse haften (Insecta: Ephemeroptera) uit de province Limburg. Natuurhist. Maandblad 74: 5-8.

Mol, A. W. M. 1987. *Caenis beskidensis* Sowa new to Belgium, with remarks on the Ephemeroptera of the river Meuse. Ent. Ber. (Amst.) 47 (4): 60-64.

Moog, O. (ed.). 1995. Fauna Aquatica Austriaca. Wasserwirtschaftskataster, Bundesministerium für Land- und Forstwirtschaft, Wien.

Müller-Liebenau, I. 1980. Die Arten der Gattung Baetis Leach der belgischen Fauna aus der Sammlung im Museum des Institut royal des Sciences Naturelles de Belgique in Brüssel. Bull. Inst. r Sci. nat. Belg. 52 (3): 1-31.

Puthz, V., 1978. Ephemeroptera in: J. Illies Ed. Limnofauna Europaea. G. Fischer Verlag, Stuttgart: 256-263.

Rabinowitz, D. 1981. Seven forms of rarity, pp. 205-217. In: H. Synge (ed.). The biological aspects of plant conservation. Wiley and Sons, New York.

Ruffieux, L. 1997 Réserves énergétiques et stratégies de reproduction chez les Ephéméroptères (Insecta, Ephemeroptera). PhD Thesis, Université de Lausanne: 1-168.

Russev, P. 1993. Review of literature and established mayfly species (Ephemeroptera) from Bulgaria. Lauterbornia 14: 71-77.

Saaristo, M. I. and E. Savolainen. 1980.The Finnish mayflies. Notul. ent. 60: 181-186.

Sartori, M. and P. Landolt. 1999. Atlas de distribution des éphémères de Suisse (Insecta, Ephemeroptera). Centre Suisse de cartographie de la faune et Société entomologique suisse, Neuchâtel (eds.). Fauna Helvetica 2: 1-212.

Sartori, M., P. Landolt and A. Zurwerra, A. 1994. Liste rouge des éphémères de Suisse (Ephemeroptera). In: Duelli, P. (Ed.). Liste rouge des espèces animales menacées de Suisse. Office fédéral de l'environnement, des forêts et du paysage: 72-74.

Schönemund, E. 1930. Eintagsfliegen oder Ephemeroptera. Dahl's Tierwelt Deutschands 19: 1-103.

Schröder, P. and P. Rey. 1991. Fliessgewässernetz Rhein und Einzugsgebiet. IFAH-Scientific Publications, Konstanz: 1-303.

Sowa, R. 1962. Material for the study of Ephemeroptera and Plecoptera in Poland. Acta Hydrobiol. 4: 205-224.

Sowa, R. 1975a. Ecology and biogeography of mayflies of running waters in the polish part of the Carpathians. 1: Distribution and quantitative analysis. Acta Hydrobiol. 17 (3): 223-297.

Sowa, R. 1975b. Ecology and biogeography of mayflies of running waters in the Polish part of the Carpathians. 2: Life cycles. Acta Hydrobiol. 17 (4): 319-353.

Sowa, R. 1992. Ephemeroptera. In: Glowacinski, Z. (Ed.), Red list of threatened animals in Poland, Polish Academy of Sciences, Krakow: 97-101

Studemann, D., P. Landolt, M. Sartori, D. Hefti and I. Tomka. 1992. Ephemeroptera. Insecta Helvetica, Fauna 9: 1-174.

Sweeney, B. W. and R. L. Vannote. 1978. Variation and the distribution of hemimetabolous aquatic insects: two thermal equilibrium hypotheses. Science 200: 444-446.

Thomas, A. and G. Masselot. 1996. Les Ephémères de la France: inventaire des espèces signalees et des espèces potentielles (Ephemeroptera). Bull. Soc. ent. France 101: 467-488.

Ujhelyi, S. 1966. The mayflies of Hungary, with the description of a new species, *Baetis pentaphlebodes* sp n. (Ephemeroptera). Acta Zool. Acad. Sc. Hung. 12: 203-210.

Weichselbaumer, P. 1997. Die Eintagsfliegen Nordtirols (Insecta: Ephemeroptera). Ber. Nat.-Med. Ver. Innsbruck 84: 321-341.

Zabric, D. and M. Sartori. 1997. First contribution to the mayfly fauna from Slovenia (Ephemeroptera). In: P. Landolt and M. Sartori (eds). Ephemeroptera and Plecoptera: Biology-Ecology-Systematics. MTL, Fribourg: 147-151.

Zimmermann, W. 1986. Neue Funde bemerkenswerter Eintagsfliegen (Ephemeroptera) aus Thüringen. Ent. Nachr. 30 (2): 69-71.

Zollhöfer, J. M. 1997. Quellen, die unbekannte Biotope erfassen, bewerten, schützen. Bristol-Stiftung, Zürich 6: 1-153.

DISTRIBUTION OF *ANACRONEURIA* SPECIES (PLECOPTERA; PERLIDAE) IN COLOMBIA

María del C. Zúñiga[1], Bill P. Stark[2], Angela M. Rojas[1], and Martha L. Baena[1]

[1] Universidad del Valle
Departamentos de Biología y Procesos Químicos y Biológicos
Apartado Aéreo 25360, Cali, Colombia
[2] Department of Biology, Mississippi College
Clinton, MS 39058, USA

ABSTRACT

Thirty six *Anacroneuria* Klapálek species are recorded for Colombia including 23 recently described. These species are distributed primarily in the Region Andina; 13 are associated with the Andina Occidental, 11 with the Andina Central and 12 with the Andina Oriental. Two species occur in the Region del Pacifico near its border with the Andina Occidental and one unidentified female is known from the Region de la Amazonia; presently no records are available from the Region de la Orinoquia or the Region del Caribe. The presence of only a few species in more than one region suggests a pattern of strong endemism within the three cordilleras.

INTRODUCTION

Although no comprehensive publications exist for Colombian Plecoptera, by 1930 twenty *Anacroneuria* and a single *Klapalekia* species had already been proposed based on Colombian specimens. Unfortunately these species were not recognizable from the descriptions and most were reported from single, often cryptic, or non-specific localities (*e.g.* "Colombia" or "Monte Socorro"). The present study is a summary of the results from the 1991-1996 field seasons in Valle del Cauca, Cauca, Nariño, and Quindio, and from the examination of material in the National Museum of Natural History, Washington, D.C. Twenty three species recently described, twelve species previously known, and one species represented by an unidentified female are included in this material but their descriptions are presented elsewhere (Stark et al., 1999) and in this paper we address only their regional distributions.

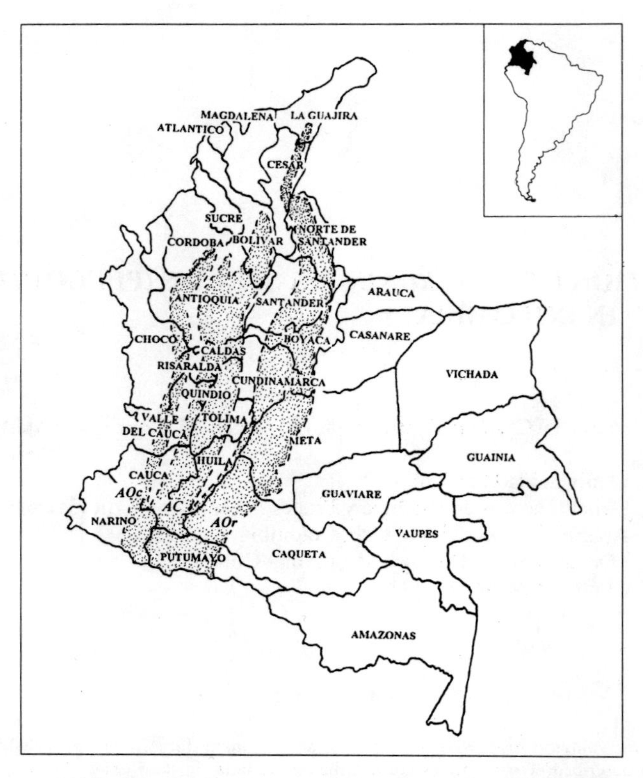

Fig. 1. Political map of Colombia with the Región Andina shaded. AOc= Andina Occidental, AC=Andina Central, AOr= Andina Oriental.

RESULTS AND DISCUSSION

The 36 *Anacroneuria* species known to us from Colombia (Table 1) are distributed principally within the three parallel cordilleras which comprise the Región Andina (Fig. 1). One unidentified female is known from the Región de la Amazonia and two species are known from the Región del Pacífico, but near the boundary of the Andina Occidental. Presently no specimens are available from the Región de la Orinoquia or the Región del Caribe but this doubtlessly is a reflection of collecting effort.

Región Andina Occidental

This region extends from southern Córdoba in the north, to Nariño in the south. Thirteen *Anacroneuria* species currently are known from this region (Table 1) and all are known from at least one site in the Departamento Valle del Cauca; two species also occur in Departamento de Nariño. Of the 13 species listed from the Andina Occidental, only *Anacroneuria anchicaya*, *A. azul*, *A. cipriano* and *A. quilla* occur in another region. The first species is known from Río Anchicayá on the western slope of the Andina Occidental and from Río Desbaratado on the western slope of Andina Central in Valle del Cauca. The second species is known from Río Azul on the western slope of the Andina Occidental, Departamento del Valle del Cauca and from two sites on the western slope of the Andina

Table 1. Colombian *Anacroneuria* species list and regional distribution. An unidentified femal is designated by number.

Species	Department	Natural Region	# Sites
1. *A. albimacula* Klapálek	Antioquia, Cundinamarca	Andina Central y Oriental	2
2. *A. angusticollis* (Enderlein)	Cundinamarca?	Andina Oriental	1
3. *A. apicalis* (Enderlein)	Cundinamarca?	Andina Oriental	1
4. *A. bifasciata* (Pictet)	?		
5. *A. bolivari* (Banks)	Antioquia, Risaralda	Andina Central	3
6. *A. farallonensis* Rojas & Baena	Valle del Cauca	Andina Occidental	1
7. *A. fenestrata* (Pictet)	?		
8. *A. paleta* Stark	Santander	Andina Oriental	1
9. *A. pallens* Klapálek	Cundinamarca	Andina Oriental	1
10. *A. pehlkei* (Enderlein)	Tolima	Andina Central	1
11. *A. schmidti* (Enderlein)	Cundinamarca	Andina Oriental	1
12. *A. vespertilio* Klapálek	Cundinamarca	Andina Oriental	1
13. *A. anchicaya* Baena & Zúñiga	Valle del Cauca	Andina Occidental y Central	2
14. *A. azul* Rojas & Baena	Cauca, Valle del Cauca	Andina Occidental y Central	3
15. *A. calima* Baena & Rojas	Valle del Cauca	Andina Occidental	1
16. *A. choachi* Stark & Zúñiga	Cundinamarca	Andina Oriental	1
17. *A. cipriano* Zúñiga & Rojas	Valle del Cauca	Andina Occidental y Pacìfica	2
18. *A. cordillera* Rojas & Zúñiga	Valle del Cauca	Andina Occidental	1
19. *A. forcipata* Rojas & Baena	Valle del Cauca	Andina Occidental	1
20. *A. guambiana* Zúñiga & Stark	Cauca	Andina Central	5
21. *A. guayaquil* Zúñiga & Rojas	Quindio	Andina Central	1
22. *A. meta* Stark & Zúñiga	Meta	Andina Oriental	1
23. *A. morena* Stark & Zúñiga	Cundinamarca	Andina Oriental	1
24. *A. oreja* Zúñiga & Stark	Valle del Cauca	Andina Occidental	1
25. *A. pacifica* Rojas & Baena	Valle del Cauca	Andina Occidental	2
26. *A. paez* Zúñiga & Stark	Cauca	Andina Central	2
27. *A. planada* Baena & Rojas	Nariño, Valle del Cauca	Andina Occidental	2
28. *A. portilla* Stark & Rojas	Cauca	Andina Central	1
29. *A. quilla* Stark & Zúñiga	Risaralda, Valle del Cauca	Andina Occidental y Central	4
30. *A. regleta* Stark & Rojas	Meta	Andina Oriental	1
31. *A. rosita* Stark & Rojas	Caquetá	Andina Oriental	1
32. *A. socapa* Stark & Zúñiga	Antioquia, Risaralda	Andina Central	2
33. *A. tejon* Baena & Stark	Nariño, Valle del Cauca	Andina Occidental	2
34. *A. undulosa* Stark & Rojas	Chocó	Pacífica	1
35. *A. valle* Zúñiga & Baena	Valle del Cauca	Andina Occidental	1
36. *A.* sp. 1	Amazonas	Amazónica	1

Central, Departamento del Cauca, whereas the latter species is known from three sites west of Cali and one locality on the western slope of the Andina Central, Departamento de Risaralda. *Anacroneuria cipriano* is known from Rio Azul (see *A. azul*) and from Rio San Cipriano near the border of the Andina Occidental in the Region del Pacifico.

Región Andina Central

This region extends from Antioquia to Nariño, lying parallel to the Andina Occidental and separated from this cordillera by a valley approximately 35 km wide at its widest point. Presently eleven *Anacroneuria* species are known from this region (Table 1) and only *A. anchicaya*, *A. azul* and *A. quilla*, as discussed above, occur in another region. *Anacroneuria pehlkei* (Enderlein) was named from Natagaima, a site on the Rio Magdalena, Departamento del Tolima, in the approximately 80 km wide valley lying between the Andina Central and Andina Oriental.

Región Andina Oriental

This region extends from Departamento Caquetá in the south to Guajira and Cesar in the northeast along the Venezuelan border. Twelve species are listed for this region (Table 1) but this includes six of the older species described from "Bogotá" or "Rio Magdalena" by Enderlein (1909) and Klapálek (1921, 1922) and *A. paleta*, a species recently described from Venezuela (Stark, 1995). No recent Colombian material of the Enderlein or Klapálek species is available but *A. angusticollis* is common in recent collections from Ecuador (Stark et al., 1999). No species reported from this region other than *A. angusticollis, A. albimacula* and *A. paleta* are currently known from localities outside the Andina Oriental . *Anacroneuria albimacula* was originally described from Bogotá and Stark et al. (1999) selected a neotype from Antioquia in the Andina Central.

Región del Pacífico

This region extends along the Pacific coast from the Panamanian border to the Mira Valley along the border with Ecuador and west of the Andina Occidental. Two species, *Anacroneuria cipriano* and *A. undulosa* are presently known from this region but from sites near the boundary with the Andina Occidental. No specimens are available from the low coastal mountains, the Serranía de Baudó or Serranía de Los Saltos, where some overlap with the mesoamerican fauna might occur.

Indefinite localities

Two species, *A. bifasciata* and *A. fenestrata* are known from Colombian specimens and have also been reported from Venezuela (Stark, 1995). Unfortunately, no recent Colombian collections of these species are available but both are expected to occur in the northern extension of the Andina Oriental and perhaps in the Region del Caribe.

ACKNOWLEDGMENT

This study was supported, in part, by the Instituto Colombiano de Investigaciones y Proyectos especiales "Francisco Jose de Caldas" - COLCIENCIAS and Universidad del Valle.

REFERENCES

Enderlein, G. 1909. Klassification der Plecopteren sowie Diagnosen neuer Gattungen und Arten. Zool. Anz. 34: 385-419.

Klapálek, F. 1921. Plécoptères nouveaux. Anns. Soc. ent. Belg. 61: 57-67, 146-150, 320-327.

Klapálek, F. 1922. Plécoptères nouveaux. Anns. Soc. ent. Belg. 62: 89-95.

Stark, B. P. 1995. New species and records of *Anacroneuria* (Klapálek) from Venezuela (Insecta, Plecoptera, Perlidae). Spixiana 68: 211-249.

Stark, B. P., M. del C. Zúñiga, A. M. Rojas and M. L. Baena. 1999. Colombian *Anacroneuria*: Descriptions of new and old species (Insecta: Plecoptera: Perlidae). Spixiana 22: 13-46.

A CLADISTIC ANALYSIS OF *AUSTROPHLEBIOIDES* AND RELATED GENERA (LEPTOPHLEBIIDAE: ATALOPHLEBIINAE)

Faye Christidis

Department of Zoology and Tropical Ecology
James Cook University
Townsville 4811, Australia

ABSTRACT

Phylogenetic relationships among the Australian genera and species of the *Meridialaris* lineage were investigated using a cladistic analysis of 35 morphological characters, to test the monophyly of this group and determine possible placement of undescribed taxa. The monophyly of a clade containing the three Australian genera of the *Meridialaris* lineage, *Austrophlebioides* Campbell and Suter, *Kirrara* Harker and *Tillyardophlebia* Dean, was strongly supported. Several undescribed species formed a monophyletic group with *Austrophlebioides pusillus* (Harker), suggesting the placement of these species in *Austrophlebioides*. The placement of several other undescribed taxa, not easily accommodated in currently recognised genera, is discussed. The outcomes from the parsimony analysis of the Australian genera agree in part with Pescador and Peters' (1980) phylogeny.

INTRODUCTION

The Leptophlebiidae is the largest mayfly family in Australia comprising 64 described species in 15 genera. All Australian genera belong to the subfamily Atalophlebiinae and are believed to be members of a Gondwanan lineage which is distributed throughout the southern hemisphere. A phylogeny was proposed by Pescador and Peters (1980) for this southern group of genera, in which they recognised five main monophyletic lineages, the *Hapsiphlebia, Penaphlebia, Nousia, Dactylophlebia* and *Meridialaris* lineages. Their phylogeny was based on morphological characters of the nymphs and adults and was constructed using traditional Hennigian methods. Character polarity was inferred by reference to a hypothetical ancestor and the possession of shared derived character states was used to define each monophyletic lineage.

The *Meridialaris* lineage is argued to be the most derived (Pescador and Peters 1980), with representative genera in Australia (*Austrophlebioides* Campbell and Suter, *Kirrara* Harker and *Tillyardophlebia* Dean), South America (*Meridialaris* Peters and Edmunds, *Massartellopsis* Demoulin and *Secochela* Pescador and Peters), New Zealand

Table 1. Data matrix used in the cladistic analysis. Characters and character states are described in Appendix 1.

Taxa	Character state
Leptophlebia cupida	0 0 0 0 0 0 0 0 0 0 0 1 0 1 0 0 0 1 2 2 0 2 0 0 0 1 1 0 1 0 1 0 0 0 0
Atalophlebia	0 0 0 2 0 0 1 0 0 1 0 1 0 0 0 0 0 1 0 2 0 3 0 0 1 1 1 0 0 0 0 0 2 0 0
Jappa	0 0 0 1 0 0 1 0 0 1 0 2 0 0 0 0 0 1 2 2 1 2 0 1 1 2 2 0 0 0 1 0 1 0 0
Ulmerophlebia	0 0 0 1 0 0 1 0 0 1 0 2 0 0 0 0 0 1 2 2 1 2 0 1 1 2 2 0 0 0 1 0 1 0 0
Atalomicria	0 0 0 2 0 0 1 0 1 1 0 1 0 0 0 0 0 1 2 2 0 2 0 0 1 1 0 0 0 1 0 0 2 1 0
Kalbaybaria	0 0 0 0 0 0 1 0 0 0 0 0 0 0 0 0 0 0 2 2 0 3 0 0 1 1 1 0 1 0 0 0 2 0 0
Bibulmena	1 1 0 2 0 0 0 0 1 0 0 2 1 1 ? 0 0 0 1 0 0 2 0 ? 1 1 2 0 1 0 0 0 2 0 0
Neboissophlebia	1 1 0 1 0 0 0 0 1 0 0 2 1 0 0 0 0 1 1 0 0 1 0 0 1 1 2 1 0 0 1 0 1 ? 0
"Genus K"	1 1 0 2 0 1 0 0 1 0 0 2 1 1 0 0 0 0 2 0 0 1 0 0 1 1 2 0 0 0 0 0 2 0 0
Garinjuga	1 1 0 2 0 1 0 0 1 1 0 2 0 0 0 1 0 0 2 0 0 2 0 0 0 0 1 0 ? 0 0 0 0 1 0
Nyungara	1 1 0 2 0 0 0 0 1 0 0 1 1 1 ? 1 0 0 1 0 0 1 0 0 0 1 1 1 0 1 0 0 1 1 0
Nousia	1 1 0 2 0 1 0 0 1 0 0 1 1 1 1 1 0 0 1 0 0 1 0 0 0 0 1 1 0 1 0 0 1 0 0
Koorrnonga	1 1 0 2 0 1 0 0 1 0 0 1 1 1 1 1 0 0 1 0 0 1 0 0 0 0 1 1 0 1 0 0 1 0 0
Kirrara procera	2 2 0 2 0 2 0 1 1 0 0 2 0 1 ? 1 1 0 0 0 1 0 1 0 0 0 2 0 1 0 1 1 0 0 0
"Northern *Kirrara*"	2 2 0 2 0 2 0 1 1 0 0 3 0 1 1 1 1 0 0 0 1 0 1 0 0 0 2 0 1 0 1 1 0 0 0
Tillyardophlebia rufosa	1 1 1 2 1 1 0 ? 1 0 1 3 1 1 1 1 1 0 0 0 0 1 0 0 0 0 0 0 1 0 1 1 0 0 1
Tillyardophlebia alpina	1 1 1 2 1 1 0 ? 1 0 1 3 1 1 1 1 1 0 0 0 0 1 0 0 0 0 0 0 1 0 1 1 0 0 1
"WT species 1"	1 1 1 2 1 1 0 ? 1 0 1 3 1 1 1 1 1 0 0 0 0 1 0 0 0 0 2 0 0 0 1 1 1 1 0
"WT species 2"	1 1 1 2 1 1 0 1 1 0 1 3 1 1 1 1 1 0 0 0 0 1 0 0 0 0 ? 0 1 0 1 1 1 0 0
Austrophlebioides pusillus	2 2 1 2 1 1 0 1 1 0 1 3 1 1 1 1 1 0 0 1 0 1 0 0 0 0 2 0 1 0 1 1 1 1 0
"Paluma"	2 2 1 2 1 1 0 1 1 0 0 3 1 1 1 1 1 0 0 1 0 1 0 0 0 0 2 0 1 0 1 1 1 1 0
"Henrietta"	2 2 1 2 1 1 0 1 1 0 0 3 1 1 1 1 1 0 0 1 0 1 0 0 0 0 2 0 1 0 1 1 1 1 0
"Daintree"	2 2 1 2 1 1 0 1 1 0 0 3 1 1 1 1 1 0 0 1 0 1 0 0 0 0 2 0 1 0 1 1 1 1 0

(*Atalophlebioides* Phillips and *Deleatidium* Eaton), Sulawesi (*Sulawesia* Peters and Edmunds), southern India (*Petersula* Sivaramakrishnan) and Madagascar (*Petersophlebia* Demoulin) (Pescador and Peters 1980, 1982, 1987; Towns and Peters 1996; Campbell and Peters 1986; Campbell and Suter 1988; Peters and Edmunds 1990; Dean 1997). Members of this lineage possess the following characters: labrum broader than clypeus, clypeus with lateral margins apically divergent, lateral margins of submentum without any setae, galea-lacinia of maxilla apically broad and ninth sternum of the adult female entire or slightly cleft.

This investigation is part of a study in progress on the systematics of the Australian genera of the *Meridialaris* lineage, namely *Austrophlebioides*, *Kirrara* and *Tillyardophlebia*. The genus *Austrophlebioides* currently comprises five described species. *Kirrara procera* Harker is presently the only species of the monotypic genus *Kirrara*, and the recently established genus, *Tillyardophlebia*, comprises two species *T. rufosa* and *T. alpina* Dean. In addition, numerous undescribed taxa are known, some of which appear to belong to recognised genera, particularly *Austrophlebioides*, while others are not easily accommodated in any of the described genera as presently defined. Placement and rank of these taxa is uncertain. The phylogenetic relationships among the Australian genera and species of the *Meridialaris* lineage are presently unknown and the monophyly of this group has not been tested using explicit cladistic parsimony analysis.

In this study a cladistic analysis based on morphological characters of nymphs and adults was used to investigate the phylogenetic relationships of the Australian genera and species of the *Meridialaris* lineage. The aims of this analysis were to test the monophyly of this group and to determine possible placement and rank of undescribed taxa. The study also provided the opportunity to compare the tree obtained from parsimony analysis of the Australian genera with Pescador and Peters' (1980) phylogeny for the Gondwanan genera of the southern hemisphere.

METHODS

Morphological Characters

A total of 35 informative morphological characters were coded from the nymphs and adults of 23 taxa. Nymphal characters were scored from various body regions including the mouthparts, thorax, abdomen and legs. Adult characters were scored from the fore and hind wings, male genitalia and female abdomen. Characters were coded as either binary or qualitative multi-state. Characters and character states used in the analysis are listed in Appendix 1. The data matrix is given in Table 1. The terminology used in this study follows that of Edmunds *et al.* (1976). Morphological characters identified by other systematists as phylogenetically informative (Pescador and Peters 1980; Towns and Peters 1980, 1996) were included in the matrix.

Taxa

The following taxa were included in the analysis:

(1) *Austrophlebioides pusillus* (Harker) and three undescribed species which appear to belong to this genus, collected from northern Queensland and informally named after their collection localities ("Paluma", "Henrietta" and "Daintree").

(2) *Kirrara procera* and an undescribed species, "Northern Kirrara", collected by Edgar Riek from northern Queensland, that may belong in *Kirrara.*

(3) Two species of *Tillyardophlebia*, *T. rufosa* and *T. alpina*, and two undescribed taxa designated "WT species 1" and "WT species 2".

The association of adults and nymphs of all undescribed species, except "Northern *Kirrara*", was by rearing. Morphological characters were coded from nymph and adult material and from published descriptions (Campbell and Peters 1986; Campbell and Suter 1988; Dean 1997).

Representative species of the remaining genera described from Australia were also included in the analysis to investigate the higher level relationships proposed by Pescador and Peters (1980). Taxa included were: *Atalophlebia* sp., *Kalbaybaria doantrangae* Campbell, *Ulmerophlebia* sp., *Jappa edmundsi* Skedros and Polhemus, and *Atalomicria sexfasciata* (Ulmer) (*Hapsiphlebia* lineage); *Garinjuga maryannae* Campbell and Suter (*Penaphlebia* lineage); *Nousia fuscula* (Tillyard), *Koorrnonga inconspicua* (Eaton) and *Nyungara bunni* Dean (*Nousia* lineage); *Bibulmena kadjina* Dean and *Neboissophlebia hamulata* Dean (affinities uncertain). A species of an undescribed genus, "Genus K", was also included. Morphological characters of these taxa were coded from examination of nymphal and adult material and from the literature (Suter 1986; Campbell and Peters 1993; Campbell and Suter 1988; Dean 1987, 1988; Campbell 1993).

Phylogenetic Analysis

Cladistic analyses were performed using PAUP 3.1.1 (Phylogenetic Analysis Using Parsimony) (Swofford 1993). The branch-and-bound search option was used with characters treated as unordered and equally weighted. A representative species from the subfamily Leptophlebiinae, *Leptophlebia cupida* (Say), was used as the outgroup. Bootstrap analysis (500 randomizations) was used to assess support for each node of the tree (Felsenstein 1985; Hillis and Bull 1993).

RESULTS

Cladistic analysis resulted in eighteen most parsimonious trees, each with a length of 91 steps, a consistency index (CI) excluding uninformative characters of 0.527 and a retention

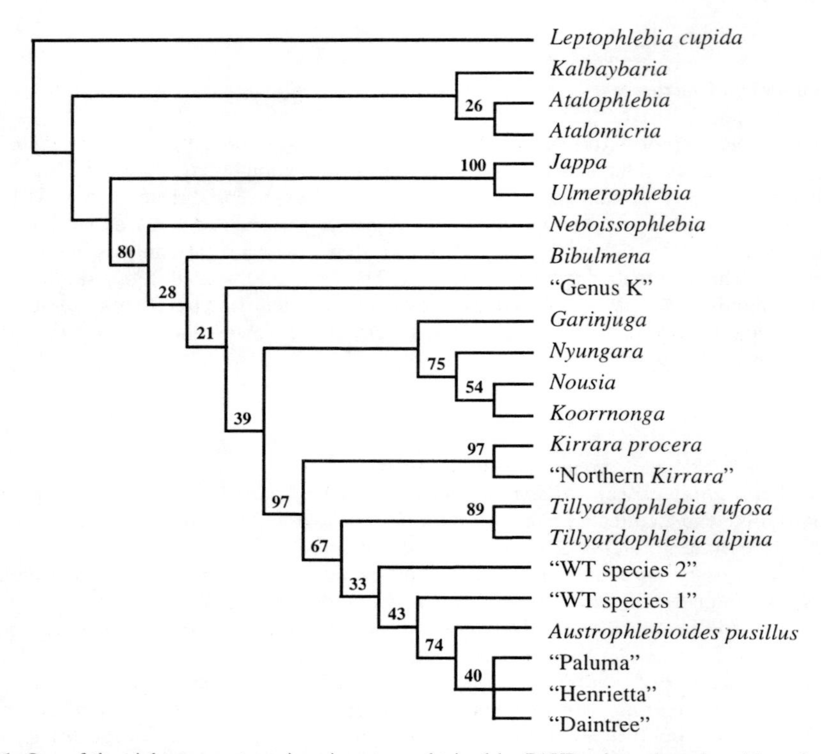

Fig. 1. One of the eighteen most parsimonious trees obtained by PAUP using a branch-and-bound search. Bootstrap values are shown above each node.

index (RI) of 0.809. One of these trees is shown in Figure 1 and a strict consensus tree is given in Figure 2.

Atalophlebia, Kalbaybaria, Atalomicria, Ulmerophlebia and *Jappa,* all members of the *Hapsiphlebia* lineage, occupy a basal position on the strict consensus tree (Fig. 2). Relationships among these genera were unresolved except for a sister group relationship between *Jappa* and *Ulmerophlebia* (Fig. 2).

All remaining ingroup taxa were placed in a separate clade (Fig. 2). Bootstrap support for this clade was 80%. *Nousia*, *Koorrnonga* and *Nyungara* formed a monophyletic group with bootstrap support of 75%. There was strong support for the monophyly of the clade containing *Kirrara*, *Tillyardophlebia* "WT species 1", "WT species 2" and *Austrophlebioides*, with a very high bootstrap value of 97% for this node. Within this clade "Northern *Kirrara*" formed a monophyletic group with *K. procera. A. pusillus* and the three undescribed species "Paluma", "Henrietta" and "Daintree" also formed a monophyletic group, as did the two *Tillyardophlebia* species. "WT species 1" and "WT species 2" were excluded from both the *Austrophlebioides* and the *Tillyardophlebia* groups.

Relationships among *Neboissophlebia*, *Bibulmena*, "Genus K" and *Garinjuga* were unresolved (Fig. 2). On some of the most parsimonious trees *Garinjuga* was placed as the sister group to the clade containing *Nousia*, *Koorrnonga* and *Nyungara* (Fig. 1), while on other most parsimonious trees (not shown) it was the sister group to the clade comprising *Kirrara*, *Tillyardophlebia* "WT species 1", "WT species 2" and *Austrophlebioides*.

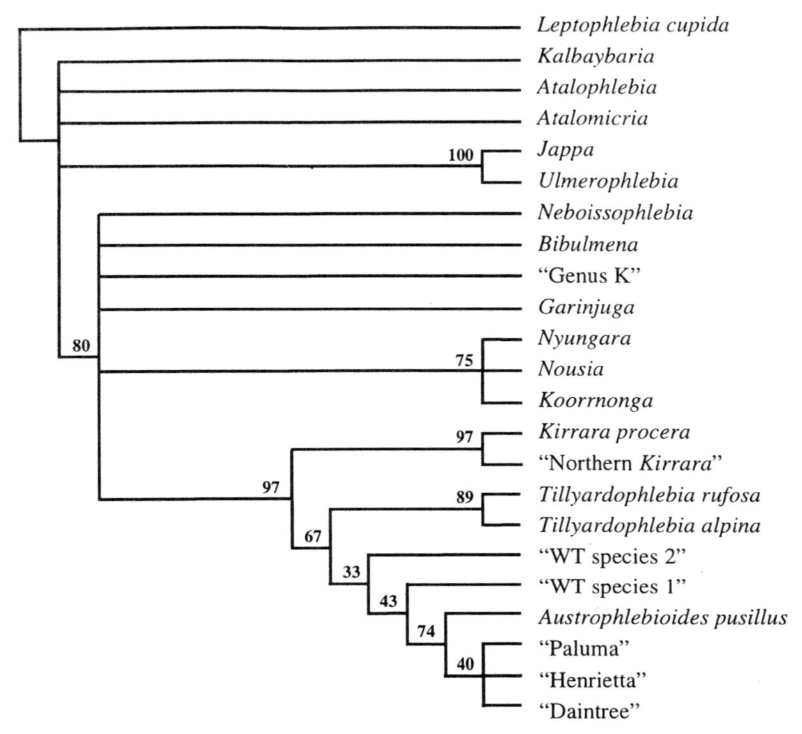

Fig. 2. Strict consensus tree of the eighteen most parsimonious trees obtained by PAUP using a branch-and-bound search. Bootstrap values are shown above each node.

DISCUSSION

Phylogenetic Relationships Among the Australian Genera of the *Meridialaris* Lineage

The findings of this study support the monophyly of a clade containing all Australian members of Pescador and Peters' *Meridialaris* lineage (*Kirrara*, *Tillyardophlebia*, *Austrophlebioides*, "WT species 1", and "WT species 2"). Character states unique to this group were the absence of hairs on the lateral margins of the submentum (character 17) and the ninth sternum of the adult female entire or only slightly cleft (character 32). These character states were identified by Pescador and Peters (1980) as synapomorphies for this group. The results indicate that "Northern *Kirrara*" and *Kirrara procera* are closely related and may be considered congeneric. The two species share many features. The nymphs of both species possess large plate-like gills with the ventral lamella of each gill greatly reduced, and mandibles with outer margin strongly right angled. Wing venation and male genitalia of the two species are also very similar. The nymphs of "Northern *Kirrara*" are easily distinguished from *K. procera* by the presence of a suction disc on the labrum and absence of median protuberances on the abdominal terga. These two species appear to be only distantly related to the other Australian members of the *Meridialaris* lineage.

Cladistic analysis supports the tentative placement of the undescribed species "Paluma", "Henrietta" and "Daintree" into *Austrophlebioides*. The three undescribed species differ from the original diagnosis given by Campbell and Suter (1988) in the absence of a series of fine ventral spines on the penes, and lack of fine setae on the outer margin of the mandible between the setal tuft and outer incisor. Parnrong and Campbell (1997) recently modified the generic description of *Austrophlebioides* to accommodate a new species, *A. marchanti*, which also lacked ventral spines on the penes. Further revision of the generic diagnosis of *Austrophlebioides* may be necessary as additional species are described.

The exclusion of "WT species 1" and "WT species 2" from both the *Austrophlebioides* and the *Tillyardophlebia* group would suggest placement of these taxa in a new genus or genera. The nymphs of "WT species 1", and "WT species 2" are very similar to those of *Tillyardophlebia*. Shared characters states include: labrum a little broader than the clypeus, anterior margin of labrum hooded and with a U-shaped emargination, and absence of a fringe of fine setae on the lateral margins of the abdominal segments. The genitalia of the adult males, however, are unlike those of *Tillyardophlebia*. They are fused along most of their length and lack the two ventral spines present at the base of the penes of *Tillyardophlebia*. It is unclear at the present time whether these two species are congeners.

Comparison with Pescador and Peters' Phylogeny

The outcomes from the parsimony analysis of the Australian genera agreed in part with Pescador and Peters' proposed phylogeny; however, relationships among the Australian genera in the present study were far less resolved. As previously discussed, a clade comprising all of the Australian genera of Pescador and Peters' *Meridialaris* lineage is strongly supported by this study. The results also support the basal position of *Atalophlebia*, *Atalomicria*, *Kalbaybaria*, *Ulmerophlebia* and *Jappa*, relative to the other genera. However, relationships among these genera were poorly resolved, with some uncertainty as to whether these genera represent a monophyletic lineage. The exception was the sister group relationships between *Jappa* and *Ulmerophlebia*. The close relationship between these two genera has previously been discussed by Tsui and Peters (1975) and Suter (1986).

The grouping of *Nousia*, *Koorrnonga* and *Nyungara* into a monophyletic clade is in agreement with the current placement of these genera (Pescador and Peters 1980; Campbell and Suter 1988; Dean 1987). The relationships among *Neboissophlebia*, *Bibulmena*, *Garinjuga* and "Genus K" were unresolved and remain problematic.

ACKNOWLEDGMENTS

I would like to thank John Dean, Peter Cranston, Richard Pearson, Alistair Cheal, Lynne van Herwerden and Brendon McKie for their constructive comments on the manuscript. Additional thanks go to Bill Peters, John Dean and Peter Cranston for making available some of the specimens used in this study. This research was supported by a grant and PhD scholarship from the Land and Water Resources Research and Development Corporation. I would also like to thank the Cooperative Research Centre for Tropical Ecology and Management, and James Cook University for travel awards enabling me to present this paper at the IX International Conference on Ephemeroptera. Finally, I would like to thank the two referees for their comments on the manuscript.

APPENDIX 1
MORPHOLOGICAL CHARACTERS USED IN THE CLADISTIC ANALYSIS

(*) Characters identified by Pescador and Peters (1980) and Towns and Peters (1980, 1996) as phylogenetically informative.

Nymph

Clypeus / labrum:
1.* Labrum width / clypeus width: (0) labrum narrower than clypeus, (1) labrum subequal to slightly wider than clypeus, (2) labrum wider than clypeus.
Clypeus:
2.* Lateral margins: (0) parallel, (1) slightly divergent anteriorly, (2) strongly divergent anteriorly.
Labrum:
3.* Median hood: (0) absent, (1) present.
4. Setae on dorsal surface of labrum: (0) scattered, (1) in 3 rows, an apical, median and basal row (2) in 1 or 2 rows, an apical plus subapical row or apical row only.
5.* Anterior margin of labrum: (0) entire or with broad emargination, (1) with narrow U-shaped emargination.
Mandible:
6.* Shape of outer margin: (0) angled, (1) rounded, (2) right angled.
7. Right outer incisor: (0) slender, parallel-sided, (1) robust, triangular.
8.* Serrations on apex of right outer incisor: (0) absent, (1) present.
9. Tuft of long setae midway along outer margin of mandible: (0) absent, (1) present.
10. Long setae along outer margin: (0) absent, (1) present.
11. Fine setae along outer margin between setal tuft and outer incisor: (0) absent, (1) present.
Maxillae:
12.* Galea-lacinia: (0) subapical pectinate setae absent, (1) with less than 15 subapical pectinate setae, (2) with 16 to 23 subapical pectinate setae, (3) usually with more than 24 subapical pectinate setae.
13. Spines or long setae present on outer margin of: (0) cardo and stipes, (1) cardo only.
14. Outer margin of cardo fringed with: (0) long setae, (1) short spine like setae.
15. Blunt seta on outer margin of stipes: (0) absent, (1) present.
Labium:
16.* Glossae: (0) not on the same plane as paraglossae, (1) on about the same plane as paraglossae.
17.* Submentum: (0) with spines or setae on lateral margins, (1) without spines or setae on lateral margins.
Thorax:
18. Setae on lateral margins of pronotum: (0) absent, (1) present.
Abdomen:
19. Postero-lateral projections on segments: (0) 2 to 9, (1) 6 to 9, (2) 7 or 8 to 9.
20*. Lateral margin of abdominal segments: (0) bare, without fringe of fine setae, (1) fringed with fine setae only, (2) fringed with thick setae or spines.
21. Row of setae on mid-dorsal region of segments: (0) absent, (1) present.
Gills:
22. Gill shape: (0) plate-like, (1) lanceolate, tapering gradually to a fine point, (2) lanceolate to ovate, tapering at about two-thirds the length of the gill and ending in a single apical filament, (3) divided.
23. Gill size: (0) lower and upper gill about the same size, (1) lower gill greatly reduced.
24. Outer margin of gill: (0) not fringed with fine setae, (1) fringed with fine setae.
Legs:
25. Tarsal claws with ventral teeth: (0) present, (1) greatly reduced to fine denticles or absent.
Caudal filaments:
26. Segments with: (0) whorls of spines, with or without short setae, (1) whorls of spines and long setae, (2) long setae only.

Adult

Fore wing:
27. Attachment of ICu1: (0) free basally, (1) attached to CuA, (2) attached to CuA-CuP cross vein.
Hind wing:
28. Length of hind wing relative to fore wing: (0) hind wing not greatly reduced, greater than 0.2 of the length of fore wing, (1) hind wing reduced, less than 0.2 of the length of fore wing.
29. Subcostal vein: (0) less than 0.9 of wing length, (1) greater than 0.9 of wing length.
30. Total number of cross veins in hind wing: (0) more than 16, (1) less than 15.
Legs:
31.* Tarsal claws: (0) similar, (1) dissimilar.
Female abdomen:
32.* 9th sternite: (0) with deep to moderate cleft, (1) very shallow cleft to entire.
Male genitalia:
33.* Penes: (0) divided except at base, (1) separated in apical $^{1}/_{6}$ to $^{2}/_{3}$, (2) fused.
34. Prominent spines on penes: (0) absent, (1) present.
35. Pair of ventral spines near base of penes: (0) absent, (1) present.

REFERENCES

Campbell, I. C. 1993. A new genus and species of leptophlebiid mayfly (Ephemeroptera: Leptophlebiidae: Atalophlebiinae) from tropical Australia. Aquat. Insects 15: 159-167.

Campbell, I. C. and W. L. Peters. 1986. Redefinition of *Kirrara* Harker with a redescription of *Kirrara procera* Harker (Ephemeroptera: Leptophlebiidae: Atalophlebiinae). Aquat. Insects 8: 71-81.

Campbell, I. C. and W. L. Peters. 1993. A revision of the Australian Ephemeroptera genus *Atalomicria* Harker (Leptophlebiidae: Atalophlebiinae). Aquat. Insects 15: 89-107.

Campbell, I. C. and P. J. Suter. 1988. Three new genera, a new subgenus and a new species of Leptophlebiidae (Ephemeroptera) from Australia. J. Aust. ent. Soc. 27: 259-273.

Dean, J. C. 1987. Two new genera of Leptophlebiidae (Insecta: Ephemeroptera) from south-western Australia. Mem. Mus. Vict. 48: 91-100.

Dean, J. C. 1988. Description of a new genus of leptophlebiid mayfly from Australia (Ephemeroptera: Leptophlebiidae: Atalophlebiinae). Proc. R. Soc. Vict. 100: 39-45.

Dean, J. C. 1997. Description of new Leptophlebiidae (Insecta: Ephemeroptera) from Australia. I. *Tillyardophlebia* gen. nov. Mem. Mus. Vict. 56: 83-89.

Edmunds G. F. (Jr.), S. L. Jensen and L. Berner. 1976. The mayflies of North and Central America. 330 pp. University of Minnesota Press, Minneapolis.

Felsenstein, J. 1985. Confidence limits on phylogenies: An approach using the bootstrap. Evolution 39: 783-791.

Hillis, D. M. and J. J. Bull. 1993. An empirical test of bootstrapping as a method for assessing confidence in phylogenetic analysis. Syst. Biol. 42: 182-192.

Parnrong, S. and I. C. Campbell. 1997. Two new species of *Austrophlebioides* Campbell and Suter (Ephemeroptera: Leptophlebiidae) from Australia, with notes on the genus. Aust. J. ent. 36: 121-127.

Pescador, M. L. and W. L. Peters. 1980. Phylogenetic relationships and zoogeography of cool-adapted Leptophlebiidae (Ephemeroptera) in Southern South America, pp. 43-56. In: J. F. Flannagan and K. E. Marshall (eds.). Advances in Ephemeroptera Biology. Plenum, New York.

Pescador, M. L. and W. L. Peters. 1982. Four new genera of Leptophlebiidae (Ephemeroptera: Atalophlebiinae) from southern South America. Aquat. Insects 4: 1-19.

Pescador, M. L. and W. L. Peters. 1987. Revision of the genera *Meridialaris* and *Massartellopsis* (Ephemeroptera: Leptophlebiidae: Atalophlebiinae) from South America. Trans. Amer. ent. Soc.: 112:147-189.

Peters, W. L. and G. F. Edmunds (Jr.). 1990. A new genus and species of Leptophlebiidae: Atalophlebiinae from the Celebes (Sulawesi) (Ephemeroptera), pp. 327-335. In: I. C. Campbell (ed.). Mayflies and Stoneflies:life histories and biology. Kluwer Academic, Dordrecht, The Netherlands.

Suter, P. J. 1986. The Ephemeroptera (Mayflies) of South Australia. Rec. S. Aust. Mus. 19: 339-397.

Swofford, D. L. 1993. Phylogenetic Analysis Using Parsimony (PAUP), version 3.1.1. Illinois Natural History Survey, Champaign.

Towns, D. R. and W. L. Peters. 1980. Phylogenetic relationships of the Leptophlebiidae of New Zealand (Ephemeroptera), pp. 57-69. In: J. F. Flannagan and K. E. Marshall (eds.). Advances in Ephemeroptera Biology. Plenum, New York.

Towns, D. R. and W. L. Peters. 1996. Leptophlebiidae. Fauna of New Zealand 36:1-143.

Tsui, P. T. P. and W. L. Peters. 1975. The comparative morphology and phylogeny of certain Gondwanian Leptophlebiidae based on the thorax, tentorium, and abdominal terga (Ephemeroptera). Trans. Amer. ent. Soc. 101: 505-595.

REDESCRIPTION AND PHYLOGENETIC RELATIONSHIPS OF *LEENTVAARIA* DEMOULIN (EPHEMEROPTERA: LEPTOPHLEBIIDAE)

Eduardo Domínguez[1], María Joze Ferreira[2], and Carolina Nieto[3]

[1] CONICET-Facultad de Ciencias Naturales, U.N.T.
Miguel Lillo 205, (4000) Tucumán, Argentina
[2] INPA-Bioecologia, Alameda Cosme Ferreira 1756
Cx. Postal 478, 69.000 Manaus, Amazonas, Brazil
[3] Facultad de Ciencias Naturales, U.N.T.
Miguel Lillo 205, (4000) Tucumán, Argentina

ABSTRACT

The genus *Leentvaaria* was established by Demoulin in 1966 for a single species, *L. palpalis*, known from a few nymphs from Surinam. Since then, there was no other report on this species.

The phylogenetic relationships of the components of the *Hermanella* generic complex were studied by Flowers and Domínguez (1991). Although *Leentvaaria* almost surely belonged to this complex, it was not included because the necessary characters were not available at that time.

Recently, we have obtained new nymphal material of *Leentvaaria* from Brazil, that allowed us to obtain the nymphal characters required to establish the relationships of *Leentvaaria* with the other components of the *Hermanella* complex.

In this study *Leentvaaria* appears as the sister group of *Needhamella* and is included within the *Hermanella* complex. The nymph of *L. palpalis* Demoulin, is redescribed based on the new material.

INTRODUCTION

The genus *Leentvaaria* was established by Demoulin (1966), for a single species, *L. palpalis*, described from 3 nymphs collected in Surinam. Since then, there was no other report of this genus. The phylogenetic relationships of the *Hermanella* generic complex, a distinctive group of leptophlebiid mayflies, were studied by Flowers and Domínguez (1991). The nymphs of this group can be characterized mainly by extremely broad mouthparts, bearing even rows of long setae, and male imagoes with modified subgenital plates. Although *Leentvaaria* almost surely belonged to this complex, it was not included in that analysis because several of the necessary characters were not available at that time. Recently, we have obtained good nymphal material from Brazil, that allowed us to study in detail the nymphs of *Leentvaaria*. Continuing with a series of papers dedicated to the systematics of this group (Domínguez and Flowers, 1989; Flowers and Domínguez, 1992; Savage and Domínguez, 1992) we analyze and propose the phylogenetic relationships of *Leentvaaria* with the other components of the *Hermanella* complex. The nymph of the genus is also redescribed, based on the new material.

MATERIALS AND METHODS

The material used in this study is deposited in the following insitutions: INPA (Instituto Nacional de Pesquisas de Amazonia, Brazil) and IFML (Instituto Fundacion Miguel Lillo, Argentina).

The phylogenetic relationships were analyzed with the aid of the computer programs Pee-Wee (Goloboff, 1993) and CLADOS (Nixon, 1992).

Genus *Leentvaaria* Demoulin

Leentvaaria Demoulin, 1966: 13.

Mature nymph: Head. Prognathous. Antennae 2.5 times length of head. Mouthparts (Figs. 1-7). *Clypeus* with lateral margins strongly concave. Maximum width of labrum 1.5-1.6 times maximum width of clypeus; length of labrum less than 0.4 maximum width, lateral margins rounded as in fig. 1, with posterolateral angulation; anteromedian emargination shallow, V-shaped dorsally, deep ventrally as in fig. 2; divided row of long dorsal setae on basal 1/5 of labrum with 25-28 setae on each side, short setae on margins. *Left mandible* (Fig. 3): outer margin angularly curved, angle sharp. *Maxillae* (Fig. 4) galea-lacinia with one long thick seta on venter, close to inner margin; subapical pectinate setae lacking; very prominent tusk on inner apical angle. Segment 1 of maxillary palpi 0.5 length of segment 2; segment 3 0.8 length of segment 2. Segment 1 with thick setae on outer margin, segment 2 with two long setae on inner apical angle, segment 3 with long setae in ordered rows. Lingua of hypopharinx with well-developed lateral processes, anterior margin with broad median V-shaped cleft; superlingua with long setae along anterior margin (Fig. 5). *Labium* (Fig. 6): Segment 1 of palpi 0.65 length of segment 2, segment 3 0.15 of segment 2. Segment 1 with a basal acute prominence; segment 2 elbowed, widened on distal 1/4 and with a dorsal row of 15-19 setae (Fig. 6); segment 3 curved, with long setae on ventral surface and external margin, short spines on inner margin (Fig. 7); glossae straight, flat, with short setae along anterior margins; paraglossae with subapical row of long setae on ventral surface. Anterolateral margins of pronotum with three large setae. Wing pads glabrous. *Legs* (Fig.8): trochanters with row of setae on apico-dorsal surface; femora with thick, long pointed setae along posterior margin and short spines along inner margin; fore and middle femora with long, fine setae along posterior margin, short spines on dorsum of middle and hind femora; tibiae with short spines along inner margin, fine setae on outer margin; tarsi with short spines on inner margin, setae on outer margin of tarsi 2 and 3. Claws hooked, narrow, ventral denticles as in fig. 9 . Gills (Fig. 10). Gills on segments 1-7, birramous, long and narrow, tapering evenly from base to apex, smaller posteriorly; trachea along median line, not branched. Posterolateral projections on abdominal segments 8 and 9. Terminal filament longer than cerci, small spines on posterior margin of each segment.

Material: BRAZIL, RR, BR 174, Km 914, Rio Paricarana, 28-X-87, Equipo Granfinale, 10 nymphs. 5 in INPA, 5 in IFML.

Distribution: Surinam and Northern Brazil.

Discussion: The type species, described by Demoulin is known only by three young nymphs. It is difficult to know if the species here used to redescribe the genus is conspecific with the type species "*L. palpalis*" and for this reason is not assigned to it. In 1992, Savage

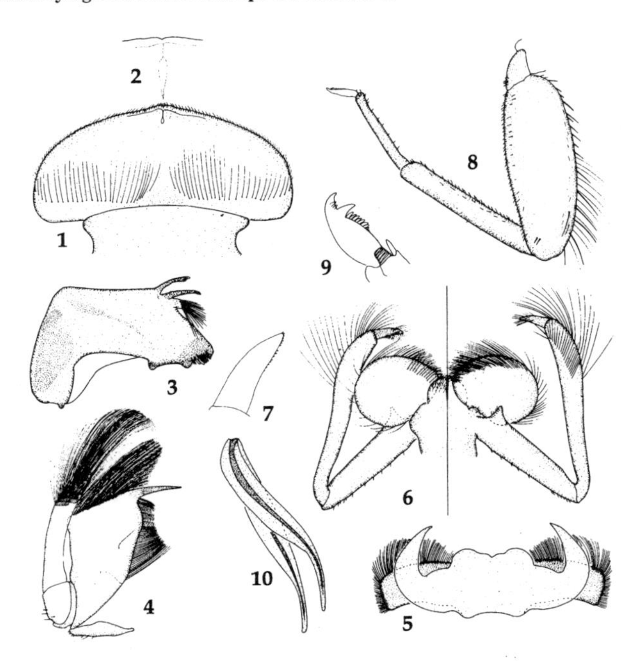

Figs. 1-10. *Leentvaaria* sp. nymph. 1, Labrum and clypeus; 2, Labrum (detail of anteromedian emargination); 3, Left mandible; 4, Maxilla; 5, Hypopharynx; 6, Labium (right, ventral view; left, dorsal view); 7, Detail of third segment of labial palpus; 8, Foreleg; 9, Tarsal claw; 10, Gill.

and Domínguez when establishing the genus *Paramaka*, raised the possibility that *P. convexa*, its type species, could represent the unknown adult of *Leentvaaria*. After comparing the abdominal pattern from the new specimens of *Leentvaaria* with *P. convexa* it appears very improbable that they could be congeneric.

The genus *Leentvaaria* belongs to the *Hermanella* generic complex (Flowers and Domínguez, 1991) based on the following synapomorphies present in the single tree obtained: 10-11, 14-16, 18 and 21-23.

The nymphs of *Leentvaaria* can be separated from the other genera of Leptophlebiidae by the following combination of characters. Labrum as wide as head, with shape and dorsal setae as in fig. 1; long setae on maxillary palpi in even rows (Fig. 4); long row of dorsal setae on segment 2 of labial palpi present; prominent tusk on inner apical margin of maxillae; segment 1 of labial palpi shorter than segment 2 (Fig. 6); enlarged subapical denticle on tarsal claws (Fig. 9).

Leentvaaria is very close to *Needhamella*, from which it can be distinguished by the size of the maxillary tusk, the shape of the gills, and the presence of a basal prominence in segment 1 of labial palpi.

CLADISTIC ANALYSIS

Characters and Coding

For the phylogenetic analysis a matrix of 57 characters (Appendix I) was compiled, including 41 nymphal and 16 adult external morphological characters. In this study 22 taxa are treated, the same that were analyzed in Flowers and Domínguez (1991), except for

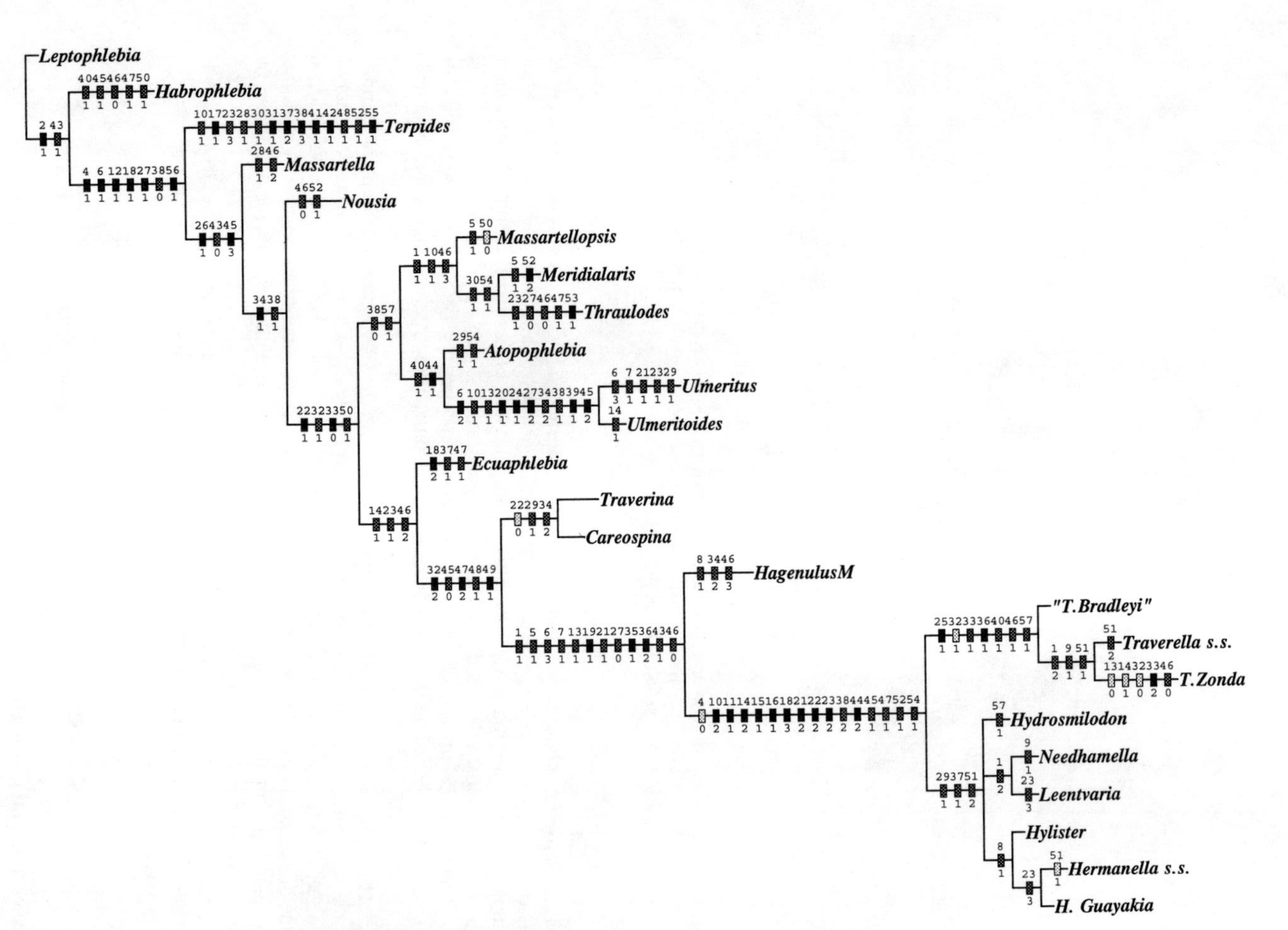

Fig. 11. Cladogram obtained from the analysis. Black boxes= apomorphies; dark grey boxes= parallelisms; light gray boxes= reversals.

Hagenulus caligatus that was not included. Most of the characters are also treated in the same way, except for a few that needed to be recoded due to new evidence. Binary characters were coded as 0 and 1. Multistate characters were assigned different numbers, and treated in two different ways: additive or non additive (see list of characters). Both programs permit the use of "Full polymorphism", coded as "*", and "Subset polymorphism" coded as "$". Characters not comparable or with no information available were assigned a missing code (?).

Character List

1. *Width of labrum/width of clypeus*: < OR = 1.1 (0); 1.2-1.4 (1); = OR > 1.5 (2)[additive].
2. *Lateral margins of labrum*: subparallel (0); rounded to angular (1).
3. *Lateral margins of labrum rounded, widest part on apical 2/3*: no (0) yes (1).
4. *Denticles on anteromedian emargination of labrum*: absent (0); present (1).
5. *Median hood in labrum*: absent (0); present, U-shaped, or V-shaped ventrally (1); cleft (2).
6. *Dorsal row of setae on labrum*: absent (0); apical (1); medial (2); basal (3)[additive].
7. *Shape of dorsal row of setae on labrum*: entire (0); divided (1).
8. *Area anterior to dorsal row of labrum covered with long setae*: absent (0); present (1).
9. *Anteromedian projection of clypeus*: absent (0); present (1).
10. *Lateral margins of clypeus*: parallel (0); divergent (1); strongly concave (2)[additive].
11. *Subapical pectinate setae on maxillae*: present (0); absent (1).
12. *Hairs or spines on brush on anterior margin of maxillae*: scattered or unevenly arranged (0); evenly arranged (1).
13. *Tusk on inner apical margin of maxillae*: absent (0); present (1).
14. *Segment 2/segment 1 of maxillary palpi*: subequal (0); 1.1 - 2 (1); >2 (2)[additive].
15. *Ordered rows of setae on segment 3 of palpi*: absent (0); present (1).
16. *Thick, blunt setae on segment 1 of maxillary palpi*: absent (0); present (1).
17. *Large non pectinated seta on inner apical margin of maxillae*: absent (0); present (1).
18. *Setae on inner margin maxillary palpi 2*: spine-like, along all margin (0); needle-like, apical 2/3 to 1/2 (1); needle-like, apical 1/5 (2); needle-like, apical corner (3); absent (4) [nonadditive].
19. *Strong setae on inner margin of palpi 3*: present (0); absent (1).
20. *Palpifer size of maxillae*: normal (0); enlarged (1).
21. *Position of articulation of palpi of maxillae*: on apical 1/2 (0); medial (1); basal (2); [additive].
22. *Shape of outer margin of mandible*: smoothly curved (0); obtuse (1); right angled (2); [additive].
23. *Setae on outer margin of mandible*: on 2/3 or more (0); on 1/2 (1); on basal 1/4 (2); absent (3); [nonadditive].
24. *Setae at base of outer incisor*: absent (0); present (1).
25. *Patch of long setae on venter of mandible*: absent (0); present (1).
26. *Shape of lingua of hypopharynx*: lateral arms lacking (0); lateral arms present (1).
27. *Long spines on labial palpi*: absent (0); on segment 3 only (1); on segment 2 and 3 (2); [nonadditive].
28. *Glossae of labium curved ventrally*: absent (0); present (1).
29. *Subapical setae row on paraglossae of labium*: absent (0); present (1).
30. *Setae or spines on submentum of labium*: present (0); absent (1).
31. *Anterolateral margins of submentum developed anteriorly*: no (0); yes (1).
32. *Segment 1/segment 2 of labial palpi*: >1.1 (0); subequal: 1.1-0.9 (1); <0.9 (2); [additive].
33. *Segment 3/segment 2 of labial palpi*: < 0.8 (0); 0.8-1.2 (1); >1.2 (2); [nonadditive].
34. *Segment 3 of labial palpi*: triangular (0); elongated (1); shortened (2); [nonadditive].
35. *Shape of labial palpi segment 2*: not elbowed (0); elbowed (1).
36. *Row of dorsal setae on palpal segment 2*: absent (0); present, <4 (1); present, many (2); [nonadditive].

37. *Denticles on tarsal claws*: subequal (0); subapical larger (1); medial larger (2); [nonadditive].
38. *Posterolateral projections on abdominal segments*: 2 or 4 to 9 (0); 5 or 6 to 9 (1); 7 or 8 to 9 (2); 3-6 and 8-9 (3); [nonadditive].
39. *Lateral margins of abdominal terga*: bare or with small spines (0); prominent setae or spines (1).
40. *Gills tracheae*: main tracheae present (0); tracheae divided basally (1).
41. *Rows of setae on base of terminal filaments*: absent (0); present (1).
42. *Dorsal portion of eyes of male on stalk*: no (0); yes (1).
43. *Fork of MA of forewings*: symmetrical (0); asymmetrical (1).
44. *Slanting cross vein above MA fork*: absent (0); present, ma symmetrical (1); present, ma asymmetrical (2); [nonadditive].
45. *Fork of MP of forewings*: symmetrical (0); slightly asymmetrical (1); asymmetrical (2); MP2 attached by cross vein (3); [nonadditive].
46. *Attachment of ICu1*: free basally (0); attached to CUA (1); attached to CUP (2); attached to both (3); [nonadditive].
47. *Shape of costal projection of hind wings*: obtuse (0); acute (1); very acute (2); [nonadditive].
48. *Vein MP of hind wings*: forked (0); unforked (1).
49. *Ending of Sc*: in wing margin (0); in cross vein or costal projection (1).
50. *Claws of a pair*: similar (0); dissimilar (1).
51. *Paired submedial projections on subgenital plate*: absent (0); broad (1); narrow (2); [additive].
52. *Lobes of penis*: completely divided (0); apical 1/2-1/4 separated (1); fused (2); [nonadditive].
53. *Forceps sockets*: separate (0); united (1).
54. *Base of penes abruptly swollen*: absent (0); present (1).
55. *Posterolateral corners of styliger plate*: not developed (0); developed (1).
56. *Styliger plate of males*: deeply cleft (0); fused (1).
57. *9th female abdominal sternite*: strongly cleft (0); entire or shallowly cleft (1).

Outgroup Selection

The genera *Leptophlebia* and *Habrophlebia* were used as outgroups, representing two different lineages of Leptophlebiinae, the sister group of the Atalophlebiinae. It was prefered to use two "real" taxa instead of an "hypothetical ancestor", with all "0" characters.

Analysis

The computer program "Pee-Wee" was used for the cladistic analysis, and "Clados" to show the character distribution in the resultant tree. Pee-Wee is a program for parsimony analysis under implied weights. It searches for trees which maximize fit across character retaining only the trees with highest total fit. Characters are given weight in inverse relation with the amount of homoplasy (extra steps) they show in every tree examined. Trees with the highest total fit (sum of character weight) are considered the best trees. Those trees resolve character conflict in favor of the characters which have less homoplasy on the trees. In this way it is possible to find trees with greatest explanatory power given the weights the characters deserve.

The options used with the program Pee-Wee were "Hold 1000; Mult *20". The command "Hold" determines the number of suboptimal trees retained in memory for the next analysis. The command "Mult" randomizes the order of the taxa in each replication, creating a weighted Wagner tree and submitting it to branch-swapping, repeating the process the number of times indicated, to find all possible "islands" present in the matrix.

RESULTS AND DISCUSSION

Only one tree was obtained with a fit 354.7 (Fig. 11). The topology of this tree is totally compatible with the consensus tree presented in Flowers and Domínguez (1991, Fig. 2), and is almost identical to the tree presented in fig. 3 in that paper (representing one of the 22 obtained in that analysis and used to illustrate the character distribution). The difference, besides the presence of *Leentvaaria*, is the resolution of the relationships of *Terpides*, *Massartella* and *Nousia*, that were unresolved in the previous analysis. Nevertheless, the relationships of the genera outside the *Hermanella* complex must be taken with caution as there are several genera that were not included.

As suspected, *Leentvaaria* belongs to the *Hermanella* complex, due to several synapomorphies (namely, characters 10, 11, 14-16, 18, 21-23 and 44, see Character list). The group of "*Traverella*", composed of the two subgenera of *Traverella*, plus "*T. Bradleyi*", that will represent a different genus, is supported by two synapomorphies: characters 25 and 36.

The *Hermanella* group does not have clear synapomorphies, but is supported by three characters that appear homoplastically: 29, 37 and 51. The relationships of ((*Hydrosmilodon*), (*Needhamella-Leentvaaria*), (*Hylister, (Hermanella s.s.-H. Guayakia)))* are not resolved but, with the characters available *Leentvaaria* appears as the sister group of *Needhamella*. Their relationship is only supported by one homoplastic character: 1(2). It is important to remember, that at present *Leentvaaria* is only known from nymphs, so all its adult characters are missing in this analysis. When the adults of this genus became available, they will allow us to test this hypothesis. It is important to stress the fact that despite the inclusion of a new taxon and the use of different programs for the analysis of the original matrix, the monophyletic groups proposed originally (Flowers and Domínguez, 1991) remained unchanged.

```
                    0    5    10   15   20   25   30   35   40   45   50 . 55
                    |    |    |    |    |    |    |    |    |    |    |    |
@Leptophlebia       -000000??000000000000000000000001000020000000100000000000
@Habrophlebia       -010000??000000000000000000000001000020100101010010000000
Hermanella s.s.     -110013110211121103102230010010020112120000121011111101010
H. Guayakia         -110013110211121103102230010010020112120000121011112101010
Needhamella         -210013101211121103102220010010020112120000121011112101010
Leentvaria          -210013100211121103102230010010020112120000???????????????
Hylister            -110013110211121103102220010010020112120000121011112101010
Traverella s.s.     -210013101211121103102220110000011111020100121111112101011
Hydrosmilodon       -110013100211121103102220010010020112120000121011112101011
T.Zonda             -210013101211011103102220110000002111020100121011111101011
"T.Bradleyi"        -110013100211121103102220110000011111020100121111110101011
Ulmeritus           -010103100101100001011111012010010200011100012100010000011
Ulmeritoides        -010102000101110001010101012000010200011100012100010000011
Atopophlebia        -010101000001000001000100011010010100000100013100010001011
Traverina           -010101000001010001000010011010020200$10000*0022111000010
Careospina          -0101010000010100010000100110100202000010000*0022111000010
Nousia              -010101000001000001000000011000001100$1000000300000010010
Massartella         -010101000001000001000000011100001000000000003200000000010
Massartellopsis     -110111000101000001000100011000010100000000003300000000011
Meridialaris        -110111000101000001000100011001010100000000003300010201011
Thraulodes          -110101000101000001000110010001010100000000003010010011011
Terpides            -010101000101000011000030001101101000230011100101000100110
HagenulusM          -1101131100011100011011100100000202120100000*0032111000010
Ecuaphlebia         -01010100000101000200011001100001010011000000321001000001?
```

Apendix I. Data matrix for the taxa used in this study. Description of characters given in text. Unknown conditions indicated by "?", subset polymorphism by "$" and full polymorphism by "*". Outgroups indicated by "@".

REFERENCES

Demoulin, G. 1966. Contribution a l'etude des Ephemeropteres du Surinam. Bull. Inst. r. Sci. nat. Belg. 42 (37): 1-22.

Domínguez, E. and R. W. Flowers. 1989. A revision of *Hermanella* and related Genera (Ephemeroptera: Leptophlebiidae: Atalophlebiinae) from Subtropical South America. Ann. ent. Soc. Amer. 82: 555-573.

Flowers, R. W. and E. Domínguez. 1991. Preliminary cladistics of the *Hermanella* complex (Ephemeroptera: Leptophlebiidae: Atalophlebiinae). Overview and Strategies on Ephemeroptera and Plecoptera. J. Alba-Tercedor and A. Sanchez-Ortega (eds.), Sandhill Crane Press, Gainesville, Florida. pp. 49-62.

Flowers, R. W. and E. Domínguez. 1992. New Genus of Leptophlebiidae (Ephemeroptera) from Central and South America. Ann. ent. Soc. Amer. 85: 655-661.

Goloboff, P. 1993. Pee-Wee (Ver. 2.8) computer program and manual distributed by the author. INSUE, Facultad de Ciencias Naturales, Universidad Nacional de Tucumán, Argentina.

Nixon, K. C. 1992. CLADOS version 1.2 manual; software and MSDOS program. Cornell University, Ithaca, N.Y.

Savage, H. M. and E. Domínguez. 1992. A new genus of Atalophlebiinae (Ephemeroptera: Leptophlebiidae) from Northern South America. Aquat. Insects 14: 243-248.

PREDACEOUS BAETIDAE IN MADAGASCAR: AN UNCOMMON AND UNSUSPECTED HIGH DIVERSITY

J. L. Gattolliat[1] and M. Sartori[2]

[1] ORSTOM, LRSAE (Lab. Rech. sur les Systèmes Aquatiques et leur environnement), BP 434, 101 Antananarivo, Madagascar
Present address: Museum of Zoology
P.O. Box 448, CH-1000 Lausanne 17, Switzerland
[2] Museum of Zoology, P.O. Box 448
CH-1000 Lausanne 17, Switzerland

ABSTRACT

Mayflies are generally considered as primary consumers; the majority of genera are either filterers or collector-gatherers. Very few feed on invertebrates. Predation can be found in different families, suggesting that the habit evolved independently in major lineages.

Three predaceous genera belonging to the family Baetidae were found in Madagascar. They feed mainly on other mayfly larvae. The diet at different stages was investigated by analysing the gut contents, and comparisons among species were performed. Besides, the study of morphology of mouthparts and legs allows us to propose apomorphies related to the predatory behaviour.

INTRODUCTION

The original nature of Madagascar is demonstrated by the extremely high diversity of some taxa and the absence of others (Paulian, 1996). This originality is the result of several factors. The long isolation of the Malagasy Plate, first from Africa (165 Ma B.P.), then from the subcontinent India (80 Ma B.P.) was the most fundamental event (Battistini, 1996). Major geodynamic events took place during the Eocene (50 Ma B.P.) with the new distribution of continental fragments derived from Gondwanaland. They induced the reorganisation of atmospheric systems and important changes on the Malagasy climate, as well as on its hydrological system (Fröhlich, 1996). As these events took place after the isolation of Madagascar, it allowed the exploitation by the present taxa of new free ecological niches. The absence of predators and more evolved taxa, at different levels of the trophic web, explains the remaining presence or the unusual success of different taxa; for example the absence of Felidae and evolved monkeys allowed the lemurs to survive and diversify (Mittermeier *et al.*, 1994); the rarity of insectivorous fishes (De Rham, 1996) led to the

Trends in Research in Ephemeroptera and Plecoptera
Edited by E. Dominguez, Kluwer Academic/Plenum Publishers, 2001

reduction of predator pressure on the aquatic invertebrate community and allowed the evolution of the flightless mayfly genus *Cheirogenesia* (Ruffieux *et al.*, 1998). On the other hand, Madagascar is constituted by very contrasted climatic areas (tropical rain forest and degraded areas on the Eastern coast; deciduous forest, very dry on the Southern and Western coasts and almost completely degraded into relatively dry grassland on the Highlands); this implies also a high diversity of aquatic environments (Chaperon *et al.*, 1993).

The Baetidae is composed of more than 60 different genera. This is the most cosmopolitan mayfly family with the Leptophlebiidae. It is present on every continent and in every kind of freshwater habitat, from the smallest pond to the waterfall. The baetid diversity is very high in Madagascar. Fifteen genera are actually known, and we estimate that seven others are present but still not mentioned. It means that almost one-third of the world baetid genera are present in Madagascar. Although the present knowledge of Malagasy baetid is still very incomplete, we have evidence that this diversity is not only high at the generic level but also at the specific one.

Most of the baetid genera are considered as collector-gatherers or collector-filterers. They are trophic generalists, extremely facultative in the type of food they consume (Brown, 1961). Among them, only 5 are recognized as carnivorous: *Centroptiloides* Lestage from Africa, *Echinobaetis* Mol from Sulawesi (formerly Celebes), *Harpagobaetis* Mol from Surinam, *Raptobaetopus* Müller-Liebenau from Eurasia and *Barnumus* McCafferty and Lugo-Ortiz from Southern Africa. All these genera are monospecific, except *Raptobaetopus* known by two allopatric species.

Analysis of the gut contents of Malagasy Baetidae allowed the discovery of three further predaceous genera: *Nesoptiloides* Demoulin, *Herbrossus* McCafferty and Lugo-Ortiz and *Guloptiloides* Gattolliat and Sartori. This diversity is unsuspected in comparison to only five other carnivorous genera occurring in the rest of the world. These three genera are strictly endemic to Madagascar. According to our present knowledge, *Nesoptiloides* and *Guloptiloides* are monospecific, *Herbrossus* being represented on the Eastern coast by three species. These three genera are closely related. Actually, the classification of the African Baetidae is still not clearly established. According to Lugo-Ortiz and McCafferty (1998), these three genera belong to the *Centroptiloides* complex, that gathers together all the African carnivorous genera (*Centroptiloides* and *Barnumus*) and also more primitive genera such as *Afroptilum* Gillies and *Dicentroptilum* Wuillot and Gillies. According to Kluge (1997), they belong to the subfamily Afroptilinae.

In this work, we want to compare the feeding behaviour between five species: *Herbrossus edmundsorum* McCafferty and Lugo-Ortiz, *H. christinae* Gattolliat and Sartori, *H. elouardi* Gattolliat and Sartori, *Nesoptiloides electroptera* (Demoulin) and *Guloptiloides gargantua* Gattolliat and Sartori. It is generally admitted that changes in mayfly diet occur during the final larval stages, either for higher-quality nutritional requirement or because larger animals can ingest larger particles (Cummins, 1973; Riaño *et al.*, 1997). Thus the diet was also compared among the different larval stages of each species.

We also observed the functional morphological evolution related to this diet. The most important adaptations have been found on the mouthparts and the legs. As the *Centroptiloides* complex presents all types of feeding behaviour, it appears particularly well-adapted to this study, as it offers, at the same locality, primitive and specialised genera.

MATERIAL AND METHODS

This study took place as a part of a general program of the ORSTOM lab: Biodiversity and Biotypology of the Malagasy Freshwaters. In this aim, more than 700 samples were taken in every types of running waters. The larvae as well as the imagoes of Ephemeroptera and Trichoptera were collected.

We chose stations where we possessed a large number of specimens of different stages. Whenever it was possible, we also chose localities where several carnivorous species were

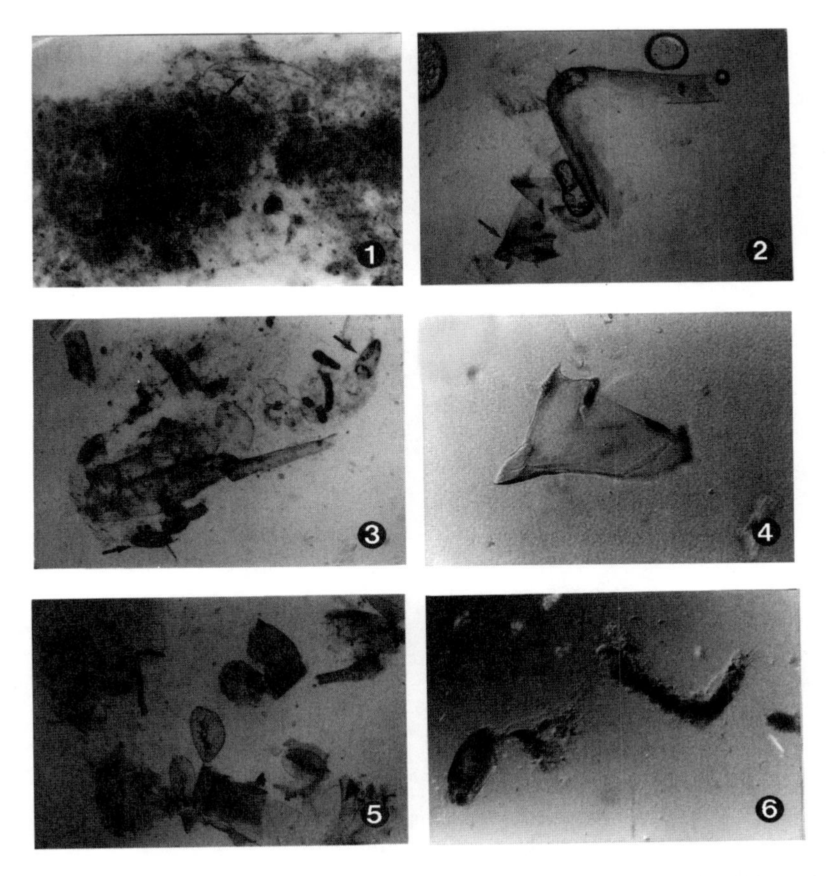

Figs. 1 to 6. Gut contents: 1: second part of the gut content of H. edmundsorum, arrow: baetid leg, enlargement x100. 2: first part of the gut content of *H. edmundsorum*, enlargement x40. 3: gut content of *H. christinae* with pieces of *Xyrodromeus*, arrow: chironomid head. 4: gut content of *H. christinae*: mandible of *Xyrodromeus*, enlargement x400. 5: gut content of *N. electroptera*: pieces of *Afroptilum*, enlargement x40. 6: gut content of *N. electroptera*: chironomid, enlargement x100.

present. We sorted the larvae according to features based on size, the degree of development of the forewing pads, and for the female, the degree of maturation of the eggs. We named the last stage LM for Matured Larva, and then we numbered the different stages in decreasing order.

LM stage: larvae with the forewings well-visible inside the pads, the forewing pads reaching at least the anterior margin of the second tergite and eggs well-separated.

LM-1 stage: forewing scarcely distinctive inside of the pads, forewing pad reaching at least the middle of the first tergite and eggs still piled up.

LM-2 stage: forewing pad reaching at least the anterior margin of the first tergite.

LM-3 stage: forewing pad reaching at least the middle of the metathorax.

Because of the differences between male and female, we could not only use size to sort larval stages. We dissected each specimen and extracted the gut. A brief description of its

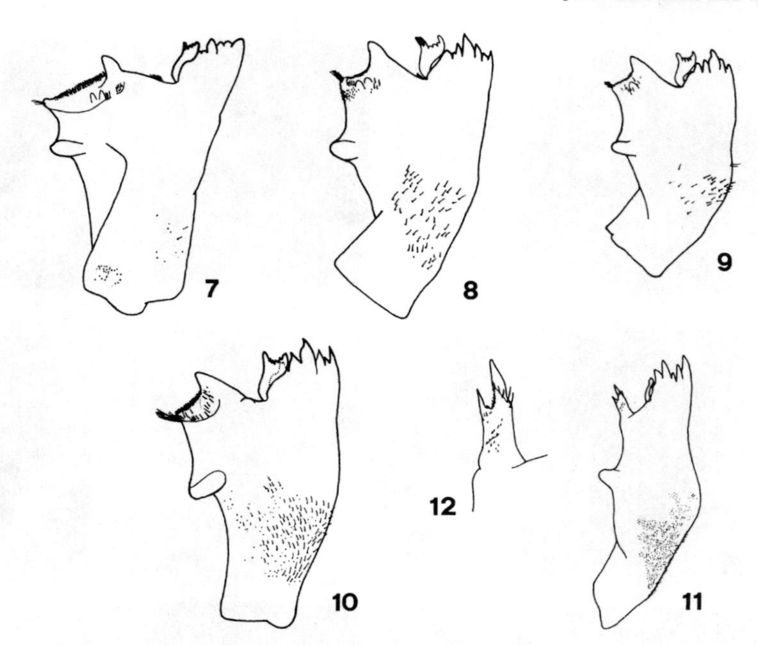

Figs. 7 to 12. Left mandible: 7: *Afroptilum* sp. 8: *H. elouardi*. 9: *H. christinae*. 10: *N. electroptera*. 11: *G. gargantua*. 12: *G. gargantua*: detail of the mola.

contents was made and the proportions of fine detritus, algae, macrophytes and prey were estimated. These values were relatively rough since we wanted to demonstrate a tendency more than an absolute number. Among the invertebrates, we tried to determinate the prey to the lowest systematic level: at the family for all the insects except to the genus for the Baetidae. The gut contents was mounted on slides in Liquid de Faure.

Mouthparts and legs of each species were mounted in Canadian Balsam after being bathed in a Creosote solution. The different parts were compared between species and genera.

RESULTS

Diet

The three species of *Herbrossus* show different foraging strategies.

H. edmundsorum is detritivore during all the larval stages, except the last one. During this stage, its diet changes abruptly, and it becomes a quasi-strict predator (figs 1,2). The prey are mainly larvae of the baetid genera *Afroptilum* and *Dicentroptilum,* and more rarely, Chironomidae larvae (fig. 23).

H. elouardi presents also a transition in the diet. During the young stages, since LM-3, the larvae feed mostly on fine detritus and macrophytes. Until LM-2, the larvae begin to feed on macroinvertebrates, mostly Simuliidae. *H. elouardi* presents a progressive transition from a detritivore to a carnivorous diet. At LM-2, *H. elouardi* feeds quasi-exclusively on Simuliidae. We did not own any data for the LM stage; all our mature specimens presented reduced and empty digestive tracts (fig. 23).

At LM-3, *H. christinae* is already partly carnivorous. Although the invertebrates, predominantly Simuliidae, represent half of the gut contents at LM-2, they are the main energetic

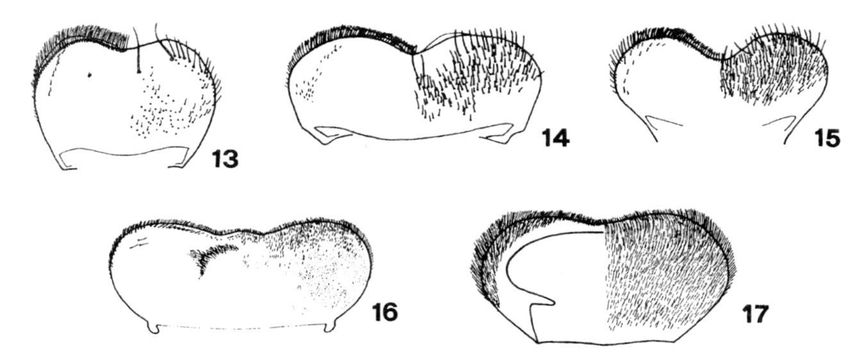

Figs. 13 to 17. Labrum (left ventral, right dorsal): 13: *Afroptilum* sp. 14: *H. elouardi*. 15: *H. christinae*. 16: *N. electroptera*. 17: *G. gargantua*.

source. The proportion of baetid larvae increases since LM-2 and becomes predominant at LM-1 (fig. 23). It is mostly constituted by larvae of the small genus *Xyrodromeus* Lugo-Ortiz and McCafferty (figs 3,4).

Nesoptiloides electroptera is a strict carnivorous species. Since the earliest stages, its gut contains quasi-exclusively invertebrates (figs 5,6). The size and diversity of the prey evolve between the different stages. At first this species feeds mostly on Diptera larvae (Chironomidae, Simuliidae and Blepharoceridae). Since LM-3, the baetid proportion becomes more and more important, with three genera: *Afroptilum*, *Dicentroptilum* and in lesser importance *Xyrodromeus*. At LM-2 and LM-1, the prey are chewed; we found only small particles of cuticles of the same prey instead of whole parts such as legs, head and abdomen.

Little information is available on the diet of *Guloptiloides gargantua*. This species is actually only known at the larval stage by three specimens. Two of them had reduced and empty digestive tracts. Only one had its gut full. Its contents were exclusively baetids of the genus *Afroptilum* (fig. 23). According to the degree of development of the hindwing pads, we assign this larva to LM-3.

Morphological Adaptation

The comparison of the mouthparts of these five species inidicated that the mandibles and the labrum are clearly the most adaptable part. The mandibles of *N. electroptera* (fig. 10), *H. edmundsorum* and *H. elouardi* (fig. 8) are the less transformed; they are stouter with more robust incisors, compared to those of *Afroptilum* (fig. 7). The evolution of the mandibles in *H. christinae* (fig. 9) shows a clear adaptation to predation: reduction of the mola and an increase of the lateral margin angulation (Gattolliat and Sartori, 1998). The molas of *G. gargantua* (fig. 12) are transformed into incisors, and the incisors themselves are completely modified for impaling the prey (Gattolliat and Sartori, 2000). This species presents the highest degree of morphological adaptation.

The labrum of the five species is conspicuously wider than long and almost completely covered dorsally with fine setae, which is not the case of those of *Dicentroptilum* and especially of *Afroptilum* (fig. 13). The three species of *Herbrossus* have a deep and broad U-shaped anteromedian emargination (fig. 14), most developed in *H. christinae* (fig. 15). The labrum of *G. gargantua* (fig. 16) is three-lobed, but the median lobe is shorter and consequently could function as an anteromedian emargination. The labrum of *N. electroptera* (fig. 17) is by far the most transformed. It is dorsoventrally extremely thick, to some extent comparable to the seal elephant nose.

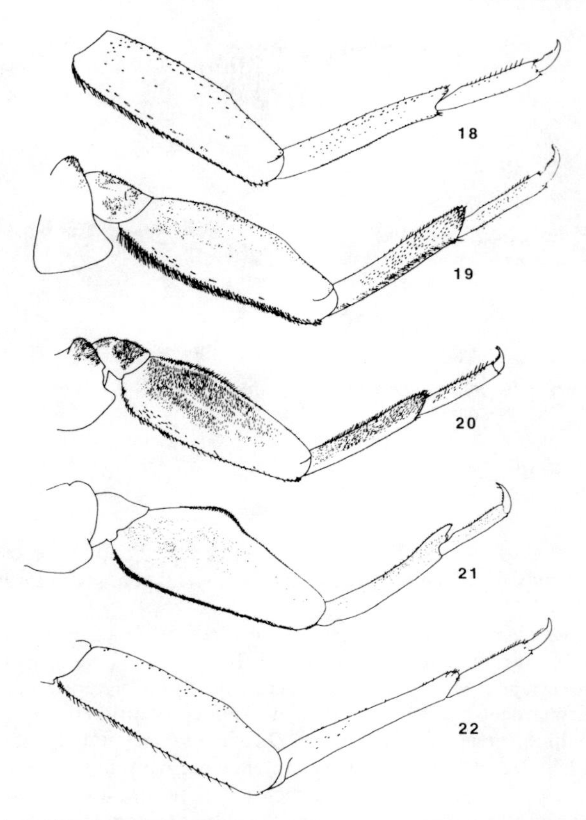

Figs. 18 to 22. Forleg: 18: *Afroptilum* sp. 19: *H. elouardi*. 20: *H. christinae*. 21: *N. electroptera*. 22: *G. gargantua*.

The forelegs of the three species of *Herbrossus* (fig. 19) have numerous, acute setae on the femora and tibiae. As for the mandibles, *H. christinae* (fig. 20) shows the most important adaptation with numerous and well-developed setae on both femora and tibiae. Apart from the numerous setae, *N. electroptera* (fig. 21) shows two apomorphies on the forelegs; femora with pronounced ventral convexity and tibiae with a conspicuous ventrodistal process (Demoulin, 1973). The second apomorphy is clearly an adaptation to catching and holding prey. The forelegs of *G. gargantua* (fig. 22) do not show any apomophy; they are similar to those of *Afroptilum* (fig. 18).

DISCUSSION

Among these five predaceous genera, we found three different strategies.

N. electroptera, *G. gargantua*, and to a lesser extent *H. christinae*, can be considered as strictly carnivorous. Since the earliest stages, they mainly or strictly feed on invertebrates. The only changes are the increase of the size and proportion between Baetidae and Diptera. The earliest stages feed mainly on Diptera, especially on Simuliidae and then switch to Ephemeroptera (mainly on Baetidae) during the next stages. This kind of diet occurs seldom among the carnivorous baetid, most of the genera feeding mainly on Diptera (Chironomidae and Simuliidae). This is the case of *Raptobaetopus* which is strictly carnivorous, feeding on

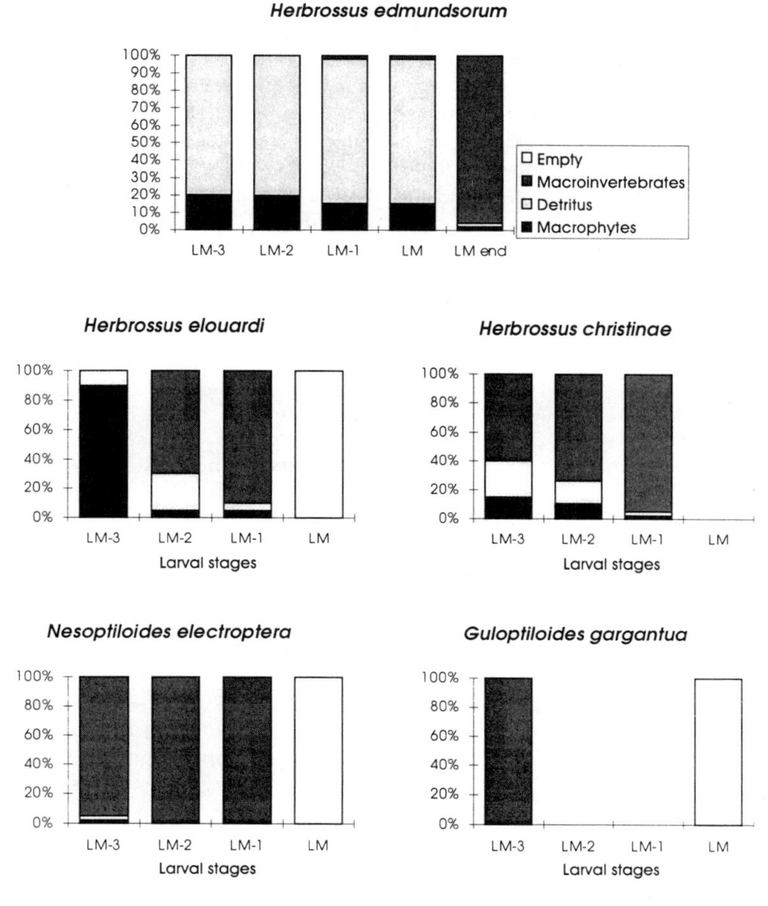

Fig. 23. Diet of the five species. Larval stages: see material and method.

Oligocheta and Diptera (Müller-Liebenau, 1978; Fontaine and Perrin, 1981), and of *Harpagobaetis gulosus* which feeds exclusively on Simuliidae (Mol, 1986). *Echinobaetis phagas* feeds mainly on Baetidae and in a less important proportion on Simuliidae and mayflies from other families (Mol, 1989). *Centroptiloides bifuscata* feeds mainly on Simuliidae and Baetidae. The proportion of Baetidae increases with the size and the larval stage. No filamentous algae were present in the digestive tract and much fine detritus was identifiable as larval remains (Agnew, 1962). This author also suggested that the type of food ingested by *C. bifasciata* depends on the availability and relative abundance of the different prey. He found a more or less direct correlation between the size of the prey eaten and the size of the larva. Prey found in the foregut of *C. bifuscata* seemed to have been little damaged by chewing, although some of the larger larvae appeared to have been merely bitten in half, probably to aid ingestion (Agnew, 1962). These two species, *C. bifuscata* and *E. phagas* appear to be the only species with a diet similar to those of *N. electroptera*, *G. gargantua* and *H. christinae*.

Among the *Centroptiloides* complex, we considered that *Afroptilum* and *Dicentroptilum* are the most primitive species presenting less adaptive structures; the mouthparts and the legs are the most plesiomorphic. In this complex, the mouthparts, particularly the mandibles, show important morphological adaptations. We can mention the large and robust mandibles with specialised armature of *Edmulmeatus* (Lugo-Ortiz and McCafferty, 1997a) and the bladelike mandibles of *Xyrodromeus* (Lugo-Ortiz and McCafferty, 1997b).

The Malagasy carnivorous species show a number of morphological adaptations. Among them, we can list: the short and broad shape of labrum with numerous sensitive setae, mandibles with stout and acute incisors, the increase of the lateral angulation, the molar reduction and its apical transformation to form a process analogous to the incisors (*G. gargantua*, fig. 23). It had been proposed that these adaptations are apomorphies clearly associated with a predatory behaviour (Lugo-Ortiz and McCafferty, 1998). The degree of specialisation of diet is directly related to the degree of adaptation of the mouthparts; for example the most adapted labrum to maintain the prey is the one of *N. electroptera* (fig. 17) and the most adapted mandibles for impaling are those of *G. gargantua* (fig. 11). *N. electroptera* appears to be the only species which chews the prey at least at the last stages. It means that its mandibles (fig. 10) are not only adapted for impaling, as in *G. gargantua*, but also for chewing.

We can find the same adaptations of the mouthparts among the other carnivorous genera of Baetidae (*Raptobaetopus*, *Harpagobaetis*, *Echinobaetis*, *Centroptiloides*), but also among three Northern American predaceous genera of Heptageniidae (McCafferty and Provonsha, 1986).

The forelegs also show significant adaptation for catching, holding and the impaling of the prey. The ventral margin of the femora and tibiae are covered with stout and acute setae. They are especially numerous in *N. electroptera* (fig. 21) and *H. christinae* (fig. 22).

The high degree of development of the femoral ventral convexity and the ventrodistal process of *N. electroptera* represent apomorphic adaptation for catching and impaling the prey. In this way, *N. electroptera* appears to have the most adapted legs for carnivorous behaviour.

H. elouardi changes its diet at the LM-2 stage (fig. 23). During the earliest stage, it can be considered as a collector-shredder, then its diet becomes more and more carnivorous, feeding mostly on Simuliidae. All the mature larvae caught had empty gut. We can presume that it must not be an important change of diet with the previous stage, maybe the proportion of Baetidae would increase. Its mouthparts and legs are less modified, allowing a more omnivorous diet.

H. edmundsorum is the most interesting case. It is a collector-shredder species during all its larval life except the second part of the last stage (fig. 23). The change between fine detritus and macrophytes to invertebrate prey is extremely fast (figs 1,2). It is difficult to find a single explanation for this behavioural change. It is obvious that invertebrates make up an energetic source much more important than detritus and macrophytes. They are the highest quality food because of both high calorific and protein contents (Cummins and Klug, 1979). Proteins represent between 50 and 85% of the total dryweight of mayflies, lipids between 10 to 25% and carbohydrates between 5 and 10% (Meyer, 1990; Ruffieux, 1997). Moreover ingested food is not equal to assimilable food; we can admit that invertebrate are easier ingested than detritus (Hawkins, 1985). This change consequently permits an important input of protein and lipid. The female imagoes need a large amount of energetic reserve for gamete production and compensation flight. But this diet change seems not directly connected with these supplementary energetic needs, since we found the same change of feeding behaviour among male and female mature larvae. It could be a determinant input if we consider that the imagoes do not eat, and permit a longer imaginal life or the attribution of more abundant reserves for flight. If the advantages of this change of diet are quite clear, we do not understand why it does not occur at earlier stages, especially if we consider that there are no morphological changes between the carnivorous and the collector-gather larval stages. The abundance of prey refutes the possible competition between the different stages; moreover, the differences among prey size avoids this competition. It had been demonstrated that addition of invertebrates to the diet of the *Clistoronia magnifica* (Trichoptera) at the last stage, highly increases the production of eggs and the rate of development, and decreases the death rate (Anderson, 1976). The carnivorous diet of *H. edmundsorum* could also be optional, but could highly increase its reproductive success.

We can state that the changes in diet as the larvae are not matched by changes in the mouthparts between the stages. Each species mouthparts are an adaptative compromise to offer the best fitness at all stages. Consequently, we can not deduce from the mouthparts the diet at each stage, but we can formulate a global diet for the entire larval life.

We noticed that four of the five species feed mainly on Baetidae, at least at one stage. The identification at the genus level is relatively easy, but a determination at a species level is impossible as long as the knowledge of the Malagasy Baetidae is so scarce. We found prey of the genera *Afroptilum*, *Dicentroptilum* and *Xyrodromeus*. *Afroptilum* and *Dicentroptilum* were by far the most abundant in the gut of *G. gargantua*, *H. edmundsorum* and *N. electroptera* (figs 2, 5, 6). *Afroptilum* and *Dicentroptilum* were also the most abundant genera in all the prospected stations. The proportion between *Afroptilum* and *Dicentroptilum* depended on the localities for the same species at the same larval stage. According to that, we can conclude that these three species are more opportunist than selective in their choice of prey.

Xyrodromeus is dominant in the gut of *H. christinae* (figs 3,4), but is relatively scarce in the gut of other species. This difference can have two explanations. The Malagasy species of *Xyrodromeus* are smaller than those of *Afroptilum* and *Dicentroptilum*; in this way, they appear more suitable prey for smaller predator such as *H. christinae*. *Xyrodromeus* present morphological adaptation to the fast flow and its bladelike mandibles seem to indicate that *Xyrodromeus* exploits a different ecological niche than *Afroptilum* and *Dicentroptilum*. The study of its gut contents shows this genus feed mostly on macrophytes and algae. As *H. christinae* is sympatric with *H. edmundsorum*, we can admit that both species exploit different ecological niches. Field observations confirm that the niche expoited by *Xyrodromeus* and *H. christinae* is more reduced.

A lot of larvae that are ready to moult are present in the gut contents. This could be explained by the reduced mobility of the larvae during this interstage.

We must note that we found no larva of caddisfly in the gut contents of our five species, even though this order is quite abundant in Eastern coast rivers (Gibon *et al.*, 1996). None of the five species fed on algae at any stage.

We can conclude that the five species show clear adaptations to predaceous behavior. We propose apomorphies directly related to the diet on the labrum, the mandibles and the forelegs. These features are more developed in the species *N. electroptera*, *G. gargantua* and *H. christinae*. It can easily be explained by their quasi-carnivorous diet from the earliest stages. At the opposite, *H. edumdsorum* and *H. elouardi* have less adapted mouthparts; this allows a more omnivorous diet, becoming carnivorous only at the last stages to supply high quality nutritional requirements.

ACKNOWLEDGMENTS

We thank the whole team of the Laboratoire de Recherche sur les Systèmes Aquatiques et leur Environnement (LRSAE) and especially its director Dr. J.-M. Elouard, for logistical assistance, great help during field work in Madagascar and for donation of specimens. We want also to thank the Swets and Zeitlinger Publishers (Aquatic Insects) for kindly authorize us to use some figures of *Guloptiloides gargantua*.

REFERENCES

Agnew, J. D. 1962. The distribution of *Centroptiloides bifasciata* (Baetidae: Ephemeroptera) in Southern Africa with ecological observation on the nymphs. Hydrobiologia 20: 367-372.

Anderson, N. H. 1976. Carnivory by an aquatic detritivore, *Clistoronia magnifica* (Trichoptera: Limnephilidae). Ecology 57: 1081-1085.

Battistini, R. 1996. Paléogéographie et variété des milieux naturels à Madagascar et dans les îles voisines: quelques données de base pour l'étude biogéographique de la "région malgache", pp. 1-17. In: W. R. Lourenço (ed.). Biogéographie de Madagascar. ORSTOM, Paris.

Brown, D. S. 1961. The food of the larvae of *Chloëon dipterum* L. and *Baetis Rhodani* (Pictet) (Insecta, Ephemeroptera). J. Anim. Ecol. 30: 55-75.

Chaperon, P., J. Danloux and L. Ferry. 1993. Fleuves et rivières de Madagascar. 874 ORSTOM, Paris.

Cummins, K. W. 1973. Trophic relations of aquatic insects. Ann. Rev. Ent. 18: 183-206.

Cummins, K. W. and M. J. Klug. 1979. Feeding ecology of stream invertebrates. Ann. Rev. Ecol. Syst. 10: 147-172.

De Rham, P. H. 1996. Poissons des eaux intérieures de Madagascar. In W. R. Lourenço (Ed.), Biogéographie de Madagascar. 423-440. ORSTOM, Paris.

Demoulin, G. 1973. Ephéméroptères de Madagascar. Bull. Inst. r. Sci. Nat. Belg. 49: 1-20.

Fontaine, J. and J. F. Perrin. 1981. Structure et fonctionnement des écosystèmes du Haut-Rhône français. XIII. *Raptobaetopus tenellus* (Albarda), nouvelle espèce pour la faune française (Ephéméroptère, Baetidae). Bull. Eco. 12: 85-94.

Fröhlich, F. 1996. La position de Madagascar dans le cadre de l'évolution géodynamique et de l'environnement de l'Océan Indien, pp. 19-26. In: W. R. Lourenço (ed.). Biogéographie de Madagascar. ORSTOM, Paris.

Gattolliat, J. L. and M. Sartori. 1998. Two new Malagasy species of *Herbrossus* (Ephemeroptera: Baetidae) with the first generic description of the adults. Ann. Limnol., Toulouse, 34 (3): 305-314.

Gattolliat, J. L. and M. Sartori. 2000. *Guloptiloides*: an extraordinary new carnivorous genus of Baetidae (Insecta: Ephemeroptera). Aquat. Insects 22 (2): 148-159.

Gibon, F. M., J. M. Elouard and M. Sartori. 1996. Spatial distribution of some aquatic insects in the Réserve Naturelle Intégrale d'Andringitra, Madagascar, pp. 109-120. In: S. M. Goodman (ed.). A floral and faunal inventory of the Eastern slopes of the Réserve Naturelle Intégrale d'Andringitra, Madagascar: with reference to elevational variation. Fieldiana. Fiel Museum of Natural History, Chicago.

Hawkins, C. P. 1985. Food habits of species of Ephemerellid mayflies (Ephemeroptera: Insecta) in streams of Oregon. Amer. Midl. Natural 113: 343-352.

Kluge, N. J. 1997. Classification and phylogeny of the Baetidae (Ephemeroptera) with description of the new species from the Upper Cretaceous resins of Taimyr, pp. 527-535. In: P. Landolt and M. Sartori (eds.). Ephemeroptera and Plecoptera. Biology-Ecology-Systematics. Mauron, Tinguely and Lachat, Fribourg (Switzerland).

Lugo-Ortiz, C. R. and W. P. McCafferty. 1997a. *Edmulmeatus grandis*: an extraordinary new genus and species of Baetidae (Ephemeroptera). Ann. Limnol. 33 (3): 191-195.

Lugo-Ortiz, C. R. and W. P. McCafferty. 1997b. New Afrotropical genus of Baetidae (Insecta: Ephemeroptera) with bladelike mandibles. Bull. Soc. Hist. Nat. Toulouse 133: 41-46.

Lugo-Ortiz, C. R. and W. P. McCafferty. 1998. The *Centroptiloides* Complex of Afrotropical small minnow mayflies (Ephemroptera: Baetidae). Ann. ent. Soc. Amer. 91 (1): 1-26.

McCafferty, W. P. and A. V. Provonsha. 1986. Comparative mouthpart Morphology and evolution of carnivorous Heptageniidae (Ephemeroptera). Aquat. Insects 8 (2): 83-89.

Meyer, E. 1990. Levels of major body compounds in nymphs of the stream mayfly *Epeorus sylvicola* (Pict.) (Ephemeroptera: Heptageniidae). Arch. Hydrobiol. 117 (4): 497-510.

Mittermeier, R. A., I. Tattersal, W. R. Konstant, D. M. Meyers and M. R. B. 1994. Lemur of Madagascar. 356 Conservation International, Washington D.C.

Mol, A. W. M. 1986. *Harpagobaetis gulosus* gen. nov., spec. nov., a new mayfly from Suriname (Ephemeroptera: Baetidae). Zool. Meded., 60 (4): 63-70.

Mol, A. W. M. 1989. *Echinobaetis phagas* gen. nov., spec. nov., a new mayfly from Sulawesi (Ephemeroptera: Baetidae). Zool. Meded., 63 (7): 61-72.

Müller-Liebenau, I. 1978. *Raptobaetopus*, eine neue carnivore Ephemeropteren-Gattung aus Malaysia (Insecta, Ephemeroptera: Baetidae). Arch. Hydriobiol., 82(1): 465-481.

Paulian, R. 1996. Réflexion sur la zoogéographie de Madagascar, pp. 219-230. In: W. R. Lourenço (ed.). Biogéographie de Madagascar. ORSTOM, Paris.

Riaño, P., A. Basaguren and J. Pozo, J. 1997. Life cycle and diet of *Ephemerella ignita* (Poda) (Ephemeroptera: Ephemerellidae) in relation to the develpmental stage, pp. 60-64. In: P. Landolt and M. Sartori (eds.). Ephemeroptera and Plecoptera. Biology - Ecology - Systematics. Mauron, Tinguely and Lachat, Fribourg (Switzerland).

Ruffieux, L. 1997. Réserves énergétiques et stratégies de reproduction chez les Ephéméroptères (Insecta, Ephemeroptera). Unpublished PhD, Univ. Lausanne, Lausanne.

Ruffieux, L., J. M. Elouard and M. Sartori. 1998. Flightlessness in mayflies and its relevance to hypotheses on the origin of insect flight. Proc. R. Soc. (265). 2135-2140. Agnew, J. D. (1962). The distribution of *Centroptiloides bifasciata* (Baetidae: Ephemeroptera) in Southern Africa with ecological observation on the nymphs. Hydrobiologia, 20. 367-372.

A NEW SPECIES OF *ATURBINA* (EPHEMEROPTERA, BAETIDAE) LUGO-ORTIZ AND MCCAFFERTY FROM URUGUAY

M. T. Gillies

Whitfeld, Hamsey, Lewes
Sussex, BN8 5TD, England

ABSTRACT

Aturbina was described as an aberrant baetid mayfly that lacked turbinate eyes in the male. The species *A. georgei* was shown to be widespread in Brazil and tropical South America. A second species, *A. beatrixae*, is now described from Departamento Maldonado, Uruguay. It differs from all known Baetidae in that the genital forceps possess two short or globular distal segments. The larva is believed to be associated with rotting logs in streams.

INTRODUCTION

In the course of a short visit to Uruguay in the summer of 1983-84, I made light-trap collections of adults of a curious pale baetid from a major afferent of the River Maldonado. The males were apparently unique for this family in lacking turbinate eyes. But despite extensive searches I was unable to find any mature male larvae that could be identified by the presence of the same distinctive feature. A number of years later, Lugo-Ortiz and McCafferty (1996) reported the presence of numerous specimens of a baetid with similar eyes in collections made from Brazil and other South American countries. Of greater interest still, they reported the capture of male baetid larvae with undivided eyes from some of the same localities. Thus, they were able to provide convincing evidence of the all-important association of the two stages, without which a satisfactory description of this curious taxon could not be made. They named the genus *Aturbina*.

Armed with this information, I re-examined the female larvae previously collected from the River Maldonado basin and found one closely resembling those described by Lugo-Ortiz and McCafferty. There were important differences, however, in both adults and nymph.

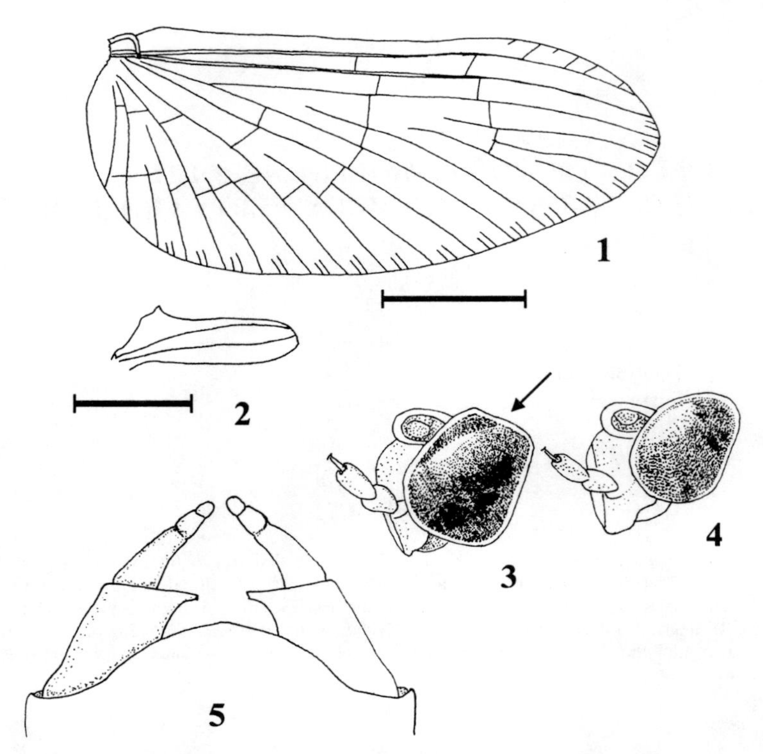

Aturbina beatrixae. **Fig. 1:** Fore wing. Bar = 1 cm. **Fig. 2:** Hind wing. Bar = 0.5 mm. **Fig. 3:** Male eye, lateral view (From an unpublished drawing by A.D. Mol). Arrow points to sulcus. **Fig. 4:** Female eye, lateral view (the same). **Fig. 5:** Male forceps.

DESCRIPTION

Aturbina beatrixae sp. n.

Male imago. (*In life*) A generally pale, colourless insect. Eyes translucent grey; thorax dove-grey with blackish-grey markings posterior to wing roots and on mesonotum; legs white, apex of fore tibia and tarsus I dark grey. Abdomen translucent colourless except for postero-lateral white patches on terga II, IV and VI; VII largely white and VIII-X entirely so; tails white.

(*In spirit*) Oculi undivided (Fig. 3), but extreme dorsal portion separated from the rest by a shallow sulcus. Thorax cream, main sutures and metanotum golden, a distinct metanotal hump present. Legs white. Fore wing (Fig. 1) with a small but conspicuous dark spot at junction of costal brace and subcosta; hind wing (Fig. 2) with two veins, main costal spur preceded basally by a rudimentary peak. Abdomen and cerci creamy white. Forceps (Fig. 5) stout and short with four segments; basal segment equal in length to other three together, the inner basal portion overlapped by a broad posterior extension of sternum IX, the inner distal portion drawn out into a blunt inwardly-directed process; second long segment dwarfed by the first segment, tapering distally; third and fourth segments rounded, the latter globular.

Male: body 4.2 - 5mm; wing 4.2 - 4.5mm

Female: body 5mm; wing 5mm.

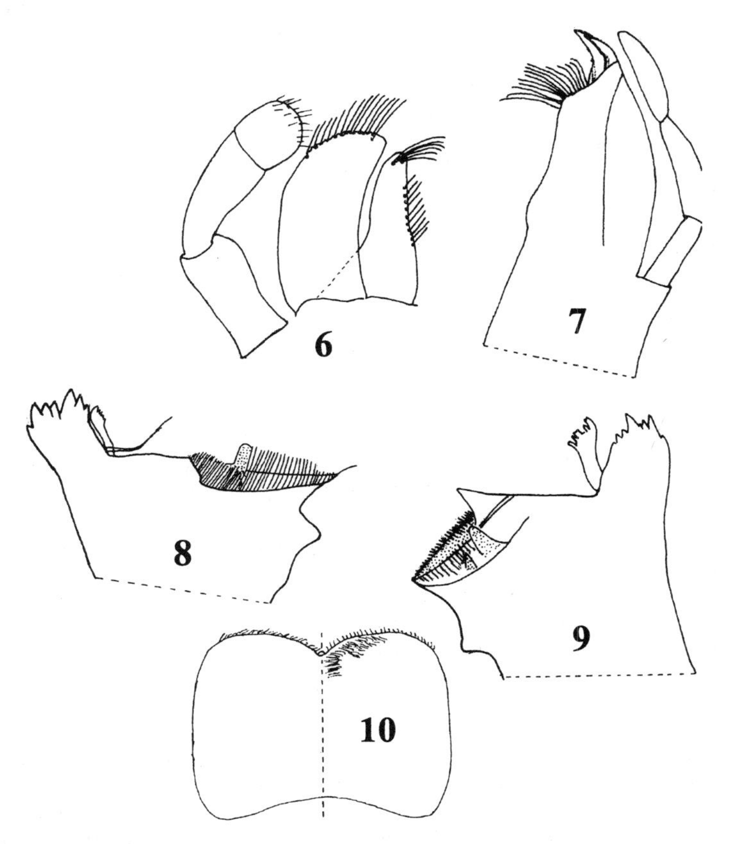

Aturbina beatrixae. Larval mouthparts. **Fig. 6:** Right mandible. **Fig. 7:** Left mandible. **Fig. 8:** Labium (palp separated). **Fig. 9:** Maxilla. **Fig. 10:** Labrum.

Larva (*assigned*). *Mouthparts* (Figs. 6 to 10): labrum with deep median notch; incisors of both mandibles fused to near apex, a conspicuous long fine seta projecting medially from base of right prostheca, this seta with barely discernible fraying in outer half; maxillary palps slender, with three segments, reaching to apex of galea-lacinia; labial palps with three segments, apical segment equal in width to second segment. Legs slender; posterior margins of femora bearing 8-15 short, spine-like setae; similar more closely spaced setae on anterior margins of tibiae and tarsi; tarsal claws with 3-4 fine teeth in middle section merging with roughened base of claw. Gills as in *A. georgei*. Cerci with medial hair fringe, terminal filament fringed on both sides.

Material. URUGUAY, Departamento Maldonado, Arroyo de las Cañas, 37 km north of Punta del Este by bridge on road from San Carlos to Aiguá, holotype male imago, 21.I.84; ibid. paratypes, 12 male, 4 female, 10 subimagoes, 4.I.84 - 21.I.84, 1 nymph, 8.II.84; Dos Hermanos, 23km north of Abra del Perdomo, 2 females, 20.I.84. Holotype and one male and female paratypes deposited in Instituto Miguel Lillo, Universidad Nacional de Tucumán, Argentina, remaining material deposited in the collection of the British Museum of Natural History, London.

Named for Mrs. Trixie Ingham of Buenos Aires, whose unstinting help in many ways made this study possible.

DISCUSSION

The adults of both sexes of *Aturbina beatrixae* generally resemble those of *A. georgei* Lugo-Ortiz and McCafferty in their overall pallor. The wing spot on the costal brace in *A. beatrixae* would seem to be distinctive. The most striking difference between the species lies in the male genitalia. While the greatly enlarged basal segment of the forceps is clearly a shared apomorphy, the presence of a globular fourth forceps segment in *A. beatrixae* immediately distinguishes it from *A. georgei*. Indeed, this last character has not been described for any other member of the Baetidae.

The single larva of *Aturbina* from the Arroyo de las Cañas is not now in very good condition. It is assumed to be that of *A. beatrixae*, but proof of this association is lacking at the present time. The mouthparts of the two species are generally similar, and the legs and gills of the two are almost identical. Most importantly, it has the same distinctive long seta arising from the base of the prostheca of the right mandible as *A. georgei*. However, in *A. beatrixae*, this hair is unbranched, at most minutely frayed.

Until much more is known about the Baetidae of South America little is to be gained by discussion of the phylogeny of *Aturbina*. While lacking turbinate eyes it could be argued that the male eye is nevertheless partially divided into two portions by the sulcus shown in Fig. 3. This sulcus is also present in the species from Surinam studied by Dr. A. D. Mol. although it is not clear from the description if it also exists in the type species *A. georgei*. This sulcus is lined with similar facets to the rest of the eye. It is not therefore homologous with the division between the turbinate and lower portions of the eye seen in other Baetidae.

The presence of a fourth forceps segment in the male is autapomorphic and is presumably not homologous with that seen, for example, in the Siphlonuridae.

Structure of the Eyes

I have been fortunate to have had the generous cooperation of Professor M.F. Land of the Centre for Neuroscience at the University of Sussex. From his examination of the eyes of *Aturbina beatrixae* he has been able to show that the eyes of both sexes are of the apposition type throughout. This is in contrast with most baetids where the turbinate region of the male eye is of the superposition type. The female has uniform facets throughout; the male, on the other hand, has a wide horizontal band of enlarged facets. These are not concentrated dorsally, as in the normal turbinate eye in the Baetidae, but are distributed round the middle region of the eye.

This has important consequences in terms of behaviour. Enlarged facets are generally found in regions of high visual acuity (Land, 1997). In the Baetidae, for instance, they are found in the dorsal turbinate eyes, where they presumably serve to locate females entering mating swarms. Professor Land points out that a medial band of large facets is also seen in empid flies that fly over the water surface on the lookout for insects struggling in the surface. He suggests that the mating dance and the pursuit of females by *Aturbina* may take place horizontally, close to a water surface, rather than in the more or less vertical swarms of most other mayflies.

Larval Habitat

The Arroyo de las Cañas flows for 30-40 km between heavily wooded banks across the undulating countryside to the north of San Carlos. Attempts to locate the larval habitat of *A. beatrixae* were only partially successful. Certain Old World baetids are known to live as inquilines in the branchial chamber of fresh water mussels (clams). These mollusks are abundant in certain reaches of the Arroyo de las Cañas but, out of more than 100 examined, none contained mayfly larvae. A more promising clue came from a collection of leaves and rotting logs brought into the lab. A single adult of *Aturbina* emerged from this brew, but the larval skin was unfortunately not recovered. A later collection of rotting wood, dredged from the bed of the river, harboured a variety of baetids including the single larva of

Aturbina described above. Thus, although the association with this habitat cannot be regarded as certain, there is enough evidence to merit further study.

Aturbina appears to be a widespread and not uncommon mayfly in South America (Lugo-Ortiz and McCafferty, 1996; A.D. Mol, in litt.). At a latitude of 34° 41´S, San Carlos is in the southernmost region of Uruguay. Winters are cool and summer heat is moderated by proximity to the Atlantic. Thus *A. beatrixae* would appear to be a species of the Temperate zone, possibly replacing *A. georgei* in this region.

ACKNOWLEDGMENTS

My indebtedness to Mrs. Trixie Ingham for hospitality and help in many ways cannot be exaggerated. I must also record my thanks to Dr. Ad Mol, for sending me details of his material of *Aturbina* from Surinam and for permission to use his drawings of the eyes. I am also most grateful to Professor Mike Land of the University of Sussex for examining and reporting on the structure of the eyes in the male.

REFERENCES

Land, M. F. 1997. Visual acuity in insects. Ann. Rev. Ent. 42: 147-177.

Lugo-Ortiz, C. R. and W. P. McCafferty. 1996. *Aturbina georgei* gen. et sp. n.: A small minnow mayfly (Ephemeroptera: Baetidae) without turbinate eyes. Aquat. Insects 18: 175-183.

A NEW GENUS OF LEPTOHYPHIDAE (INSECTA: EPHEMEROPTERA)

Carlos Molineri

INSUE (CONICET) - Fundación Miguel Lillo
Miguel Lillo 252
(4000) San Miguel de Tucumán, Tucumán, Argentina

ABSTRACT

A new genus of Leptohyphidae, *Yaurina* gen. nov., is established. Males have very characteristic genitalia. Nymphs of this new genus are superficially similar to *Leptohyphes*, but can be distinguished from them by the form and length of the femoral spines, number and arrangement of denticles on the tarsal claws and abdominal gill structure among others characters. Three species are described, *Y. yuta* and *Y. mota* from male, female and nymphs from Argentina, and *Y. yapa* from a male subimago from Ecuador. Some nymphs previously assigned to *Leptohyphes* probably belong to this new genus.

INTRODUCTION

The family Leptohyphidae is presently composed of 8 genera: *Leptohyphes, Tricorythodes, Leptohyphodes, Tricorythopsis, Haplohyphes, Coryphorus, Cotopaxi* and *Allenhyphes*.

The most common South American genus is *Leptohyphes*, with the majority of its species described from nymphs only. The adult stage is poorly known, and the male genitalia is seldom used in the systematic of the genus. In South America, there are only a few species with both the adult and nymphal stages known: *L. maculatus*, *L. setosus* and *L. petersi* (Allen, 1967), and those species described from imagos only have little variation in the male genitalia (*peterseni* type: Traver, 1958a: 497 and Traver, 1958b). The male genitalia of *Leptohyphes indicator* Needham and Murphy (1924: 33) is very atypical for the genus (Domínguez et al., 1994: 99) and represents a distinct group that will be discussed in another paper.

In this paper is described a new genus with male genitalia characterized by the presence of two large spine-like appendages arising from the base of the penes (Figs. 1-6, 8). This character and others of the penes, forceps and styliger plate distinguish this genus from other genera of the family. Nymphs of the new genus have some of the diagnostic characters of *Leptohyphes*: rather ovoid operculate gills, a transverse row of spines on the fore femora, and hind wing pads present in males but absent in females. These characters are of little systematic value to separate different genera because they show a gradual range of variation, i.e. operculate gills vary from ovoid to triangular, and dorsal spines of the fore femora range from short and blunt to long and thin (setae-like) without discrete morphological gaps. Moreover, the presence or absence of hind wings in females is variable in different species or populations, thus reducing the use of these characters for generic determination.

As a result, some nymphs previously described in *Leptohyphes* may represent other genera when adults can be associated, as is the case of the recently described genus *Allenhyphes* Hofmann and Sartori (in Hofmann et al., 1999); the type species was formerly known only from nymphs as *Leptohyphes flinti* Allen. The nymphs here described as *Yaurina* gen. nov. do not agree with any of the descriptions of the *Leptohyphes* species in the literature. Diagnostic characters are given in the generic description and discussion.

Yaurina Gen. nov. (Figs. 1-30)

Type species: *Yaurina yuta* sp. nov.
Species included: *Y. yuta*, *Y. mota* and *Y. yapa*.
Distribution: Argentina and Ecuador.

Male imago: Length: body, 3.0-3.2 mm; fore wings, 2.9-3.3 mm; hind wings, 0.40-0.50 mm. *Head*: compound eyes blackish separated by a distance 3 times diameter of an eye, median ocellus 2/5 of diameter of lateral ocelli. Compound eyes blackish. Ocelli whitish basally ringed with black. Antennae: scape 1/2-1/3 length of pedicel, flagellum 3-4 times length of scape and pedicel together. *Thorax*: mesonotum with heavy sclerotized bands on anterolateral margins of mesoscutum or between these sclerites and mesonotal protuberance (fore mesonotal transverse invagination, Kluge, 1992); mesoscutellum with two membranous filaments 7 times longer than wide. Fore wings (Fig. 12): vein MP2 united basally to veins CuA and MP1 by a cross vein, base of vein IMP ending freely in wing membrane or united to vein MP2 by a cross vein, vein ICu1 united basally to veins CuA and CuP by a cross vein; fore wings fringed on hind margin. Hind wings (Figs. 10-11) present, total length of hind wings 0.13-0.15 total length of fore wings; with a pair of weak longitudinal veins and fringed posterior margin; costal projection 0.50-0.55 of wing length and curved anteriorly. Legs: fore and middle femora of similar length, hind femora 1.4 times length of fore femora; fore tibiae 1.84 times length of middle tibiae and 1.62 times length of hind tibiae; fore tarsi 2.05 times length of middle tarsi and 1.56 times length of hind tarsi. Fore tarsal claws of a pair similar, blunt; middle- and hind tarsal claws of a pair dissimilar, one blunt, paddle-like and other apically hooked. *Genitalia*: rear margin of styliger plate extended posteriorly as in Figs. 1-4. Forceps (Figs. 1-4) three-segmented, segment 1 short and stout, segment 2 inserted on apical inner margin of segment 1; segment 2 with a slightly bulbous base and 3 times length of segment 1; segment 3 small and globular. Forceps extends perpendicularly to styliger plate (Fig. 4). Penes (Figs. 5, 6, 8) long and slender, completely fused except on small apical furrow, penes with a pair of long spine-like appendages on ventral side, appendages with wide base and thinner toward apex, dorsally or dorsolaterally curved. Penes excavated laterally (Fig. 4) and ventrally; more or less quadrangular in transversal view (Figs. 7, 9). Cerci 2 1/2 times length of body, terminal filament 3 1/2 times length of body and bearing short setae.

Female imago: Length: body, 3.7-4.5 mm; fore wings, 4.3-4.5 mm. *Head*. As in male. *Thorax* as in male. Legs: in all legs, tarsal claws of a pair dissimilar, one blunt, paddle-like, other apically hooked. Wings: fore wings as in Fig. 13. Hind wings absent. *Abdomen*: ninth sternum with rounded posterior margin, very slightly emarginated apically. Cerci slightly longer than body, terminal filament 1 1/2 times the length of body.

Mature nymph (Fig. 14): Body length (without cerci), 3.1-4.5 mm. *Head*: hypognathous. Antennae 1.25 times width of head. Mouthparts (Figs. 20-25): clypeus with convergent lateral margins, anterior margin as wide as labrum; width of labrum (Figs. 20-21) 1.8-1.9 times maximum length, lateral margins straight, anteromedian emargination broad and shallow; long setae present on dorsum of lateral and anterior margins (Fig. 20), with submedian longitudinal rows of setae ventrally as in Fig. 21. Mandibles as in Figs. 24-25. Maxillae (Fig. 22) with galea-lacinia completely fused except on apical furrow; inner

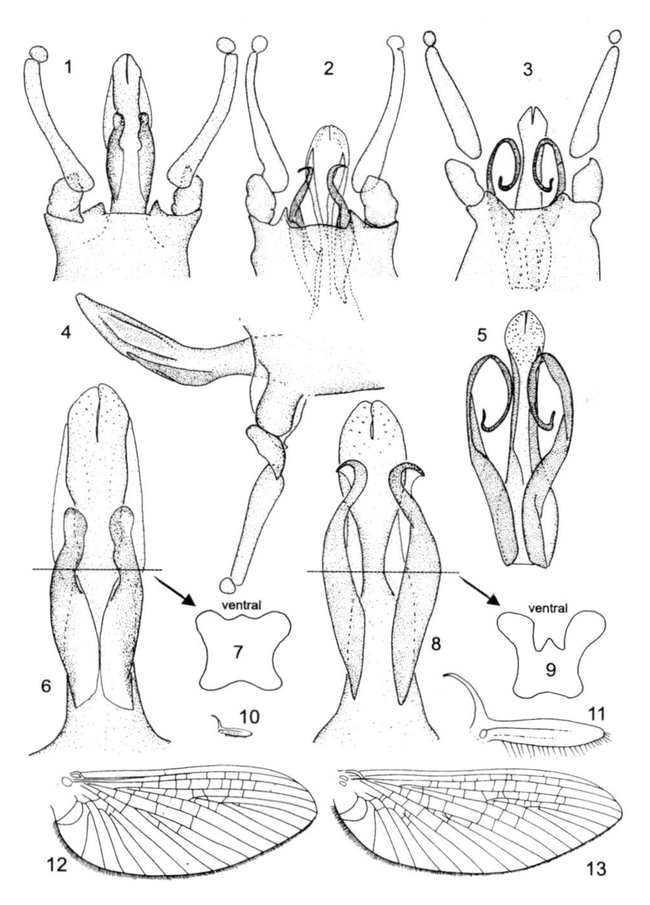

Figs. 1-13. *Yaurina* gen. nov. *Yaurina mota* sp. nov.: 1, male genitalia, v.v.; 4, same, lateral v.; 6, penes, enlarged, v.v.; 7, transversal view of penes; 10, male hind wing; 11, same, enlarged; 12, male fore wing; 13, female fore wing. *Yaurina yuta* sp. nov: 2, male genitalia, v.v.; 8, penes, enlarged, v.v.; 9, transversal view of penes. *Yaurina yapa* sp. nov.: 3, male genitalia, v.v.; 5, detail of penes, v.v.

margin near apex with row of 6-7 spines, a second row of 5-6 long spines at midpoint; outer margin with 5 long setae at basal half and numerous long curved setae at apex; maxillary palpi 2-segmented, segment 1 half length of segment 2. Lingua of hypopharynx rectangular with anterior margin slightly concave; superlinguae elongated and oval with setae on anterior margin. Labium (Fig. 23) with maximum width of submentum 3.2 times maximum width of mentum; labial palpi 3-segmented, segment 1 2.26 times length of segment 2, segment 2 1.25 times length of segment 3; outer margin of segment 1 and 2 of palpi and paraglossae with long setae, shorter setae on glossae; glossae and paraglossae fused almost completely, with rounded margins. *Thorax* with anterior margin of pro- and mesonotum with short denticles; base of coxae with long setae. Legs (Figs. 15-17): femora of fore legs with a transverse row of spines extending as in Fig. 17, spines long and flattened distally as in Fig. 18; tibiae with a double row of thick setae on anterior margin; tarsi with a row of setae on anterior margin; tarsal claw of all legs acute and slightly curved apically, with a row of 5-6 blunt marginal denticles at middle and a palisade of 6-7 submarginal slender denticles on distal half (Fig. 19). Middle and hind femora (Figs. 16 and 15 respectively) with spines on posterior margin, spines similar to those on dorsum of fore femora; middle and hind tibiae with a double row of thick setae on anterior margin and a single row on posterior margin, middle tibiae with an additional row of setae on ventral side; middle and hind tarsi with a ventral row of setae; proportions of middle femora 1.2-1.3 times length of fore femora, hind femora 1.6-1.7 times length of fore femora; middle tibiae 1.3 times length of fore tibiae and hind tibiae 1.9 times length of fore tibiae; length of fore femora 2.6 times maximum width. Gills (Figs. 26-30): gills on segments II-VI, gills II-V formed by three lamellae (Figs. 26b-29), gills VI by a single one (Fig. 30). Dorsal lamellae of gills II operculate elongate-ovoid (Figs. 26a, 26b) covering remaining gills almost completely; gills II with a ventral pair of lamellae of different shape and position (Fig. 26b), inferior one placed perpendicularly to dorsal lamella (operculate part of the gill) and elongated, forming (together with lateral projections of abdominal terga III-VI) the base of a cavity where remaining gills lie (the roof of the cavity being formed by operculate lamella); ventral upper lamella small and elongated as in Fig. 26b. Lamellae of gills III-VI ovoid, smaller posteriorly (Figs. 27-30). Abdominal segments III-VII laterally expanded, lateral margins of VII narrower. Small posterolateral spines present on abdominal segments II-VII, smaller on VII. Posterior margins of terga I-IX with short denticles. Terminal filament slightly longer than cerci, 0.8 times length of body, with whorls of spines at articulations.

Etymology: "Yaurina", from a Quechua voice for "fishing hook", because of the form of the spine-like appendages of penes.

Discussion: Yaurina gen. nov. is distinguished principally by adult characters and is distinct from all described species of Leptohyphidae known from male imagos. However, because the male imago of the type species *Leptohyphes eximius* Eaton was unknown, it became necessary to rear *L. eximius* to confirm that it was not *Yaurina*. This description will be published separately. For purposes of this paper, I can report that the male genitalia of *L. eximius* is clearly of the type conventionally attributed to *Leptohyphes*: the *peterseni*-type of Traver (1958a,b) and that the nymph is conespecific with that described by Kluge (1992).

It is more difficult to distinguish the nymph of *Yaurina* from other species described only from nymphs, and several known species may eventually prove to belong to *Yaurina*, as for example *Leptohyphes* sp. A of Traver (1944: 15), *L. tinctus* Allen (1973), *L. viriosus* Allen (1973), and *L. spinosus* Allen and Roback (1969). Other nymphs which may be confused with *Yaurina* are *Tricorythodes* sp. of Demoulin (1966: 20) and *T. sierramaestrae* Kluge and Naranjo (1990), but they can be distinguished from *Yaurina* by the characters discussed below. Also, the nymphs of *Allenhyphes flinti* are similar, but they can be distinguished by the number of denticles on the tarsal claws and the presence of bipectinate setae on the labrum, labium, and hind tibiae. Among the three subgenera established for *Tricorythodes* nymphs by Allen and Murvosh (1987), only *T.* (*Homoleptohyphes)* has obovate opercular gills, and it can be separated from *Yaurina* by the form of the legs, fore femoral bands of spines, and the tarsal claws.

Kluge (1992) listed thoracic characters that distinguish *Leptohyphes* from *Tricorythodes*. Some of them are: form of "posterior scutal protuberances (PSP)" (divergent in *Leptohyphes*, convergent in *Tricorythodes*), "transverse mesonotal suture" (present in *Leptohyphes*, absent in *Tricorythodes*), and "inferior dorsoventral suture of lateropostnotum" (forming a straight line with the "superior dorsoventral suture" in *Tricorythodes* but not in *Leptohyphes*). In *Yaurina*, the PSP are almost parallel, the "transverse mesonotal suture" is absent or slightly marked and the "inferior dorsoventral suture of lateropostnotum" is also slightly marked but is otherwise similar to *Leptohyphes*.

Yaurina can be separated from the other genera of the family, including all the taxa discussed above, by the following combination of characters. In the male imagos: 1) hind wings present (Figs. 10-11); 2) length of costal projection 0.50-0.55 of hind wings length (Figs. 10-11); 3) forceps 3-segmented (Figs. 1-4); 4) styliger plate extended posteriorly at base of forceps as in Figs. 1-4; 5) penes completely fused except on small apical furrow (Figs. 5, 6, 8); and 6) penes with a pair of long ventral spine-like appendages arising from the base (Figs. 5, 6, 8). In the nymphs: 1) gills present on abdominal segments II-VI (Figs. 26-30); 2) gills on abdominal segment II formed by an operculate dorsal lamella, elongated and oval in shape,

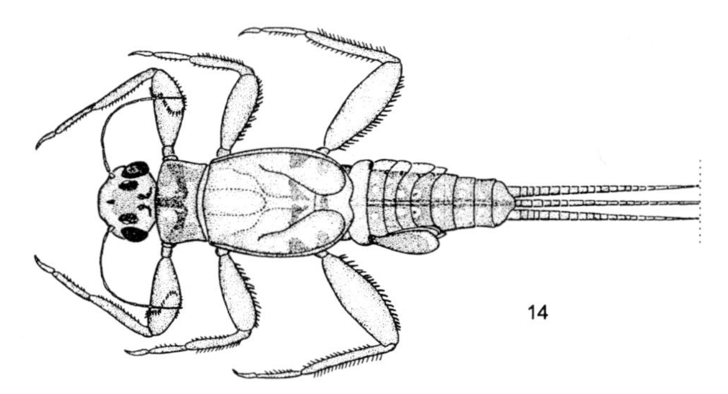

Fig. 14. *Yaurina* gen. nov. *Yaurina yuta* nymph, dorsal view.

and with the inferior ventral lamella perpendicular to it (Fig. 26b); 3) long, thin and distally flattened spines on the fore femora (Fig. 18); 4) maxillary palpi 2-segmented (Fig. 22); 5) maximum width of submentum 3.2 times maximum width of mentum (Fig. 23); and 6) tarsal claws with a row of 5-7 blunt marginal denticles at middle and an apical palisade of 4-7 submarginal slender denticles (Fig. 19).

Yaurina yuta sp. nov. (Figs. 2, 8, 9, 14-25)

Male imago (in alcohol): Length: body, 3-3.2 mm; fore wings, 3.1-3.3 mm; hind wings, 0.40-0.45 mm. General coloration light yellowish-orange, abdomen whitish translucent. *Head*: whitish, widely shaded with gray except ventrally. Antennae light yellowish-white. *Thorax*: dorsal portion of pronotum whitish shaded with gray, mediolateral zones yellowish-white with black margins and a black sublateral irregular band. Mesonotum light yellowish-orange, turning whitish-yellow at middle and yellowish-orange between mesonotal protuberance and mesoscutum; shaded diffusely with gray; mesoscutellum yellowish translucent; mesopleura whitish-yellow, shaded with gray anterior to insertion of wings, base of coxae grayish; lateral sclerites of meso- and metathoracic sterna yellowish-orange, median sterna whitish shaded with gray. Metanotum light yellowish-orange slightly shaded with gray. Legs yellowish-white, darker on margin except fore tibiae and fore tarsi whitish translucent, diffusely shaded with gray on all coxae and fore tibiae and fore tarsi. Wings (similar to Figs. 10-12): membrane of fore wings hyaline slightly tinged with yellow, darker in C and Sc sectors; longitudinal veins yellowish-orange, transverse veins diffusely tinged with yellow turning darker toward fore margin. Membrane of hind wings slightly yellowish, costal projection darker, longitudinal vein hyaline. *Abdomen*: whitish translucent shaded with gray on terga I-X almost evenly, except on a longitudinal median line on II-IX and a pair of semicircular anterosubmedian marks on III-IX. Sterna diffusely shaded with gray except on semicircular submedian marks on anterior margin of II-VIII. *Genitalia* (Figs. 2, 8, 9): styliger plate whitish except on posterior margin between forceps, yellowish; slightly shaded with gray; basal segment of forceps yellowish-white, second and third segments whitish translucent; penes (Fig. 8, 9) whitish translucent except lateral spines light yellowish-orange; spines shorter than penes, with apex curved outward as in Fig. 8. Cerci whitish translucent.

Female imago: Unknown.

Female subimago (in alcohol): Length: body, 3.7-4.5 mm; fore wings, 4.3 mm. General coloration light whitish-brown, abdomen yellowish. *Head* as in male except antenna: scape and pedicel whitish, flagellum shaded with light brown. *Thorax*: pronotum whitish-yellow, shaded extensively with gray except on sublateral zones; propleura and prosternum whitish diffusely washed with gray. Mesonotum as in male. Legs: fore legs yellowish-white shaded with brown; middle and hind legs whitish. Fore wings as in male. Hind wings absent.

Abdomen: whitish translucent shaded with gray almost completely except on a medio-longitudinal line on terga VI-IX, sterna as in male; eggs yellowish. Terminal filament whitish translucent [cerci broken off and lost].

Mature nymph (in alcohol) (Fig. 14): Body length 3.4 mm, cerci length 1.8 mm (male); body length of female larvae 4.3-4.5 mm. General coloration pale light yellowish-brown. *Head*: whitish-yellow shaded diffusely with gray between and behind lateral ocelli, around base of and between antennae, and on irregular marks near posterior margin. Compound eyes blackish. Ocelli whitish rounded with black. Antennae whitish washed with brown on flagellum. Mouthparts (Figs. 20-25) yellowish except incisive and molar surfaces of mandibles orangeish-brown and labrum yellowish-brown. *Thorax*: pronotum pale yellowish shaded with gray as in male imago, pleura and sternum whitish-yellow. Mesonotum pale yellowish shaded with gray on mesoscutellum, pleura and sternum whitish-yellow shaded with gray at base of coxae and sternum. Metanotum whitish-yellow shaded with gray at base of hind wingpads. Legs (Figs. 15-17). Whitish-yellow with yellowish spines, tarsal claws with yellowish apex. *Abdomen* yellowish, shaded with gray as in male except below gills, whitish. Gills: gills II yellowish-white tinged with gray almost completely except on distal and lateral margins (similar to Fig. 26a), remaining gills whitish washed with gray at base. Cerci whitish-yellow with whorls of setae at articulations.

Material: Holotype male imago, ARGENTINA, Salta, Estancia Jakúlica (near Parque Nac. Baritú), Arroyo de la Casa, 25-II-1989, E. Domínguez col.. Paratypes: 37 male imagos same data as holotype, 2 reared male subimagos from ARGENTINA, Salta, Río Piedras, Ruta 18 to Isla de Cañas, 25-XII-1997, C. Molineri Col.; 2 reared females subimagos and 1 nymph same data as holotype except date and collector: 26 to 28-XII-1997, C. Molineri Col. Holotype and paratypes deposited in the Entomological Collections of Instituto-Fundación Miguel Lillo, Tucumán, Argentina, except 5 paratypes male imagos in Florida A and M University, Tallahassee, Florida, USA and 5 paratypes male imagos in Musée de Zoologie, Palais de Rumine, Place dela Riponne 6, Lausanne, Switzerland.

Life cycle associations: Association between nymphs and adults was made by rearing nymphs of both sexes.

Etymology: "Yuta", from a Quechua voice meaning "shortened", because the spine-like appendages do not reach the apex of penes.

Discussion: Males of *Yaurina yuta* can be distinguished from the other males of the genus because the spine-like appendages of penes are slightly shorter than penes (0.68 times the total length of penes) and their apices are acute and laterally curved (Fig. 8). The nymphs here described as *Y. yuta* and *Y. mota* are not distinguishable at this time.

Yaurina mota sp. nov. (Figs. 1, 4, 6-7, 10-13)

Male imago (in alcohol): Length: body, 3.2-3.6 mm; fore wings, 3.6-3.8 mm; hind wings, 0.5 mm. General color whitish light brown. *Head* yellowish-white suffused with light brown almost completely except on frons (between antennae and median ocellus), on a paler transverse line between lateral ocelli, and on a pair of submedian small circular sclerites on the occipute. Antennae whitish translucent. *Thorax*. Pronotum almost completely shaded with brown, except on three paler circular marks at each side; propleurae and prosternum whitish washed with light brown. Mesonotum yellowish light brown, darker on anterolateral corners and FMI (fore mesonotal transverse invagination); mesopleurae and sternum paler. Metanotum whitish tinged with light brown medially and shaded with gray at wing base. Wings. Membrane of wings hyaline, slightly tinged with brown. Fore wings (Fig.12): longitudinal veins grayish-brown, cross veins light brown. Hind wings as in Figs. 10-11. Legs yellowish-white, tinged with light brown on fore legs, middle and hind legs paler. *Abdomen* whitish translucent tinged with brownish gray dorsally, darker on a pair of submedian longitudinal lines, median line between them paler. *Genitalia* (Figs. 1, 4, 6, 7): styliger plate whitish, forceps and penes yellowish-white except spine like appendages of penes, orangeish. Caudal filaments whitish.

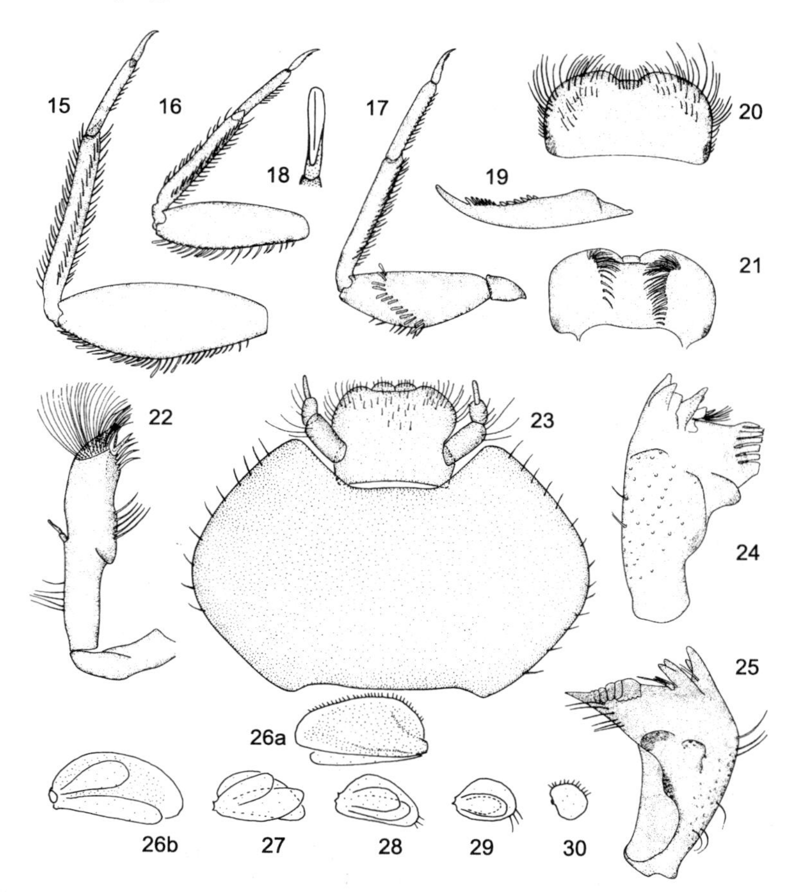

Figs. 15-30. *Yaurina* gen. nov., nymph. *Yaurina yuta* sp. nov.: 15, hind leg; 16, middle leg; 17, fore leg; 18, detail of femoral spine; 19, fore tarsal claw, enlarged; 20, labrum, d.v.; 21, labrum, v.v.; 22, right maxilla, d.v.; 23, labium, v.v.; 24, left mandible, d.v.; 25, right mandible, d.v. *Yaurina mota* sp nov., abdominal gills: 26a, gill II, d.v.; 26b, gill II, v.v.; 27 gill III, v.v.; 28, gill IV, v.v.; 29, gill V, v.v.; 30, gill VI, v.v.

Female imago (in alcohol): Length: body, 3.7-4.0 mm; fore wings (Fig. 13), 4.3-4.5 mm. As in male imago except abdomen yellowish when full of eggs.

Mature nymph (in alcohol): Length: body, 3.0-3.2 mm; cerci 2.4 mm; terminal filament, 2.7 mm. General color yellowish-white to yellowish light brown with gray markings. *Head* yellowish light brown shaded with gray between antennae and a pair of oblique bands extended from posterior margin of each eye to lateral ocelli; occipute with a slightly marked semicircular band. Antennae and mouthparts yellowish translucent. *Thorax* yellowish with gray markings, ventrally paler. Legs yellowish-white, spines yellowish. *Abdomen* yellowish-white, turning yellowish toward rear segments; shaded with gray dorsally, ventrally paler. Gills (Figs. 26-30): opercular gills translucent yellowish-brown, margins hyaline, remaining gills whitish. Caudal filaments yellowish white.

Material: Holotype male imago from ARGENTINA, Jujuy, El Carmen, Camino de Cornisa Km 1658, S 24° 27' 17"-W 65° 17' 48", A° Las Lanzas, 1250 m, 12-III-2000, Domínguez and Molineri Cols. Paratypes: 32 male imagos, 26 female imagos, 1 reared male subimago with its nymphal cuticle, and 10 nymphs, same data as holotype. Holotype and paratypes deposited in the Entomological Collections of Instituto-Fundación Miguel Lillo,

Tucumán, Argentina, except 5 paratypes male imagos and 5 paratypes female imagos in Florida A&M University, Tallahassee, Florida, USA.

Life cycle associations: Female and male adults were captured at the same time and share color patterns and wing venation. Nymphs and adults were associated by a reared nymph.

The adults were captured with a light trap just before sunrise. Swarms occur at early morning and last just a few minutes.

Etymology: From the Quechua voice "mota", that means "short and blunt", for the form and relative length of the spine-like appendages of penes.

Discussion: Males of *Y. mota* can be distinguished from the other members of the genus by the relatively shorter spine-like appendages of penes, that are apically blunt (length of spines 0.54 length of penes) as in Fig. 6. At this time it is impossible to distinguish the nymphs of this species from the other of the genus.

***Yaurina yapa* sp. nov. (Figs. 3, 5)**

Male subimago (in alcohol, wings and genitalia on slides): Length: body, 3.0 mm; fore wings, 2.9 mm; hind wings, 0.4 mm. General coloration yellowish, abdomen whitish. *Head*: yellowish-white shaded with gray behind lateral ocelli and around antennae. Antennae. Scape and pedicel whitish, flagellum yellowish-white. *Thorax*: pronotum whitish translucent diffusely shaded with gray, stronger on anterior margin and on a sublateral oblique band. Mesonotum yellowish with anterolateral margins of mesoscutum orangeish-brown; mesopleura and mesosternum yellowish-white diffusely shaded with gray. Metanotum yellowish, shaded with gray on base of coxae. Fore wings similar to that in Fig. 12; veins C, Sc and R1 yellowish, remaining veins whitish. Hind wings similar to that in Fig. 10. *Abdomen*: whitish except segments IX-X yellowish-white, diffusely shaded with gray. *Genitalia* (Figs. 3, 5): styliger plate yellowish-white, forceps whitish, penes whitish translucent except spine-like appendages light yellowish-orange. Appendages of penes (Fig. 6) long and coiled distally, total length surpassing that of penes.

Female and nymph: Unknown.

Material: Holotype male subimago from ECUADOR, Past. Puyo, 30-I-1976, black light, Spangler et al. Cols. In National Museum of Natural History, Smithsonian Institution, Washington D.C., USA.

Etymology: "Yapa", from a Quechua voice for "something more" or "aggregate" because the spine-like appendages are longer than penes.

Discussion: Although only a single subimago is known, the characteristic male genitalia of this species permits rapid identification. On *Y. mota* and *Y. yuta*, the relative length of the spine-like appendages do not change between the subimaginal and imaginal stages, suggesting that imaginal genitalia of *Y. yapa* should not differ from the subimaginal one, here described. None of the *Leptohyphes* nymphs described from Ecuador by Mayo (1968) or Wang et al. (1998) have the diagnostic characters of *Yaurina*, suggesting that the nymph of *Y. yapa* has not been described yet. Males of *Y. yapa* can be distinguished from the other males of the genus because the spine like appendages of penes are very long (if extended, longer than penes) and curled dorsally as in Figs. 3 and 5 (subimaginal cuticle omitted in the figures).

ACKNOWLEDGMENTS

I wish to thank E. Domínguez, W. L. and J. G. Peters and J. M. Elouard for valuable comments on this manuscript. I also want to thank C. Hofmann and M. Sartori for the loan of *Allenhyphes* material used for comparison. This manuscript was completed while the author was supported by a fellowship from the Argentine National Council of Scientific Research (CONICET) and this support is greatly acknowledged.

REFERENCES

Allen, R. K. 1967. New Species of New World Leptohyphinae (Ephemeroptera: Tricorythidae). Canad. Ent., 99 (4): 350-375.

Allen, R. K. 1973. New Species of *Leptohyphes* Eaton (Ephemeroptera: Tricorythidae). Pan-Pac. Ent., 49 (4): 363-372.

Allen, R. K. and C. M. Murvosh. 1987. Mayflies (Ephemeroptera: Tricorythidae) of the Southwestern United States and Northern Mexico. Ann. ent. Soc. Amer., 80 (1): 35-40.

Allen, R. K. and S. S. Roback. 1969. New Species and Records of New World Leptohyphinae (Ephemeroptera: Tricorythidae). J. Kans. ent. Soc., 42 (4): 372-379.

Demoulin, G. 1966. Contribution a l'Etude des Ephemeropteres du Surinam. Bull. Inst. r. Sci. Nat. Belg., 42 (37): 1-22.

Domínguez, E., M. D. Hubbard and M. L. Pescador. 1994. Los Ephemeroptera en Argentina. Fauna de Agua Dulce de la Republica Argentina, 33 (1): 142 pp.

Hofmann, C., M. Sartori and A. Thomas. 1999. Les Ephéméroptères (Ephemeroptera) de la Guadeloupe (petites Antilles françaises). Mém. Soc. Vaud. Sc. Nat., 20 (1):1-96.

Kluge, N. Yu. and J. C. Naranjo. 1990. Mayflies of the family Leptohyphidae (Ephemeroptera) of Cuba. Rev. Ent. URSS, 69 (3): 564-578.

Kluge, N. Ju. 1992. Redescription of *Leptohyphes eximius* Eaton and diagnoses of the genera *Leptohyphes* and *Tricorythodes* based on the structure of pterothorax (Ephemeroptera: Tricorythidae, Leptohyphinae). Opusc. Zool. Flumin., 98: 1-16.

Mayo, V. K. 1968. Some New Mayflies of the Subfamily Leptohyphinae (Ephemeroptera: Tricorythidae). Pan-Pac. Ent. 44 (4): 301-308.

Needham J. G. and H. E. Murphy. 1924. Neotropical Mayflies. Bull. Lloyd Lib. 24, Ent. Ser. No. 4: 1-79.

Traver, J. R. 1944. Notes on Brazilian Mayflies. Bol. Mus. nac., Zool. 22: 2-53.

Traver, J. R. 1958a. The Subfamily Leptohyphinae (Ephemeroptera: Tricorythidae). Ann. ent. Soc. Amer., 51 (5): 491-503.

Traver, J. R. 1958b. Some Mexican and Costa Rican Mayflies. Bull. Brooklyn ent. Soc., 53 (4): 81-89.

Wang, T. Q., R. W. Sites and W. P. McCafferty. 1998. Two new species of *Leptohyphes* (Ephemeroptera: Leptohyphidae) from Ecuador. Florida Ent. 81: 68-75.

THE NYMPH OF *SECOCHELA ILLIESI* (EPHEMEROPTERA: LEPTOPHLEBIIDAE: ATALOPHLEBIINAE) FROM SOUTH AMERICA

Manuel L. Pescador[1] and Tom Gonser[2]

[1] Entomology, Center for Water Quality
Florida A&M University
Tallahassee, FL 32307-4100, USA
[2] EAWAG, Limnological Center
Kastanienbaum, Switzerland

ABSTRACT

The nymph of *Secochela illiesi,* the only species presently known in the genus, is described. The characters that distinguish the nymph of *Secochela* from other genera of Leptophlebiidae are discussed. The nymph shares the derived characters that define the *Meridialaris* lineage, which includes genera from South America, Australia, and New Zealand.

INTRODUCTION

Pescador and Peters (1982) established the genus *Secochela* for the species *Secochela illiesi* based on male and female imagos, male subimago, and egg. The nymph of this monotypic genus is herein described for the first time. The specimens were collected by one of us (TG), while conducting research on the ecology and biology of mayflies in Chile (Gonser,1990). Nymphs were associated with adults by rearing.

This paper represents an on-going taxonomic documentation of the leptophlebiid fauna of southern South America (Peters and Edmunds,1972; Pescador and Peters,1980a, b, 1982, 1985, 1987, 1990, 1991; Pescador, 1997; Dominguez and Pescador,1983; Dominguez and Flowers,1989)). The techniques of Pescador and Peters (1980a) were followed to prepare structures for taxonomic description of the nymph.

Secochela Pescador and Peters

Genus E Pescador and Peters, 1980b
Secochela Pescador and Peters, 1982:1

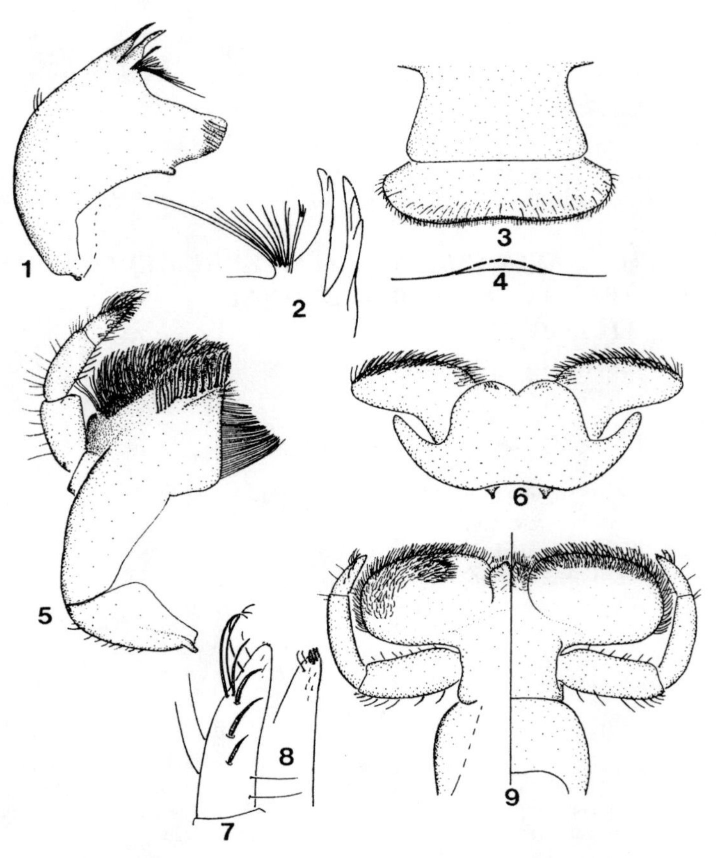

Figs. 1-9. *Secochela illiesi,* mature nymph: 1, left mandible; 2, right mandibular incisor; 3, clypeus and labrum; 4, enlarged anteromedian emargination of labrum; 5, right maxilla, ventral; 6, hypopharynx; 7and 8, segment 3 of labial palpi, dorsal view and ventral view; 9, labium, dorsal (left), ventral (right).

Mature Nymph: Head prognathous. Antennae 1 1/3 times as long as head, flagellum with clusters of fine subapical hairs usually in-groups of 3's and 4's (Fig. 15). Mouthparts (Figs. 1-9): Clypeus distinctly narrower than labrum, lateral margins apically divergent (Fig. 3). Length of labrum approximately 1/3 times maximum width, angularly curved laterally (Fig. 3); anteromedian emargination broad, U-shaped with undefined indentations (Fig. 4). Outer margin of mandibles smoothly curved, glabrous except for a thin submedian hair tuft (Fig. 1); outer incisors with minute apical serrations (Fig. 2); prostheca well-developed (Fig. 4). Galea-laciniae of maxillae apically broad with 15-17 pectinate setae (Fig. 5); segments 1-2 of maxillary palpi subequal in length, segment 3 slightly shorter than either segment 1 or 2, with long thick hair (Fig. 5); inner margin of segments 2 and 3 with spinous setae. Hypopharyngeal lingua with well-developed lateral processes (Fig. 6), anterior margin moderately cleft; superlinguae with thick hair along anterior margin, lateral margin smoothly curved (Fig. 6). Segments 1 and 2 of labial palpi subequal in length, segment 3 approximately ½ length of either segment 1 or 2; segment 3 with a row of well-developed dorsal setae (Fig. 7) and with few apical denticles (Fig. 8); glossae straight and flat, paraglossae ventral to glossae (Fig. 9); lateral margins of submentum glabrous (Fig. 9). Pronotum with a row of short anterolateral spines. Legs (Figs. 10-12). Maximal width of tibiae approximately 1 ½ x

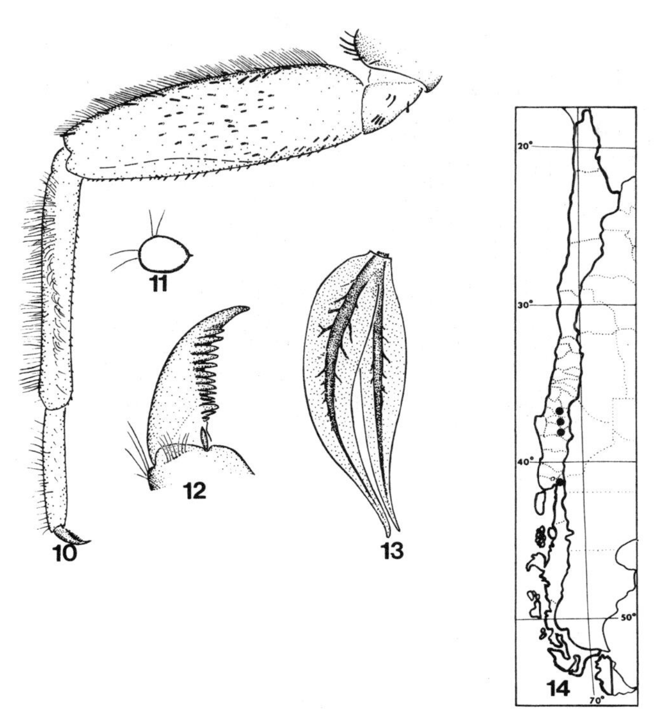

Figs. 10-14. *Secochela illiesi,* mature nymph: 10, fore leg; 11, cross section of tibia of fore leg; 12, fore claw; 13, gill IV; 14, geographical distribution of *S. illiesi.*

maximal width of tarsi, tibiae of prothoracic legs in cross section slightly oval (Fig. 11); femora with fine hair and prominent spoon-shaped-setae (Fig. 19); dorsum of tibiae with submarginal row of spatulate setae (Fig. 17), apex of setae with minute papillae (Fig. 18); tarsal claws with moderately large and unequal-sized basal denticles, remaining denticles progressively larger apically (Fig. 12). Gills (Fig. 13): gills on abdominal segments I-VII alike; dorsal and ventral lamellae slender, gradually tapered apically (Fig. 13); main trachea with weakly developed branches (Fig. 13). Posterolateral spines on abdominal segments II-IX, progressively longer posteriorly (Fig. 21); abdominal terga with few scattered filamentous dorsal setae, with fringe of lateral setae, and needle-like posterior spines (Figs. 20-21). Caudal filaments: terminal filament approximately 1/3 longer than cerci; surface of caudal filaments granulate, segments with sharp-pointed apical spines (Fig. 16).

Discussion: Pescador and Peters (1982) had previously discussed the adult and egg characters that distinguish *Secochela* from the other genera of Leptophlebiidae. The nymph of *Secochela* can be distinguished from all other genera of Leptophlebiidae by the following combination of characters: (1) clypeus is distinctly narrower than labrum and the lateral margins are apically divergent (Fig. 3); (2) length of labrum is approximately 1/3 times maximum width and angularly curved laterally (Fig. 3); (3) outer margin of mandibles is smoothly curved and glabrous except for a thin submedian hair tuft (Fig. 1); (4) lateral margins of submentum are glabrous (Fig. 9); (5) dorsum of tibiae has submarginal dorsal row of spatulate setae (Fig. 17), apex of setae with minute papillae (Fig. 18); (6) tarsal claws have unequal-sized basal denticles and the remaining denticles are progressively larger apically (Fig. 12); (7) posterolateral spines occur on abdominal segments II-IX; (9) abdominal gills I-VII are alike, both dorsal and ventral lamellae are slender and gradually tapered apically

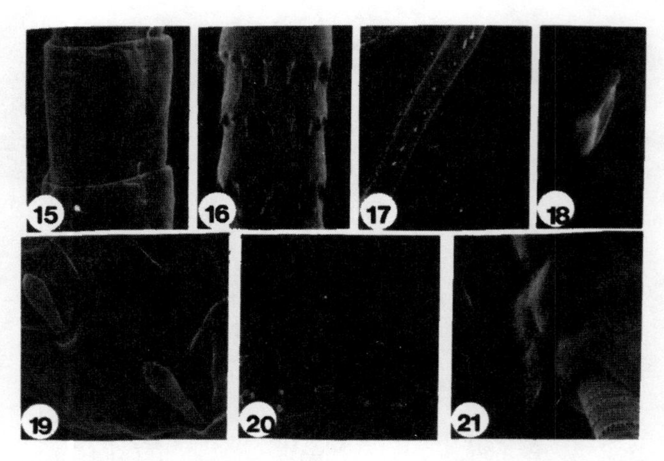

Figs. 15-21. Scanning electron micrographs of *Secochela illiesi,* mature nymph: 15, antenna (1000X); 16, caudal filament (400X); 17, tibial setae and hair, dorsal (100X); 18, enlarged tibial setae (2000X); 19, femoral setae and hair, dorsal (1000X); 20 and 21, abdominal terga (600X, 150X).

(Fig. 13); and (10) caudal filaments have granulate surface and sharp-pointed apical spines (Fig. 16).

Pescador and Peters (1980b, 1982) have indicated that *Secochela* belongs to the *Meridialaris* lineage that includes genera from Chile and southern Argentina [*Meridialaris* and *Masssartellopsis* (Pescador and Peters, 1980, 1982, 1987)], Australia [*Austrophlebioides* (Campbell and Suter, 1988)] and New Zealand [*Atalophlebioides and Deleatidium* (Towns and Peters, 1978, 1996)] The adults of *Secochela* according to Pescador and Peters (1982) can be distinguished from the other genera of the lineage by the shape and armatures of the penis lobes, and the type of tarsal claws. The *Secochela* nymph shares the same derived characters that define the *Meridialaris* lineage which include the strongly divergent lateral margins of the clypeus (Fig. 3), the labrum distinctly broader than the clypeus (Fig. 3), the apically broad galea-lacinia of maxillae with 15 or more subapical pectinate setae (Fig. 5), and the glabrous lateral margins of the labial submentum (Fig. 9). The nymph, however, can be distinguished from the other genera of the *Meridialaris* lineage by the following characters: (1). outer margin of mandibles is smoothly curved and glabrous except for a thin submedian hair tuft (Fig. 1), (2) femora have prominent spoon-shaped setae (Fig. 19); (3) dorsum of tibiae has submarginal row of spatulate setae (Fig. 17), apex of setae with minute papillae (Fig. 18); and (4) tarsal claws have unequal-sized basal denticles and the remaining denticles are progressively larger apically (Fig. 12).

Secochela illiesi Pescador and Peters

Secochela illiesi Pescador and Peters, 1982:8 (male, female, egg).

Mature Nymph (in alcohol): Body length 6.0-7.0 mm. Dorsum of head dull yellow, lightly washed with dark brown between ocelli extending to anterior base of compound eyes, forming a narrow irregular transverse band. Antennae yellow, scape faintly washed with black. Female eyes black; upper portion of male eyes orange-yellow, lower portion black. Median ocellus approximately 2x smaller than lateral ocelli. Mouthparts: labrum yellow, posterior margin thinly lined with orange-brown; dorsum of mandible yellow, faintly washed with dark brown along margins, venter pale. Maxillae yellow, stipes and basal half of outer margin of galea-lacinia faintly washed with dark brown; apical hair of galea-lacinia brownish; inner margin of segment 2 of maxillary palpi with series of 5-7 spinous setae (Fig. 5). Labium pale yellow, marginal setae brownish-yellow. Thorax: nota yellow, median and margins of pronotum, mesoscutellum and anterolateral corners of mesonotum washed with dark brown. Pleura yellow, pleural sutures and areas around base of legs washed with dark brown. Legs yellow, subcoxae and coxae dorsally washed with dark brown; tarsal claws with 14-16 denticles (Fig. 12). Abdomen: terga yellow, faintly washed with dark brown, progressively darker towards lateral margins and posterior margins; posterolateral spines on abdominal segments VIII and IX slightly incurved apically (Fig. 20). Sterna yellow; brownish abdominal ganglia externally visible, particularly pronounced on fused ganglia on sterna VII-VIII. Gills with short tracheal branches (Fig. 13). Gills grayish-white, tracheae dark brown. Caudal filaments: yellow with 2-3 narrow, prominent orange-brown annulations near base.

Geographical Distribution (Fig.14): *Secochela illiesi* has only been collected in Chile. Pescador and Peters (1982) reported the species from *Llanquihue Prov.*, Bío *Bío Prov.*, *Linares Prov.*, *Malleco Prov.,* and *Nuble Prov.* The specimens for this study included 4 nymphs that were collected from *Los Lagos Prov.*, Río Quinchila. 9.III.85, T. Gonser. The specimens are deposited in the collections of Florida A&M University (2 nymphs) and Universidad Nacional de Tucumán, Argentina (2 nymphs).

ACKNOWLEDGMENTS

We would like to thank Janice G. Peters for the illustrations and to Kim Riddle for her assistance with the electron microscope. Janice G. and William L. Peters, Florida A&M University read and offered valuable comments on the manuscript. This research was supported by a research grant (FLAX 91004) from CSREES-USDA to Florida A&M University.

REFERENCES

Campbell, I. C. and P. J. Suter. 1988. Three new genera, a new subgenus and a new species of Leptophlebiidae (Ephemeroptera) from Australia. J. Aust. ent. Soc. 27: 259-273.

Domínguez, E. and M. L. Pescador. 1983. A new species of *Penaphlebia* (Ephemeroptera: Leptophlebiidae) from Argentina. Ent. News 94:21-24.

Domínguez, E. and R. W. Flowers. 1989. A revision of *Hermanella* and related genera (Ephemeroptera: Leptophlebiidae: Atalophlebiinae) from subtropical South America. Ann. ent. Soc. Amer. 82: 555-573.

Pescador, M. L. 1997. *Gonserellus*: A new genus of Leptophlebiidae (Ephemeroptera) from southern South America. Aquat. Insects 19: 237-242.

Pescador, M. L. and W. L. Peters. 1980a. Two new genera of cool-adapted Leptophlebiidae (Ephemeroptera) from southern South America. Ann. ent. Soc. Amer. 73: 332-338

Pescador, M. L. and W. L. Peters. 1980b. Phylogenetic relationships and zoogeography of cool-adapted Leptophlebiidae (Ephemeroptera) from southern South America, pp. 43-56. In: J. F. Flannagan and K. E. Marshall (eds.). Advances in Ephemeroptera Biology. New York, Plenum.

Pescador, M. L. and W. L. Peters. 1982. Four new genera of Leptophlebiidae (Ephemeroptera: Atalophlebiinae) from southern South America. Aquat. Insects 4: 1-19.

Pescador, M. L. and W. L. Peters. 1985. Biosystematics of the genus *Nousia* from southern South America (Ephemeroptera: Leptophlebiidae: Atalophlebiinae). J. Kans. ent. Soc. 58: 91-123.

Pescador, M. L. and W. L. Peters. 1987. Revision of the genera *Meridialaris* and *Massartellopsis* (Ephemeroptera: Leptophlebiidae: Atalophlebiinae) from southern South America. Trans. Amer. ent. Soc. 112: 147-189.

Pescador, M. L. and W. L. Peters. 1990. Biosystematics of the genus *Massartella* Lestage (Ephemeroptera: Leptophlebiidae: Atalophlebiinae) from South America. Aquat. Insects 12: 145-160.

Pescador, M. L. and W. L. Peters. 1991. Biosystematics of the genus *Penaphlebia* (Ephemeroptera: Leptophlebiidae: Atalophlebiinae) from South America. Trans. Amer. ent. Soc., 117: 1-38.

Peters, W. L. and G. F. Edmunds (Jr.). 1972. A revision of the generic classification of certain Leptophlebiidae from southern South America (Ephemeroptera). Ann. ent. Soc. Amer. 65: 1398-1414.

Towns, D. R. and W. L. Peters. 1978. A revision of genus *Atalophlebioides* (Ephemeroptera: Leptophlebiidae). N. Z. J. Zool. 5: 607-614.

Towns, D. R. and W. L. Peters. 1996. Fauna of New Zealand: Leptophlebiidae (Insecta: Ephemeroptera). Manaaki Whenua Press, Lincoln Canterbury, New Zealand. No. 36, 143 pp.

THE IDENTITY OF *HAGENULOPSIS MINUTA* SPIETH (LEPTOPHLEBIIDAE: ATALOPHLEBIINAE)

William L. Peters[1] and Eduardo Domínguez[2]

[1] Entomology, Orr Drive
Florida A&M University, Tallahassee
FL 32307, USA
[2] CONICET - Facultad de Ciencias Naturales, UNT
(4000) San Miguel de Tucumán, Argentina

ABSTRACT

Additional characters of the male imago of *Hagenulopsis minuta* Spieth are described from material collected in Northern South America and the nymph is described based on reared specimens. Distributional limits are given for the species. The original species epithet *minutus* is changed to *minuta as* "-opsis" is feminine (Article 30a, Code). The subgenus Borinquena (Australphlebia) Peters known from Dominica and St. Lucia is synonymized with *Hagenulopsis* and the type species becomes *Hagenulopsis traverae comb. n.*

INTRODUCTION

Ulmer (1920) established *Hagenulopsis* for *H. diptera* based on male imagos and subimagos and female subimagos collected in Isabella region, Humboldt District, Santa Catarina State, Brazil. Later Spieth (1943) described a second species *H. minutus* based on 1 female imago collected from the Marowijne River in Surinam.

While studying the original material of *Choroterpes emersoni* Needham and Murphy (all collected at a single locality in Guyana), Traver (1946) indicated that the female description of *C. emersoni* by Needham and Murphy (1924) was in reality the female description of *H. minutus*. Further, the original material labeled *C. emersoni* contained 2 male imagos of *H. minutus* which Traver described in 1946. The species *emersoni* was later transferred to *Miroculitus* by Savage and Peters (1983).

Traver (1944) described the probable nymph of *Hagenulopsis* as "*Hagenulopsis*-ally" based on 2 nymphs from Minas Gerais State, Brazil. Peters (1969) confirmed the identity of these nymphs as *Hagenulopsis* based on unpublished reared material. The 2 nymphs described by Traver (1944) are probably the nymphs of *H. diptera* based on their distribution in southern Brazil and the clouded cross veins in the developing wing pads.

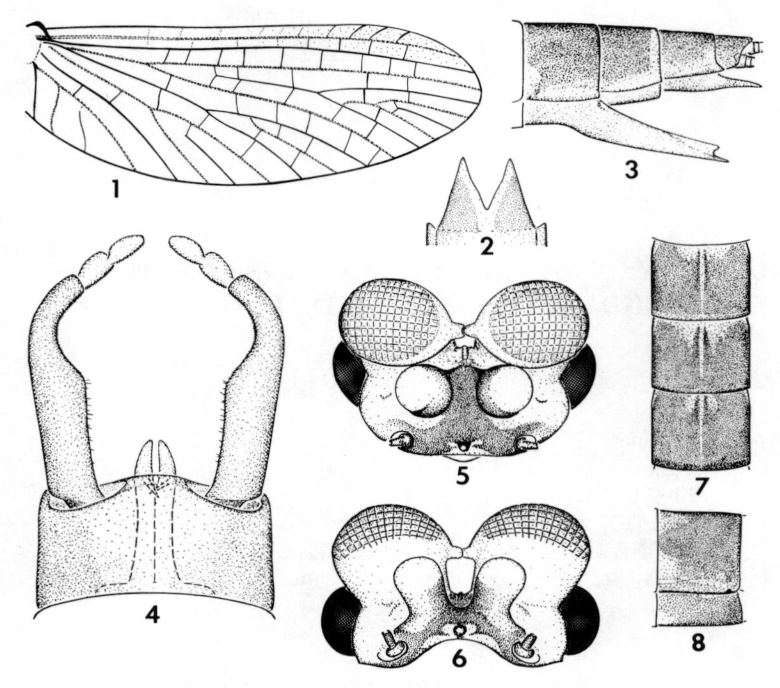

Fig. 1-8. *Hagenulopsis minuta* imagos: 1, ♂ fore wing; 2, 9th sternum of ♀; 3, lateral view of abdominal segments 6-10 of ♀; 4, ♂ genitalia, ventral view; 5-6, head of ♂ (5 dorsal, 6 frontal, without flagellum); 7, abdominal terga 5-7; 8, lateral view of abdominal segment 6.

In this paper the species epithet *minutus* is changed to *minuta* as "-opsis" is feminine (Article 30a, Code), additional characters of the male imagos of *H. minuta* are described, and the nymph of *H. minuta* is described based on reared material. The subgenus *Borinquena (Australphlebia)* Peters known from Dominica and St. Lucia is synonymized with *Hagenulopsis*.

Hagenulopsis minuta Spieth, 1943 (Fig. 1-20)

Hagenulopsis minutus Spieth, 1943, 1233:10; Traver, 1946, 17:427-428.

Hagenulopsis sp., Edmunds *et al.*, 1976, fig. 212-213, 376.

Hagenulopsis (in discussion of *Choroterpes emersoni*), Savage and Peters, 1983, 108:569.

Male imago (in alcohol): Length: body 3.5-3.8 mm; fore wings 3.9-4.0 mm. Eyes (Fig. 5-6): upper portion separated basally, meeting medially with well developed eye bridge on inner margin of upper portion; facets of turbinate portion large, 14-16 complete facets in longest row, reddish-brown; facets of lower portion black, small. Head reddish-brown, carinae darker. Scape and pedicel darker, washed with blackish-brown, flagellum paler. Lateral ocelli enlarged (Fig. 6). Abdomen (Fig. 7-8): translucent, reddish-brown, terga and sterna heavily washed with blackish-brown as in Fig. 7-8. Genitalia (Fig. 4): styliger plate and forceps segment 1 heavily washed with blackish-brown, forceps segment 2 and 3 paler; penes pale. Caudal filaments pale, articulations of basal 3 segments darker.

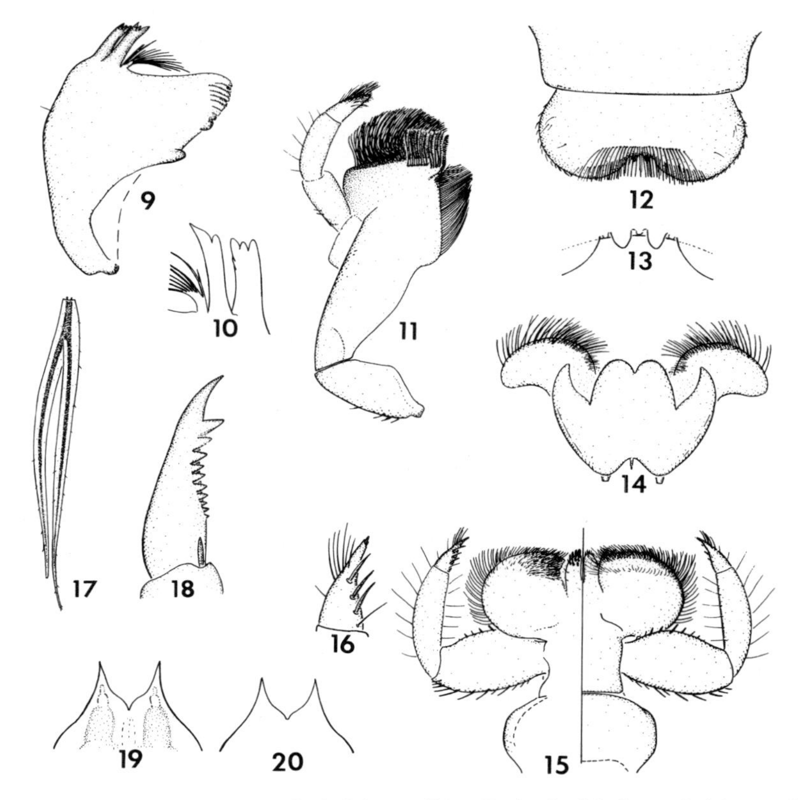

Fig. 9-20. *Hagenulopsis minuta*, nymph: 9, left mandible; 10, detail of incisors of right mandible; 11, maxilla (ventral); 12, labrum and clypeus; 13, ventral detail of margin of labrum; 14, hypopharynx; 15, labium (dorsum on left, venter on right); 16, detail of 3rd segment of labial palp; 17, gill 4; 18, fore claw; 19-20, apex of sternum 9 (19 mature ♂, 20 partially grown ♀).

Legs (description from Savage and Peters 1983 and additional material from Cerro de Neblina): ratios of segments in prothoracic legs, 0.46-0.50: 1.00 (1.57-1.59 mm): 0.03: 0.30-0.35: 0.21-0.27: 0.15-0.17: 0.06-0.07. Legs pale, except coxae, trochanters and prothoracic femora heavily washed with blackish-brown, apex of prothoracic tibiae blackish-brown, mesothoracic femora with an indistinct, subbasal and apical blackish-brown band, metathoracic femora darker with wide dark blackish-brown subbasal and apical bands; claws of a pair dissimilar, one apically hooked, other obtuse, pad-like.

Male subimago (in alcohol): Eyes as in ♂ imago, except bridge only developed as a protuberance on inner margin of stalks and not meeting on vertex of head.

Mature nymph (in alcohol): Body length 3.8-4.2 mm. Eyes of ♀ black, upper portion of eyes of ♂ reddish-brown, lower portion black. Dorsum of head light brown, area between base of antennae to lateral ocelli washed heavily with blackish-brown. Scape and pedicel washed with blackish-brown, flagellum pale. Mouthparts (Fig. 9-16): anteromedian emargination of labrum cleft with 2 large denticles and remnants of 3 small denticles (fig. 12-13); left mandible as in Fig. 9; hypopharynx as in Fig. 14; maxillae as in Fig. 11, segment 1 of palpi 1-1/3 length of segment 2, segment 3 2/5-3/5 length of segment 2; labium as in Fig. 14, segment 1 of palpi a little shorter than segment 2, segment 3 a little less than 1/2 length of segment 2. Thorax: brown, pronotum and anterolateral areas of mesonotum uniformly washed with blackish-brown. Legs: brown, blackish-brown marks as in ♂ and ♀ imagos, except all marks paler (especially on prothoracic femora). Abdomen: brown, terga 1-9 uniformly washed with dark blackish-brown, except for paler median longitudinal line as in ♂ imago; distinct posterolateral projections progressively larger posteriorly on terga 6-9 (may be weakly indicated on terga 3-5); sterna marked as in ♂ and ♀ imagos; apex of 9th sternum divided into pair of acute projections (Fig. 19-20). Gills (Fig. 17): membrane grayish, translucent; tracheae black, unbranched. Caudal filaments brown, basal articulations darker as in ♂ and ♀ imagos.

Type locality: SURINAM: Marowijne River, August, 1939, Geijskes coll.

Deposition of Type: Holotype ♀ imago, deposited in American Museum of Natural History (currently holotype can not be found in museum).

Specimens examined: SURINAM: Brokopondo Dist., Kreek on N. edge of Brokopondo, 90 m, 27-XII-1968, W.L. and J.G. Peters, nymphs; BRAZIL: Pará State: Akahe Creek nr. Tiriyos Mission, nr. Brazil-Surinam border, 15-27-III-1962, E.J. Fittkau, 2 ♂ imagos, 2 ♀ imagos, ♂ and ♀ subimagos, reared ♀ subimago, nymphs; Okueima Creek, nr. Brazil-Surinam border, 18-IV-1962, E.J. Fittkau, nymphs.

Additional material examined: BRAZIL: Amazonas State: Rio Maurauia, 3 day's trip above S. Antonio Mission, NW of Taparuquara, 24-25-I, 28-I-1963, E.J. Fittkau, ♂ imago, ♂ subimago, nymphs; mountain stream II, nr. Rio Maurauia, other data as given above, 27-I-1963, nymphs; VENEZUELA: T.F. Amazonas, Cerro de Neblina: Base Camp, 0°50'N, 66°10'W, rapids of Rio Baria, 12-II-1985, P.J. and P.M. Spangler, R. Faitoute, nymphs; same locality, 4-12-II-1984, D. Davis and T. McCabe, ♂ imagos; same locality, Camp VII, 1800 m, 0°50'N, 66°68'W, small stream, 30-I-10-II-1985, P.J. and P.M. Spangler, R. Faitoute, ♂ imago; Agua Blanca, 0°49'N, 66°08'W, 160 m, 20-21-III-1984, O. Flint and J. Louton, ♂ and ♀ subimagos.

Variations: Spieth (1943) noted the small size of *H. minuta* based on 1 ♀ imago. Both the body and wings of the holotype are 3.0 mm. Traver (1946) indicated that the body length of the ♂ imago is 3.5 mm and the wings are about 4.0 mm. Specimens we examined from Surinam and Pará State, Brazil, are all within the range of measurements given in the description, but specimens from Venezuela are larger. Traver (1946) noted that on some ♂ imagos the basal and middle terga possess a pale crescentic area on the anteromedian margin. Male specimens we examined have such a mark on terga 2-6. The mark on specimens from all sites in Brazil is indistinct, while the mark on specimens from Venezuela is very distinct. Dark marks are more extensive on specimens from Venezuela and the Amazonas localities. Number of facets in the dorsal portion of ♂ eyes differ somewhat by locality, the smallest (10 facets in longest row) being found on the single ♂ from Venezuela, Cerro de Neblina, at 1800 m.

DISCUSSION

Spieth (1943) differentiated *H. minuta* from *H. diptera* by the small size and the reduction of the cross veins in cells C and Sc of the fore wings; however, the 2 species can be differentiated by several other characteristics. In imagos of *H. minuta*: (1) well developed bridge between the stalks of the upper portion of the ♂ eyes (Fig. 5-6); (2) absence of blackish clouds (fringes) around cross veins in the fore wings of ♂ and ♀ imagos (Fig. 1); (3) indistinct subbasal and apical blackish-brown bands on the mesothoracic femora in ♂ and ♀ imagos; and (4) darker annulations at articulations of only the basal 3 caudal filaments of ♂ and ♀ imagos. Although the nymph of *H. diptera* is not known with certainty, it is clear that the color pattern of mature nymphs is the same as that of imagos, so that the lack of wing markings or apical markings on the caudal filaments should serve to distinguish the nymph of *H. minuta.*

We recognize the ♂ parts on the allotype slide of *Choroterpes emersoni as* those of *Hagenulopsis minuta* as suggested by Traver (1946) and Savage and Peters (1983) because of the color pattern of legs and caudal filaments and the fact that there is some evidence of a protuberance between the eyes on the slide. For illustrations in Edmunds *et al.* (1976), 3 are definitely *H. minuta* (212-213, 376) and 2 are probably *H. minuta* (160, 218); other figures may represent other species except for Fig. 164, an atypical claw which is not characteristic of any known species of *Hagenulopsis*. All figures of imagos in this paper were drawn from the Akahe Creek locality (Pará, Brazil) and those of the nymph from the Brokopondo locality (Surinam), except that detail of the submentum and ventral detail of the labrum were added from Akahe Creek specimens because of better condition on the slides.

Based on collections of *Hagenulopsis* at Florida A&M University, the genus occurs from Rio Grande do Sul State, Brazil, to Surinam, across the Guiana Shield to Peru and Venezuela and north to Honduras. Savage (1987) listed *Hagenulopsis* as a representative of the "North Andean, Central American Genera"; however, based on the known distribution of the genus it is a member of the "Guiana and Brazilian Shields Associated Genera". This generic distribution includes 4 described and several undescribed species, and *Hagenulopsis* can be differentiated from other Neotropical genera of the Leptophlebiidae by the following combination of characters. In the imagos: (1) vein ICu_1 attached at base to vein CuP in the fore wings (Fig. 1); (2) hind wings absent (Fig. 1); and (3) inner angle of forceps segment 1 located about 1/2 distance from base (Fig. 4). In the nymph: (1) abdominal gills slender, deeply forked with tracheae unbranched (Fig. 17); (2) ninth sternum deeply forked with apices acute (Fig. 19-20); and (3) apical 1/2 of segment 3 of labial palpi constricted (Fig. 16).

Kluge (1994) suggested that the nymph of *Borinquena carmencita* redescribed by Peters (1971) could not be the true nymph of *B. carmencita* but must represent another species. Traver (1938) designated only imagos from the Luquillo Mountains in the type series, but cited nymphs from other areas in Puerto Rico. The nymphal description and illustrations in Peters (1971) were made from specimens collected at the Hicaco River, Rio Blanco, March 7, 1935, by J. Needham and J. Garcia-Diaz. It now appears that Traver (1938) was looking at more than 1 species, and the nymphal description of Peters (1971) is not *Borinquena*, but is *Hagenulopsis*. The nymph of *Borinquena* as described by Traver (1938) and Kluge (1994) can be differentiated from *Hagenulopsis* by the characters cited above.

Peters (1971) established *Borinquena (Australphlebia)* for *B. (A.) traverae* from Domínica and an undescribed species from St. Lucia. Based on the definition of *Hagenulopsis* given above, *Australphlebia* is a junior synonym of *Hagenulopsis* and the type species of *Australphlebia* becomes *Hagenulopsis traverae* comb. n. McCafferty (1985) noted 2 undescribed species of *Borinquena* (*Australphlebia*) from Costa Rica, which were recently described by Lugo-Ortiz and McCafferty (1996) as *Hagenulopsis ingens* and *H. ramosa* based upon nymphs.

ACKNOWLEDGMENTS

We would like to thank the following persons who loaned or donated material for this study: Drs E. J. Fittkau, Zoologische Sammlung des Bayerischen Staates, München; O.S. Flint, Jr., National Museum of Natural History, Washington, D.C.; and Mr. E. R. Hoebeke, Cornell University, Ithaca, NY. We sincerely thank Mrs. J. G. Peters for the preparation of illustrations. Figures 3-4 and 14 were originally published in Edmunds *et al.* 1976 and are used with permission of G. F. Edmunds, Jr.

REFERENCES

Edmunds, G. F. (Jr.), S. L. Jensen and L. Berner. 1976. The Mayflies of North and Central America. Univ. Minnesota Press, St. Paul.

Kluge, N. Yu. 1994. A revision of Leptophlebiidae of Cuba (Ephemeroptera). Zoosyst. Rossica, 1993, 2:247-285.

Lugo-Ortiz, C. and W. P. McCafferty. 1996. New species of Leptophlebiidae (Ephemeroptera) from Mexico and Central America. Ann. Limnol. 32: 3-18.

McCafferty, W. P. 1985. New records of Ephemeroptera from Middle America. Int. Q. Ent. 1: 9-11.
Peters, W. L. 1969. *Askola froehlichi* a new genus and species from Southern Brazil (Leptophlebiidae: Ephemeroptera). Florida Ent. 52: 253-258.
Peters, W. L. 1971. A revision of the Leptophlebiidae of the West Indies. Smithson. Contr. Zool. 62:1-48.
Savage, H. M. 1987. Biogeographic classification of the Neotropical Leptophlebiidae (Ephemeroptera) based upon geological centers of ancestral origin and ecology. Stud. Neotrop. Fauna Environ. 22: 199-222.
Savage, H. M. and W. L. Peters. 1983. Systematics of *Miroculis* and related genera from Northern South America (Ephemeroptera: Leptophlebiidae). Trans. Amer. ent. Soc., 1982, 108: 491-600.
Spieth, H. 1943. Taxonomic studies on the Ephemeroptera. III. Some interesting ephemerids from Surinam and other Neotropical localities. Am. Mus. Novit. 1244: 1-13.
Traver, J. R. 1938. The mayflies of Puerto Rico. J. Agri. Univ. Puerto Rico 22: 5-42, 3 plates.
Traver, J. R. 1944. Notes on Brazilian mayflies. Bol. Mus. Nac., Rio de J., N.S., Zool. 22: 2-52.
Traver, J. R. 1946. Notes on Neotropical mayflies. Part I. Family Baetidae, Subfamily Leptophlebiinae. Rev. ent., Rio de J., 17: 418-430.
Ulmer, G. 1920. Neue Ephemeropteren. Arch. Naturg. 85 (A): 1-80.

PLATYBAETIS GAGADJUENSIS, A NEW SPECIES FROM NORTHERN AUSTRALIA (EPHEMEROPTERA : BAETIDAE)

Phillip J. Suter

Department of Environmental Management
and Ecology La Trobe University
Albury/Wodonga Campus
PO Box 821, Wodonga, Victoria, Australia 3689

ABSTRACT

A new species of *Platybaetis* is described from nymphal and subimago material collected from Kakadu National Park, Northern Territory. These dorso-ventrally flattened mayfly nymphs are very small and lack a number of characters expressed by the four described species from south east Asia. A subimago was reared and verified Müeller-Liebenau's recognition that *Platybaetis* was closely related to a "*Pseudocloeon*" type adult.

Field collections from the South Alligator River in Kakadu National Park suggested that this species has a distinct diurnal activity which reduces the chance of collection of this small nymph.

The Australian Monitoring River Health Initiative collections from northern Queensland also included nymphs of this species and extends the known distribution across the tropical north of the Australian continent.

INTRODUCTION

The genus *Platybaetis* was described by Müller-Liebenau (1980a) to include a dorsoventrally flattened nymph in the family Baetidae. In her paper, Müller-Liebenau described *P. edmundsi* from the Philippines, and included a second species *P. eunoi* which was described by Ueno (1955) as "Baetis sp 2" from Nepal. In subsequent papers Müller-Liebenau (1980b and 1984) described two more species, *P. bishopi* from Malaysia and *P. probus* from Sabah. This wide distribution in the oriental region has been extended with specimens of this genus being recorded in Australia by Suter (1992) from the Northern Territory. This species has remained undescribed until this paper.

In 1992 the Monitoring River Health Initiative (MRHI) was established as part of the National Rivers Health Programme in Australia with the aim to assess the health of Australian rivers by developing a series of predictive models for geographical regions throughout Australia. Approximately 1500 sites were chosen as being pristine or at worst having mini-

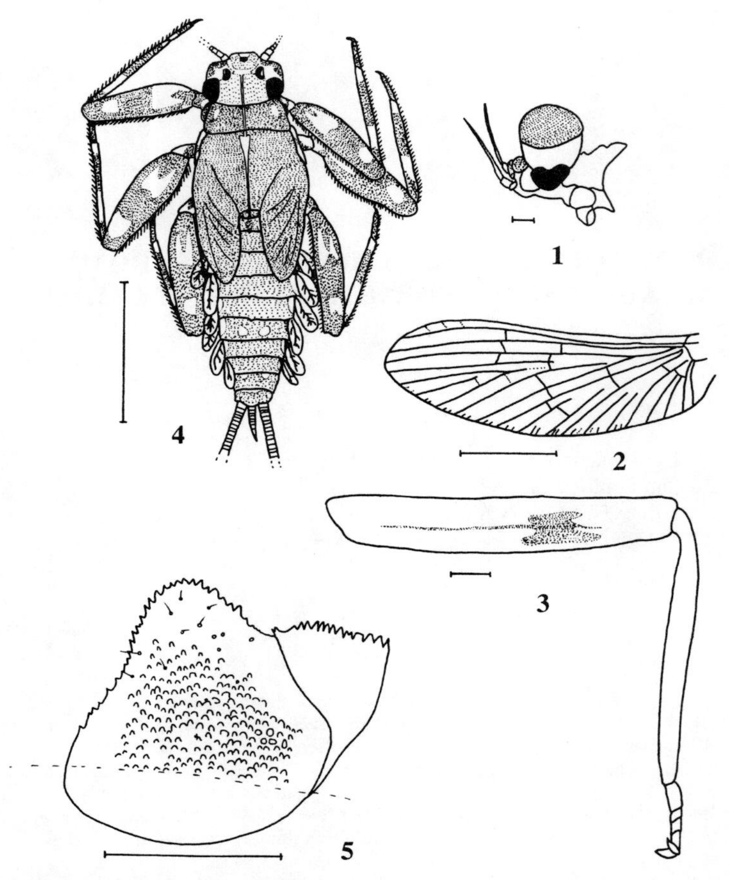

Figs. 1-5. Characteristics of male subimago and nymph of *Platybaetis gagadjuensis*. 1) Lateral view of turbinate eye of male subimago. 2) Forewing of male subimago. 3) Foreleg of male subimago. 4) Whole nymph. 5) Paraproct.
Scale Lines: Figures 1, 3 and 5 = 0.1mm, Figures 2 and 4 = 1.0mm.

mal human impact, and macroinvertebrate samples were taken from each of these sites in Autumn and Spring in 1994 -1996. The predictive models were constructed using family level taxonomic discrimination. Currently the First National Assessment of River Health (FNARH) is being undertaken at a further 1-2000 sites which have been modified by human activities (eg mining, effluent disposal, channel modification, land clearance, hydrological change etc.).

These initiatives have provided a large material base for taxonomic study over the whole continent. I was involved in a taxonomic project under the MRHI which was to provide illustrated keys to voucher species of the Baetidae and Caenidae. Similar such projects were included for the other Australian mayfly families. Through this work additional *Platybaetis* material has been collected from northern Australia, particularly in Queensland. It is likely that more specimens will be forthcoming as FNARH continues.

The subimago and nymph of the Australian species are described and the nymph is compared with the other four described species.

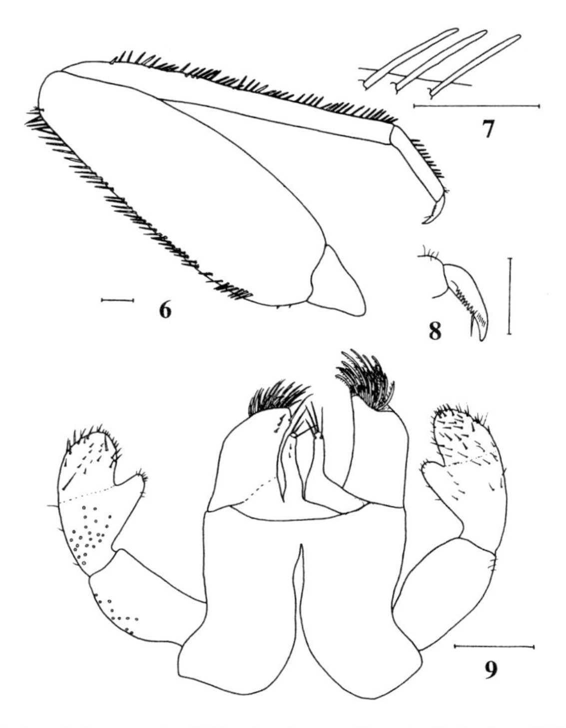

Figs. 6-9. Characteristics of the nymph of *Platybaetis gagadjuensis*. 6) foreleg. 7) Detail of setae lining femora, tibiae and tarsi of legs. 8) detail of tarsal claw of foreleg. 9) Labium.
Scale Lines = 0.1mm.

METHODS

Nymphs were collected using a hand-held dip net with mesh pore size of 250mm held downstream of disturbed substrate, or by hand picking of nymphs clinging to the under-surface of rocks. Mature nymphs from the South Alligator River were kept in 1L plastic rearing containers which had mesh sides and fitted into a polystyrene float so that it could remain in the river and maintain flow and adequate oxygen. Other specimens were preserved in 75% ethanol.

Nymphs were dissected and mounted on slides in polyvinyl lacto-phenol mounting medium. Illustrations were prepared with the aid of a camera lucida. Mouthparts were viewed ventrally (except labrum) and the labium is illustrated with the ventral surface shown on the right hand side of the illustration and the dorsal surface on the left.

***Platybaetis gagadjuensis* sp. nov.**

Baetidae Genus A sp1 Suter 1992
Platybaetis Dean and Suter 1996
Platybaetis sp 1 Suter 1997

Holotype: 1 nymph dissected and mounted on two slides in polyvinyl lactophenol mountant. Collected from Stewart Ck at timber crossing, Queensland, MRHI site 1121028 17°39' S, 145°58' E 10 July 1997 by Department of Primary Industries (DPI).

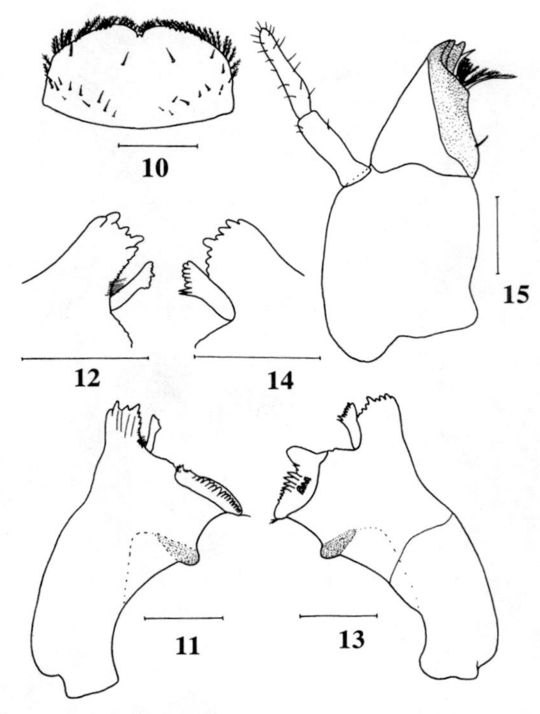

Figs. 10-15. Mouthparts of the nymph of *Platybaetis gagadjuensis*. 10) Labrum. 11) Ventral view of right mandible. 12) Detail of incisors and prostheca of right mandible. 13) Ventral view of left mandible. 14) Detail of incisors and prostheca of left mandible. 15) Maxilla.
Scale Lines = 0.1mm.

Paratypes: 6 nymphs from the type locality as above; 1 nymph dissected and mounted on two slides in polyvinyl lactophenol mountant collected from South Alligator River above Fisher Ck confluence NT. 13°34' S, 132°34' E, April 1989, coll. P. Suter, A. Wells, P. Cranston; 1 subimago and cast skin in ethanol collected from South Alligator River above Fisher Ck confluence NT. 13°34' S, 132°34' E, reared 21 April 1989, coll. P. Suter, A. Wells, P. Cranston. 3 nymphs Annan River at Main Rd, Qld. MRHI site 107003A 15°42' S, 145°12' E, 29 June 1997, coll. DPI; 2 nymphs South Johnstone River upstream of Central Mill, Qld. MRHI site 112101B 17°37' S, 145°59' E, 26 October 1994 coll. DPI; 1 nymph Herbert River at Long Pocket, Qld. MRHI site 1160098 18°31' S, 146°02' E, 25 June 1997 coll. DPI; 2 nymphs McLeod River at McLeod, Qld. MRHI site 919013A 16°30' S, 145°00' E, 24 June 1997 coll. DPI; 1 nymph Meunga Ck at Ellerbeck Rd., Qld. MRHI site 1140015 18°14' S, 145°56' E 2 July 1997 coll. DPI; 2 nymphs North Johnstone River at Tung Oil, Qld. MRHI site 114004A 17°33' S, 145°56' E, 8 July 1997 coll. DPI; 1 nymph Herbert River at Abergowie, Qld. MRHI site 116006B 18°30' S, 145°55' E, 25 June 1997 coll. DPI; 1 nymph Murray River at Leichhardt, Qld. MRHI site 1140010 18°01' S, 145°56' E, 4 July 1997 coll. DPI.

All types are placed in the Australian National Insect Collection, CSIRO Canberra.

Other Material Examined: Northern Territory: 3 nymphs South Alligator River above Coronation Hill 13°36' S, 132°37' E, May 1988, April 1989, coll. P. Suter, A. Wells, P. Cranston; 4 nymphs South Alligator River at Gimbat OSS Field Station 13°35' S, 132°36' E, April 1989, coll. P. Suter, A. Wells, P. Cranston; 4 nymphs and 1 subimago South Alligator

River above Fisher Ck confluence 13°34' S, 132°34' E, April 1989, coll. P. Suter, A. Wells, P. Cranston.

Imago: unknown.

Subimago (Figs. 1-3): associated by rearing on 21 April 1989.

Body length approx. 3mm; forewings (Fig. 2) with reduced crossvein system, length 3x longer than wide; marginal intercalaries paired; pterostigma with 3 very faint crossveins; hind wings absent. Forelegs (Fig. 3) with a black marking on femora, other segments pale; 4 tarsal segments, leg ratios 1.19: 1.00: 0.11: 0.04: 0.04: 0.08 (tibial length 0.76mm); tarsal claws dissimilar; turbinate eyes round dorsally (Fig.1).

Nymph (Figs. 4-15): Body length: 2.8-4.0mm. Cerci length approx. 2mm, terminal filament reduced to <16 segments, 0.4mm. Body (Fig. 4) colour grey-brown with light markings. Head prognathous, broader than long with very small labrum, lacking a deep incision in posterior margin. Pronotum dark with two light lateral patches. Mesonotum dark with a light triangular central marking. Hind wings absent. Legs appear banded, long with femora flattened; villopore absent; femora dark with two light patches, tibiae dark with light portion centrally and apically, fore tarsi light basally, dark apically, mid and hind tarsi all dark; femora, tibiae and tarsi lined with long blunt setae (Figs. 6 and 7); surface of femora rugose; tarsal claws long with 8-10 teeth and a long apical setule (Fig. 8); sement ratios:

Fore leg: 1.00: 0.97: 0.27 (Femur Length 0.8-1.1mm)

Mid leg: 1.00: 0.90: 0.25 (Femur Length 0.9-1.2mm)

Hind leg 1.00: 0.82: 0.23 (Femur Length 0.9-1.2mm)

Femur length to width ratio, fore leg 2.75, mid leg 3.3, hind leg 3.8.

Abdomen dark with light patches on tergites 6 and 9; small tubercle present on posterior margin of tergites 1-5, not apparent in small specimens; paraprocts with numerous small mesial and apical spines, surface with numerous tubercles (Fig. 5); gills present on segments 2-7, lamellae single and plate-like, margins serrated and ciliated.

Mouthparts: labrum (Fig. 10) lined with pinnate setae, deep central concavity present; right mandible (Fig. 11) with fused incisors with at least 7 apical teeth (8 clear teeth in newly moulted nymphs), inner margin rugose and with basal hairs present, prostheca robust, broadest apically with 8-10 apical teeth (Fig. 12); left mandible (Fig 13) with incisors fused, with approx. 8 apical teeth and 3 small inner teeth, prostheca robust with 6-8 apical teeth and 2 spines (Fig. 14), molar region with large broad tooth on inner margin, smaller teeth centrally. Maxillae (Fig. 15) with a single setule near medial hump, two segmented palp approximately as long as galeolacinia, basal segment 0.8x segment 2. Labium (Fig. 9) with narrow glossae which are shorter than the broad paraglossae; labial palpi three segmented, second segment well developed lobe on mesial margin; segment ratios, 1.00: 0.82: 0.51 (0.14mm), basal segment length 1.6x width.

Etymology: Named after the Gagadju country in which Kakadu National Park is situated and from where this species was first recorded.

DISCUSSION

This species is placed in *Platybaetis* because it possesses the following characters; nymph in both sexes is dorso-ventrally flattened; head is prognathous, is broader than long with very small labrum; hind wings absent; legs with femora flattened; femora, tibiae and tarsi lined with long blunt setae.

Platybaetis gagadjuensis is distinguished from the four other species in the genus by the size, less than 5mm in length, gills present on abdominal segment 2-7, head lacking a deep incision in posterior margin; prosthecae of mandibles large and robust; right mandible with hair fringe on inner margin of incisors; glossae narrow and distinctly shorter than the broad paraglossae; second segment of labial palp with well developed lobe on mesial margin.

Waltz and McCafferty (1987) noted that *Platybaetis* and other closely related genera are "...distinguished from the larvae of all other baetids by their synapomorphic possession of a ventral femoral patch" (Waltz and McCafferty 1987:553). However, *P. gagadjuensis* does not possess this femoral patch, but is in strong agreement with other generic characters of the genus. It is suggested that the absence of this femoral patch in this species may be due to the very small size of the fully mature nymph and that its expression has not developed. Similarly the absence of a gill on the first abdominal segment may also be related to the reduced size of this species.

The presence of this species in Northern Australia extends the known distribution of this genus in the oriental region being recorded from Nepal (*P. uenoi*), Malaysia (*P. bishopi*), The Philippines (*P. edmundsi*) and Sabah, E. Malaysia (*P. probus*).

The nymphs of *P. gagadjuensis* were collected from riffle zones in the rivers, principally from on rocks in moderately flowing water. Observations made in the South Alligator River in 1989 suggested that the nymphs have a distinctive period of activity in the morning. Nymphs were all collected before midday in the South Alligator River, but never in the afternoon. A similar observation has been made with specimens collected in Queensland. All specimens were collected before 2.30pm eastern standard time. The absence of this nymph in samples collected later in the afternoon may mean that its distribution is more widespread, but it has been missed due to time of collection. In addition with the MRHI samples only being identified to Family some specimens may have been incorrectly identified as juvenile Leptophlebiids, although the quality assurance program has not detected this mis-identification.

Only one subimago was attracted to a UV light trap in the South Alligator River. The reared specimen died as a subimago, but the characteristics of the wing venation is consistent with the *Pseudocloeon* group. Müller-Liebenau (1980a) considered *Platybaetis* closely related to *Pseudocloeon*, but with the difficulties associated with the adult taxonomy no further conclusions can be drawn.

ACKNOWLEDGMENTS

I would like to thank Peter Dostine and Chris Humphrey who provided the opportunity to collect in Kakadu National Park, Alice Wells and Peter Cranston for the assistance in the field in Kakadu; Dianne Conrick from the Queensland Department of Primary Industries for providing the MRHI material. Funds from the Land and Water Resources Research and Development Corporation through the taxonomic support program of the MRHI are gratefully acknowledged. This work also includes work carried out as part of the research program of the Supervising Scientist for the Alligator Rivers Region.

REFERENCES

Dean, J. C. and P. J. Suter. 1996. Mayfly Nymphs of Australia a Guide to Genera. CRC Freshwater Ecology Identification Guide No. 7.

Müller-Liebenau, I. 1980a. *Jubabaetis* Gen.N. and *Platybaetis* Gen. N., Two New Genera of the Family Baetidae from the Oriental Region, pp. 103-114. In: J. F. Flannagan and K. E. Marshall (eds.). Advances in Ephemeroptera Biology. Plenum Press New York.

Müller-Liebenau, I. 1980b. A New Species of the Genus *Platybaetis* Müller-Liebenau 1980, *P. bishopi* sp.n., from Malaysia (Insecta, Ephemeroptera). Gewäss. Abwäss. 66/67: 95-101.

Müller-Liebenau, I. 1984. Baetidae from Sabah (East Malaysia) (Ephemeroptera), pp. 85-89. In: V. Landa *et al.* (eds.). Proc. Ivth International Conference Ephemeroptera; CSAV.

Suter, P. J. 1992. Taxonomic key to the Ephemeroptera (Mayflies) of the Alligator Rivers Region, Northern Territory. Supervising Scientist of the Alligator Rivers Region Open File Record 96.

Suter, P. J. 1997. Preliminary Guide to the Identification of Nymphs of Australian Baetid Mayflies (Insecta: Ephemeroptera) found in Flowing Waters. CRC Freshwater Ecology Identification Guide No. 14.

Ueno, M. 1955. Mayfly Nymphs. In: fauna and Flora of Nepal Himalaya. Scientific Results of the Japanese Expedition to Nepal Himalaya 1952-1953. 1: 301-316. Contribution No. 150 Otsu Hydrobiological Station, Kyoto University.

Waltz, R. D. and W. P. McCafferty. 1987. Systematics of *Pseudocloeon, Acentrella, Baetiella* and *Liebebiella*, New Genus (Ephemeroptera: Baetidae). J. N. Y. ent. Soc. 95: 553-568.

DESCRIPTION OF THE MALE IMAGO OF *AMELETUS PRIMITIVUS* TRAVER (EPHEMEROPTERA: AMELETIDAE) WITH NOTES ON ITS RELATIONSHIP WITH OTHER *AMELETUS* SPECIES

Jacek Zloty

Division of Ecology, Department of Biological Sciences
The University of Calgary
Calgary, Alberta, Canada T2N 1N4

ABSTRACT

The male imago of *Ameletus primitivus* (Ephemeroptera: Ameletidae) is described and illustrated for the first time from material collected from northern India. Diagnostic features are provided and comparison is made with adults and larvae of other *Ameletus* species.

INTRODUCTION

Ameletus primitivus was described by Traver (1939) from the female imago, female subimago and the larva from northern India. The persistent mouth-parts (labial and maxillary palps) of the winged stages are unique in the Ephemeroptera and led to the suggestion that the species is one of the most primitive of all mayflies (Traver, 1939). Lestage (1940) created a new genus *Paleoameletus* for *A. primitivus* and Edmunds and Traver (1954) synonymised *Paleoameletus* with *Ameletus*, without giving any explanation. Recently, I located a single male imago (in the collection of George F. Edmunds, Jr.; recently transferred to Purdue University, West Lafayette, Indiana) that was collected from the same area as Traver's (1939) types. The specimen has clearly visible mouth parts, and other morphological characters (e.g. size, wing coloration, emergence time) are also in accordance with Traver's (1939) description. There seems little reason to doubt that this male is *A. primitivus*. In this paper, I describe and illustrate the male imago of *A. primitivus*, and I also discuss its relationship with other *Ameletus* species in both adult and larval stages. Information on the interpretation of morphological characters used in the description may be found in Zloty (1996a) and Zloty and Pritchard (1997).

Description of the Male Imago of *Ameletus primitivus*

Ameletus primitivus Traver, 1939: 46; Lestage, 1940: 124; Edmunds and Traver, 1954: 237.

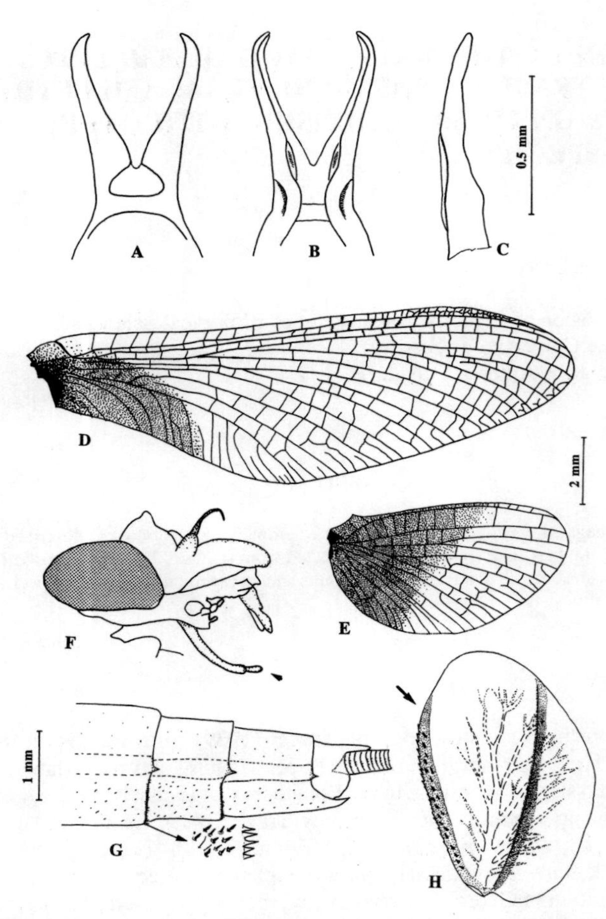

Fig. 1. Morphological characters of *Ameletus primitivus.* A- dorsal view of penes, B- ventral view of penes, C- right lateral view of penes, D- right fore wing, E- right hind wing, F- adult head in lateral view (after Traver 1939), G- larval abdominal segments 7-10, H- gill of abdominal segment 4 (modified from Traver 1939).

Type material: Female holotype, female subimago and larval paratypes: Kyam, North India, 22 June, 1932, Yale North Indian Expedition; in Purdue University, Indiana (examined).

Description: Male imago (in alcohol): body length 16 mm, fore wings 16 mm. Head dark brown; upper portion of compound eyes gray, lower portion brown. Labial and maxillary palps present (Fig. 1F). Pronotum, mesonotum, scutellum and infrascutellum dark brown; thoracic pleura dark brown with light brown between plates, at the centre of the katepimeron, and at the base of the legs and wings. Wings transparent with brown veins and light brown cross-veins and with pronounced suffusion at the base (Fig. 1D-E). First abdominal tergite brown; tergites 2-9 brown to dark brown with light brown along anterior margins; tergite 10 light brown. Abdominal sternite 1 light brown; sternites 2-8 opaque white with some light brown laterally. Ganglionic markings on sternites 2-8. Genitalia (Figs. 1A-C) with elongated lateral lobes (length of lateral lobes about 2X the width of the penes; as in Fig. 1A), ventral plates absent (Fig. 1C). Caudal filaments yellow.

DISCUSSION

The presence of labial and maxillary palps in adult mayflies is currently known only in this species. It is uncertain whether two other species (*A. alexandrae* Brodsky and *A. asiacentralis* Soldán) that are closely related to *A. primitivus* (Zloty, 1996b), also have these mouth-parts in the winged stages. Penes lack ventral plates (Fig. 1C), a characteristic shared only with *A. inopinatus* Eaton (widely distributed in Asia, Europe and northern part of North America) and *A. velox* Dodds (known only from western North America). However, the apical ends of the lateral lobes of the penes are straight in *A. primitivus,* are inwardly twisted in *A. inopinatus* (Zloty, 1996a: Fig. 9C) and are curved mesally in *A. velox* (Zloty, 1996a: Fig. 10A). Larvae of *A. primitivus* have numerous small spine-like bristles on the abdominal sternites and tergites (Fig. 1G), and on the dorsal surface of the lateral band on the abdominal gills (Fig. 1H). These characteristics are shared with three other Asian species, *A. alexandrae*, *A. asiacentralis* (Soldán, 1978: Plate II, Figs. 16-18) and an undescribed species from Talas River, Kirgizstan (in N.J. Kluge collection). Only one Nearctic species, *A. edmundsi* Zloty, has numerous spine-like bristles on the abdominal sternites and tergites, but none of the North American species has spines on the dorsal surface of the lateral band on the abdominal gills. Eggs of *A. primitivus* are ovoid (similar to Fig. 32B in Zloty and Pritchard, 1997).

Distribution - Northern India; June-July. The single male, from which the above description is made, was collected at Lhamo Tso Lake, extreme NE Sikkim, elevation 5,090 m, 25 June 1959, by F. Schmid.

ACKNOWLEDGMENTS

I thank G.F. Edmunds, Jr., N.J. Kluge and W.P. McCafferty for allowing examination of *Ameletus* specimens in their collections. I also thank Gordon Pritchard for his critical comments and support during the study. The work was supported by a grant from the Natural Sciences and Engineering Research Council of Canada to G. Pritchard.

REFERENCES

Edmunds, G. F. (Jr.) and J. R. Traver. 1954. An outline of a reclassification of the Ephemeroptera. Proc. ent. Soc. Wash. 56: 236-240.

Lestage, J. A. 1940. Contribution à l'étude des Ephéméroptères XXIV. Un cas de non-agnathisme chez l'adulte de *Paleoameletus primitivus* Trav. de l'Himalaya. Bull. Ann. Soc. ent. Belg. 80: 118-124.

Soldán, T. 1978. *Ameletus asiacentralis* sp. n. from Uzbekistan, with notes on *A. alexandrae* (Ephemeroptera, Siphlonuridae). Acta ent. Bohemoslov. 75: 379-382.

Traver, J. R. 1939. Himalayan mayflies (Ephemeroptera). Ann. Mag. Nat. Hist. 4: 32-56.

Zloty, J. 1996a. A revision of the Nearctic *Ameletus* mayflies based on adult males, with descriptions of seven new species (Ephemeroptera: Ameletidae). Canad. Ent. 128: 293-346.

Zloty, J. 1996b. Systematics of Nearctic *Ameletus* mayflies (Ephemeroptera: Ameletidae). Ph. D. diss., University of Calgary, Calgary, AB. 231 pp.

Zloty, J. and G. Pritchard. 1997. Larvae and adults of *Ameletus* mayflies (Ephemeroptera: Ameletidae) from Alberta. Canad. Ent. 129: 251-289.

PROGRESS ON TAXONOMIC STUDY OF THE FAMILY PERLIDAE FROM CHINA

Du Yuzhou[1,2] and He Junhua[1]

[1] Department of Plant Protection, Agricultural College
Zhejiang University, Hangzhou 310029, China
[2] Present Address: Department of Plant Protection, Agricultural College
Yangzhou University, Yangzhou, Jiangsu 225009, China

ABSTRACT

A review of taxonomic research on Chinese Perlidae shows that twelve foreign entomologists and six Chinese scholars have recorded 165 species of the family from China. In this paper the history, current status, biodiversity, biogeography and phylogeny of Chinese Perlidae is reviewed and some proposals for future research are given.

HISTORY OF TAXONOMIC RESEARCH ON CHINESE PERLIDAE

In 1909, Günther Enderlein described the first Chinese species of the family Perlidae, *Hemacroneuria violacea* Enderlein 1909 = *Acroneuria violacea* (Enderlein), from Wuyi Mountain, Fujian Province. Franz Klapálek, the founder of Plecopterology, described 27 species of Perlidae from China in 1912-1923. In this pre-1940's period, Longinos Navás (1911-1936) added 14 perlid species, Nathan Banks (1920-1940) described 15, H. Okamoto (1912) described three Taiwanese species and Peter W. Claassen, in cooperation with Wu Chenfu, described 10 species in 1934. The pioneering works authored by these researchers provide a foundation on which subsequent taxonomic research of Chinese perlids is based. More recently, Peter Zwick, Ignac Sivec, Bill P. Stark and others have described Chinese perlids bringing the total number of species described by twelve foreign entomologists given in Table 1, to 69. Four of these species known from China have type localities in other countries.

Prof. Chu Yanting, one of two Chinese pioneers of stonefly research, published the first taxonomic paper on the family Perlidae of China in 1928 and in that paper *Neoperla sinensis* Chu became the earliest perlid species named by a Chinese scholar. Prof. Chu published four papers on stoneflies and described five perlid species from Hangzhou, China in 1928-1929, before changing his research interests to the taxonomy of Chinese fish. Prof. Wu Chenfu, pioneer and founder of Chinese stonefly research, studied with James G. Needham at Cornell University in the early 1920's and received his Ph.D. degree in 1922. After returning to China he published three taxonomic papers on Chinese Nemouridae in1926-1929 before enlarging his studies to the entire plecopteran fauna of China. During the 1930's and 1940's he published 20 papers on stoneflies including his monograph, published in 1938, in which many new species were described. His work on stoneflies was interrupted then until 1962 when he completed a study of stoneflies collected in Yunnan. His final paper appeared in 1973, after his death in 1972. His life work includes 27 stonefly publications in which 148 new Chinese species from seven families are described; ten of these are co-described with Prof. Claassen. Seventy three of these species from 15 genera are from the family Perlidae. Unfortunately, most of the type specimens of Wu species were lost during World War II (Wu, 1962)

Following Dr. Wu's death in 1972, research on stoneflies in China stopped until recent studies of Prof. Yang Chikuen and Dr. Yang Ding in which 14 species of Chinese Perlidae are described; Dr. Du Yuzhou and Prof. Chou Io also recently described three new species which brings the number of species described by Chinese scholars to 95 (Table 1). Together, the twelve foreign entomologists and six Chinese scholars have recorded 165 species of Perlidae from China.

Table 1. Describers and the number of chinese Perlidae species described in each genera

	Klapáleck	Banks	Navás	Okamoto	Stark	Zwick	Mclachlan	Enderlein	Illies	Sivec et Zwick	Sivec et Zhiltzova	Wu	Wu et Claassen	Yang et Yang	Chu	Du et Chou	Total
Claassenia	3	1	1									4	1				10
Agnetina	1	2	1 (+1)								1	6				3	15
Etrocorema												1					1
Kamimuria	7		1									19			1		28
Neoperlops		2										1					3
Oyamia		1					(1)										2
Paragnetina	5	1	3									1	1	1			12
Togoperla	2		1														3
Tyloperla			1	1								2	1	1			6
Chinoperla		1										2					3
Furcaperla												1		1			2
Neoperla	4	3	2	2	2 (+1)	1 (+1)			1	1		12	4	9	3		46
Phanoperla		1															1
Acroneuria								1				4	1	1			7
Brahmana												1					1
Gibosia	2											2			1		5
Kiotina		3										4	1				8
Perlesta												2					2
Sinacroneuria			1											1			2
Mesoperla	1																1
Incertae sedis	2		2									2	1				7
Total	27	15	13 (+1)	3	2 (+1)	1 (+1)	(1)	1	1	1	1	64	10	14	5	3	165

( ) The type locality is not from China.

CURRENT RESEARCH STATUS

Chinese Perlidae Diversity

Perlidae is the most speciose stonefly family in China. Du (1995) examined numerous collections and published records and found 231 species in 21 genera in the Chinese perlid fauna. This includes 67 undescribed species and one undescribed genus. Presently the Chinese perlid species comprise about 35% of the world Perlidae making China the pre-eminent area of world perlid diversity. *Neoperla* with 79 species and *Kamimuria* with 42 species are two of the dominant perlid genera in China. Because much of the Oriental region of China is poorly studied we anticipate this survey of Chinese perlid species is far from complete.

Distribution of Chinese Perlidae

Only a preliminary analysis of distribution is possible based on current records. At present, three genera (*Furcaperla, Sinacroneuria* and *Mesoperla*) are known only from the South of China and Taiwan. The total perlid fauna includes nine strictly Oriental genus, one Palaearctic genus, four Oriental-Palaearctic genera, one Oriental-Nearctic genus, four Oriental-Palaearctic-Nearctic genera and one Oriental-Palaearctic-Nearctic-Afrotropical genus. These results show the Chinese perlid fauna is dominated by Oriental genera (Table 2).

Figure 1 identifies major zoogeographic regions of China which we use to carry out a preliminary analysis of known perlid distributions. Central China (Region VI) contains the greates perlid diversity with only 3 genera (*Etrocorema, Oyamia,* and *Mesoperla*) unrepor-

Table 2. Distribution species and genera of the family Perlidae in China

	Paleartic region										Oriental region									
	I			II		III			IV		V		VI		VII				NR	AR
Genera	IA	IB	IC	IIA	IIB	IIIA	IIIB	IIIC	IVA	IVB	VA	VB	VIA	VIB	VIIA	VIIB	VIIC	VIID		
Claassenia		2	1		2						5		4	4	1				+	
Agnetina							1				3		3	4	7	1		1	+	
Etrocorema																1	1			
Kamimuria		3			8	1					4	3	12	19	2	4		3		
Neoperlops											1		1	3	2					
Oyamia		2																		
Paragnetina						1					6		3	3	1			1	+	
Togoperla													3	5						
Tyloperla											1		2		2			2		
Chinoperla													1				2			
Furcaperla													2							
Neoperla		1		2		1					4	4	20	25	4	6	6	9	+	+
Phanoperla													1			3	1			
Acroneuria				1							4		3	3			1		+	
Brahmana											1		1	3						
Gibosia				1							1		3	1	2			1		
Kiotina											1		5		1			1		
Mesoperla																		1		
Perlesta					1?						1		2						+	
Sinacroneuria													2							

NR: Neartic region; AR: Afrotropical region.
The distributions of *Kiotina* and *Sinacroneuria* in Japan belong to Palaeartic region.

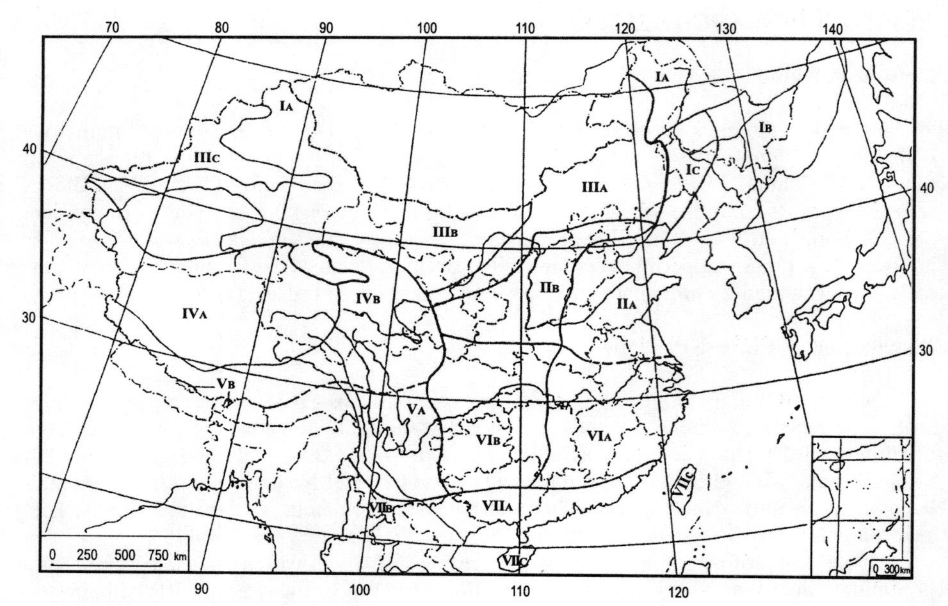

Fig. 1. Map of zoogeographical regions designing of China.
(Modified from Editorial Committee of "Chinese Natural Geography", Academia Sinica, 1979)
Northeast China region; IA- Daxingan mountain and the Altai subregion; IB- Changbai mountain subregion; IC- Liaosong plain subregion; II-North China region; IIA- Huanghuai plain subregion; IIB- Huangtu plateau subregion; III- Meng-Xin region; IIIA- Eastern prairie subregion; IIIB- Western desert subregion; IIIC-Tianshan subregion; IV- Qing-Zang region; IVA-Qiangtang plateau subregion; IVB- Qinghai Zangnan subregion; V- Southwest China region; VA- Southwestern mountainous district subrejion; VB- Himalayas subregion; VI- Central China region; VIA- Eastern hills and plain subregion; VIB- Western mountainous district subregion; VII- South China region; VIIA- Guangdong-Fujian coast subregion; VIIB- South Fujian mountainous district subregion; VIIC- Hainan Island subregion; VIID- Taiwan subregion.

ted from this region (Table 2). In addition, the species are more diverse in this region, especially those of *Neoperla* and *Kamimuria.* Therefore we suggest the Central China region as the distribution center for Chinese Perlidae. Southwest China (Region V) contains 12 genera but only *Kamimuria* and *Neoperla* are known in the Himalayas Subregion (VB) (Table 2). South China (Region VII) includes subregions along the coast of Fujian and Guangdong (VIIA), mountains of South Fujian (VIIB), Hainan Island (VIIC) and Taiwan (VIID). Fourteen genera are known from the South China Region; eight occur on Taiwan and five on Hainan Island. In North China (Region II), six genera are known, in Northeast China (Region I), four genera and in Meng-Xin (Region III), four genera. Presently no perlids are known from Qing-Zang (Region IV) where additional study is needed.

PHYLOGENY

Characters extracted from Sivec et. al (1988) were used with the phylogenetic program Hennig86 (version 1.5) to evaluate the phylogenetic relationships among 18 genera of Perlinae. The results are similar to those obtained by Sivec et. al (1988).

DISCUSSION AND PROSPECTS

Early studies of Chinese perlids by Klapálek, Navás, Wu and others provide an important base on which to begin the study of these insects but they generally lack the detail in penis characters, eggs and larvae needed. Species recorded in these studies require review, especially those based solely on females. These reviews are sometimes made difficult by loss of type specimens and other collections. It is important, now, to continue faunal investigations which will lead to material needed for phylogenetic and biogeographic analysis, and, at the same time, begin the study of stonefly biology and ecology which have been neglected in China. Present difficulties in pursuing these studies are due to the lack of research funds and researchers.

ACKNOWLEDGMENTS

This work was supported by the National Natural Science Foundation of China (N° 39560017).

REFERENCES

Banks, N. 1920. New neuropteroid insects. Bull. Mus. comp. Zool. Harv., 64: 314-325.
Banks, N. 1937a. Perlidae. In: Neuropteroid insects from Formosa. Philipp. J. Sci., 62: 269-275.
Banks, N. 1937b. Philippine neuropteroid insects. Philipp. J. Sci., 63: 134-137.
Banks, N. 1938. New Malayan neropteroid insects. J. F. M. S. Mus., 18: 221-223.
Banks, N. 1939. New genera and species of neuropteroid insects. Bull. Mus. comp. Zool. Harv., 85:439-504.
Banks, N. 1940. Report on certain groups of neuropteroid insects from Szechwan, China. Proc. U. S. nat. Mus., 88: 173-220.
Chu, Y. T. 1928. Descriptions of a new genus and three new species of stone-flies from Hangchow. China J., 9 (4): 194-198.
Chu, Y. T. 1929. Descriptions of four new species and one new genus of stone-flies in the family Perlidae from Hangchow. China J., 10 (2): 88-92.
Claassen, P. W. 1940. A catalogue of the Plecoptera of the world. Mem. Agr. Exp. Sta., 232: 147.
Du, Yuzhou. 1995. A taxonomic study on the family Perlidae from China (Plecoptera: Perloidea). Ph. D. thesis, Northwestern Agricultural University, pp.1-254.
Du, Yuzhou. 1998. Two new record species of genus *Neoperla* Needham (Plecoptera: Perlidae) from China. J. Zhejiang agric. Univ., 24 (4): 392-394.
Du, Yuzhou and Io Chou. 1998. Taxonomic study of the genus *Agnetina* Klapálek (Plecoptera: Perlidae: Perlinae) from China. Entomotaxonomia, 20 (2): 100-110.
Editorial Committee of "Chinese Natural Geography", Academia Sinica 1979. Chinese Natural Geography-Zoogeography. Science Press, Beijing, pp. 1- 150.
Enderlein, G. 1909. Klassification der Plecopteren sowie diagnosen neuer gattungen und arten. Zool. Anz., 34: 385-419.
Hennig, W. 1965. Phylogenetic systematics. Ann. Rev. Ent., 10: 97-116.
Hynes, H. B. N. 1987. Biogeography and Origin of the North American Stoneflies (Plecoptera). Mem. ent. Soc. Canada, 144: 31-37.
Illies J. 1965. Phylogeny and Zoogeography of the Plecoptera. Ann. Rev. Ent., 10: 117-140.
Illies, J. 1966. Katalog der rezenten Plecoptera. Tierreich, 82: i-xxx, 1-632.
Klapálek, Fr. 1912a. Plecopterorum genus: *Kamimuria* Klp. Casopis Ceské Spolec Ent., 9: 84-110.
Klapálek, Fr. 1912b. Plecoptera. In H. Sauter's Formosa-Ausbeute. Ent. Mitt., 1: 342-351.
Klapálek, Fr. 1913. Plecoptera II. In: H. Sauter's Formosa-Ausbeute. Suppl. Ent., 2: 112-123.
Klapálek, Fr. 1916. Subfamilia Acroneuriinae Klp.. Casopis Ceské Spolec Ent., 13: 45-84.
Klapálek, Fr. 1921. Pleecopteres nouveaux. Ann. Soc. ent. Belg., 61: 57-67, 146-150, 320-327.
Klapálek, Fr. 1923. Plecopteres II. Fam. Perlidae. Collections zool. du Baron Edm. de Selys Longchamps, Bruxelles, 4 (2): 1-193.
Liang, Aiping. 1993. On the phylogenetic program Hennig 86 (Version 1.5). Acta Zootaxon. Sin. 18 (4): 499-502.
Navás, R. P. L. 1911. Fam. Perlidae, pp. 11: 111-112. In: Névroptères nouveaux de l'extrème Orient. Rev. ent. URSS (Moskau).

Navás, R. P. L. 1912. Famille Perlidae, pp. 12: 417-418. In: Quelques névroptères de la Sibèrie méridionale-orientale. Rev. ent. URSS (Moskau).

Navás, R. P. L. 1919. Plecoptera, pp. 9: 186-189. In: Névroptères de l'Indo-Chine. Insecta, Rennes.

Navás, R. P. L. 1922a. Plecópteros, pp. 20: 49-50. In: Insectos exóticos. Brotéria, Sér. Zool. (Lisboa).

Navás, R. P. L 1922b. Plecópteros, pp. 17 (15): 6-9. In: Insectos nuevos o poco conocidos. Mem. R. Acad. (Barcelona).

Navás, R. P. L. 1922c. Plecópteros, pp. 7:33 -44. In: Algunos insectos del Museo de Paris. Rev. Acad. Cienc. Zaragoza.

Navás, R. P. L. 1926. Plecópteros, pp. 23: 103-112. In: Algunos insectos del Museo de Paris. Broteria. Ser. Zool., Lisboa.

Navás, R. P. L. 1929. Plecópteros, pp. 12:75-83. In: Insectos del Museo de Hamburgo. Bol. Soc. ent. Esp.(Zaragoza).

Navás, R. P. L. 1933a. Plécoptères, pp. 1 (9): 15-16. In: Névroptères et insectes voisins (Chine et pays environnants). Notes Ent. Chin., Musée Heude (Shanghai).

Navás, R. P. L. 1933b. Insecta orientalia. Mem. Acad. Sci. Nuovi Lincei, 17: 81-85.

Navás, R. P. L. 1936. Schwedisch-chinesische wissenschaftliche expedition nach den nordwestlichen Provinzen Chinas, Plecoptera. Ark. Zool., Uppsala, 27A: 1-11.

Okamoto, H. 1912. Erster Beitrag zur Kenntnis der japanischen Plecoptera. Trans. Sappora Nat. Hist. Soc., 4: 105-170.

Sivec, I. 1984. Redescription of *Neoperla klapaleki* Banks holotype from Taiwan/Formosa (Plecoptera: Perlidae). Biol. Vestn. 32: 105-108.

Sivec, I. and P. Zwick. 1987. Some *Neoperla* from Taiwan (Plecoptera: Perlidae). Beitr. Ent. 37: 391-406.

Sivec, I. and P. Zwick. 1989. Addition to the knowledge of genus *Chinoperla* (Plecoptera: Perlidae). Aquat. Insects, 11 (1): 11-16.

Sivec, I. and A. Zhiltzova. 1996. Description of *Neoperla ussurica* sp.n. from the Russian Far East (Plecoptera: Perlidae). Acta Ent. Sloven., 4 (1): 13-18.

Sivec, I., B. P. Stark and S. Uchida. 1988. Synopsis of the world genera of Perlinae(Plcoptera: Perlidae). Scopolia, 16: 1-66.

Sivec, I., Ping-Shih Yang and Chi-Zeng Lee. 1997. Name lists of insects in Taiwan-Plecoptera. Chinese J. Ent., 17 (3): 188-193.

Stark, B. P. and A. R. Gaufin. 1976. The Nearctic genera of Perlidae (Plecoptera). Misc. Pub. ent. Soc. Amer., 10: 1-79.

Stark, B. P. 1987. Records and descriptions of oriental Neoperlini (Plecoptera: Perlidae). Aquat. Insects, 9: 45-50.

Stark, B. P. and I. Sivec. 1991. Descriptions of oriental Perlini (Plecoptera: Perlidae). Aquat. Insects, 13 (3): 151-160.

Uchida, S. 1987. The lectotype of *Neoperla formosana* Okamoto (Plecoptera: Perlidae). Aquat. Insects, 9 (3): 159-160.

Wu, C. F. 1934. A homonym of a plecopterous genus. Ann. ent. Soc. Amer., 27: 256.

Wu, C. F. 1935a. New species of stoneflies from cast and south China. Peking Nat. Hist. Bull., 9: 227-243.

Wu, C. F. 1935b. Catalogus insectorum Sinensium. Fan Mem. Inst. Biol., 1: 299-315.

Wu, C. F. 1935c. Aquatic insects of China. Article XXII. Two new species of stoneflies from Kwangsi (Order Plecoptera). Peking Nat. Hist. Bull., 10: 61-62.

Wu, C. F. 1935d. Supplementum primum catalogi insectorum Sinensium. Peking Nat. Hist. Bull., 10: 63-88.

Wu, C. F. 1936-38: The stoneflies of China (Order Plecoptera). Peking Nat. Hist. Bull., 11: 49-82, 163-189; 11: 297-307, 441-443; 12: 57-70, 127-166 (1937); 12: 225-252, 319-351; 13: 53-87.

Wu, C. F. 1938. Plecopterorum sinensium. A monograph of stoneflies of China (Order: Plecoptera), pp.1-225.

Wu, C. F. 1939-40. First supplement to the stoneflies of China (Order Plecoptera). Peking Nat. Hist. Bull., 14 (2): 153-157.

Wu, C. F. 1940-41. Second supplement to the stoneflies of China (Order Plecoptera). Peking Nat. Hist. Bull., 14 (4): 331-333.

Wu, C. F. 1947-49. Third-sixth supplements to the stoneflies of China (Order Plecoptera). Peking Nat. Hist. Bull. 16 (3-4): 265-272 (1947-48); 17 (1): 75-80 (1948-49); 17 (2): 145-150 (1948-49); 17 (4): 251-256 (1949).

Wu, C. F. 1962. Results of the zoologico-botanical expedition to Southwest China, 1955-1957 (Plecoptera). Acta Ent. Sin., Suppl., 11:139-153.

Wu, C. F. 1973. New species of Chinese stoneflies (Order Plecoptera). Acta Ent. Sin., 16(2):97-118.

Wu, C. F. and P.W. Claassen 1934. New species of China stoneflies. Peking Nat. Hist. Bull., 9: 111-129.

Yang, Chikun and Ding Yang. 1990. New and little known species of Plecoptera from Guizhou province (I). Guizhou Sci., 8 (4): 1-4.

Yang, Chikun and Ding Yang. 1991a. New and little known species of Plecoptera from Guizhou province (II). Guizhou Sci., 9 (1): 48-50.

Yang, Chikun and Ding Yang. 1993. One new species of the genus *Tyloperla* Sivec et Stark from Guangxi (Plecoptera:Perlidae). J. Guangxi Acad. Sci., 9 (1): 61-62.

Yang, Chikun and Ding Yang. 1994. A new genus and new species of Plecoptera from East China (Perlidae:Acroneuriinae). Ent. J. East China, 4 (1): 1-2.

Yang, Ding and Chikun Yang. 1991. One new species of the genus *Furcaperla* from Jiangxi (Plecoptera: Perlidae). Acta Agric. Univ. Jiangxiensis, 13 (1): 16-18.

Yang, Ding and Chikun Yang. 1992 Plecoptera: Perlidae. In: Insects of Wuling Mountains Area, Southwestern China (Ed. Huang Fusheng), Science Press, Beijing, pp. 62-65.

Yang, Ding and Chikun Yang. 1993. New and little known species of Plecoptera from Guizhou province(III). Entomotaxonomia, 15 (4): 235-238.

Yang, Ding and Chikun Yang. 1995a. Three new species of Plecoptera from Hainan Province. Acta Agric. Univ. Pekinensis, 21 (2): 223-225.

Yang, Ding and Chikun Yang. 1995b. Plecoptera: Perlidae. In: Insects of Baishanzu Mountain, Eastern China (ed. Wu Hong). China Forestry Press, Beijing, pp. 59-60.

Yang, Ding and Chikun Yang. 1996. Four new species of Plecoptera from Nei Mongol. J. China Agricult. Univ., 1 (5): 115-118.

Yang, Xingke and Hongguo Sun. 1991 Catalogue of the insect typ specimens preserved in the insect collections of the Institute of Zoology, Academia Sinica. Agricultural Press, Beijing, pp. 5-9.

Zhang, Shimei. 1996. Synopsis of insect geography. Jiangxi Science and Technology Press, Nanchang, pp. 1-102.

Zwick, P. 1973a. Die Plecopteren-Arten Enderliens (Insecta); revision der typen. Ann. Zool., Warszawa, 31 (16): 471-4507.

Zwick, P. 1973b. Insecta:Plecoptera, phylogenetisches system und katalog. Das Tiereich, 94: 1-465.

Zwick, P. 1982a. A revision of the stonefly genus *Phanoperla* (Plecoptera: Perlidae). Syst. Ent., 7: 87-126.

Zwick, P. 1982b. Contribution to the knowledge of *Chinoperla* (Plecoptera: Perlidae: Neoperlini). Aquat. Insects. 4: 167-170.

Zwick, P. 1984a. Notes on the genus *Agnetina* (=*Phasganophora*) (Plecoptera: Perlidae). Aquat. Insects, 6: 71-79.

Zwick, P. 1984b. The genera *Tetropina* and *Neoperlops* (Plecoptera: Perlidae). Aquat. Insects, 6: 169-176.

Zwick, P. 1988a. Notes on Plecoptera (16). *Tylopyge* Klapálek: a synonym of *Paragnetina* Klapálek. Aquat. Insects, 10: 201-203.

Zwick, P. 1988b. Species of *Neoperla* from the South East Asian mainland (Plecoptera: Perlidae). Ent. Scand., 18: 393-407.

Zwick, P. and I. Sivec. 1980. Beiträge zur Kenntnis der Plecoptera des Himalaja. Ent. Basiliens., 5:59-138.

GUARANYPERLA, A NEW GENUS IN THE GRIPOPTERYGIDAE (PLECOPTERA)

Claudio G. Froehlich[1]

Dept. of Biology, Univ. of São Paulo
14040-901 Ribeirão Preto
São Paulo, Brazil

ABSTRACT

Guaranyperla g.n. from Brazil is described. The genus is defined by some autapomorphies, especially of the nymphs, viz., the expanded paranota and a body covering of vesicular hairs, both unique in the family. The type species is *G. guapiara* sp.n.; other described species are *G. beckeri* sp.n. and *G. nitens* sp.n.; a few unassociated nymphs are also considered. The genus occurs in southeastern Brazil.

INTRODUCTION

The family Gripopterygidae, comprising five subfamilies (McLellan, 1977), is South Gondwanian, occurring in the Australian Region and in southern South America. One subfamily, the Gripopteryginae, has a few genera that extend into tropical South America. In Brazil, they are represented by three genera, *Paragripopteryx, Gripopteryx*, and *Tupiperla*. The northern limits of the latter two are the highlands of Central Brazil. Here a fourth genus in the same subfamily is described.

Guaranyperla gen. nov.

Type species: Guaranyperla guapiara sp.nov.

Diagnosis: A genus in the Gripopteryginae with the following autapomorphies: In the nymph, thoracic segments with broad paranota; body covering of vesicular hairs. In the adult, pronotum relatively broad with squarish or projecting anterior corners. Additional characteristics: Legs with ventral femoral spine. Wings with Rs fork reduced or missing; costal crossveins present or absent; pterostigmatic crossveins present. male genitalia without epiproct; tergite 10 extension short and ending in two teeth.

Remarks: Guaranyperla shares with *Tupiperla* a ventral femoral spine. Compared with the latter genus, adults have a relatively broad pronotum with squarish or pointed anterior

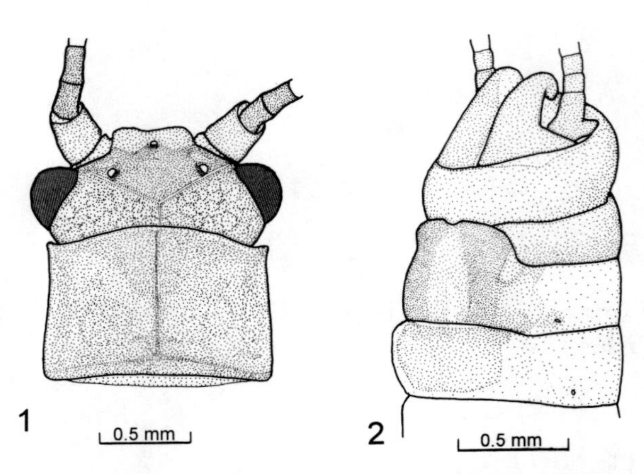

Figs. 1-2. *Guaranyperla guapiara* sp.n.; adult female. Fig. 1, head and pronotum, dorsal view; Fig. 2, terminalia, ventro-lateral view.

corners. The wings of one species (*G. nitens*) show an evident tessellated pattern, as do many *Tupiperla*, but not the other two (*G. guapiara* and *G. beckeri*). An apparent trend in the genus is the reduction or loss of the Rs fork. Concerning some cross-veins, *G. nitens* has a few costal ones, absent in the other two; these cross-veins occur in *Gripopteryx*. All species show pterostigmatic cross-veins, in some cases reduced to only one in one wing; these are always absent in *Tupiperla*.

The single known male lacks an epiproct, a condition shared with *Tupiperla* but that is found also in a few *Gripopteryx* and *Paragripopteryx*; in these cases probably a convergence with *Tupiperla*. The T10 extension is very short, ending in two teeth, a condition prevalent in *Paragripopteryx*. *Tupiperla* has a longer T10 extension.

Nymphs present the most striking apomorphies, listed in the diagnosis. The prothoracic paranota extend forward to the sides of the head (Figs. 3, 13). The dorsal side of the body is densely covered by the vesicular hairs, the ventral side has a sparser cover; the covering extend also to the appendages, excepting their extremities. Among the vesicular ones, usual filiform hairs are interspersed. The shape of the hairs varies with the location (Figs. 4-6). *Tupiperla* has also cover hairs standing on elevated sockets, but they have only dilated bases, the apices being thin; sometimes the apices are bent, giving the hair a retort shape; hairs at tergal margins may be broad and parallel-sided.

Distribution: The genus is known from southeastern Brazil (Fig. 14).

Etymology: After the Guarany group of Tupy Indians.

Guaranyperla guapiara sp. nov. (Figs. 1-6)

Material: Holotype: adult ♀ and nymphal exuviae, BRAZIL, SÃO PAULO, Parque Estadual Intervales, Rio das Mortes, 540 m, nymph collected 06/VIII/1997, adult emerged 12/VIII, C.G. Froehlich and V. Ribeiro col. Paratypes: Same locality, 1 ♀ nymph, 8-10/08/1991, L.G. Oliveira; same data as holotype, 1 adult ♀, at light, C.G. Froehlich, A.S. Melo and V. Ribeiro; 8 nymphs, one ♂ and 7 ♀♀, of which 2 adults emerged in cages (8/VIII and 15/VIII). Paranapiacaba Biological Station, 1 nymph, 05/VII/1963, C.G. Froehlich

Numerical data: Holotype, ♀: Forewing length, 10.6 mm; antenna length, 12.0 mm; 15 cercomeres. Paratypes, ♀♀ (N=3): forewing length, 10.3 - 11.0 mm; antenna length, 10.6 - 12.6 mm; 14-18 cercomeres, median 15 cercomeres. Nymphs, last instar: Male, head width, 1.13

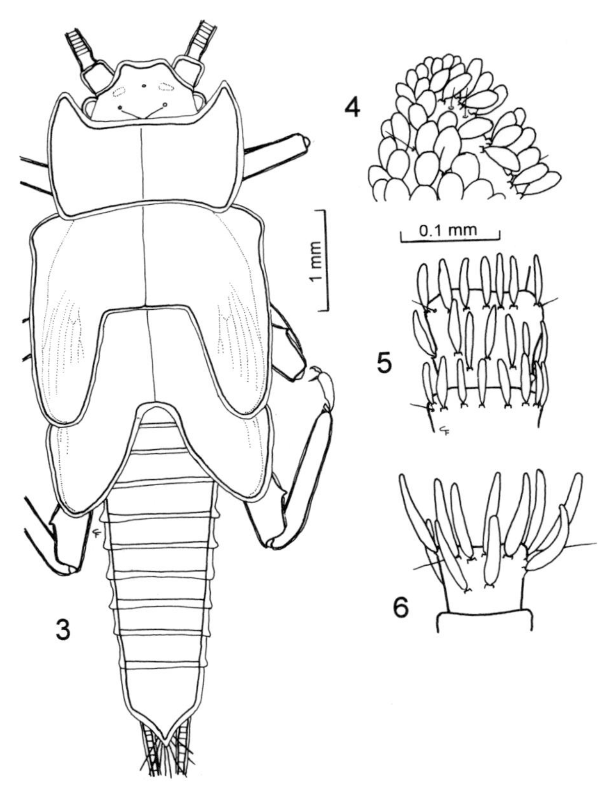

Figs. 3-6. *Guaranyperla guapiara* sp.n.; nymph. Fig. 3, nymph, dorsal view; Figs. 4-6, vesicular hairs; Fig. 4, on edge of prothoracic paranotum; Fig. 5, on antenna; Fig. 6, on cercus.

mm; body length, 6.7 mm, antenna length, 5.0 mm, cercus length, 2.7 mm. Females (N=6): head width, 1.21 - 1.41 mm, mean = 1.30 mm; body length, 5.7 - 9.0 mm, mean = 7.80 mm; antenna length, 5.7 - 6.7 mm, mean = 6.2; cercus length, 2.4 - 3.0 mm, mean = 2.8 mm.

Description: Male: Unknown. Female: General colour dark brown, almost black in life. Wings dark, with lighter line between R and M; distal cross-veins behind R thin. Centre of cells in forewings may be lighter; subcostal and pterostigmatic cell often darker. Pronotum a bit narrower than to about as wide as head, its anterior corners projecting a little (Fig. 1). Sternum 7, a single broad plate posteriorly; subgenital plate dark with median lighter area, apical margin with shallow emargination (Fig.2). Nymph brown, thorax segments with paranota, the prothoracic ones extending forward to sides of head (Fig.3). Body cover of numerous vesicular hairs which stand on elevated bases; the shape of the hairs differ in different parts of the body (Figs. 4-6).

Distribution: The species is known from the Intervales State Park (around 24° 19'S, 48° 25'W; Fig. 14, no. 1), and the Paranapiacaba Biological Station (23° 47'S, 46° 20'W; ib.,no. 2), São Paulo State.

Etymology: Guapiara, the name of a town near Intervales State Park.

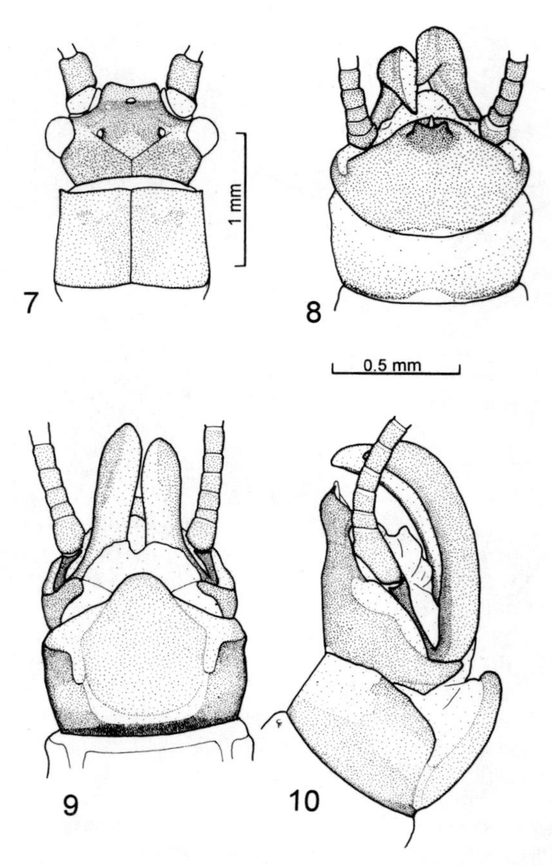

Figs. 7-10. *Guaranyperla beckeri* sp.n.; male. Fig. 7, head and pronotum, dorsal view; Figs. 8-10, terminalia in dorsal, ventral and lateral views.

Guaranyperla beckeri sp. nov. (Figs. 7-10)

Material: Holotype ♂: BRAZIL, MINAS GERAIS, Poços de Caldas, Morro do Ferro, 20/X/1963, J. Becker col.

Numerical data: Forewing length, 8.4 mm; antenna length, 8.6 mm; 17-18 cercomeres.

Description: Male. General colour light to medium brown. Pronotum squarish, with slightly produced anterior corners (Fig. 7). Forewing pattern inconspicuous, subcostal and pterostigmatic cells darker. In both pairs of wings Rs simple; 1 pterostigmatic cross-vein in right hind wing only. Tergite 10 extension very short, ending in a pair of teeth. Paraprocts relatively thick, simple, curving upward to level of T10 extension. Sclerotized epiproct absent (Figs. 8-10).

Female and nymph. Unknown.

Remarks: This species is considered to belong to *Guaranyperla* by the shape of the pronotum and also by the unforked Rs vein. Compared with *Tupiperla* it has a very short T10 extension and shares the lack of an epiproct.

Distribution: Known only from the type locality, Poços de Caldas, Minas Gerais State (Fig. 14, no. 3).

Etymology: Dedicated to the collector, Joahnn Becker.

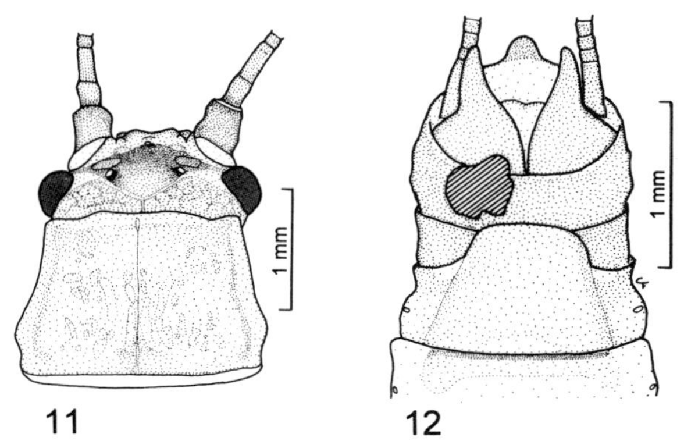

Figs. 11-12. *Guaranyperla nitens* sp.n.; female. Fig. 11, head and pronotum, dorsal view; Fig. 12, terminalia, ventral view. The lined structure is a scar.

Guaranyperla nitens sp. nov. (Figs. 11-12)

Material: Holotype, ♀: BRAZIL, SÃO PAULO, Campos do Jordão State Park, 30.X.1986. C.G. Froehlich.

Numerical data: Forewing length, 13.2. mm; antenna length, 13.5 mm; 25 cercomeres.

Description: General colour brown to dark brown; head darkest at frons; parietalia with an irregular reticulate pattern (Fig. 11). Forewings with a distinct pattern of dark borders to cross-veins and along some veins. Hind wings rather uniform brown, in life with a marked blue iridescence. In both pairs of wings, distal cross-veins behind Rs very pale.

Pronotum a little narrower than head anteriorly, corners squarish; posteriorly it widens becoming broader than head. Wings with relatively narrow tips. In both fore- and hindwings 1-3 costal and 3 pterostigmatic cross-veins. Rs fork very short or absent.

St7 with a transverse slightly sclerotized bar along posterior border. Subgenital plate trapezoid, light brown, apex broadly rounded, covering St9 (Fig. 12). Paraprocts simple, apices rounded in side view. T10 terminates in a rounded knob.

Remarks: This species is included in *Guaranyperla* because of the shape of the pronotum and by the condition of the Rs fork. The presence of costal and several pterostigmatic cross-veins are characters otherwise known only in *Gripopteryx*.

Distribution: Known only from the type locality, Campos do Jordão, São Paulo State Fig. 14, no. 4).

Etymology: Latin, *nitens* meaning shining, in reference to the hind wings.

UNASSOCIATED NYMPHS

Guaranyperla spec. A (Fig. 13)

Material: BRAZIL, SÃO PAULO, Campos do Jordão State Park, 2 nymphs, 18.V.1993, S.A. Vanin col.

These nymphs differ from those of *G. guapiara* in having a very long posterior extension of T10 (Fig. 13). Both are ca. 7 mm long and probably young, for no developing wings can be seen inside the meso- and metathoracic paranota. Having been collected in the same stream at which *G. nitens* was collected, they could belong to this species (Fig. 14, no. 4).

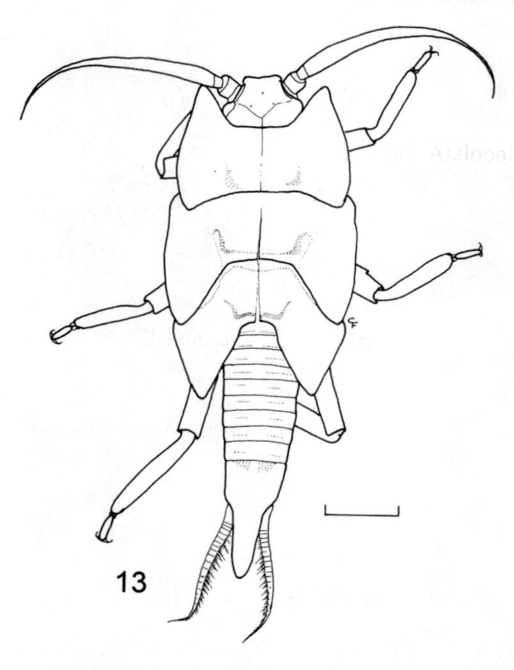

Fig. 13. *Guaranyperla* sp. A, nymph, dorsal view.

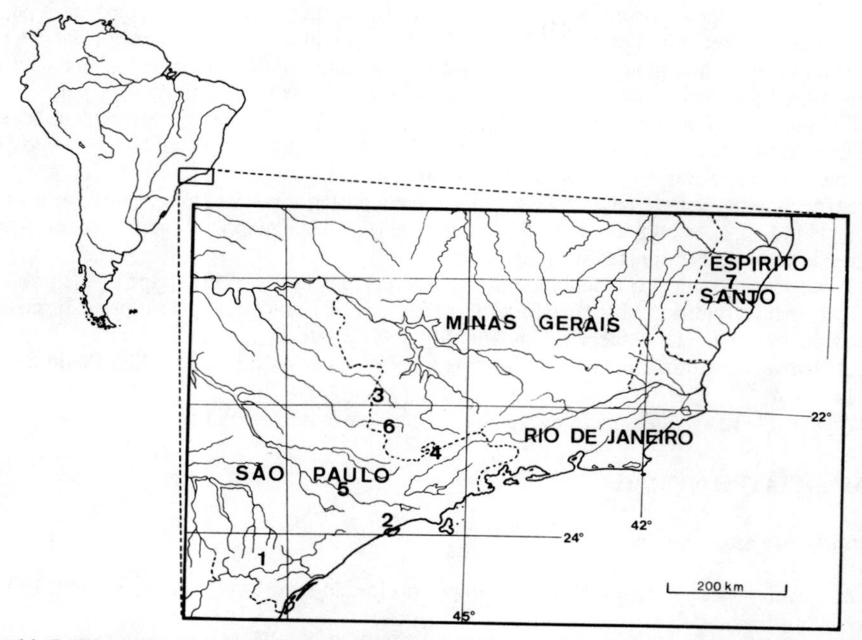

Fig. 14. Outline map of southeastern Brazil, showing localities, as follows: 1, Intervales State Park, SP (*G. guapiara*); 2, Paranapiacaba Biological Station, SP (*G. guapiara*); 3, Poços de Caldas, MG (*G. beckeri*); 4, Campos do Jordão State Park, SP (*G. nitens* and *G.* sp. A); 5, Serra do Japi, SP (*G.* sp.); 6, Ouro Fino, MG (*G.* sp.); 7, Santa Teresa, ES (*G.* sp.).

Guaranyperla spp.

Material: BRAZIL, SÃO PAULO, Jundiaí, Serra do Japi (Fig. 14, no. 5), 3 nymphs, 1995; 1 nymph, 25-26.IX.1955; 4 nymphs, 1.IV.1997, all A.S. Melo col. MINAS GERAIS, Ouro Fino (Fig. 14, no. 6), Córrego Boaventura, 1 nymph, 21.VIII.1940, H. Kleerekoper col (NMNH). ESPÍRITO SANTO, Santa Tereza (Fig. 14, no. 7), 1 nymph, 23.IV.1977, O.S. Flint, Jr., col. (NMNH).

These nymphs have a pointed T10 extension as *G. guapiara* but differ in details and probably belong to more than one species.

ACKNOWLEDGMENTS

To CNPq (Brazilian Council for Scientific and Technological Development) for research fellowship no. 301247/96-0. To O.S. Flint, Jr. for help with the material of the National Museum of Natural History (NMNH), Smithsonian Institution, Washington, DC, USA.

To an anonymous reviewer for improvements in the text.

REFERENCE

McLellan, I. D., 1977. New alpine and southern Plecoptera from New Zealand, and a new classification of the Gripopterygidae. N. Z. J. Zool. 4: 119-147.

DESCRIPTION OF THE NYMPH OF *KEMPNYIA TIJUCANA* DORVILLÉ AND FROEHLICH (PLECOPTERA, PERLIDAE), WITH NOTES ON ITS DEVELOPMENT AND BIOLOGY

Luís Fernando M. Dorvillé[1] and Cláudio G. Froehlich[2]

[1] Universidade Federal do Rio de Janeiro
Instituto de Biologia, Departamento de Zoologia
Caixa Postal 68044, CEP 21944-970
Rio de Janeiro, RJ, Brazil
[2] Universidade de São Paulo
Av. dos Bandeirantes s/n, CEP 14040-901
Ribeirão Preto, SP, Brazil

ABSTRACT

Although very common in montane areas of the southeastern and southern regions of Brazil, the nymphs of *Kempnyia* have never been thoroughly described. In this work the nymph of *Kempnyia tijucana* is described and data on its development and biology are provided.

INTRODUCTION

The described species of *Kempnyia* Klapálek are found in the southern and southeastern regions of Brazil. Their nymphs are oligostenothermal insects occuring mainly in montane streams (Froehlich, 1981), especially in areas occupied by the Atlantic Forest. They are easily distinguished from the other two perlid genera found in these regions (*Anacroneuria* Klapálek and *Macrogynoplax* Enderlein) by the characters recorded by Froehlich (1984b). However, few nymphs from these three genera have been thoroughly described up to now. The few exceptions include *Anacroneuria aroucana* Kimmins (Hynes, 1948), from Trinidad and Tobago, later redescribed by Stark (1994); and *A.blanca*, *A. shamatari*, *A. caraca*, *A. cruza* (Stark, 1995), from Venezuela. The nymph of *A. wipukupa*, from Arizona (USA), was briefly characterized by Baumann and Olson (1984), and those from *A. maritza* and *A. uatsi* (Stark, 1998), from Costa Rica. A general description of the nymphs of the genus was made by Needham and Broughton (1927). Short descriptions of *Macrogynoplax* nymphs were also made by Froehlich (1984a) and Stark and Zwick (1989), but nothing on the immatures of *Kempnyia*.

Table I. Frequency of *Kempnyia tijucana* in the samples taken from Rio da Fazenda, Rio de Janeiro, Brazil. LR - litter in riffle areas; ST - stone; LP - litter in pool areas; AS - sand.

August				November				February				May			
LR	ST	LP	A	LR	ST	LP	A	LR	ST	LP	A	LR	ST	LP	A
3	1	1	0	3	0	1	0	3	0	0	0	6	1	0	0
2	3	1	0	1	3	0	0	3	1	0	0	3	0	0	0
4	0	0	0	2	0	0	0	1	1	0	0	0	0	0	0
5	0	0	0	0	2	0	0	3	1	0	0	0	0	0	0
5	0	0	0	0	1	0	0	26	0	0	0	0	0	0	0

In this work the nymph of *Kempnyia tijucana* Dorvillé and Froehlich is described and some aspects of its development and biology are provided. All the specimens were collected at the Parque Nacional da Tijuca (National Park of Tijuca), Rio da Fazenda (1st order stream, altitude 400 m), in the city of Rio de Janeiro. This small stony stream has a mean width of 2 m and depths ranging from 10 cm to 30 cm. Four mature nymphs of this species (two males and two females) were deposited in each of the following institutions: Museu Nacional/UFRJ (National Museum of the Federal University of Rio de Janeiro); Museu de Zoologia da Universidade de São Paulo (Museum of Zoology of the University of São Paulo), and in the National Museum of Natural History, Smithsonian Institution, Washington D.C.

MATERIALS AND METHODS

The nymphs were quantitatively collected by means of a Surber sampler (350 mm mesh) in the four most frequent microhabitats present in the stream: 1) litter in pool areas; 2) stones; 3) litter held in riffle areas; and 4) sand. Five random samples were taken from each one of them, comprising 20 samples per month. The material was collected in 13-VIII-94 (winter), 20-XI-94 (spring), 19-II-95 (summer), and 25-V-95 (autumn). Additional material was obtained from qualitative collections conducted with a sieve in the same site. All specimens were preserved in 80% alcohol.

Each specimen was measured under a stereoscopic microscope with the aid of an eyepiece micrometer scale. Two measures were registered: head width just behind the compound eyes and maximum length of the mesonotum from its base to the apex of the wing pads. The measures of head width were also employed to estimate the total number of instars, assuming *K. tijucana* follows the Dyar's equation in its growth: $\log y = \log a + (x-1) \log b$ (where "a" is the size of the first instar, "y" the size of instar x, "x" is the instar, and "b" is the rate of growth). According to this rule, the plot of mean head widths (in log scale) against instar numbers should result in a straight line. Growth rate "b" was obtained from the mean head width values of the three last instars, which are more easily distinguished, with the following number of specimens for each instar: two from the last instar (Ad – 1); 19 from the Ad – 2; and 31 from Ad – 3. First instar nymphs were recognised because they had only three cercomeres, as Harper (1979) pointed out for the Systellognatha studied by him.

The sexes were determined for all specimens whenever possible and the presence or not of the following characters analysed: median ocellus, paired ocelli, distinct M-line, white spots in the nota, and the post-frontal line projecting between the paired ocelli.

Gut contents of 20 specimens were examined. The guts were extracted, shredded on a slide with glycerin, and examined under a microscope. Also, the area occupied by each of the following items was measured: detritus, plant fragments, animals, algae, and fungi.

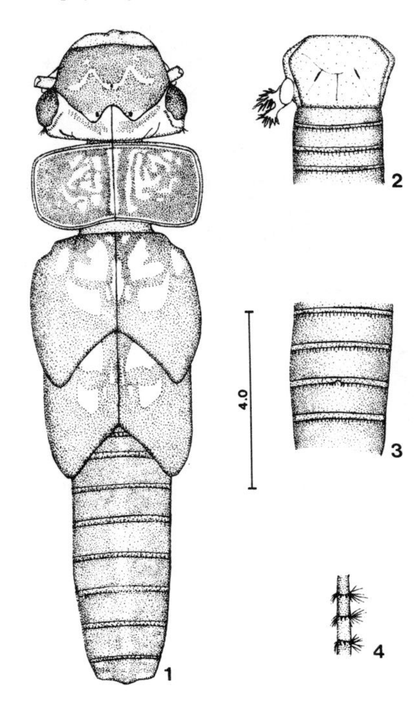

Figs. 1-4. *Kempnyia tijucana*, last instar nymph: 1. general aspect in dorsal view (appendages not drawn); 2. metathorax and abdomen in ventral view, showing gill SC3 and the position where gill III is placed in other species of *Kempnyia*; 3. abdominal sterna VII – IX; 4. Median cercomeres.

RESULTS

Collections (Table I)

The quantitative collections provided 88 nymphs, with 71 (80.68%) in litter held in riffle areas, followed by 14 (15.91%) in stones, and three (3.41%) in litter deposited in pool areas (Table I). No specimen of *K. tijucana* was found in the sand. The qualitative collections yielded 119 nymphs.

Material: Quantitative collections. 5 ♂♂, 3 dd, 17 not determined, 13.VIII.94 (Dorvillé, L.F.M. and Nessimian, J.L.); 4 ♂♂, 4 ♀♀, 6 not determined, 20.XI.94 (Dorvillé, L.F.M. and Nessimian, J.L.); 11 ♂♂, 5 ♀♀, 23 not determined, 19.II.95 (Dorvillé, L.F.M. and Weber, L.N.); 3 ♂♂, 1 ♀, 6 not determined, 25.V.95 (Dorvillé, L.F.M. and Sachsse, R.). Qualitative collections. 23 ♂♂, 24 ♀♀, 1 not determined, 07.X.90 (Dorvillé, L.F.M. and Silva, E.R.); 36 ♂♂, 35 ♀♀, 10.I.91 (Dorvillé, L.F.M.).

Nymph Description (Figs. 1 – 5)

General colour ochre (dark), with light areas yellow-ochraceous; pale yellow ventrally.

Head wide, moderately flattened dorso-ventrally, without occipital ridge. Compound eyes without white markings. Median ocellus always present, never reduced; paired ocelli

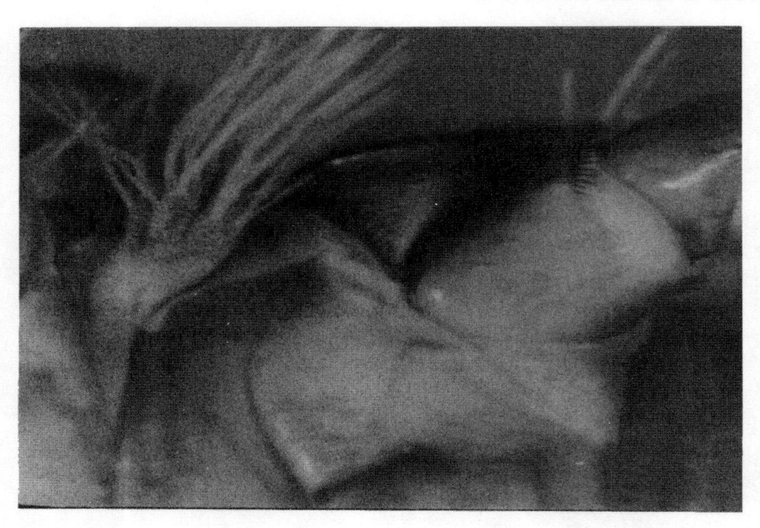

Fig. 5. *Kempnyia tijucana*, area between the metaesternum and abdomen, showing the shape of gill II.

distant from each other about two thirds of their distance to the nearest compound eye. Post-frontal line with a well developed curve between the paired ocelli (reaching the apex of the paired ocelli or beyond) occupying all the space between them (Fig. 1). Head pattern divided basically in two portions: an ochre frons and a lighter parietalia. The former is interrupted by a distinct M-line, two small spots in front of it, and a pair of dorsal callosities, all yellow-ochraceous. Parietalia with two darker bands which touch post-frontal line laterally. Moreover, slender dark lines branch from the posterior border of the head capsule. Labrum lighter than the frons, as well as the antenna and palpi. Mandibles yellow ochraceous, except for its teeth and external border, which are dark brown. Lacinia yellow ochraceous but with the apex ochre. Mouthparts as described by Froehlich (1984b).

Thorax: Pronotum quadrangular, with posterior lateral borders narrow, and with a weak developed groove. Nota ochre, with several nearly symmetrical light spots. Legs dark, with conspicuous light coloured femoral line. Tarsi with 1st and 2nd articles much smaller than the 3rd and claws developed and toothed, as in *A. aroucana* (Hynes, 1948).

Gills: Each side of the thorax with three tufts of filamentous gills present between the pro- and the mesothorax, and between the meso- and metathorax (gills I and II), as well as two tufts of supra-coxal gills on each side of the pro-, meso-, and metathorax (gills SC1, SC2, and SC3). No gill is found between the metathorax and the abdomen (gills III), as seen in Fig. 2. Gill filaments originating directly from the central part of the gill (Fig. 5). These gills have no distinct bristles on their bases. Paraproct gills present.

Abdomen: Tergites dark, with diffuse lighter marks in the central portion of each segment. The last tergite also presents four light areas, two basal and two apical. Sternites pale yellow, the last ones darker. Segment II divided into tergum and sternum, segments III – X forming complete rings. Paraprocts dark. The sexes of the nymphs are distinguishable by the aspect of sternum VIII. In females it shows a bristleless notch in the central portion (Fig. 3), while in the males the border of this segment is not interrupted and presents a complete row of bristles. In younger nymphs this character is not easily observed. Cerci dark, median cercomeres with a length two-three times their width, very hairy and with tufts and spines (Fig. 4). The same pattern is found in the last cercomeres, which are more slender. At the base there are no conspicuous spines.

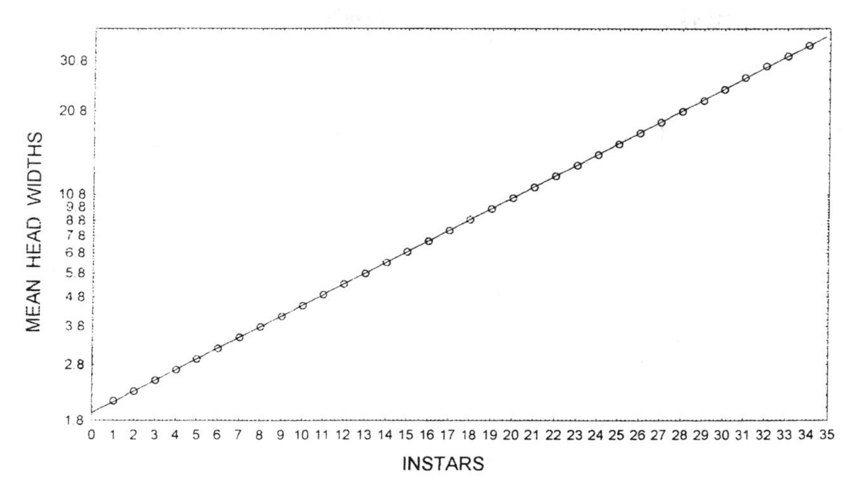

Fig. 6. *Kempnyia tijucana*, growth rate based on the mean head width values of the first and the last three instars. Ordinate in 1/10 mm (log scale).

Development and Biology (Figs. 6 –7, Table II)

The growth rate of *K. tijucana* seems to follow the Dyar's rule as depicted in Fig. 6, which also estimates as 34 the maximum number of nymphal instars for this species.

The relation between head width and length of mesonotum is displayed in Fig. 7, where is seen that male nymphs with head widths larger than 2.50 mm (about instar XXX) tend to have longer wing pads.

The first modification recorded during the development of young nymphs is the presence of light spots in the nota, found in nymphs with 0.33 mm head width (instars VI or VII) or larger. Although diffuse in the beginning, these spots become increasingly clear in older nymphs. The next change is the curve of the post-frontal line between the paired ocelli (still absent), occuring in nymphs from 0.71 mm head width on (about instar XV). This curve is well developed only in specimens with a head width larger than 1.00 mm (about instar XIX). Paired ocelli are observed in nymphs with 1.13 mm head width or larger (about instar XXI). The M-line and median ocellus are recorded in nymphs with 1.31 mm head width or larger (instars XXII or XXIII), although some older ones do not present these characters (five out of 156 nymphs without median ocellus and two without M-line). Sexual dimorphism was evident from head widths of 1.50 mm on (about instar XXIV).

Gut content analysis showed a very eclectic diet, which is mainly carnivorous, with the following items: Plecoptera and Ephemeroptera (*Pseudocloeon* sp., *Miroculis* sp.?) nymphs; Trichoptera larvae; Diptera larvae (Chironomidae – Tanytarsini, other Chironominae, and Tanypodinae; Ceratopogonidae – Forcipomiinae and Ceratopogoninae; Simuliidae); adults of Diptera, Hymenoptera and Coleoptera; Crustacea; Acarina; Collembola; Oligochaeta; Nematoda, and items that may have been consumed by the prey (plant fragments; detritus; algae, and fungi). Two nymphs have not shown any matter in their guts and so were excluded from the following comparisons. Among food items, those present in most of the nymphs were algae (72.2%) and Chironomidae larvae (66.7%), followed by fungi (44.4%), plant fragments (33.3%), and Ceratopogonidae larvae (27.8%). Regarding the amount of material ingested by each nymph, the areas occupied by the five food types reveal a great dominance of animal matter in the diet of this species (Table II). In this way, just three nymphs have not shown any animal matter in their guts, while 13 had at least 80% of the ingested material in this category, and the remaining two specimens 60%.

Table II. Percentage of the area occupied by the main food items found in the gut contents of specimens of *Kempnyia tijucana*

Specimens	Animals	Detritus	Fungi	Algae	Plant fragments
1	90.0%	6.5%	—	1.0%	2.5%
2	93.0%	5.0%	1.0%	1.0%	—
3	92.0%	8.0%	—	—	—
4	94.0%	5.0%	0.5%	0.5%	—
5	92.0%	6.0%	1.0%	1.0%	—
6	65.8%	16.4%	0.3%	—	8.2%
7	85.0%	12.0%	—	3.0%	—
8	95.0%	4.5%	0.5%	—	—
9	—	—	1.2%	21.7%	77.1%
10	95.0%	4.0%	0.5%	0.5%	—
11	95.5%	4.0%	—	0.5%	—
12	—	—	—	32.0%	68.0%
13	80.0%	7.0%	—	3.0%	10.0%
14	99.0%	1.0%	—	—	—
15	91.5%	6.0%	—	—	2.5%
16	93.0%	5.0%	—	2.0%	—
17	—	90.0%	—	10.0%	—
18	70.0%	24.0%	—	5.0%	1.0%

DISCUSSION

This species was described by Dorvillé and Froehlich (1997) from 16 adults (seven males and nine females) obtained from reared nymphs. Although no other *Kempnyia* nymph has been described, the comparison of *K. tijucana* with other species of the genus provides some important characters to distinguish this species from the others. First of all, the absence of gill III is unique in the genus. Froehlich (1984b) observed this character in the gill patterns of all the five species of *Kempnyia* he studied in order to separate the southeastern perlid genera. Other useful constant characters, although not exclusive, are: degree of development of the post-frontal line; shape of pronotum and its posterior lateral borders; degree of development of the pronotal groove; shape of thoracic gills; and presence of paraproct gills. It is important to remark that the shape of the post-frontal line between the paired ocelli was assigned by Froehlich (1988) as a good character to separate some adults of *Kempnyia*, some of them (*K. colossica*, *K. guassu*, and *K. brasiliensis*) showing a very weak curve between the ocelli. Later studies may prove this character to be common both to nymphs and adults.

The nymphs of *K. tijucana* seem to be directly influenced by water velocity. About 97% of the specimens were collected in fast flow areas (litter in riffles and stones), suggesting the association of this species to habitats with a high oxygen concentration. The absence of this species from sand can be explained by its carnivore habits and the low number of macroinvertebrates found in these samples. This paucity of invertebrates in sand was reported by Ward (1992), and is due mainly to the instability of the substrate.

The number of nymphal instars is quite variable, with values such as 23 for *Neoperla* (Perlidae), 12 (males) and 13 (females) for *Pteronarcys* (Pteronarcyidae), 16 for *Nemoura trispinosa* (Nemouridae), 14 for *Amphinemura decemceta* (Nemouridae), and 14-16 for *Capnia bifrons* (Capniidae) (Hynes, 1976). Froehlich (1969) estimated in 13 the number of

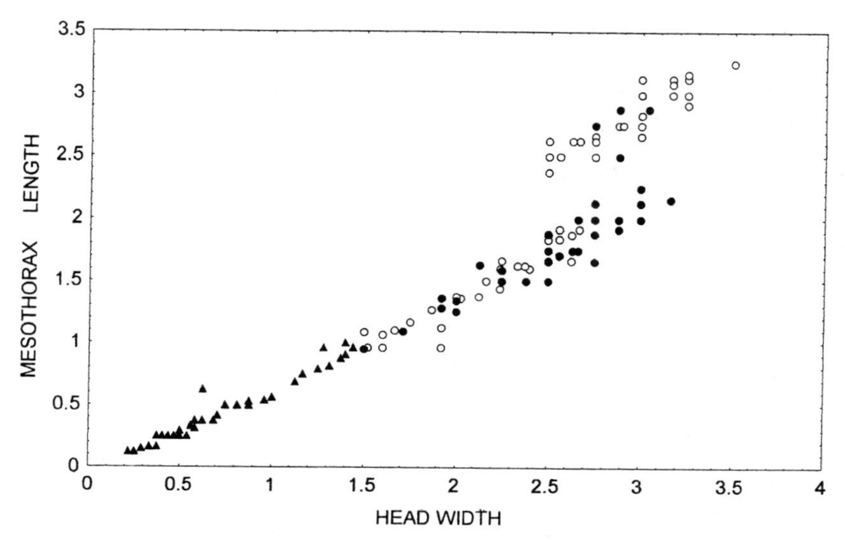

Fig. 7. *Kempnyia tijucana*, relation between head width and mesonotum length (in mm). Triangles indicate nymphs too young to be sexed, empty circles male nymphs, and black circles female nymphs.

nymphal instars for another neotropical species, *Paragripopteryx anga* (Gripopterygidae). Pennak (1978) recorded a variation of 12 to 36 instars for the members of this order. The number of instars (34) assigned to *K. tijucana* in this study should be regarded as a rough estimate for both sexes. The number of collected specimens is very low (especially mature females) and besides that, the number of instars within a species varies a lot, depending on factors such as temperature, food and stress conditions.

According to Beer-Stiller and Zwick (1995), growth and development of hemimetabolous insects are frequently described as a series of molts, followed by a continuous and gradual change. However, both authors report that this is not true for some species of Plecoptera. Plots of wing lengths against head widths of *Protonemura auberti* and *P. intricata* (Nemouridae) form discrete groups, corresponding to ante-penultimate, penultimate, and last instar nymphs. Hence, the wing pad length/head width ratio can be used to separate the three last instars of these species, independently of the nymph size, which certainly is not the case in *K. tijucana* (Fig. 7). In this species, although the three last instars are more easily recognised, they do not form discrete groups but exhibit a great change in form (mesonotum length) as a function of size (head width). In this way, the growth of this species seems to be gradual and continuous, without detection of distinct morphological stages. Also, differently than the species studied by Beer-Stiller and Zwick (1995), the males and females of *K. tijucana* separate into two distinct groups when mesonotum length is plotted against head width (Fig. 7).

Gut contents show a great variety in the food items consumed, although the diet of *K. tijucana* is mainly carnivorous, as previously registered for the family Perlidae (Hynes, 1941, and Harper and Stewart, 1984) and especially for *Kempnyia* (Froehlich and Oliveira, 1997). However, this is the first detailed report of the animal items consumed by a species of this genus. In the laboratory, the authors observed that recently emerged adults of *K. tijucana* were devoured by nymphs of the same species if not removed to a separate vial.

ACKNOWLEDGMENTS

The authors are grateful to CNPQ (Brazilian Research Council) for the grant that supported this work; to Prof. Elidiomar Ribeiro da Silva (UNIRIO), Luiz Norberto Weber and Richard Sachsse, for the great help in the field work; to Prof. Jorge Luiz Nessimian (UFRJ) and Prof. Alcimar do Lago Carvalho (UFRJ), for critical review of the manuscript.

REFERENCES

Baumann, R. W. and C. A. Olson. 1984. Confirmation of the stonefly genus *Anacroneuria* (Plecoptera: Perlidae) from the Nearctic region with description of a new species from Arizona. Southwest. Natur. 29: 489-492.

Beer-Stiller, A. and P. Zwick. 1995. Biometric studies of some stoneflies and a mayfly (Plecoptera and Ephemeroptera). Hydrobiologia 299: 169-178.

Dorvillé, L. F. M. and C. G. Froehlich. 1997. *Kempnyia tijucana* sp. n. from Southeastern Brazil. Aquat. Insects 19: 177-181.

Froehlich, C. G. 1969. Studies on Brazilian Plecoptera 1. Some Gripopterygidae from the Biological Station of Paranapiacaba, State of São Paulo. Beitr. Neotrop. Fauna 6: 17-39.

Froehlich, C. G. 1981. Plecoptera, pp. 86-88. In: S. H. Hurlbert, G. Rodríguez and N. D. Santos (eds.). Aquatic Biota of Tropical South America. Part I, Arthropoda. San Diego State University, San Diego.

Froehlich, C. G. 1984a. Brazilian Plecoptera 3. *Macrogynoplax veneranda* sp n. (Perlidae: Acroneuriinae). Ann. Limnol. 20: 39-42.

Froehlich, C. G. 1984b. Brazilian Plecoptera 4. Nymphs of perlid genera from southeastern Brazil. Ann. Limnol. 20: 43-48.

Froehlich, C. G. 1988. Brazilian Plecoptera 5. Old and New species of *Kempnyia* (Perlidae). Aquat. Insects 10: 153-170.

Froehlich, C. G. and L. G. Oliveira. 1997. Ephemeroptera and Plecoptera nymphs from riffles in low-order streams in southeastern Brazil, pp. 180-185. In: P. Landolt and M. Sartori (eds.). Ephemeroptera and Plecoptera: Biology-Ecology-Systematics, Fribourg.

Harper, P. P. 1979. Observations on the early instars of stoneflies. Limnol. Schr. Gew. Abw. 64: 18-28.

Harper, P. P. and K. W. Stewart. 1984. Plecoptera, pp. 182-230. In: R. W. Merritt and K. W. Cummins. Aquatic Insects of North America. Kendall/Hunt, Dubuque. 722p.

Hynes, H. B. N. 1941. The taxonomy and biology of nymphs of British Plecoptera with notes on the adults and eggs. Trans. R. ent. Soc. Lond. 91: 459-557.

Hynes, H. B. N. 1948. The nymph of *Anacroneuria aroucana* Kimmins (Plecoptera, Perlidae). Proc. R. ent. Soc. Lond. (A) 23: 105-110.

Hynes, H. B. N. 1976. Biology of Plecoptera. Ann. Rev. Ent. 21: 135-153.

Needham, J. G. and E. Broughton. 1927. Central American stoneflies, with descriptions of new species (Plecoptera) J. N. Y. ent. Soc. 35: 109-121.

Pennak, R. W. 1978. Freshwater Invertebrates of the United States (2nd edition). John Wiley and Sons, New York. 803 p.

Stark, B. P. 1994. *Anacroneuria* of Trinidad and Tobago (Plecoptera, Perlidae). Aquat. Insects 16: 171-175.

Stark, B. P. 1995. New species and records of *Anacroneuria* (Klapálek) from Venezuela. Spixiana 18: 211-249.

Stark, B. P. 1998. The *Anacroneuria* of Costa Rica and Panama (Insecta: Plecoptera: Perlidae). Proc. biol. Soc. Wash. 111: 551-603.

Stark, B. P. and P. Zwick. 1989. New species of *Macrogynoplax* from Venezuela and Surinam (Plecoptera: Perlidae). Aquat. Insects 11: 247-255.

Ward, J. V. (1992): Aquatic Insect Ecology. 1. Biology and Habitat. John Wiley and Sons, New York. 438 p.

SPHAERONEMOURA, A NEW GENUS OF THE AMPHINEMURINAE (NEMOURIDAE, PLECOPTERA) FROM ASIA

Takao Shimizu[1] and Ignac Sivec[2]

[1] Aqua Restoration Research Center
Kasada, Kawashima-cho
Gifu, 501-6021, Japan
[2] Slovene Museum of Natural History
Presernova 20, P. P. 290
Slo 1001, Liubljana, Slovenia

ABSTRACT

The new genus, *Sphaeronemoura*, is erected for two new species and four already named species confirmed by bibliographic references. They are: type species, *S. plutonis* (Banks, 1937) comb. n., Taiwan, transferred from *Nemoura*, *S. paraproctalis* (Aubert, 1967), comb. n., Assam, transferred from *Protonemura*, *S. elephas* (Zwick 1974), comb. n., Taiwan, transferred from *Nemoura*, *S. hamistyla* (Wu, 1962), comb. n., China, transferred from *Mesonemoura*, *S. formosana*, sp. n. from Taiwan and *S. inthanonica*, sp. n. from Thailand. This genus is distinguished from related genera by its two simple cervical gills on either side and by the swelling segments of the nymphal cerci.

INTRODUCTION

Baumann (1975) divided the family Nemouridae into two subfamilies, the Nemourinae and the Amphinemurinae, and erected two Oriental genera, *Mesonemoura* and *Indonemoura* in the latter subfamily. Zwick and Sivec (1980), subsequently, recorded a strange nymph with two simple cervical gills on either side and swelling segments of the nymphal cerci from the Himalayas. The nymphal characteristic had never been observed in any nemourid genera, except the gills of several *Zapada*. Then Sivec (1981) identified the nymph with *paraproctalis* Aubert, 1967, and transferred it back from *Mesonemoura* to *Protonemura*, since the species has elongate cervical gills.

Incidentally, the authors each obtained additional nymphs and adults related to *paraproctalis* from the Oriental Region. The males are characterized by the elongated apex of the epiproct which is similar to those of *Mesonemoura*, but its females and nymphs are very much different from the latter. They are considered to be excluded from any genera of

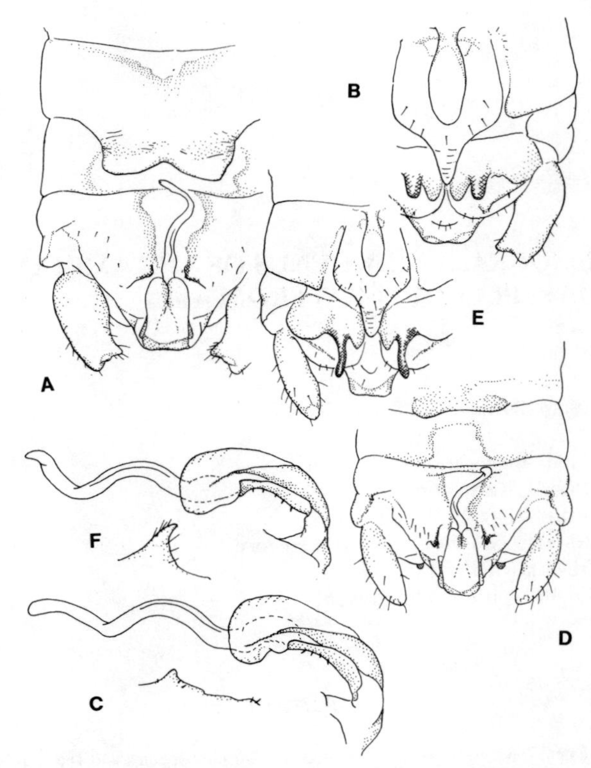

Fig. 1. A-C, *S. plutonis* (Banks): (A and B) male terminalia, dorsal and ventral view; (C) epiproct and the elevation of tergite 10, lateral view. D-F, *S. formosana* sp. n.: (D and E) male terminalia, dorsal and ventral view; (F) epiproct and the elevation of tergite 10, lateral view.

the Nemouridae, and the new genus, *Sphaeronemoura*, is erected. In this paper, two Taiwanese species and one Thai species are treated. One of the Taiwanese species is merged with *Nemoura plutonis* Banks, 1937 and is designated as the genotype. The other two species are described as new to science.

The genus *Sphaeronemoura* presently includes the four species and two additional species confirmed by bibliographic references: 1) *S. elephas* (Zwick 1974: p. 77, figs. 6-8; from Taiwan), comb. n., transferred from *Nemoura*, and 2) *S. hamistyla* (Wu, 1962: p. 142, figs. 13-16; from Southwest China), comb. n. transferred from *Mesonemoura*. A key to species is provided for adults. The relationship within the Amphinemurinae is discussed.

DESCRIPTION

Sphaeronemoura Gen. nov.

Etymology: A Greek noun '*sphaira*' meaning globe plus the generic name *Nemoura*; the name refers to the uniquely bulged segments of the nymphal cerci.

Type species: Nemoura plutonis Banks, 1937: 274.

Diagnosis: All the species have the following characteristics: (1) nymphal cerci globularly swelling in several segments, and the last segments forming bell-like shapes; (2)

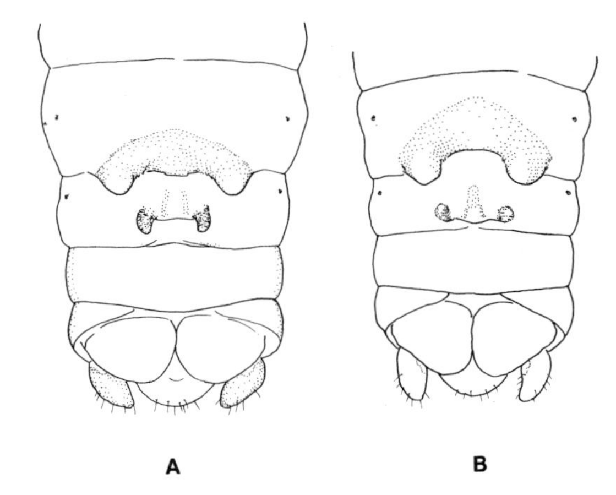

Fig. 2. Female terminalia, ventral view: A, *S. plutonis* (Banks); B, *S. formosana* sp. n.

cervical gills sausage-like in shape, one simple gill located on either side of each cervical sclerite, like some *Zapada* species; (3) male paraproct less developed, bearing a small blunt projection on the median lobe; (4) male epiproct bearing a long stout flagellum at the apex; (5) female pregenital plate widely hollowed in the middle of the hind margin. A characteristic (4) is also observed also in *Mesonemoura*, but the male of the latter generally differs in having a setose median lobe of the paraproct and the insertion of the apex of the epiproct to the modified male tergite 9.

Male: Tergite 8 more or less asymmetrically protruded posteriorly, while uncertain in a few species. Tergite 10 with longitudinal membranous area below epiproct, and bearing a small acute hump on either side near the epiproct. Paraprocts trilobed: inner lobe small and simple; median lobe strongly sclerotized and forming a blunt projection; outer lobe mostly membranous. Epiproct asymmetrical, bearing elongated flagellum, which is curved in all species, except *S. elephas* which has a straight one; dorsal hyaline portion weakly scaled; ventral sclerite broad and subquadrangular, bearing several setae laterally in basal portion, and tapered anteriorly.

Female: Pregenital plate emarginate anteriorly in the middle of hind margin except in *S. inthanonica*, which has rounded plate with a pair of papilla-shaped protuberances near the middle of hind margin. Subgenital plate reduced to small lateral sclerites by the genital opening, except for *S. paraproctalis*, which has a large subrectangular plate. Vaginal lobes reduced.

Nymph: Cerci forming uniquely bulb-shaped segments, which increase in diameter from the base to the terminal segments; each segment with a whorl of long stout bristles and a tuft of short setae between the bristles.

Discussion. This new genus is certainly a member of the subfamily Amphinemurinae by having a paraproct that is divided into three lobes. The subfamily Amphinemurinae is divided into two groups. The first includes *Amphinemura* and *Malenka* which are discriminated from the other four genera by having branched cervical gills (Baumann, 1975). The other four genera, *Sphaeronemoura*, *Protonemura*, *Indonemoura* and *Mesonemoura*, look similar to one another in having a sclerotized projection (tigellus) on the median lobe of the paraproct, even though the tigellus becomes indistinct in some species as in *Mesonemoura*. The females of the other three genera, except *Sphaeronemoura*, are characterized by having sternite 8 with a developed median lobe overlapping on the vaginal lobes as Baumann (1975)

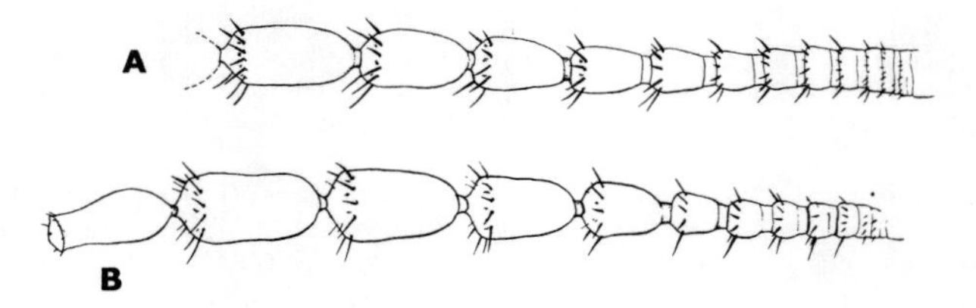

Fig. 3. Nymphal cerci of *Sphaeronemoura plutonis* (Banks) (A) and *S. formosana* sp. n. (B).

said. Therefore this new genus is regarded not to be closely related to *Mesonemoura*, even though the male terminalia are apparently quite similar.

Remarks. Though no additional species can be transferred to this new genus with certainty, the following two species now placed in *Mesonemoura* may be attributed to this genus: *Nemoura gradicauda* Wu (1973, p. 114, fig. 36-38) because of the feature of tergite 10 (4), and *Nemoura multispira* Wu (1973, p. 114, fig. 46-49) because of the female pregenital feature (5).

The genus *Mesonemoura* is defined by the following general points: 1) male tergite protruded on the 9th segment, and 2) the flat flagellum of the male epiproct. Comparative morphology on the features of gills and female terminalia will provide a better definition on the genus, though the appearances are uncertain in some species at present.

KEY TO THE SPECIES OF *SPHAERONEMOURA*

Males:

1. Epiproct with a straight flagellum at the apex .... *S. elephas* (Zwick)
   Epiproct with curvy flagellum 2
2. Legs brown with wide yellow band .... *S. inthanonica* sp. n.
   Legs uniformly brown 3
3. Tergite 10 unmodified .... *S. paraproctalis* (Aubert)
   Tergite 10 with a pointed hump on either side near epiproct 4
4. Tergite 8 unmodified .... *S. hamistyla* (Wu)
   Tergite 8 modified, protruded posteriorly 5
5. Tergite 8 distinctively extending posteriorly; rounded projection of paraproct short .... *S. plutonis* (Banks)
   Tergite 8 weakly extending posteriorly; rounded projection of paraproct long .... *S. formosana* sp. n.

Females: excluding *S. elephas*

1. Pregenital plate subcircular with two protuberances .... *S. inthanonica* sp. n.
   Pregenital plate deeply concave in the middle of hind margin 2
2. Subgenital plate mostly sclerotized .... *S. paraproctalis* (Aubert)
   Subgenital plate mostly membranous except for either side of hind margin 3
3. Subgenital plate extending posteriorly on either side .... *S. plutonis* (Banks)
   Subgenital plate not extending posteriorly on either side .... *S. formosana* sp. n./*S. hamistyla* (Wu)

Sphaeronemoura plutonis (Banks, 1937) (Figs. 1A-C, 2A, 3A)

Nemoura plutonis Banks, 1937: 274.

Specimens examined: Taiwan - ♀ (Syntype; MCZ, no. 20191); 1 ♂ 8 ♀♀ 8 nymphs (LBM), Taichung Prefecture, Wuling Farm, Hsuehshan-hsi, 1.v.1994, T. Shimizu; 1 ♀, Nantou

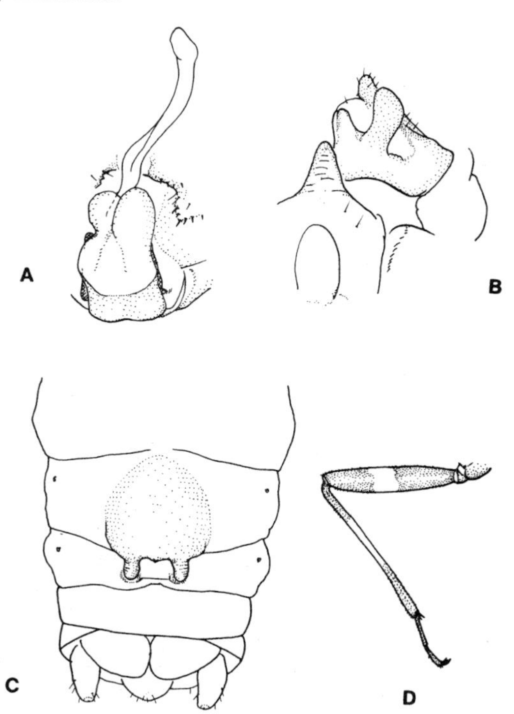

Fig. 4. *S. inthanonica* sp. n. A, Epiproct and the neighbor region, dorsal view; B, male terminalia, ventral view; C, hind leg; D, female terminalia, ventral view.

Pref., Hohuan-hsi near Lishan, 30.iv.1994, T. Shimizu; 1 ♀, Taichung Pref., Kukhan, 11.iv.1991, T. Kishimoto.

Uniformly dark brown to blackish brown. Wings fully infuscated and brown, pterostigma darker. Antennae, head, thorax and legs blackish brown.

Male: Body ca. 9.0 mm long; forewing ca. 10.5 mm long. Sternite 9: vesicle long, oval and stout; subgenital plate subpentagonal with elongated blunt corner. Tergite 8 posteromedially expanded and forming a wide overhang with a weak notch at the middle of hind margin. Tergite 9 broadly membranous medially. Tergite 10 with vertical membranous area medially and a small point situated on either side near the base of epiproct. Paraprocts broadly membranous in posterior portions: inner lobe simple; median lobe broad at base and gently projected posteriorly to flat blunt projection; outer lobe linearly and slightly sclerotized, several small setae scattered at the apex. Cerci truncate and weakly bent inward, each apex with a bulge on inner and outer side. Epiproct with a long flagellum, which is grooved dorsomedially and curved; ventral sclerite wide in basal portion and bearing several setae sporadically, then extending and tapered anteriorly to inner side.

Female: Body 9.0-10.0 mm long; forewing 11.5-12.0 mm long. Pregenital plate large and transverse, roundly protruded on either side of hind margin, which is widely hollowed in the middle of hind margin. Subgenital plate rudimentary, mostly membranous, bearing papilla-shaped projection on either side. Cerci short pineal or sometimes slightly modified like that of male.

Nymph: Cerci illustrated in Fig. 3A.

Remarks: The second author borrowed and studied the type. The dry pinned female was relaxed by a KOH solution on the abdominal tip. This female is identical by having two sets of finger -like projections on sternite 8.

Sphaeronemoura formosana sp. n. (Figs. 1D-F, 2B, 3B)

Holotype: 1 ♂ (LBM), Taiwan, Nantou Prefecture, Nanshan-hsi, 25.iv.1994. *Paratypes:* Taiwan: 5 ♂♂ 7 ♀♀ (LBM), collected with holotype; 1 ♂, Nantou Pref., Shihtzutou-hsi, Saide Waterfall, 28.iv.1994; 1 ♂ 2 ♀♀, Kuanin Waterfall, 22.iv.1994; 1 ♀, Nanshan-hsi, 16.v.1983, M. Hasegawa; 2 ♀♀ 7 nymphs, Nantou Pref., Lushan, Mahebo-hsi, 1,300m, 24.iv.1994; 4 ♂♂ 2 ♀♀ 27 nymphs, Taichung Pref., Chihpen, 3.v.1994.

Uniformly dark brown to blackish brown. Wings fully infuscated and brown, pterostigma darker. Antennae, head, thorax and legs blackish brown.

Male: Body 7.5-8.0 mm long; forewing 8.0-8.5 mm long. Sternite 9: vesicle long, oval in shape; subgenital plate subpentagonal, and elongated posteromedially. Tergite 8 often slightly expanded posteriorly in the middle of hind margin, but nearly unmodified. Tergite 9 membranous in median portion. Tergite 10 with vertical membranous area medially, bearing an angulate point situated on either side. Paraprocts broadly membranous in posterior portions: inner lobe simple; median lobe broadly sclerotized at the base, and bearing long blunt projection; outer lobe linearly and slightly sclerotized along outer margin. Cerci long conical, slightly curved inward. Epiproct with a long flagellum, which is grooved dorsomedially; ventral sclerite parallel sided in basal portion but gently notched on either side at the middle, bearing several setae sporadically, then extending and tapered anteriorly to inner side.

Female: Body 8.5-10.0 mm long; forewing 10.0-10.5 mm long. Pregenital plate expanded anteromedially, roundly bulged on either side of hind margin with a large round indentation in the middle. Subgenital plate mostly membranous and rudimentary, bearing a small sclerotized area on posterolateral margins, anteromesal, and ventral inner side of vulva. Cerci simply conical.

Nymph: Cerci illustrated in Fig. 3B.

Etymology: The specific epithet refers to the area where the type was collected, Taiwan (Formosa).

Sphaeronemoura inthanonica sp. n. (Figs. 4A-D)

Holotype: 1 ♂ (LBM), Thailand: Chiang Mai Province, Doi Inthanon, 10.x.1992, H. Maruyama. *Paratype:* Thailand: 1 ♀ (LBM), collected with holotype.

Wings blackish brown. General color blackish brown; antennae, head and thorax blackish. Legs blackish brown except for wide yellowish rings in the middle of femora and hind tibiae.

Male: The male specimen is not in a good condition; it is slightly discolored and partly damaged at the left side of the abdomen. Forewing ca. 9.5 mm long. Sternite 9: vesicle oval in shape; subgenital plate pentagonal, posteromedially extending. Tergite 8 widely expanded posteriorly on mesal hind margin. Tergite 10 with membranous field along the midline below epiproct, bearing weak elevation situated on either side near epiproct. Paraprocts broadly membranous in posterior portions: inner lobe simple; median lobe sclerotized at the base, bearing stout rounded projection; outer lobe linearly sclerotized along outer margin, apex slightly roundly bulged. Cerci long, conical, slightly curved inward. Epiproct with a long curvy flagellum, grooved dorsomedially, asymmetrical also on the body, which is depressed on the left side and bends to the right; ventral sclerite almost parallel-sided in basal half, then tapered to apex.

Female: Body 10.0 mm long; forewing 11.5 mm long. Pregenital plate suboval and enlarged, with a pair of papilla-shaped projections on sides of hind margin. Subgenital plate rudimentary, mostly membranous except for lateral sclerotized areas. Cerci simply conical.

Etymology: The specific epithet refers to the type locality, Mt. Doi Inthanon.

ACKNOWLEDGMENTS

We would like to thank Dr. Richard W. Baumann for his information on *Mesonemoura* and comments on the manuscript. Thanks are also due to curators of (MCZ) and (LBM).

REFERENCES

Aubert, J. 1967. Les Nemouridae de l'Assam (Plécoptères). Mitt. schweiz. ent. Ges., 39: 209-253.
Banks, N. 1937. Neuropteroid insects from Formosa. Philipp. J. Sci., 62: 269-275.
Baumann, R. W. 1975. Revision of the stonefly family Nemouridae (Plecoptera): a study of the world fauna at the generic level. Smithson. Contr. Zool., (211): 3+74.
Sivec, I. 1981. Some notes about Nemouridae larvae (Plecoptera) from Nepal. Ent. Basil., 6: 108-119.
Wu, C. F. 1962. Results of the zoologico-botanical expedition to Southwest China, 1955-1957. Acta ent. Sin., 11: 139-153.
Wu, C. F. 1973. New species of Chinese stoneflies (Order Plecoptera). Acta ent. Sin., 16: 97-126.
Zwick, P. 1974. Zwei neue Nemouridae (Ins., Plecoptera) aus dem Fernen Orten. Neuv. Rev. Ent., 4: 75-78.
Zwick, P. and I. Sivec. 1980. Beiträge zur Kenntnis der Plecoptera des Himalaja. Ent. Basil., 5: 59-138.

STONEFLIES OF TAIWAN WITHIN THE ORIENTAL STONEFLY FAUNA DIVERSITY

Ignac Sivec[1] and Ping-Shih Yang[2]

[1] Slovenian Museum of Natural History
Presernova 20, POB290, 1001 Ljubljana, Slovenia
[2] Lab. Insect Conservation, Dept. of Plant Pathology and Entomology
National Taiwan University, Taipei, Taiwan, R. O. C.

ABSTRACT

Unlike some other insect groups such as Coleoptera and Lepidoptera, the advanced study of Plecoptera was begun rather late. Still the taxonomic knowledge of European and North American stoneflies is among the best of all insects. Biodiversity estimates for other areas including much of the Oriental region, however, are far from complete. The inventory of Taiwan Plecoptera, initiated by Okamoto (1912) and Klapálek (1912), had until recently reached 31 species, but a few intense collecting trips has revealed a much richer Taiwanese stonefly diversity than expected. In this study, an overview of the stonefly fauna from different Asian countries is presented and comparison is made with the holarctic fauna.

INTRODUCTION

Unlike some other insect groups such as Coleoptera and Lepidoptera, the advanced study of Plecoptera began rather late. The first illustrations of stoneflies originate from Hoefnagel (1592) and the first description from 1683, however serious studies were initiated only at the beginning of this century. The first records of stoneflies from Taiwan were published independently in 1912 by Okamoto and in 1912 and 1913 by Klapálek. The majority of species originate from Sauter's collecting and only a few were described by other authors like Banks, Navas, Kawai, Zwick and Sivec and Shimizu. The reported Taiwan stonefly fauna (Sivec, I., Ping-Shih Yang 1997) has now reached 31 species, but a few intensive collecting trips in recent years has revealed a much richer stonefly diversity than expected.

Table 1. Stonefly species number of the most common families in different areas

	Taiwan	Asia[1]	Europe[2]	N. America[3]
Peltoperlidae	2	24	0	20
Perlidae	18	270	21	82
Nemouridae	7	241	173	64
Leuctridae	2	63	137	55
Chloroperlidae	0	29	30	81
Perlodidae	0	33	76	124
Taeniopterygidae	0	20	47	35
Capniidae	0	36	30	151
Total	29	716	514	612

[1] Not all Asian countries are included; Chinese stoneflies are not included.
[2] After unpublished list of European stoneflies according to Aubert.
[3] After "A Check List of the Insects of North America", Stark (1997).

LIST OF SPECIES

Styloperlidae

Cerconychia brunnea Klapálek, 1913
Cerconychia livida Klapálek, 1913

Peltoperlidae

Cryptoperla formosana (Okamoto, 1912)
Peltoperla formosana Klapálek, 1913

Perlidae

Agnetina aequalis (Banks, 1937)
Tyloperla formosana (Okamoto, 1912)
Tyloperla sauteri (Navas, 1929)
Paragnetina planidorsa Klapálek, 1913
Kamimuria formosana Klapálek, 1921
Kamimuria lepida Klapálek, 1913
Neoperla cavaleriei (Navas, 1922)
Neoperla costalis (Klapálek, 1913)
Neoperla formosana Okamoto, 1912
Neoperla klapaleki Banks, 1937
Neoperla sauteri Klapálek, 1912
Neoperla signatalis Banks, 1937
Neoperla taihorinensis Klapálek, 1913
Neoperla taiwanica Sivec and Zwick, 1987
Neoperla uniformis Banks, 1937
Gibosia lucida (Klapálek, 1913)
Kiotina collaris (Banks, 1937)
Mesoperlina crucigera Klapálek, 1913

Table 2. Stoneflies species diversity for the different Asian countries

	Peltoperlidae	Perlidae	Nemouridae	Leuctridae	Capniidae	Perlodidae	Others
Taiwan	2	18	7	2	0	0	2
Japan	7	44	105	50	34	27	44
Korea	0	1	5	1	0	1	1
Philippines	1	21	0	3	0	0	0
Borneo	1	37	1	0	0	0	0
Indonesia	0	31	5	0	0	0	0
Malaysia	1	11	2	3	0	0	0
Thailand	6	21	2	0	0	0	0
Cambodia	0	0	0	0	0	0	0
Vietnam	1	17	2	0	0	0	0
Laos	0	2	0	0	0	0	0
Mjanmar	0	3	0	0	0	0	0
Buthan	1	5	17	2	0	0	3
Bangladesh	0	2	0	0	0	0	0
Sri Lanka	0	10	0	0	0	0	0
India	7	31	64	1	5	2	2
Nepal	1	16	31	3	7	3	2

Nemouridae

Amphinemura flavicollis Klapálek, 1912
Amphinemura flavinotus Shimizu, 1997
Illiesonemoura bispinosa (Kawai, 1968)
Nemoura brevilobata (Klapálek, 1912)
Nemoura formosana Shimizu, 1997
Nemoura elephas Zwick, 1982
Nemoura plutonis Banks, 1937

Leuctridae

Rhopalopsole dentata Klapálek, 1912
Rhopalopsole subnigra Okamoto, 1922

Practically all museum material was collected by light trapping or while collecting other insect groups. This is the main reason for the high number of Perlidae represented among museum material. After only a few recent collecting trips by stonefly specialists, about 20 additional and new species will be added to the Taiwan stonefly fauna. Still, as in other Southeast Asian islands, the stonefly fauna is poor in comparison to the mainland, but practically all the species reported for Taiwan are endemic. Most undescribed species belong to Nemouridae and Leuctridae. It is questionable if representatives of Capniidae, Perlodidae, Taeniopterygidae and Chloroperlidae occur on the island.

No previous Taiwanese stonefly collections from high mountain seeps and springs had been made. Possible reasons for their absence in museums are the small size of seepage species, their less attractive appearance, or their winter and early spring emergence.

DIVERSITY OF ASIAN STONEFLIES

Asian stonefly diversity is much greater in comparison to that of Europe or North America. This is true despite the fact that except for Japan, our knowledge of the area is extremely poor, thus it is only a small fragment that we know. Chinese stonefly species are not included in this comparison, although over 200 species have been described from China. Unfortunately, the majority of types were lost or destroyed and the identity of several recently described species is questionable.

If such brief collecting activities in Taiwan can add significant numbers of additional and new species to the Asian stonefly fauna, it seems likely that intensive future research will certainly result in hundreds of new species in areas like China. However, negative human impact is more and more evident in Asia and it is an open question how many new species we will be able to recognize and describe before they become extinct. Already most type localities for species described in the early part of the century are polluted, changed or completely destroyed.

ACKNOWLEDGMENTS

We would like to thank Bill Stark for his linguistic help and Takao Shimizu to provide unpublished data.

REFERENCES

Hoefnagel, G. 1592. Archetypa studiaque Patris Georgii Hoefnagelii Jacobus F: genio duce ab ipso scalpta omnibus philomusis ainice D: ac perbenigne communicat. Francofurti ad Moenum.

Klapálek, F. 1912. H. Sauter's Formosa - Ausbeute. Plecoptera. Ent. Mitt. 1: 342-351.

Klapálek, F. 1913. H. Sauter's Formosa - Ausbeute. Plecoptera II. Suppl. Ent. 2: 112-127.

Okamoto, H. 1912. Erster Beitrag zur Kenntnis der japanischen Plecopteren. Trans. Sapporo nat. Hist. Soc. 40: 105-170.

Sivec, I., Ping-Shih Yang 1997: Name Lists of Insects in Taiwan. Plecoptera. Chinese J. Ent. 17: 188-194.

Stark, B. 1997. Plecoptera from Nomina Insecta Nearctica. A Check List of the Insects of North America. Volume 4: Non-Holometabolous Orders, pp. 641-664. In: R. W. Poole and P. Gentili (eds.). Entomological Information Services.

A SYNOPSIS OF NEOTROPICAL PERLIDAE (PLECOPTERA)

Bill P. Stark

Department of Biology
Mississippi College
Clinton, Mississippi 39058, USA

ABSTRACT

The Neotropical perlid fauna currently includes about 280 species placed in ten genera. *Anacroneuria*, the dominant and most diverse genus throughout much of the Neotropics, is largely replaced in southern South America by *Inconeuria*, *Kempnyella*, *Kempnyia*, *Nigroperla*, and *Pictetoperla*. *Enderleina* and *Macrogynoplax* are modestly diverse genera found sporadically around the fringe of the Amazon Basin while *Onychoplax* and *Klapalekia* are monotypic genera known from holotypes. A review of the distribution, distinctive morphological characters and current status is presented for each genus.

INTRODUCTION

Perlid stoneflies are an abundant and sometimes dominant component of Neotropical streams. Currently in the Neotropics, ten perlid genera and approximately 280 species are considered valid, however the fauna includes some poorly known genera and numerous *nomina dubia* at the species level (Illies,1966).

Recent studies of *Anacroneuria* (Stark,1994,1995,1998,1999, Stark and Sivec,1998, Stark et al.1999), *Enderleina* (Stark,1989), *Kempnyia* (Froehlich,1984a,1988), *Klapalekia* (Stark,1991) and *Macrogynoplax* (Froehlich,1984b, Stark,1996, Stark and Zwick, 1989) have greatly improved our knowledge of these groups but none of these studies include generic keys. Illies (1964) gave the last comprehensive overview of Neotropical perlids but his account included only *Anacroneuria* and the four Patagonian genera (*Inconeuria*, *Kempnyella*, *Nigroperla*, *Pictetoperla*). More recently, Froehlich (1984c) offered a key to the nymphs of perlid genera found in southeastern Brazil (*Anacroneuria*, *Kempnyia*, *Macrogynoplax*).

This study, based on specimens obtained from the California Academy of Sciences, San Francisco (CAS), Carnegie Museum of Natural History, Pittsburgh (CMNH), Limnologische Flussstation Schlitz des Max-Planck-Instituts für Limnologie, Schlitz (LFS), and the National Museum of Natural History, Washington (USNM), presents the first generic key for Neotropical Perlidae, and a checklist of known species.

Trends in Research in Ephemeroptera and Plecoptera
Edited by E. Dominguez, Kluwer Academic/Plenum Publishers, 2001

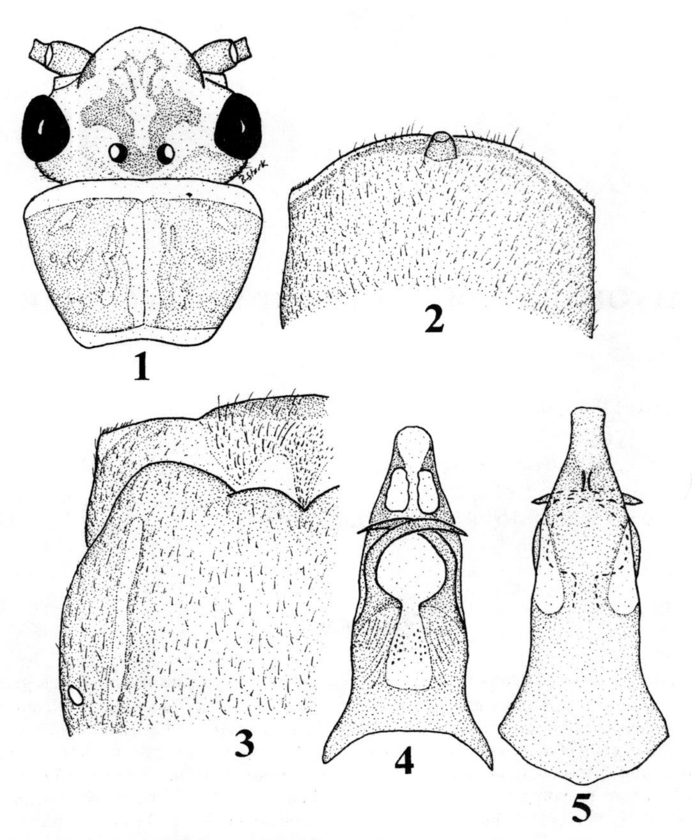

Figs. 1-5. *Anacroneuria* sp. A, Misiones, Argentina. 1. Head and pronotum. 2. Male sternum 9. 3. Female sterna 8, 9 (partial). 4. Aedeagus ventral. 5. Aedeagus dorsal.

Gill designations, adapted from Shepard and Stewart (1983) include the following:

AT = Gills located on the anterior margin of the second (AT2) or third (AT3) thoracic segments.

ASC = Gills located anteriorly, and dorsal to, the first (ASC1), second (ASC2) or third (ASC3) coxa.

PSC = Gills located posteriorly, and dorsal to, the first(PSC1), second (PSC2) or third (PSC3) coxa.

PT = Gills located on the posterior margin of the third (PT3) thoracic segment.

KEYS TO GENERA

Males (*Klapalekia* and *Onychoplax* unknown)

1. Two ocelli 2
   Three ocelli, anterior ocellus sometimes minute 6
2. Posteromesal margin of S9 scarcely produced (Fig. 2); hammer often thimble shaped, when present .... *Anacroneuria*

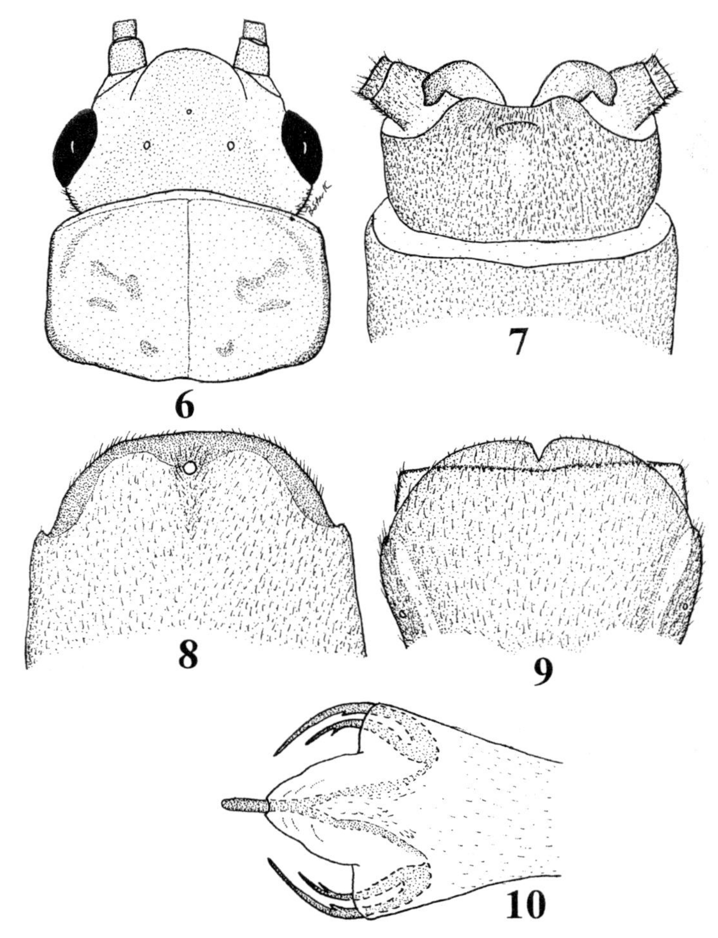

Figs. 6-10. *Enderleina* spp. 6. *E. flinti* head and pronotum. 7. *E. yano* male terga 9, 10. 8. *E. yano* male sternum 9. 9. *E. yano* female sterna 8, 9. 10. *E. yano* aedeagus dorsal.

Posteromesal margin of S9 strongly produced (Fig. 35); hammer a low callus 3

3. Body color green in life, pale white in alcohol; posteromesal ommatidial rows usually unpigmented (Fig. 33) .... *Macrogynoplax*
   Body color pale to dark brown; ommatidia normally pigmented 4
4. A short longitudinal ridge extends from behind ocellar triangle to hind margin of head (Fig. 25); aedeagus usually with a pair of hooks (Fig. 29) .... *Kempnyia* (in part)
   Post ocellar area without longitudinal ridge; aedeagus without hooks (Fig. 49) 5
5. Ocellar area with at least some dark pigment; mesal area of aedeagus with a narrow transverse band of overlapping scale-like spines (Fig. 49).......... *Pictetoperla* (in part)
   Ocellar area pale; mesal area of aedeagus with a partially sclerotized tube covered with patches of spines (Fig. 16) .... *Inconeuria*
6. Hammer small and circular (Fig. 8); pronotum orange, wings dark brown to purple .... *Enderleina*
   Hammer large and oval or elongate (Fig. 28);pronotum dark brown or black often patterned with yellow, wings pale to dark brown 7

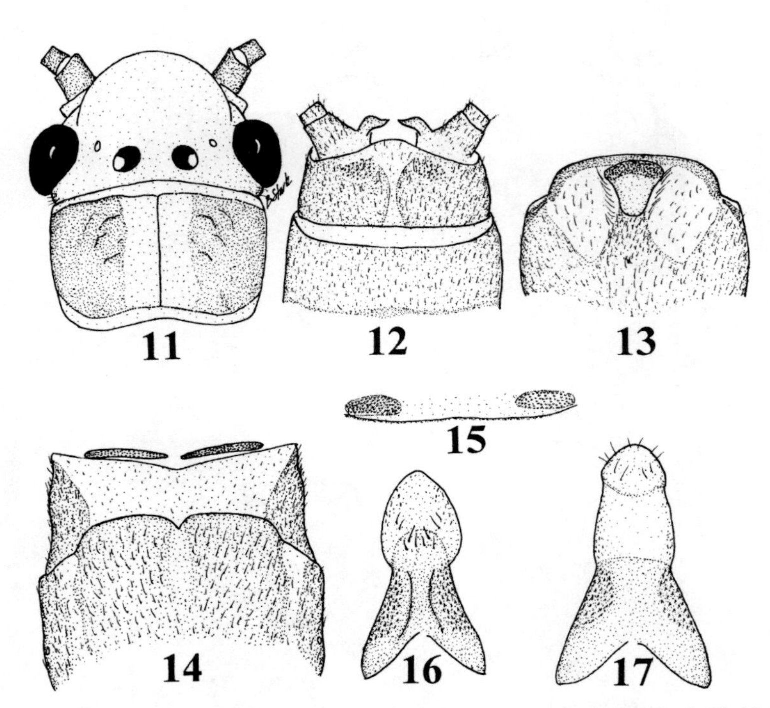

Figs. 11-17. *Inconeuria* spp. 11. *I. porteri* head and pronotum. 12. *I. porteri* male terga 9, 10. 13. *I. porteri* male sternum 9. 14. *I. porteri* female sterna 8, 9. 15. *I. marcapatica* holotype female microtrichia bands in sternum 9 intersegmental membrane. 16. *I. porteri* aedeagus ventral. 17. *I. porteri* aedeagus dorsal.

7. A short longitudinal ridge extends from behind ocellar triangle to hind margin of head (Fig. 25); aedeagus usually with a pair of hooks (Fig. 29) .... *Kempnyia* (in part)
   Post ocellar area without longitudinal ridge; aedeagus without hooks (Fig. 43) 8
8. Head and pronotum black; subapical area of aedeagus a sclerotized tube (Fig. 43).................... *Nigroperla*
   Head and pronotum pale, or patterned with pale pigment; sclerotized portion of aedeagus not forming a tube 9
9. Aedeagal bands composed of densely packed regular spine rows (Fig. 22); area anterior to aedeagal bands partially sclerotized .... *Kempnyella*
   Aedeagal bands composed of large spines in irregular rows (Fig. 49); area forward of bands membranous .... *Pictetoperla* (in part)

Females (*Onychoplax* not included)

1. Two ocelli 2
   Three ocelli, anterior ocellus sometimes minute 7
2. Subgenital plate poorly developed, broadly emarginate (Fig. 32); egg collar short and wide ...... *Klapalekia*
   Subgenital plate well developed (Figs. 3,9); egg collar a small button-like disc, or absent 3
3. A short longitudinal ridge extends from behind ocellar triangle to hind margin of head (Fig.25) .... *Kempnyia* (in part)
   Post ocellar area without longitudinal ridge 4
4. Post-intersegmental area of S9 with a sclerite or pair of sclerites covered with microtrichia (Fig.14)..... 5

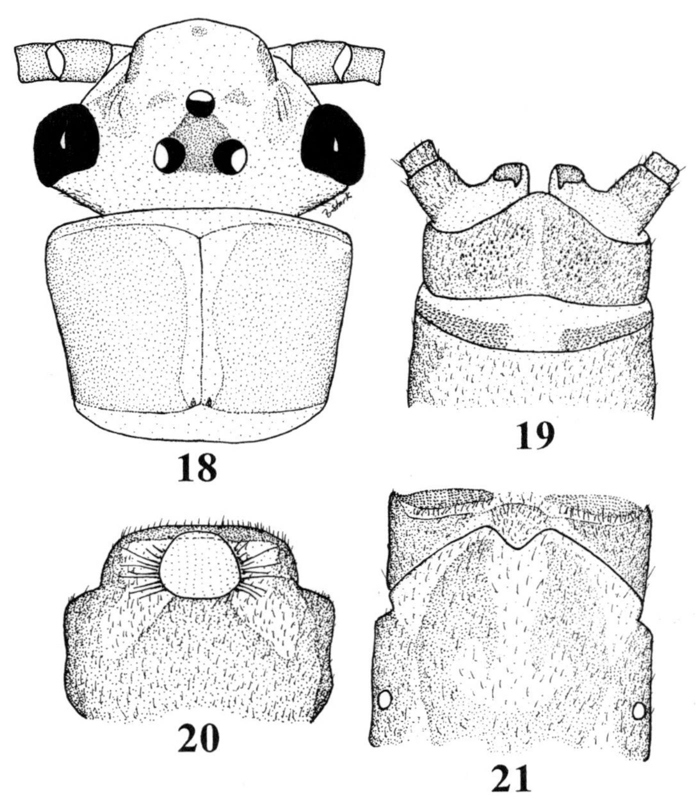

Figs. 18-21. *Kempnyella genualis*. 18. Head and pronotum 19. Male terga 9, 10. 20. Male sternum 9. 21. Female sterna 8, 9.

Post-intersegmental area of S9 without microtrichia bands or patches 6

5. Ocellar area covered with dark pigment .... *Pictetoperla* (in part)
   Ocellar area pale .... *Inconeuria*
6. Body color green in life, pale white in alcohol; marginal posteromesal ommatidial rows usually unpigmented (Fig. 33); S9 without mesal setal patch or sclerite .... *Macrogynoplax*
   Body color pale to dark brown; ommatidia normally pigmented; S9 with a mesal setal patch or sclerite .... *Anacroneuria*
7. Posterior margin of S9 with a narrow dark sclerite usually bearing a row of slender spines (Fig. 9); head and pronotum orange, wings dark brown to purple .... *Enderleina*
   Posterior margin of S9 without a narrow dark sclerite; head and pronotum brown patterned with yellow, wings pale to dark brown 8
8. A short longitudinal ridge extends from behind ocellar triangle to hind margin of head .... *Kempnyia* (in part)
   Post ocellar area without longitudinal ridge 9
9. Subgenital plate margin with a narrow, V-shaped notch (Fig. 21) 10
   Subgenital plate margin entire or emarginate (Fig. 40) 11
10. Intersegmental microtrichia patches separated on sternum 9; S9 with a small mesal sclerite covered with fine setae(Fig. 21); pronotum almost entirely dark brown .... *Kempnyella* (in part)

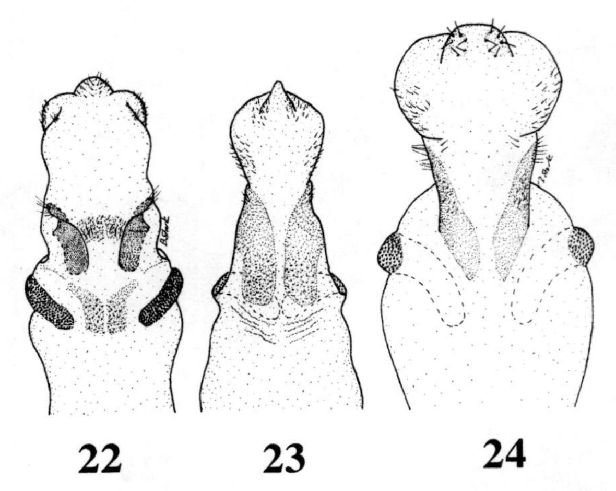

Figs. 22-24. *Kempnyella* aedeagi. 22. *K. genualis* ventral. 23. *K. genualis* dorsal. 24. *K. walperi* dorsal.

Intersegmental microtrichia patches joined on sternum 9; S9 without a mesal sclerite; pronotum with a broad pale band, or almost entirely pale .... *Pictetoperla* (in part)

11. Head and pronotum almost entirely black; femora conspicuously banded; forewing greater than 30 mm in length .... *Nigroperla*

    Head and pronotum with pale bands; femora without conspicuous bands; forewing less than 28 mm in length .... *Kempnyella* (in part)

Nymphs (*Klapalekia*, *Nigroperla* and *Onychoplax* unknown)

1. Anal gills absent 2
   Anal gills present 4
2. PSC2 gills absent .... *Anacroneuria*
   PSC2 gills present 3
3. AT2, AT3 gills with triple trunks .... *Pictetoperla* (in part)
   AT2, AT3 gills with double trunks .... *Kempnyia*
4. Two ocelli 5
   Three ocelli 6
5. Fore femora swollen, raptorial in appearance .... *Macrogynoplax*
   Fore femora normal in appearance .... *Inconeuria*
6. Ocelli minute and widely spaced; cerci without dense swimming fringe .... *Enderleina*
   Ocelli normal; at least basal cercal segments with swimming fringe .... *Kempnyella*

Anacroneuria Klapálek, 1909 (Figs. 1-5)

Type species. Anacroneuria albimacula Klapálek.

Distribution. Arizona [USA] to Brazil and Argentina.

Material. See Stark (1995), Stark (1998), and Stark and Sivec (1998).

Diagnosis. Biocellate. Adult habitus pale to dark, often brown patterned with yellow or yellow patterned with brown or black.

Size variable, forewing length 7-30 mm. Hammer variable, typically thimble shaped with small diameter, absent in a few species. Mesoapical area of sternum 9 only slightly prolonged. Aedeagus a sclerotized tube with broad base, narrow apex and a ventral pair of large opposable hooks. Female subgenital plate covers most of sternum 9; apex bilobed or quadrilobed. Sternum 9 usually with a mesal setose sclerite; posterior margin often with a narrow transverse sclerite. Nymphal thoracic gills ASC [1], PSC [1] and AT [2,3] with double trunks, PT [3] with triple trunks; ASC [2,3] PSC [2,3] and anal gills absent. Egg spindle shaped with button collar. Chorion typically smooth. Posterior pole sometimes prolonged into a spine.

Species list. Over 230 species are included. A current species list is available from the author.

Discussion. Recent and ongoing studies of *Anacroneuria* in Mexico (Kondratieff and Baumann, unpublished), Mesoamerica (Harper, 1992, Stark,1998), Colombia (Rojas and Baena,1993, Stark et al., 1999, Zúñiga et al., unpublished), Venezuela-Guyana-Suriname (Stark,1995, Stark,1999), Trinidad-Tobago (Stark,1994), Ecuador (Stark, unpublished), and Bolivia-Peru (Stark and Sivec,1998) suggest the number of species in this genus will likely exceed 300. No comprehensive key is available, but regional keys for male specimens have been developed for Costa Rica-Panama (Stark,1998), Venezuela (Stark,1995), Colombia (Stark et al., 1999), and Bolivia-Peru (Stark and Sivec,1998). Much work remains, particularly in the association of males with females and in the rearing and association of nymphal specimens.

Enderleina Jewett, 1960 (Figs. 6-10)

Type species. Enderleina preclara Jewett.

Distribution. Known from Brazil and Venezuela.

Material. E. flinti Stark: Holotype ♀, Venezuela, Cerro de la Neblina, Camp XI, 1450 m, 27 February 1985, P. J. Spangler (USNM). *E. yano* Stark: Holotype ♂, paratype ♀, Venezuela, Cerro de la Neblina, Camp IV, 760 m, 15-18 March 1984, O. S. Flint (USNM).

Diagnosis. Triocellate, ocelli small. Pronotum orange, wings dark brown to purple. Antennae and cerci long. Male forewing length ca. 12 mm; female forewings 17-18 mm. Hammer a small, low circular callus. Apex of sternum 9 moderately produced. Aedeagus heavily armed with a mesoapical tongue-like sclerite and a lateral pair of bifid, toothed spines. Female subgenital plate large, completely covering sternum 9, and notched apically. Posterior margin of sternum 9 with an irregular row of slender, anteriorly directed spines. Nymphal cercal segments with whorls of long slender and short stout setae; hair fringe absent. Thoracic gills ASC [1,2,3], PSC [1,2,3], AT [2,3] and PT [3] with single trunks; anal gills with multiple trunks. Egg spindle shaped or oval with button collar. Chorion thin and smooth. Micropyles in oval pits.

Species list. E. bonita Stark; *E. flinti* Stark; *E. froehlichi* Ribeira-Ferreira; *E. preclara* Jewett; *E. yano* Stark.

Discussion No key exists for the six members of this rare genus. Only three male specimens, representing three different species (*E. preclara, E. froehlichi, E. yano*) have been collected and the *E. preclara* specimen is apparently lost (Stark,1989). Females are available for only three species (*E. bonita, E. flinti, E. yano*) and these are distinguished on the basis of color pattern, subgenital plate structure, and in the form of the sclerotized posterior margin of sternum 9 (Stark,1989).

Inconeuria Klapálek, 1916 (Figs. 11-17, 51-52)

Type species. Inconeuria marcapatica Klapálek.

Distribution. Presently known from Chile and Peru.

Material. Inconeuria marcapatica Klapálek: Holotype ♀, Peru, Marcapata (Natural History Museum, Prague). *Inconeuria porteri* (Navas): 1 ♀, Chile, Pucatrihue, Osorno, February 1980, L. E. Pena (CMNH). 1 ♂, Chile, Chiloe, Huequetrumao, 22 km N Quellon, 26-28 December 1981, L. E. Pena (USNM). 1 ♂, 1 ♀, Chile, Llanquihue, Salto Chamiza, Correntoso, 100 m, 19 January 1987, C. M. Flint, O. S. Flint (USNM). 2 ♂, 1 ♀, Chile, Cautin, Fundo el Coique, 500 m, 27 km NE Villarica, 28 February-3 March 1979, D. Davis, M. Da-

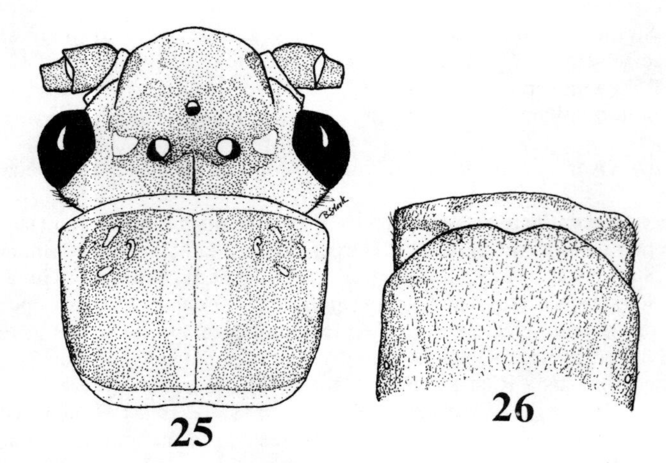

Figs. 25-26. *Kempnyia* sp. A, Rio Marumbi, Paraná State, Brazil. 25. Head and pronotum. 26. Female sterna 8, 9.

vis, B. Akerbergs (USNM). 5 ♂, Chile, Nuble, Alto Trequalemu, 500 m,20 km SE Chovellen, 26-27 January 1979, D. Davis, M. Davis, B. Akerbergs (USNM). 1 ♂, 1 ♀, Chile, Valdivia, 8 mi E Rio Bueno, 15 January 1951, Ross, Michelbacher (CAS).

Diagnosis. Biocellate. Adult head yellow; pronotum brown on lateral third, pale mesally. Male forewing length 16-20 mm, female forewings 22-24 mm. Hammer subtriangular, apex of sternum 9 prolonged. Tergum 10 with membranous mesal band and a pair of midlateral spinule patches. Apical section of aedeagus partially sclerotized; sclerites bearing ventrolateral patches of apically directed spines. Membranous apex bearing a sparse cluster of thick setae and fine scattered setae. Female subgenital plate small, covering ca. half of sternum 9; apex with a shallow notch. Mesal field of sternum 9 weakly sclerotized; prominent microtrichia bands in intersegmental membrane. Nymphal thoracic gills ASC [1,2,3], PSC [1,2,3], AT [3] and PT [3] with single trunks; AT [2] with triple trunks; anal gills present. Egg spindle shaped but much wider at collar; collar button-like, small. Chorion of posterior pole irregular, covered with low, coarse tubercles. Micropyles small, difficult to observe among tubercles.

Species list. I. marcapatica (Klapálek); *I. porteri* (Navas).

Discussion. The females of *I. marcapatica* and *I. porteri* are distinguished by the shapes of the respective microtrichia patches of sternum 9. The holotype of *I. marcapatica* bears short, oval, widely separated patches, whereas in *I. porteri* these are long, slender and almost meet on the midline of sternum 9. The meager nymphal characters were extracted from the same small specimens used by Illies (1964). The nymphs generally resemble *Anacroneuria* specimens but with anal gills.

Kempnyella Illies, 1964 (Figs. 18-24, 57-58)

Type species. Kempnyella genualis (Navas).

Distribution. Argentina and Chile.

Material. Kempnyella genualis (Navas): 1 ♀, Argentina, Neuquen, Rio Agrco, N Zapala, 9-11 December 1983, L. E. Pena (USNM). 3 nymphs, Argentina, Neuquen, Rio Malleo, 17 km N Junin de los Andes, 2 February 1987, C. M. Flint, O. S. Flint (USNM). 1 ♂, Argentina, El Bolson, 13 February 1961, Kovacs (CAS). 3 ♂, 10 ♀, Chile, Linares, Puente Malcho, 600 m, near Longavi River, 13-15 January 1979, D. Davis, M. Davis, B. Akerbergs (USNM). 1 ♂, Chile, L. Chapo, Llanquihue, January 1980, L. E. Pena (CMNH). 2 ♂, Chile,

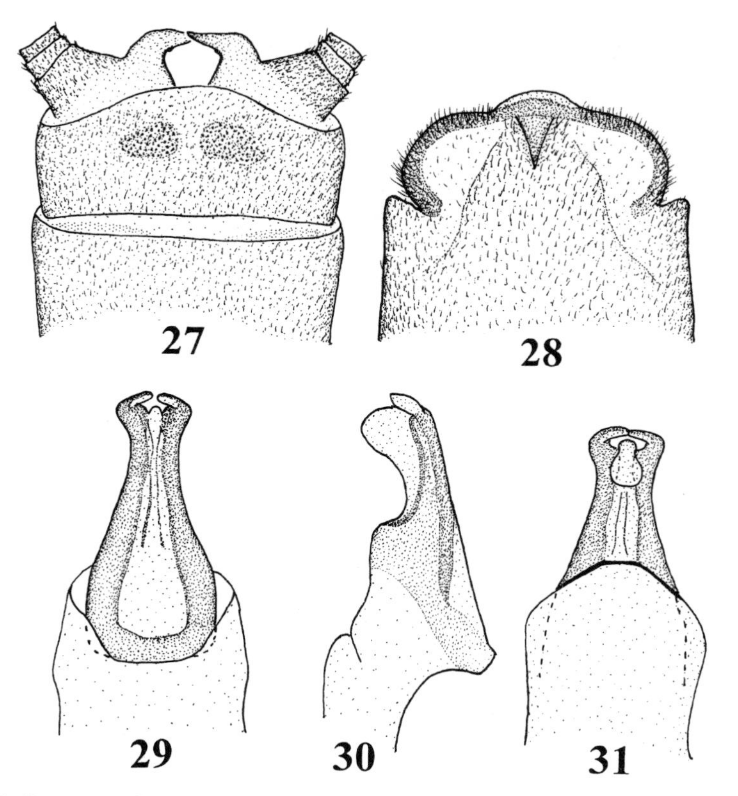

Figs. 27-31. *Kempnyia collosica*. 27. Male terga 9, 10. 28. Male sternum 9. 29. Aedeagus dorsal. 30. Aedeagus lateral. 31. Aedeagus ventral.

Cautin, Fundo el Coique, 500 m, 27 km NE Villarica, 28 February-3 March 1979, D. Davis, M. Davis, B. Akerbergs (USNM). 6 ♂, 2 ♀, Chile, Aisen, Lago Risopatron, 17 km N Puyuhuapi, 24 January 1987, C. M. Flint, O. S. Flint (USNM). 1 ♂, Chile, Rio Limay, 9 March 1958 (LFS). *Kempnyella walperi* Illies: 1 ♂, Chile, Pucatrihue, Osorno, February 1980, L. E. Pena (CMNH). 1 ♂, Chile, Villarica, Cautin, March 1979, L. E. Pena (CMNH). 2 ♂, Chile, Cautin, Fundo el Coique, 500 m, 27 km NE Villarica, 28 February-3 March 1979, D. Davis, M. Davis, B. Akerbergs (USNM). Holotype ♀, Chile, Valdivia, ca. 800 m, Bergbach der Kuestenkordillere, im Fundo Walper, Punucapa, 16 February 1958 (LFS).

Diagnosis. Triocellate. Adult head yellow or brown patterned with darker brown areas; pronotum mostly brown with pale median stripe. Male forewing length 19-24 mm, female forewings 25-29 mm. Hammer a large circular disc; mesoapical area of sternum 9 produced into a prominent truncate lobe. Membranous aedeagal base with a pair of prominent bands of thick spines interrupted on midventral and middorsal lines. Subapical section of aedeagus with sclerotized lateral plates and scattered setae. Intersegmental membrane of tergum 9 with microtrichial patches.

Female subgenital plate covers most of sternum 9; apex with or without mesal notch. Intersegmental membrane of sternum 9 armed with a pair of prominent microtrichial patches. Nymphal femoral and tibial hair fringe well developed, cercal fringes developed basally but sparse on mid and apical segments. Thoracic gills ASC [1] with double trunks, PSC [1,2,3],

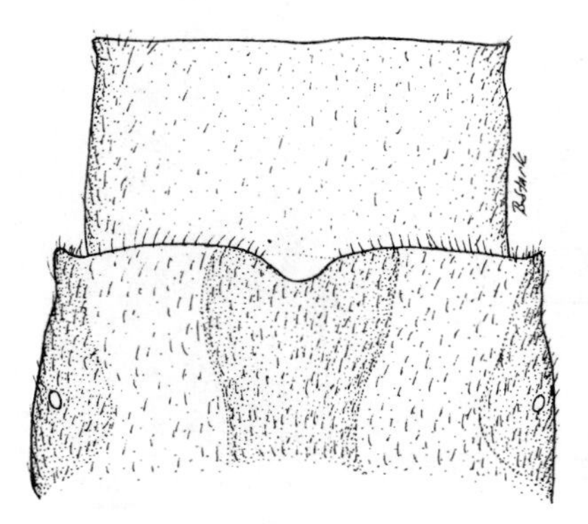

Fig. 32. *Klapalekia* female sterna 8, 9.

ASC [2,3], and PT [3] trunks single, AT [2,3] trunks triple; anal gills with multiple trunks. Egg spindle shaped with collar absent. Chorion smooth except anterior pole covered with follicle cell impressions; anterior pole encircled by opercular ring. Micropyles set in oval pits.

Species list. K. genualis (Navas)= *Pictetoperla brundini* Illies, new synonymy; *K. walperi* Illies.

Discussion. The female holotype of *P. brundini* is identical in subgenital plate and sternal 9 structure to *K. genualis* and is therefore placed as a junior synonym of that species. Female *Pictetoperla* specimens have the microtrichia joined into a single band on the intersegmental membrane of sternum 9 rather than separated into a pair of bands as in the *Kempnyella* species.

Several male *Kempnyella* specimens were found which differ in aedeagal features from typical *K. genualis*; these are presumed to be males of *K. walperi.* The aedeagal apex of these specimens has a pair of stout setal clusters, and the venter lacks the obscure mesal sclerite found between the lateral sclerites of most *K. genualis* specimens. A few male specimens determined as *K. genualis* by Illies also lack the mesal sclerite but are otherwise indistinguishable from other *K. genualis* specimens.

Kempnyia Klapálek, 1916 (Figs. 25-31, 59)

Type species. Kempnyia klugii (Pictet).

Distribution. Brazil.

Material. Kempnyia colossica (Navas): 1 ♂, Brazil, Paraná, Rio Marumbi, Marumbi, 15-16 February 1969, W. L. Peters, J. G. Peters (USNM). *Kempnyia guassu* Froehlich: 1 ♀, Brazil, Rio de Janeiro, 27 April 1974 (LFS). *Kempnyia* sp. A: 11 ♀, Brazil, Paraná, Rio Marumbi, Marumbi, 15-16 February 1969, W. L. Peters, J. G. Peters (USNM). *Kempnyia* sp. B: 2 ♀, Brazil, Paraná, Rio Marumbi, Marumbi, 15-16 February 1969, W. L. Peters, J. G. Peters (USNM).

Diagnosis. Biocellate or triocellate. Adult head and pronotum entirely yellow or yellow patterned with brown. Male forewing length 8-24 mm, female forewings 10-35 mm. Hammer usually longer than wide, often subtriangular. Mesoapical area of sternum 9 greatly prolonged. Aedeagus partially sclerotized and typically bearing a pair of short hooks. Tergum 10 membranous around posteromesal margin and between patches of peg setae. Female subgenital plate large, usually covering sternum 9, margin typically with shallow notch. Intersegmental membrane of sternum 9 with a broad band of microtrichia or fine setae. Nymphal thoracic gills ASC [1,2,3] and AT [2,3] with double trunks, PSC [1,2,3] and PT [3] with single trunks; anal gills present. Egg spindle shaped with button collar. Chorion smooth.

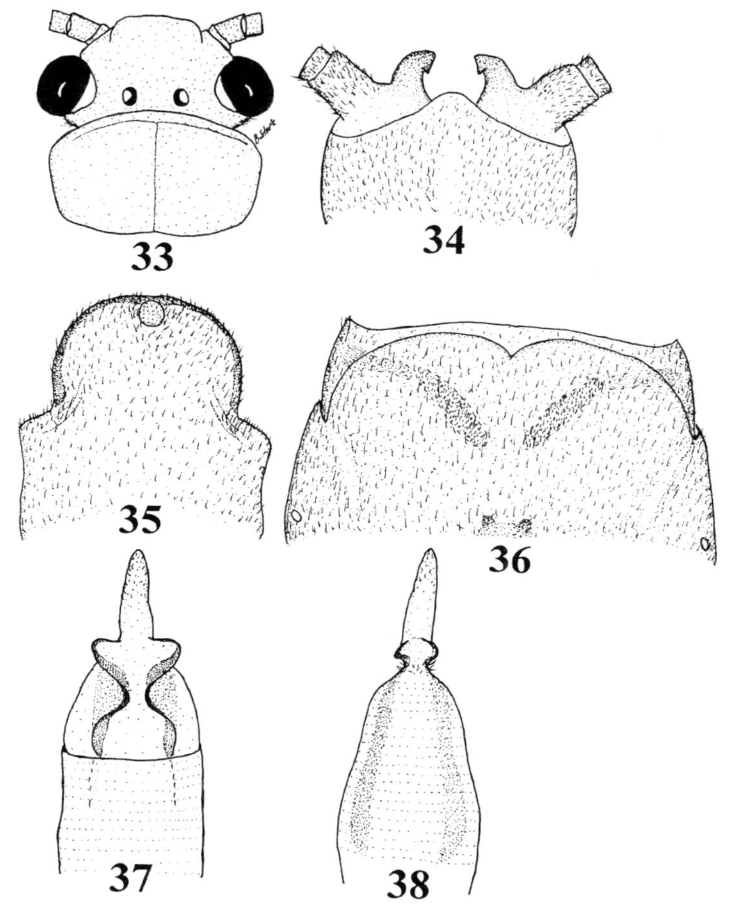

Figs. 33-38. *Macrogynoplax spangleri*. 33. Head and pronotum. 34. Male terga 9, 10. 35. Male sternum 9. 36. Female sterna 8, 9. 37. Aedeagus dorsal. 38. Aedeagus ventral.

Species list. *K. alterosarum* Froehlich; *K. auberti* Froehlich; *K. barbiellinii* (Navas), new comb.(Froehlich i.l.); *K. brasilica* (Navas), new comb.(Froehlich i.l.); *K. brasiliensis* (Pictet); *K. colossica* (Navas); *K. flava* Klapálek; *K. gracilenta* (Enderlein); *K. guassu* Froehlich; *K. jatim* Froehlich,; *K. klugii* (Pictet); *K. mirim* Froehlich; *K. neotropica* (Jacobson & Bianchi); *K. obtusa* Klapálek; *K. petersorum* Froehlich; *K. petropolitana* (Navas); *K. reichardti* Froehlich; *K. remota* (Banks), new comb. (Froehlich i.l.); *K. reticulata* Klapálek; *K. sazimai* Froehlich; *K. serrana* (Navas); *K. sordida* Klapálek; *K. tamoya* Froehlich; *K. taunayi* (Navas), sp. prop. (Froehlich i.l.); *K. tenebrosa* Klapálek; *K. tijucana* Dorville & Froehlich; *K. umbrina* Froehlich; *K. vaini* Froehlich; *K. varipes* Klapálek.

Discussion. Several new and a few of the older species have been described by Froehlich (1984a, 1988, 1996) but presently no keys for the 29 species listed above have been published.

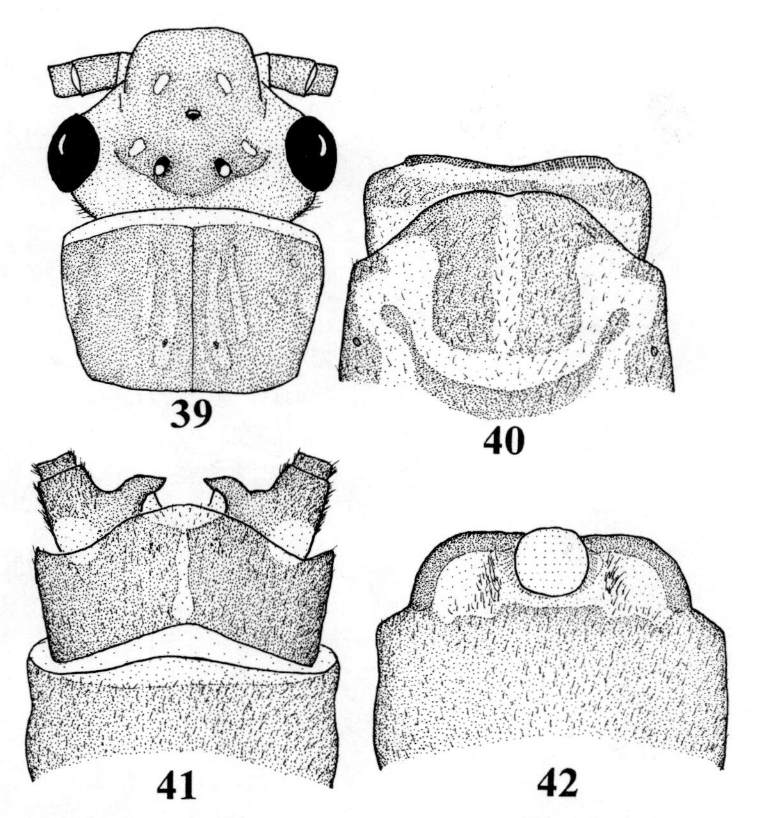

Figs. 39-42. *Nigroperla costalis.* 39. Head and pronotum. 40. Female sterna 8, 9. 41. Male terga 9, 10. 42. Male sternum 9.

Klapalekia Claassen, 1936 (Fig. 32)

Type species. Klapalekia augustibraueri (Klapálek).

Distribution. Known only from the holotype from Bogota.

Material. K. augustibraueri (Klapálek): Holotype ♀, Colombia, Bogota (National Museum, Prague).

Diagnosis. Biocellate. Adult head yellow with darker ocellar area. Pronotum brown with narrow, pale median stripe. Female forewing length ca. 18 mm. Subgenital plate scarcely produced, barely reaching anterior margin of sternum 9; posteromesal margin of plate shallowly excavated. Sternum 9 unmodified. Egg outline oval, collar short and stalked. Chorion punctate throughout; punctations enclosed in shallow follicle cell impressions. Male and nymph unknown.

Species list. K. augustibraueri (Klapálek).

Macrogynoplax Enderlein, 1909 (Figs. 33-38, 60)

Type species. Macrogynoplax guyanensis Enderlein.

Distribution. Presently known from Brazil, Guyana, Suriname, Peru, and Venezuela.

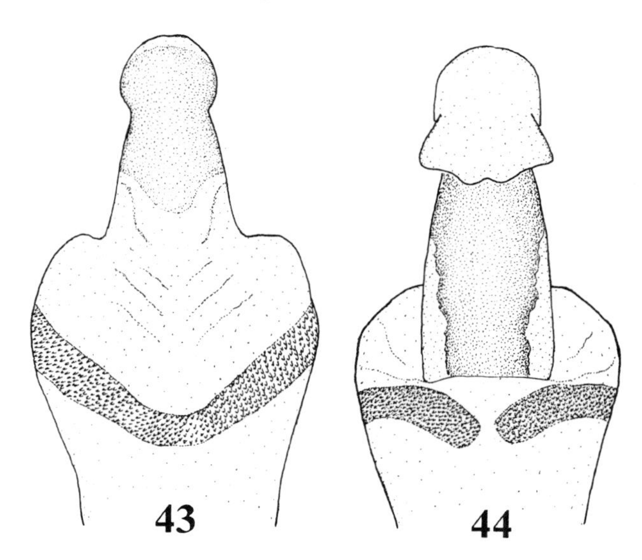

Figs. 43-44. *Nigroperla costalis* aedeagi. 43. Ventral. 44. Dorsal.

Material. M. flinti Stark: Holotype ♂, Guyana, Aramatani Creek, Dubulay Ranch, 15-18 April 1995, O. S. Flint (USNM). *M. kanuku* Stark: Holotype ♂, paratype ♀, Guyana, Kumu River, Kanuku Mountains, 28-30 April 1995, O. S. Flint (USNM). *M. neblina* Stark: Holotype ♂, paratype ♀, Venezuela, Cerro de la Neblina, Camp VII, 1800 m, 30 January-10 February 1985, P. J. Spangler, P. M. Spangler, R. Faitoute (USNM). *M. spangleri* Stark: Paratype ♂ and ♀, Venezuela, Cerro de la Neblina, Basecamp, 140 m, 20-24 March 1984, O. S. Flint, J. Louton (USNM). *M. truncata* Stark: Holotype ♂, paratype ♀, Peru, Rio Nanay, 25 km SW Iquitos, 120 m, 10-17 January 1980, J. B. Heppner (USNM). *M. yupanqui* Stark: Holotype ♂, Peru, Madre de Dios, Rio Tambopata, 30 km SW Maldonado, 290 m, 21-25 October 1979. J. B. Heppner (USNM).

Diagnosis. Biocellate. Adult body color green in life. Eyes usually with rows of unpigmented ommatidia along inner margins. Male forewing length 9-21 mm, female forewings 12-23 mm. Hammer a small, low callus, circular to rectangular in outline. Apex of sternum 9 prolonged into large mesal lobe bearing hammer. Paraprocts weakly to moderately sclerotized, often with an anterior spine. Aedeagus partially sclerotized, with or without hooks. Female subgenital plate covers at least half of sternum 9. Posterior margin often emarginate but sometimes rounded or truncate. Sternum 9 sclerotized lateral margins extend under subgenital plate as slender bars. Nymphal forelegs with short thick femora and long slender tibiae. Maxillary palpi long and slender. Thoracic gills ASC [1,2,3] and PSC [1,2] with single trunks, PSC [3] trunk double, AT [2,3] trunks triple; anal gills with multiple trunks. Egg small, spindle shaped with button collar. Chorion thin and smooth.

Discussion. Stark (1996) presents a key to the males of this genus.

Species list. M. flinti Stark; *M. geijskesii* Zwick; *M. guyanensis* Enderlein; *M. kanuku* Stark; *M. neblina* Stark; *M. spangleri* Stark; *M. truncata* Stark; *M. veneranda* Froehlich; *M. yupanqui* Stark.

Nigroperla Illies, 1964 (Figs. 39-44, 61-62)

Type species. Nigroperla costalis Illies.

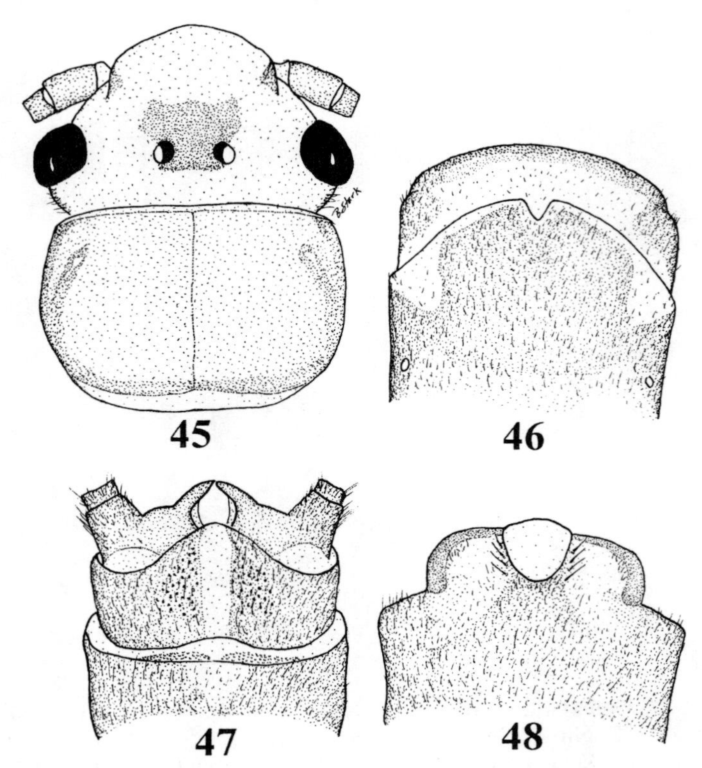

Figs. 45-48. *Pictetoperla gayi.* 45. Head and pronotum. 46. Female sterna 8, 9. 47. Male terga 9, 10. Male sternum 9.

Distribution. Known only from Chile.

Material. N. costalis Illies: 1 ♂, Chile, Arauco, Rio Llinco, Nahuelbuta, 13 Februa 1952, G. Monsalve (LFS). 4 ♀, Chile, Malleco, Nahuelbuta, 1 February 1979, L. E. Pe (CMNH). 1 ♂, Chile, Concepcion, Peuco, 1956, E. P. Reed (CAS). 1 ♂, Chile, Malleco, R Manzanares, 2 January 1966, Flint, Cekalovic (USNM). 1 ♀, Chile, Puyen, February 19 (USNM).

Diagnosis. Adult head and pronotum black with scattered pale areas. Femo conspicuously banded; base and apical half black, midbasal third yellow. Male forewi length 23-28 mm, female forewings 33-34 mm. Hammer a large circular bulb; mesoapic area of sternum 9 produced into a prominent truncate lobe. Membranous aedeagal base wi prominent band of thick spines, continuous ventrally but interrupted on dorsum. Subapic section of aedeagus a partially sclerotized tube; apex membranous. Female subgenital pla covers most of sternum 9; apex rounded with mesal emargination. Plate with a mes membranous band and large membranous areas basally and laterally. Posterior margin sternum 9 with a prominent band of intersegmental microtrichia; microtrichia ba interrupted on meson. Egg spindle shaped with button collar. Posterior pole strong attenuated below micropylar line. Micropyles set on oval plates. Nymph unknown.

Species list. N. costalis Illies.

Discussion. The aedeagal figures in Illies (1964) show only the distinctive sclerotized tu and therefore miss the band of aedeagal spines which allies this genus closely with *Kempnyella.*

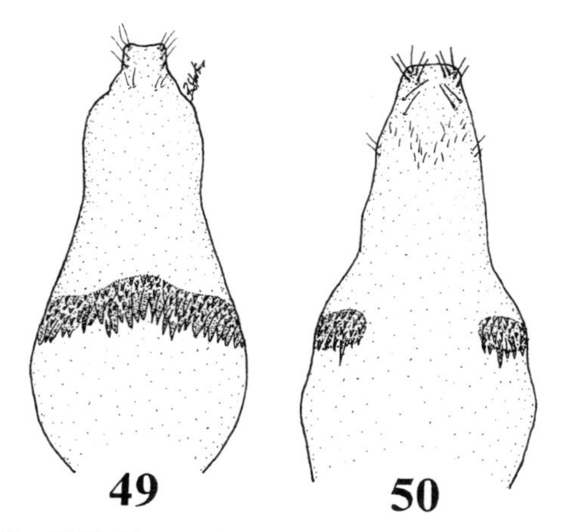

Figs. 49-50. *Pictetoperla gayi* aedeagi. 49. Ventral. 50. Dorsal.

Onychoplax Klapálek, 1916

Type species. Onychoplax limbatella Klapálek.

Distribution. Klapálek (1916) listed the holotype questionably from Brazil.

Material. None.

Diagnosis (modified from Klapálek 1916). Biocellate. Adult head mostly yellow but with a dark area in the center. Pronotum brown with dark borders. Female forewing length ca. 11 mm. Subgenital plate small, slightly emarginate. Male, nymph and egg unknown.

Species list. O. limbatella Klapálek.

Pictetoperla Illies, 1964 (Figs. 45-50, 53-56)

Type species. Pictetoperla gayi (Pictet 1841).

Distribution. Argentina and Chile.

Material. Pictetoperla gayi (Pictet): 1 ♂, Argentina, Neuquen, tributary Arroyo Trompul, W San Martin de los Andes, 23 February 1978, O. S. Flint (USNM). 1 ♂, Chile, Nuble, Rio Chillan near Recinto, 6 March 1968, O. S. Flint, L. E. Pena (USNM). 2 ♂, Chile, Las Trancas, Curico, 1 March 1979, L. E. Pena (CMNH). 1 ♀, Chile, Dalcahue, 10-20 February 1957 (LFS). 1 ♀, Chile, Cautin, Rio Llanquihue, 600 m, 11 February 1958, Illies, Besch (LFS). *Pictetoperla repanda* (Banks): 1 ♀, Argentina, R. N. Case, Mallin, Ahogado, El Bolson, 9 February 1974, O. S. Flint (USNM). 1 ♂, El Bolson, December 1959, A. Kovacs (CAS). 2 ♂, 1 ♀,Chile, Maule, Rio Teno, 800 m, ca. 49 km E Curico, 25-27 November 1981, D. Davis (USNM).

Diagnosis. Triocellate, anterior ocellus sometimes minute. Adult head and pronotum uniformly yellow brown or strongly patterned with dark brown areas. Male forewing length 23-30 mm, female forewings 28-35 mm. Hammer a large oval or subtriangular callus, mesoapical area of sternum 9 produced into a prominent truncate lobe. Membranous aedeagal base with a continuous, or narrowly interrupted band of large, irregular, overlapping spines; area apical to band membranous with a few large spine-like setae. Intersegmental membrane of tergum 9 with patches of microtrichia. Tergum 10 divided by a mesal membranous band; midlateral areas covered with spinules. Female subgenital plate covers most of sternum 9; apex of plate with a shallow V-shaped notch. Intersegmental membrane of sternum 9 covered

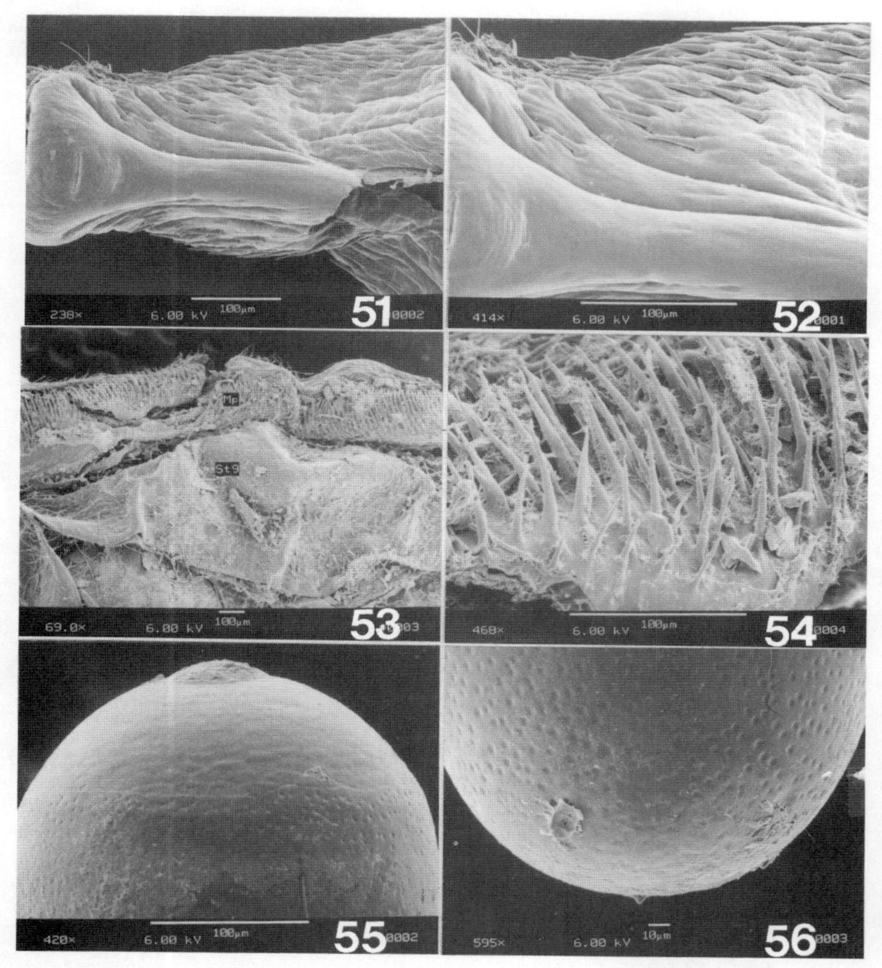

Figs. 51-56. SEM micrographs of *Inconeuria* and *Pictetoperla* structures. 51. *I. porteri* aedeagus. 52. *I. porteri* aedeagus. 53. *P. repanda* female sternum 9 [St9] and intersegmental microtrichia [Mp]. 54. *P. repanda* intersegmental microtrichia. 55. *P. gayi* egg, anterior pole and collar. 56. *P. gayi* egg, posterior pole.

with a continuous microtrichia band. Nymphal femoral, tibial and cercal hair fringes well developed. Thoracic gills ASC [1] and PSC [1,2,3] with single trunks, AT [2,3] trunks triple, PT [3} trunks double; ASC [2,3] and anal gills absent. Egg spindle shaped with button collar. Chorion smooth or covered with shallow pits. Micropyles set in shallow depressions.

Species list. P. gayi (Pictet); *P. repanda* (Banks).

Discussion. With *P. brundini* placed as a synonym of *Kempnyella genualis* (see above), adults of the remaining species of *Pictetoperla* are readily distinguished on the basis of color pattern (Illies, 1964) but the two species are also distinguished by details of the basal band of male aedeagal spines. In *P. repanda*, the band is almost continuous across the dorsum whereas in *P. gayi* the band is broadly separated on the dorsum.

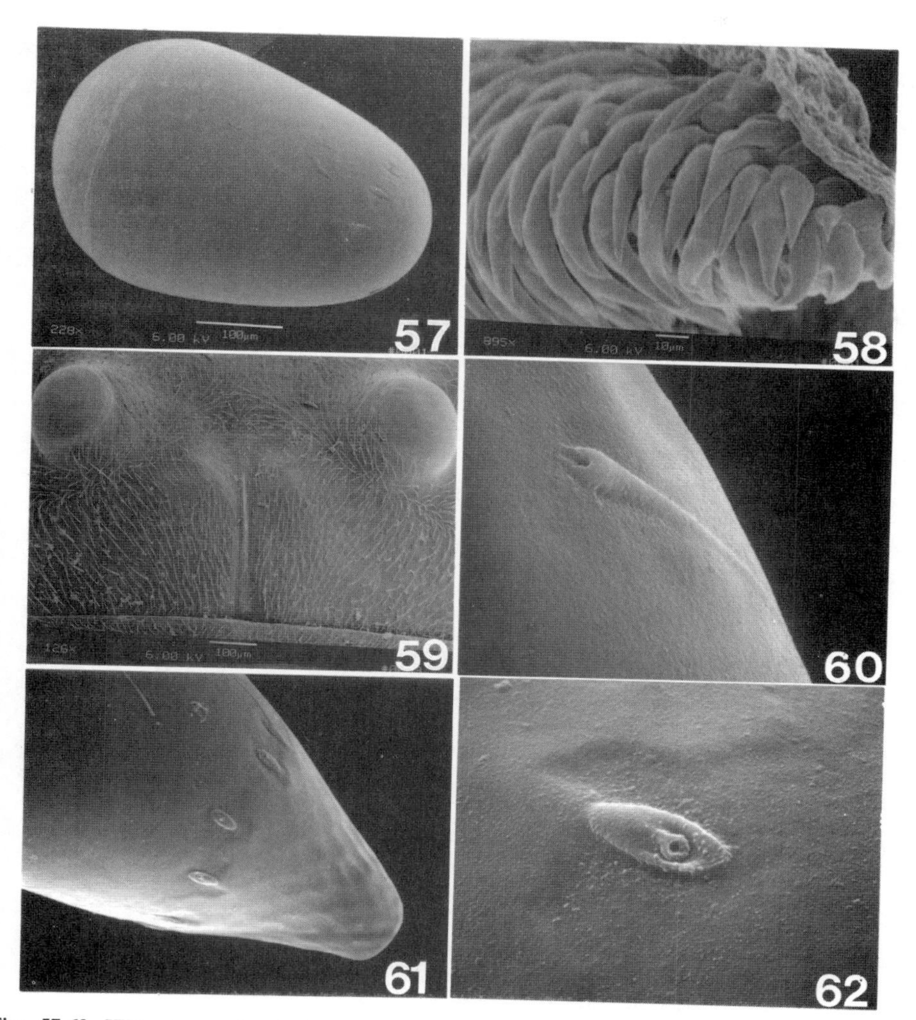

Figs. 57-62. SEM micrographs of *Kempnyella*, *Kempnyia*, *Macrogynoplax* and *Nigroperla* structures. 57. *Kempnyella walperi* egg from holotype. 58. *K. genualis* spines from aedeagal band. 59. *Kempnyia* sp. A, postocular occipital ridge. 60. *M. spangleri* egg micropyle. 61. *N. costalis* egg, posterior pole. 62. *N. costalis* micropyle.

ACKNOWLEDGMENTS

I thank O.S. Flint and N. Adams (National Museum of Natural History), John Rawlins (Carnegie Museum of Natural History), P. Zwick (Limnologische Flussstation Schlitz des Max-Planck-Instituts für Limnologie) and N. Penny (California Academy of Science) for arranging the loan of material used in this study.

C.G. Froehlich provided a sketch of the *Onychoplax limbatella* subgenital plate.

REFERENCES

Banks, N. 1920. New neuropteroid insects. Bull. Mus. comp. Zool. Harv. 64: 314-325.

Claassen, P. W. 1936. New names for stoneflies (Plecoptera). Ann. ent. Soc. Amer. 29:622-623.

Dorville, F. F. M. and C. G. Froehlich. 1997. *Kempnyia tijucana* sp.n. from Southeastern Brazil (Plecoptera, Perlidae). Aquat. Insects 19: 177-181.

Enderlein, G. 1909. Klassification der Plecopteren sowie Diagnosen neuer Gattungen und Arten. Zool. Anz. 34:385-419.

Froehlich, C. G. 1984a. Brazilian Plecoptera 2. Species of the *serrana*-group of *Kempnyia* (Plecoptera). Aquat. Insects 6: 137-147.

Froehlich, C. G. 1984b. Brazilian Plecoptera 3. *Macrogynoplax veneranda* sp.n. (Perlidae: Acroneuriinae). Ann. Limnol. 20: 39-42.

Froehlich, C. G. 1984c. Brazilian Plecoptera 4. Nymphs of perlid genera from southeastern Brazil. Ann. Limnol. 20: 43-48.

Froehlich, C. G. 1988. Brazilian Plecoptera 5. Old and new species of *Kempnyia* (Perlidae). Aquat. Insects 10: 153-170.

Froehlich, C. G. 1996. Two new species of *Kempnyia* from southern Brazil (Plecoptera: Perlidae). Bull. Soc. ent. Suisse 69: 117-120.

Harper, P. P. 1992. Stoneflies of Panama (Plecoptera), pp. 114-121. In: Quintero and Aiello (eds) Insects of Panama and Mesoamerica, selected studies. Oxford University Press, Oxford.

Illies, J. 1964. Südamerikanische Perlidae (Plecoptera), besonders aus Chile und Argentinien. Beitr. Neotrop. Fauna 3: 207-233.

Illies, J. 1966. Katalog der rezenten Plecoptera. Das Tierreich 82: I-xxx, 1-632.

Jewett, S. G. 1960. Notes and descriptions concerning Brazilian stoneflies (Plecoptera). Arch. Mus. nac., Rio de J. 50: 167-183.

Klapálek, F. 1909. Vorläufiger Bericht über exotische Plecopteren. Wien. ent. Ztg. 28: 215-232.

Klapálek, F. 1916. Subfamilia Acroneuriinae Klp. Cas. Ceské Spol. Ent. 13: 451-84.

Pictet, F.J. 1841. Histoire naturelle generale et particuliere des insectes Neuropteres. Famille des Perlides. 1. Partie: 1-423.

Rojas, A. M. and M. L. Baena. 1993. *Anacroneuria farallonensis* (Plecoptera: Perlidae) una nueva especies para Colombia. Bol. Mus. Ent. Univ. Valle 1: 23-28.

Stark, B. P. 1989. The genus *Enderleina* (Plecoptera: Perlidae). Aquat. Insects 11: 153-160.

Stark, B. P. 1991. Redescription of *Klapalekia augustibraueri* (Klapálek) (Plecoptera: Perlidae). Aquat. Insects 13: 189-192.

Stark, B. P. 1994. *Anacroneuria* of Trinidad and Tobago (Plecoptera: Perlidae). Aquat. Insects 16: 171-175.

Stark, B. P. 1995. New species and records of *Anacroneuria* (Klapálek) from Venezuela. Spixiana 68: 211-249.

Stark, B. P. 1996. New species of *Macrogynoplax* (Insecta: Plecoptera: Perlidae) from Peru and Guyana. Proc. biol. Soc. Wash. 109: 318-325.

Stark, B. P. 1998. The *Anacroneuria* of Costa Rica and Panama (Insecta: Plecoptera: Perlidae). Proc. biol. Soc. Wash. 111: 551-603.

Stark, B. P. 1999. *Anacroneuria* from northeastern South America (Insecta: Plecoptera: Perlidae). Proc. biol. Soc. Wash. 112: 70-93.

Stark, B. P. and I. Sivec. 1998. *Anacroneuria* of Peru and Bolivia (Plecoptera: Perlidae). Scopolia 40: 1-64.

Stark, B. P. and P. Zwick. 1989. New species of *Macrogynoplax* from Venezuela and Surinam (Plecoptera: Perlidae). Aquat. Insects 11: 247-255.

Stark, B. P., M. C. Zúñiga, A. M. Rojas and M. L. Baena. 1999. Colombian *Anacroneuria*: Descriptions of new and old species (Insecta: Plecoptera: Perlidae). Spixiana 22: 13-46.

PLECOPTERA FAUNA OF CAPNIIDAE OF RUSSIA AND ADJACENT TERRITORIES (WITHIN THE LIMITS OF THE FORMER USSR)

Lidija A. Zhiltzova

Zoological Institute, Russian Academy of Sciences
St. Petersburg, 199034 Russia

ABSTRACT

Ten genera and 60 species of Capniidae were found in the examined territory of the former USSR. I establish six well-defined species groups of *Capnia*, differing by the shape of male epiproct and some other characters: 1) "*atra*" group - 10 species, 2) "*vidua*" group - 2 species, 3) "*japonica*" group - 2 species, 4) "*bifrons*" group - 4 species, 5) "*cordata*" group - 4 species from Middle Asia and 6 species from Himalaya, 6) "*pedestris*" group - 7 species from Middle Asia. Six species of *Capnia* still remain unassigned to any of the groups. Geographic distribution of genera and species is discussed. Six groups of generic distribution and 15 groups of specific distribution are established.

INTRODUCTION

The present report is based on many years of investigation by me (from 1953 to present) on the Plecoptera fauna of Russia and adjacent territories (within the limits of the former USSR). Plecoptera was collected during numerous expeditions of the author to the Caucasus, the Ukrainian Carpathians, the Urals, the north and middle regions of the European part of Russia, in Middle Asia, East Siberia and in the Russian Far East (including Sakhalin and South Kurile islands). Besides my own collections I studied the general collections of the Zoological Institute in St. Petersburg, of the Zoological Museum of Moscow University, the Biology and Pedology Institute of the Far Eastern Branch of the Academy of Sciences of Russia, of the Reservation "Stolby" (Krasnojarsk), of the Reservation Sikhote-Alin (Primorye), of the Komy branch of the Academy of Sciences of Russia, and of many other scientific institutions and individuals. Eight families, 70 genera and about 350 species of Plecoptera are presently known in the fauna of Russia and adjacent territories. The Capniidae are discussed below.

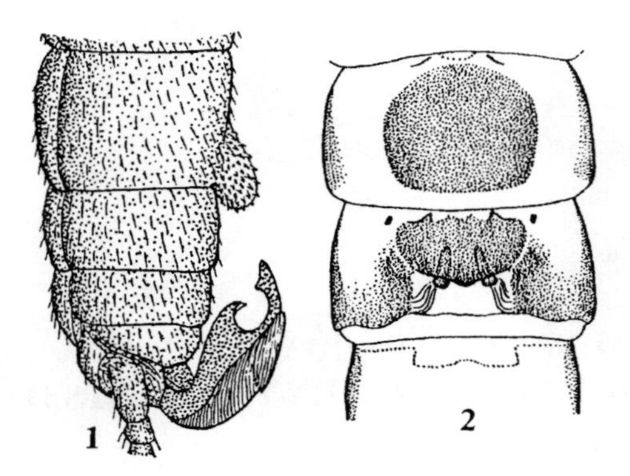

Figs. 1-2. *Capnia pygmaea* Zett. (the "*atra*" group): 1 - male abdomen, laterally; 2 - female abdomen, ventrally.

RESULTS AND DISCUSSION

I recognize 10 genera and 60 species and subspecies of Capniidae in Russia and adjacent territories. This family ranks second place in number of species, after the family Nemouridae (92 species). 1 genus (*Baikaloperla*) and 31 species were described by the author as new to science (*Capnia*-19 species, *Mesocapnia* - 1, *Baikaloperla* - 2, *Takagripopteryx* - 1, *Capniella* - 1, *Isocapnia* - 3, *Paracapnia* - 3, *Capnioneura* - 1 species).

The genus *Capnia* has the largest number of species in the investigated territory (35 species). The most important diagnostic characters, used by modern authors and by me in keys and descriptions of new species of *Capnia*, are: peculiarities of male epiproct shape; number and shape of processes on tergites 6-9 of abdomen; the shape of the vesicle on male sternite 9, or its absence; male brachyptery; modifications of female sternites 7 and 8, especially the shape of the subgenital plate. Most morphological structures used as diagnostic characters display a wide variability within the limits of *Capnia*. This makes it possible to divide genus *Capnia* into some groups of morphologically close species. I divide the species of *Capnia* of Russia and adjacent territories into six well defined groups, namely:

1) "*atra*" group (Figs 1-2), characterized by male epiproct divided longitudinally into ventral and dorsal halves, and the presence of a dark tubercle on tergite 7. I include in this group *Capnia atra, C. ahngeri, C. aligera, C. pygmaea, C. zaicevi, C. nearctica, C. kurnakovi, C. alternata, C. nigra, C. khubsugulica*;

2) "*vidua*" group (Figs 3-4), holds two species - *Capnia vidua* and *C. potikhae*; it is similar to the first group, but characterized by a longitudinally divided epiproct and the presence of a pair of tubercles on tergite 6, and an single tubercle on tergite 7;

3) "*japonica*" group (Figs 5-6), characterized by the cylindrical male epiproct which is not longitudinally divided, and the presence of a conical tubercle on male tergite 7; the *japonica* group holds two species of the fauna of the Russian Far East (*Capnia sidimiensis, C. iturupiensis*), and several Japanese species (*C. japonica* and some other);

4) "*bifrons*" group (Figs 7-8), characterized by arch-shaped curved epiproct of equal width over all its length, and not divided longitudinally; tergite 9 of male with unpaired tubercle. I include in this group *Capnia bifrons, C. sevanica, C. tuberculata, C. turkestanica;*

5) "*cordata*" group (Figs 9-10), characterized by elongated longitudinally undivided epiproct, raised hind margin of tergite 9, presence of vesicle on male sternite 9, special shape of female subgenital plate, narrowed posteriorly, with unpaired lobe of hind margin. I include in this group *Capnia prolongata, C. badakhshanica* and Himalayan species (*C. cordata, C. montana, C. tibetana, C. hingstoni* and possibly, *C. manii, C. montivaga*). Provisionally, I include here *C. ansobiensis* and *C. shugnanica*, known as females only. The membership in the group will be more definitely assessed after the description of males.

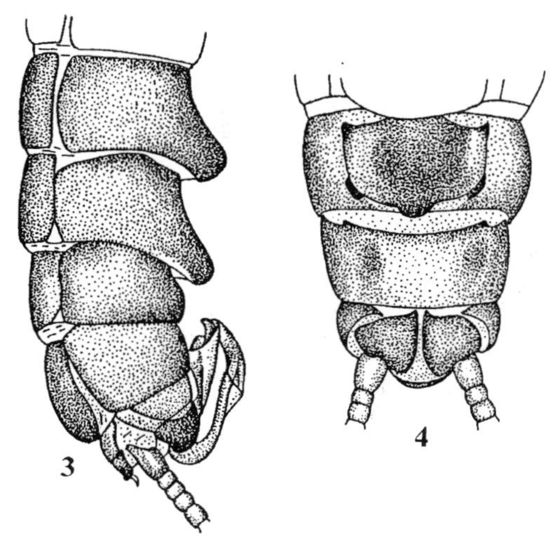

Figs. 3-4. *Capnia vidua* Klap. (the "*vidua*" group): 3 - male abdomen, laterally; 4 -female abdomen, ventrally.

6) "*pedestris*" group (Figs 11-13) established by Zwick and Sivec (1980) for several Himalayan and Middle Asian species. Main characters of the group are: tergite 9 of male with two elongate processes; vesicle on male sternite 9 lacking in most species; males are mostly apterous; sternite 8 of female complicatedly modified. I include in this group *Capnia pedestris, C. bimaculata, C. bicuspidata, C. longicauda, C. singularis, C. hamifera, C. jankowskajae*, also the Himalayan *C. bifida* and *C. femina*. Zwick and Sivec (1980) include here also Caucasian *C. ahrensi*. However this species is morphologically and geographically isolated. I consider its inclusion in the group as preliminary.

Relations of a number of species are not yet clear and I cannot include them in any of the above groups. These species are: *Capnia levanidovae, C. kolymensis, C. lepnevae, C. bargusinica, C. rara* and *C. tshukotica;* the latter is known only as female. Possibly, each of these species is the representative of a separate group.

The remaining genera are represented in the fauna of Russia and adjacent territories by small numbers of species and cannot be divided into groups within each genus. Among the palaearctic species of *Isocapnia, I. kudia* is morphologically isolated, differing from other species by the male epiproct shape and bigger size of body.

Peculiarities of the geographical distribution of the genera and species of Capniidae are of interest (Table 1). All the examined genera are divided into six groups according to their geographical distribution:

1) Holarctic group: *Capnia*. This genus is widely distributed in the European part of Russia (6 species), in the Asian part of Russia (8 species), in the Caucasus (4 species) and in Middle Asia (12 species). Only 3 species of *Capnia* are transpalaearctic, the rest are limited to smaller regions.

2) Nearctic-East-Palaearctic group: *Paracapnia, Isocapnia, Eucapnopsis, Mesocapnia*;

3) West-Palaearctic group: *Capnopsis, Capnioneura*; the monotypic genus *Capnopsis* is represented by a separate subspecies in the European part of Russia and in the Caucasus; *Capnioneura* is represented by one endemic species in the Caucasus;

4) East-Palaearctic group: *Capniella*;

5) Palaearchearctic[1] (Manchuro-Chinese) group: *Takagripopteryx*;

---

[1] Term by Semenov-Tian-Shanskij (1936)

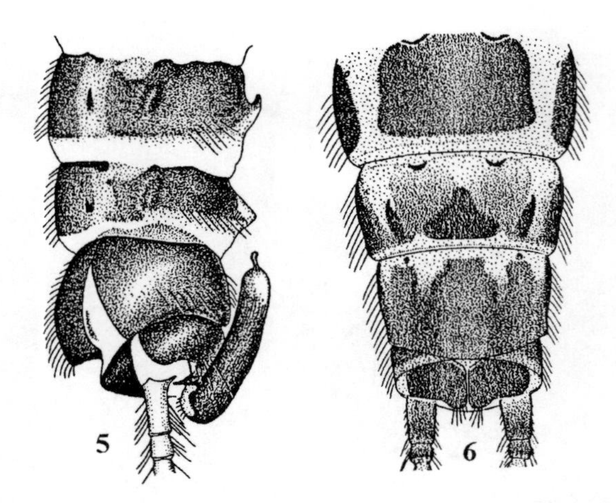

Figs. 5-6. *Capnia sidimiensis* Zhiltz. (the "*japonica*" group): 5 - male abdomen, laterally; 6- female abdomen, ventrally.

6) Endemic of the lake Baikal: *Baikaloperla*.

The faunas of the European and Asian part of Russia are quite different in generic composition. Mainly two genera - *Capnia and Capnopsis* - are represented in the European part. One more genus - *Mesocapnia*-is distributed in the Asian part but reaches the north of the European part by the border of distribution of one single species only, *Mesocapnia variabilis*.

In the Asian part of Russia, 8 genera of Capniidae are represented. Among them, 5 genera are shared with the Nearctic: *Capnia, Mesocapnia, Paracapnia, Isocapnia, Eucapnopsis*. The genus *Capniella* is restricted to Siberia and the Russian Far East; *Takagripopteryx* is an island genus. *Baikaloperla* is endemic of Lake Baikal. Thus only 3 genera are common to the faunas of the European and Asian parts of Russia; the fauna of the Asian part is richer in both genera (8) and species (40).

With regard to geographical distribution, species of Capniidae are divided into two complexes: Holarctic (3 species) and Palaearctic (57 species). The holarctic complex includes two groups:

1) Circumpolar: *Mesocapnia variabilis*;

1) Siberian – North American (*Eucapnopsis brevicauda, Capnia nearctica*).

The Palaearctic complex is divided into 13 groups:

1) Transpalaeartic group (3 species): *Capnia nigra, C. atra, C. pygmaea;*

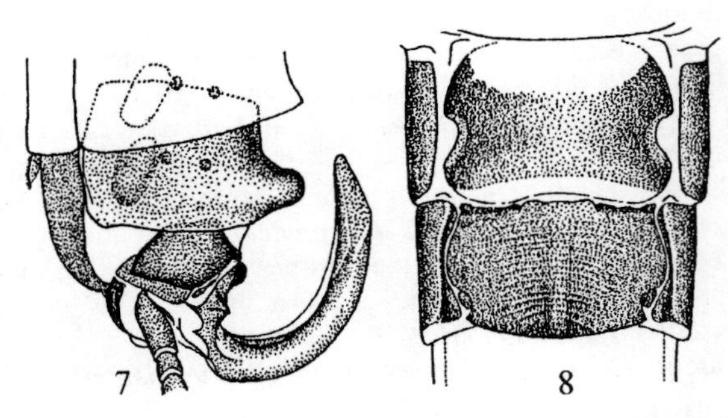

Figs. 7-8. *Capnia tuberculata* Zhiltz. (the "*bifrons*" group); 7 - male abdomen, laterally; 8 - female abdomen, ventrally.

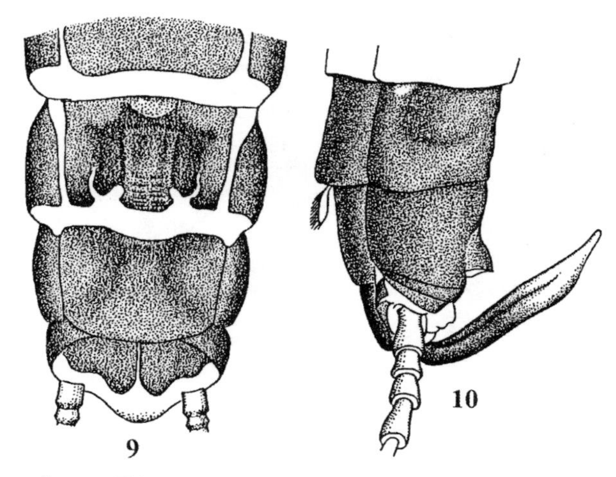

Figs. 9-10. *Capnia prolongata* Zhiltz. (the "*cordata*" group); 9 - female abdomen, ventrally; 10 - male abdomen, laterally.

2) Arctic Euro-Asian (1 species): *Capnia zaicevi*;
3) West-Palaearctic (3 species): *Capnia bifrons, C. vidua, Capnopsis schilleri*;
4) Caucasian-Asia Minor (3 species): *Capnia ahrensi, C. sevanica, C. tuberculata*;
5) Caucasian endemics (1 species and 1 subspecies): *Capnioneura caucasica, Capnopsis schilleri archaica*;

Table 1. Zoogeographic groups of the family Capniidae (Plecoptera) of Russia and adjacent Territories. Numbers indicate number of species in corresponding group.

Genera	Circumpolar	Siberian-North American	Transpalaearctic	Arctic Euro-Asian	West-Palaearctic	Caucasian-Asian Minor	Caucasian Endemics	Central Asian	Middle Asian Endemics	East-Palaearctic	Endemics of Sajan and Altai	Okhotskian-Amurian	Endemics of North-East Asia	Baikalian Endemics	Palaearchearctic
Capnia		1	3	1	2	3			12	5	1		4		3
Mesocapnia	1										1		1		
Capnopsis					1		1								
Paracapnia															3
Eucapnopsis		1						1							
Capnioneura							1								
Isocapnia								1		3		2			
Baikaloperla														2	
Takagripopteryx															3
Capniella										1	1				1
Total species	1		3	1	3	3	2	2	12	9	3	2	5	2	10

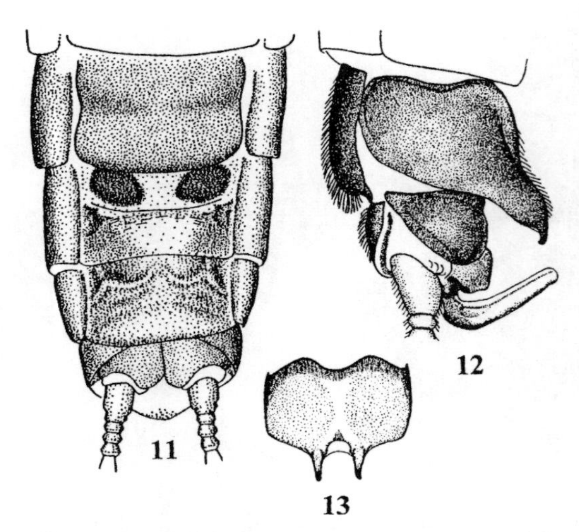

Figs. 11-13. *Capnia bimaculata* Zhiltz. (the "*pedestris*" group): 11 - female abdomen, ventrally; 12 - male abdomen, laterally; 13 - tergite 9 of male abdomen, dorsally.

6) Central Asian (2 species): *Capnia pedestris, Eucapnopsis stigmatica transversa*;

7) Endemics of Middle Asia (12 species): *Capnia turkestanica, C. prolongata, C. badakhshanica, C. bimaculata, C. bicuspidata, C. longicauda, C. singularis, C. hamifera, C. jankowskajae, C. ansobiensis, C. shugnanica, Isocapnia aptera;*

8) Widespread East-Palaearctic (9 species): *Capnia aligera, C. ahngeri, C. bargusinica, C. lepnevae, C. rara, Isocapnia guentheri, I. kudia, I. sibirica, Capniella nodosa;*

Table 2. List of Stoneflies of the Family Capniidae of Russia and Adjacent Territories

Capnia atra Mort.
C. ahngeri Kop.
C. aligera Zap.-Dulk.
C. pygmaea Zett.
C. zaicevi Klap.
C. nearctica Banks
C. kurnakovi Zhiltz.
C. alternata Zap.- Dulk.
C. nigra (Pict.)
C. khubsugulica Zhiltz. et Varikh.
C. vidua Klap.
C. potikhae Zhiltz.
C. sidimiensis Zhiltz.
C. iturupiensis Zhiltz.
C. bifrons (Newm.)
C. sevanica Zhiltz.
C. tuberculata Zhiltz.
C. turkestanica Kimm.
C. prolongata Zhiltz.
C. badakhshanica Zhiltz.
C. shugnanica Zhiltz.
C. ansobiensis Zhiltz.
C. pedestris Kimm.
C. bimaculata Zhiltz.
C. longicauda Zhiltz.
C. singularis Zhiltz.
C. hamifera Zhiltz.
C. bicuspidata Zhiltz.
C. jankowskajae Zhiltz.
C. ahrensi Zhiltz.
C. levanidovae Kaw.
C. kolymensis Zhiltz.
C. lepnevae Zap.-Dulk.
C. bargusinica Zap.-Dulk.
C. rara Zap.-Dulk.
C. tshukotica Zhiltz. et Lev.
Mesocapnia altaica (Zap.-Dulk.)
M. gorodkovi Zhiltz. et Baum.
M. variabilis (Klap.)
Capnopsis schilleri schilleri (Rostock)
Cs. schilleri archaica Zw.
Paracapnia khorensis Zhiltz.
P. sikhotensis Zhiltz.
P. recta Zhiltz.
Eucapnopsis stigmatica transversa Aub.
E. brevicauda (Claass.)
Capnioneura caucasica Zhiltz.
Isocapnia sibirica (Zap.-Dulk.)
I. orientalis Zhiltz.
I. guentheri (Joost)
I. arcuata Zhiltz.
I. kudia Ricker
I. aptera Zhiltz.
Baikaloperla elongata Zap.-Dulk. *et* Zhiltz.
B. kozhovi Zap.-Dulk. *et* Zhiltz.
Takagripopteryx nigra Okam.
T. imamurai Kohno
T. zhuikovae Zhiltz.
Capniella nodosa Klap.
Ca. ghilarovi Zhiltz.
Ca. endemica (Zap.-Dulk.), comb. nov.

9) Endemics of Sajan and Altai (3 species) : *Capnia alternata, Capniella endemica, Mesocapnia altaica;*

10) Okhotskian-Amurian (2 species): *Isocapnia arcuata, I. orientalis*;

11) Endemics of North-East Asia (5 species): *Capnia kurnakovi, C. kolymensis, C. levanidovae, C. tshukotica, Mesocapnia gorodkovi*;

12) Endemics of the Lake Baikal (2 species): *Baikaloperla elongata, B.kozhovi*;

13) Palaearchearctic (Manchuro-Chinese) group (10 species); among them mainland species (6): *Paracapnia khorensis, P. sikhotensis, P. recta, Capnia sidimiensis, C. potikhae, Capniella ghilarovi*; and island species (4): *Capnia iturupiensis, Takagripopteryx imamurai, T. nigra, T. zhuikovae.*

On the whole, representatives of four groups prevail in the fauna of Russia and adjacent territories, namely East-Palaearctic (9 species), Palaearchearctic (10), Central - and Middle-Asian (14), endemics of the North-East (5); they represent more then half of the species. It is characteristic that the number of widespread species is small: 1 Holarctic, 2 Siberian-North American and 3 Transpalaearctic. The percentage of local endemics is rather high, especially in mountain regions (more then 1/3 of species). The mountain regions of Middle Asia are the richest in endemic species.

REFERENCES

Semenov-Tian-Shankij, A. P. Boundaries and zoogeographical divisions of the Palaearctic Region for terrestrial animals of the basis of the geographical distribution of Coleoptera. Moscov-Leningrad, 1936. 16 p. [in Russian.]

Zwick, P., Sivec I. Beitrage zur Kenntnis der Plecoptera des Himalaja. Ent. Basil., 5, 1980: 59-138.

FINE STRUCTURE OF THE MALPIGHIAN TUBULES OF MAYFLY NYMPHS, *BAETIS RHODANI* AND *ECDYONURUS VENOSUS* (EPHEMEROPTERA)

Elda Gaino and Manuela Rebora

Dipartimento di Biologia Animale ed Ecologia
Via Elce di Sotto-06123 Perugia, Italy

ABSTRACT

The Malpighian tubules (MT) of *Baetis rhodani* and *Ecdyonurus venosus* were studied by light, scanning and transmission electron microscopy. Special attention was paid to the ultrastructural organization of the constitutive cells, which in the secretory regions typically have a microvillated luminal border. Cells rest on a basal lamina, show varying degrees of basal infoldings and a large number of mitochondria. Cells of the collecting ducts show vesicled cytoplasm and have a sponge-like appearance. In *B. rhodani* MT empty individually into the alimentary canal whereas in *E. venosus* they enter through intermediate trunks. Numerous MT proper are attached to each common trunk, mainly in its distal portion. In this species MT are more complex owing to regional specializations. The distal region coils to form an asymmetrical plate, to one side of which the narrow collecting duct adheres before emerging right in the middle of the opposite side. Each duct extends to a certain distance and ends in the trunk. Different cell types are present in the trunk: a ring of cells bordering the duct/trunk junction, cells with apical infoldings and a luminal cuticle intima, and cells whose long thin microvilli almost fill the cavity. The relationship between structural features and known physiological function are discussed.

INTRODUCTION

The comparative anatomy and arrangement of the Malpighian tubules (MT) of Ephemeroptera larvae have been the object of intense scrutiny by Landa (1969), Landa *et al.* (1980), Landa and Soldán (1985), on account of the phylogenetic relevance of these structures. According to these authors, in primitive Ephemeroptera the plesiomorphic arrangement of the MT results from a large number of tubules attached individually to the alimentary canal. The further acquisition is represented by the intermediate stage of buds and their ensuing elongation to form common trunks. Trunks are variable in number and shape, and show a general tendency to reduce the number of their pairs. Numerous interstages characterize mayfly MT system, as does the differentiation of the tubules from straight units to the acquisition of dis-

Trends in Research in Ephemeroptera and Plecoptera
Edited by E. Dominguez, Kluwer Academic/Plenum Publishers, 2001

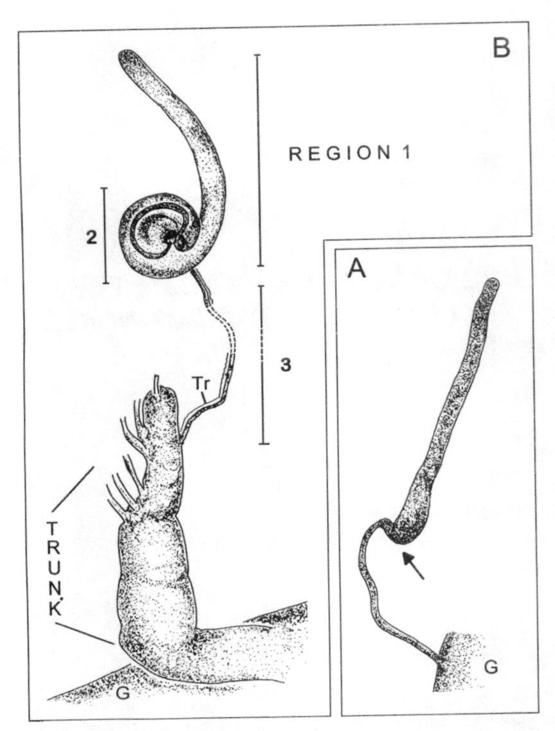

Fig. 1. Sketch of a distinct unit of the Malpighian tubules in *Baetis rhodani* (A) and *Ecdyonurus venosus* (B). In *B. rhodani* the distal portion curves (arrow) at the junction with the proximal portion. Each tubule enters the gut (G) individually. In *E. venosus* each unit achieves a more complex diversification and opens into the gut (G) through an intermediate trunk. Tr= tracheal supply.

coidal plates in the distal region. In short, two main features have to be considered in the anagenesis of the MT: the shape of each tubule and its emptying into the alimentary canal.

Some information on the morphological and histological organization of MT has been provided for several Italian mayfly species by Grandi (1950). The ultrastructural investigation of Nicholls (1983) on the heptageniid *Ecdyonurus dispar* showed a relationship between regional specialization to the function of cells in each tract.

The present paper describes the ultrastructural organization of Malpighian tubules in the nymphs of *Baetis rhodani* and *Ecdyonurus venosus*.

MATERIAL AND METHODS

Mature nymphs of *Baetis rhodani* and *Ecdyonurus venosus* were collected from the Lemme stream in Alessandria, Piedmont, Italy. The gut and Malpighian tubules were dissected out under a stereo-microscope. The arrangement of the MT was observed by light and scanning electron microscopy (SEM).

Dissected material was fixed in Karnovsky's medium (1965) using cacodylate buffer, pH 7.2, for 1 hour, rinsed several times in the same buffer and postfixed in 1% osmium tetroxide for 1 hour.

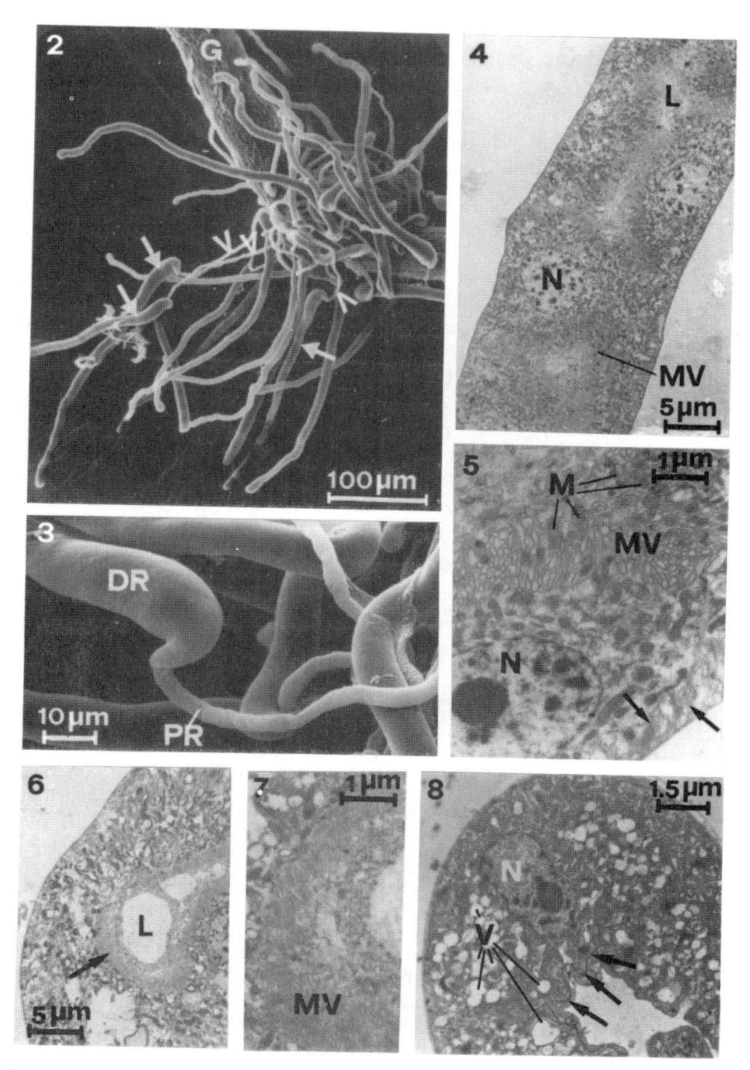

Figs. 2-8. Malpighian tubules (MT) of the nymph of *Baetis rhodani* under SEM (2,3) and TEM (4-8). 2- General view of the MT showing their arrangement at their junction with the gut (G). Note the broad (arrows) and the thin (open arrowheads) portions; 3- Detail of a single MT showing the junction between the broad distal region (DR) and the thin proximal one (PR); 4- Closely aligned cells making up the distal region of the MT. L= lumen. MV= microvilli. N= nucleus; 5- Detail of a cell with infoldings (arrows) and microvilli (MV) that include mitochondria (M). N= nucleus; 6- Cross-section of the tubule in the curved region close to the junction with the thin collecting duct. Note the lumen (L) with the microvillated border (arrow); 7- Zoomed view of the microvillated luminal border. Note the absence of mitochondria inside microvilli (MV); 8- Cross-section of the thin collecting duct with cells full of vesicles (V). N= nucleus. Note the interdigitations of the cells joined by extensive junctions (arrows).

For SEM, the specimens were dehydrated in a graded ethanol series, critical-point dried, mounted on stubs with silver conducting paint and sputter-coated with gold-palladium in a Balzers Union Evaporator. Specimens were observed with a Philips EM 515 scanning electron microscope.

For TEM, ethanol processed tissue was embedded in Epon-Araldite mixture resin. Ultrathin sections, obtained with a Reichert ultramicrotome, were collected on formvar-coated copper grids, stained with uranyl acetate and lead citrate. Sections were examined under a Philips EM 400 electron microscope.

Immuno-labelling techniques were used to visualize both cytoskeletal actin and nuclei. F-actin was detected by a phalloidin-staining procedure. This consisted of treatment for ten minutes in a mixture of fixative (Triton 0.1%), after fixation of the tissue in a 4% solution of paraformaldehyde for 30 minutes. Afterwards, the material was washed in PBS buffer and stained with phalloidin-rhodamine in a dark oven at 37 °C for 30 minutes.

The fluorescence of the nuclei was detected with 2.8 mM 4', 6-diamino-2-phenylindole 2HCl (DAPI, Sigma), a DNA specific fluorochrome, dissolved in Tris buffer (stock solution) to a final saturating concentration. After a final wash in PBS, differently treated MT were mounted on slides in 60-70% glycerol in PBS. Various controls were performed. Stained slides were examined under a Leika microscope, equipped for epifluorescence in the rhodamine mode. Photographs were taken with a Kodak Ektacrome 400 ASA colour slide film. The terminology in the paper follows that used by Landa and Soldán (1985).

RESULTS

Whatever the organization, the Malpighian tubule system extends out of the alimentary canal at the transition from the midgut to the hindgut. Each tubule floats in the hemolymph and is invested with tracheae. Two main models are considered (see the sketch of figure 1): (a) tubules attached individually to the alimentary canal (Fig. 1A) and (b) tubules with discoidal plate and common trunk (Fig. 1B).

Tubules Attached Individually to the Alimentary Canal

In *Baetis rhodani*, tubules are arranged around the alimentary canal (Fig. 2). Each tubule consists of two regions: a distal portion and a thin proximal one (Fig. 1A). At the junction between the two, the distal portion enlarges and curves (Fig. 3) leading into the proximal portion, or thin collecting duct (Fig. 3).

Ultrathin sections of the distal part show that the epithelial wall is made up of microvillated cells on a thin basal lamina facing an eccentric lumen (Fig. 4). The polytene nucleus occupies the expanded part of the cell body. Malpighian cells have a fusiform shape, each of them extends between the neighbouring ones (Fig. 4). The cell apex protrudes towards the lumen and has numerous microvilli filled with long, thin mitochondria (Fig. 5). The region adjacent to the microvilli also includes mitochondria of varying shape and length. At the basal edge of these cells the plasma membrane forms short infoldings (Fig. 5). Towards the junction with the thin collecting duct, the lumen becomes larger (Fig. 6) and the apical microvillated border is devoid of mitochondria (Fig. 7). The thin collecting duct results from cells delimiting a large lumen (Fig. 8). These cells lie on a basal lamina, have few mitochondria and the cytoplasm is filled with vesicles varying in diameters and containing flocculent material. The adjacent cell membranes are greatly interdigitated and joined by extensive junctions (Fig. 8). Unlike the distal portion of the tubule, the cells lack both microvilli and plasma membrane infoldings (Fig. 8).

Tubules with Discoidal Plate and Common Trunk

Ecdyonurus venosus can represent a typical model where trunks are located around the alimentary canal (Fig. 9) and may branch at their bases. Numerous MT enter each trunk, mainly in its distal portion (Fig. 10).

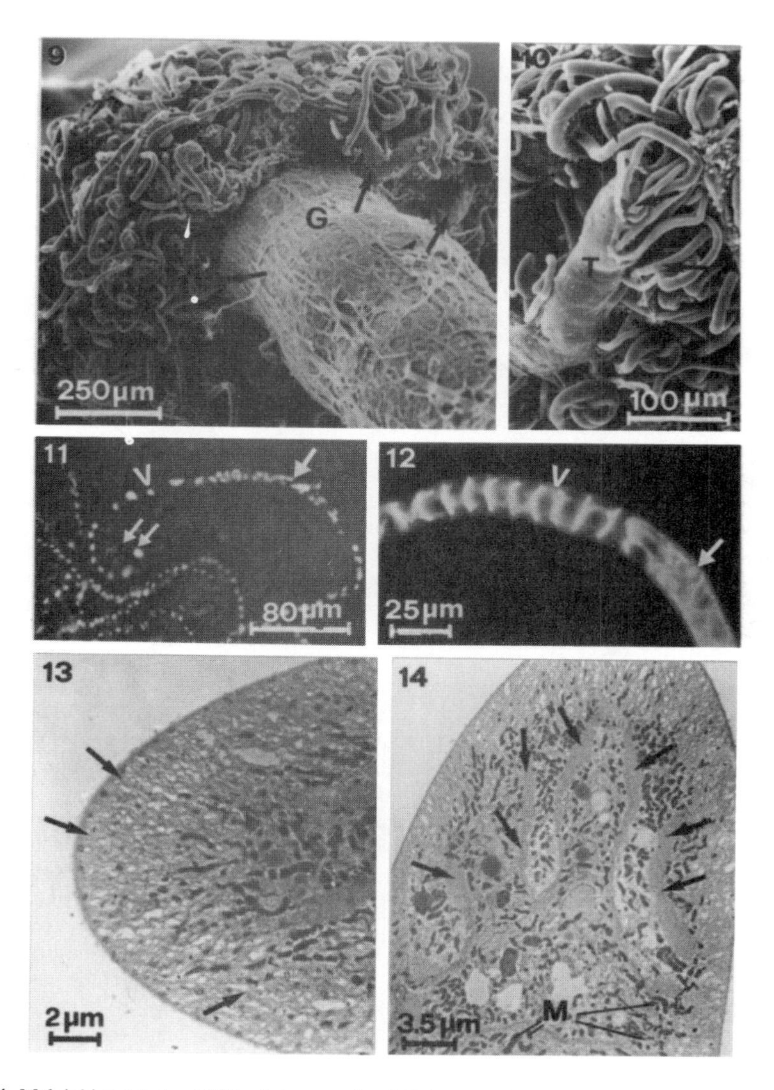

Figs. 9-14. Malpighian tubules (MT) of the nymph of *Ecdyonurus venosus* under SEM (9,10), light microscopy (11-12) and TEM (13-14). 9- General view of the arrangement of MT showing their junction to the gut (G) by the intermediate trunks (arrows); 10- Detail showing the MT (arrows) at their entry to the trunk (T) in its distal portion; 11-MT whole-mount labelled with DAPI staining. The sequence of large polytene nuclei along the length of the MT reveals a distal tract with packed cells (arrow) and a proximal one with extended cells (open arrowhead). Note the coiled region (double arrow); 12-MT whole-mount with F-actin labelling. Note the marked difference between the distal tract (arrow) and the proximal one with actin bundles arranged perpendicularly to the longitudinal MT axis (open arrowhead); 13- Distal blind portion of MT showing the deep infoldings (arrows). 14- Distal blind portion of the tubule showing cellular borders with short microvilli (arrows) that almost completely occlude the lumen. M= mitochondria.

Figure 1B shows several histological distinct regions. The primary most distal region (region 1) leads to a next coiled region (region 2), which acquires the shape of a discoidal plate and leads to a thin collecting duct (region 3). Each duct connects a single plate to a common trunk.

The arrangement and number of the cells in a Malpighian tubule is obtained by DAPI technique (see material and method section), which reveals a sequence of large polytene nuclei along the length of a Malpighian tubule (Fig. 11). This technique coupled with the rodamine-labelled phalloidin staining method and detergent extraction (see material and method section), shows that the primary tubule (region 1) includes two tracts of about equal length (Fig. 12). The blind distal end terminates in a solid rod of tightly packed cells that, after staining for cytoskeletal filamentous actin with rhodamine labelled phalloidine, show a diffuse fluorescence (Fig. 12). This tract leads abruptly into the next tract, whose cells are more elongated and show periphery actin bundles. These bundles are oriented perpendicularly along the axis of the MT (Fig. 12). TEM observations confirm a distinct organization of the two tracts. In the blind distal end, cells rest on a thin basal lamina and show some elaboration of the infoldings (Figs. 13,14). Luminal cells have a border of short microvilli that almost completely fill the lumen (Fig. 14). The majority of mitochondria tend to gather along the luminal side of the cells (Fig. 14).

At a point where the distal end meets the next tract, the lumen is not occluded and mitochondria are sparsely distributed in the infoldings (Fig. 15). Nuclei are located near the apical border of the cells. Moving towards the coiled region, basal infoldings are more extended (Fig. 16). They are regularly spaced and oriented at right angles to the basal lamina. Mitochondria accumulate along the deep infoldings (Fig. 16). The lumen widens and the cell apex of the bordering cells shows an extensive microvillar border of uniform length (Fig. 16).

The coiled region (region 2) is the most complex part and gives rise to a discoidal plate (Figs. 17,18). The discoidal plate has two different sides, one showing a thin raised C-shaped tubular structure (Fig. 17) and the other containing a single emerging thin duct (Fig. 18). The latter is usually associated with a trachea (Fig. 18). Such an organization is the result of an abrupt narrowing of the primary tubule at the end of the coil from which the thin collecting duct departs (region 3). The initial part of the collecting duct adheres by a fibrous sheet to one side of the discoidal plate, and assumes a raised C-shaped configuration (Fig. 17). The collecting duct is kept adherent to the plate by a fibrous sheet (Fig. 17). This duct then emerges in the middle from the opposite side of the discoidal plate (Fig. 18) and, after extending to a certain distance, empties into the common trunk. This arrangement of the duct results in an asymmetry on the sides of the plate. The discoidal plate (about 50 μm in diameter) contains several cavities delimited by large cells (about 5 cells). Ultrastructurally, two cell types are recognizable. The cells located on the upper side of the plate just below the C-shaped collecting duct differ from the remaining ones in many respect. The cells in the upper side have remarkably electron-dense cytoplasm filled with mitochondria that penetrate into the infoldings of the basal plasma membrane (Fig. 19). The remaining cells at the lower end have electron-translucent cytoplasm and a few mitochondria. The apical border of both cell types shows facing microvilli that delimit cavity lumina (Fig. 19). The lumen of the collecting duct (region 3) tends to enlarge towards the trunk and is bordered by cells whose cytoplasm accumulates a large number of vesicles in such a way that it acquires a sponge-like appearance (Fig. 20).

Semithin sections show the junction between trunks and alimentary canal at the transition from the midgut to the hindgut (Fig. 21). Three different cell types are present in the common trunk. In the most distal part, the junction with the collecting duct is bordered by a ring of cells (Fig. 22). The rest of the trunk is constituted of two very distinct cell types delimiting an internal cavity. Type-1 cells are elongated and show deep infoldings along their basal region. They exhibit a more electron-dense cytoplasm around the nucleus whereas the apical region bordering the cavity greatly enlarges to form a lamina from which long microvilli originate (Fig. 23). Microvilli are numerous and form a network in the lumen (Figs. 23,24). Type-2 cells are larger than the previous ones and have apical plasma membrane folds and are bordered by a cuticle intima (Figs. 23,24). The microvilli of the type-1 cells

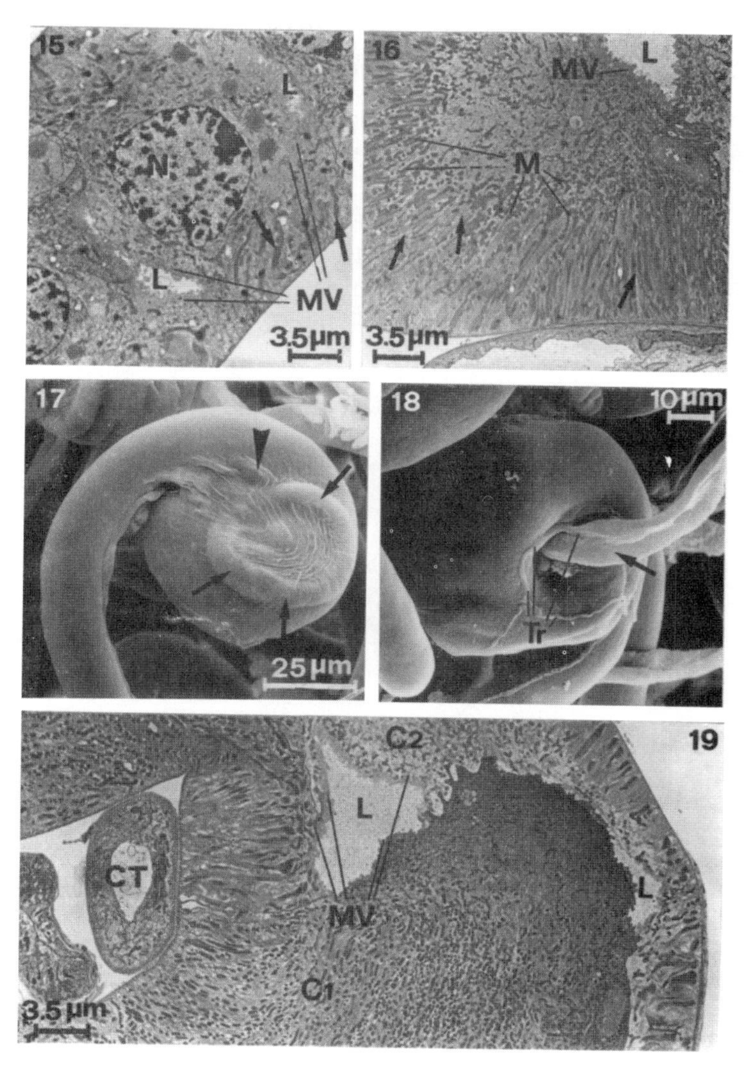

Figs. 15-19. Malpighian tubules (MT) of the nymph of *Ecdyonurus venosus* under TEM (15,16,19) and SEM (17,18). 15- Longitudinal section of the MT at the transition between region 1 and 2. Note the lumen (L) between microvillated cells (MV) and the mitochondria in the infoldings (arrows). N= nucleus; 16- Cross-section of the tubule close to the coiled region whose cells have very extended basal infoldings (arrows) rich in mitochondria (M). L= lumen bordered by microvilli (MV); 17- Side of the discoidal plate showing the thin raised C-shaped arrangement of the thin collecting duct (arrows). The fibrous sheet (arrowhead) covers the duct; 18- Side of the discoidal plate showing the centrally emerging duct (arrow). Note the associated trachea (Tr); 19- Cross-section of the coiled region showing electron-dense (C1) and electron-translucent (C2) cells whose facing microvilli (MV) delimit cavity lumina (L). CT= section of the thin collecting duct.

may reach the cuticular border of the type-2 cells (Fig. 24). The distribution of the type-1 and type-2 cells may vary along the length of the trunk.

Phalloidin-labelling shows spiral muscle arising from and continuous with the muscle layer of the gut.

DISCUSSION

Physiology and ultrastructure of the Malpighian tubules have been investigated in a wide range of insect species (Wigglesworth and Salpeter, 1962; Berridge and Oschman, 1969; Sohal, 1974; Green, 1979, 1980; Ryerse, 1979; Meyran, 1982; Alkassis and Schoeller-Raccaud, 1984; Satmary and Bradley, 1984; Bradley, 1985; Hazelton *et al.*, 1988; Kukel and Komnick, 1989; Dallai *et al.*, 1991; Kapoor, 1994).

Malpighian tubules are regarded as typical transporting epithelia. They are involved in secretion and reabsorption processes which in some instances occur along successive segments (Maddrell, 1977) and lead to the formation of the primary urine.

Variations between species and between segments have emerged from comparative studies of cell structure, organizational details, membrane specialization and mitochondrial populations of the constitutive elements (Wigglesworth and Salpeter, 1962; Eichelberg and Wessing, 1975; Wall *et al.*, 1975). In addition, the ultrastructural features may yield information on the activity of the various segments. For instance, the preferential location of mitochondria at the apical or basal cellular pole seems to express a different function of the cells along the tubules (Meyran, 1982). In this regard, mitochondria recruitment from cell cytoplasm and insertion into the microvilli has been experimentally obtained by stimulating a rapid fluid transport (Bradley and Satir, 1977). Likewise, the extension of basal infoldings expresses the cell's ability to create concentration gradients (Berridge and Oschman, 1969; Oschman and Berridge, 1971), and the number of these invaginations seems to be correlated to the activity of the tubule.

MT cell membranes are involved in ion transport (see review in Nicolson, 1993); microvilli and basal infolds are the localization of the ion pump on which fluid secretion depends (Maddrell, 1977; Pannabecker *et al.*, 1992).

In spite of the different physiological activity performed by MT in the various insect species examined, there is a surprising morphological similarity in the primary cells: extensive membrane infoldings along their basal region, microvillar apical border and a remarkable number of mitochondria. The present investigation visualise that in their basic morphology the cells of the distal part of the tubules of *B. rhodani* and *E. venosus* conform to this general description. Other well-known features in insects, such as polyploidization and regional differentiation, were highlighted in *E.venosus* by MT whole-mounts utilising DAPI and F-actin labelling techniques respectively. Cytoskeletal actin patterns provided a clear insight into the differentiation of the successive segments in region 1 of *E. venosus* MT, thus confirming the close relationship between actin filament arrangement and cell shape (Kukel and Komnick, 1989; Meulemans and De Loof, 1990,1992).

Extreme regional specializations have been observed in *Ecdyonurus dispar* by Nicholls (1983), who suggested that in the distal part of the MT the fluid flow takes place via the paracellular route at the interspace between lateral cells membranes. Our TEM images of *E. venosus* do not show similar specialization, thus highlighting the controversy regarding fluid movement via this route (Diamond, 1979; Møllgard and Rostegaard, 1981).

Like *E. dispar*, the MT of *E. venosus* are divided into distinct regions and, in comparison with those of *B. rhodani*, they have a more complex organization, as proved by the ultrastructural investigation. Even though in both species the general morphological features of the cells in the solid terminal segment of the MT differ from those of the luminated lower segment, there is no doubt that the model of *B. rhodani* is far simpler. Indeed, in this latter species, the distal portion of each MT curves at its junction with the proximal duct, without however reaching the regional specialization of the discoidal plate of

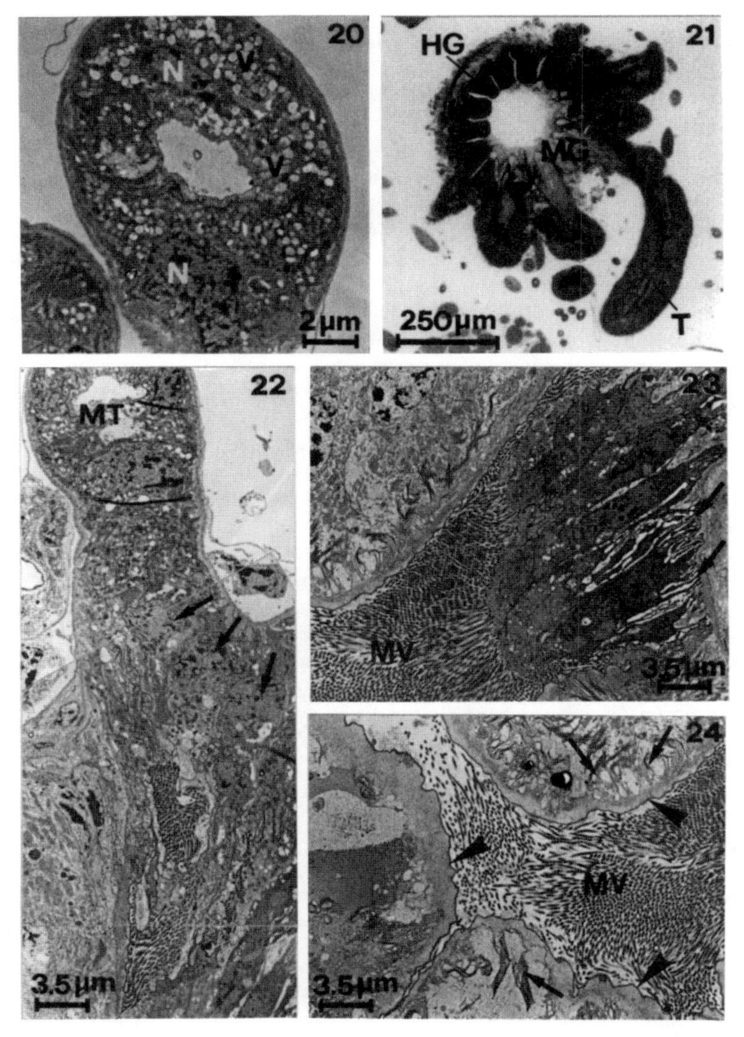

Figs. 20-24. Malpighian tubules (MT) of the nymph of *Ecdyonurus venosus* under TEM (Figs. 20, 22-24) and light microscope (Fig. 21). 20- Cross-section of a collecting duct bordered by cells full of vesicles (V). N= nucleus; 21- Semi-thin section showing the transition from the midgut (MG) to the hindgut (HG). T= trunk; 22- MT proper at its junction with a trunk. Note a ring of cells (arrows); 23- Elongated type-1 cells of the trunk showing deep infoldings along the basal region (arrows). Note the electron-dense cytoplasm around the nucleus and their long microvilli (MV); 24- Type-2 cells with folded apical plasma membrane (arrows) and cuticle intima (arrowheads). MV= microvilli of type-1 cells.

E. venosus. In the discoidal plate, two different cell types are easily recognizable on the basis of their morphology. The reason for such a difference is so far obscure, as is the occurrence of the thin collecting duct adherent to one side of the plate, and must await a more detailed physiological study. We can speculate that the dominant feature consisting of electron-dense cytoplasm, deep basal membrane infoldings and associated mitochondria may reflect the involvement of this cell type in the removal of ions from the primary urine, thus recalling special structures supporting this function in other aquatic insects (Wall and Oschman, 1975). Indeed, regional specializations of the tubules have been observed in various insect groups and correlated to a variety of reabsorptive processes able to modify the primary urine (Irvine, 1969; Maddrell and Phillips, 1975; Wall *et al*., 1975).

The thin collecting duct of the MT of *B. rhodani* and *E. venosus* appears similar in both species and its organization is coherent with the function of a simple conducting device. However, the constitutive cells, characterized by a rich amount of vesicles with flocculent material, recall mucosubstances described in the cells of MT collecting ducts of a dragonfly (Kukel and Komnick, 1989). We cannot exclude that these vesicles might have included some components dissolved by the fixation procedure. Indeed, it is rather surprising that the mineral concretions normally found in the MT of insects were not evidenced inside cells, nor did we encounter profiles indicating extrusion of such concretions into the lumen.

The most striking feature of *E. venosus* is the presence of the common trunk into which collecting ducts open separately. The trunk is more than a simple bridge between MT proper and the hindgut, as it consists of different cell types, as pointed out in *E. dispar* by Nicholls (1983). In particular, the cells with long microvilli and deep basal infoldings may be extensively modified midgut elements. The cells with the apical cuticular intima show the general profile of the hindgut cells. As a consequence, it seems acceptable that cells from different stem lines cooperate in building up the trunk. The function of these different cell types is obscure. We can speculate that mayflies need to eliminate excess fluid from the hemolymph owing to a continuous dilution in the fresh-water environment. These microvillated cells may compensate for loss of ions as is presumed to occur in the discoidal plate. An active ion uptake together with a reabsorption of substances by the cells bordered by the cuticle may allow the insect to adjust its hemolymph content, thereby recalling the hindgut-cell activity.

In conclusion, as emerges from the present study, the MT of *B. rhodani* and *E. venosus* express two different levels of complexity and may be extreme models among numerous interstages of increasing diversification.

REFERENCES

Alkassis, W. and J. Schoeller-Raccaud. 1984. Ultrastructure of the Malpighian tubules of blow fly larva, *Calliphora erythrocephala* Meigen (Diptera: Calliphoridae). Int. J. Insect Morphol. Embryol. 13: 215-231.

Berridge, M. J. and J. L. Oschman. 1969. A structural basis for fluid secretion by Malpighian tubules. Tiss. Cell 1: 247-272.

Bradley, T. J. 1985. The excretory system: structure and physiology, pp. 421-465. In: G. A. Kerkut and L. I. Gilbert (eds.). Comprehensive Insect Physiology, Biochemistry and Pharmacology. Pergamon Press, Oxford.

Bradley, T. J. and P. Satir. 1977. Microvillar beating and mitochondrial migration in Malpighian tubules. J. Cell Biol. 75: 255a.

Dallai, R., G. Del Bene and D. Marchini. 1991. The ultrastructure of Malpighian tubules and hindgut of *Frankliniella occidentalis* (Pergande) (Thysanoptera: Thripidae). Int. J. Insect Morphol. Embryol. 20: 223-233.

Diamond, J. M. 1979. Osmotic water flow in leaky epithelia. J. Memb. Biol. 51: 195-216.

Eichelberg, D. and A. Wessing. 1975. Morphology of the Malpighian tubules in insects. Fortschr. Zool. 23: 124-147.

Grandi, M. 1950. Contributi allo studio degli "Efemeroidei" italiani. XIV. Morfologia ed istologia dell'apparato digerente degli stadi preimmaginali, subimmaginali ed immaginali di vari generi e specie. Boll. Ist. Ent. Univ. Bologna 18: 58-92.
Green, L. F. B. 1979. Regional specialization in the Malpighian tubules of the New Zealand glow-worm *Arachnocampa luminosa* (Diptera: Mycetophilidae). The structure and function of type I and II cells. Tiss. Cell 11: 673-702.
Green, L. F. B. 1980. Cryptonephric Malpighian tubules system in a dipteran larva, the New Zealand glow-worm, *Arachnocampa luminosa* (Diptera: Mycetophilidae): A structural study. Tiss. Cell 12: 141-151.
Hazelton, S. R., S. W. Parker and J. H. Spring. 1988. Excretion in the house cricket (*Acheta domesticus*): Fine structure of the Malpighian tubules. Tiss. Cell 20: 443-460.
Irvine, H. B. 1969. Sodium and potassium secretion by isolated insect Malpighian tubules. Amer. J. Physiol. 217: 1520-1527.
Kapoor, N. N. 1994. A study on the Malpighian tubules of the plecopteran nymph *Paragnetina media* (Walker) (Plecoptera: Perlidae) by light, scanning electron, and transmission electron microscopy. Can. J. Zool. 72: 1566-1575.
Karnovsky, M. S. 1965. A formaldehyde-glutaraldehyde fixative of high osmolality for use in electron microscopy. J. Cell Biol. 27: 137-138.
Kukel, S. and H. Komnick. 1989. Development, cytology, lipid storage and motility of the Malpighian tubules of the nymphal dragonfly, *Aeshna cyanea* (Müller) (Odonata: Aeshnidae). Int. J. Insect Morphol. Embryol. 18: 119-134.
Landa, V. 1969. Comparative anatomy of mayfly larvae (Ephemeroptera). Acta ent. Bohemoslov. 66: 289-316.
Landa, V. and T. Soldán. 1985. Phylogeny and higher classification of the order *Ephemeroptera*: a discussion from the comparative anatomical point of view. Academia Nak. Cesk., pp.1-121.
Landa, V., T. Soldán and W. L. Peters. 1980. Comparative anatomy of larvae of the family Leptophlebiidae (Ephemeroptera) based on ventral nerve cord, alimentary canal, malpighian tubules, gonads and tracheal system. Acta ent. Bohemoslov. 77: 169-195.
Maddrell, S. H. P. 1977. Insect Malpighian tubules, pp. 541-569. In: B. L. Gupta, R. B. Moreton, J. L. Oscheman and B. J. Wall (eds.). Transport of Ions and Water in Animals, Academic Press, New York, London.
Maddrell, S. H. P. and J. E. Phillips. 1975. Secretion of hypo-osmotic fluid by the lower Malpighian tubules of *Rhodnius prolixus*. J. Exp. Biol. 62: 671-683.
Meulemans, W. and A. De Loof. 1990. Cytoskeletal F-actin patterns in whole-mounted insect Malpighian tubules. Tiss. Cell 22: 283-290.
Meulemans, W. and A. De Loof. 1992. Changes in cytoskeletal actin patterns in the Malpighian tubules of the fleshfly, *Sarcophaga bullata* (Parker) (Diptera: Calliphoridae), during metamorphosis. Int. J. Insect Morphol. Embryol. 21: 1-16.
Meyran, J. C. 1982. Comparative study of the segmental specializations in the Malpighian tubules of *Blattella germanica* (L.) (Dictyoptera: Blatellidae) and *Tenebrio molitor* (L.) (Coleoptera: Tenebrionidae). Int. J. Insect Morphol. Embryol. 11: 79-98.
Møllgard, M. and J. Rostegaard. 1981. Morphological aspects of transepithelial transport with special reference to the endoplasmic reticulum, pp. 209-231. In: S. O. Schulz (ed.). Ion Transport by Epithelia. Raven Press, New York.
Nicholls, S. P. 1983. Ultrastructural evidence for paracellular fluid flown in the Malpighian tubules of a larval mayfly. Tiss. Cell 15: 627-637.
Nicolson, S. W. 1993. The ionic basis of flluid secretion in insect Malpighian tubules: Advances in the last ten years. J. Insect Physiol. 39: 451-458.
Oschman, J. L. and M. J. Berridge, 1971. The structural basis of fluid secretion. Fed. Proc. Fed. Amer. Soc. Exp. Biol. 30: 49-56.
Pannabecker, T. L., D. J. Aneshansley and K. W. Beyenbach. 1992. Unique electrophysiological effects of dinitrophenol in Malpighian tubules. Am. J. Physiol. 263: R609-R614.
Ryerse, J. S. 1979. Developmental changes in Malpighian tubule cell structure. Tiss. Cell 11: 533-551.
Satmary, W. M. and T. J. Bradley. 1984. The distribution of cell types in the Malpighian tubules of *Aedes taeniorhynchus* (Wiedemann) (Diptera: Culicidae). Int. J. Insect Morphol. Embryol. 13: 209-214.
Sohal, R. S. 1974. Fine structure of the Malpighian tubules in the housefly, *Musca domestica*. Tiss. Cell 6: 719-728.
Wall, B. L. and J. L. Oschman. 1975. Structure and function of the rectum in insects. Fortschr. Zool. 23: 193-222.
Wall, B. L., J. L. Oschman and B. A. Schmidt. 1975. Morphology and function of Malpighian tubules and associated structures in the cockroach, *Periplaneta americana*. J. Morphol. 146: 265-306.
Wigglesworth, V. and M. M. Salpeter. 1962. Histology of the Malpighian tubules in *Rhodnius prolixus* Stål (Hemiptera). J. Int. Physiol. 8: 299-307.

ULTRASTRUCTURAL STUDIES ON THE DEVELOPMENT OF THE GREGARINE *ENTEROCYSTIS RACOVITZAI* IN THE GUT OF *BAETIS RHODANI* (EPHEMEROPTERA, BAETIDAE)

Elda Gaino and Manuela Rebora

Dipartimento di Biologia Animale ed Ecologia
Via Elce di Sotto 06123 Perugia, Italy

ABSTRACT

Different developmental phases have been observed in the gregarine *Enterocystis racovitzai* living in *Baetis rhodani* (Insecta: Ephemeroptera): the young gregarine in the midgut cell epithelium; the syzygy in the midgut lumen and the cyst in the terminal portion of the midgut. Gregarines in mature syzygy exhibit characters useful for their identification. Even though the growing phases are difficult to identify, we have tentatively assigned to this species the observed single gamonts and individuals in pairs. Among the ultrastructural features, epicyte modifications and the presence of gametes in formation are described. A synopsis on the occurrence of various species of *Enterocystis* gregarines associated with Ephemeroptera is also reported.

INTRODUCTION

The genus *Enterocystis* was created by Zwetkow (1926) for *E. ensis*, a gregarine parasite of the larval *Caenis*. Since then, various species of gregarines attributed to *Enterocystis* have been collected from several species of Ephemeroptera (see reviews in Geus, 1969; Arvy and Peters, 1973; Peters and Arvy, 1979).

The representatives of the genus *Enterocystis* are characterised by the intracellular development of the young gregarine, the lack of a septum between proto- and deutomerite, and by an early pairing of the gamonts to form syzygy (Desportes, 1966). The lack of a real septum subdividing the gregarine cell has been also observed in *Gamocystis ephemerae*, a feature reported by Rühl (1976) as a valid trait to reconsider the taxonomic position of this species. Desportes (1963), on the basis of some similarities with other gregarines belonging to *Enterocystis*, proposed to attribute this species to this last genus. The doubtful allocation of *G. ephemerae* has been previously pointed out by Codreanu (1940).

So far, gregarines belonging to *Enterocystis* are exclusive to Ephemeroptera. They have been mainly collected from Baetidae, as reported by Desportes (1966), who sketched some distinct morphological features of the various parasitic species which are useful for specific attribution. Among Baetidae, larvae of *Baetis rhodani* constitute the preferential host for several gregarines, such as *E. ensis*, *E. fungoides* and *E. racovitzai* (Desportes, 1963; 1964; and see table I herein).

The ultrastructure of the syzygy of *Enterocystis* is known for *E. fungoides* (Desportes, 1974) and we have recently described some ultrastructural aspects of *E. racovitzai* (Gaino and Rebora, 1998).

The purpose of this study is to present additional data on the ultrastructure of *E. racovitzai* during its development.

MATERIAL AND METHODS

Nymphs of *Baetis rhodani* (Pictet), which were collected in the Lemme stream (Voltaggio, Piedmont-Italy, from 18 October 1996 to 10 April 1998) contained gregarines belonging to *Enterocystis racovitzai* Codreanu, 1940. Among 48 examined insects, gregarine were found in 10 individuals. These gregarines were found both intracellularly and in syzygies in the alimentary canal. A single mayfly with dark wing-pads showed a mature gametocyst in the terminal tract of the midgut.

Gregarines removed from the midgut were observed *in vivo* under both light and interference contrast microscopes.

For ultrastructural investigation, selected material was differently fixed: (a) 1 hour in Karnovsky's medium (1965); (b) 1 hour in glutaraldehyde diluted to 2% in Na-cacodylate buffer (0.2 M). After fixation, specimens were repeatedly rinsed in the same buffer, postfixed in 1% osmium tetroxide for 1 hour at 4°C, and then dehydrated in a graded ethanol series.

For observations under the scanning electron microscope (SEM), the samples were critical-point dried using a CO_2 Pabisch CPD 750 apparatus, mounted on stubs with silver conducting paint, coated with gold-palladium in a Balzers Union Evaporator, and observed under a Philips EM 515 at an accelerating voltage of 18 kv.

For observations under the transmission electron microscope (TEM), the material was embedded through propylene oxide in Epon-Araldite. Thin sections were obtained using a Reichert ultramicrotome, stained with uranyl acetate and lead citrate, and were examined under a Philips EM 400 T.

RESULTS

Frantzius (1848) described the first gregarine, *Zygocystis ephemerae*, in the mayfly *Ephemera vulgata*. The taxonomic position of this gregarine has repeatedly changed as time went by (review in Arvy and Peters, 1973), because the species was then redescribed as *Gamocystis francisci* by Schneider (1882) and changed to *Gamocystis ephemerae* by Labbé (1899, in Codreanu, 1940), to *Enterocystis* (*Gamocystis*) *ephemerae* by Codreanu (1940) and back to *Gamocystis ephemerae* by Geus (1969). A new species was described by Geus (1969), *G. cloeonis*, a parasite of *Cloeon* sp.

A synopsis of the presence of *Enterocystis* gregarines in the gut lumen of mayflies from various sampling areas is reported in Table I. This synopsis does not include other gregarines whose attribution to *Enterocystis* is doubtful (such as *Gamocystis ephemerae*). It emerges that Baetidae during their aquatic life cycle are the main host for the parasites. On occasion, the presence of tentatively identified enterocystids has been reported in the larvae of *Ephemerella*, *Leptophlebia*, *Caenis* and *Siphlonurus* mayflies (Shtein, 1960).

The absence of a species-specific relationship between host and parasite makes a specific attribution difficult at the early stage of gregarine development in the gut lumen, where the individuals join together to form syzygy. Better identification is possible as the

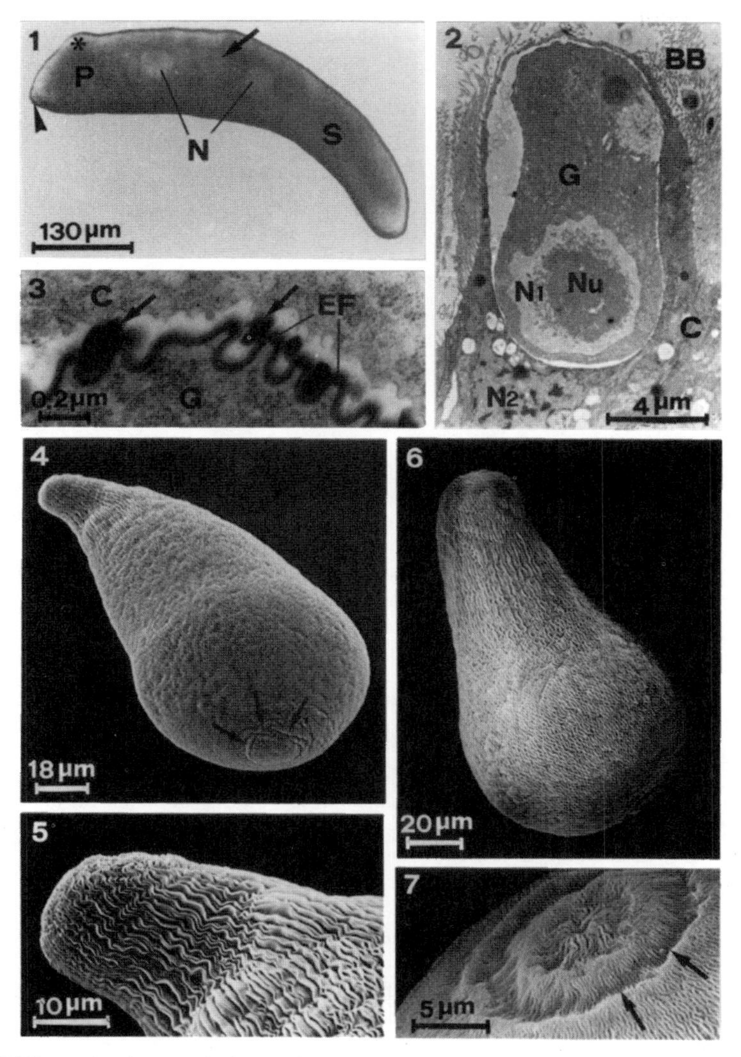

Figs. 1-7. Different developmental phases of *Enterocystis racovitzai* from the midgut of *Baetis rhodani* under interference contrast microscope (1), TEM (2,3), SEM (4-7). 1- Mature syzygy. Note the nucleus (N) of the primite (P) and of the satellite (S), the septum (arrow), the apical protuberance (arrowhead) and the lateral lobe (asterisk); 2- Intracellular young gregarine (G) which protrudes towards the gut lumen. C= midgut cell with its brush border (BB); N_1= nucleus of the gregarine and nucleolus (Nu); N_2= nucleus of the midgut cell; 3- Epicyte folds (EF) of the intracellular young gregarine (G). Note the electron-dense material (arrows) among the epicyte folds. C= midgut cell; 4- Pear-shaped young gregarine with its slightly uplifted apical region (arrows); 5- Magnification of the "tail" of the pear-shaped young gregarine; 6- Pear-shaped gregarine in a more advanced phase of growth; 7- Detail of the apical protuberance of the pear-shaped gregarine. Note the circular deep groove (arrows).

Table 1. *Enterocystis* gregarines in the gut lumen of mayflies from various sampling areas.

Gregarine	Ephemeroptera	Locality	Author
Enterocystis ensis Zwetkow, 1926	*Caenis sp.*	Russia (Peterhof)	Zwetkow, 1926
	Cloeon sp.	Russia	Bobyleva, 1963
	Baetis rhodani	France (East Pyrenees)	Desportes, 1964
E. fungoides Codreanu, 1940	*Baetis rhodani*	Rumania	Codreanu, 1940
	Baetis rhodani	France	Desportes, 1963
	Baetis rhodani	France (East Pyrenees)	Desportes, 1964
	Baetis vernus	Rumania	Codreanu, 1940
	Baetidae	France	Desportes, 1974
E. grassei Desportes, 1963	*Baetis vernus*	France	Desportes, 1963
	Baetis tenax	France	Desportes, 1963
	Heptagenia flava	France	Desportes, 1966
	Ecdyonurus sp.	France	Desportes, 1966
	Epeorus torrentium	France	Desportes, 1966
E. palmata Codreanu, 1940	*Baetis buceratus*	Rumania	Codreanu, 1940
E. racovitzai Codreanu, 1940	*Baetis vernus*	Rumania	Codreanu, 1940
	Baetis rhodani	France	Desportes, 1963
	Baetis rhodani	France (East Pyrenees)	Desportes, 1964
	Baetis rhodani	Italy	Gaino & Rebora, 1998
E. rhithrogenae Codreanu, 1940	*Rhithrogena semicolorata*	Rumania	Codreanu, 1940
Enterocystis sp.	*Baetis rhodani*	Rumania	Codreanu, 1940

maturation proceeds and the paired gregarines differentiate characteristics useful for a proper diagnosis. In this regard, the syzygy of *E. racovitzai* consists of a primite showing a lateral short extension, an apical protuberance, and a satellite (Fig. 1).

The onset of the gregarine development takes place inside the cells of the midgut. The first intracellular phase that can be identified as a gregarine under TEM is represented by an individual that shows a slightly elongated shape and is included into a vacuole (Fig. 2). The presence of the parasite deforms the apical part of the epithelial cell that markedly protrudes towards the gut lumen. The cytoplasm of the gregarine is fairly homogeneous and does not contain special inclusions, and the nucleus is located in the region opposite to the brush border of the host cell. Irregular epicyte folds are evident and electron-dense granules occur between them (Fig. 3).

The youngest gregarine found in the gut lumen before pairing is pear-shaped (Fig. 4). Under SEM its body surface reveals the differentiation of longitudinal epicyte folds that converge into a slightly uplifted apical region. The posterior region extends into a short "tail" along which the epicyte folds are looser than on the rest of the gregarine's surface (Fig. 5). Other pear-shaped gregarines in a more advanced phase of growth (Fig. 6) show modifications of both apical and posterior regions. These changes result in a better definition of the apical uplifted area, consisting of a central protuberance delimited by a deep groove (Fig. 7) and in the disappearance of the tail (Fig. 6).

Gamonts in pairs show a remarkable variability in morphology during the course of their growth in the gut lumen. The gregarines growing in pairs showed a clear dimorphism

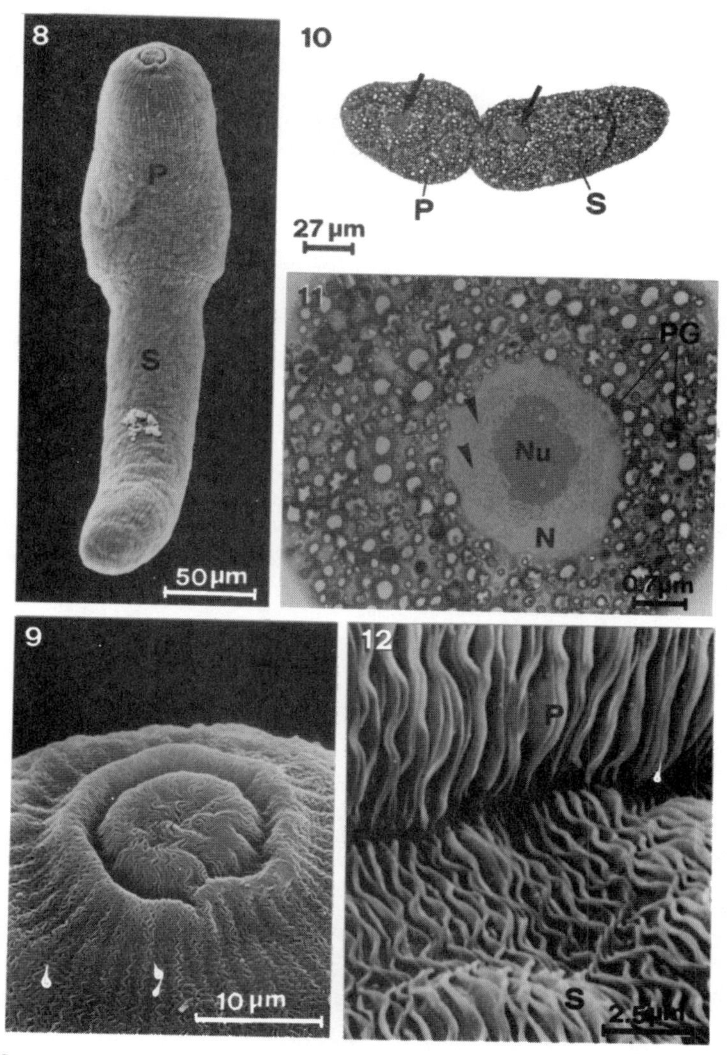

Figs. 8-12. Syzygy of *Enterocystis racovitzai* from the midgut of *Baetis rhodani* under SEM (8,9,12) and TEM (10,11). 8- Syzygy. Note that the primite (P) is shorter than the satellite (S). 9- Detail of the apical uplifted area of the syzygy in the Fig. 8; 10- Syzygy in longitudinal section. The position of the nuclei is indicated by arrows. P= primite; S= satellite; 11- Section of a syzygy showing the nucleus (N) of the primite. Note the nucleolus (Nu) and the scattered filaments (arrowheads); PG= paraglycogen granules; 12- Contact region between primite (P) and satellite (S) showing the different arrangement of the epicyte folds in the early syzygy.

with a primite shorter than the satellite (Fig. 8). Typically, the satellite tends to narrow as the gregarine maturation proceeds (Fig. 8). Concomitantly, the apical uplifted area acquires a sucker-like appearance (Fig. 9).

The position of the nuclei in the primite and in the satellite is clearly visible under TEM in a longitudinal section (Fig. 10). The nucleus of each individual shows an irregular border, and contains a nucleolus with an electron-dense region and scattered filaments (Fig. 11).

In young gregarines, dimorphism is also indicated by a different arrangement of the epicyte folds: looser and taller in the primite than in the satellite (Fig. 12). This feature tends to disappear during syzygy maturation.

Epicyte folds emerge from the thin peripheral ectocyte that bounds the endocyte (Fig. 13). The ectocyte includes vacuoles, electron-dense inclusions and fibrillar components (Fig. 14). The paraglycogen granules, which represent typical storage material, accumulate in the endocyte (Fig. 13).

In cross-sections, the epicytes are delimited by a multilayered border that presents a striated appearance in the apical part of the epicyte folds (Fig. 15). Electron-dense granules are occasionally found adherent to the cytomembranes (Fig. 15) and are usually interposed between consecutive folds (Fig. 14).

In pairing gamonts, observed under SEM, the epicytes are modified along their contact region to form a narrow band (Fig. 16). Longitudinal sections of the contact area confirm the drastic change in the epicyte fold pattern: the cell membranes of the associated individuals run in parallel fashion showing a close adhesion (Fig. 17).

A single gametocyst has been found in the terminal part of the midgut of a mature nymph of *Baetis rhodani*, which was about to emerge. The cyst has a spherical shape (about 140 μm in diameter) (Fig. 18) and almost completely occupies the gut lumen (Figs. 18, 19). The cyst is delimited by a wall (1.8 μm in thickness) (Fig. 20) with a multilayered organization (Fig. 21). Below the envelope the surface of the gamonts appears remarkably modified. Indeed, the epicyte folds drastically reduce in length up to their final disappearance (Fig. 20). In contrast, modified epicyte folds appear along the contact area, thus making more evident the boundary between the pairs (Fig. 22). Mucous material accumulates in the space leading to the contact area between the pairing gamonts. Spherical cells of 2.5-3 μm, characterised by a homogeneous cytoplasm and a wide nucleus, appear among the abundant paraglycogen granules that tend to shadow these minute cells. Their morphology is consistent with their identification as developing gametes (Fig. 23).

No further developmental steps, or other cyst formation, have been observed in spite of our efforts.

DISCUSSION

The main feature distinguishing Gregarinidae from Enterocystidae, the latter which includes only the genus *Enterocystis*, is the lack of a septum between proto- and deutomerite in enterocystids. Therefore, it is quite surprising that *Gamocystis*, in spite of its lack of septate gregarines (Rühl, 1976), is still included in Gregarinidae, thus making the lack of a septum not an exclusive feature of the genus *Enterocystis*.

The present study on *E. racovitzai* confirms previous observations (Codreanu, 1940; Desportes, 1963; Geus, 1969) that paired gamonts constitute the most common phase of development of this parasite. Indeed, intracellular gregarines and single individuals can be observed only rarely. The joined partners grow together and, except for the epicyte folds, their sexual dimorphism increases with maturation, thereby emphasizing the morphological differences between primite and satellite. Primite undergoes a major diversification consisting of lateral expansions and an apical protrusion. Our SEM images showed that this apical protrusion is already present in single individuals living in the midgut lumen and could represent a marker useful for identifying this gamont.

The fine organization of *E. racovitzai* conforms broadly to that described for other gregarines regarding the epicyte structure and the remarkable amount of paraglycogen

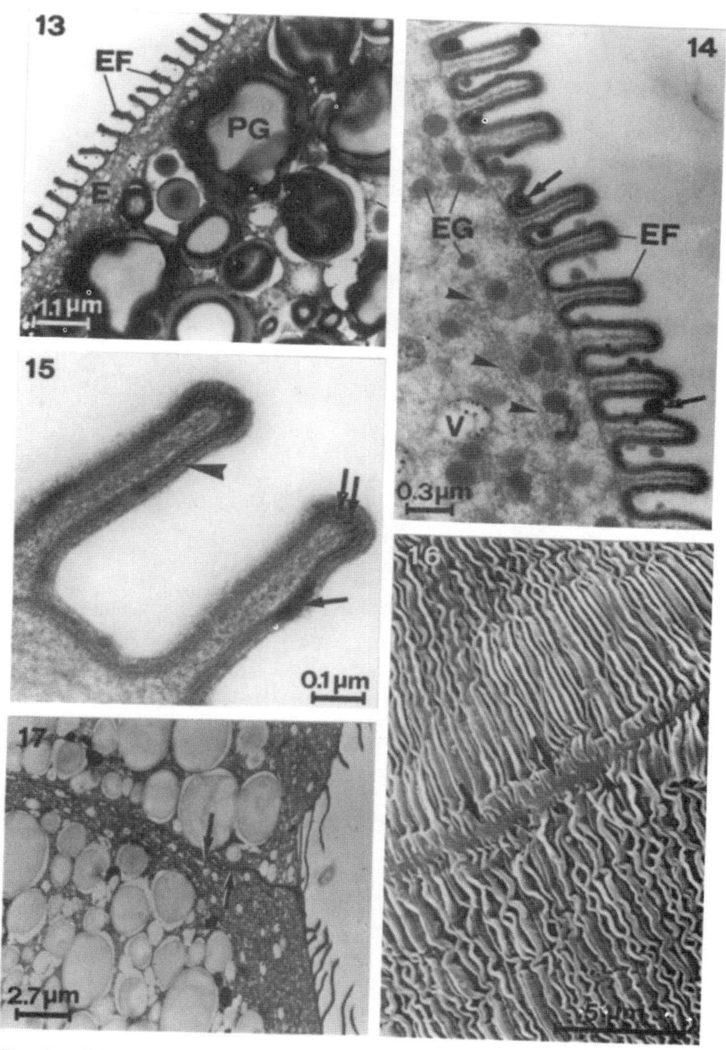

Figs. 13-17. Details of the mature syzygy of *Enterocystis racovitzai* from the midgut of *Baetis rhodani* under TEM (13-15,17) and SEM (16). 13- Sequence of epicyte folds (EF) emerging from the ectocyte (E) that bounds the endocyte-rich area of paraglycogen granules (PG); 14- Accumulation of extracellular electron-dense material (arrows) between epicyte folds (EF) emerging from the endocyte including vacuoles (V), electron-dense granules (EG) and fibrillar components (arrowheads); 15- Epicyte folds delimited by a multilayered border (arrowhead) with a striated appearance (arrows). Note the electron-dense granules adherent to the cytomembranes (arrow); 16- Contact region in the mature syzygy. Note the narrow band (arrows) between the two gamonts. 17- Longitudinal sections of the contact area showing the close adhesion (arrows) between the two gamonts.

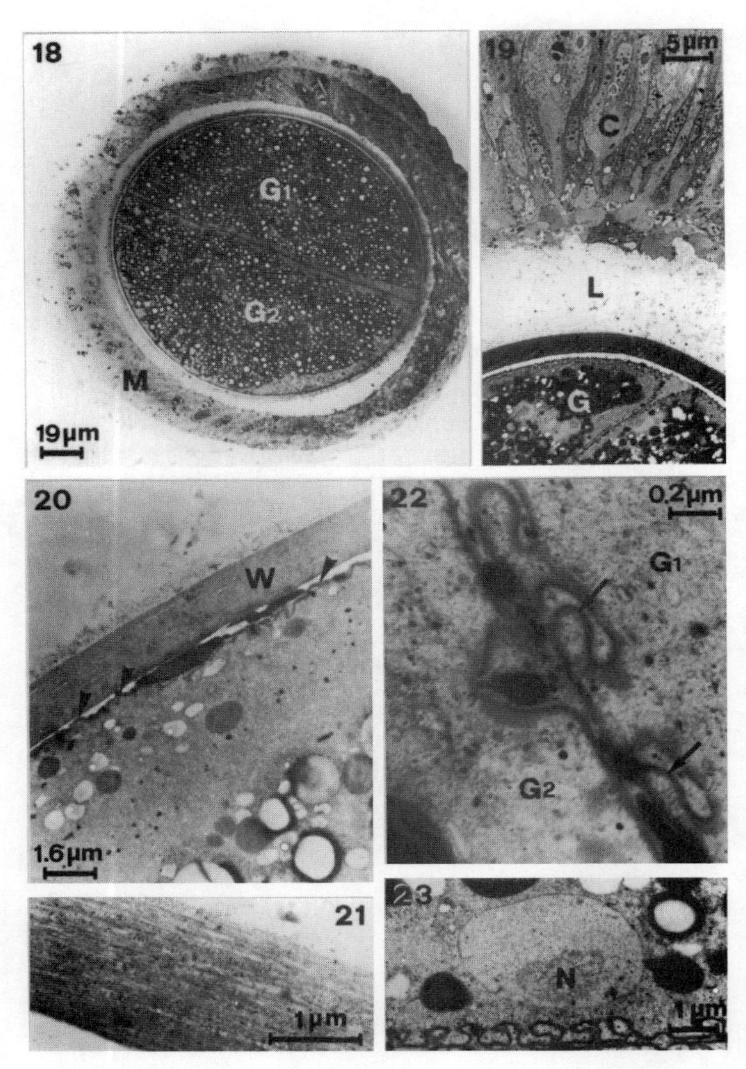

Figs 18-23. Gametocyst of *Enterocystis racovitzai* in the midgut of *Baetis rhodani* under TEM. 18- Gametocyst in the midgut (M) showing gamonts in association (G1-G2); 19- Epithelial cells (C) facing the gametocyst (G). L= midgut lumen; 20- Gametocyst wall (W). Note the remarkably reduced epicyte folds (arrowheads); 21- Detail of the gametocyst wall showing its multilayered organization; 22- Modified epicyte folds (arrows) along the contact area between gamonts (G1-G2); 23- A gamete in formation. N= nucleus.

granules necessary for parasite development (Vegni Talluri and Dallai, 1985). Another common feature is represented by the extrusion of electron-dense material through the epicytes, a feature already shown by Desportes (1974) in *E. fungoides*. On the basis of the resemblance of our images with those of authors who used histochemical techniques, we believe that this material is composed of mucopolysaccharides. As stressed for other gregarines, this material is associated with the life cycle of the parasite, and it is important for gliding movements (King *et al.*, 1982; Dallai and Vegni Talluri, 1983; Vegni Talluri and Dallai, 1983, Ghazali *et al.* 1989; Ghazali and Schrével, 1993), cyst wall differentiation and, presumably, the homogeneous matrix we observed between the gamonts inside the cyst.

SEM images of young syzygy showed that epicyte folds of the primite are slightly longer than those of the satellite, thus emphasizing that ectocyte organization could be the expression of an early sexual dimorphism. As observed by Desportes (1974), the epicyte folds disappear along the contact zone and the opposite membranes tend to run in a parallel fashion. In *E. racovitzai,* they are more closely associated, a feature consistent with the occurrence of a specialization strengthening the joining of partners. Indeed, the representatives of Enterocystidae are peculiar among gregarines for their early pairing, a feature that needs a stable connection between the partners during their further growth. The mechanical role of cell junction in gamont association has been stressed by Dallai and Vegni Talluri (1988), who described the presence of a septate junction in the syzygy of *Gregarina polymorpha*.

A certain level of plasticity in the shape of the epicyte folds along the contact region is suggested by the comparison between the syzygy not included in the cyst and the syzygy wrapped in the cyst wall. Indeed, in the latter, the folds, even though reduced in length, become evident again, thereby making the joining between the partners more evident. This feature contrasts remarkably with the rest of the ectocyte surface where the folds are no longer visible.

We believe that the modifications of the epicytes reflect their function in the life cycle of the parasite, while the gamonts are growing in the gut lumen or are included in the cyst. The need for firm adhesion is mainly related to the gliding movement that is accomplished by vertical undulation of the epicyte folds (Walker *et al.*, 1979; MacKenzie and Walker, 1983), which ceases at the level of the contact region to avoid distortions (Dallai and Vegni Talluri, 1988).

According to previous observations (Desportes, 1963; Codreanu and Codreanu-Balcescu, 1979), cyst differentiation takes place at the end of the aquatic portion of the mayfly life cycle, in such a way that the cyst, with gametes in formation, is released before the emergence of the insect. This represents a successful strategy allowing the parasite to infect other individuals.

The attribution of the gregarines dissected from *Baetis rhodani* to *E. racovitzai* is mainly based on the comparison between the drawings reported in the literature for this species and for other gregarines that are parasites of Ephemeroptera (see references in Table 1). The mature syzygy offers a good opportunity for specific attribution, while the growing phases are difficult to determine. In this report, the latter are referred to *E. racovitzai* in consideration of the fact that so far only this gregarine has been found in the gut lumen of our specimens of *B. rhodani.*

In conclusion, our finding of *E. racovitzai* in Italian representatives of *B. rhodani* allowed us to identify some ultrastructural morphological traits that are useful for taxonomic purposes and to increase the knowledge on the interaction between host and parasite.

REFERENCES

Arvy, L. and W. L. Peters. 1973. Phorésies, biocoenoses et thanatocoenoses chez les Ephéméroptères, pp. 254-312. In: W. Peters and J. Peters (eds.). Proceedings of the 1st International Conference on Ephemeroptera. E.J. Brill, Leiden.

Bobyleva, N. N. 1963. Recherches histochimiques sur les differénts stades du cycle vital de la grégarine *Enterocystis ensis*, vivant dans les larves des Ephéméroptères du genre *Cloeon*. Inst. Citol. Sbor. Rab. U.R.S.S. 3: 35-43.

Codreanu, M. 1940. Sur quatre grégarines nouvelles du genre *Enterocystis*, parasites des éphémères torrenticoles. Arch. Zool. Exp. Gén. (Notes et Revue) 81:113-122
Codreanu, R. and D. Codreanu-Balcescu. 1979. Remarques critiques sur le parasites et leur effects chez les éphéméroptères, pp. 227-243. In: K. Pasternak and R. Sowa (eds.). Proceedings of the 2nd International Conference on Ephemeroptera. Polska Akademia Nauk., Krakow.
Dallai, R. and M. Vegni Talluri. 1983. Freeze-fracture study of the Gregarine trophozoite: I. The top of the epicyte folds. Boll. Zool. 50: 235-244.
Dallai, R. and M. Vegni Talluri. 1988. Evidence for septate junctions in the syzygy of the protozoon *Gregarina polymorpha* (Protozoa, Apicomplexa). J. Cell Sci. 89: 217-224.
Desportes, I. 1963. Quelques grégarines parasites d'Insects aquatiques de France. Ann. Parasit. Hum. Comp. 38: 341-377.
Desportes, I. 1964. Sur la présence d' *Enterocystis ensis* Zwetkow (Eugregarina, Enterocystidae) chez un Ephéméroptère des Pyrénées-Orientales. Vie et Milieu, Suppl. 17: 103-106.
Desportes, I. 1966. Révision des grégarines de la famille des Enterocystidae (Eugregarinidae) parasites des larves d'Ephéméroptères. Protistologica 2: 141-144.
Desportes, I. 1974. Ultrastructure et evolution nucléaire des trophozoites d'une grégarine d'Ephéméroptère: *Enterocystis fungoides* M. Codreanu. J. Protozool. 21: 83-94.
Frantzius, A. von, 1848. Einige nachträgliche Bemerkungen über Gregarinen. Arch. Naturg. 14:188-197.
Gaino, E. and M. Rebora, 1998. Contribution to the study of *Enterocystis racovitzai,* a gregarine parasite of *Baetis rhodani* (Ephemeroptera, Baetidae). Acta Protozool., 37: 125-131.
Geus, A. 1969. Sporentierchen, Sporozoa, die Gregarinida der land-und süsswasserbewohnenden Arthropoden Mitteleuropas, pp. 1-608. In: F. Dahl, M. Dahl and F. Peus (eds.). Die Tierwelt Deutschlands. VEB Gustav Fiscer, Jena, 57.
Ghazali, M., M. Philippe, A. Deguercy, P. Gounon, J. M. Gallo and J. Schrével. 1989. Actin and spectrin-like (M_r=260-240.000) proteins in Gregarines. Biol. Cell. 67: 173-184.
Ghazali, M. and J. Schrével, 1993. Myosin-like protein (Mr 175,000) in *Gregarina blaberae*. J. Euk. Microbiol. 40: 345-354.
Karnovsky, M. S. 1965. A formaldehyde-glutaraldehyde fixative of high osmolality for use in electron microscopy. J. Cell Biol. 27: 137A-138A.
King, C. A., K. Lee, L. Cooper and T. M. Preston. 1982. Studies on gregarine gliding. J. Protozool., 29: 637-638
MacKenzie, C. and M. H. Walker, 1983. Substrate contact, mucus and Eugregarine gliding. J. Protozool. 30: 3-8
Peters, W. L. and L. Arvy. 1979. Phoresis, biocoenosis et thanatocoenosis in the Ephemeroptera, pp. 245-263. In: K. Pasternak and R. Sowa (eds.). Proceedings of the 2nd International Conference on Ephemeroptera. Polska Akad. Nauk., Krakow.
Rühl, H. 1976. Uber die Bedeutung spezifischer Ultrastrukturen einiger Gamonten verschiedener Gregarinenfamilien der Ordnung Eugregarinida (Doflein, 1901) als systematische Merkmale. Arch. Protistenkd. 118: 353-363.
Schneider, A. 1882. Contribution a l'étude des grégarines. Arch. Zool. Exp. Gen. 10: 423-450.
Shtein, G. A. 1960. Gregarines des arthropodes aquatiques dans les lacs de Karélie. Zool. Zh. 39: 1535-1544.
Vegni Talluri, M. and R. Dallai. 1983. Freeze-fracture study of the gregarine trophozoite: II. Evidence of "rosette" organization on cytomembranes in relation with micropore structure. Boll. Zool. 50: 245-255.
Vegni Talluri, M. and R. Dallai. 1985. Freeze-fracture study of the gregarine trophozoite: III. Cytoplasmatic differentations in *Lepismatophila thermobiae*. Boll. Zool. 52: 247-258.
Walker M. H., C. MacKenzie, S. P. Bainbridge and C. Orme. 1979. Study of the structure and gliding movement of *Gregarine garnhami.* J. Protozool. 24: 566-574.
Zwetkow, W. N. 1926. Eine neue Gregarinengattung, *Enterocystis ensis*, aus den Larven einer Eintagsfliege. Arch. Russ. Protist. Moskau 5: 45-55.

MORPHOLOGY OF THE HAMMER AND DRUMMING SIGNALS IN TWO STONEFLY SPECIES, *GIBOSIA HAGIENSIS* (OKAMOTO) AND *KIOTINA PICTETII* (KLAPÁLEK) (PERLIDAE)

Satoko Hanada[1] and Hiroki Maruyama[2]

[1] Department of Biological Science, Faculty of Science
Nara Women's University
Kitauoya-nishimachi, Nara 630-8506, Japan
[2] 1-25-5-200 Mukonoso-higashi, Amagasaki-shi
Hyogo Pref., 661-0032, Japan

ABSTRACT

Drumming signals of males in *Gibosia hagiensis* (Okamoto) and *Kiotina pictetii* (Klapálek) were recorded in the laboratory. Males of the species had a specialized "hammer" on the ventral abdomen, which is used for signal production. The ultrastructure of the hammer was described using a scanning electron microscope. *G. hagiensis* had a hammer with a scaly surface and produced monophasic signals by rubbing the substrate with the hammer. The hammer of *K. pictetii* was oval in shape and had a ridged surface with scattered short sensilla basiconica. They produced monophasic signals by tapping the substrate with the hammer.

INTRODUCTION

Drumming, the vibrational and sexual communication system in Northern Hemisphere stoneflies, is diversified in methods of signal production, quantitative characteristics of signals and exchange patterns between males and females (Stewart, *et al.*, 1988; Maketon, *et al.*, 1988 a, b; Stewart, *et al.*, 1991). Drumming signals are generally produced by striking the substrate with the postero-ventral portion of the abdomen (Rupprecht, 1968). Although both sexes produce low-frequency vibrational signals, males are more diversified in methods of signal production by the abdominal structures specialized for drumming named as "vesicle", "lobe", "knob" or "hammer", and various species-specific signals (Rupprecht, 1976; Stewart, *et al.*, 1991).

Males of two Japanese perlid species, *Gibosia hagiensis* (Okamoto) and *Kiotina pictetii* (Klapálek) were studied in the present study. *G. hagiensis* is distributed in Hokkaido, Honshu and Kyushu, and *K. pictetii* in Honshu and Shikoku (Uchida, 1990;

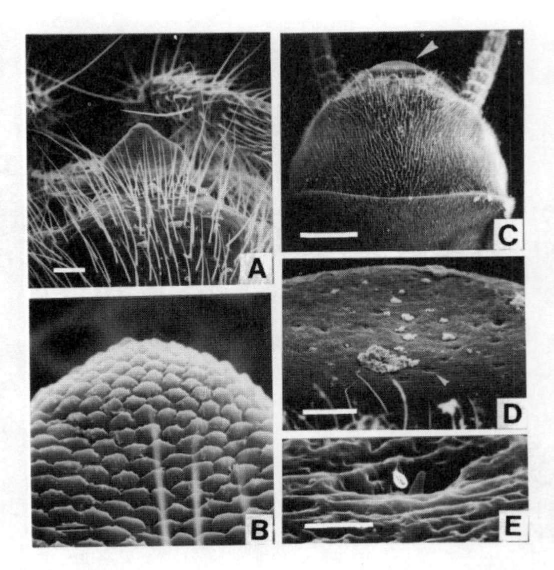

Fig. 1. Morphology of terminal abdomen (A, C), ventral hammer structure (B, D) and sensilla basiconia of hammer (E). A and B: *Gibosia hagiensis*, C, D and E: *Kiotina pictetii*. Scales: C: 500 μm; A, D: 50 μm; B, E: 5 μm. Arrow in C and D indicates the hammer and a pit with sensilla basiconia, respectively. See materials and methods for details of the specimens examined.

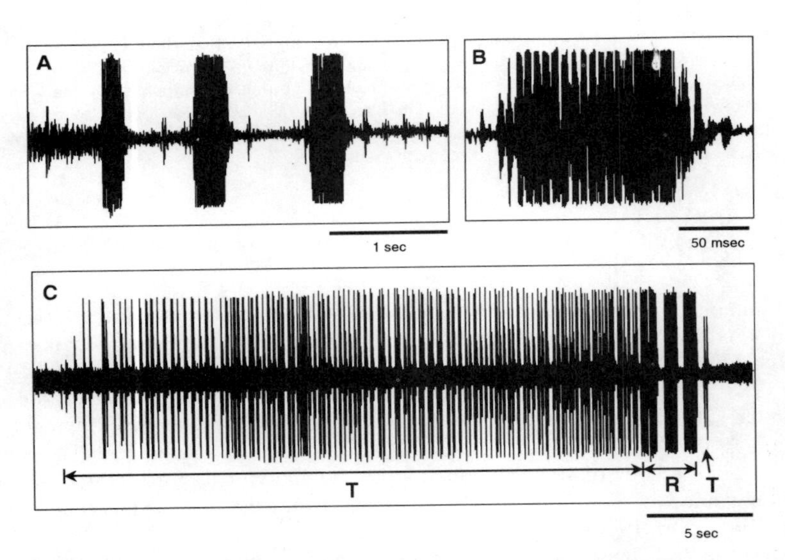

Fig. 2. Drumming signals of male *Gibosia hagiensis*. A: a male call consisted of 3 rubs; B: the second rub in A was magnified; C: an irregular male call consisted of tapping (118 beats and 2 beats) and rubbing (3 rubs) parts. T and R in C indicate the tapping part and the rubbing part, respectively.

Maruyama, *et al.*, 1994 a, b). In this paper, we describe the ultra-structure of the "hammer" specialized for drumming using a scanning electron microscope and present the quantitative and descriptive characteristics of drumming signals of the males.

MATERIALS AND METHODS

Morphological Observation

The following specimens were examined: (1) *Gibosia hagiensis* (Okamoto), 1 male, Nishihora, Takasu-mura, Gifu Pref., Japan, 20-VII-1997, H. Maruyama and (2) *Kiotina pictetii* (Klapálek), 1 male, Mioguchi, Higashi-yoshino-mura, Yoshino-gun, Nara Pref., Japan, 24-IV-1992, S. Hanada.

The terminal abdomens were removed, cleaned in an ultrasonic cleaner and dehydrated in ethanol series, acetone and isoamylacetate. The terminalia were dried with a critical point dryer Hitachi HCP-2 (Hitachi Koki Co., Ltd., Ibaragi), attached to aluminum stubs ventral side up, and coated with gold using an ion coater Eiko IB-3 (Eiko, Ibaragi). Mounted specimens were examined and photographed with the SEM, Hitachi S-430 (Hitachi Ltd., Tokyo).

Recording of Drumming Signals

Wild adults of the following species were collected and used in behavioural observations and recording signals: (1) *G. hagiensis*, 3 males, Nishihora, Takasu-mura, Gifu Pref., Japan, 20-VII-1997 and (2) *K. pictetii*, 1 male, Mioguchi, Higashi-yoshino-mura, Yoshino-gun, Nara Pref., Japan, 24-IV-1992; 1 male, Mio, Higashi-yoshino-mura, Yoshino-gun, Nara Pref., Japan, 26-IV-1992.

The recording apparatus consisted of a recording chamber (6 X 6 X 4H cm) made of paper with a clear plastic cover, a contact microphone (Cony Electronics Service, CN-555, Tokyo) and a cassette tape recorder (Aiwa, HS-J20). The adults were kept individually in plastic cylinders with wet cottons, and active drummers in the cylinders were put into the chamber for behavioural observation and recording signals. Signals were recorded at room temperatures of 29.0-31.0 °C in *G. hagiensis* and of 18.0-19.0 °C in *K. pictetii*. Recorded signals were measured and analyzed using SoundHandle Ver.1.03 (1994-95 by Dale Veeneman) with a Macintosh computer.

RESULTS

Gibosia hagiensis. Males of *G. hagiensis* have an oval hammer attached near the anterior margin of the 9th sternum (Figure 1 A). The hammer has a scaly surface (Figure 1B).

Males began drumming spontaneously under the solitary condition. They stopped, trembled their abdomen slightly, keeping it detached from the ground and then rubbed the ventral surface of their hammer against the substrate. The typical single male call is shown in Figure 2 A and B. Each rub was like a prolonged percussion stroke , during which the ventral surface of the hammer was dragged across the substrate, producing a squeaking sound (Figure 2 B). Males sequentially rubbed their hammer from one to three times (n=12, mean ± SD=2.7 ± 0.7) at the mean interval of 705.2 msec (n=20, SD=99.1) (Table 1). Each rub duration was 242.9 msec on average (n=32, SD=56.7).

One male produced two irregular signals. The male continued tapping his hammer on the substrate for a long time. One signal consisted of 187 beats with the mean interval of 272.9 msec (n=185, SD=77.8) and lasted for ca. 51 sec. In another case, the male produced 118 beats with the mean interval of 261.1 ± 83.7 msec (n=117) for ca. 31 sec and then rubbed the hammer against the substrate three times followed by two beats at the end (Figure 2 C).

Kiotina pictetii. Males of *K. pictetii* have a broadly oval membranous hammer on the anterior margin of the 9th sternum (Figure 1C). The hammer has a surface with irregularly ridged and scattered pits (Figure 1D). There are short sensilla basiconica in the pits (Figure 1E).

Males made the signals by tapping the hammer on the substrate while bending their abdomen like an arch. Figure 3 shows an example of the single male call. Males tapped the hammer from 7 to 20 times (n=8, mean ± SD=14.4 ± 4.8) with irregular intervals (n=105, mean ± SD=272.3 ± 480.0 msec, range=44 - 3892) and then started their call with constant intervals. The male call was monophasic, consisted of 70 mode beats (n=8, mean ± SD=68.3 ± 9.2) with the mean beat interval of 84.3 msec (n=538, SD=11.1) (Table 1) and had the mean signal duration of 5.7 sec (n=8, SD=0.8).

DISCUSSION

Male drumming signals of eighteen species belonging to the subfamily, Acroneuriinae in Perlidae, have been reported including the two species in this study. The drumming of *Gibosia hagiensis* and *Kiotina pictetii* are the first reports for the genera. All males of the eighteen species in Acroneuriinae except for *Perlesta placida* have the hammers for signal production on their ventral abdomen. The morphological features of the hammers such as shape, form of surface and presence of modified receptors on the surface, are related to the methods of drumming (rubbing or tapping) and the characteristics of drumming signals (Stewart, *et al.*, 1991). Only the species having the hammers with ridged or papillar surfaces produce derived rubbing signals (Stewart, *et al.*, 1991). Males of *G. hagiensis* have a hammer with a scaly surface, which is consistent with its rubbing signals. On the other hand, *K. pictetii* having the hammer with a ridged and scattered pits on the surface and bearing sensilla basiconia in the pits has retained the tapping method.

The two species showed specific drumming behaviours and signals among the species of Acroneuriinae. *G. hagiensis* produced signals consisting of 1-3 rubs. Generally, the signals by rubbing tend to consist of a small number of rubs (ex. *Carineuria californica*, 1 rub; *Doroneuria baumanni*, 2 mode rubs) (Maketon, *et al.*, 1984; Stewart, *et al.*, 1982, 1991) and the case of *G. hagiensis* are consistent with this tendency. Males of the species were always observed to tremble just before rubbing. It is possible that the tremble induces an exact frequency of the following rub and is important for the start of a male call. However, one male of *G. hagiensis* produced two irregular signals, one by tapping for a long time and another by combination of tap and rub. The long-time tapping appears to be the modification or failure of the trembling.

Males of *K. pictetii* produced signals consisting of a large number of beats by tapping. The mode or mean beat number of the species is the largest among the males of

Table 1. Characteristics of male drumming signals of *G. hagiensis* and *K. pictetii*

Species	N	Method of signal production	Beat or rub number / signal $\overline{X}$ ± SD (mode, range)	Beat or rub interval [msec] $\overline{X}$ ± SD (range)
Gibosia hagiensis	12	Rubbing	2.7 ± 0.7 (3, 1-3)	705.2 ± 99.1 (555-935)
Kiotina picteii	8	Tapping	68.3 ± 9.2 (70, 54-83)	84.3 ± 11.1 (44-131)

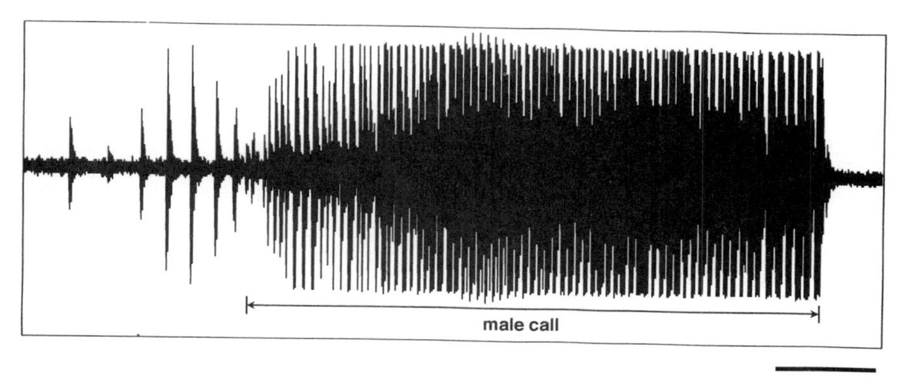

Fig. 3. A drumming signal of *Kiotina pictetii* male.

perlid species whose drumming signals have been reported (Stewart, *et al.*, 1991). Tapping with irregular intervals just before a male call may be necessary for making beats with constant intervals during long-time signaling.

ACKNOWLEDGMENTS

Thanks are due to Prof. T. Oishi and Dr. Y. Isobe, Nara Women's University, for reading an earlier draft of the text and their valuable suggestions.

REFERENCES

Maketon, M. and K. W. Stewart. 1984. Further studies of drumming behavior of North American Perlidae (Plecoptera). Ann. ent. Soc. Amer. 77: 770-778.

Maketon, M. and K. W. Stewart. 1988a. Patterns and evolution of drumming behavior in the stonefly families Perlidae and Peltoperlidae. Aquat. Insects 10: 77-98.

Maketon, M., K. W. Stewart, B. C. Kondratieff and R. F. Kirchner. 1988b. New descriptions of drumming and evolution of the behavior in North American Perlodidae (Plecoptera). J. Kans. ent. Soc. 61: 161-168.

Maruyama, A. I. Saleh and H., M. Takai. 1994a. Adult stoneflies (Insecta, Plecoptera) collected from Mt. Tsurugi, Tokushima Prefecture, Shikoku. Bull. Tokushima Museum 4: 91-96.

Maruyama, H., M. Takai and T. Kawasawa. 1994b. Stonefly fauna of Kochi Prefecture, Shikoku, Japan (2). Gensei 65: 13-16. (In Japanese, with English summary).

Rupprecht, R. 1968. Das Trommeln der Plecopteren. Z. vergl. Physiol. 59: 38-71.

Rupprecht, R. 1976. Strukture und Funktion der Bauchblase und des Hammers von Plecopteren. Zool. Jb. Anat. Bd. 95: 9-80.

Stewart, K. W. and M. Maketon. 1991. Structure used by Nearctic stoneflies (Plecoptera) for drumming, and their relationship to behavioral pattern diversity. Aquat. Insects 13: 33-53.

Stewart, K. W., S. W. Szczytko, B. P. Stark and D. D. Zeigler. 1982. Drumming behavior of six North American Perlidae (Plecoptera) species. Ann. ent. Soc. Amer. 75: 549-554.

Stewart, K. W., S. W. Szczytko and M. Maketon. 1988. Drumming as a behavioral line of evidence for delineating species in the genera *Isoperla*, *Pteronarcys*, and *Taeniopteryx* (Plecoptera). Ann. ent. Soc. Amer. 81: 689-699.

Uchida, S. 1990. A revision of the Japanese Perlidae (Insecta: Plecoptera) with special reference to their phylogeny. D. Sc. thesis, Tokyo Metropolitan University.

FINE STRUCTURE OF THE GUT OF A STONEFLY NYMPH, *PARAGNETINA MEDIA* (WALKER) (PLECOPTERA: PERLIDAE)

N. N. Kapoor

Concordia University, Department of Biology
1455 de Maisonneuve Blvd. W., Montreal
Quebec H3G 1M8, Canada

ABSTRACT

The present study concerns the ultrastructural survey of the gut of a stonefly nymph, *Paragnetina media*. The cuticular lining of the foregut is highly folded. Mircrofibers in the cuticle of the foregut form a supporting string-like structure in each cuticular fold. The wall is invested with circular and visceral muscles. The luminal surface of the midgut is regularly folded and covered with microvilli. Eight gastric caeca from the midgut extend forward to cover the proventriculus. Each gastric caecum has lobular luminal surface with short microvilli. The hindgut shows ultrastructural features of an osmoregularory organ.

INTRODUCTION

Nymphs of *Paragnetina media*, a predatory stonefly has provided interesting material for several morphological and physiological studies (Kapoor and Zachariah 1973; Kapoor 1974, 1978, 1979, 1994, 1997). My interest in ionoregulation prompted the present study since a detailed morphological information on the gut of plecopteran is lacking. Such information is essential to understand aspects of feeding and water-electrolyte balance in plecopteran nymphs. This paper gives a survey of the fine structure of the gut.

MATERIALS AND METHODS

Nymphs of *Paragnetina media* in their final instar were used for this study. The animals were dissected in 2.5% glutaraldehyde in phosphate buffer 0.15 M pH 7.2. The guts, free from contents were divided into appropriated sections and processed for scanning (SEM) and transmission (TEM) electron microscopy by methods described earlier (Kapoor 1994).

Trends in Research in Ephemeroptera and Plecoptera
Edited by E. Dominguez, Kluwer Academic/Plenum Publishers, 2001

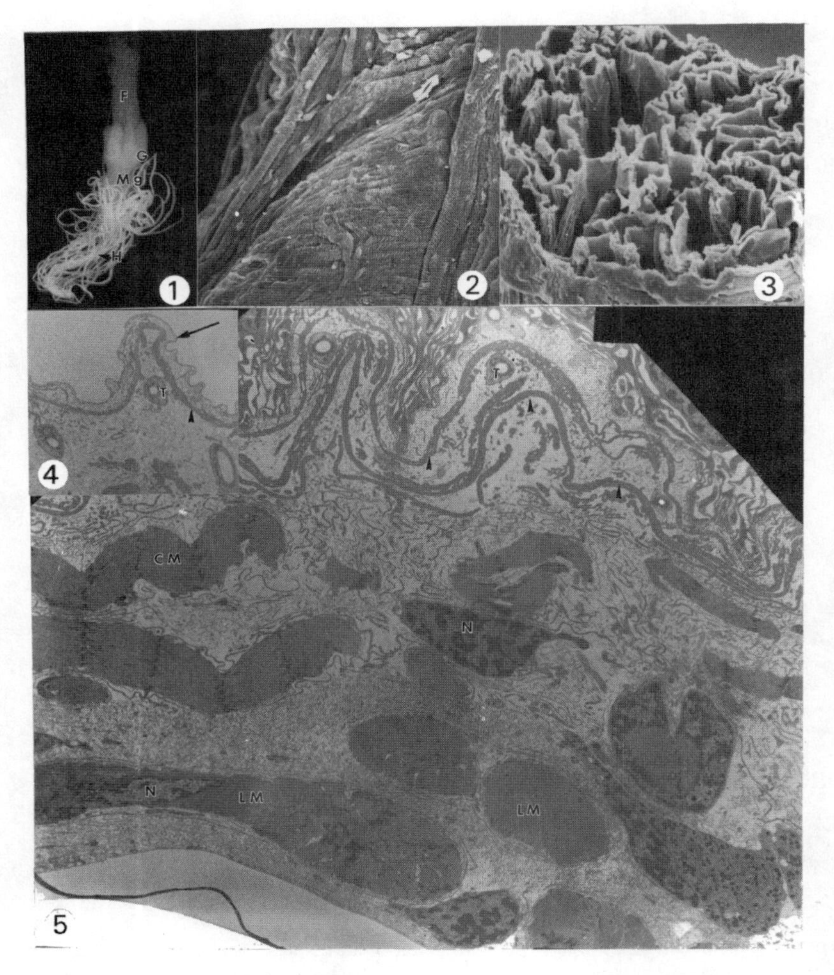

Figs. 1-5. 1) A light micrograph of the gut F, foregut; G, gastric caeca; Mg, midgut; H, hindgut. A cluster of Malpighian tubules (MT) originates at the junction of midgut and hindgut. X 4
2) The muscular investment of the circular and longitudinal muscle fibres of the foregut. X 300
3) Transversely fractured foregut by SEM, revealing the lumen of the foregut filled with luminal folds. X 400
4) Electron micrograph of the foregut showing irregular surface and a fold of the intima (arrow). T, tracheole; arrowhead pointing to string formed by microfibres. X 4,000
5) A longitudinal section of the foregut wall. CM, circular muscle; LM longitudinal muscle; N, nucleus; T, tracheole. Arrow heads point to strings of microfibres in the cuticular intima. X 3,800

RESULTS AND DISCUSSION

The alimentary canal of the *Paragnetina media* comprises three structurally and embryologically distinct regions: foregut, midgut and hindgut (Fig. 1). These three regions of the gut are covered with visceral muscle fibers, and this muscular investment is well developed around the foremost region of the foregut, the pharynx (Fig. 2). The foregut and hindgut are lined with a cuticle that shows surface specialization of folds (Fig. 3). These folds bear arrays of spines (Kapoor 1997). The intervening midgut is endodermal, and lacks cuticular intima, but the food is enclosed in a peritrophic membrane secreted by the epithelial cells of the midgut. This membrane was removed during cleaning the gut contents.

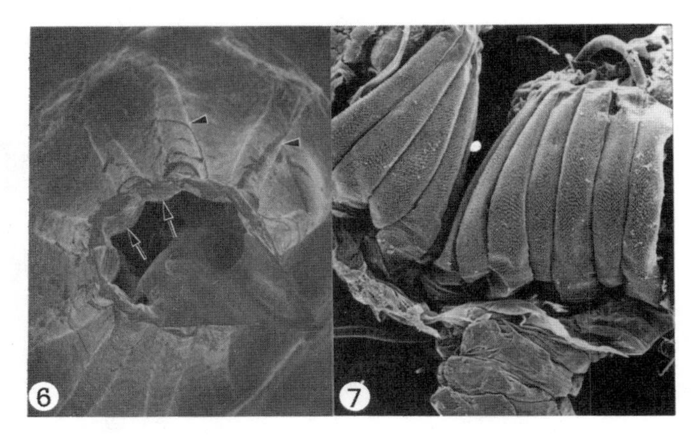

Figs. 6-7. 6) Scanning electron micrograph of the proventriculus showing ridges on the surface, strengthening the wall of the proventriculus (arrowheads). The posterior end of the proventriculus is cut open to reveal inner folds of the intima bearing cuticular spines (arrows). X 70
7) Proventriculus is cut open to show intima folds with spines. X 40

Foregut

Electron micrographs of the thin sections of the foregut illustrate a cuticular fold (Fig. 4, arrow) with an irregular surface. The cuticular intima is richly supplied with tracheoles and microfibres. Microfibres form string-like structure that extends into each cuticular fold (Figs. 4 and 5, arrowheads). Packing arrangement of fibrils or microfibrils in the matrix of cuticle is variable in insects. They are generally seen in longitudinal sections as dark rows of lines and dots. In obliquely cut section, the fibrils appear as rows of fibrous arcs or lamellae in different anthropod cuticle (reviewed by Filshie, 1982). In *Paragnetina media*, a string-like fibrillar orientation seems to support delicate folds of the cuticular lining of the foregut. The visceral muscle fibres run circularly and longitudinally around the tubular gut (Fig. 2). In a transverse section of the gut (Fig. 5), there is a narrow band of contractile fibre in a longitudinal profile, and thus runs circularly (CM) around the gut. Z-bands divide the narow band of contractile tissue into sarcomeres, but other bands are indistinct. Outside the circular muscle fibers are running parallel with the foregut, filled with contractile material that do not distinguish into separate fibrils (LM). These visceral fibers perform slow peristaltic contractions that propel the food material forward within the gut lumen. The proventriculus is a distinct part of the foregut that leads into the midgut. Its outer surface has ridges (Fig. 6, arrowheads) that support the intervening areas of the luminal folds. These folds are heavily sclerotised and bear an armature of tooth-like spines (Figs. 6, 7). Structural details of the proventiculus are given in a recent publication (Kapoor 1997).

Midgut

A rosette of eight gastric caeca originates from the midgut and extends forward to cover the proventriculus (Fig. 1, 8). The inner surface of each caecum has longitudinal folds. Each fold is constricted at places along its length and gives a lobular appearance (Figs. 8,9). Longitudinal and cross sections of folds show short microvilli lining the lumen of each lobe (Figs. 9, 10 arrowheads).

The lumen side of the midgut wall is regularly folded and covered with microvilli (Fig. 11). Columnar epithelial cells lining the wall have apposed lateral cell membrane linked by a smooth type of desmosomes near the apical ends of luminal borders (Fig. 12 arrowheads,

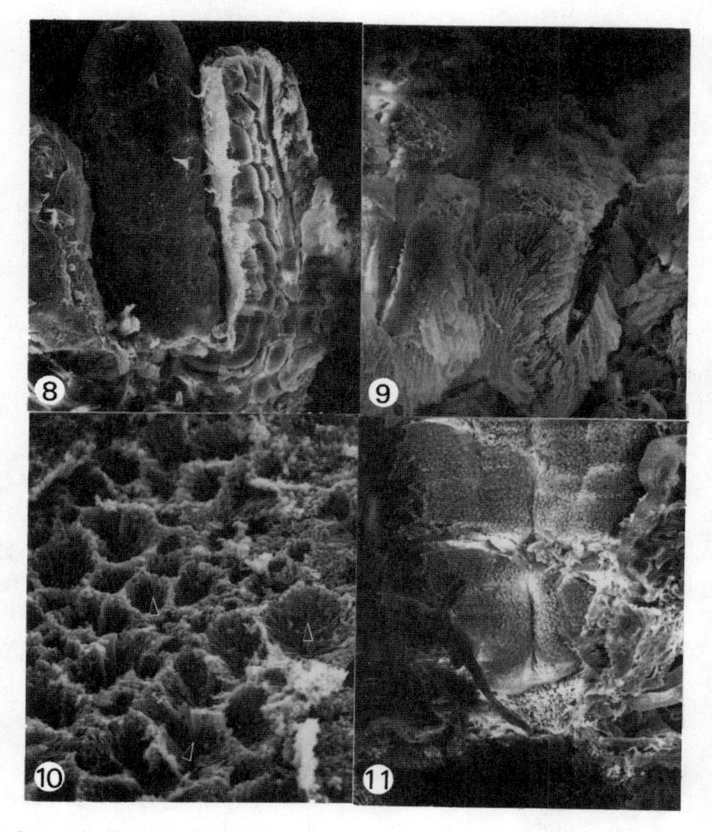

Figs. 8-11. 8) Scanning electron micrograph of three gastric caeca. One caecum is longitudinally cut to show lobular folds. X 100
9) A longitudinal section of the lobes to show arrangement of microvilli. X 400
10) Transversely fractured inner surface of the gastric caecum to show arrangement of microvilli in each lobe (arrowheads). X 3,500
11) Surface of the midgut showing tufts of microvilli. X 170

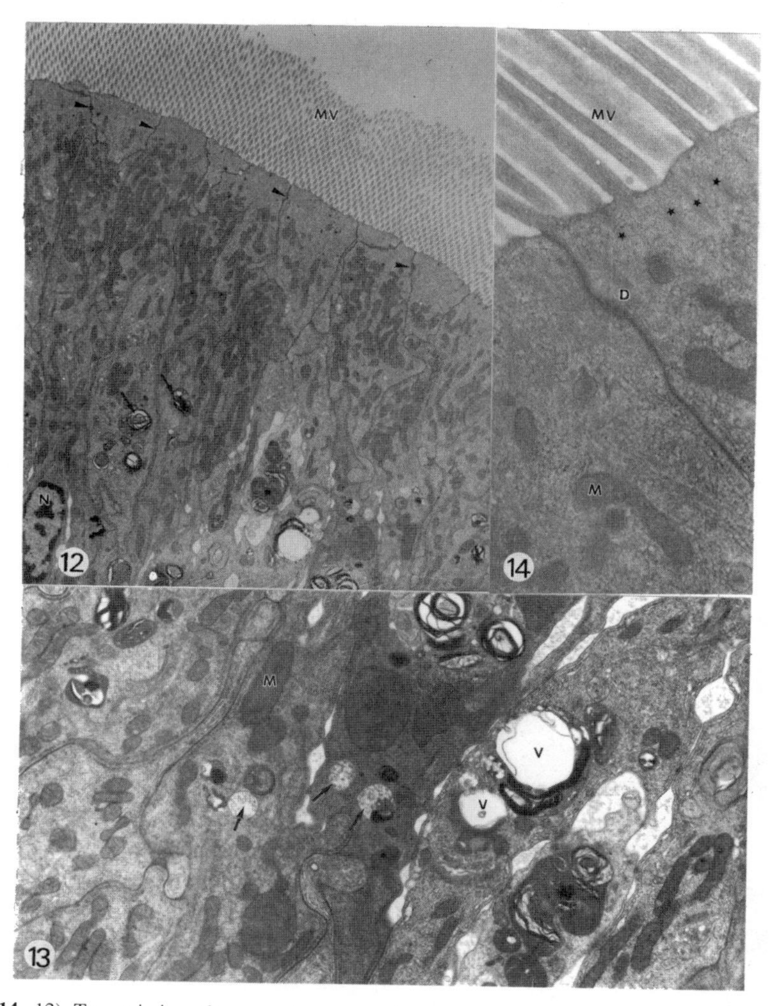

Figs. 12-14. 12) Transmission electron micrograph of the midgut. Microvilli (MV) are cut obliquely. Arrowheads show desmosomes near the microvillar end of the columnar epithelial cells. Arrows show accumulation of electron dense secretory products in vacuoles. N. nucleus. X 5,000
13) A magnified portion of the columnar epithelial cells showing empty vacuoles (V) and some of them filled with granules (arrows). The cytoplasm is filled with mitochondria (M) and granular endoplasmic reticulum. X 11,666
14) The apical membrane is extended at a regular interval along microvilli (MV), projecting from the cell surface into the gut lumen. The villi contain fine fibrils oriented longitudinally in each villus, which continue for a short distance into the cytoplasm (*). D, desmosome. X 30,000

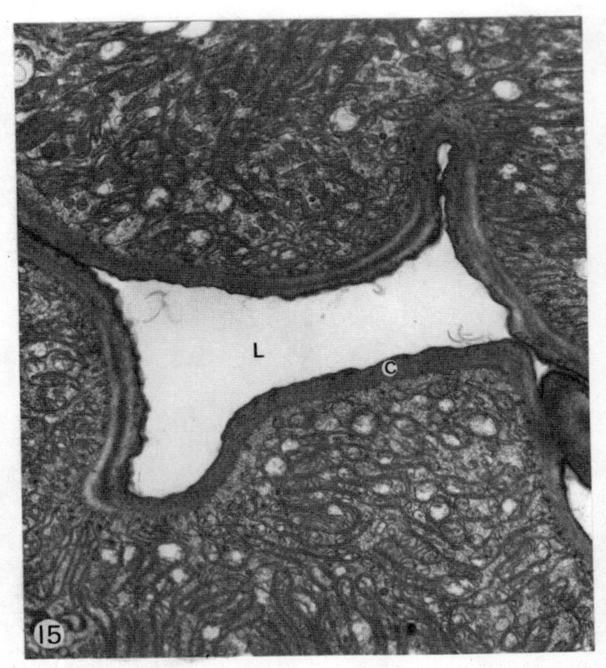

Fig. 15. A cross section of the anterior hindgut. An extensive network of invagination of the plasma membrane fills each fold of the gut wall. The cytoplasm is full of vacuoles and mitochondria. L, lumen; C, cuticle. X 15,000

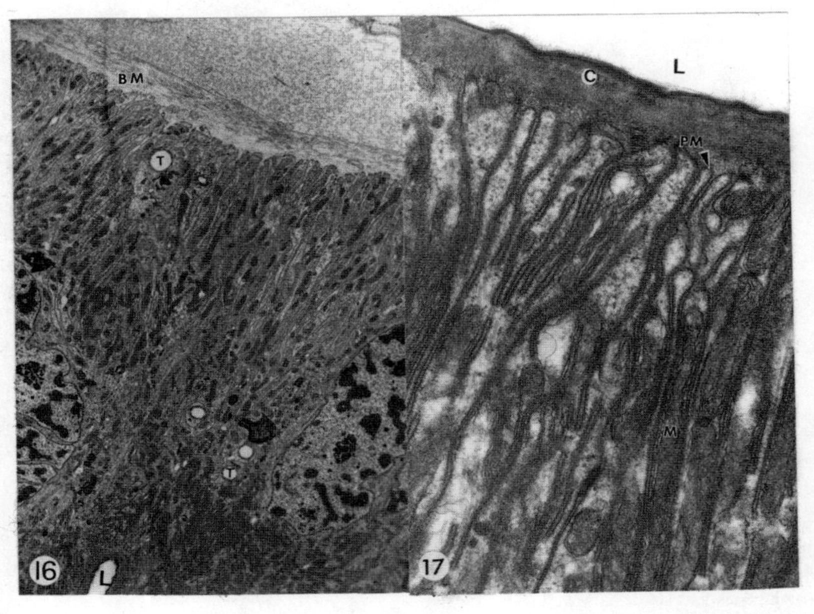

Figs. 16-17. 16) A longitudinal section of the hindgut showing extensive infoldings of the basal membrane (MB). Mitochondria in close association with infoldings are profusely distributed in the cytoplasm. Thin basement membrane contains collagen fibres. L, lumen; T, tracheoles. X 5,000.
17) A magnified view of the luminal side (L) of the posterior hindgut showing plasma membrane (PM) infoldings associated with mitochondria (M). C, cuticle. X 37,500

14, D). Such junctions are common in the midgut of arthropods and presumed to be associated with cell renewal, and may be involved in permeability barrier for certain ions and molecules between cells (Lane, 1982). The microvilli originate as extensions of the plasma membrane at regular intervals and are supported by bundles of filaments that extends into the cytoplasm of the cell (Fig. 14, MV,*). The cytoplasm of the epithelial cell is packed with mitochondria, free ribosomes rough endoplasmic reticulum and vacuoles. Some vacuoles are filled with electron dense material (arrows, Fig. 12) while others contain secretary granules (arrows, Fig. 13).

The midgut and caeca are the principal site of digestion and absorption or nutrient in the insect alimentary canal. They are linked to water uptake and to an active transport of sodium (Berridge, 1970). Close resemblance of ultrastructural features of the midgut and caeca of the *Paragnetina media* nymphs to that of other insects suggests that these two regions may have osmoregulatory function.

Hindgut

The Malpighian tubules about 100 in number arise from the gut at the junction of the mid-and hindgut (Fig. 1). A detailed morphology of these tubules has been described elsewhere (Kapoor 1994). Figures 15, 16 and 17 show morphological features of the hindgut. A basement membrane bounds the epithelial cells (BM) and the basal lamina is thrown into long folds. The cytoplasm of the cell has a rich supply of tracheoles (T) and mitochondria (M). On the luminal side, the gut wall is thrown into folds lined with a thin cuticle. Plasma membrane forms an extensive network of membranes. This region is highly vacuolated and has a high concentration of mitochondria.

In *Paragnetina media*, it is presumed that Na+, K+ ATPase is located in the basolateral membranes of the hindgut. Changes in the activity of this enzyme is associated with ionic activity in the rectum (Kapoor, 1980). Present observations show that infoldings of the plasma membrane form an association with mitochondria in such a way that mitochondria are brought into a close relationship to an area of the cell surface. Such an arrangement is related to an active-transport mechanism in the hindgut. Future histochemical and physiological studies will be required to elaborate structural-functional relationships in various regions of the gut of plecopteran nymphs.

REFERENCES

Berridge, M. J. 1970. A structural analysis of intestinal absorption, pp. 135-151. In: A. C. Neville (ed.). Insect Ultrastructure. Blackwell Scientific Publications, Oxford, England.

Filshie, B. K. 1982. Fine structure of the cuticle of insect and other arthropods, pp. 281-312. In: R. C. King and H. Akai (eds.). Insect Ultrastructure vol, 1. Plenum Press, New York, U.S.A.

Kapoor, N. N. 1974. Some studies on the respiration of stonefly nymph, *Paragnetina media* (Walker). Hydrobiologia 44: 37-41.

Kapoor, N. N. 1978. Effect of salinity on the osmoregulatory cells in the tracheal gills of the stonefly nymph, Paragnetina media (Plecoptera: Perlidae). Can. J. Zool. 56: 2194-2197.

Kapoor, N. N. 1979. Osmotic regulation and salinity tolerance of the stonefly nymph, *Paragnetina media*. J. Insect Physiol. 25: 17-20.

Kapoor, N. N. 1980. Relationship between gill Na^+, K^+- ATPase activity and osmotic stress in the plecopteran nymph, *Paragnetina media*. J. Exp. Zool. 213: 213-218.

Kapoor, N. N. 1994. A study of the Malpighian tubules of the plecopteran nymph *Paragnetinna media* (Walker) (Plecoptera: Perlidae) by light, scanning electron, and transmission electron microscopy. Can. J. Zool. 72: 1566-1575.

Kapoor, N. N. 1997. A scanning electron microscopic study of the cuticular lining of the gut of a stonefly nymph, *Paragnetina media* (Walker) (Plecoptera: Perlidae). In: P. Landolt and M. Sartori (eds.). Ephemeroptera and Plecoptera: Biology-Ecology-Systematics. MTL, Fribourg, Switzerland.

Kapoor, N. N. and K. Zachariah. 1973. A study of specialized cells of the tracheal gills of *Paragnetina media* (Plecoptera). Can. J. Zool. 51: 983-986

Lane, N. J. 1982. Insect intracellular junctions: their structure and development, pp. 402-433. In: R. C. King and H. Akai (eds.). Insect Ultrastructure Vol. 1. Plenum Press, New York, U.S.A.

CONTRIBUTION TO THE EGG MORPHOLOGY OF THE SPANISH STONEFLIES (PLECOPTERA) AN OVERVIEW IN THE SEGURA RIVER BASIN (S.E. OF SPAIN)

N. A. Ubero-Pascal[1], A. Soler[1], and M. A. Puig[2]

[1] Departamento de Biología Animal
Universidad de Murcia, Campus de Espinardo
30100 Murcia, Spain
[2] Centro de Estudios Avanzados de Blanes (C.S.I.C.)
Camino de Santa Bárbara
17300 Blanes, Gerona, Spain

ABSTRACT

The eggs of several species of both Systellognatha and Euholognatha from the Spanish plecopteran fauna are observed using by SEM. We show the first observations and descriptions of the temporal external membrane which enveloping the plecopteran eggs. In systellognathan species, we studied in detail the eggs morphological features of *Eoperla ochracea* (Kolbe, 1885), and we considered doubtful the presence of collar in the egg of *Perla marginata* (Panzer, 1799). In euholognathan species, we show the first observations of the eggs in Nemouroidea. New eggs information, such as micropyles or chorionic sculpture, from *Capnioneura* Ris 1905, *Leuctra* Stephens 1836, and *Nemoura* Latreille 1796 are appointed.

INTRODUCTION

Most of the stonefly eggs present several morphological features with taxonomical validity that permits their specific determination. In some cases, the eggs can be used to identify females lacking specific sexual features (Berthelemy, 1964). The egg morphological knowledge has contributed yet, or it can contribute, to establish the phylogenetic relationship in certain taxonomic groups (Sivec *et al.*, 1988; Stark and Stewart, 1981; Stark and Szczytko, 1988; Starkand Nelson, 1994; Isobe, 1997).

The egg morphology of European systellognathan species, like in other continents, is better known than euholognathan species. Superficial descriptions by means of light microscope of euholognathan species may found in some european works (Hynnes, 1941; Brinck, 1949; Degrange, 1957; Zwick; 1973). Although several eggs of european species,

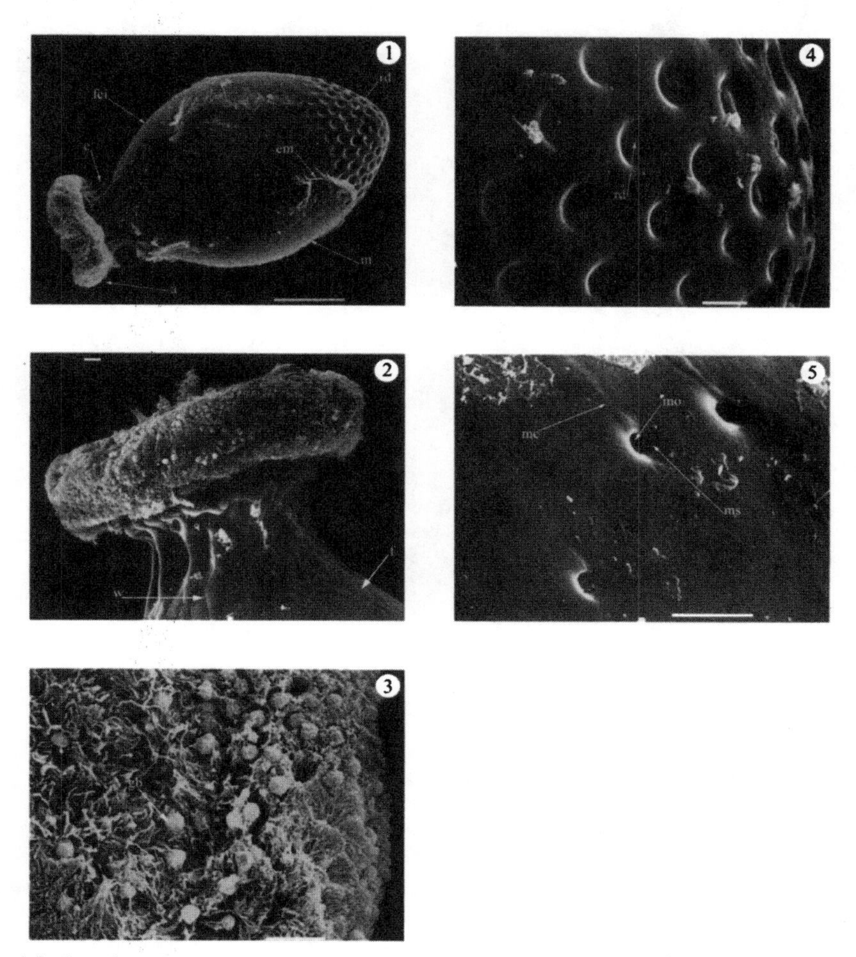

Figs. 1-5. *Eoperla ochracea.*
1: egg, 2: detail of collar, 3: detail of anchor plate, 4: posterior pole of egg, 5: detail of micropyle. a: anchor, c: collar, em: external membrane, f: floor, fci: follicle cell impression, gb: globular bodies, m: micropyle, mc: micropylar canal, mo: micropylar opening, ms: micropylar sperm guide, rd: round depression sculpture, w: wall. Scale: 1: 100 μm, 2-5: 10 μm.

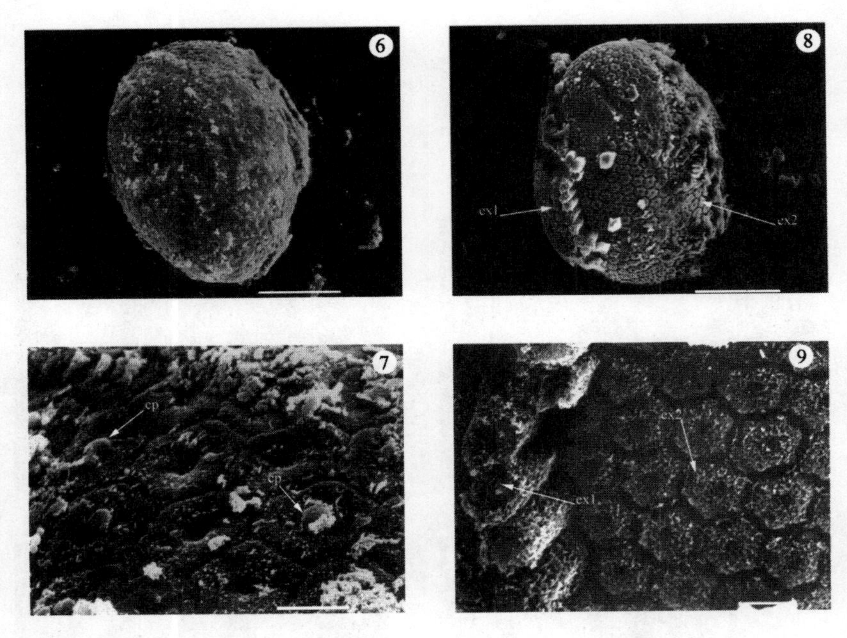

Figs. 6-9. *Perla marginata.*
6: egg, 7: detail of outer exochorion, 8: egg partly covered with the outer exochorion, 9: detail of the inner exochorion. ex_1: outer exochorion, ex_2: inner exochorion, gb: globular bodies. Scale: 6 and 8: 100 µm, 7 and 9: 10 µm.

floors is uniform (Fig. 2). In some cases the FCI can not be observed because a fine external temporal membrane (Fig.1) hides it.

Perla marginata (Panzer, 1799) (Perlidae Latreille, 1802)

Egg with oval-shape (Figs. 6 and 8). Mean size: 300 µm length and 237 µm width. The color is brown clear. The egg of *Perla marginata*, against to the observations of Isobe (1997), doesn't present collar (Figs. 6 and 8). Sivec (pers. comm.), who has reviewed the european species of *Perla* Geoffroy 1762, has observed this feature, as well as any misidentified of this species.

The micropyles haven't been observed.

Degrange (1957) noted that the egg has an external membrane with a regular hexagonal reticulation and on the center of each hexagonal unit appears a hemispheric refringent corpuscle. Stark and Szczytko (1988) pointed that the hexagonal reticulation is an impression of the follicular cells over the exochorion surface very common in systellognathan species and, in most species, the floor of each hexagonal unit is associated with globular bodies. These globular bodies are similar to find us in this species The external membrane features noted by Degrange (1957) are showed in Fig. 7, but in accord with Stark and Szczytko (1988) we think that these features are from the exochorionic layer, and not from the external membrane. Against of Degrange (1957) the corpuscles aren't hemispheric but spherical. If possible, like in *Eoperla ochracea* (Fig. 1), that the external temporal layer in *Perla marginata* could be very thin and transparent, and Degrange (1957) really observed the exochorionic features over the temporal external membrane.

Degrange (1957) pointed that the chorion of this species is constituted of three layers (one endochorionic layer and two exochorionic layers), and the outer chorionic layer has an hexagonal reticulation with a central perforation too. The two exochorionic layers are showed in the Fig. 8, and the differences between these two layers may see in the Figs. 7 and 9. The surface of both exochorionic layers has a similar hexagonal aspect; but the outer exochorion is more compact than the inner exochorion, and the separation among hexagonal units is more grooves in the inner than the outer exochorion. The reticulate aspect of the outer exochorionic layer is formed by hexagonal prismatic units essembles and not is a simply impression of follicle cells (Fig.9). Degrange (1957) had noted that the outer exochorion was formed by central perforated hexagonal plate. The central perforation observed in the hexagonal units of outer exochorion continued in each hexagonal unit of the inner exochorion.

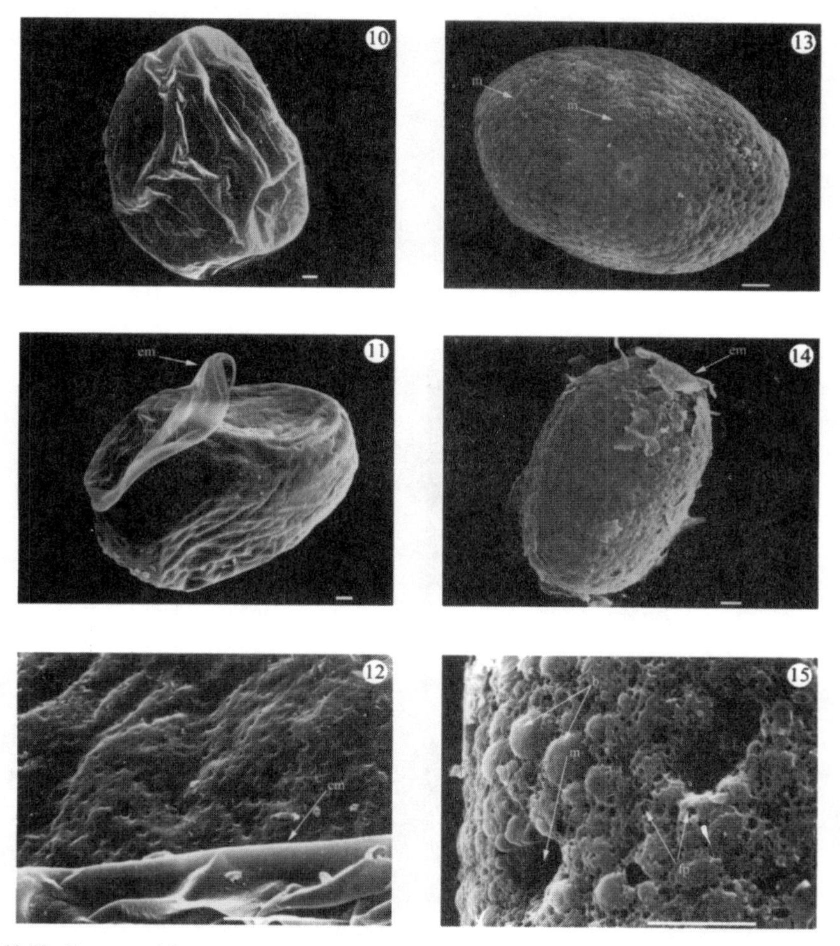

Figs. 10-12. *Nemoura fulviceps* and **Figs. 13-15.** *Capnioneura mitis*.
10: egg covered with the external membrane, 13: egg without the external membrane and micropyle, 11 and 14: egg with rest of external membrane, 12 and 15: detail of exochorionic sculpture. b: bubble-like process, em: external membrane, fp: fine pores, m: micropyle. Scale: 10-14: 10 μm, 15: 1 μm.

Nemoura fulviceps Klapalek, 1902 (Nemouridae Newman, 1853)

Egg with oval-shape, although it may present dents (Figs. 10 and 11). Mean size: 181 μm length and 128 μm width. The color is clear yellow-brown. The egg doesn't present collar. The egg doesn't present any attachment structures clearly distinguishable, although appears a thick external membrane hiding the exochorionic sculpture (Fig. 10 and 11). It doesn't appear any clear micropyles. The exochorionic surface sculpture is formed by thin pores of relatively size (Fig. 12).

Capnioneura mitis Despax, 1932 (Capniidae Klapalek, 1905)

Egg with oval-shape (Fig. 13). Mean size: 112 μm length and 77 μm width. The color is clear yellow. The egg doesn't present collar. The egg doesn't present any attachment structures clearly distinguishable. External membrane, when present, is sculptured by slight creases. The sculpture of exochorion is very regular, and it is slightly reticulate with bubble-like process and pores of different sizes (Figs. 14 and 15). The micropyles are big pores distributed to all the surface of the egg (Fig. 15).

Leuctra cazorlana Aubert, 1962 (Leuctridae Klapalek, 1905)

Egg with oval-shape, almost round (Fig. 16). Mean size: 124 μm length and 97 μm width. The color is yellow and clear. The egg doesn't present collar. The egg doesn't present any attachment structures clearly distinguishable, and neither the external membrane. The

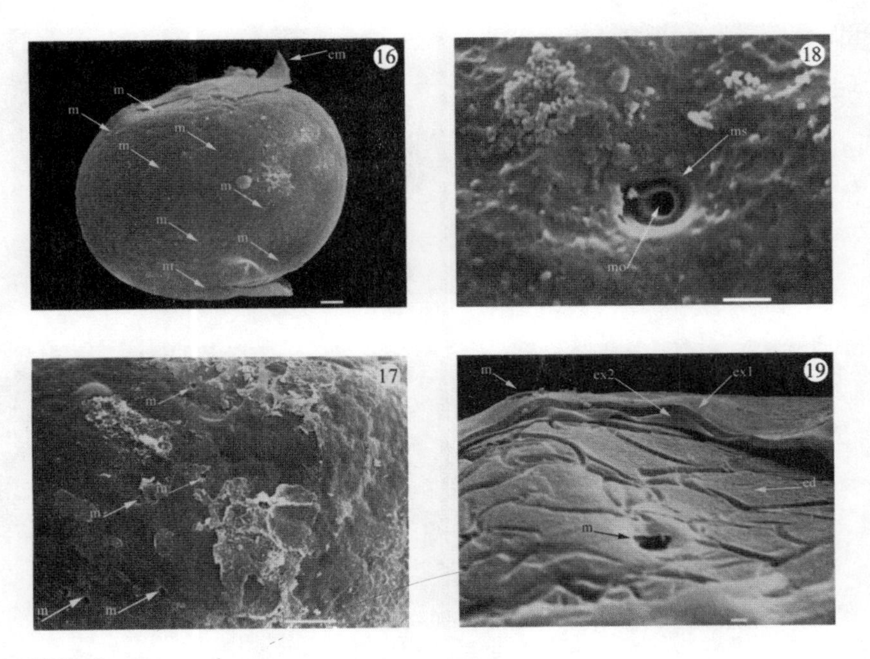

Figs. 16-19. *Leuctra cazorlana.*
16: Egg, 17: micropyle disposition, 18: micropyle, 19: details of the chorionic layers. ed: endochorion, em: external membrane, ex_1: outer exochorion, ex_2: inner exochorion, m: micropyles, mo: micropylar opening, ms: micropylar sperm guide. Scale: 16 and 17: 10 μm, 18 and 19: 1 μm.

micropyles are clearly evident (Figs. 16 and 17) and they are close to one pole. The number of micropyles is variable (from 12 to 14). Their disposition in the egg is in zigzag (Fig. 17). The sperm guide is a deep round depression of the outer layer of the exochorion and in its center appears the micropylar opening (Fig. 18).

The chorion is composed of at least by three layers, two exochorionic layers and probably one endochorionic layer. The outer exochorionic layer is thicker than the inner one (Fig. 19). The exochorion surface sculpture is slightly creased and the surface of the endochorion presents several irregular grooves (Fig. 17).

Leuctra geniculata Stephens, 1835 (Leuctridae Klapalek, 1905)

Egg with oval-shape, almost round, although it may present dents (Fig. 20). Mean size: 133 μm length and 107 μm width. The color is clear yellow-brown. The egg doesn't present

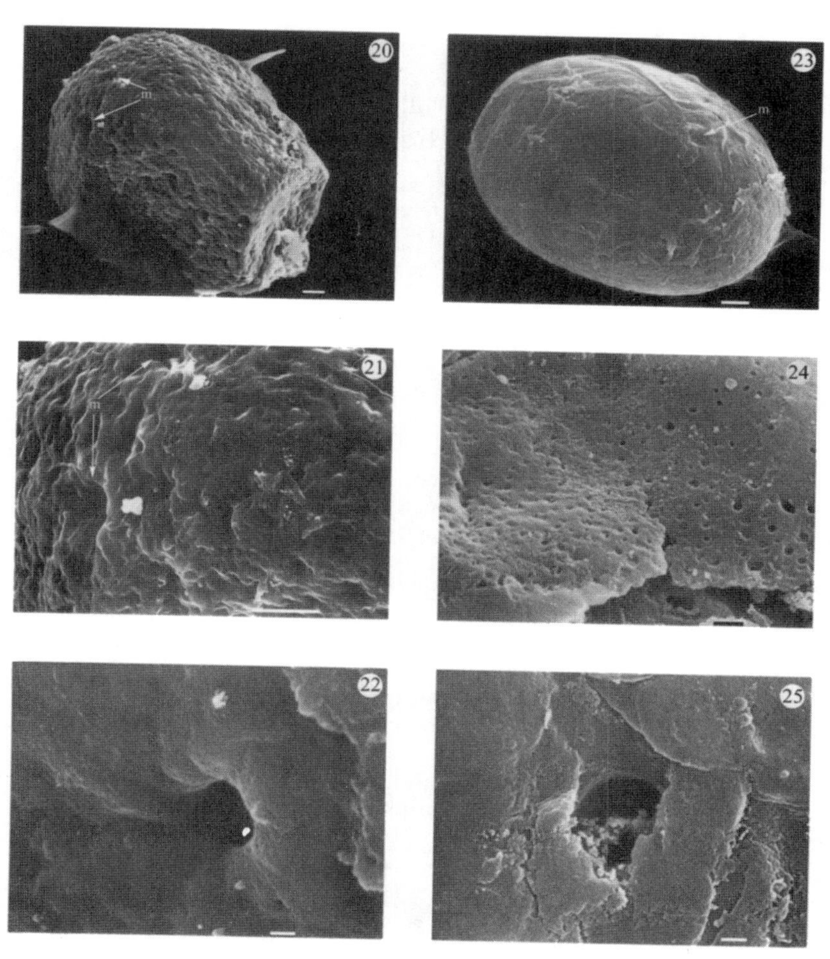

Figs. 20-22. *Leuctra geniculata*, and **Figs. 23-25.** *Leuctra iliberis*.
20 and 23: egg covered with the external membrane, 21: Chorionic sculpture covered by the external membrane, 22: Micropyle, 24: Exochorionic sculpture without the external membrane, 25: micropyle. m: micropyle. Scale: 20, 21 and 23: 10 μm, 22, 24 and 25: 1 μm.

collar. The egg doesn't present any attachment structure clearly distinguishable, although the temporal external membrane appears like a slight layer very close to the exochorion, which permitted to distinguish the exochorionic sculpture (Fig. 21). The sculpture of exochorionic surface is very irregular, with many creąses and pores of different size. The micropyles are large pores that don't present any accessory structures (Fig. 22).

***Leuctra iliberis* (Sánchez-Ortega y Alba-Tercedor, 1988) (Leuctridae Klapalek, 1905)**

Egg with oval-shape, (Fig. 23). Mean size: 112 µm length and 79 µm width. The color is clear yellow. The egg doesn't present collar. The egg doesn't present any attachment structures clearly distinguishable, although appears the external membrane by which the exochorionic sculpture is hidden (Fig. 25). The exochorionic surface presents a reticulated sculpture (Fig. 24). The micropyle may be observed with difficulty in the Fig. 25, and it is a round perforation without any distinguishable accessory structures.

CONCLUSIONS

It is well know in Systhellognatha that the egg chorionic features are important to identify the females, to solve taxonomic mistakes, and to understand the phylogenetic relationships. Furthermore, Isobe (1997) also points the importance for these goals of attachment structure, that is, anchor. However, external membrane has not been observed yet, because the membrane is temporal and interferes to observe the chorionic structures. In this study, the interesting features of the external membrane, particularly in Euhollognatha, are observed in Spanish species.

The euholognathan eggs are poorly known, perhaps because they haven't showy structures and because the external membrane usually hides the specific features using a light microscope. Although in many case, the external membrane may show a typical ornamentation using SEM, such as in *Capnioneura mitis*.

If compared the features of euholognathan eggs to that of the systellognathan eggs, we may observed several basic differences. In general, the sygtellognathan eggs have clearly attachment structures, a common hexagonal reticulation, micropyles with ornamentation, and a very thin external temporal membrane. Normally, the euholognathan eggs lack of clearly attachment structures, hexagonal reticulation, and micropyles with ornamentation, but they have a thick temporal external membrane and the chorionic surface present a heterogeneous ornamentation.

We have found that the eggs of systellognathan species are attached directly, each one, on the rivers substrate, while in euholognathan species all eggs together are fixed on the river substrate into gelatinous mass (egg-laying). Perhaps, this observation could to explain the lack of anchor in euholognathan eggs. The temporal external membrane, in these cases, would functioned to mix well the eggs and the gelatinous mass.

The egg terminology proposed by Stark and Szczytko (1988) for the systellognathan eggs can be used for euholognathan eggs too. Although, the pole orientation proposed by these authors can not applied in the euholognathan eggs, due to absence of both collar and anchor. Otherwise, we propose to use the terms micropylar opening, micropylar canal and micropylar sperm guide (Koss, 1968; Koss and Edmunds, 1974) in case that these features can be distinguishable in plecopteran eggs, such as in *Eoperla ochracea*.

ACKNOWLEDGMENTS

We wish to thank Dr. I. Sivec for his comments about the species *Perla marginata*. We are grateful to Dr. B. Stark for his interest in this work and animate us to publish it. Finally, we wish to thank the referee for his comments about the useful of this paper to understand the order Plecoptera and his orientations in the new redaction.

REFERENCES

Berthelemy, C. 1964. Interet taxonomique des oeufs chez les *Perlodes* europeens (Plecopteres). Bull. Soc. Hist. nat. Toulouse, 99 (3-4): 529-537.

Brinck, P. 1949. Studies on Swedish stoneflies (Plecoptera). Opusc. Ent., 11: 1-250.

Degrange, C. H. 1957. L'oeuf et le mode d'eclosion de quelques plecopteres. Trav. Lab. Hydrobiol. Grenoble, 48/49: 37-49.

Hynes, H. B. N. 1941. The taxonomy and ecology of the nymphs of British Plecoptera with notes on the adults and eggs. Trans. R. ent. Soc. London, 91: 459-555.

Isobe, Y. 1997. Anchors of stonefly eggs, pp. 349-361. In: P. Landolt and M. Sartori (eds.). Ephemeroptera and Plecoptera. Biology-Ecology-Systematic. Mauron+Tinguely and Lachat. Fribourg.

Koss, R. W. 1968. Morphology and taxonomic use of Ephemeroptera eggs. Ann. ent. Soc. Amer., 61: 696-721.

Koss, R. W. and G. F. Edmunds (Jr.). 1974. Ephemeroptera eggs and their contribution to phylogenetic studies of the order. Zool. J. Linn. Soc., 55: 267-349.

Sivec, I., B. P. Stark and S. Uchida. 1988. Sypnosis of the world genera of Perlinae (Plecoptera: Perlidae). Scopolia, 16: 1-66.

Stark, B. P., M. González del Tánago and W. Szczytko. 1986. Systematic studies on western paleartic Perlodini (Plecoptera: Perlodidae). Aquat. Insects, 8 (2): 91-98.

Stark, B. P. and C. R. Nelson. 1994. Systematics, phylogeny and zoogeogrphy of genus *Yoraperla* (Plecoptera: Peltoperlidae). Ent. Scand. 25: 241-273.

Stark, B. P. and K. W. Stewart. 1981. The neartic genera of Peltoperlidae (Plecoptera). J. Kans. ent. Soc. 54 (2): 285-311.

Stark, B. P. and S. W. Szczytko. 1988. Egg morphology and phylogeny in Arcynopterygini (Plecoptera: Perlodidae). J. Kans. ent. Soc. 61 (2): 143-160.

Tierno de Figueroa, J. M., J. M. Luzón-Ortega and A. Sánchez-Ortega. 1998. Imaginal biology of the stonefly *Hemimelaena flaviventris* (Pictet, 1841) (Plecoptera: Perlodidae). Ann. Zool. Fennici, 35 (4): 225-230.

Zwick, P. 1973. Insecta Plecoptera. Phylogenetisches System und Katalog. Das Tierreich 94: I-XXXII, 1-465.

INDEX